Engineering Materials

This series provides topical information on innovative, structural and functional materials and composites with applications in optical, electrical, mechanical, civil, aeronautical, medical, bio- and nano-engineering. The individual volumes are complete, comprehensive monographs covering the structure, properties, manufacturing process and applications of these materials. This multidisciplinary series is devoted to professionals, students and all those interested in the latest developments in the Materials Science field, that look for a carefully selected collection of high quality review articles on their respective field of expertise.

Indexed at Compendex (2021)

Peter M. Schweizer

Premetered Coating Methods

Attractiveness and Limitations

Springer

Peter M. Schweizer
Schweizer Coating Consulting GmbH
Wünnewil, Fribourg, Switzerland

ISSN 1612-1317 ISSN 1868-1212 (electronic)
Engineering Materials
ISBN 978-3-031-04182-2 ISBN 978-3-031-04180-8 (eBook)
https://doi.org/10.1007/978-3-031-04180-8

This Springer imprint is published by the registered company Springer Nature Switzerland AG
The registered company address is: Gewerbestrasse 11, 6330 Cham, Switzerland

This book is dedicated to Ken Ruschak with whom I worked at Kodak from 1981–1986, and who shared his invaluable research in premetered coating technology with the coating community.

Preface

Motivation for and Purpose of this Book

The basis of my experience (know-how) and knowledge (know-why) in coating technology is from the photographic industry, where premetered coating methods were prevalent. The level of understanding in this sector was rather high owing to the complex multilayered products and their demanding requirements for product uniformity and process productivity. In the meantime, mainly due to digital photography, this industry has dramatically faded away, and those researchers, who had acquired much knowledge and experience in premetered coating methods, are now in the process of retiring. Sadly, this means that their wisdom and experience will be lost to public access. At the same time, my continuous involvement as a teacher in short courses indicates to me that the need for good, thorough knowledge of coating processes is only growing, as new coated products are continually emerging. Products such as lithium-ion batteries or organic photovoltaic films are just two examples of what young people entering this industry using premetered coating methods will later be responsible for efficiently manufacturing. In addition, the chapters about premetered coating methods in the 1997 book *Liquid Film Coating* were contributed by people who either did not have much practical experience or who worked for the photographic industry. The latter group of contributors was not allowed to report freely about their knowledge because this industry was very competitive. Consequently, much understanding about premetered coating process was not made available to the public at that time.

The driving motivation for writing this book is my concern that the basic knowledge about premetered coating methods, needed by today's engineers for today's emerging products, could soon be lost. The retirement of many experts whose knowledge in this field was obtained in the heyday of the photographic industry will soon be unavailable for conferences, courses, and consultations. Coupled together with the fact that most up-to-date information is published on a large variety of platforms will make it increasingly difficult for the industrial engineer to learn about these processes. With this book, I wish to address the lack of a comprehensive overview and preserve this knowledge by summarizing from experts, publications, and my own experience in the field of premetered coating methods. Resulting from my concerns is this new book entitled *Premetered Coating Methods*.

By comparison, *Liquid Film Coating* was a complete summary of what was known publicly about the wide scope of coating technology up until about 1996. Each chapter was written by one or two recognized experts in their respective fields and included an extensive list of references. In contrast, this book only focuses on premetered coating methods presenting many aspects of this technology in much more detail. I did not attempt to address every publication that has appeared since 1996. Instead, based on my practical experience of more than 35 years, I emphasize what I believe is important for achieving highly uniform coated films at a desired production rate. In that sense, the book is a reflection of my personal views about premetered coating methods.

Premetered Coating Methods is primarily addressed to coating engineers working for industrial coating companies. The book describes and explains all process-technological aspects from preparing the coating fluid until it is applied to a moving substrate in form of a uniform film and ready to be dried or cured. Coating engineers should understand these process-technological aspects in terms of the underlying physical concepts. Therefore, the purpose of the book is also to provide tools to the coating engineers, often in form of analytical equations, that allows the engineers to quantify the technology for their specific operating conditions, and, more importantly, to optimize all aspects of the coating process for those conditions. There is no better way of operating a coating process than by tuning the values of the relevant parameters such that the desired output in terms of product uniformity and process productivity is optimal. Analytical equations, which are promoted in this book, cannot always accurately model every detail of a particular situation, but they can nicely visualize the underlying physical concept of that situation. Specifically, they show how a parameter of interest depends on other relevant process parameters, and they show in which direction a given parameter must be changed in order to influence a given process in a desirable way. This approach allows the coating engineer to get the essential aspects of a process right. Fine-tuning the details and optimizing the process must then be supported by experiments, preferably first on a pilot machine, and later on the production coater. The concept of this approach is the basis of this book.

This book is a mixture of reporting results previously published by other experts and presenting results from my own unpublished theoretical and experimental investigations. Personally, I prefer to not only see the results of an investigation but also understand the process leading to the results. I have followed this view on many occasions throughout the book. I am aware that some readers will appreciate this approach and the added details, and I hope that other readers will bear with me.

Outline of the Book

This book is divided into three parts.

Part I starts by re-capturing the history of the R&D work in coating technology carried out and disseminated across the world, and by comparing premetered to self-metered coating processes. It continues with basic information that is needed for studying coating methods, i.e., mass balance, flow rates, and physical fluid properties. The main focus of Part I is on the basic flow fields that are found in premetered coating methods, i.e., pipe, duct, slot, film, curtain, and boundary layer flow, as well as creeping flow after coating. These flows have been amply discussed in other books. Nevertheless, it is vital for coating engineers to soundly understand these basic concepts, which is why a comprehensive summary of these flows is included here. Next, the importance of achieving a minimum wall shear stress in the confined flow fields is discussed. Furthermore, it is illustrated how the wall shear stress controls the residence time behavior of the fluid delivery system including the slot die. Part I closes with a presentation on the concept of hydrodynamic assist of dynamic wetting. In my opinion, imposing a minimum wall shear stress and exploiting the hydrodynamic assist are responsible for making premetered coating methods superior to self-metered processes in terms of product uniformity and process productivity.

Part II addresses general properties that are common to all three premetered coating methods. The first chapter is on fluid preparation and delivery systems, followed by a discussion of the nominal thickness of the coated film, as well as the thickness uniformity in machine and cross-web direction. In this context, a large portion of Part II is devoted to the design of slot dies. Next, the attractive feature of simultaneously applying several different fluid layers is reviewed, the question of interlayer mixing is examined, and guidelines for designing single- and multilayer films are suggested. Part II closes by presenting examples of economic considerations related to premetered coating methods.

Part III summarizes specific information about slot, slide, and curtain coating. Specifically, preferred process configurations and typical equipment are described, the operating windows with various coating defects that mark the boundaries of the window are discussed, operational issues are addressed, and ways for optimizing the processes are suggested. For slot coating, the specific features of tensioned-web coating, double-sided coating, as well as stripe and intermittent coating are reviewed.

Past, Present, and Future of Premetered Coating Methods

As explained in Chap. 2 of this book, there is no doubt in my mind that premetered coating methods are superior to self-metered methods in terms of product uniformity and process productivity, and that curtain coating is the best method with regard to these attributes. This does not mean, however, that curtain coating is a suitable process for every coating application. Moreover, and disappointingly, the current reality in the converting industry does not reflect this opinion. A survey during the period 2008–2016 of all projects involving coating technology at Polytype Converting revealed that multiple smooth roll processes were used in 39% of these projects, 42% used gravure coating, 2% used blade coating, and only 17 % of the project used premetered coating methods, in particular multilayer curtain coating and single-layer slot coating. The average coating speed of these projects was 381 m/min and the average coating width was 1525 mm. Discussions with other machine suppliers at that time showed a similar picture throughout the converting industry. These numbers, therefore, provide an image of the state of the art of the converting industry in the decade 2010–2020.

Why did the converting industry at large not turn to premetered coating methods more enthusiastically despite the high-performance level that was reached in the photographic industry? One important reason is the fact that the well-endowed photographic companies owned and developed both the process and the product, i.e., they employed scores of well-educated chemists who were qualified to develop films and papers with ever-improving functional properties, and they also employed scores of chemical engineers who had pilot machines available to develop new coating processes and improve their performance. Moreover, chemists and chemical engineers applied the concept of parallel engineering, i.e., chemists did not develop new products without considering boundary conditions posted by process engineers and related to the efficient manufacturing of such products.

In contrast, many of the converting companies today do not own pilot coating facilities to enable them to improve their understanding of the manufacturing process. They only own their products, but not their processes, meaning they have to buy the process and the associated machines from a suitable supplier. Consequently, the discipline of process engineering is only modestly staffed at best. Moreover, most of the converting companies buy the least expensive manufacturing machines from external suppliers rather than the best-performing manufacturing processes. This situation is accentuated by the fact that most suppliers of coating machines focus on the design and the fabrication of the machines rather than on the coating and drying processes that run on these machines. This view is incorrect because a converter needs well-performing processes in the first place; the associated machines are merely the housings of the process. During the commissioning of new coating equipment at customer sites while working at TSE and Polytype Converting, I observed time and again that machine-related problems were normally few and could easily be resolved, but that process-related issues often were numerous and much more difficult to resolve.

I am convinced that the converting industry, as well as the associated technology suppliers, need more process engineers who thoroughly understand their manufacturing processes, and I hope that this book will contribute to improving this situation by explaining and spreading the good news about premetered coating methods.

Wünnewil, Switzerland Peter M. Schweizer

Acknowledgements

First and foremost, I want to thank Ernst Meier, former CEO of Polytype Converting AG, who allowed me to report on most of the R&D work I carried out during my 15 years at the company. Without that permission, I could not have written this book. The same thanks go to Marcel Koestinger and Bruno Kaeser, the most experienced and valuable coating technicians one can wish for. They were and still are responsible for operating the Polytype Pilot Facility, and they supported me in all the experimental studies I carried out on those coating machines.

I am also grateful to Maick Nielsen, CEO of TSE Troller AG, for providing me with drawings and photos that show design solutions to flow problems inside and outside of slot dies.

A special thank you goes to Alfred Roulier and Hans Bebié. Fred holds a Ph.D. in Physics from the University of Bern. He was also my boss when I worked at ILFORD in Switzerland. Hans is a professor emeritus from the Physics Department at the University of Bern. Hans and Fred share their hobby of finding all sorts of physical problems that affect the everyday life of ordinary people, using their skills in mathematic methods for solving the problems, and communicating the results in a way that is understandable to ordinary people. When asked, Hans and Fred immediately agreed to help me with my book project by solving several flow problems of power-law fluids inside slot dies. Sadly, Hans passed away before the book was published.

I thank my wife, Sue. Being American, she improved the English language of my manuscript, which is for the benefit of the readers of the book. She also endured many hours I spent in my home office writing this book during the corona pandemic.

Last but not least, my appreciation goes to Willi Schabel, Head of the Thin Film Technology Laboratory at the Karlsruhe Institute of Technology, and organizer of the Short Course Coating and Drying of Thin Films, where I am part of the teaching team. Willi's encouragement prompted me to write this book.

Contents

About the Author

Peter M. Schweizer After graduating from ETH Zurich with a diploma in Mechanical Engineering and a Ph.D. in Process Engineering, I was fortunate and grateful to have spent my entire professional career in the coating industry. It started in 1979 with a postdoctoral position with Skip Scriven at the University of Minnesota. Following was a 16-year period when I worked in the photographic industry, i.e., on the user side of coating technology. First, I spent 6 years in the Coating Flow Research Group at Kodak in Rochester, New York. Then, I moved back to Switzerland, where I worked for 10 years at ILFORD in Fribourg. My responsibilities included the development of a visualization method for coating flows, the study of the stability of coating flows, the optimization of die design, and the implementation of various R&D findings into the industrial manufacturing environment. In both companies, the technological focus was on premetered coating methods, in particular, on slide and curtain coating because these methods allowed the many layers of photographic films and papers to be coated simultaneously in one pass, i.e., with very high process productivity, and with very high uniformity in each layer, therefore with high product quality.

For the second part of my career, I switched to the supplier side of coating technology. For the first 4 years, I was a partner at TSE Troller Schweizer Engineering, a company affiliated with Troller AG, a well-known manufacturer of slot dies and multilayer slide dies for premetered coating methods. More importantly, I later worked for 15 years at Polytype Converting AG in Fribourg, a major manufacturer of coating and drying machines for the converting industry. My main responsibility was the introduction of premetered and simultaneous multilayer coating processes to the portfolio and the customers of Polytype. To that extent, I was involved in the further development of various aspects of these technologies, such as designing and optimizing edge guides and the suction baffle for curtain coating, experimentally determining the onset of ribbing lines in slide coating, preventing ribbing lines in slot coating, introducing tensioned-web, simultaneous double-sided, strip and intermittent coating for slot coating, etc. All these activities were carried out in conjunction with running numerous coating trials for customers in Polytype's pilot coating center. This experience was immensely valuable to me. Unlike in the photographic industry, where the focus was limited to one product group with a relatively narrow range of coating parameters, there I was exposed to the entire world of coated products with a much wider range of relevant coating parameters. This environment was not only more difficult and challenging than the photographic industry but also more rewarding.

In 2006, I received the John Tallmadge Award for significant contributions to the understanding and improvement of liquid coating process technology from the International Society of Coating Science and Technology, and I am the president of the European Coating Society since 2018. Now in my retirement, I continue to work as an independent consultant in coating technology.

Definition of Symbols

Most of the symbols are defined in the text where they first appear. The following list defines the symbols most commonly used in the book. The list is not complete, and several symbols have more than one meaning. Apart from a few well-marked exceptions, SI units are used throughout the book.

Letters

a	Yasuda parameter in Carreau-Yasuda equation, long side of rectangle, square side
a_i	constants in regression equation
A	area
A_0	initial cross-section of inner cavity, annual production volume
A_e	end cross-section of inner cavity
A_{dry}	dry coat weight
A_{gross}	gross lot size
A_{wet}	wet coat weight
b	shear-thinning parameter in Carreau-Yasuda equation, short side of rectangle
b_i	constants in regression equation
c_i	constants in regression equation
dA_{dry}	variation of dry coat weight
dH_{wet}	variation of wet film thickness
$d\mu$	variation of viscosity
dU	variation of web speed
dV*	variation of volumetric flow rate
dw_1	variation of inner slot height
dw_2	variation of outer slot height
C_C	product unit costs associated with the coating process
C_{CM}	hourly cost of the coating machine

$C_{M,i}$	specific raw material costs of a given fluid
C_{RM}	annual costs for raw materials
CD	depth of outer cavity
CL	length of outer cavity
C_V	concentration by volume
$C_{V,solid}$	solids concentration by volume
$C_{V,solvent}$	solvent concentration by volume
C_W	concentration by weight
$C_{W,solid}$	solids concentration by weight
$C_{W,solvent}$	solvent concentration by weight
d	constant in regression equation
$d_{b,0}$	bubble diameter at ambient pressure P_0
d_b	bubble diameter at vacuum pressure p
D	pipe diameter
DF	damping factor of outer cavity
e	constant in regression equation
e_c	extension rate in curtain flow
E	modulus of elasticity
$E(t_r)$	residence time spectrum
f	constant in regression equation, step height between adjacent die plates, factor for calculating boundary layer thickness
g	acceleration of gravity
H	film thickness
H_0	uniform film thickness
H_c	curtain thickness
H_{cr}	capillary rise
H_{dry}	dry film thickness
H_δ	thickness of boundary layer
H_{edge}	height of coated edge
H_{gap}	coating gap, gap between suction baffle and substrate
H_{porous}	porous film thickness
H_{waist}	thickness of bead waist inside coating
H_{wet}	wet film thickness
i_i	constant in regression equation
I	moment of inertia
$J(t_r)$	transfer function
k	wave number
k_i	constant in regression equation
ℓ	relative wetting line position
l_i	constant in regression equation
L	length of die slot, die lip, pipe
L_{bar}	length of die bar
L_c	length of curtain
L_δ	length of boundary layer
L_{dryer}	length of dryer

L_m position of upstream meniscus
L_w working length
n power-law index, number of coating passes, revolutions per minute
m consistency
$m^*_{solvent}$ solvent mass flow rate/unit area
M visibility of defect, momentum
M_c momentum of curtain
M^* mass flow rate
M^*_{solid} solids mass flow rate
$M^*_{solvent}$ solvent mass flow rate
OS offset between adjacent die plates
P pressure
P_0 initial pressure
P_i equilibrium vapor pressure
P_{load} pressure-driven load on die bar
P_L pressure after length L
P_v vacuum pressure
Q volumetric flow rate/width
Q_0 uniform volumetric flow rate/width
Q_c critical volumetric flow rate/width
Q_δ average volumetric flow rate/width inside boundary layer
Q_{eff} effective volumetric flow rate/width
Q_{min} minimum volumetric flow rate/width
r radius
r^2 measure for the quality of a regression equation
R pipe radius
R_c corner radius
s dimensionless group in equation 5.4.15
S spreading coefficient, wetted perimeter
SMU specific machine utilization
t time
t_b bubble rise time
t_c curtain fall time
t_{np} non-productive time
$\bar{t}_r$ mean residence time
T temperature
T_{bar} thickness of die bar
U coating speed, web speed
U_i coating speed of pass i
U_{max} maximum coating speed
U_{min} minimum coating speed
$\overline{U}$ average coating speed
v_b bubble rise velocity
v_d disturbance velocity in curtain

v_z	velocity in z-direction
$v_{z,average}$	average velocity in z-direction
$v_{z,max}$	maximum velocity in z-direction
V	velocity
V_c	local curtain velocity
$V_{c,0}$	initial curtain velocity
V_{in}^*	flow rate entering the die at the inlet port
V_{out}^*	flow rate exiting the die at the end of the distribution cavity
V*	volumetric flow rate
w	slot height
W	die width, coating width, cavity length
W_{max}	maximum coating width
W_{min}	minimum coating width
x	coordinate
x_i	molar concentration of component i in liquid phase
y	coordinate
y_i	molar concentration of component i in gas phase
y_{max}	maximum deflection of die bar
z	coordinate

Dimensionless Numbers

Bo	Bond number
Ca	capillary number
Ca_c	critical capillary number
Ca_{film}	capillary number of film flow
Fr	Froude number
Oh	Ohnesorge number
Po	property number
Re	Reynolds number
Re_c	critical Reynolds number
Re_s	stretched Reynolds number
Sto	Stokes number
We	Weber number
We_c	critical Weber number

Greek Letters

α	application angle, half Mach angle, diffuser angle, upstream angle of outer cavity, die tilt angle, by-pass factor

β slide angle, angle between suction baffle and curtain, isentropic exponent, measure for cavity end cross-section
β_0 deterministic contributions to the cross profile after inner distribution system
β_1 total deterministic contribution to the cross profile
γ shear rate, impingement angle, angle of downstream die lip
γ_o wall shear rate
$\gamma_{average}$ average shear rate
γ_c critical shear rate
γ_Γ critical shear rate related to yield stress
γ_{uc} upper critical shear rate
Γ inertia shape factor
δ lip angle, V_c/U, change-over time
Δ difference
ΔP pressure drop, bead vacuum
ε porosity, lip angle of slide die
ε_{max} total cross profile
φ $\delta \sin\alpha$, taper index of inner cavity
Φ location where wall shear stress in inner cavity drops below its initial value
λ viscous shape factor
μ liquid viscosity
μ_0 low-shear viscosity
μ_∞ high-shear viscosity
μ_e equivalent Newtonian viscosity
ω angular velocity
ρ_g gas density
ρ_l liquid density
ρ_{liquid} liquid density
ρ_{porous} porous layer density
ρ_{solid} solid density
$\rho_{solvent}$ solvent density
σ liquid surface tension
σ_{LG} surface tension at liquid-gas interface
σ_{SG} surface energy at solid-gas interface
σ_{SL} surface energy at solid-liquid interface
σ^d disperse part of surface tension
σ^p polar part of surface tension
θ static contact angle
τ shear stress
τ_0 wall shear stress
$\tau_{0,min}$ minimum wall shear stress
τ_Y yield stress
ψ dynamic contact angle
ζ shape factor of cross-flow in outer cavity

Part I
R&D in Premetered Coating Technology

Abstract

Part I begins by reviewing the history of R&D activities in coating technology and how the attained knowledge was disseminated throughout the coating industry. It continues by comparing premetered with self-metered coating methods and by highlighting the attractive features of the three premetered application processes. Various equations based on the law of mass conservation are then provided that show how parameters that are relevant for fluid flow simulations such as the wet film thickness and the flow rate depend on the dry thickness or the dry coat weight of the coated film, which are relevant for the function of the final product.
Physical fluid properties including the density, the viscosity (rheological properties), the surface tension, the static contact angle, and wetting properties are then discussed. It is also shown how these material properties can be altered by changing the solids concentration or the temperature of the coating fluid.

Knowing the flow rate and the physical fluid properties is a prerequisite for simulating the basic flows that appear in premetered coating processes. Specifically, Chap. 5 summarizes how pipe, duct, slot, film, curtain, and boundary layer flows as well as flow after coating can be modeled for Newtonian and shear-thinning flow behavior. In addition, the stability of some of these flows is addressed, and advice is provided on how vortices, which are an unwanted feature in many of these flows can be prevented. Finally, several examples show how the geometry of the boundaries or the operating conditions can be optimized to maximize the uniformity or the performance of these flows.

Part I closes by presenting the fluid mechanical concepts of wall shear stress and hydrodynamic assist. Wall shear stress is relevant for designing slot dies and fluid delivery systems because it affects the residence time behavior of such confined flows and it determines how easily they contaminate and can be cleaned. Hydrodynamic assist is essential for dynamic wetting, and exploiting this feature is one of the major advantages of premetered over self-metered coating methods.

Chapter 1
Introduction

Abstract The introduction to this book re-captures the history of the R&D work in coating technology carried out and disseminated across the world. It prepares a platform for this book, which adds another piece to the vast puzzle of information about coating technologies.

1.1 History of R&D in Coating Technology

Liquid film coating, i.e., the application of a thin and uniform liquid film onto a moving substrate, is a complex and multi-disciplinary technology. It involves fluid conditioning and delivery, fluid dynamics of confined and free surface flows, static and dynamic wetting, interfacial phenomena, rheology, process control, optimization, and other disciplines such as calculus and numerical mathematics.

To date and to my knowledge, no school in the world offers a comprehensive curriculum in coating technology. A suitable basic education for entering coating technology is chemical engineering or process technology. In the 1970s Prof. L. E. (Skip) Scriven at the University of Minnesota was the first one to offer a high-level graduate program in coating and drying technology. Several years later, a few professors at schools in Europe and Asia also started graduate-level research in coating and drying. Today, some of the Ph.D. graduates from these early professors became professors themselves and continue to investigate coating and drying issues.

In parallel, also starting in about the seventies of the last century, all major companies of the photographic industry, which was thriving at that time, began to build up internal coating flow research groups to improve the performance of their main manufacturing process. These efforts were particularly important because the quality requirements of photographic films and papers were very demanding.

P. M. Schweizer, *Premetered Coating Methods*, Engineering Materials,
https://doi.org/10.1007/978-3-031-04180-8_1

1.2 History of Dissemination of Knowledge in Coating Technology

In 1982, Ed Gutoff from Polaroid and Ken Ruschak from Kodak took it upon themselves to organize the first international conference on coating and drying technology. This conference continues to be organized bi-annually up to the present. During the first years, it operated under the umbrella of the American Institute of Chemical Engineers (AIChE), and since 1998 under the patronage of the International Society of Coating Science and Technology (ISCST). In 1995, a similar bi-annual conference was first organized in Europe by Mike Savage and Phil Gaskell from the Leeds University. Today this event is still ongoing under the helm of the European Coating Society (ECS). Similar regular conferences also take place in Asia.

The main purpose of such conferences is to give researchers from both academia and R&D Labs of larger coating companies the opportunity to present their latest findings, and to offer an international networking platform. These findings describe the forefront of science in coating and drying technology. However, this is not always valuable for those coating engineers representing smaller converting companies who are working to solve everyday problems occurring in an industrial manufacturing environment.

In 1990, for the first time, Ed Gutoff from Polaroid and Ed Cohen from DuPont organized a 2-day short course in coating and drying technology in conjunction with the AIChE coating symposium mentioned above. Since then, this type of short course has been offered every year both by various universities in the US, Europe, and Asia and by professional organizations such as AIChE, ISCST, ECS, TAPPI, and AIMCAL. Over time, the need for such courses has not diminished as exemplified by continued good attendance. The latest of these courses entitled *Coating and Drying of Thin Films* was organized for the 13th time in 2022 by Prof. Wilhelm Schabel at the Karlsruhe Institute of Technology in Germany. While the teaching staff was recruited from academia and industry, some 75% of the attendees came from industrial coating companies, thus manifesting the desire of industrialists to learn the details of their manufacturing technologies. I am fortunate to have been a teacher for such courses in the US and Europe from the beginning up to the present time.

By the end of the 1980s some 15 years of intensive research in coating and drying had passed. Results were presented at coating conferences and published in scientific journals, but were not readily available in a comprehensive way. As a consequence, Ed Gutoff and Ed Cohen published the book entitled *Modern Coating and Drying Technology* (Cohen & Gutoff, 1992). The content of this book was based on their short course as mentioned above. Moreover, the book was geared towards engineers, chemists and physicists new to coating and drying who were eager to learn about these technologies. Three years later, the same authors added another volume entitled *Coating and Drying Defects* (Gutoff & Cohen, 1995).

Driven by the same motivation Stefan Kistler and I published the book entitled *Liquid Film Coating* (Kistler & Schweizer, 1997). As editors we were fortunate to engage some 30 well-known and respected coating experts from academia and industry across the world. This volume was more comprehensive and its content was on a higher scientific level than the aforementioned *Modern Coating and Drying Technology*. In the ensuing time many additional findings regarding coating have been published, although again in a scattered way. Nevertheless, most of the content in *Liquid Film Coating* still holds true today.

An interesting book entitled *Making KODAK Film* was published by Shanebrook (2016). The author spent his entire 35-year career at Kodak. He was involved in the development and manufacturing of many of the professional photographic film products. In the book, Robert Shanebrook provides detailed insights into the history of making films, thereby covering the complete life cycle of light-sensitive films and papers. Specifically, the topics discussed by Robert include film base making, emulsion making, coating and drying, finishing, quality control, and more.

References

Cohen, E., & Gutoff, E. (Eds.). (1992). *Modern coating and drying technology*. VCH Publishers.

Cohen, E., & Gutoff, E. (1995). *Coating and drying defects—Troubleshooting, operating, problems*. Wiley.

Kistler, S. F., & Schweizer, P. M. (Eds.). (1997). *Liquid film coating—scientific principles and technological implications*. Chapman & Hall.

Shanebrook, R. L. (2016). *Making KODAK film* (2nd ed.). ISBN 978-0-615-41825-4. Self-published. www.makingkodakfilm.com

Chapter 2
Premetered Versus Selfmetered Coating Methods

Abstract The class of premetered coating methods including slot, slide, and curtain coating is compared to the class of selfmetered methods, which comprises some 50 variations of roll and blade coating processes. Process limitations and application ranges for relevant operating parameters are listed, and the ease of implementing process-enhancing concepts in terms of product uniformity and process productivity are discussed. It is concluded that premetered coating methods are preferred over selfmetered processes.

2.1 Premetered Coating Methods

All in all, more than 50 different coating methods are known for applying a thin liquid film onto a moving substrate. In the scope of this book, all these methods have been grouped into only two classes, namely premetered and selfmetered processes. This makes it easier to point out and discuss the differences between these two groups.

The term *premetered* refers to the fact that the flow rate that is pumped to the coating station in order to obtain a specified thickness of the coated film is accurately measured (metered) before (pre) coating. Consequently, the physical law of mass conservation applies (what goes in must come out), which allows the wet film thickness to be calculated according to

$$H_{wet} = \frac{V^*}{WU} = \frac{Q}{U} \tag{2.1}$$

H_{wet} wet film thickness [m],
V^* volumetric flow rate [m^3/s],
Q volumetric flow rate/width [m^2/s],
U coating speed [m/s],
W coating width [m].

Equation 2.1 visualizes one of the attractions of premetered coating methods, namely that the wet film thickness only depends on the flow rate, the web speed, and the coating width, but not, what-so-ever, on any physical fluid properties such as the

P. M. Schweizer, *Premetered Coating Methods*, Engineering Materials,
https://doi.org/10.1007/978-3-031-04180-8_2

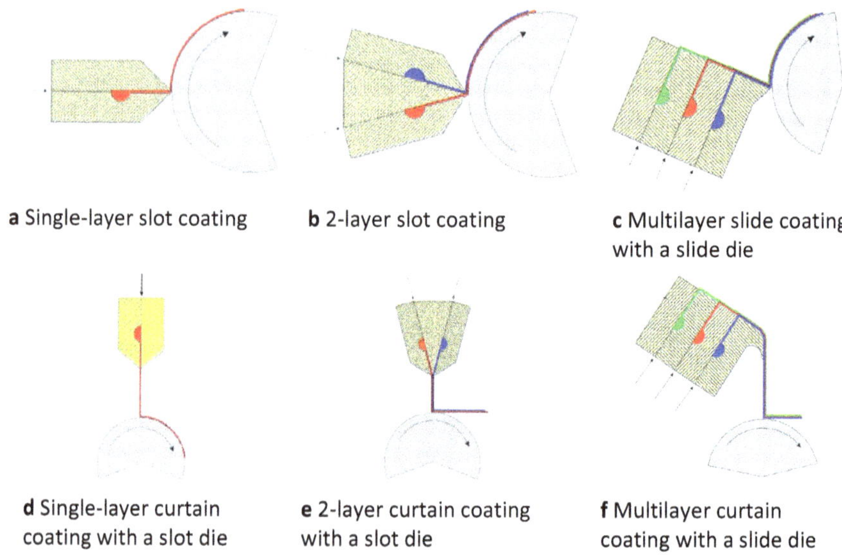

Fig. 2.1 Schematic diagrams of premetered coating methods

viscosity, or any of the geometrical process parameters such as the slot height, the coating gap, or the curtain height.

Premetered coating methods comprise only three different processes called slot, slide, and curtain coating. These schematic process configurations are shown in Fig. 2.1.

A common and distinctive feature of all these processes is that they use dies for distributing the coating fluid in the cross-web direction. Dies are classified as single or multilayer slot dies (Fig. 2.1a, b, d, e) or single or multilayer slide dies (Fig. 2.1c, f). The word *slide* refers to the inclined plane on top of the die, where, upon exiting from the die slot, the liquids first form a single-layer or multilayer film before being coated onto the substrate. The concepts for designing the internal geometry, which controls the fluid distribution process, are the same for both slot and slide dies, see Sect. 9.4.

Premetered coating methods were invented by the photographic industry. In particular, Eastman Kodak received broad patents for slot coating in 1954 and for multilayer slide coating in 1956. The major technological and economic advantages of slide over slot coating are the larger number of layers that can be coated simultaneously, and the fact that the coating fluid does not have to flow through a narrow gap formed between the die lip and the substrate surface. Then, in 1970, Kodak again received a landmark patent for multilayer curtain coating. This process can be considered as the next technological step forward from slide coating, with the big difference being the much larger distance between the die lip and the substrate surface, and consequently the much higher momentum of the impinging curtain.

It is no surprise that the photographic industry invented simultaneous multilayer coating methods because photographic color papers contain 6–8 layers, and some color films contain more than a dozen layers. In fact, a finished instant photographic product from Polaroid has more than 20 layers. Being able to coat at least some of these layers in one pass drastically increased the productivity of the manufacturing process and hence lowered the costs of those products, see also Chap. 15. I heard that Polaroid coated 19 layers at the same time in a production environment, and an engineer from Fuji Photo Film informed me that he used the slide coating method for applying 25 layers simultaneously, although only for a test in a laboratory environment.

Today, slot coating is most often operated in the single-layer mode, although 2 or perhaps 3 layers can also be coated simultaneously. Four or more layers are not possible because the die design concept shown in Fig. 2.1 does not provide enough room for implementing more than 3 slots. Slide coating was and still is very popular throughout the photographic industry, particularly during the seventies and eighties, when curtain coating was protected and reserved for Kodak by the comprehensive patent mentioned above. Outside the photographic industry, slide coating has not found much interest to date, perhaps owing to the relatively small operating window, which is dominated by the ribbing line instability, see Sect. 18.5.1. During the lifetime of the Kodak patent, ILFORD in Marly, Switzerland, was the only company having a license agreement that allowed them to use multilayer curtain coating. Thereafter, curtain coating was widely used by other photographic companies. Moreover, the paper industry, driven by the paper machine manufacturer Voith, was venturing toward curtain coating, particularly for high-speed single-layer thermal paper applications. Later, at the end of the 1990s, Bachofen and Meier was the first equipment supplier to introduce curtain coating to the converting industry, beginning with single-layer adhesive applications. At the beginning of the twenty-first century, other suppliers such as Polytype Converting and Kroenert started to spread this process across the converting industry for multilayer applications such as high-grade thermal paper, ink jet imaging, and pressure-sensitive adhesives (PSA).

2.2 Selfmetered Coating Methods

Selfmetered coating methods include roll, blade, and dip coating, see Fig. 2.2. The term *selfmetered* expresses the fact that, in a first step, too much liquid is applied to the substrate, and then the excess fluid must be removed (metered) by a suitable device to obtain the desired film thickness. In other words, the film thickness is controlled by the coating process. Moreover, the film thickness depends on physical fluid properties and relevant geometrical process parameters, such as the gap between adjacent rollers, the gap between the blade tip and the substrate surface, or the cell volume of a patterned roller surface.

Roll coating comes in many different configurations. The distinctive features are:

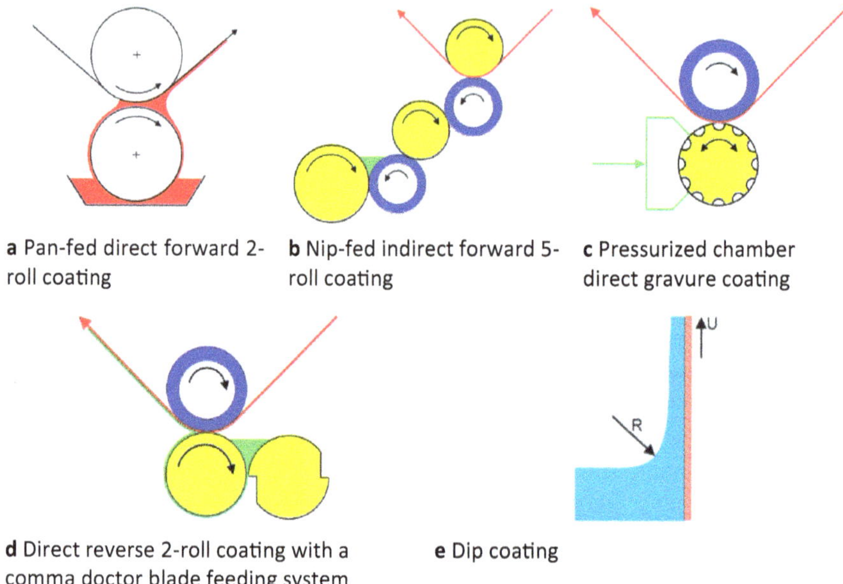

Fig. 2.2 Schematic diagrams of selfmetered coating methods

- Number of rollers: 2–6.
- Roller material: steel, rubber-covered.
- Roller surface: smooth, engraved.
- Application mode: direct, indirect.
- Film transfer to a substrate: forward, reverse.
- Liquid feeding: pan, nip, pressurized chamber.
- Liquid metering: gap (nip), gravure volume, pressure in closed chamber of gravure coater.

Most of the published studies on roll coating were carried out with a pair of steel rolls. This configuration was suitable for investigating the basic flow fields across a nip. In industry, however, pairs of steel rolls are very rarely used because the minimum gap must be larger than about 30 μm to avoid collision between rigid bodies and to reduce the sensitivity of the film thickness to mechanical gap non-uniformities. Consequently, very thin wet films cannot be coated with steel-steel roller pairs. Instead, most industrial roll coating processes use steel-rubber combinations, which allow wet films of <1 μm to be coated, and which allow splices to pass through the nip without any problems.

Blade coating comes with flexible or rigid blades. The latter is sometimes called knife coating or knife over roll coating. In sketch 2.2d two rigid knifes are incorporated into a cylindrical body. Rotating this body allows the knife to be exchanged quickly in case of contamination or mechanical damage. Since the knives are not straight like blades but curved like comas, this configuration is called coma coater.

Moreover, the coma coater is depicted in the indirect mode, which prevents the splice from having to move through the metering gap, and allows for intermittent coating by temporarily separating the two moving rolls.

Dip coating is an old, simple, and slow method, but it is still used today, for example for coating metal coils. Sometimes, dip coating is combined with an air jet, which serves as a metering device, and allows thinner films at higher speeds to be coated.

The books *Modern Coating and Drying Technology* (Cohen & Gutoff, 1992) and *Liquid Film Coating* (Kistler & Schweizer, 1997) provide an overview of both premetered and selfmetered coating methods.

2.3 Process Limitations

If more than 50 coating processes are known, which one then is the best choice for manufacturing a given product? The answer is not trivial at all because our theoretical understanding and the practical experience of these processes is often insufficient for judging the process performance relative to a new set of operating conditions and physical fluid properties. For example, the admissible range of a given process parameter, say the coating speed, depends on many other relevant process parameters, such as the viscosity, the wet film thickness, the substrate properties, etc. Most of the time, formal relationships between these parameters are not known because theoretical calculations are difficult if not impossible to carry out, and experimental data are sparse in the public domain, not the least owing to a lack of suitable pilot machines throughout the converting industry.

The most important criterion for selecting a suitable coating method is the admissible wet thickness of the coated film. Furthermore, criteria relating to product uniformity and process productivity must be considered in parallel.

Product uniformity criteria include:

- Film thickness uniformity in the machine direction.
- Film thickness uniformity in the cross-web direction.
- Coating defects:
  - longitudinal or transverse lines and streaks;
  - ribbing lines;
  - air bubbles imported by the fluid or generated by wetting failure (air entrainment).
- Influence of substrate properties:
  - thickness variation;
  - porosity;
  - mottle.

Process productivity criteria include:

- Production time:
  - coating speed, U;
  - coating width, W.
- Downtime, t_{np}:
  - ease of cleaning;
  - ease of changing coating width.
- Size of coating lot, A_{lot}.
- Simultaneous multilayer capability:
  - Number of coating passes, n.

Sometimes, the strict imposition of one given criterion, for example, the absence of air bubbles, or the capability of coating beyond a minimum speed, allows one to eliminate several coating methods right away. Most of the time, however, choosing the best coating method for a given application is difficult. In any case, before investing in a new method, coating trials should be carried out on a suitable pilot machine. This in itself may be difficult too because many of the available pilot facilities are not capable of running the process at the envisioned operating conditions of the industrial manufacturing process. Often, for example, the available speed and/or drying capacity of the pilot line is not in sync with the desired coating speed of the industrial machine.

Despite the difficulties described above and the impossibility of condensing the task of selecting the best coating method into one small table or diagram, Table 2.1 might help start a meaningful discussion on this issue.

When reading Table 2.1 the following explanations must be considered:

- None of the numbers are precise but merely express orders of magnitude. The best value of a given process parameter usually depends on the values of several other relevant process parameters, and formal relationships between these parameters are not always known.
- If, for example, the upper viscosity is indicated with a value of 10,000 mPas, then values of, for example, 8,000 or 20,000 are included in this upper limit.
- All viscosity values refer to the low shear viscosity. However, since no shear rate value is associated with these viscosity values, they merely indicated the fluidity of a given liquid. Honey-like fluids have a high low shear viscosity, while low viscosities point toward solvent-like liquids of low solids concentration.
- A viscosity range of 1–10,000 mPas means that very low-viscosity fluids and very high-viscosity fluids can be coated with this method. If the lower viscosity limit is not 1 but 10 mPas means that very low-viscosity fluids such as pure solvents should not be coated with this method.
- Curtain coating is the only method with requirements for the surface tension and the volumetric flow rate/width.
- Ranges for the film thickness are difficult to specify without considering other circumstances. For example, liquid film coating methods deposit wet films. Hence,

Table 2.1 Limitations of coating processes

	Slot	Slide	Curtain	Gravure	Rigid gap and knife	Elasto-hydro-dynamic roll and knife
μ [mPas]	1–10,000	1–5,000	10–10,000	1–500	10–10,000	10–10,000
σ [mN/m]	–	–	<40	–	–	–
U [m/min]	1–1,000	1–400	100–2,000	10–1,000	1–500	10–1,000
H_{wet} [mm]	10–>100	50–>100	10–>100	0.3–50	20–>100	0.1–50
H_{dry} [mm]	<1–>100	<1–>100	<0.1–>100	<0.1–50	<1–>100	<0.1–50
# Layers [−]	1–3	1–>10	1–>10	1	1	1
Q [cm²/s]	–	–	>1.0	–	–	–
Lines + streaks	Critical	Less critical	Not critical	Less critical	Less critical	Not critical
Bubbles + dirt	Critical	Critical	Critical	Critical	Critical	Critical
Ribbing lines	Yes	Yes	No	Yes	Yes	Yes
Dry or wet edges	Dry	Dry	Both	Both	Both	Both
Reactive chemicals	On-line injection	On-line injection	On-line injection	Not	Not	Not

the resulting dry film thickness depends on the solids concentration of the fluid. Moreover, slot coating, for example, has two operating modes. In the capillary mode where surface tension forces dominate viscous forces, wet films of <10 μm can be applied, but only at unattractive process productivities. For more details see Sect. 17.5. In addition, using the simultaneous multilayer capability of curtain coating, very thin individual dry layers on the order of 10 nm can be achieved, either in combination with other functional layers or with the help of a dummy layer. For more details see Chap. 12.

- More meaningful and more precise operating limitations of slot, slide, and curtain coating are presented in Chaps. 17, 18, and 19, respectively.

2.4 Attractiveness of Premetered Coating Methods

Selecting the best coating method for a given application involves comparing the available processes and evaluating their performance relative to uniformity and productivity criteria as well as to product unit costs and investment costs. Economic aspects of coating processes are discussed in more detail in Chap. 15. Additional information about this topic can be found in Schweizer (1997).

With regard to uniformity and productivity criteria, there is no doubt in my mind that premetered coating methods are superior to selfmetered processes, and that curtain coating is the absolute best method of all.

Beyond the wide speed and thickness ranges indicated in Table 2.1, and beyond the unique capability of laying down several layers at the same time, premetered coating methods are better suited than any other coating method for implementing performance-enhancing criteria such as.

- Minimum wall shear stress, which is beneficial regarding:
  - Preventing (minimizing) contamination of flow fields.
  - Preventing (minimizing) cleaning of flow fields.
  - Minimizing fluid change-over times.
  - Preventing vortices, which in turn improves the uniformity of the coated film.

High wall shear stress can be achieved by designing flow fields with appropriate flow boundaries (small flow cross-sections, gentle changes in flow direction). For more details see Chap. 6.

- Hydrodynamic assist to dynamic wetting, which
  - Allows the coating process to run at high speeds without air entrainment.
  - Improves the uniformity of the coated film.

Hydrodynamic assist can most effectively be implemented in curtain coating, which explains the high web speeds that can be attained with this process. In fact, on their pilot machine in Samstagern, Switzerland, Trinseo (formerly Dow Europe) in 2006 achieved a coating speed of >2,000 m/min when applying a single layer of paper color. For more details see Sect. 5.9.6, Chap. 7 and 12.

- Premetering the coating fluid, which
  - Results in a film thickness non-uniformity in the machine direction of <1%. For more details see Sect. 9.3.
  - Prevents problems related to recycling of excess fluid such as
    - air entrainment,
    - evaporation of solvents,
    - replenishing of solvents,
    - foam formation,
    - crust formation.
  - Allows easy handling of reactive chemicals by online injection of one of the reactive components. For more details see Sect. 8.2.
- Distributing the coating fluid in the cross-web direction with dies, which
  - Results in film thickness non-uniformities in the cross-web direction of <1%, at least for dedicated applications. For more details see Sect. 9.4.

The above list of attractive features of premetered coating methods is long and impressive. However, for the moment being, these features are merely unsupported claims. The following chapters, therefore, provide information, discussions, and experience to substantiate the claims. In any case, while some products such as those using very low-viscosity coating fluids, or very thin films applied at high web speeds may better be coated with selfmetered methods, premetered coating methods should be preferred over alternative methods whenever possible.

References

Cohen, E., & Gutoff, E. (Eds.). (1992). *Modern coating and drying technology*. VCH Publishers.

Kistler, S. F., & Schweizer, P. M. (Eds.). (1997). *Liquid film coating—scientific principles and technological implications*. Chapman & Hall.

Schweizer, P. M. (1997). Minimization of product unit costs, Example on page 761 ff. in Chapter 15 entitled Control and optimization of coating processes. In S. F. Kistler & P. M. Schweizer (Eds.), *Liquid film coating*. Chapman & Hall.

Chapter 3
Mass Balance and Flow Rates

Abstract The physical law of mass conservation is basic for premetered coating methods. This chapter summarizes correlations between various thickness measures of the coated film and related flow rates that are required for obtaining coated films with a desired thickness. Specifically, the wet thickness of the coated film, which is the relevant film dimension from a coating and fluid mechanics point of view, as well as relevant flow rates are expressed in terms of the dry thickness of the coated film, which is the relevant dimension from a product performance point of view. Alternatively, thickness and flow rate parameters are expressed in terms of the dry coat weight, which is more widely used in the converting industry.

3.1 Wet Versus Dry Film Thickness

The thickness of a coated layer, as well as related dependent variables, can be expressed in terms of volume-based or mass-based independent process parameters. From a product performance point of view, the relevant coating dimension is the **dry thickness** $\mathbf{H_{dry}}$ measured in units of [m], or, more widely used in the converting industry, the mass-based **dry coat weight** $\mathbf{A_{dry}}$ measured in units of [g/m^2] because the performance of the coated layer is defined by these parameters. In contrast, the volume-based **wet film thickness** $\mathbf{H_{wet}}$ measured in units of [m] is the relevant parameter from a coating process point of view. When studying any given coating flow, therefore, the wet thickness of the coated film must be expressed in terms of the dry coat weight, and these two parameters are related to each other by the solids concentration and the liquid density. Details of this correlation are given in Sect. 3.3.

3.2 Flow Rates

As already mentioned in Chap. 2, premetered coating methods satisfy the physical law of mass conservation. This is a big advantage over selfmetered processes because the wet thickness of the coated film is always correct from the very beginning of the

P. M. Schweizer, *Premetered Coating Methods*, Engineering Materials,
https://doi.org/10.1007/978-3-031-04180-8_3

coating, as long as the flow rate pumped from the supply vessel to the coating station is correctly calculated and controlled at that value during a production run.

The volumetric flow rate V^* [m^3/s] or the mass flow rate M^* [kg/s] are important parameters when discussing the amount of liquid that is required for alimenting the coating process. However, when discussing the performance of a coating process, the flow rate is irrelevant. Instead, the relevant parameter is the volumetric flow rate/width Q expressed in units of [m^2/s], or more conveniently in units of [cm^2/s]. The flow fields of successful coating processes are two-dimensional, except for narrow regions along both edges of the coated substrate. Therefore, the coating width must be an irrelevant parameter, particularly when discussing and comparing the performance of different coating methods. This statement is in line with the first law of coating, which states:

Anything can be done to the coating liquid except that, which changes in the cross-web direction.

Note that the volumetric flow rate is used when the flow rate is measured with a flow meter based on the concept of magnetic induction. In contrast, the mass flow rate is used when the flow rate is measured with a Coriolis flow meter.

3.3 Correlations Between Volume-Based and Mass-Based Parameters

The purpose of this section is to summarize the relevant formulas that describe the correlations between the various flow rate and thickness parameters mentioned in Sects. 3.1 and 3.2. Moreover, all these correlations are expressed either in terms of the dry film thickness H_{dry} or the dry coat weight A_{dry}, such that they can readily be used depending on the local language in a given coating company.

It is suggested that an EXCEL file containing all the equations is prepared in such a way that, when providing application-specific input information, all resulting thickness and flow rate parameters are immediately and correctly calculated.

All symbols used in the following equations are defined in the *List of Symbols* at the beginning of the book.

3.3.1 Units

SI units, i.e., meter, kilogram, and second [m, kg, s] are used throughout this book, particularly when dealing with equations. However, many companies in the converting and coating industry are still expressing some of the parameters in non-SI units. The most prominent example is the coating speed, which is often expressed in units of [m/min]. Consequently, this speed unit is used in this book when denoting a speed axis in a diagram. However, when using the coating speed or other parameters in equations in an EXCEL file, input values given in non-SI units are always first transformed into values with SI units to prevent mistakes.

The same principle applies when using abbreviations for powers of 10 to obtain comfortable parameter values on the order of 1. Examples include gram [g], millimeter [mm] or kilopascal [kPa]. Such abbreviations do not change the nature of the units. However, when working with such values in an EXCEL file, it is suggested to first convert these values into true SI units.

3.3.2 Correlations Based on Dry Coat Weight A_{dry}

$$H_{dry} = \frac{A_{dry}}{\rho_{solid}} \tag{3.3.1}$$

$$A_{wet} = \frac{A_{dry}}{C_{W,solid}} \tag{3.3.2}$$

$$H_{wet} = \frac{A_{wet}}{\rho_{liquid}} = \frac{A_{dry}}{C_{W,solid}\rho_{liquid}} \tag{3.3.3}$$

$$Q = H_{wet}U = \frac{A_{dry}U}{C_{W,solid}\rho_{liquid}} \tag{3.3.4}$$

$$M^* = \frac{A_{dry}UW}{C_{W,solid}} \tag{3.3.5}$$

$$M^*_{solvent} = M^*\left(1 - C_{W,solid}\right) = A_{dry}UW\left(\frac{1}{C_{W,solid}} - 1\right) \tag{3.3.6}$$

$$M^*_{solid} = M^*C_{W,solid} = A_{dry}UW \tag{3.3.7}$$

$$V^* = QW = \frac{M^*}{\rho_{liquid}} = \frac{A_{dry}UW}{C_{W,solid}\rho_{liquid}} \tag{3.3.8}$$

3.3.3 Correlations Based on Dry Film Thickness H_{dry}

$$A_{dry} = H_{dry}\rho_{solid} \tag{3.3.9}$$

$$H_{wet} = \frac{H_{dry}\rho_{solid}}{C_{W,solid}\rho_{liquid}} \tag{3.3.10}$$

$$A_{wet} = \frac{H_{dry}\rho_{solid}}{C_{W,solid}} \tag{3.3.11}$$

$$Q = H_{wet}U = \frac{H_{dry}\rho_{solid}U}{C_{W,solid}\rho_{liquid}} \tag{3.3.12}$$

$$M^* = \frac{H_{dry}\rho_{solid}UW}{C_{W,solid}} \tag{3.3.13}$$

$$M^*_{solvent} = H_{dry}\rho_{solid}UW\left(\frac{1}{C_{W,solid}} - 1\right) \tag{3.3.14}$$

$$M^*_{solid} = H_{dry}\rho_{solid}UW \tag{3.3.15}$$

$$V^* = \frac{H_{dry}\rho_{solid}UW}{C_{W,solid}\rho_{liquid}} \tag{3.3.16}$$

3.3.4 *Correlations Based on Porous Film Thickness H_{porous}*

With regard to thickness, wet films are coated in the form of solutions (pure liquids), emulsions (liquid/liquid systems) or suspensions (liquid/solid systems), or perhaps combinations thereof, see Fig. 3.1. When dried, the first two systems tend to give compact solid films, while dried suspensions may result in compact or non-compact,

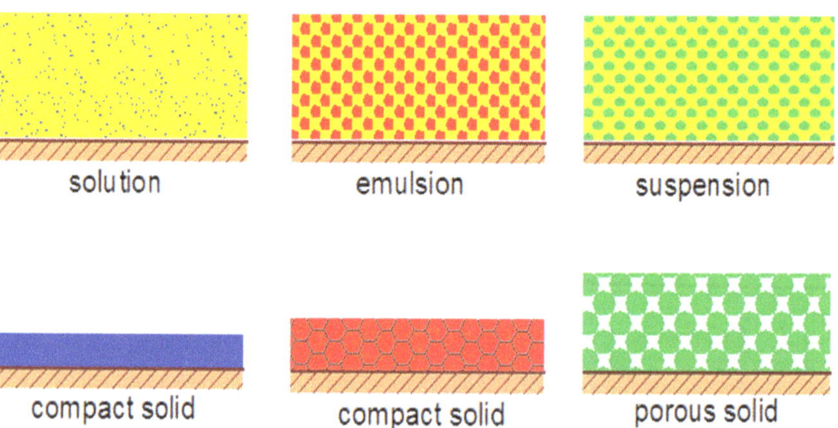

Fig. 3.1 Wet and dry film thickness of various coating fluids

i.e., porous, films. If the solid material is not densely packed as shown in the bottom middle sketch in Fig. 3.1, but if the solid material is porous as shown in the bottom right sketch, then some of the above equations must be modified as follows to account for the porosity of the material. Again, the law of mass conservation is used for deriving the following correlations. Moreover, they only consider the concentration by weight, as this is the concentration of choice in most industrial applications.

The porosity ε is defined as the ratio of void to total volume:

$$\epsilon = \frac{V_{void}}{V_{total}} \tag{3.3.17}$$

Consequently, we can derive the following expressions for the void and the solid volumes:

$$V_{void} = \varepsilon V_{total} \tag{3.3.18}$$

$$V_{solid} = (1 - \varepsilon) V_{total} \tag{3.3.19}$$

Moreover, the density of the porous material is given by

$$\rho_{porous} = \frac{M_{solid}}{V_{total}} = \frac{\rho_{solid} V_{solid}}{V_{total}} = (1 - \varepsilon)\rho_{solid} \tag{3.3.20}$$

and the thickness of the porous layer is given by

$$H_{porous} = \frac{H_{dry}}{(1 - \varepsilon)} \tag{3.3.21}$$

Now, obviously, $H_{porous}\ \rho_{porous} = H_{dry}\ \rho_{solid}$, and therefore Eqs. 3.3.9–3.3.16 must be re-written as follows:

$$A_{dry} = H_{porous} \rho_{porous} \tag{3.3.22}$$

$$H_{wet} = \frac{H_{porous} \rho_{porous}}{C_{W,solid} \rho_{liquid}} \tag{3.3.23}$$

$$A_{wet} = \frac{H_{porous} \rho_{porous}}{C_{W,solid}} \tag{3.3.24}$$

$$Q = H_{wet} U = \frac{H_{porous} \rho_{porous} U}{C_{W,solid} \rho_{liquid}} \tag{3.3.25}$$

$$M^* = \frac{H_{porous} \rho_{porous} U W}{C_{W,solid}} \tag{3.3.26}$$

$$M^*_{solvent} = H_{porous}\rho_{porous}UW\left(\frac{1}{C_{W,solid}} - 1\right) \tag{3.3.27}$$

$$M^*_{solid} = H_{porous}\rho_{porous}UW \tag{3.3.28}$$

$$V^* = \frac{H_{porous}\rho_{porous}UW}{C_{W,solid}\rho_{liquid}} \tag{3.3.29}$$

3.3.5 Operating Requirements for the Flow Rate

As explained in Sect. 2.1 and Eq. 2.1, the specified wet thickness H_{wet} or wet coat weight A_{wet} of the coated film is obtained for all premetered coating methods by calculating the required volumetric or mass flow rate, see Sects. 3.3.2, 3.3.3 and 3.3.4. As shown by the corresponding equations, these flow rates depend on several operating parameters, notably the solids concentration by weight, the liquid density, the solid density, the porous density, and the web speed. Therefore, it is important to emphasize that these parameters must be determined accurately, and they must be kept constant throughout a given coating campaign if the goal is to obtain a precise value of the specified coating thickness.

Information about the solids concentration is given in Sect. 4.2, and information about densities can be found in Sect. 4.3.

3.3.6 Maximum Web Speed Resulting from Drying Limitation

The solvent flow rate per unit dryer area is called the specific solvent flow rate, and it is defined as follows, see Eq. 3.3.6:

$$m^*_{solvent} = \frac{M^*_{solvent}}{WL_{dryer}} = \frac{A_{dry}U}{L_{dryer}}\left(\frac{1}{C_{W,solid}} - 1\right) \tag{3.3.30}$$

L_{dryer} = length of dryer.

$m^*_{solvent}$ is sometimes considered as a material property, which characterizes the maximum drying load to which a given product can be exposed. Therefore, even if more drying capacity is installed on a given coating machine, the maximum specific solvent flow rate $m^*_{solvent,max}$ must not be exceeded to prevent drying-related quality degradations in the coated film.

Table 3.1 shows experience-based examples of maximum specific solvent flow rates for selected aqueous coating fluids.

Table 3.1 Maximum specific solvent flow rates for aqueous coating fluids

Fluid	Substrate	A_{dry} [g/m^2]	$C_{W,solid}$ [–]	$m^*_{solvent,max}$ [kg/h m^2]
PSA	Paper	20	0.60	20
PSA	PET Film	20	0.60	10
Thermal paper	Paper	3	0.30	12

Consequently, an associated **maximum coating speed** can be derived from Eq. 3.3.30.

$$U_{max} = \frac{m^*_{solvent,max} L_{dryer}}{A_{dry}\left(\frac{1}{C_{W,solid}} - 1\right)} \tag{3.3.31}$$

Conversely, an expression for a **minimum dryer length** can be obtained by rearranging Eq. 3.3.31:

$$L_{dryer,min} = \frac{A_{dry} U}{m^*_{solvent,max}}\left(\frac{1}{C_{W,solid}} - 1\right) \tag{3.3.32}$$

Still, conversely, an expression for a **minimum solids concentration** can be derived by re-arranging Eq. 3.3.31:

$$C_{W,solid,min} = \frac{1}{\frac{m^*_{solvent,max} L_{dryer}}{A_{dry} U} + 1} \tag{3.3.33}$$

Using the specific solvent flow rate to formulate a purely evaporation-based drying limitation is a simplification that is helpful for certain applications as explained above. In reality, drying of thin, coated films is a complex process involving simultaneous heat and mass transfer. While the underlying differential equations are known and can be solved numerically, accurately predicting the drying performance for any given applications is often not possible because relevant material properties are not known. Specifically, it is difficult to measure the diffusion coefficient as a function of solvent concentration and temperature. Mamaliga et al. (2004), Schabel et al. (2007), and Siebel et al. (2015) have developed suitable measurement methods. In practice, however, suitable material data of this kind are often not available to industrial coating companies.

Describing in detail the drying process for thin films goes beyond the scope of this book. Perez and Carvalho (2007) as well as Scharfer (2019) show how elaborate numerical methods can be used for simulating the performance of industrial multi-zone dryers.

References

Mamaliga, I., Schabel, W., & Kind, M. (2004). Measurements of sorption isotherms and diffusion coefficients by means of a magnetic suspension balance. *Chemical Engineering and Processing—Process Intensification, 43*(6), 753–763.

Perez, E. B., & Carvalho, M. S. (2007). Drying of thin films of polymer solutions coated over impermeable substrates. *Heat Transfer Engineering, 28*(6), 559–566.

Schabel, W., Scharfer, P., Kind, M., & Mamaliga, I. (2007). Sorption and diffusion measurements in ternary polymer–solvent–solvent systems by means of a magnetic suspension balance—Experimental methods and correlations with a modified Flory-Huggins and free-volume theory. *Chemical Engineering Science, 62*(8), 2254–2266.

Scharfer, P. (2019). Simulation and design of industrial thin film dryers. In *11th short course in coating and drying of thin films*. Karlsruhe Institute of Technology. May 13–15, 2019.

Siebel, D., Scharfer, P., & Schabel, W. (2015). Direct determination of diffusion coefficients in polymer solvent systems from Raman spectroscopic data. In *Book of Abstracts, 11th European Coating Symposium*, Eindhoven, NL, September 9–11, 2015.

Chapter 4
Physical Fluid Properties

Abstract Coating involves fluid mechanics, and understanding coating requires understanding the various basic flow fields that may exist in a given coating method. The performance of a given flow depends on three classes of parameters, namely operating conditions such as flow rate, physical fluid properties such as viscosity, and geometric dimensions such as pipe diameter. This chapter discusses physical fluid properties that are relevant for coating flows, including density, viscosity, surface tension, contact angle, wettability of substrates, as well as their dependence upon the solids concentration and temperature. Regarding the viscosity, equations for modeling the dependence upon the shear rate are presented. Finally, methods and techniques for measuring the fluid properties are reviewed.

4.1 Introduction

Understanding physical fluid properties is important because they influence the behavior of the basic flows that are encountered in premetered coating processes. All fluid properties depend on the temperature, some of them strongly, others only weakly. Fluid properties also depend on the solids concentration, if the coating fluid contains one or several solvents. Temperature and solids concentration, therefore, are interesting parameters because they allow us to alter physical fluid properties and hence influence the behavior of premetered coating processes without changing the chemical composition of the fluid. Often, therefore, the solids concentration is controlled by the process engineer and not by the formulation chemist. Moreover, the solids concentration, i.e., the amount of solvent in the coating fluid, determines temporary material properties that can be changed to optimize the coating process. After solidification, i.e., after evaporation of the solvent, the final material properties of the coated film only depend on the chemical composition.

Care must be taken when altering the solids concentration because it will change the requirements for the subsequent drying process.

Before discussing physical fluid properties in more detail, particularly as they relate to premetered coating methods, it is suitable to clarify issues related to the solids concentration.

P. M. Schweizer, *Premetered Coating Methods*, Engineering Materials,
https://doi.org/10.1007/978-3-031-04180-8_4

4.2 Solids Concentration

Consider a unit volume of fluid as shown in Fig. 4.1, and assume that this fluid consists of a solvent and a solid material. Generally speaking, the solvent portion of this unit volume may consist of several different solvents, and the solid portion may consist of several different solids.

4.2.1 Solids Concentration by Volume

The **concentration by volume** of any of the solid or liquid components i is given by Eq. 4.2.1:

$$C_{V,i} = \frac{V_i}{\sum_{i=1}^{k} V_i} = \frac{M_i/\rho_i}{\sum_{i=1}^{k} M_i/\rho_i} \tag{4.2.1}$$

In the example of Fig. 4.1, k = 2, $C_{V,1} = C_{V,solvent} = 0.7/(0.7 + 0.3) = 0.7$, and $C_{V,2} = C_{V,solid} = 0.3/(0.7 + 0.3) = 0.3$.

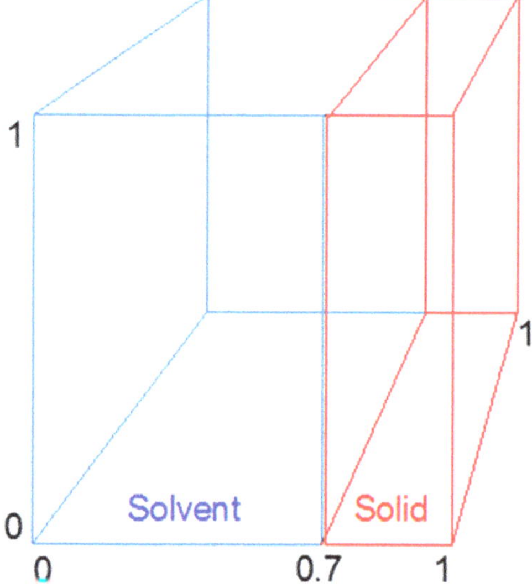

Fig. 4.1 Cubic unit volume consisting of 70% solvent and 30% solid

4.2.2 Solids Concentration by Weight

In contrast, the **concentration by weight**, or the **concentration by mass**, of any of the solid or liquid components i is given by Eq. 4.2.2:

$$C_{W,i} = \frac{M_i}{\sum_{i=1}^{k} M_i} = \frac{\rho_i V_i}{\sum_{i=1}^{k} \rho_i V_i} = \frac{\rho_i C_{V,i}}{\sum_{i=1}^{k} \rho_i C_{V,i}} \tag{4.2.2}$$

Assuming that the density of the solvent in Fig. 4.1 is 900 kg/m^3, and the density of the solid is 1600 kg/m^3, then k = 2, $C_{W,1} = C_{W,solvent} = 0.7 * 900/(0.7 * 900 + 0.3 * 1'600) = 0.568$, and $C_{W,2} = C_{W,solid} = 0.3 * 1'600/(0.7 * 900 + 0.3 * 1'600) = 0.432$.

Note that regardless of whether the concentration is measured by volume or weight, the sum of the concentrations of all components of a given fluid always equals unity according to Eq. 4.2.3.

$$\sum_{i=1}^{k} C_{V,i} = \sum_{i=1}^{k} C_{W,i} = 1 \tag{4.2.3}$$

For a fluid consisting of two components, i.e., solvent and solid matter, the conversion between the solids concentration by weight and the solids concentration by volume are given by Eqs. 4.2.4 and 4.2.5, respectively.

$$C_{V,solid} = \frac{1}{1 + \frac{\rho_{solid}}{\rho_{solvent}}\left(\frac{1}{C_{W,solid}} - 1\right)} \tag{4.2.4}$$

$$C_{W,solid} = \frac{1}{1 + \frac{\rho_{solvent}}{\rho_{solid}}\left(\frac{1}{C_{V,solid}} - 1\right)} \tag{4.2.5}$$

Also note that, strictly speaking, Eqs. 4.2.1–4.2.5 only apply in the absence of excess volume of mixing. Excess volume is a property of a mixture, which quantifies the non-ideal behavior of real mixtures. It is defined as the difference between the value of the volume in a real mixture and the value that would exist in an ideal mixture under the same conditions (Wikipedia, 2022).

As a general guideline, the solids concentration should be as high as possible to minimize the operating and investment costs of the drying process, and as low as necessary to avoid operational problems of the fluid delivery and coating processes such as degassing or air entrainment, which become more cumbersome as the viscosity increases.

Sometimes, particularly during the development of new product formulations, molar concentrations are used to quantify the amounts of the various chemical components. In our experience, however, the solids concentration by weight is the preferred property in an industrial coating plant because weighting is an easy and readily available process, either by using a table-top balance for small amounts, or by mounting larger vessels onto load cells.

4.2.3 Measuring the Solids Concentration by Weight

In industrial or pilot coating environments, it is often convenient to measure the solids concentration by weight. Measuring this parameter also increases the confidence that the calculated value is correct and that no mistakes were made during the preparation of the coating fluid.

Measuring the concentration must be carried out with great care to avoid unwanted effects from the environment, such as evaporation of solvent from the sample or resorption back into the probe. The following operating procedure guarantees reliable results.

1. Take two small, round aluminum trays, stack them together and weigh them; result: weight G_1.
2. Separate the trays.
3. Use a spoon and transfer a small amount of coating fluid into the lower tray.
4. Place the upper tray onto the lower tray and bend both tray edges downward; the filling and sealing of the trays must be carried out quickly to prevent evaporation of solvent, particularly if the coating fluid contains low-boiling point solvents.
5. Weigh both trays and the liquid; result: weight G_2.
6. Separate the trays and place them in the oven; for aqueous fluids: oven temperature = 120 °C, oven time = 10 min.
7. Remove both trays from the oven, quickly stack them together and bend both tray edges downward to prevent that the dried probe resorbs moisture from the environment.
8. Wait a few seconds until the aluminum trays have cooled down.
9. Weigh both trays with the solid material; result: weight G_3.
10. Calculate the solids concentration according to

$$C_{w,solid} = \frac{G_3 - G_1}{G_2 - G_1} \tag{4.2.6}$$

Example: $G_1 = 4.1866$ g, $G_2 = 4.6730$ g, $G_3 = 4.3108$ g; $C_{w,solid} = 0.255$.

4.3 Densities

The fluid or liquid density is a physical material property, which defines the ratio of mass to volume of a given quantity of fluid:

$$\rho_{liquid} = \frac{M_{liquid}}{V_{liquid}} \tag{4.3.1}$$

Consequently, the density of any given coating fluid can easily be determined experimentally by measuring (i.e., weighing) the mass of a known volume of fluid. A graded 100 ml glass cylinder works well for this purpose.

Density values of most water-based coating fluids are in the range of 1′000–1′300 kg/m^3, while densities of solvent-based fluids are lower, i.e., in the range of 800–1′100 kg/m^3. From a coating process point of view, the density can be changed by altering the solids concentration. However, owing to the rather narrow range of density values, changing the density typically has little effect on the performance of the coating process. Moreover, the effect of temperature on the density is usually negligible as is exemplified by density values of water at ambient pressure, which decreases from 998.4 kg/m^3 at 20 °C to 983.2 kg/m^3 at 60 °C, i.e., by 1.5% (VDI-Wärmeatlas, 1974).

Alternatively, for a liquid containing solvent and solid matter, either in dissolved (solution) or dispersed (suspension or emulsion) form, the liquid density can be calculated based on the density values of the solvent(s) and the solid(s). The appropriate equations to be used for calculating the liquid density depend on whether the solids concentration is measured by volume or by weight (or mass), see above.

4.3.1 Liquid Density Based on Concentration by Volume

According to Eq. 4.3.1 and Fig. 4.1, the liquid density of a 2-component system consisting of a solvent and a solid material is equal to

$$\rho_{liquid} = \frac{M_{liquid}}{V_{liquid}} = \frac{\rho_{solvent} V_{solvent} + \rho_{solid} V_{solid}}{V_{solvent} + V_{solid}} \tag{4.3.2}$$

with $V_{solvent} = C_{V,solvent}(V_{solvent} + V_{solid})$, and $V_{solid} = C_{V,solid}(V_{solvent} + V_{solid})$. Therefore, the density of a fluid consisting of k chemical components can be calculated according to Eq. 4.3.3:

$$\rho_{liquid} = \sum_{i=1}^{k} \rho_i C_{V,i} \tag{4.3.3}$$

4.3.2 Liquid Density Based on Concentration by Mass

Alternatively, but also according to Eq. 4.3.1 and Fig. 4.1, the liquid density of a 2-component system consisting of a solvent and a solid material is equal to

$$\rho_{liquid} = \frac{M_{solvent} + M_{solid}}{\frac{M_{solvent}}{\rho_{solvent}} + \frac{M_{solid}}{\rho_{solid}}} \tag{4.3.4}$$

with $M_{solvent} = C_{M,solvent}(M_{solvent} + M_{solid})$, and $M_{solid} = C_{M,solid}(M_{solvent} + M_{solid})$. Therefore, the density of a fluid consisting of k chemical components can be calculated according to Eq. 4.3.5:

$$\rho_{liquid} = \frac{1}{\sum_{i=1}^{k} \frac{C_{W,i}}{\rho_i}} \tag{4.3.5}$$

For the various concentration values and the values of the solvent and solid density given in the example in Sect. 4.2, Eqs. 4.3.3 and 4.3.5 both yield a liquid density value of 1′110.0 kg/m^3.

4.3.3 Solid Density

So far, the liquid density was calculated based on the solvent and solid densities of a given fluid. In practice, however, the density values of the solid components of a given fluid are often not known, and it is often difficult to measure such properties. In contrast, it is easy to measure the liquid density with the help of a pycnometer, or with a graded glass cylinder as explained above. In addition, the density of many solvents can be found in readily available databases. Therefore, it is often convenient to calculate the density of the solid component as a function of the solvent and liquid densities. For a 2-component liquid consisting of a solvent and a solid material, the density of the solid material can be calculated with Eqs. 4.3.6 or 4.3.7, depending on whether the concentration by volume or by weight is known:

$$\rho_{solid} = \rho_{solvent} + \frac{(\rho_{liquid} - \rho_{solvent})}{C_{V,solid}} \tag{4.3.6}$$

$$\rho_{solid} = \frac{1}{\frac{1}{C_{W,solid}}\left(\frac{1}{\rho_{liquid}} - \frac{1}{\rho_{solvent}}\right) + \frac{1}{\rho_{solvent}}} \tag{4.3.7}$$

Using the density and concentration values from the example above, Eqs. 4.3.6 and 4.3.7 both yield a solid density value of 1′600 kg/m^3.

4.4 Rheological Properties

Rheology is the science that investigates how materials, mostly fluids, flow and deform when stressed. Many books and papers have been written about this topic, and suitable examples include Bird et al. (1977), Macosko (1994), or Glass and Prud'homme (1997).

The rheological properties of a coating fluid are material properties that depend on the chemical composition, the structure of the molecules, the temperature, the solids concentration, and most importantly, the type of deformation that is imposed on the fluid while it is flowing through the coating process. Larson (1999) provides a comprehensive review of how rheological material properties are affected by the structure of the material. Moreover, rheological properties cannot be predicted accurately by theoretical models and hence must be measured with proper devices and across an appropriate range of deformation rates.

All premetered coating methods comprise several characteristic flow fields that ultimately govern the quality and uniformity of the coated film. Examples include pipe flow in fluid conditioning and delivery systems (1), flow in distribution cavities (2) and narrow slots (3) inside slot-type and slide-type dies, film flow on an inclined plane of slide-type dies, flow in falling liquid curtain (5), and flow inside a boundary layer associated with dynamic wetting (6), see Fig. 4.2. The flow field in the bead of a slide coater (7), i.e., between the die lip and the moving substrate, is similar to the flow field of an impinging curtain (6) because it is also characterized by a boundary layer. The big difference is the magnitude of the velocity of the impinging liquid stream, i.e., curtain or film flow. The flow field in the bead of a slot coater (8) is more complex because the boundary layer flow is embedded in a slot flow, which in itself is a combination of pressure-driven Hagen-Poiseuille flow and Couette flow.

Pipe-, cavity-, slot- and film flows are shear flows and hence the **shear viscosity** is the relevant rheological property, at least for purely viscous fluids. As is explained in more detail in Chap. 5, each characteristic flow field has its characteristic range of shear rates. Consequently, any given fluid flowing through a coating process such as the one shown in Fig. 4.2a changes its viscosity in response to the shear rate present at a given location.

In contrast, and apart from short transition zones between film or slot flow and curtain flow at the beginning of the curtain as well as between curtain flow and boundary layer flow at the end of the curtain, uniform curtain flow is a plug flow, i.e., it is a shear-free flow. Therefore, the concept of shear viscosity does not apply. Instead, curtain flow is an extensional flow and hence the **extensional** or **elongational viscosity** is the relevant rheological property. The extensional viscosity depends on the extension rate, see Sect. 5.8.2. In addition, the ratio of elongational to shear viscosity is called Trouton ratio (Bird et al., 1977; Macosko, 1994). For polymer melts at low strain rates ($<1\ s^{-1}$), the value of the Trouton ratio is 3, meaning that the extensional viscosity is three times as high as the shear viscosity for such flow conditions. Note, however, that these considerations do not apply to the curtain flow. On the one hand, Eq. 5.8.5 shows that the extension rate for most industrial curtains

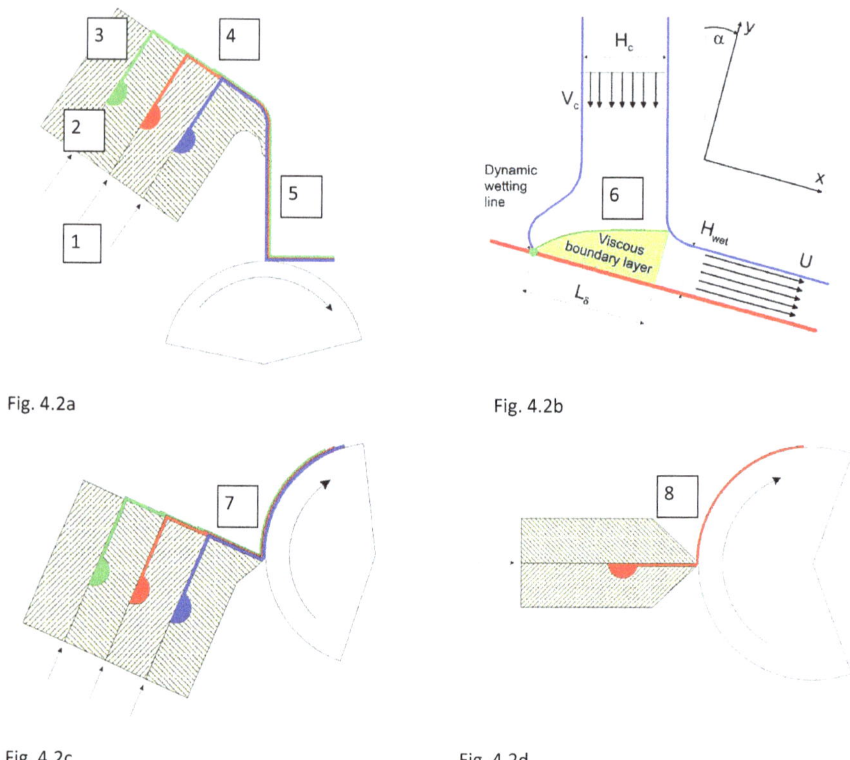

Fig. 4.2 Characteristic flow fields of premetered coating flows

is in the range of 5–100 s^{-1}. On the other hand, viscosity measurements of polymer melts indicate that the elongational viscosity remains constant with increasing extension rate, while the shear viscosity drops with increasing shear rate. As a result, the Trouton ratio increases to values of >3 with increasing strain rate ratio (Bird et al., 1977; Macosko, 1994).

Finally, boundary layer flow associated with dynamic wetting is a combination of shear and extensional flow and thus more complex. However, when experimentally observing the behavior of an impinging curtain (see Fig. 4.2b) and comparing it with theoretical predictions based on a viscous boundary layer model, the agreement is rather good, at least for low solids content polymeric solutions and dispersion such aqueous gelatin-based photographic formulations, fairly concentrated aqueous acrylic emulsions such as pressure-sensitive adhesives, and others. For many industrial coating fluids, therefore, it seems that the shear viscosity is the relevant rheological property for modeling the behavior of the characteristic flow fields of premetered coating processes.

Macosko (1994, page 109) states that viscoelasticity, i.e., a time-dependent rheological response to deformation, is typical of all polymeric materials. This is a disturbing situation because it applies to almost all industrial coating fluids. On the one hand, viscoelastic material properties are more difficult to measure than purely viscous properties such as the shear viscosity. On the other hand, modeling of viscoelastic flow behavior requires high-level mathematical tools and cannot be accomplished with simple models. Fortunately, however, and as stated above, investigating the behavior of coating processes with purely viscous models is often reasonably successful on an engineering level when compared with experimental results. This then indicates that the level of viscoelasticity of many industrial coating fluids is not very high and thus can be neglected. Based on these considerations the focus in this book is on modeling coating flows with purely viscous models and the shear viscosity.

4.4.1 *Shear Viscosity*

Experience shows that, in many coating companies across the world, the viscosity is expressed in units of seconds. This approach may be suitable for monitoring quality aspects in production, for example, if the viscosity of the next batch is the same as it was for the previous one. Using a time unit for the viscosity implies that it was determined with a viscometer, which measured some sort of characteristic time, such as the efflux time from a normalized cup with a normalized orifice, for example, a 4 mm DIN cup, or an inclined tube filled with coating fluid, where a solid ball is rolling past two markers. Moreover, viscosity measured with simple devices such as those described above implies that the viscosity was measured for unknown but low shear rates.

While this approach is simple and convenient, it is not suitable for discussing the features and the performance of various coating flows and processes on an engineering or even scientific level. Firstly, the correct SI unit for the viscosity is Pascal second [Pas = Ns/m^2 = kg/ms]. Secondly, most industrial coating fluids are characterized by a shear rate-dependent viscosity. In addition, the characteristic flow fields encountered in premetered coating methods cover a wide range of shear rates, extending over several decades, see Chap. 5 and Schweizer (2016).

Consequently, shear-dependent viscosity data and knowledge of the characteristic shear rate of the flow field under consideration are required. Discussing the influence of the viscosity on the performance of a coating process now becomes a more cumbersome and complicated issue. Hence, the purpose of this chapter is to revisit well-known information, which explains how to better handle viscosity when it is related to coating flows.

Experience confirms that most industrial coating fluids show similar rheological behavior, i.e., for low and high shear rates the viscosity remains constant (Newtonian behavior), while it decreases with increasing shear rate for an intermediate range of shear rates (shear thinning behavior). Figures 4.3 and 4.4 present exemplary viscosity data of three different fluids, namely PSA, a 60% pressure-sensitive adhesive at 22 °C; PVA, a 9% aqueous polyvinyl alcohol solution at 25 °C; and LIB, a 40% anode slurry

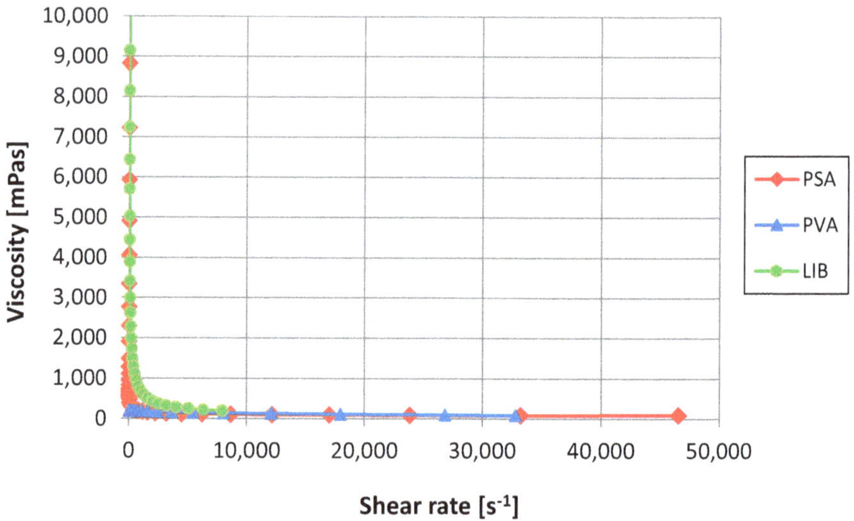

Fig. 4.3 Linear plot of shear viscosity of three different fluids

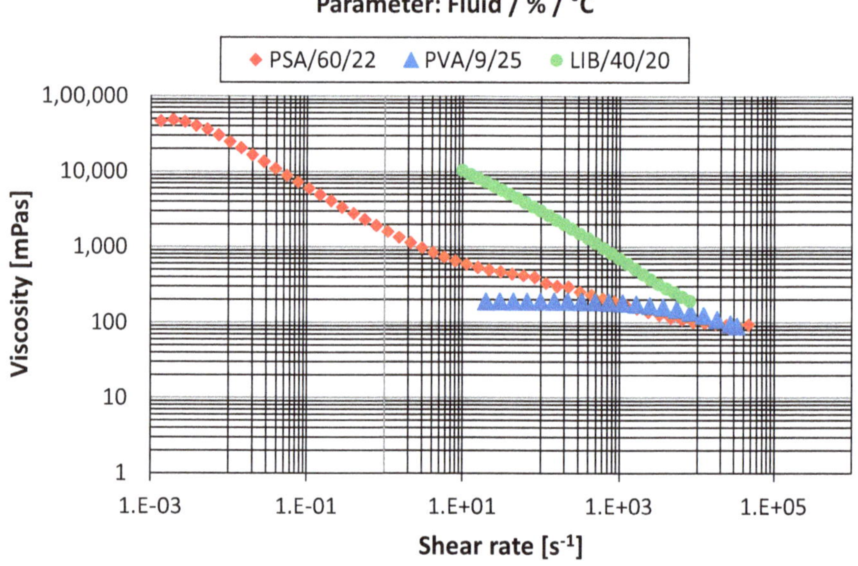

Fig. 4.4 Double-logarithmic plot of shear viscosity of three different fluids

for a Li-ion battery at 20 °C. When the data are plotted in a diagram with both the shear rate and the viscosity axes having a linear scale as in Fig. 4.3, then not much difference between the three fluids can be detected. However, when the same data are plotted in a double-logarithmic diagram, then the characteristic rheological behavior of the three fluids as well as their rheological differences can be visualized as shown in Fig. 4.4.

In particular, the PSA data covers a wide enough shear rate range, such that all three viscosity ranges can be seen. The PVA data only reveals the upper Newtonian plateau and the shear thinning range, while the LIB data only reveals the shear thinning behavior. Obviously, the shear rate range for measuring the PVA and LIB viscosities could have and should have been extended to lower and higher values to obtain information about the upper and lower Newtonian plateau. Note, however, that it is not acceptable to resort to extrapolation from viscosity data measured in a limited range of shear rates to obtain viscosity information at shear rates below and above the measured range. One example is taken from Willenbacher (2019) and shown in Fig. 4.5, where the viscosity η of certain dispersions does not remain constant beyond the shear-thinning range but increases again for sufficiently high shear rates. This behavior is called shear thickening.

Materials characterized by a yield stress are another example. They are referred to as *plastic materials*, and they show no deformation up to a certain level of stress. Beyond this yield stress, the material flows readily with a Newtonian or shear-thinning behavior (Macosko, 1994). Examples of plastic materials include slurries for electrodes of Li ion batteries or food substances like mayonnaise and ketchup.

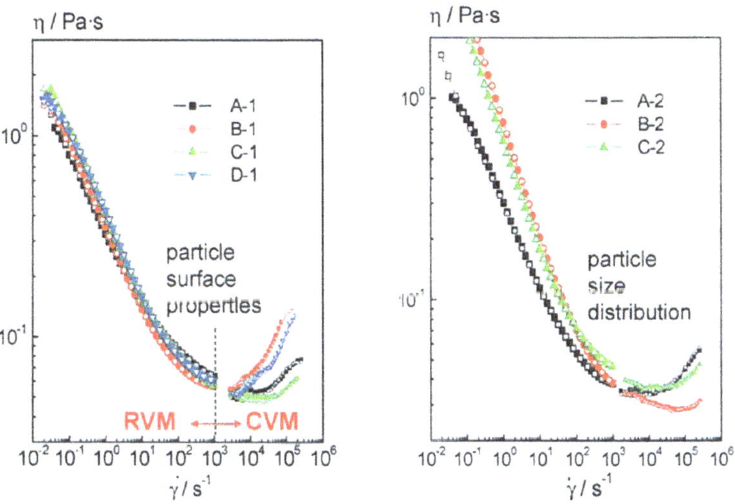

Fig. 4.5 High shear viscosity of dispersion, reproduced with permission from Willenbacher (2019)

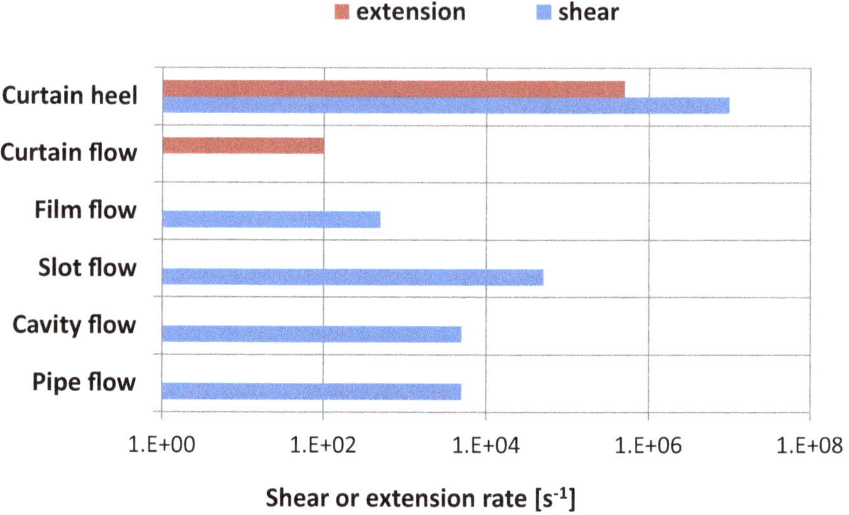

Fig. 4.6 Shear rate ranges of characteristic coating flow fields

4.4.2 Shear Rate Ranges of Coating Flows

Figure 4.6 shows the shear and extension rate ranges of the characteristic flow fields that are encountered in premetered coating methods. Both ranges are very wide, i.e., extending over several decades up to values of 10^6 s^{-1} and beyond for high-speed curtain coating. This presents a big challenge for rheological measurement devices. In particular, at least two different devices with different measurement concepts are required for covering such a wide shear rate range. The challenge is even bigger for measuring high extension rates.

Useful formulas for calculating shear and extension rates of coating flows are given in Chap. 5.

4.4.3 Measurement Devices

Acquiring viscosity data as a function of the shear rate such as shown in Fig. 4.4 requires sophisticated and costly rheological equipment. Rotational devices and capillary viscometers are necessary. Rotational devices are torque-limited, meaning that highly viscous fluids cannot be measured at high shear rates. Therefore, several viscometers with different sensors may have to be used, or they may have to be combined with capillary devices to cover a wide range of shear rates. Malvern Instruments (www.malvern.com), Anton Paar (www.anton-paar.com), and Brookfield (www.brookfieldengineering.com) are good examples of equipment manufacturers. Extensional material properties can be measured with the capillary breakup extensional rheometer (Caber), which can be found, for example, at ThermoFischer Scientific (Haake) (www.thermofisher.com). Simple but widely used devices such as the DIN or Ford cup, the rolling ball viscometer, or gravity-driven flow in a capillary glass tube are not suitable because the shear rate cannot be varied, or at best only in a limited range at low shear rate values. Instead of buying rheological equipment, it may be possible to obtain good viscosity data from the supplier of the coating fluid, or it may be possible to contract a nearby university for carrying out the measurements. It is important to remember that shear-rate dependent viscosity data must be taken for every relevant temperature and solids concentration of the coating fluid.

4.4.4 Rheological Models

Most viscosity data of industrial coating fluids show the same characteristic behavior regarding the dependency upon the shear rate. Hence, they can easily be modeled with fitting mathematical equations. Many such rheological models, or constitutive equations, have been proposed in the literature, see for example Bird et al. (1977) or Macosko (1994) for an overview. One equation that is quite performant in representing the shear-thinning viscosity behavior of many coating fluids is the 5-parameter **Carreau-Yasuda** model according to Eq. 4.4.1, see Macosko (1994, page 86).

$$\frac{\mu - \mu_\infty}{\mu_0 - \mu_\infty} = \left[1 + \left(\frac{\gamma}{\gamma_c}\right)^a\right]^{\frac{b-1}{a}} \tag{4.4.1}$$

Here, μ is the viscosity, μ_0 is the constant (Newtonian) viscosity at low shear rates, μ_∞ is the constant (Newtonian) viscosity at high shear rates, γ is the shear rate, γ_c is the critical shear rate, above which the viscosity starts to decrease with increasing shear rate, a is the Yasuda parameter, which describes the curvature of the transition between the upper Newtonian plateau and the shear thinning region, and b is a measure for the slope of the shear-thinning region.

The 5 parameters of the Carreau-Yasuda model, i.e., μ_0, μ_∞, γ_c, a and b, are unique material properties that depend on the solids concentration and the temperature. The Carreau-Yasuda parameters can easily be determined from viscosity measurements by a trial-and-error approach in an EXCEL environment, i.e., by varying the parameter values until the viscosity data and the model line match well. A visual agreement between the viscosity data and the Carreau-Yasuda equation is sufficient for engineering applications. Figure 4.7 shows shear-dependent viscosity data (symbols) and the corresponding approximation with the Carreau-Yasuda model (solid lines) for four arbitrarily chosen coating fluids. The resulting Carreau-Yasuda parameters are listed in Table 4.1.

As is revealed in Table 4.1 all Carreau-Yasuda parameters vary considerably from one fluid to the other, some of them over several decades. For shear-thinning fluids,

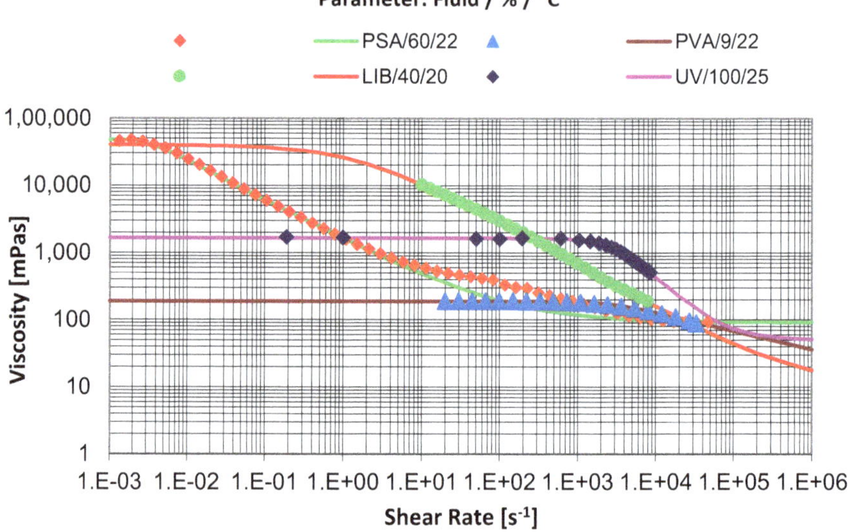

Fig. 4.7 Shear rate-dependent viscosity data of 4 fluids and approximation of data with Carreau-Yasuda model

Table 4.1 Carreau-Yasuda parameters for fluids in Fig. 4.7

Fluid	μ_0 [mPas]	μ_∞ [mPas]	γ_c [s^{-1}]	a [−]	b [−]
PSA	45′000	75	0.0025	5.0	0.45
PVA	188	1	2′600	2.0	0.72
LIB	40′000	10	1.5	0.8	0.37
UV	1′650	50	3′300	2.0	−0.23

the parameter b is <1, and many industrial coating fluids have b-values in the range of 0.2–0.9. The UV fluid, i.e., a UV-curing and hence solventless material, is a special fluid characterized by a very strong shear-thinning behavior such that the parameter b has a negative value of −0.23. This behavior is rare but real. It was observed only 3 times during 15 years of running coating trials on the Polytype Converting pilot facility for many different applications across the entire converting industry.

Note that μ_0, μ_∞, γ_c, and a had to be assumed for the LiB fluid due to lack of data. The same is true for the parameter μ_∞ of the PVA and UV fluids. This assumption would result in uncertainty of the predicted performance if, for example, these Carreau-Yasuda parameters would be used for designing the internal geometry of a slot die. Moreover, the match between the data and the Carreau-Yasuda curve for the PSA fluid is quite good, except for the shear rate range between 10 and 10′000 s^{-1}. Re-measuring the viscosity would indicate if this deviation is an artifact, or if it is a true representation of the material property. In the latter case, and from an engineering point of view, the Carreau-Yasuda model still is a viable tool because finding an alternative mathematical equation that could accurately mimic the somewhat unusual viscosity behavior in that shear rate range would be too difficult if not impossible.

The beauty of the Carreau-Yasuda model is its capability of well-modeling the shear rate-dependent viscosity behavior of many coating fluids. The drawback on the other hand is the complexity of the model Eq. 4.4.1. In particular, using this model in combination with simple equations that describe the characteristic flow fields of premetered coating processes is difficult. As will be shown in Chap. 5, using the mathematically much simpler **power law** model might be a way for overcoming this discrepancy. According to the power law, the viscosity depends on the shear rate as shown in Eq. 4.4.2, see for example Macosko (1994, page 85).

$$\mu = m\gamma^{n-1} \tag{4.4.2}$$

Here, n is the dimensionless power law index, and m is a material constant called consistency with unit [Pa s^n]. Both n and m depend on the solids concentration and the temperature. Bird et al. (1977) explains that the physical significance of the power law index n and the parameter b in the Carreau-Yasuda equation is identical, namely (n − 1) or (b − 1) is the slope of the straight line, i.e., the power law or shear-thinning region, when both the viscosity and the shear rate are plotted logarithmically, and when the power law line is placed tangent to the Carreau curve at its inflection point.

Figure 4.8 shows how well the Carreau-Yasuda model and the power law model match up with measured viscosity data, here for the example of the PSA fluid. Obviously, the Carreau-Yasuda model is the winner of this contest. In the shear thinning region the power law model matches the data reasonably well. However,

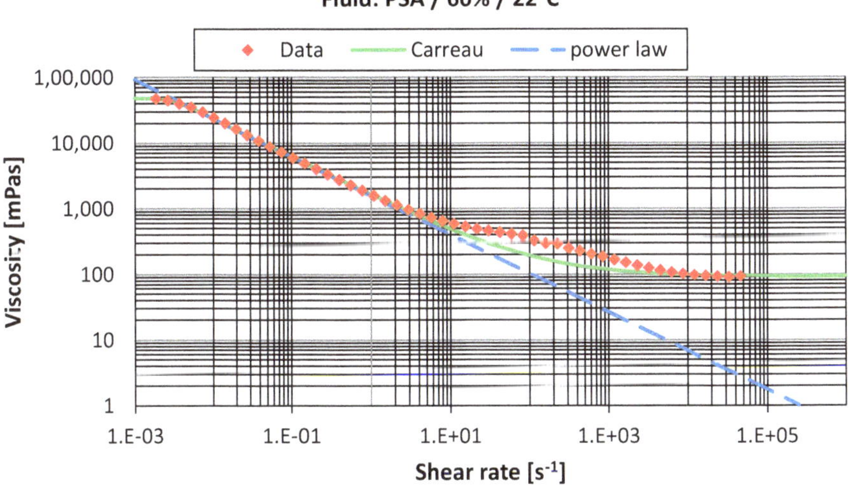

Fig. 4.8 Shear rate-dependent viscosity data of the PSA fluid (red symbols); approximation of data with Carreau-Yasuda model (solid green line) and power law model (dashed blue line)

viscosity predictions with the power law model for both the low shear and high shear Newtonian plateau are completely off and cannot be used for modeling the characteristic flow fields of coating processes.

The accuracy of viscosity predictions with the power law model is only adequate for the shear rate range of $\gamma_c < \gamma < \gamma_{uc}$. γ_{uc} is the upper critical shear rate, which marks the point where the power law viscosity equals the high shear Newtonian viscosity μ_∞, and it can be calculated according to:

$$\gamma_{uc} = \left(\frac{\mu_\infty}{m}\right)^{\frac{1}{n-1}} \tag{4.4.3}$$

One possibility for improving viscosity predictions is to use the **truncated power law** model (Bird et al., 1977; Ruschak, 2008). The key here is that the viscosity is constant at μ_0 for $\gamma < \gamma_c$, that it is calculated with the power law model according to Eq. 4.4.2 for $\gamma_c < \gamma < \gamma_{uc}$, and that it is constant again at μ_∞ for $\gamma > \gamma_{uc}$. However, Ruschak and Weinstein (2014) argued that viscosity predictions with the truncated power law model are inaccurate if the shear rate of the flow field under consideration is near the critical or upper critical shear rate of the "true" viscosity curve, which is described by the Carreau-Yasuda model. To overcome this disadvantage, Ruschak and Weinstein proposed a **local power law** model (LPL). From an engineering point of view, this approach is convincing and optimal because it combines a realistic approximation of the shear rate-dependent viscosity (Carreau-Yasuda model) over the entire relevant shear rate range with the power law model, which is efficient and manageable (i.e., low computational load) for one-dimensional flows such as pipe, cavity and slot flows that are essential, for example, for the design of the internal geometry of slot dies.

In particular, and as explained in detail by Ruschak and Weinstein (2014), constructing a local power law (LPL) model involves knowing the characteristic shear rate, i.e., typically the wall shear rate, of the flow under consideration, calculating the corresponding viscosity with the Carreau-Yasuda Eq. 4.4.1, calculating the local power law index n with Eq. 4.4.4, and calculating the local consistency parameter m with Eq. 4.4.2.

$$n = 1 - (1 - b)\left(1 - \frac{\mu_\infty}{\mu}\right)\left[1 - \left(\frac{\mu - \mu_\infty}{\mu_0 - \mu_\infty}\right)^{a/(1-b)}\right] \tag{4.4.4}$$

The following example illustrates how the power law index and the consistency change as a function of the shear rate if the local power law model is used. Specifically, using the viscosity data and the Carreau-Yasuda parameters of the PSA fluid given in Table 4.1 and assuming shear rate values in the upper transition zone, at the inflection point of the Carreau-Yasuda curve, and in the lower transition zone, results in local power law parameters as shown in Table 4.2 and as depicted in Fig. 4.9.

Figure 4.10 shows how the local power law index changes as a function of the shear rate and as a function of the viscosity, which is described by the Carreau-Yasuda

Table 4.2 Local power law parameters for PSA fluid and assumed shear rates

Shear rate zone	Assumed shear rate $[s^{-1}]$	Local power law index [−]	Local consistency [Pa s^n]
Upper transition	0.003	0.71	8′417.9
Inflection point of Carreau curve	0.022	0.41	1′565.3
Lower transition	500	0.82	398.6

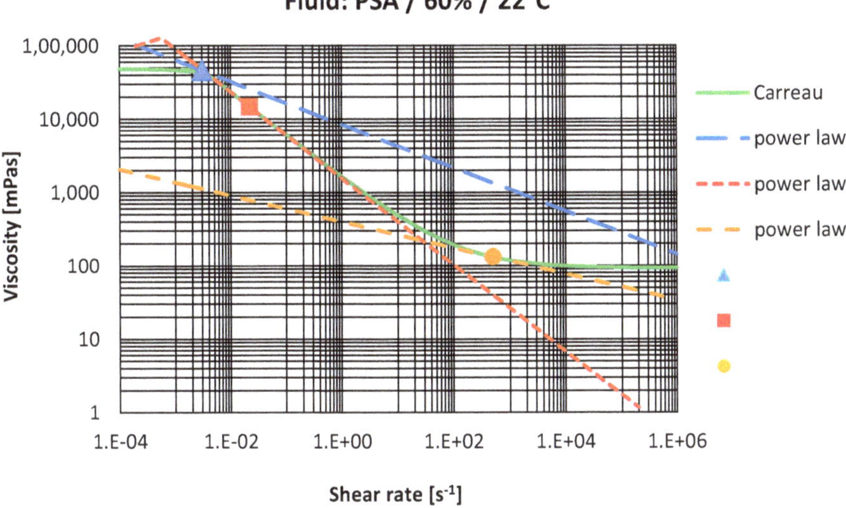

Fig. 4.9 Carreau-Yasuda model and 3 local power law models for PSA fluid

model. The viscosity data are for the PSA fluid. This graph also nicely shows that the local power law model is a smooth and continuous function that covers the low shear, the shear thinning and the high shear range for the viscosity.

Beyond the Carreau-Yasuda and power law models, other rheological models might be useful for characterizing coating fluids. One example is the **truncated Carreau-Yasuda** model, which may be applied if the high shear Newtonian plateau is not known due to a lack of viscosity data at sufficiently high shear rates. In this case, the high shear rate viscosity μ_∞ is set equal to zero and Eq. 4.4.1 mutates to

$$\mu = \mu_0 \left[1 + \left(\frac{\gamma}{\gamma_c} \right)^a \right]^{\frac{b-1}{a}} \tag{4.4.5}$$

Some coating fluids such as slurries for making electrodes of Li-ion batteries may exhibit a material property called yield stress τ_Y, i.e., such materials behave like a solid plastic as long as the applied stress is below the yield stress. However,

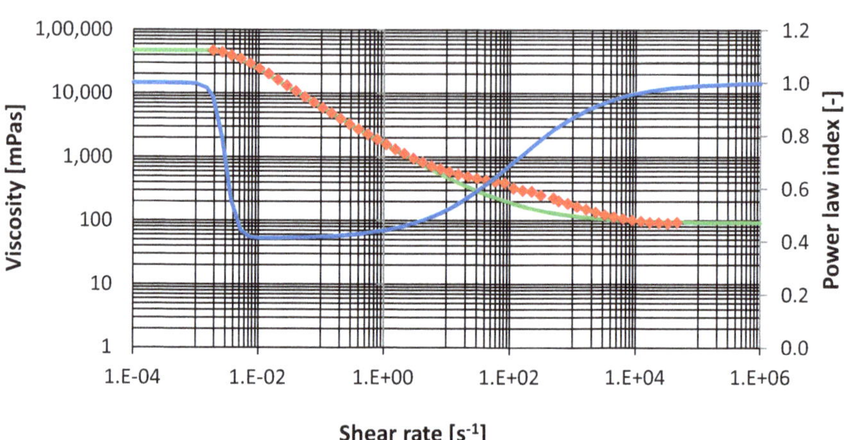

Fig. 4.10 Viscosity and local power law index as a function of the shear rate

for shear stresses above the yield stress, such materials behave like a liquid and flow readily (Bird et al., 1977; Macosko, 1994). In other words, for shear stresses $<\tau_Y$ the viscosity is infinite, but for stresses $>\tau_Y$ the viscosity becomes finite. This situation can be approximated with the **Bingham-Carreau-Yasuda** model (Nelson et al., 1990; Secor, 1997), where the viscosity increases sharply upon decreasing the shear rate beyond a critical value near γ_Y according to

$$\mu = \mu_\infty + \left(\mu_0 - \mu_\infty + \tau_Y \frac{1 - e^{-\gamma/\gamma_Y}}{\gamma}\right)\left[1 + \left(\frac{\gamma}{\gamma_c}\right)^a\right]^{\frac{b-1}{a}} \tag{4.4.6}$$

Equation 4.4.6 is graphically depicted in Fig. 4.11 and the various viscosity regimes can easily be identified.

The rheological behavior of materials with a yield stress is sometimes represented in a diagram where the shear stress is plotted as a function of the shear rate. Associated but simple rheological models for Newtonian flow behavior were developed by Bingham (1916) and Casson (1959), see also Macosko (1994) and Bird et al. (1977). While this representation serves very well to visualize the concept of a yield stress and the change from a plastic to a fluid material, it does not directly show how the viscosity depends on the shear rate because the viscosity is equal to the slope of the curve shear stress versus shear rate. However, as is shown in Chap. 5, modelling the behavior of the characteristic flows of premetered coating methods requires knowledge of how the viscosity depends on the shear rate. For this reason, we prefer working with the Bingham-Carreau-Yasuda equation rather than with the Bingham model or the Casson model when dealing with materials with a yield stress.

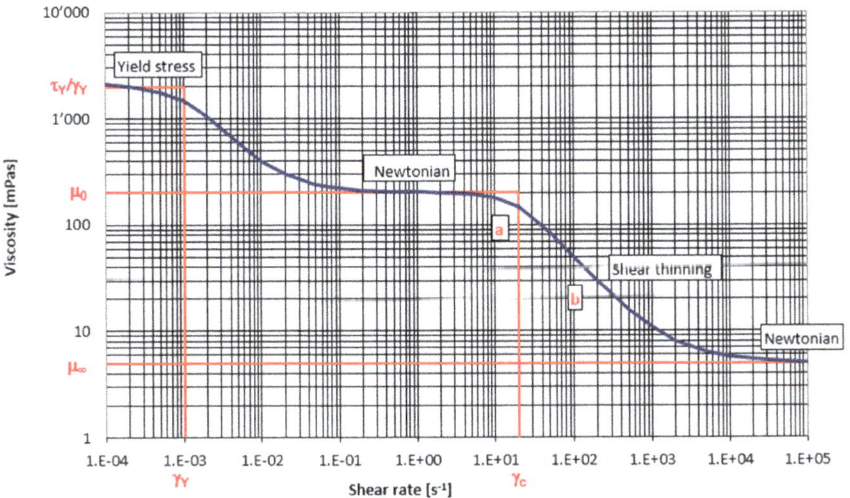

Fig. 4.11 Bingham-Carreau-Yasuda model; $\tau_Y = 2$ mPa, $\gamma_F = 0.001$ s^{-1}, $\mu_0 = 200$ mPas, $\mu_\infty =$ 5 mPas, $\gamma_c = 20$ s^{-1}, a = 2.0, b = 0.1

4.4.5 Concentration Dependence of Viscosity

The viscosity not only depends on the shear rate but also on the solids concentration of the fluid. The solids concentration is an interesting parameter for optimizing coating processes because it is under the control of the process engineer and not of the formulation chemist, and because it provides a temporary means during the manufacturing process for manipulating the fluid properties without interfering with the chemical material composition and hence with the intended product function. Whether the fluid contains more or less solvent is irrelevant for the final product because, after coating and drying, practically all solvent is removed from the material. One drawback of playing with the solvent concentration is that the amount of solvent in the coating fluid influences the length and thus the cost of a new dryer, or it limits the maximum coating speed for an existing oven.

The feature of optimizing a coating process by changing the solids concentration does not exist, or at best in only a limited amount, for solventless fluids, i.e., monomers that solidify (polymerize) under the influence of radiation energy such as UV light or electron beam. Adding a reactive thinner fluid changes the viscosity, but such materials affect the chemical composition of the final coated film and so are not desirable, or only in limited amounts.

Using an aqueous polyvinyl alcohol (PVA) solution as model fluid, Fig. 4.12 shows viscosity data and associated Carreau-Yasuda curves for 3 different solids concentrations. The data were acquired with a Bohlin rheometer at ILFORD Imaging Switzerland GmbH. Owing to the limitation of the cone-and-plate rheometer, the data is incomplete as it does not reveal the high shear Newtonian plateau. Moreover, as the

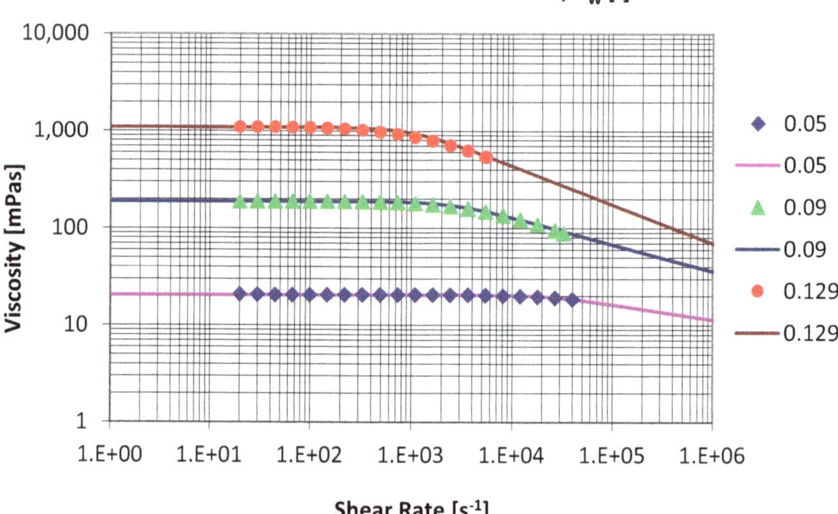

Fig. 4.12 Viscosity data and Carreau-Yasuda model for polyvinyl alcohol Mowiol 15–79 at T = 22 °C and at 3 different solids concentrations

transition between Newtonian and shear thinning behavior takes place at rather high critical shear rates, the quantification of the power law index for low solids concentrations is associated with uncertainty. Nevertheless, the data clearly shows that at least the low shear rate viscosity, the critical shear rate, and the power law index vary as the solids concentration is changed. The resulting values of the Carreau-Yasuda parameters are listed in Table 4.3. Note that the Yasuda parameter was constant at 2.0 for all tested concentrations and the high shear rate viscosity was assumed to be 1 mPas.

The dependence of these 3 rheological parameters upon the solids concentration can easily be determined by a trial-and-error approach involving the following steps:

- Plot each parameter as a function of the concentration
- Vary the scale of both diagram axes between linear, logarithmic, or a combination thereof until the data come to lie on a simple curve such as a straight line
- Select a different but simple mathematical function such as linear, low-grade polynomial, logarithmic, exponential and perform a regression analysis between the data and these functions
- Select the function that gives the best fit, i.e., the function with the highest r^2 value

Table 4.3 Carreau-Yasuda parameters as a function of the solids concentration for Mowiol 15–79

	$C_W = 0.05$	$C_W = 0.09$	$C_W = 0.129$
μ_0 [mPas]	20.56	188	1′100
γ_c [s^{-1}]	25′000	2′600	1′000
b [−]	0.83	0.72	0.60

Applying the above procedure to the PVA data produced results as shown in Fig. 4.13.

$$\mu_0 = a_0 e^{a_1 C_w} \tag{4.4.7}$$

$$\begin{aligned} \mathrm{a}_0 &= 1.766 \\ \mathrm{a}_1 &= 50.397 \\ \mathrm{r}^2 &- 0.997 \end{aligned}$$

$$\gamma_c = a_0 {C_w}^{a_1} \tag{4.4.8}$$

$$\begin{aligned} \mathrm{a}_0 &= 0.781 \\ \mathrm{a}_1 &= -3.440 \\ \mathrm{r}^2 &= 0.991 \end{aligned}$$

$$b = a_0 + a_1 C_w + a_2 {C_w}^2 \tag{4.4.9}$$

$$\begin{aligned} \mathrm{a}_0 &= 0.998 \\ \mathrm{a}_1 &= -3.376 \\ \mathrm{a}_2 &= 2.360 \\ \mathrm{r}^2 &= 0.998 \end{aligned}$$

Note that varying the scale of all diagram axes resulted in straight lines for all 3 data sets and consequently in simple equations for the rheological parameters as shown in Fig. 4.13. For the power law index, a simpler linear equation would have been sufficient as well, but a second-order polynomial function produced a slightly higher r^2 value.

Inserting Eqs. 4.4.7, 4.4.8, and 4.4.9 into 4.4.1 allows one now to conveniently calculate the PVA viscosity as a function of the shear rate and the solids concentration. Here, the regression analysis was carried out with a self-developed program, but EXCEL from Microsoft or other mathematical programs such as Mathematica (www.wolfram.com) or Maple (www.maplesoft.com) offer similar capabilities.

4.4.6 Temperature Dependence of Viscosity

The viscosity not only depends on the shear rate and the solids concentration but also the temperature of the fluid. Working at Polytype Converting AG, Beer (2020) used a Bohlin rheometer for measuring the viscosity of polyvinyl alcohol Mowiol 18–88 as a function of shear rate, solids concentration, and temperature. He used a

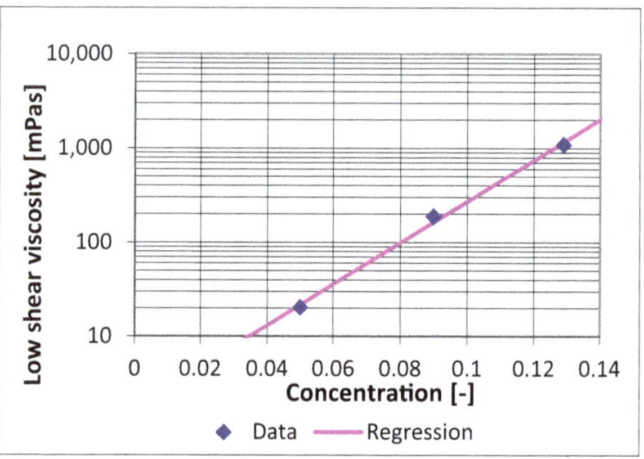

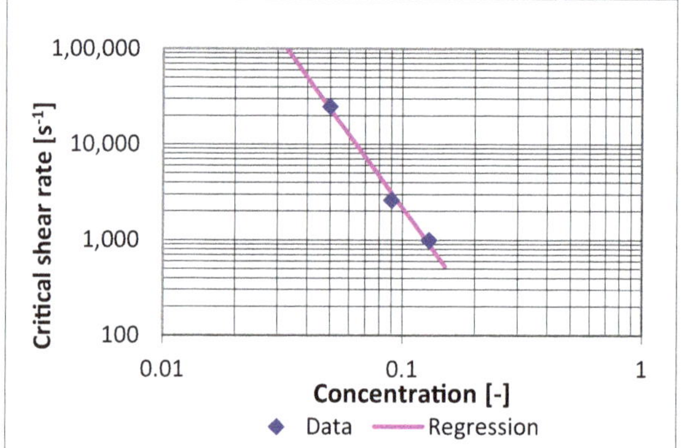

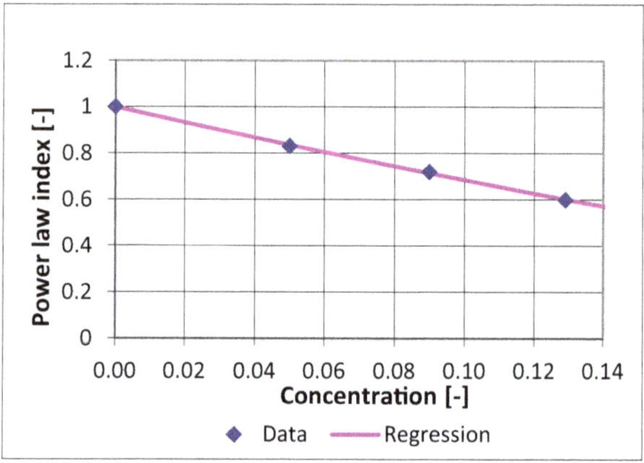

Fig. 4.13 Regression analysis on PVA viscosity data and Carreau-Yasuda parameters as a function of the solids concentration

plate-to-plate geometry and covered a shear rate range of 1–24′000 s^{-1}. The results are shown in Fig. 4.14.

The parameters of the Carreau-Yasuda equation are listed in Table 4.4.

Beer reported that the measurements were difficult to carry out, particularly at T = 50 °C where the fluid tended to dry up quickly in the rheometer. Also, the maximum shear rate available on the rheometer was too close to the critical shear rate, particularly for T = 50 °C, thus preventing sufficient viscosity data to be taken in the shear-thinning region. Consequently, the viscosity data are plagued by missing data, by scatter, and by questionable viscosity values as visualized by some of the Carreau-Yasuda parameters in Table 4.4 and, for example, by some of the measured viscosity values in Fig. 4.14c, where the viscosity drops unrealistically fast for T = 50 °C and $\gamma > 10'000\ s^{-1}$. As a result, only the low shear viscosity μ_0 could be evaluated properly as a function of the solids concentration and the temperature.

Applying the same procedure as presented in Sect. 4.4.5, the following regression equations were generated for the low shear viscosity:

$$\mu_0 = a_0 e^{a_1 C_w} \tag{4.4.10}$$

with

$$a_0 = b_0 T^{b_1} \tag{4.4.11}$$

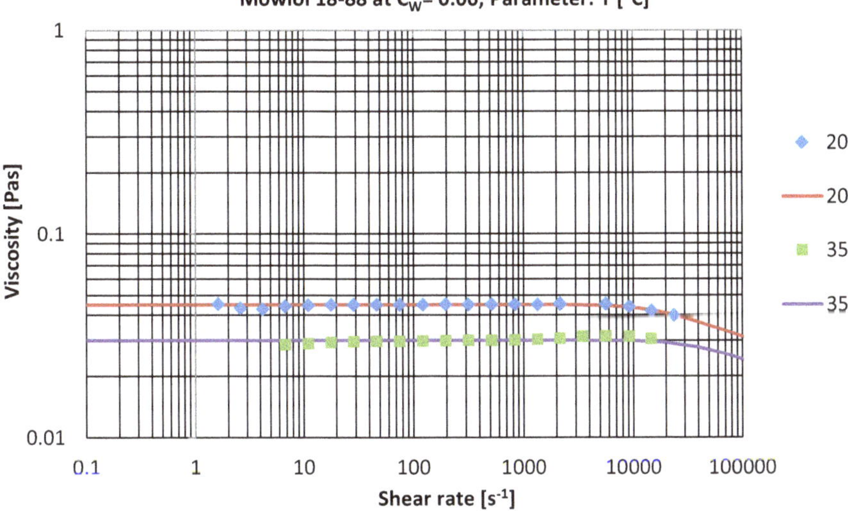

Fig. 4.14 **a** Viscosity of Mowiol 18–88 as a function of shear rate and temperature for a solids concentration of 6%. **b** Viscosity of Mowiol 18–88 as a function of shear rate and temperature for a solids concentration of 9%. **c** Viscosity of Mowiol 18–88 as a function of shear rate and temperature for a solids concentration of 12%

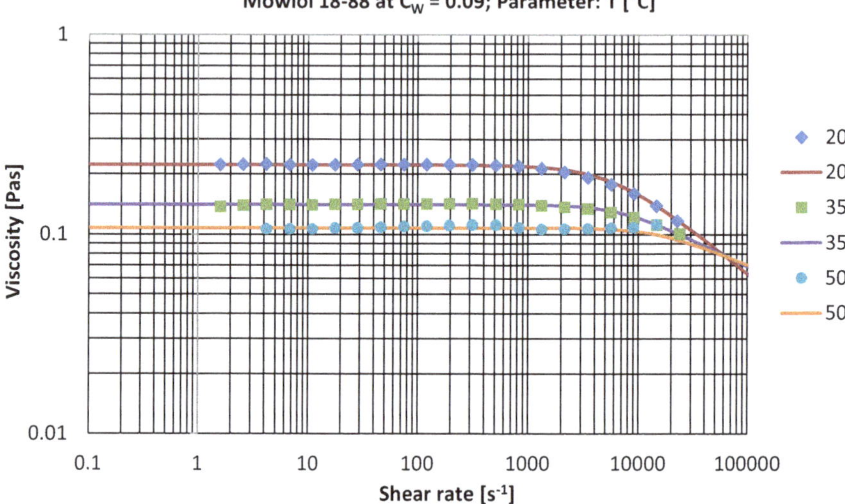

Fig. 4.14 (continued)

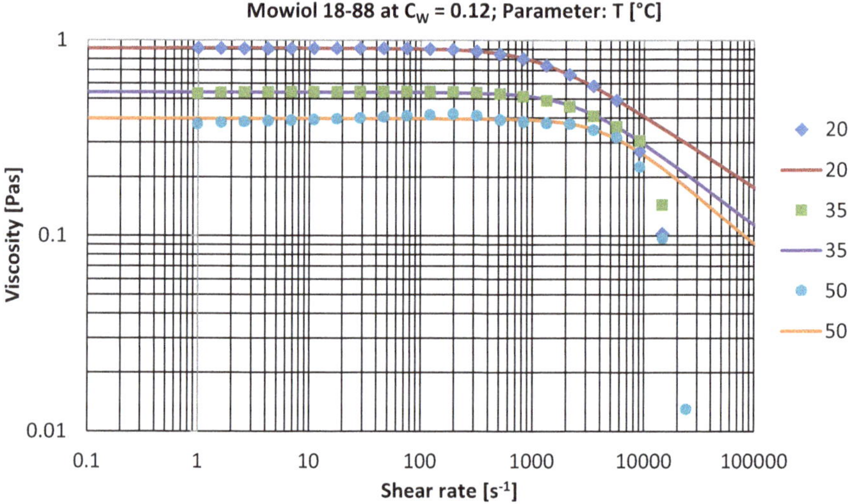

Fig. 4.14 (continued)

and

$$a_1 = b_0 e^{b_1 T} \tag{4.4.12}$$

- For a_0 : $b_0 = 51.9788281, \quad b_1 = -1.04366051, \quad r^2 = 0.999$

Table 4.4 Carreau-Yasuda parameters as a function of the solids concentration and the temperature for Mowiol 18–88

Carreau-Yasuda parameters as a function of the solids concentration and the temperature for Mowiol 18-88; C_w = 0.06

$C_w = 0.06$	T = 20 °C	T = 35 °C	T = 50 °C
μ_0 [mPas]	45	30	–
γ_c [s^{-1}]	16′500	40′000	–
a [−]	2.0	2.0	–
b [−]	0.73	–	–

Carreau-Yasuda parameters as a function of the solids concentration and the temperature for Mowiol 18-88; C_w = 0.09

$C_w = 0.09$	T = 20 °C	T = 35 °C	T = 50 °C
μ_0 [mPas]	223	141	108
γ_c [s^{-1}]	6′500	7′200	15′000
a [−]	1.5	2.0	2.0
b [−]	0.5	0.7	–

Carreau-Yasuda parameters as a function of the solids concentration and the temperature for Mowiol 18-88; C_w = 0.12

$C_w = 0.12$	T = 20 °C	T = 35 °C	T = 50 °C
μ_0 [mPas]	911	541	397
γ_c [s^{-1}]	1′200	2′700	5′000
a [−]	1.5	1.6	1.8
b [−]	0.62	0.55	0.48

- For a_1 : $b_0 = 49.1398524, \quad b_1 = 0.00093628, \quad r^2 = 0.977$

Equations 4.4.10–4.4.12 were used to produce the graph in Fig. 4.15, which visualizes the dependence of the low shear viscosity μ_0 of Mowiol 18–88 upon the solids concentration and the temperature. The round dots in the graph are the measured viscosity values. The approximation of these values with the regression equations is quite good as documented by high r^2-values and by a maximum deviation between measured and calculated values of about 16%.

The above data demonstrate that increasing the fluid temperature is an interesting means for affecting the viscosity without interfering with the product formulation. In particular, increasing the temperature of Mowiol 18–88 from 20 to 50 °C decreases the low shear viscosity by more than 50% for all of the tested solids concentrations. Note, however, that the sensitivity of the low shear viscosity to changes in temperature or solids concentration is a material-specific property that cannot be predicted but must be determined experimentally for every coating fluid. The same is true for the other material parameters of the Carreau-Yasuda equation.

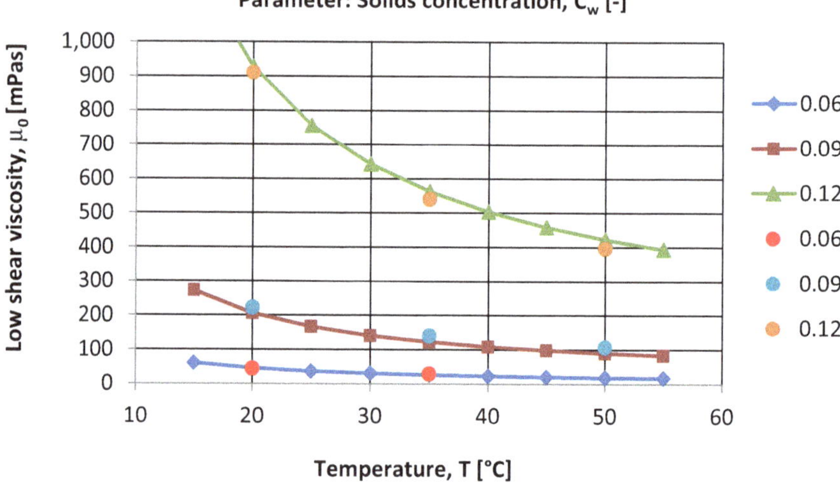

Fig. 4.15 Low shear viscosity of Mowiol 18–88 as a function of the temperature and the solids concentration

4.4.7 Viscoelastic Properties

Sometimes, particularly during the first pilot coating trials of a new product, unusual phenomena may be observed that cannot be explained by modeling the various characteristic coating flows with simple analytical approaches based solely on the shear-dependent viscosity as presented above in this chapter. In such moments, it might be helpful to investigate to what extent the coating fluid is characterized by viscoelastic properties. This is best accomplished with experimental methods.

Specifically, viscoelastic material properties can be measured by exposing the liquid to a unidirectional unsteady shear flow, such as the flow between two parallel plates where one plate is at rest and the other is oscillating with an angular frequency ω. This kind of flow is conveniently implemented in a parallel plate rotational rheometer.

Bird et al. (1977) explain that, in a small-amplitude oscillatory shear experiment with a Newtonian fluid, the shear stress is in phase with the shear rate, and normal stresses do not exist. For polymeric liquids, however, the shear stress oscillates with frequency ω, but is out of phase with the shear rate, and the normal stresses oscillate with a frequency of 2ω relative to a nonzero mean value. Bird et al. (1977) further state that the measured data are often represented in terms of the real (η') and imaginary (η'') parts of the complex viscosity η^* according to

$$\eta^* = \eta' - i\eta'' \tag{4.4.13}$$

Here, η is a non-Newtonian shear rate-dependent viscosity, and the real part η', which is the dynamic viscosity, is thought of as the viscous contribution associated with energy dissipation, while the imaginary part η'' is thought of as the elastic contribution associated with energy storage. Since the amplitude of the oscillation in the experiment is small, the resulting velocity profiles are nearly linear. This is why η' and η'' are usually called linear viscoelastic properties. Note that for Newtonian fluids, $\eta' = \mu_0$ and $\eta'' = 0$.

Instead of using the complex viscosity, the results from oscillatory experiments are often expressed in terms of two dynamic moduli, Specifically, $G' = \omega\eta''$ is called the in-phase storage or elastic modulus, and $G'' = \omega\eta'$ is the out-of-phase viscous or loss modulus, see Bird et al. (1977) and Macosko (1994). Conveniently, G' and G'' can be determined experimentally with many of the rotational rheometers. For Newtonian fluids, $G' = 0$. Therefore, the more G' is different from 0, the more viscoelastic properties can be attributed to a given liquid.

Modeling the flow of viscoelastic materials is difficult and goes beyond the scope of this book. One reason for the difficulties is the presence of time-dependent terms in the flow equation. Early attempts to form models for linear viscoelastic fluids go back to Maxwell, who combined the shear stress of a Newtonian liquid, which is proportional to the viscosity, with the shear stress of a Hookean solid, which is proportional to the elastic modulus (Bird et al. (1977). This reference also discusses other suitable models for viscoelastic materials.

Lacking the ability to easily model viscoelastic effects makes it difficult to know how and where the performance of premetered coating methods is affected and possibly degraded by such material properties. As shown in Chap. 5, most characteristic flow fields are rectilinear, steady, and they certainly are free of oscillating shear rates. However, transitional regions between successive flow fields may be locations where viscoelastic effects are substantial. Examples are the T-junction in a center-fed slot die where the feed port meets the (inner) distribution cavity, and the transition from the cavity to the slot flow, see Sect. 8.2.4. In a well-designed coating process, all flow fields downstream of the distribution cavity are nearly rectilinear, and viscoelastic effects are expected to be equal at any location across the coating width and therefore acceptable, for example at the exit of the slot flow of a multilayer slide die. Another such example is the flow of a liquid curtain. Becerra and Carvalho (2011) showed that the minimum flow rate/width needed to establish a stable curtain decreases considerably with increasing viscoelasticity, see also Sect. 5.8.5. Beyond the curtain, flow fields with highly curved free surfaces such as the beads in slide and slot coating or the area of an impinging curtain, as well as accelerating boundary layers that start at a dynamic wetting line may exhibit viscoelastic effects. If such effects are unwanted because they degrade the performance of the coating process, then relief may be obtained by heating or diluting the coating fluid.

4.5 Surface Tension

Surface tension is a material property that acts in free liquid surfaces, particularly if these surfaces are curved. Each of the premetered coating methods has two such

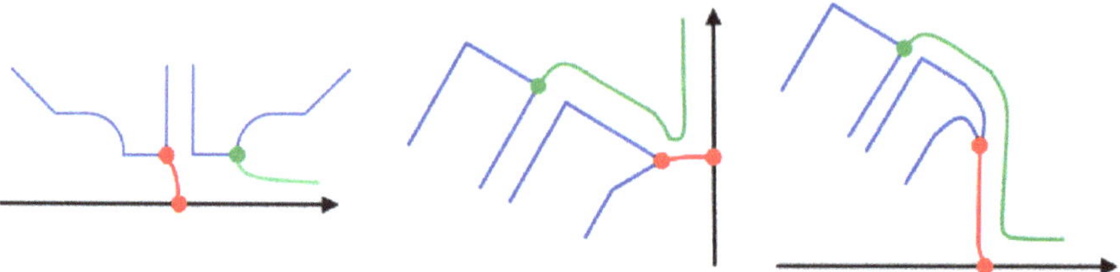

Fig. 4.16 Free surfaces and wetting lines of premetered coating methods

surfaces as schematically depicted in Fig. 4.16. One of them (green line) starts at a static wetting line located on the die and ultimately turns into the free surface of the coated wet film (film-forming meniscus). The other one (red line) also starts at a static wetting line located on the die but ends at the dynamic wetting line located on the moving substrate, i.e., at the point where coating takes place (wetting meniscus).

Curtain coating has the largest free surfaces, particularly if simultaneous multilayer applications are involved. In addition, the curtain, which typically is several centimeters long, has two free surfaces. In contrast, the curved free surfaces in slot coating are very short, on the order of the wet film thickness. In slide coating, the free surface of the liquid film on the incline may be long while the wetting meniscus is very short like in slot coating.

The subject of surface tension as it applies to coating processes is complicated. Tricot (1997) wrote a comprehensive overview on this topic. Specifically, he discussed the role of surfactants, properties of surfactants, static and dynamic surface tension, theoretical considerations, measurement techniques, and practical applications including the diffusional behavior of 12 commercial surfactants.

The purpose of this chapter is to re-visit some of the surfactant and surface tension issues as they apply to premetered coating methods.

4.5.1 Surface Tension of Coating Fluids

Aqueous coating fluids tend to have a rather high surface tension, often in the range of 50–60 mN/m because water at room temperature has a surface tension of about 72 mN/m, which is the highest value of all liquids (except liquid metals). This high surface tension value is often not a problem from a product formulation point of view, and so many product formulations do not require surfactants. Other formulations contain surfactants based on fluid-making requirements, for example for stabilizing an emulsion. In yet other formulations surfactants may not be desirable, if they negatively affect the performance of the coated film, for example in printing applications.

However, as will be seen throughout this book, high surface tension values may affect the performance of coating processes. One example is curtain coating, which, for many applications, requires a surface tension of <40 mN/m, see Table 2.1. Consequently, most formulations of aqueous fluids must include a sufficient amount of a suitable surfactant to reach such a low value.

In contrast to aqueous liquids, coating fluids based on organic solvents inherently have a low surface tension, typically in the range of 20–40 mN/m because most organic solvents have such values. Consequently, adding surfactants is typically not necessary for solvent-based formulations. The same applies to solventless fluids. Experience shows that such fluids have surface tension values in the range of 15–35 mN/m.

4.5.2 *Static or Dynamic Surface Tension*

Lowering the surface tension of aqueous fluids can be achieved in two ways, either by adding one or several surfactants or by adding an organic solvent. If the solvent is miscible with water, such as alcohols, then water and the solvent form a pure solution and the resulting surface tension has a constant (static) value that depends on the mixing ratio, see Fig. 4.17.

Using an organic solvent for modifying the surface tension of a coating fluid has the advantage that the solvent will evaporate during drying and hence will not be present in the final coated film. The disadvantage is that significant solvent quantities may be necessary to lower the tension to values of, say, less than 40 mN/m. This then could turn into an ecological and financial issue because the manufacturing plant may have to be modified to become a solvent-proof facility.

If, on the other hand, a surfactant is added to the coating fluid for lowering its surface tension, then the surfactant molecules are evenly dissolved and distributed

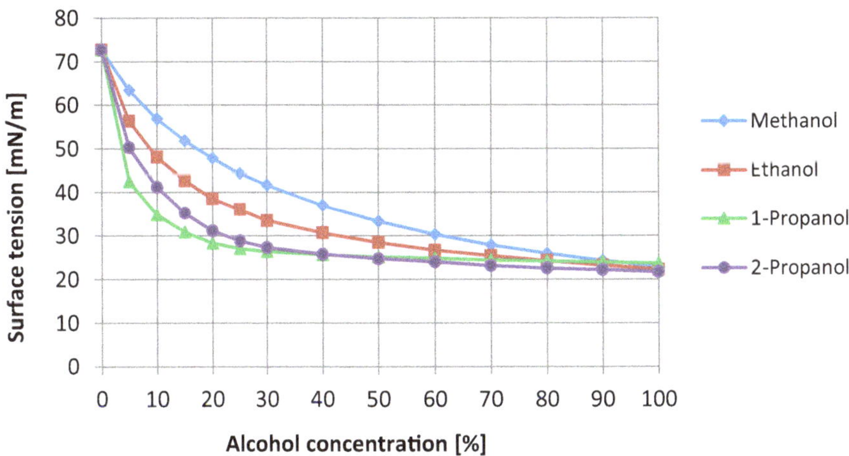

Fig. 4.17 Surface tension of various water-alcohol mixtures at 20 °C. Data taken from CRC (2019)

throughout the liquid while it is flowing from the supply vessel to the coating process, i.e., while the transporting flow field is confined by pipes and the die, etc. Moreover, the surfactant molecules are inactive during that phase. However, as soon as the liquid exits the die slot, one or two free liquid surfaces are formed (see Figs. 2.1 and 4.16), and the surfactant molecules become active, i.e., they first must move to the free liquid surface before they can lower the surface tension. Therefore, the time it takes surfactant molecules to reach a free surface is controlled by a diffusion process, which depends on the molecular structure and the concentration of the surfactant as well as on flow properties such as the viscosity, the thickness, and the speed, i.e., the age of the free surface of the flowing film. As a result, the surface tension of an aqueous coating fluid containing surfactants is time-dependent or dynamic. Tricot (1997) provides a more detailed explanation of this situation.

Figure 4.18 shows a typical example of this time-dependent behavior, here for the liquid pair of distilled water and the surfactant Surfynol 2502 from Air Products. The surface tension decreases with increasing age of the free surface, and it decreases with increasing surfactant concentration. As indicated by the data, the surface tension levels off for sufficiently large surface ages, and that constant surface tension value corresponds to the static surface tension.

To achieve low surface tension values of, say, <40 mN/m at very short surface ages of, say, <100 ms, an issue that is important for the stability of liquid curtains (see Sects. 5.8.5 and 12.3), sufficiently high surfactant concentrations are necessary,

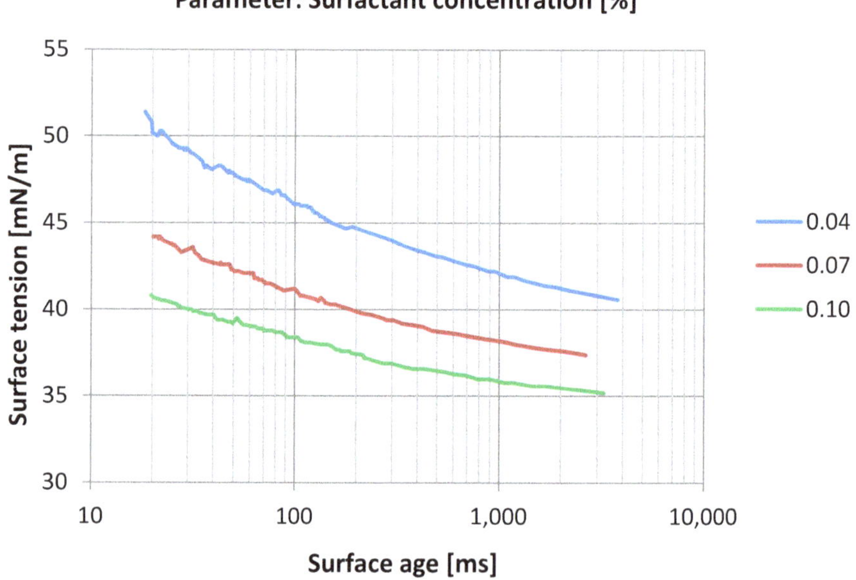

Fig. 4.18 Surface tension of distilled water and Surfynol 2502 at 25 °C; data produced with the maximum bubble pressure method by ILFORD Imaging Switzerland GmbH, Marly, Switzerland

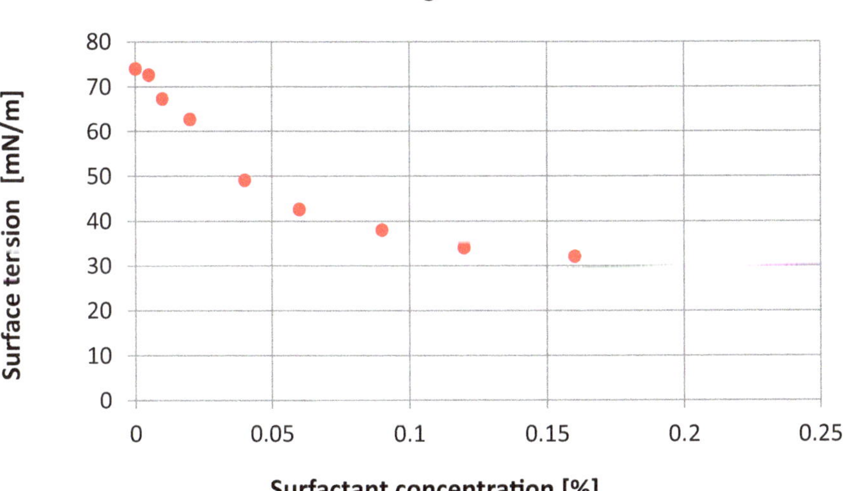

Fig. 4.19 Concentration dependence of dynamic surface tension for distilled water and Surfynol PSA 336 at 25 °C; data produced with the maximum bubble pressure method by ILFORD Imaging Switzerland GmbH, Marly, Switzerland

and fast diffusing surfactants should be preferred over slowly diffusing ones (Tricot, 1997).

The typical dependence of the dynamic surface tension upon the surfactant concentration is better shown in Fig. 4.19 for the liquid pair of distilled water and surfactant Surfynol PSA 336 from Air Products. All data were taken at a surface age of 100 ms.

Figure 4.20 shows the temperature dependence of surface tension for the example of water. Typical temperatures for liquid film coating processes are between 20 and 60 °C. In this temperature range, the surface tension of water drops by less than 10%, meaning that temperature is not an important parameter for altering the surface tension of coating fluids.

4.5.3 Methods for Measuring Surface Tension

Tricot (1997) discussed several suitable methods for measuring surface tension. The du Nouÿ ring or the Wilhelmy plate methods are well-known principles for determining the static surface tension. Regarding the dynamic surface tension, the maximum bubble pressure method is also well known and convenient. Krüss is an established equipment supplier (www.kruss-scientific.com). This technique allows measuring the surface tension for surface ages as low as 20 ms, which is excellent but only accurate for low shear rate viscosities of about <150 mPas. Tricot (1997)

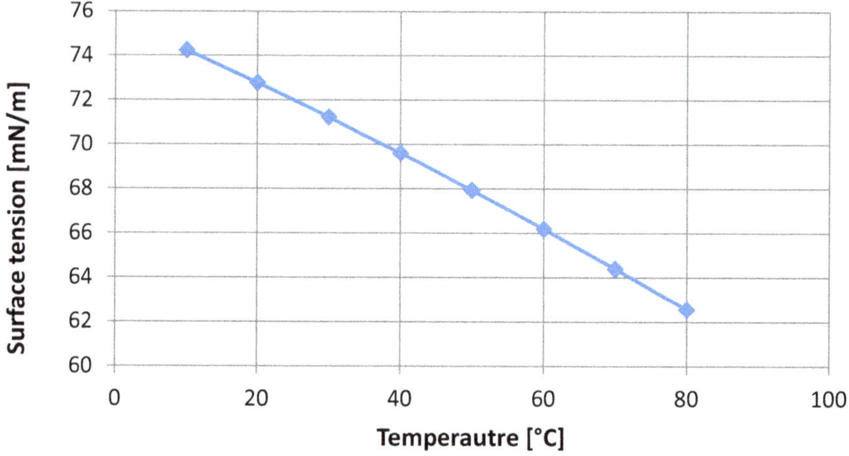

Fig. 4.20 Temperature dependence of surface tension for water. Data taken from CRC (2019)

has overcome this drawback by combining the Wilhelmy plate methods with a film flow on a long (1 m) inclined plane. He was able to measure the surface tension of surface ages down to about 10 ms. This approach is not without drawbacks either because it cannot be bought commercially, and it requires a die for generating the liquid film on top of the inclined plane.

With regard to curtain coating the **Mach angle method** is a simple way for measuring the surface tension directly where it counts, namely in the falling liquid curtain, see Brown (1961), Antoniades et al. (1980), and Tricot (1997).

If the surface of a flowing liquid membrane (liquid curtain) is disturbed in a single point, for example by a needle, then a standing wave, a so-called Mach wave, is produced in the curtain, see Fig. 4.21. The vertex of the wave coincides with the

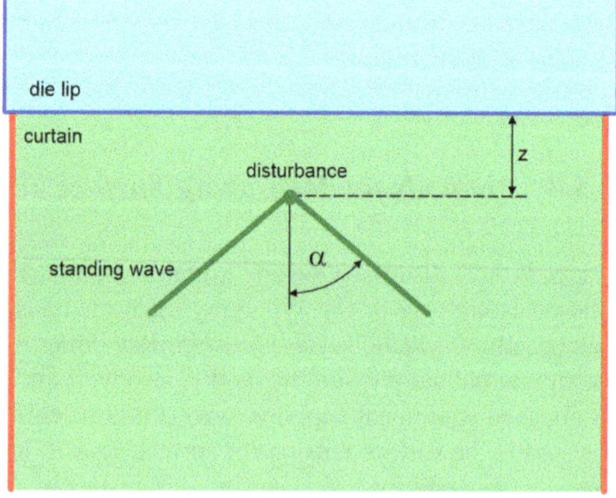

Fig. 4.21 Schematic diagram of a curtain with a point disturbance and a Mach wave

point of disturbance, and the angle of the wavefront depends on several relevant process parameters, among them the surface tension of the liquid at the point of disturbance. Therefore, the local surface tension in the curtain can be determined experimentally by measuring the angle of the Mach wave.

Equation 4.5.1 shows the dependence of the surface tension σ upon other relevant process parameters:

$$\sigma = \frac{\rho Q V_c \sin^2(\alpha)}{2} \tag{4.5.1}$$

ρ is the liquid density, Q is the volumetric flow rate/width, V_c is the local velocity of the curtain, and α is the half Mach angle.

According to Brown (1961), the local curtain velocity can be calculated by using Eq. 4.5.2.

$$V_c = \sqrt{V_{c,0}^2 + 2g(z - z_0)} \tag{4.5.2}$$

The curtain velocity depends on the point of interest z, where z = 0 coincides with the origin of the curtain, i.e., with the location of the lip of the die. $V_{c,0}$ is the initial velocity of the curtain flow, which depends on the type of die (slot or slide) in use, g is the acceleration due to gravity, and z_0 is a measure for the extent of the transitional flow between stationary film or slot flow on or inside the die and stationary curtain flow away from the die. For a slide die, the initial velocity is set equal to the average velocity of the film flow on the die slide according to Eq. 4.5.3.

$$V_{c,0}\left(\frac{\rho g \sin(\beta) Q^2}{3\mu}\right)^{1/3} \tag{4.5.3}$$

β is the angle of the inclined plane or die slide relative to horizontal taken directly at the die lip, and μ is the shear viscosity at the shear rate of the film flow at the die lip.

For a slot die, the initial velocity is set equal to the average velocity of the slot flow at the die exit according to Eq. 4.5.4.

$$V_{c,0} = \frac{Q}{w} \tag{4.5.4}$$

Here, w is the height of the die slot measured at the slot exit. More details about film, slot, and curtain flows are presented in Chap. 5.

The length z_0 depends on the density, the local shear-rate-dependent viscosity μ, and the gravitational acceleration g according to Eq. 4.5.5.

$$z_0 = 2\left(\frac{4\mu}{\rho}\right)^{2/3} (g)^{-1/3} \tag{4.5.5}$$

Equations 4.5.1–4.5.5 are best incorporated into an EXCEL sheet, such that the surface tension in the curtain can easily be calculated by considering the parameters described above.

Measuring half of the Mach angle is easily accomplished by using a special angle measurement device developed by Polytype Converting AG, see Fig. 4.22. A small pointed needle is placed in the center of the circular measurement device. The needle has to be positioned such that it just touches the front side of the flowing curtain. Disturbing the curtain in this way produces and visualizes the standing Mach wave. Now, the inner ring of the measurement device has to be rotated such that the hairline (wire) through the center of the device aligns perfectly with the left front of the Mach wave, and such that half of the Mach angle can be read in units of [degree] from the scale on the outer stationary ring of the device.

When touching the curtain with the needle point, care has to be taken to not break the curtain. Owing to the extra weight of the edge bead along a broken wavefront, the Mach angle of a ruptured wave (triangular hole in the curtain) is not the same as the Mach angle of a wave in a stable and coherent curtain.

The horizontal arm holding the ring of the measurement device is attached to a vertical bar, into which a millimeter-scale is integrated. With the help of this scale, the location of the needle point can be set and read. Apart from the value of the Mach angle, the location of the needlepoint, as well as the volumetric flow rate generated by the pump, and the width of the curtain (width of the die), must be transferred into the EXCEL file.

For certain operating conditions, the front of the standing Mach wave is only barely or even not at all visible. This is particularly true, if the needle point is positioned just a short distance below the die lip. For such circumstances, it is recommended to move the point of disturbance further downstream in the curtain until the Mach angle becomes clearly visible.

Since the surface tension depends on the volumetric flow rate/width and the local velocity in the curtain, i.e., on the location of the needle point, these two parameters

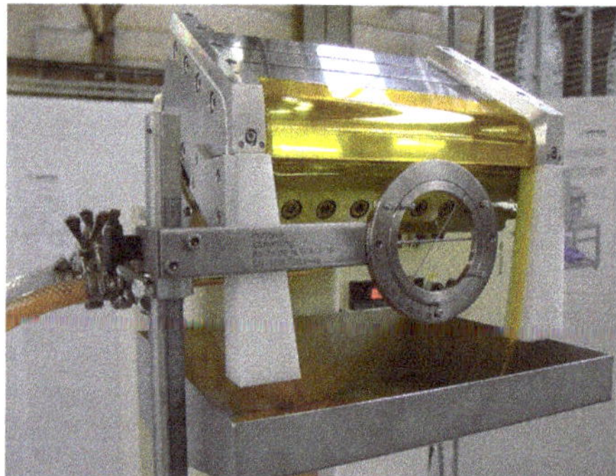

Fig. 4.22 Mach angle measurement device developed by Polytype Converting AG; photos reproduced with permission from Polytype Converting AG

can easily be changed to determine their effect on the surface tension. However, if the goal is to compare the surface tensions of different liquids, then it is usually sufficient to measure the surface tension for a constant flow rate and at a constant location of the needle point.

Exemplary surface tension data taken at Polytype Converting AG with the Mach angle method in a running curtain are presented in Figs. 4.23 and 4.24. The fluid in Fig. 4.23 was polyvinyl alcohol (PVA, Mowiol 15–79) dissolved in a 30:70 ratio ethanol/water mixture. The solids concentration was 10.35% and the low shear viscosity was 350 mPas at 20 °C.

Varying the volumetric flow rate/width did not change the surface tension, at least not for the tested flow rate range. This behavior was observed many times for many different coating fluids when a slide die as shown in Fig. 4.22 was used for producing the curtain. For this configuration, the flow rate/width only affects the initial curtain velocity according to Eq. 4.29. Experience has shown, however, that the initial curtain velocity only contributes about 10% of the final curtain speed. Therefore, changing the flow rate is not expected to greatly influence the resulting surface tension. Moreover, changing the measurement point inside the curtain from 50 to 100 to 150 mm from the die lip did not change the surface tension either. As explained above, a water-alcohol mixture is a pure solution with inherently constant surface tension, and hence such a coating fluid is not expected to show any dynamic behavior with respect to surface tension.

Figure 4.24 shows results from a similar experiment. The difference here is that the PVA was dissolved in pure water and the low surface tension was obtained by adding 0.2% of the surfactant Surfynol PSA 336. The PVA concentration was 9.55% and the low shear viscosity was 332 mPas at 20 °C.

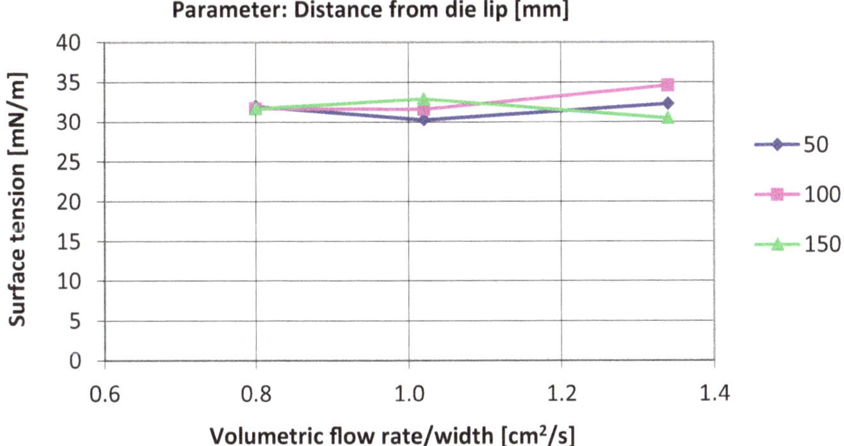

Fig. 4.23 Surface tension data taken with the Mach angle method in a running curtain of polyvinyl alcohol (PVA, Mowiol 15-79) dissolved in a 30:70 ratio ethanol/water mixture

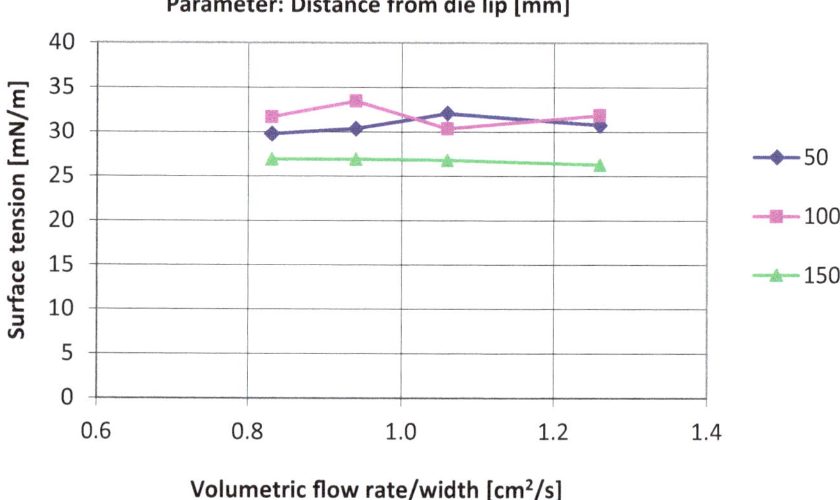

Fig. 4.24 Surface tension data taken with the Mach angle method in a running curtain of aqueous polyvinyl alcohol (PVA, Mowiol 15-79) containing 0.2% of the surfactant Surfynol PSA 336

While changing the flow rate again did not affect the surface tension, changing the measurement location, i.e., changing the surface age did change the surface tension and visualized a dynamic behavior. The flow in a falling curtain is extensional, meaning that fresh surface is generated all along its trajectory. A freshly generated surface is poor on surfactant molecules compared to an "old" surface, and therefore the surface tension there is expected to be high. However, the data suggests that, when the free surface is sufficiently old, i.e., 150 mm below the die lip, the diffusion of surfactant molecules from the bulk of the fluid to its free surface is stronger than the depletion effect from the generation of new surface such that the surface tension at the bottom of the curtain is slightly lower than at the top. This effect could be amplified or weakened by increasing or decreasing the surfactant concentration in the fluid.

4.5.4 Requirements for Surfactants

Surface tension and hence surfactants affect many aspects of coating flows, and they may be crucial in determining the uniformity of the coated film. Experience shows that various quality-related aspects lead to conflicting requirements for the optimum type(s) and concentration(s) of surfactants. Unfortunately, not all of these requirements are based on a sound scientific understanding of the subject matter. As a result, clear guidelines for determining optimum surfactant properties (type, concentration, etc.) do not exist.

The following list summarizes issues that are controlled by surface tension. These issues are discussed in more detail later in the book.

- Multilayer film flow on an inclined plane.
- Multilayer curtain flow.
- Curtain trajectory and curtain deflection.
- Curtain thinning near the edge guides.
- Wetting and spreading.
- Resistance to repellencies.
- Resistance of a free liquid surface to ambient pressure disturbances.
- Leveling and flow after coating.

4.5.5 Selection of Surfactants

Many commercial surfactants are available, and they all differ in molecular and ionic structure. Finding the right surfactant for a given application is not a trivial task. Experience helps, and most of the time a trial-and-error approach is necessary. Ideally, surface tension should be in control of the coating engineer and not of the formulation chemist, but support from the formulation specialist is necessary and helpful for finding the best surfactant.

One of the important selection criteria is the speed at which surfactant molecules diffuse from the bulk of the liquid to its free surface. Fast diffusing surfactants are particularly important for curtain coating because the time the liquid spends in the curtain is very short, i.e., on the order of 100 ms, see Chap. 5.8.3.

Tricot (1997) quantified the diffusional behavior of 12 surfactants, many of which were successfully used in photographic applications.

Fluorinated surfactants have a high surface tension lowering capability. However, many of them are slowly diffusing (Tricot, 1997). Moreover, fluorine poses an environmental concern, and hence this type of surfactant is no longer popular.

During the past 20 years or so Polytype Converting AG generated a lot of experience with surfactants, mostly from running many single and multilayer curtain coating trials with customers covering the whole spectrum of the converting industry. During such trials, or in-house studies with model fluids such as aqueous PVA solutions, good results were obtained with surfactants from the Surfynol family by Air Products. Type 2502 was a suitable general-purpose surfactant, while type PSA 336 was optimized for applications with pressure-sensitive adhesives.

4.6 Contact Angle and Wetting

Wetting occurs under static or dynamic conditions. At the coating point of premetered processes, i.e., at the location where the impinging liquid displaces air from the substrate surface, wetting is a dynamic process, see Fig. 4.16. After coating and before drying, however, wetting, or, as the case may be, de-wetting, may also take place under more or less static conditions. A well-known example is the making of adhesive tapes and labels, where edge withdrawal is observed when the adhesive is applied to a siliconized substrate.

Describing wetting processes, and in particular dynamic wetting situations, by physical principles is very difficult. Many researchers across the globe are still working on improving the fundamental understanding of wetting. Theoretical and reliable models for predicting the onset of wetting failure, which is an operating boundary of every coating process, do not yet exist, specifically models that are practical for industrial coating engineers.

Blake and Ruschak (1997) have written a comprehensive review of wetting. They addressed static menisci and stationary wetting lines, dynamic menisci and wetting lines, as well as dynamic wetting failure. More recently, Vandre et al. (2012) showed experimentally and theoretically that wetting failure can be delayed to higher substrate speeds by confining the wetting meniscus. Such meniscus confinement is present in the bead of slot and slide coating as depicted in Fig. 4.16. In addition, Blake et al. (2015) developed a model based on molecular dynamics to support experimental observations suggesting that dynamic wetting failure can be postponed by generating hydrodynamic assist. In particular, they found that "an intense shear stress in the vicinity of the contact line can assist surface tension forces in promoting dynamic wetting, thus reducing the velocity-dependence of the contact angle".

On a less scientific level, experience shows that wetting failure is not often observed in industrial coating applications. Running many premetered coating trials for more than 15 years on the pilot facility of Polytype Converting only rarely led to the onset of air entrainment, even though coating speeds were as high as 1′000 m/min, for example when using curtain coating for applying pressure-sensitive adhesives. It is believed that this result is based on optimizing the coating processes by maximizing the hydrodynamic assist. More details on this topic are presented in Chap. 7. Therefore, the focus in the chapter at hand is on methods and procedures for measuring static wetting properties, namely the static contact angle, the spreading coefficient, and the wetting envelope. These methods are not often encountered in industrial coating companies, but they could help analyze and resolve certain coating problems.

4.6.1 Static Contact Angle and Spreading Coefficient

As depicted in Figs. 4.25 and 4.26, every wetting situation is characterized by a solid substrate surface, a liquid, which is bound by the substrate and a meniscus, i.e., a curved free surface, a gas, which is bound by the substrate and the liquid, a wetting line (also called a 3-phase line), where the substrate, the liquid, and the gas meet, and a contact angle θ, which is measured through the liquid from the solid surface to the tangent of the liquid–gas interface at the static contact line, and which is a measure for the shape of the meniscus at the wetting line.

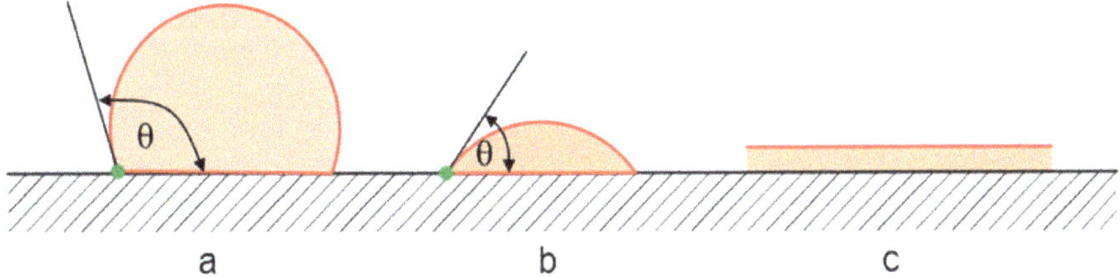

Fig. 4.25 Wetting of a solid substrate by a liquid

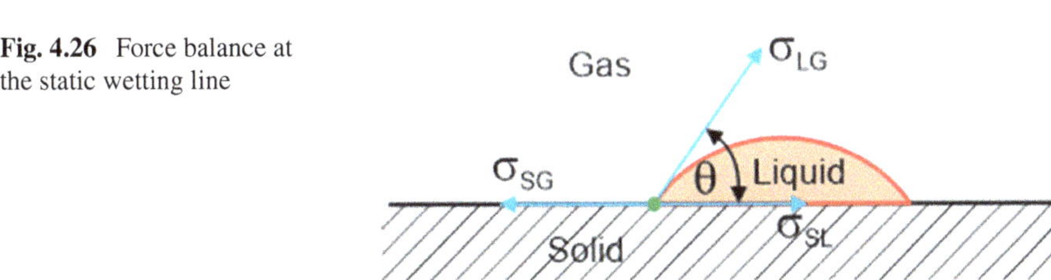

Fig. 4.26 Force balance at the static wetting line

The static contact angle is a material property, which characterizes the wetting properties of a liquid on a solid substrate. Consider a small drop of a given liquid being deposited onto a horizontal solid substrate. Depending on the capability of this liquid to wet this solid material, the drop will assume a different shape as shown in Fig. 4.25. For the sake of simplicity, it is assumed that the drop will always have a circular cross-section, which is true if the drop volume is sufficiently small.

If the liquid wets the solid not at all or only purely, the drop assumes a bulging form (bigger than a half-circle) as shown in Fig. 4.25a. If the liquid wets the solid reasonably well, the drop shape is equal to a segment of a circle (smaller than a half circle), see Fig. 4.25b. For perfect wetting, the liquid drop spreads spontaneously over the solid surface until a homogeneous film is formed, see Fig. 4.25c.

For no wetting at all, $\theta = 180°$. An example of poor wetting is given by a water drop on a Teflon surface, such as a frying pan. In contrast, excellent wetting is observed, if a drop of an organic solvent, such as acetone or ethanol, is placed onto a clean glass surface. For perfect wetting, $\theta = 0°$.

The static contact angle is related to the interfacial tension σ of the various interfaces present at the contact line. A theoretical model representing this force balance is given by Young's equation as follows, see Fig. 4.26. The indices S, L, and G stand for solid, liquid, and gas.

$$\sigma_{LG}\cos(\theta) = \sigma_{SG} - \sigma_{SL} \tag{4.6.1}$$

For perfect wetting or complete spreading as mentioned above, $\theta = 0°$, $\cos\theta = 1$, and Eq. 4.6.1 can be rewritten as

$$\sigma_{LG} \leq \sigma_{SG} - \sigma_{SL} \tag{4.6.2}$$

The difference between the two terms in Eq. 4.6.2 is the driving force of spreading, which is called the spreading coefficient S:

$$S = \sigma_{SG} - \sigma_{SL} - \sigma_{LG} \tag{4.6.3}$$

S must be positive for spontaneous spreading. Tricot (1997) writes that S is difficult to calculate for spreading on a solid surface because the solid–gas and solid–liquid interfacial tensions σ_{SG} and σ_{SL} are difficult to measure. If, however, spreading of the liquid occurs on a second liquid, σ_{SG} in Eq. 4.6.3 must be replaced by σ_{L2G}, the second liquid surface tension, and σ_{SL} by σ_{L1L2} the liquid–liquid interfacial tension.

This latter situation is relevant for simultaneous multilayer coatings, and for the flow of a liquid curtain along the curtain edge guide because the top layer must be able to spread over the layer just below, or the curtain must be able to spread on the auxiliary fluid of the edge guide. Since miscible fluids cannot sustain interfacial tensions, spontaneous spreading of the top layer will occur (positive S in Eq. 4.6.3), if its surface tension is lower than the surface tension of the layer below, or if the surface tension of the curtain is lower than the one of the edge fluid.

4.6.2 Measurement of the Contact Angle

The contact angle is a material property, which cannot be calculated from first principles or by knowing other chemical or physical properties of the material pair involved. Consequently, the contact angle must be determined experimentally, and Krüss is an example of a suitable supplier of adequate equipment and associated software (www.kruss-scientific.com).

On the other hand, it is not too difficult to experiment without a costly device. To that extent, a small drop of the liquid to be tested is dispensed by a syringe onto the substrate sample to be tested. Taking a photograph of this drop with a macro lens allows then various geometrical parameters to be extracted, by which the contact angle can be calculated, see Fig. 4.27.

The relevant geometrical parameters, which must be extracted from the shape of the drop, are depicted in Fig. 4.28. Moreover, assuming that the drop has the shape of a segment of a circle, then the contact angle can be calculated with the help of the

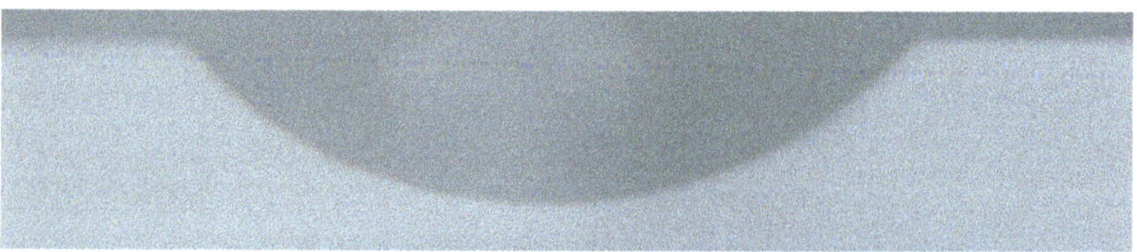

Fig. 4.27 Photograph of a liquid drop (water) on a solid substrate (steel); note that the orientation of the drop was inverted by the optical system

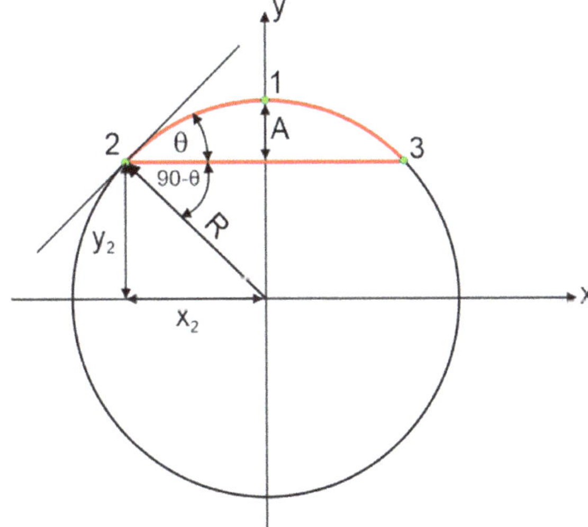

Fig. 4.28 Geometrical parameters for determining the static contact angle

following equations:

$$x_2^2 + y_2^2 = R^2 \tag{4.6.4}$$

$$y_2 + A = R \tag{4.6.5}$$

$$y_2 = \frac{x_2^2 - A^2}{2A} \tag{4.6.6}$$

$$R = \frac{x_2^2 + A^2}{2A} \tag{4.6.7}$$

$$\sin(90 - \theta) = \frac{y_2}{R} \tag{4.6.8}$$

$$\theta = 90^\circ \quad arc\sin\left(\frac{x_2^2 - A^2}{x_2^2 + A^2}\right) \tag{4.6.9}$$

The parameters x_2 and A can easily be extracted from the photograph of a drop as explained below and shown in Fig. 4.29. Equation 4.6.9 is valid for A < R and A > R. However, the calculated contact angle values become less accurate for A > R because the drop shape is no longer circular owing to the large drop volume and the large drop weight.

Experimental procedure:

1. Take a printout of a photograph of a drop on a substrate as shown in Fig. 4.29.

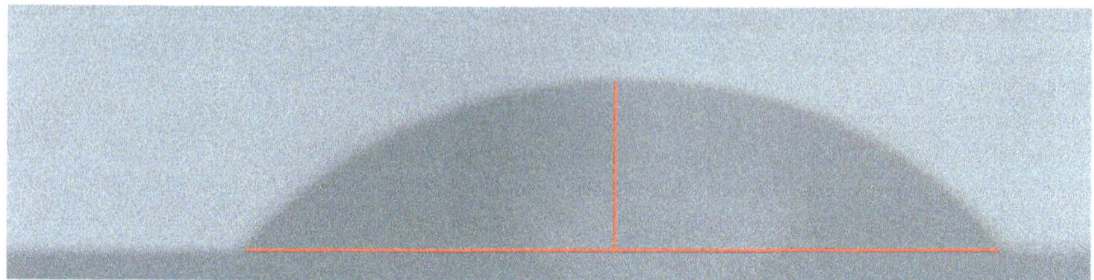

Fig. 4.29 Photograph of a liquid drop (water) on a solid substrate (steel)

2. Draw a straight line through points 2 and 3 corresponding to the location of the substrate surface.
3. Measure the distance between points 2 and 3. Half of that distance corresponds to the value of x_2.
4. Draw a perpendicular line through the midpoint of the line between points 2 and 3.
5. Measure the distance between the midpoint and point 1, which corresponds to the value of A.
6. Use Eq. 4.6.9 for calculating the contact angle θ in units of [°].

An example of carrying out the above procedure is shown with the help of Fig. 4.29, where the following values were measured and calculated:

$x_2 = 85$ mm; $A = 38$ mm; $\theta = 48.2°$

In addition, the following observation related to the performance of a curtain edge guide was made: water having a surface tension of about 72 mN/m and flowing down a vertical stainless steel plate (curtain edge guide) with a surface energy of about 55 mN/m could not form a coherent film on that steel plate. In other words, a contact angle of about 48° did not allow the water to spread sufficiently. However, adding 5.9% of iso-propanol to the water lowered the surface tension of the mixture to 50.6 mN/m. Now, the resulting contact angle was 43.6°, and the surface tension of the liquid was lower than the surface energy of the solid substrate. Consequently, sufficient spreading was possible.

4.6.3 *Wettability of Substrates*

Whether or not a given liquid can wet a given solid substrate is a fundamental question in coating. Knowledge about the wettability of a solid substrate, therefore, is important. A comprehensive summary of the physical concepts that describe wetting of solid surfaces can be found on the website of Krüss at www.kruss-scientific.com.

Accordingly, the surface free energy of a solid is closely related to its wettability, and good wettability requires a high surface free energy. It is well known in the coating industry that the surface free energy of a solid substrate, for example of a plastic material such as PET film, can be increased by a plasma, flame, or corona treatment.

As mentioned in Sect. 4.6.1 the contact angle is also a measure for the wettability of a solid. The contact angle is related to the surface free energy through Young's Eq. 4.6.1. In fact, Young's equation describes a force balance at the three-phase line between the surface tension of the liquid σ_{LG}, the interfacial tension between solid and liquid σ_{SL}, and the surface free energy of the solid substrate σ_{SG}.

The question then arises if it is possible to quantify the wettability of a solid substrate, i.e., if it is possible to quantify the surface free energy by measuring the contact angle, which, as shown above, is easy to measure. If the contact angle is measured with a liquid of known surface tension, then two terms in Young's equation remain unknown. Consequently, the surface free energy can be calculated if the interfacial tension between the liquid and the solid is first determined.

Looking at Fig. 4.26 the liquid and solid would form two different surfaces if no interaction existed between the solid and liquid phases. Therefore, the contact angle would be 180°. Since cos(180) = −1, and according to Young's equation, the interfacial tension between solid and liquid would be equal to the sum of the surface free energy and the surface tension of the liquid.

In reality, however, interactions always exist, and the interfacial tension must be reduced by the energy contributions from the interactions according to

$$\sigma_{SL} = \sigma_{SG} + \sigma_{LG} - (effect\ of\ interactions) \tag{4.6.10}$$

On a molecular level, the surface or interfacial tension can be decomposed into several components (Blake, 1984; Tricot, 1997). The most important, i.e., long-range, interaction forces are provided by disperse and poplar forces. Hence, surface and interfacial tensions can be modeled as the sum of a disperse and a polar part according to

$$\sigma = \sigma^d + \sigma^p \tag{4.6.11}$$

Based on this idea, Fowkes (1964) and later Owens and Wendt (1969), Rabel (1971), and Kaelble (1979) developed a model, which accounts for the solid–liquid interaction. They assumed that the interactions can be viewed as the geometric mean of a disperse part σ^d and a non-disperse part σ^{nd} (Fowkes), while Owens et al. defined the non-disperse part to be the polar part σ^p of the surface tension or surface free energy. Thus, Eq. 4.6.10 can be re-written as Eq. 4.6.12, which is named after the authors Owens, Wendt, Rabel, and Kaelble, and which is known as the OWRK model.

$$\sigma_{SL} = \sigma_{SG} + \sigma_{LG} - 2\left[\sqrt{\left(\sigma_{SG}^d \sigma_{LG}^d\right)} + \sqrt{\left(\sigma_{SG}^p \sigma_{LG}^p\right)}\right] \tag{4.6.12}$$

Furthermore, Owens et al. developed the following method, which allows the disperse and polar parts of the surface free energy to be determined by simply measuring the contact angles of two liquids of known surface tension and their disperse and polar parts.

Inserting Eq. 4.6.12 into Young's Eq. 4.6.1 yields

$$\sigma_{LG}(\cos(\theta) + 1) = 2\left[\sqrt{\left(\sigma_{SG}^{d}\sigma_{LG}^{d}\right)} + \sqrt{\left(\sigma_{SG}^{p}\sigma_{LG}^{p}\right)}\right] \tag{4.6.13}$$

Then, dividing Eq. 4.6.13 by $2\sqrt{\sigma_{LG}^{d}}$ produces a linear equation of the form

$$y = x\sqrt{\sigma_{SG}^{d}} + \sqrt{\sigma_{SG}^{d}} \tag{4.6.14}$$

Equation 4.6.14 is graphically shown in Fig. 4.30. The OWRK method calls for taking at least two liquids with known surface tensions including their disperse and polar components. Then, a drop of each liquid is to be dispensed onto the substrate surface to be evaluated to measure the resulting contact angle. This allows two points to be plotted in the graph shown in Fig. 4.30, and a straight line can be drawn through these two points. Consequently, the disperse and the polar part of the surface free energy can be read from the graph, and the surface free energy of the substrate can be determined with Eq. 4.6.11.

Suitable liquids for determining the surface free energy as well as the disperse and polar parts for a given substrate are water and di-iodomethane. Water is much more polar than di-iodomethane while the disperse part of di-iodomethane is larger than the one of water. The surface tensions and the disperse and polar parts of these two liquids are listed in Table 4.5.

More values of disperse and polar surface tension components for liquids can be found, for example, in Kopczynska and Ehrenstein (2017, https://www.lkt.tf.fau.de/files/2017/06/Oberflaechenspannung.pdf).

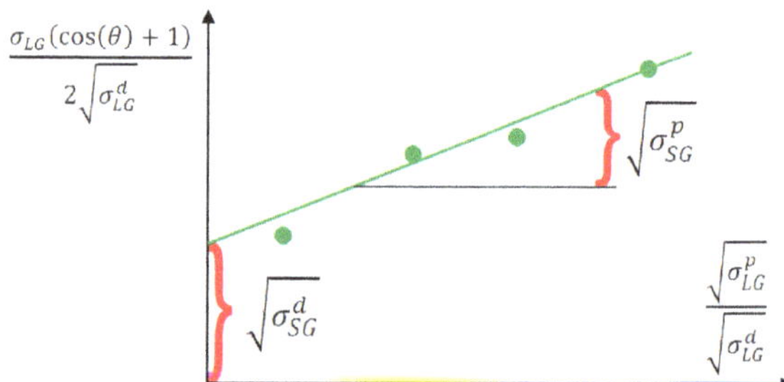

Fig. 4.30 Graphical representation of the OWRK method for determining the disperse and the polar parts of the surface free energy of a solid substrate

Table 4.5 Surface tensions and their disperse and polar parts of water and di-iodomethane

Liquid	Surface tension [mN/m]	Disperse part [mN/m]	Polar part [mN/m]
Water	72.8	21.8	51.0
Di-iodomethane	50.8	50.8	0.0

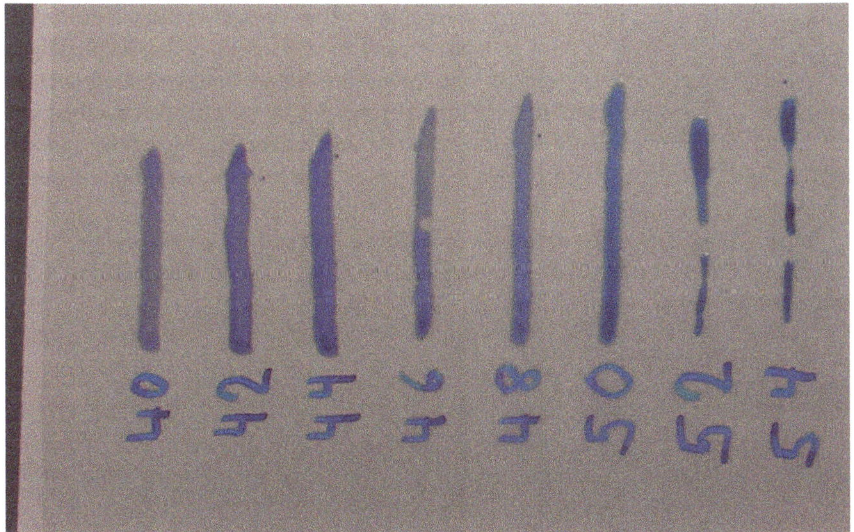

Fig. 4.31 Stripes of test inks with different surface tension on a plastic substrate

Alternatively, it is well known that the wettability of a substrate can also be determined with a set of test inks, also known as dyne pens, each of which has a slightly different surface tension. The goal is to find the ink that does not contract anymore but that is just able to spread upon coating the substrate with a stripe of ink. The so-called "three second rule" says that the right ink is found if spreading can be maintained for at least three seconds. It is then assumed that the surface tension of that ink equals the surface energy of the substrate. In the example of Fig. 4.31 the ink having a surface tension of 44 mN/m seems to be the right one. Using dyne pens is much simpler than working with the OWRK method discussed above. However, care must be taken to not contaminate the dyne pens. ISO 8296 defines a procedure for properly using dyne pens. Also, following the operating instructions provided by dyne pen manufacturers is helpful. Examples of such suppliers include Dyne Pens from Dyne Testing Ltd. (www.dynepens.co.uk) and *ACCU DYNE TEST*™ from DIVERSIFIED Enterprises (www.accudynetest.com).

4.6.4 Wetting Envelope

A wetting envelope is the wettability profile of a solid surface (www.kruss-scientific.com). The wetting envelope allows one to predict if a given liquid will completely wet a given substrate. Creating a wetting envelope builds on the OWRK method because it requires the determination of the surface free energy and its polar and disperse parts for a given substrate. The wetting envelope is a curve that is plotted in a graph with the disperse part of the surface tension or surface energy on the x-axis, and the polar part on the y-axis. If these two parameters for a given liquid are located inside the wetting envelope of a given substrate, then this liquid can completely wet that substrate. On the other hand, if these liquid properties lie outside the envelope, then the liquid will only partially wet the substrate and the wettability of the substrate by the liquid decreases with increasing distance of the liquid properties from the envelope.

In particular, complete wetting or spreading is given if the contact angle is zero, i.e., if cos(θ) is 1, see Fig. 4.25. Consequently, Eq. 4.6.13 combined with Eq. 4.6.11 can be re-written as

$$\sigma_{LG} = \sqrt{\sigma_{SG}^d \sigma_{LG}^d} + \sqrt{\sigma_{SG}^p \sigma_{LG}^p} = \sigma_{LG}^d + \sigma_{LG}^p \tag{4.6.15}$$

The wetting envelope is obtained by solving the right side of Eq. 4.6.15 for σ_{LG}^p and by calculating σ_{LG}^p as a function of σ_{LG}^d for the given substrate values of σ_{SG}^d and σ_{SG}^p. After some algebra σ_{LG}^p is given by

$$\sigma_{LG}^p = \frac{a + \sqrt{a^2 - 4b}}{2} \tag{4.6.16}$$

with

$$a = 2\left(\sqrt{\sigma_{SG}^d \sigma_{LG}^d} - \sigma_{LG}^d\right) + \sigma_{SG}^p \tag{4.6.17a}$$

and

$$b = \left(\sqrt{\sigma_{SG}^d \sigma_{LG}^d} - \sigma_{LG}^d\right)^2 \tag{4.6.17b}$$

The result is shown in Fig. 4.32 for a PTFE substrate and the liquids water, n-hexane, and toluene. n-hexane just barely lies inside the wetting envelope and therefore can wet the PTFE substrate. In contrast, toluene may be able to partially wet the substrate, while water has no chance whatsoever to wet PTFE.

Figure 4.33 shows the wetting envelopes of different substrates. Steel has a higher polar part than the other surfaces shown, while PMMA has a higher disperse part than the other substrates.

Values of disperse and polar surface energy components for different substrates can be found, for example, in Kopczynska and Ehrenstein (2017, https://www.lkt.tf.fau.de/files/2017/06/Oberflaechenspannung.pdf).

Figure 4.34 shows the effect of corona treatment on the wetting envelope of two plastic films. In particular, the corona treatment increases the polar part of the surface free energy of the substrate, thus enlarging the wetting envelope and hence allowing more liquids to better wet the substrate surface.

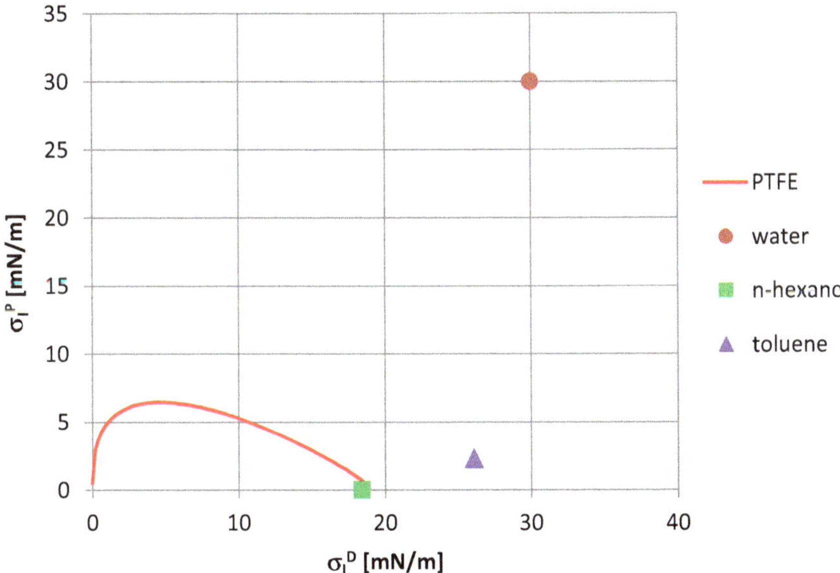

Fig. 4.32 Wetting envelope for water, n-hexane and toluene on a PTFE substrate

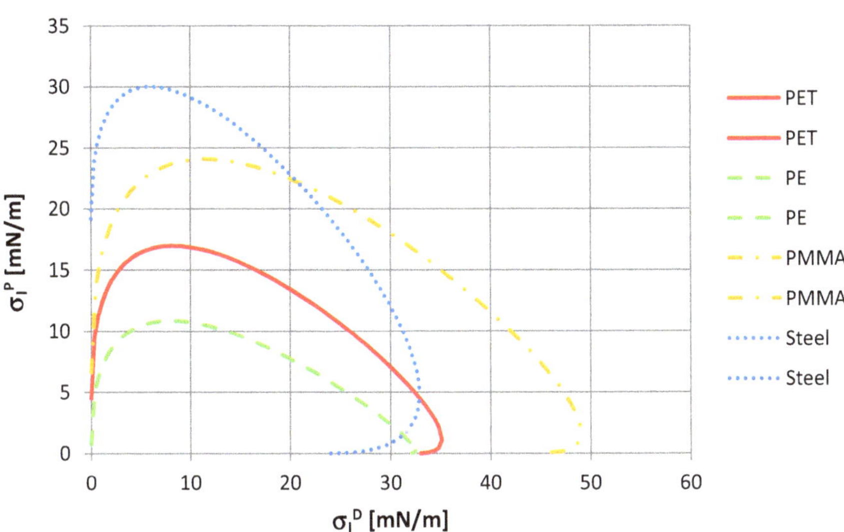

Fig. 4.33 Wetting envelope for different substrates. *Source* G. Gugler, iPrint Institute, University of Applied Sciences and Arts of Western Switzerland, Fribourg, Switzerland, reprinted with permission from G. Gugler

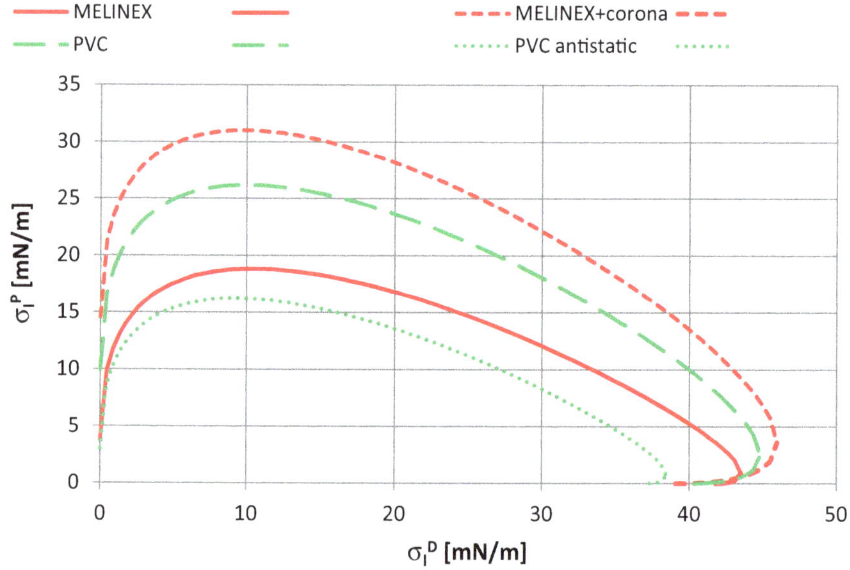

Fig. 4.34 Wetting envelope for MELINEX (voided PET) films without and with corona treatment and for PVC films with and without antistatic properties. *Source* G. Gugler, iPrint Institute, University of Applied Sciences and Arts of Western Switzerland, Fribourg, Switzerland, reprinted with permission from G. Gugler

Note that values for the polar and disperse surface tension or surface energy components of liquids and substrates vary depending on the source of the data. This variation is based on how the measurements were carried out, on how pure the liquids were, e.g., degree of desalination of water, and on the type and quantity of additives contained in substrates.

Even though coating is dynamic wetting it is important to attend to issues related to static wetting such as described in this chapter. Neglecting to assure that

- the surface to be coated is clean with regard to dirt or any other kind of contamination and homogeneous in terms of surface free energy, surface charge, and roughness and
- the requirements for the surface tension and the surface energy of the coating fluid and the substrate as well as the contact angle between these two materials are satisfied as described above

may result in insufficient adhesion between the coated film and the substrate, and it may generate coating defects such as longitudinal diffuse bands, mottle, de-wetting in the form of repellencies, edge withdrawal between coating and drying, and gross air entrainment at the dynamic wetting line.

References

Antoniades, M. G., Godwin, R., & Lin, S. P. (1980). A new method of measuring dynamic surface tension. *Journal of Colloid and Interface Science, 77*(2), 583.

Becerra, M., & Carvalho, M. S. (2011). Stability of viscoelastic liquid curtain. *Chemical Engineering and Processing: Process Intensification., 50*(5–6), 445–449.

Beer, R. (2020). Rheology of Mowiol 18–88. *Personal Communication.*

Bingham, E. C. (1916). *U.S. Bureau of Standards Bulletin, 13*, 309–353.

Bird, R. B., Armstrong, R. C., & Hassager, O. (1977). *Dynamics of polymeric liquids: Fluid mechanics.* Wiley.

Blake, T. D. (1984). Wetting. In T. F. Tadros (Ed.), *Surfactants.* Academic Press.

Blake, T. D., & Ruschak, K. J. (1997). Wetting: Static and dynamic contact lines. Chapter 3. In S. F. Kistler & P. M. Schweizer (Eds.), *Liquid film coating.* Chapman & Hall.

Blake, T. D., Fernandez-Toledano, J. C., Doyen, G., & de Coninck, J. (2015). Forced wetting and hydrodynamic assist. *Physics of Fluids, 27.*

Brown, D. R. (1961). A study of the behavior of a thin sheet of moving liquid. *Journal of Fluid Mechanics., 10*, 297.

Casson, N. (1959). In C. C. Mill (Ed.), *Rheology of disperse systems* (p. 84). Pergamon Press.

CRC Handbook of Chemistry and Physics (2019). 100th Edition. CRC Press.

Fowkes, F. M. (1964). Attractive forces at interfaces. *Industrial and Engineering Chemistry, 56*(12), 40–52.

Glass, J. E., & Prud'homme, R. K. (1997). Coating rheology: Component influence on the rheological response and performance of water-borne coatings in roll applications. Chapter 5. In S. F. Kistler & P. M. Schweizer (Eds.), *Liquid film coating.* Chapman & Hall.

Kaelble, D. H. (1979). Dispersion-polar surface tension properties of organic solids. *Journal of Adhesion, 2*, 66–81.

Kopczynska, A., & Ehrenstein, G. W. (2017). *Oberflächenspannung von Kunststoffen: Messmethoden am LKT.* Lehrstuhl für Kunststofftechnik, Friedrich-Alexander University Erlangen-Nürnberg, Germany.

Larson, R. G. (1999). *The structure and rheology of complex fluids.* Oxford University Press.

Macosko, C. W. (1994). *Rheology: Principles, measurements, and applications.* VCH Publishers Inc.

Nelson, N. K., Kistler, S. F., & Olmsted, R. D. (1990). A generalized viscous model for the steady shear rheology of magnetic dispersion. Paper at 62nd Annual Meeting of the Rheological Society. October 21 – 25, 1990. Santa Fe, NM.

Owens, D., & Wendt, R. (1969). Estimation of the surface free energy of polymers. *Journal of Applied Polymer Science, 13*, 1741–1747.

Rabel, W. (1971). Einige Aspekte der Benetzungstheorie und ihre Anwendung auf die Untersuchung und Veränderung der Oberflächeneigenschaften von Polymeren. *Farbe Und Lack., 77*(10), 997–1005.

Ruschak, K. J. (2008). Die Design. Lecture at Short Course in Coating and Drying Technology. International Society of Coating Science and Technology. September 6–7, 2008. Marina del Rey, California.

Ruschak, K. J., & Weinstein, S. J. (2014). A local power law approximation to a smooth viscosity curve with application to flow in conduits and coating dies. *Polymer Engineering and Science*, 2301 – 2309.

Schweizer, P. M. (2016). Viscosity versus rheology. *Converting Quarterly*, Quarter 3.

Secor, R. B. (1997). Analysis and design of internal coating die cavities. Chapter 10. In S. F. Kistler & P. M. Schweizer (Eds.), *Liquid film coating.* Chapman & Hall.

Tricot, Y. M. (1997). Surfactants: Static and dynamic surface tension. Chapter 4. In S. F. Kistler & P. M. Schweizer (Eds.) *Liquid film coating.* Chapman & Hall.

VDI-Wärmeatlas. (1974). Second edition. VDI-Verlag GmbH.

Vandre, E., Carvalho, M. S., & Kumar, S. (2012). Delaying the onset of dynamic wetting failure through meniscus confinement. *Journal of Fluid Mechanics., 707*, 496–520.
Wikipedia. (2022). Excess property. Accessed on February 10, 2022.
Willenbacher, N. (2019). Rheology of coating fluids. 11th Short Course in Coating and Drying of Thin Films. Karlsruhe Institute of Technology, 13–15 May 2019.

Chapter 5
Basic Flows of Premetered Coating Methods

Abstract Premetered coating methods comprise various basic flow fields including pipe, duct, slot, film, curtain, and boundary layer flow, as well as creeping flows that occur in the coated film before solidification. This chapter reviews theoretical models for analyzing and optimizing these flows. The focus is on analytical models because they are very well suited for visualizing the underlying physical concepts that control the flows. Also important is the use of the local power law model proposed by Ruschak and Weinstein (Polym Eng Sci 2301–2309, 2014), which, in combination with analytical flow equations and modest computational efforts, allows an accurate modeling of a complex behavior of the shear viscosity as described by the Carreau-Yasuda equation. In this context, various strategies for solving flow problems are proposed. Finally, for each basic flow field, potential locations for flow separation are visualized and ways for preventing the onset of vortices are suggested.

5.1 Introduction

The basic flow fields of premetered coating methods include pipe flow, duct flow, slot flow, film flow, curtain flow, boundary layer flow, and combinations thereof. To optimize the performance of these coating processes it is important to understand the details of all these basic flows. In particular, knowing the pressure drop in pipe flow upstream and downstream of the pump is necessary for designing a well-functioning fluid delivery system; knowing the pressure drop and the wall shear stress in duct and slot flows is essential for optimizing the internal geometry of dies, knowing the thickness of single and multilayer films on the inclined plane of slide dies is important for optimizing the height of the edge guides on both sides of the films, knowing the impingement speed of a falling liquid curtain and the size of the heel in the curtain impingement zone is paramount for preventing the onset of air entrainment. In addition, knowing the shear rate, particularly the wall shear rate of all these flows, is crucial for determining the prevailing viscosity of a given flow.

Knowing these flows and predicting their performance for a given set of operating conditions requires theoretical models, and various such models have long been

P. M. Schweizer, *Premetered Coating Methods*, Engineering Materials,
https://doi.org/10.1007/978-3-031-04180-8_5

discussed extensively in the literature. Bird et al. (1960, 1977), Liu (1983), and Ruschak and Weinstein (2014) are examples of suitable references.

Most of these basic flows, as they relate to premetered coating methods, can be considered as one-dimensional, meaning that the flow equations are averaged over the flow cross-section (Ruschak & Weinstein, 2014). In reality, however, many of the flows are three-dimensional. Examples of three-dimensional flow fields include the transition of the flow in the distribution cavity and the slot in a die (see Sect. 9.4 for more details), boundary layer effects in slot, film, and curtain flows, as well as vortices. Correctly modeling three-dimensional flows requires solving the equations of motion (Navier–Stokes equations) in three dimensions, and that in itself is a non-trivial task requiring powerful computers and well educated people with knowledge in fluid dynamics of confined and free surface flows, rheology and numerical mathematics. Most industrial coating companies do not employ such people. From an engineering point of view, however, exact three-dimensional solutions of flow problems are usually not necessary. It is more important to obtain decent ball-park solutions owing to the fact that industrial coating equipment is typically used for more than one application, meaning that the equipment has to be designed such that it performs optimally at best for one set of operating conditions, and that it performs reasonably well for several other sets of operating conditions. In that context, therefore, the assumption of one-dimensional flows is acceptable.

The purpose of this chapter is to

- summarize well known theoretical models for simulating the performance of the basic flows of premetered coating methods;
- focus on analytical and not numerical models because analytical models are better suited for visualizing the physical concepts that control these basic flows;
- focus on Newtonian (constant viscosity) and shear thinning flow behavior;
- discuss strategies for selecting approximate models when the flows become too complicated for simple analytical models such as multilayer film flow;
- provide suggestions for optimizing the various flow fields i.e., for designing flow fields that contain as many desirable and as few undesirable features as possible.

5.2 Vortices in Coating Flows

Before analyzing each basic flow field, it is meaningful to address the concept of vortices or flow separation, which is present in all of the basic flows. Vortices, or eddies, are areas of flow separation in a given flow field, i.e., areas where the main flow cannot follow the boundaries of the flow because its inertia is too high for a given form of the flow boundary, or because the flow boundaries are ill-designed for the inertia of the flow.

As will be shown in the following chapters, the characteristic flow fields of premetered coating processes provide many potential locations for vortices. Figure 5.1 shows a nice example of a vortex located in the upper meniscus of the coating bead in slide coating. Vortices are either attached to solid flow boundaries or to free surfaces

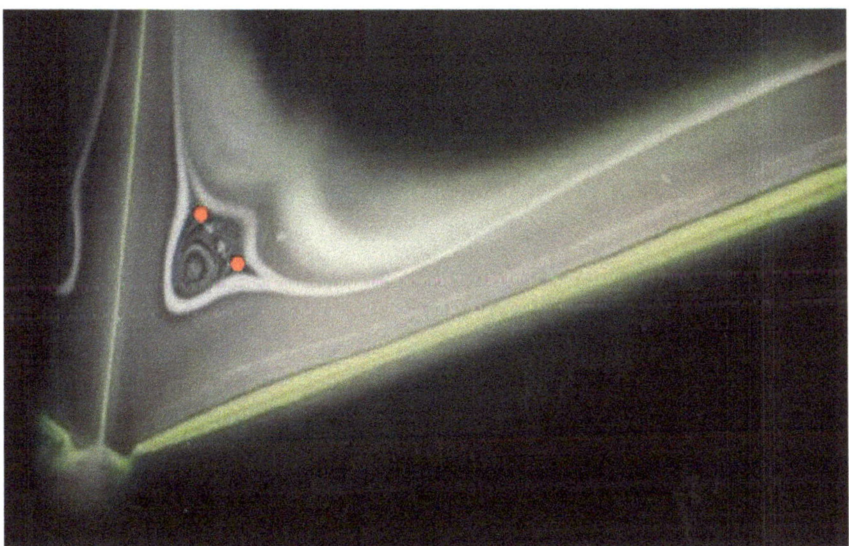

Fig. 5.1 Vortex in the coating bead of slide coating; some reflections of the flow cross-section are seen in the surface of the upper meniscus and in the substrate surface; photo taken from Schweizer (1988) and reprinted with permission from Cambridge University Press

of film flows. Rarely, a vortex may be completely contained inside a flow field without touching the flow boundary. The beginning and the end of a vortex along the flow boundary are marked by two stagnation points, where the velocity and the velocity gradient are zero as indicated by red dots in Fig. 5.1. The streamline connecting the two stagnation points across the flow field is called stagnation line. Consequently, the vortex stays put in the flow field, even though all fluid approaching the vortex, flowing around it and leaving it is flowing in the downstream direction. Moreover, the velocity of the fluid approaching a stagnation point changes from one direction to the opposite direction upon crossing the point. Therefore, all liquid inside a vortex rotates, and the sense of rotation is determined by the direction of the outer flow. At the inception of a vortex both stagnation points coincide at the same location, and the cross-section of the vortex is zero. Upon changing the flow conditions, however, the stagnation points move apart and the cross-section of the vortex increases.

Owing to the finite width of coating flows, and hence owing to the presence of side walls confining that width, the flow inside vortices is 3-dimensional and not 2-dimensional as suggested by the streamlines in Fig. 5.1. Flow visualization experiments by Schweizer (1988, 1997a, 1997b) revealed inflow and outflow into and out of vortices. The resulting flow in cross-web direction may be appreciable, at least in the vicinity of the confining side walls. In some cases, the flow inside a vortex may even become unstable by forming more or less regular cells along its axis all across the coating width. The example in Fig. 5.2 shows the nip feed of a 5-roll coater, by which highly viscous silicone is applied. Visualization of the cross-web cell structure in the vortex is made possible by micro-bubbles, which resulted from

Fig. 5.2 3-dimensional vortex structure in the cross-web direction in the nip feed of a 5-roll coater; photo reprinted with permission from Polytype Converting AG

massive air entrainment at the dynamic wetting line located in the free surface of the nip feed.

Experience and experimental evidence based on flow visualization show that vortices are an undesirable flow feature in any kind of coating flow because they are a major source for defects in the coated film, such as lines and streaks, cloudiness, diffuse and longitudinal bands. As explained by Schweizer (1997a), the mechanism for streak formation requires inhomogeneities in the liquid, and a location in the flow field, where these inhomogeneities can get trapped or stuck. Particles in the generic sense, i.e., solid particles, gas bubbles, and nicks in surfaces or edges of the flow boundaries qualify as inhomogeneities. Vortices can be considered as the main particle trap. A streak is formed, if a gas bubble, for example, gets stuck in a vortex and stays there for a while. Vortices, which are located in the free surface of the coating flow, such as the one shown in Fig. 5.1, are particularly destined to form streaks.

This statement is supported by an experiment where small particles were dropped onto the surface of a liquid film approaching the bead of a slide coater, and where the vortex in the upper meniscus of the bead was present as in Fig. 5.1. We observed that many of the particles got stuck in the vortex and the coated film was full of lines. However, when the flow conditions were continuously changed such that the cross-section of the vortex decreased, all particles were swept downstream when the vortex disappeared. At the same time, all lines disappeared in the coated film, and a particle was found at the end of each line.

As described above, vortices in coating flows bear the potential for diminishing the uniformity of the coated film, and hence they must be avoided. In the following chapters, possible locations for vortices in the various basic flow fields are presented and means for preventing these vortices are discussed.

5.3 Strategies for Solving Fluid Flow Problems

As mentioned in the introduction of this chapter, solving the full Navier–Stokes equations in 3 dimensions in combination with a suitable constitutive equation that relates flow properties such as the shear rate to physical material properties such as the viscosity is the most accurate but also the most complex and expensive way for obtaining solutions to flow problems. Coating engineers who are interested in using numerical approaches for solving coating flow problems will find a comprehensive summary on this topic in Christodoulou et al. (1997). Correctly handling the a priori unknown shape and location of free flow boundaries, and particularly incorporating a method for handling the dynamic wetting line are considerable challenges when using numerical methods. Back in the eighties and nineties of the last century, many of the numerical codes were written by Ph.D. students who solved one or two specific coating flow problems. Consequently, these codes were not available to the public. Nowadays, commercial programs are available that use the finite element or the finite volume methods for modeling the flows.

Solving simpler problems of confined flows, such as calculating shape factors (see Sect. 5.5) can easily be accomplished with the software package *Mathematica* from Wolfram (www.wolfram.com/mathematica). Flows with free surfaces can be handled by the package *Ansys Fluent* from Simu Tech Group (www.simutechgroup.com). The program *FLOW-3D* from Flow Science Inc. has been used for solving several coating flow problems and can handle dynamic wetting lines (www.flow3d.com). In any case, before buying a specific software package, it is recommended to clarify with the software supplier that the program can handle all the physical challenges of the intended coating applications such as free surfaces, dynamic wetting lines, etc.

Fortunately, numerical approaches are rarely justified when dealing with the characteristic flows of premetered coating processes as applied in industry. Instead, several simpler approaches are known for obtaining approximate solutions and the question is which of them is best for a given situation. In particular, the following solution strategies are available:

1. Newtonian flow equations with a constant viscosity, typically the low shear viscosity, or a low viscosity measured with a simple rheometer that operates at a low and often unknown shear rate such as orifice flow out of a cup.
2. Newtonian flow equations with a shear-dependent viscosity according to the Carreau-Yasuda model in combination with a suitable shear rate such as the wall shear rate or the average shear rate across the coating gap.
3. Power law flow equations with the corresponding shear-dependent viscosity according to Eq. (4.4.2) in combination with an appropriate shear rate such as the wall shear rate or the average shear rate across the coating gap. Here, the power law viscosity model is applied globally by setting the power law index n in Eq. (4.4.2) equal to the parameter b in the Carreau-Yasuda Eq. (4.4.1).
4. Power law flow equations with the corresponding shear-dependent viscosity according to Eq. (4.4.2) in combination with an appropriate shear rate such as the wall shear rate or the average shear rate across the coating gap. Here,

the power law viscosity model is applied locally (called local power law or LPL model) in accordance with Ruschak and Weinstein (2014) by determining accurate values for the material properties m and n over the entire shear rate range as described by viscosity data and the Carreau-Yasuda model.

5. Power law flow equations with the corresponding shear-dependent viscosity according to Eq. (4.4.2). Here, the power law viscosity model is applied locally (LPL model) in accordance with Ruschak and Weinstein (2014) by determining accurate values for the material properties m and n over the entire shear rate range as described by viscosity data and the Carreau-Yasuda model, and, if available, by calculating an equivalent viscosity and an equivalent shear rate.

The local power law model with an equivalent shear rate and viscosity as proposed by Ruschak and Weinstein (2014) is the best strategy because the mathematical difficulties are manageable, usually with analytical equations and in combination with an iteration, and the computational efforts are moderate compared to numerical solutions of the full Navier–Stokes equations. Moreover, comparing these two approaches, Ruschak and Weinstein obtained differences of less than about 2% when calculating the pressure gradient for pipe flow. Note, however, that the analytical equations mentioned here may contain parameters such as shape factors that must be determined with numerical methods, see Sects. 5.5.1.2, 5.5.1.3, and 5.5.2.2.

Nevertheless, both Newtonian and power law solutions are presented in this chapter for the characteristic flow fields of premetered coating methods. Including Newtonian flow equations is justified owing to their excellent ability to explain and visualize the underlying physical concept of a given flow. In addition, analytical power law solutions are not available for all the flow fields considered, e.g., boundary layer flow and multilayer film flow. Consequently, many of the results presented throughout this book are obtained with Newtonian flow simulations.

Note that most of the flow fields presented in Sect. 5.4–5.10 have been analyzed, discussed, and published in many text books and scientific papers many years ago. In spite of this fact, we decided to include these results and some of the derivations in the book at hand. Consequently, this comprehensive summary of flow equations prevents coating engineers working in industrial coating companies from consulting multiple resources.

5.4 Pipe Flow

5.4.1 Constant Viscosity (Newtonian Flow Behavior)

The velocity and shear stress profiles for one-dimensional flow in a straight pipe of constant cross-section are depicted in Fig. 5.3. The derivation and the solution of the equation of motion for this simple flow can be found, for example, in Bird et al. (1960). The velocity profile is calculated with

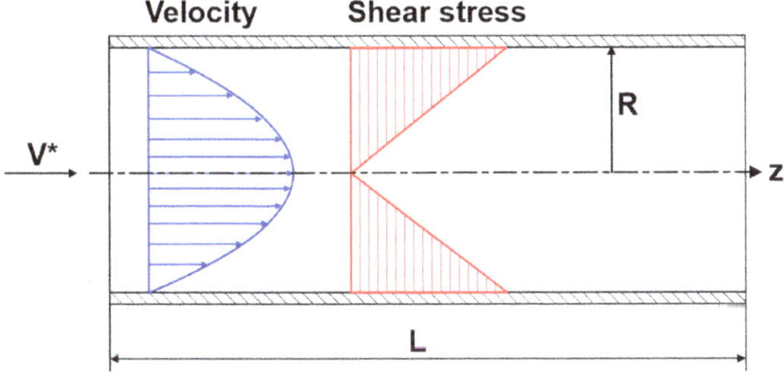

Fig. 5.3 Velocity and shear stress profiles for Newtonian pipe flow

$$v_z = \frac{(P_0 - P_L)R^2}{4\mu L}\left[1 - \left(\frac{r}{R}\right)^2\right] \tag{5.4.1}$$

P_0 is the pressure at the beginning of the pipe at z = 0, and P_L is the pressure at the end of the pipe at z = L. Moreover, P equals the sum of the static pressure and the gravitational force according to $P = p + \rho gh$, where h is the distance in the direction opposite to gravity from any reference plane. Whether or not the gravitational term ρgh can be neglected depends on the particular flow situation.

As can be seen from Eq. (5.4.1) and as is depicted in Fig. 5.3 the form of the velocity profile is parabolic and the shear stress profile is linear. Moreover, the velocity depends on the pressure drop of the pipe flow. In premetered coating methods, however, the pressure drop is unknown a priori. In contrast, the volumetric flow rate required for achieving a desired film thickness is known according to Eqs. (3.3.8) or (3.3.16). The flow rate for pipe flow can be calculated by integrating the velocity over the pipe cross-section, and the result is

$$V^* = \frac{\pi(P_0 - P_L)R^4}{8\mu L} \tag{5.4.2}$$

Equation (5.4.2) is the so called flow rate-pressure drop relationship for pipe flow or, in honor of the scientists who first developed this equation, the Hagen-Poiseuille law. Combining Eqs. (5.4.1) and (5.4.2) allows the velocity profile to be expressed in terms of known quantities according to

$$v_z = \frac{2V^*}{\pi R^2}\left[1 - \left(\frac{r}{R}\right)^2\right] \tag{5.4.3}$$

For premetered coating processes, therefore, the driving force for pipe flow is the required volumetric flow rate, and the pressure drop is a resulting consequence. Moreover, the pump for delivering this flow rate must be dimensioned such that it can overcome the resulting pressure drop.

The pressure drop for Newtonian pipe flow can be calculated by rearranging Eq. (5.4.2):

$$\frac{\Delta P}{L} = \frac{(P_0 - P_L)}{L} = \frac{8\mu V^*}{\pi R^4} \tag{5.4.4}$$

The maximum velocity is found in the center of the pipe at r = 0 and its value is

$$v_{z,max} = \frac{2V^*}{\pi R^2} \tag{5.4.5}$$

The average velocity is obtained by dividing the volumetric flow rate by the cross-sectional area of the pipe according to

$$v_{z,average} = \frac{V^*}{\pi R^2} \tag{5.4.6}$$

Note that the average velocity equals half of the maximum velocity.

The shear rate γ of pipe flow is equal to the velocity gradient, and it is calculated by taking the first derivative of the velocity profile. Taking the shear rate to be positive the result is shown by Eq. (5.4.7).

$$\gamma = \frac{4V^* r}{\pi R^4} \tag{5.4.7}$$

The shear rate is zero in the center of the pipe, and its maximum value is found at the pipe wall. Hence the wall shear rate is

$$\gamma_0 = \frac{4V^*}{\pi R^3} \tag{5.4.8}$$

Note that the wall shear rate is the appropriate shear rate value for determining the corresponding viscosity according to the Carreau-Yasuda model.

Newton's law of viscosity states that the shear stress is proportional to the viscosity and the velocity gradient (Bird et al., 1960). Consequently, the wall shear stress τ_0 can be calculated as

$$\tau_0 = -\mu \frac{dv_z}{dr} = \frac{4\mu V^*}{\pi R^3} \tag{5.4.9}$$

Equations (5.4.8) and (5.4.9) are interesting results because they show that, for a given set of operating conditions, i.e., for given values of the viscosity and the volumetric flow rate, the wall shear rate and the wall shear stress can be designed and controlled by the geometry of the flow boundary, i.e., the by the pipe radius. Normally, geometrical parameters are in control of the process engineer and not of the formulation chemist, meaning that process optimization can be carried out by manipulating geometrical parameters and not by parameters that affect the chemical product formulation. This issue is important when optimizing the fluid delivery system (see Sect. 8.1) and the internal die geometry as is explained in details in Sects. 6.3 and 8.2.4.

Finally, the Reynolds number Re for pipe flow, i.e., the dimensionless ratio of inertia to viscous force, is calculated by taking the pipe diameter as the characteristic length and the average velocity as the characteristic velocity according to

$$Re = \frac{2\rho V^*}{\pi \mu R} \tag{5.4.10}$$

Pipe flow is said to be laminar for Re < 2,100 (Bird et al., 1960). With regard to premetered coating methods it is important that all basic flows remain laminar. It is therefore the responsibility of the process engineer to assure laminar flows by choosing adequate values for geometrical parameters, e.g., pipe diameters, for given sets of operating conditions.

5.4.2 *Shear Rate Dependent Viscosity (Shear Thinning Flow Behavior)*

Flow equations for constant viscosities as presented in the previous chapter are only adequate if the shear rate-dependent viscosity remains constant up to reasonably high shear rates, and if the shear rate generated by the flow field of interest remains below the critical shear rate defined by the Carreau-Yasuda model. Referring to Fig. 4.6 this may be the case for a fluid such as a low-concentration aqueous PVA solution having a critical shear rate of 2,600 s^{-1}. Many industrial coating fluids have much lower critical shear rates, and two examples are also shown in Fig. 4.6. For such fluids, flow equations that can handle shear-dependent viscosities are more appropriate. From a mathematical point of view, and as already mentioned in Sect. 4.4.4, the simplest such model is the so called power law model (see Eq. 4.4.2), particularly if it is applied with the concept of the local power law approximation as proposed by Ruschak and Weinstein (2014). In this chapter, therefore, flow equations similar to those in Sect. 5.4.1 are presented, but now with the incorporation of the rheological power law model.

The derivation of the flow equations can be found, for example, in Bird et al. (1960), and the relevant results are as follows.

The velocity profile is described by

$$v_z = \left(\frac{(P_0 - P_L)R}{2mL}\right)^{1/n} \frac{R}{\left(\frac{1}{n}+1\right)}\left[1 - \left(\frac{r}{R}\right)^{\left(\frac{1}{n}+1\right)}\right] \tag{5.4.11}$$

The flow rate-pressure drop relationship is given by Eq. (5.4.12) and the pressure gradient is obtained by inverting Eq. (5.4.12).

$$V^* = \frac{\pi R^3}{\left(\frac{1}{n}+3\right)}\left[\frac{(P_0 - P_L)R}{2mL}\right]^{1/n} \tag{5.4.12}$$

$$\frac{(P_0 - P_L)}{L} = \frac{2m}{R}\left[\frac{V^*}{\pi R^3}\left(\frac{1}{n}+3\right)\right]^n \tag{5.4.13}$$

Inserting Eq. (5.4.13) into Eq. (5.4.11) yields the velocity profile in terms of known quantities.

$$v_z = \frac{(1+3n)}{(1+n)}\frac{V^*}{\pi R^2}\left[1 - \left(\frac{r}{R}\right)^{\frac{1}{n}+1}\right] \tag{5.4.14}$$

The dependence of the dimensionless velocity profile upon the power law index n is graphically shown in Fig. 5.4. Here the velocity is normalized with the average velocity, which is the same as shown in Eq. (5.4.6).

For Newtonian fluids, i.e., for n = 1.0, the velocity profile is parabolic and the maximum velocity is twice as high as the average velocity, which is in agreement with Eq. (5.4.5). As n decreases the wall shear rate increases, which decreases the velocity near the center of the pipe and increases this property near the pipe wall. In addition, the parabolic velocity profile changes into a boundary layer-like profile. In fact, the flow field approaches plug flow as n approaches 0. Note that if n = 1 in Eq. (5.4.14), then Eq. (5.4.3) describing the velocity profile for Newtonian fluids is recovered.

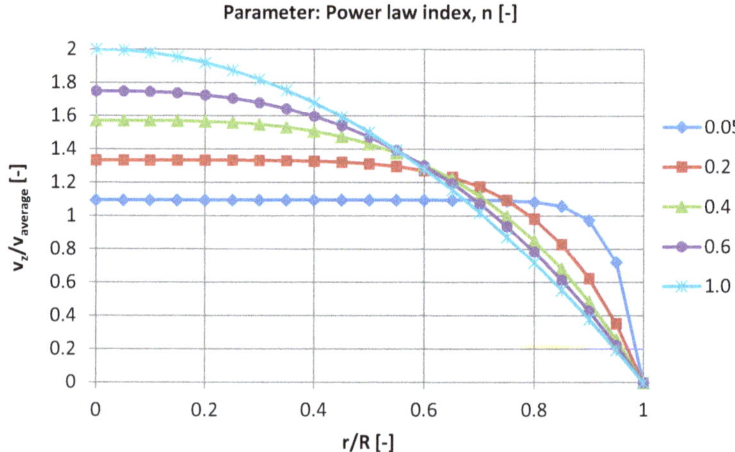

Fig. 5.4 Velocity profile of pipe flow for power law fluids

The shear rate γ of pipe flow is obtained by taking the first derivative of the velocity profile, and the relevant (positive) wall shear rate results by setting r = R as follows.

$$\gamma_0 = \frac{V^*}{\pi R^3}\left(\frac{1}{n} + 3\right) \tag{5.4.15}$$

If n = 1.0 the wall shear rate for Newtonian fluids according to Eq. (5.4.8) is recovered. The wall shear stress is given by Eq. (5.4.16).

$$\tau_0 = m\left[\frac{V^*}{\pi R^3}\left(\frac{1}{n} + 3\right)\right]^n \tag{5.4.16}$$

Assuming the same definition for the Reynolds number as used for Newtonian pipe flow, the Reynolds number for shear thinning pipe flow is

$$Re = \frac{2\rho}{m}\left(\frac{V^*}{\pi}\right)^{(2-n)} R^{(3n-4)}\left(\frac{1}{n} + 3\right)^{(1-n)} \tag{5.4.17}$$

As mentioned above, flow equations based on the power law model are only valid for the straight line portion of the viscosity curve (see Fig. 4.8), but they are inaccurate and therefore inacceptable for both the low shear and high shear Newtonian plateau as well as for both transition regions between constant and shear-dependent viscosity. To overcome this limitation without departing from the reasonably simple mathematics of the power law model, Ruschak and Weinstein (2014) introduced a **local power law model (LPL)** by constructing a tangent to the viscosity curve at the so called equivalent Newtonian viscosity μ_e for the power law model. The equivalent Newtonian viscosity was first proposed by Metzner and Reed (1955) as that giving the same wall shear stress and pressure gradient as the non-Newtonian liquid. Ruschak and Weinstein derived μ_e to be

$$\mu_e = m\left(\frac{3n + 1}{4n}\right)^n\left(\frac{8U}{D}\right)^{n-1} \tag{5.4.18}$$

where D is the pipe diameter and U the average velocity of the pipe flow. Using Eq. (5.4.6) for the average velocity, Eq. (5.4.18) can be re-written as

$$\mu_e = m\left(\frac{3n + 1}{4n}\right)^n\left(\frac{4V^*}{\pi R^3}\right)^{n-1} \tag{5.4.19}$$

The shear rate γ_e corresponding to μ_e is obtained by solving the Carreau-Yasuda Eq. (4.4.1) for γ_e according to

$$\gamma_e = \gamma_c \left\{ \left[\frac{\mu_e - \mu_\infty}{\mu_0 - \mu_\infty} \right]^{-1/(1-b)} - 1 \right\}^{1/a} \tag{5.4.20}$$

For solving a pipe flow problem, the unknown parameters m, n, μ_e and γ_e must first be determined. The solution procedure requires iteration by assuming a value for μ_e and by using Eqs. (5.4.20), (4.4.4), (4.4.2), and (5.4.19) for calculating γ_e, n, m and μ_e, where all equations are evaluated at μ_e and γ_e, respectively. The procedure must be repeated until the assumed and the calculated values for μ_e are equal. Ruschak and Weinstein recommend the Newton–Raphson iterative scheme, which works well even in the constant viscosity regions, but the goal seek function in EXCEL also serves the purpose well. Once these four parameters are known the pressure drop and the wall shear stress can be calculated with Eqs. (5.4.13) and (5.4.16).

5.4.3 Comparison of Flow Models for Solving Pipe Flow Problems

A representative flow situation was constructed in order to compare the fife solution strategies presented in Sect. 5.3. In particular, the pressure drop along a pipe with a decreasing cross-section was calculated. This type of flow is encountered when designing the distribution chamber of a slot die, see Sects. 6.3.1 and 8.2.4. In addition, a Carreau-Yasuda viscosity curve and a flow rate were chosen. The geometrical and rheological parameters as well as the flow rate were chosen such that the resulting shear rates covered a portion of the Newtonian low shear plateau and a portion of the shear thinning region, see Fig. 5.5.

Since the pipe diameter changes in flow direction, so does the shear rate and hence the viscosity according to the Carreau-Yasuda curve. The pressure drop along the pipe was calculated by dividing the pipe into several short sections, and local values for the shear rate and viscosity were calculated and kept constant over each section in order to calculate the local pressure drop. The overall pressure drop was then obtained by adding the local values. The accuracy of this approach increases with increasing number of short pipe sections.

Results are shown in Fig. 5.6 for a small flow rate generating shear rates between 384 and 25,409 s^{-1}, see red, solid, vertical lines in Fig. 5.5.

All models predict a higher pressure drop than the Ruschak–Weinstein LPL model (KJR + SJW). The global power law model with a constant power law index generates an excessively high pressure drop owing to the much higher viscosity at low shear rates when compared to the Carreau-Yasuda viscosity, see Carreau and power law curves in Fig. 5.5. The Newtonian model produces an excessively high pressure drop for higher shear rates, when the constant Newtonian viscosity remains much higher than the decreasing viscosity according to the power law model. The Newtonian model with a variable viscosity according to the Carreau-Yasuda viscosity curve comes closest to the KJR + SJW model. For the flow parameters chosen in this example the pressure drop of the Newtonian model is about 11% higher than the one predicted with the local power law model.

Increasing the flow rate by a factor of 7 such that the resulting shear rates lay between 2,686 and 172,209 s^{-1} (see green, dashed, vertical lines in Fig. 5.5) changed the pressure drops as shown in Fig. 5.7.

Now, the Newtonian model is way off because most of the shear rates of the flow are now located in the shear-thinning region of the viscosity curve. The power law model with a constant power law index is still off owing to the excessively high viscosities at the beginning of the flow field. Again, the Newtonian model with a variable viscosity comes closest to the KJR + SJW model, but the deviation of the overall pressure drop is now about 45%, which is substantial even for engineering approximations.

In summary, the local power law model according to Ruschak and Weinstein (2014) is always recommended for calculating properties of pipe flows. Whether or not other approximate models are acceptable strongly depends on the location of the characteristic shear rate range of the flow field under consideration relative to the shear rate range that characterizes the shear-thinning region of the Carreau-Yasuda viscosity curve.

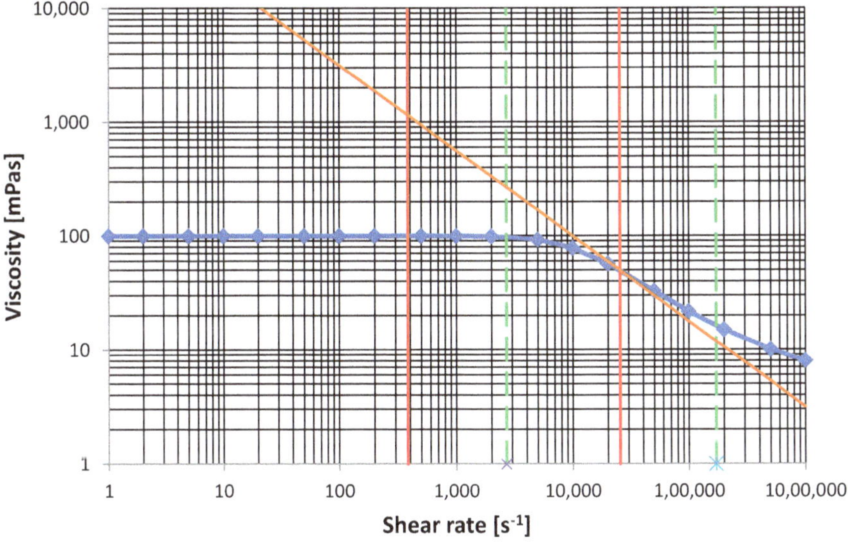

Fig. 5.5 Carreau-Yasuda viscosity curve and characteristic shear rate ranges of a pipe flow situation

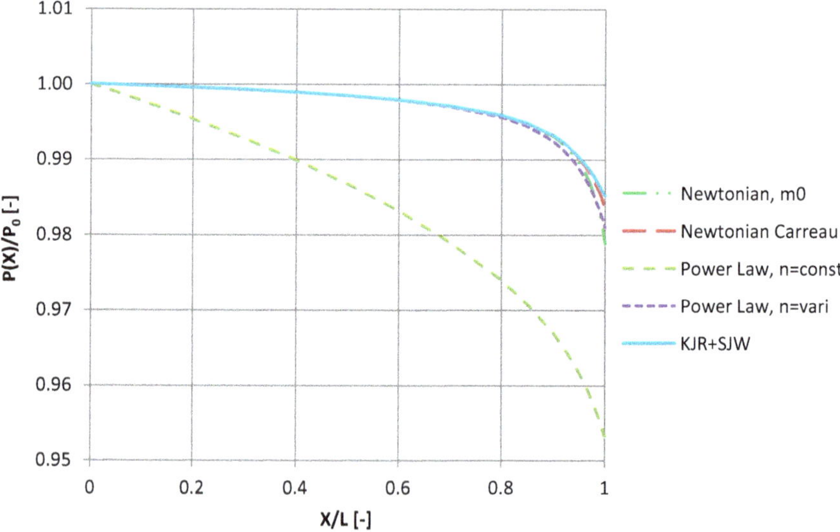

Fig. 5.6 Normalized pressure drop for flow in a pipe with decreasing cross-section, low flow rate. Parameter: Type of flow model

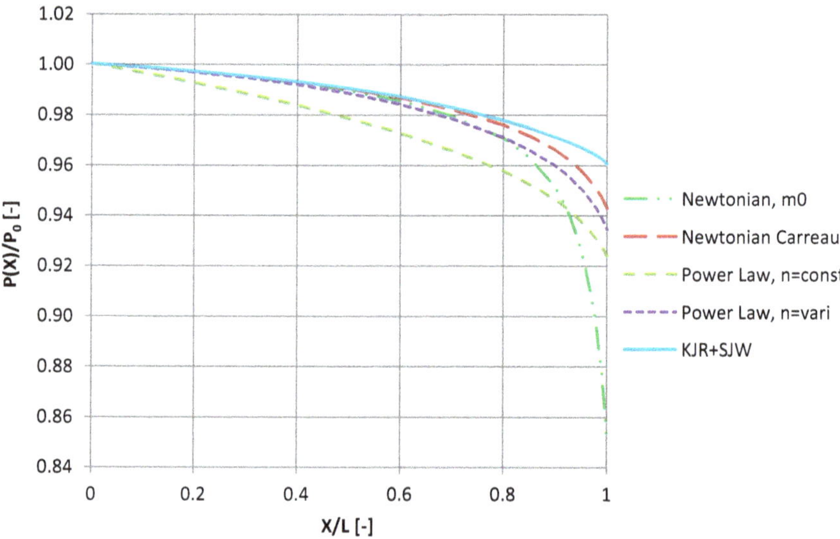

Fig. 5.7 Normalized pressure drop for flow in a pipe with decreasing cross-section, high flow rate. Parameter: Type of flow model

5.4.4 Vortices in Pipe Flow

Pipe flow is central to designing fluid delivery systems for premetered coating processes. Unfortunately, experience shows that it is difficult to design good fluid delivery systems that are free of undesirable features. The main reason is the fact that the elements that must be incorporated into a fluid delivery system such as a pump, a flow meter, a filter, a pulsation dampener, various vales, etc. all have different diameters. Consequently, the pipe diameter must be increased and decreased several times between the fluid supply vessel and the entrance to the die. Of particular concern are locations of increasing diameter because they bear the potential for flow separation, which must be avoided.

Flow details in conical diffusers have been studied extensively in the past, but most of the work focused on turbulent flows. Abramowitz (1949) investigated the flow of a viscous fluid in a diverging channel with straight walls (symmetric diffuser) and a parabolic entrance velocity profile, and he found that flow separation in such devices depends on the Reynolds number. In particular, a vortex can be avoided if the diffuser angle α is small enough according to

$$\alpha Re \leq 24.7 \qquad (5.4.21)$$

Here, Re is the Reynolds number for pipe flow according to Eq. (5.4.10). Equation (5.4.21) is graphically shown in Fig. 5.8, where the maximum diffuser angle α is plotted as a function of the Reynolds number and the diffuser geometry, i.e., for symmetrical diffusers as in pipe flow (Abramowitz constant = 24.7) and for unsymmetrical diffusers such as in slot flow where one wall is flat and the other one is diverging (Abramowitz constant = 12.35).

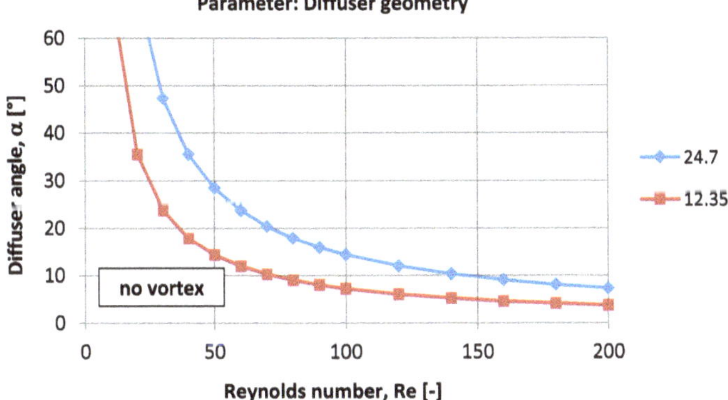

Fig. 5.8 Onset of flow separation in diffuser flow; symmetrical diffuser: constant = 24.7; unsymmetrical diffuser: constant = 12.35

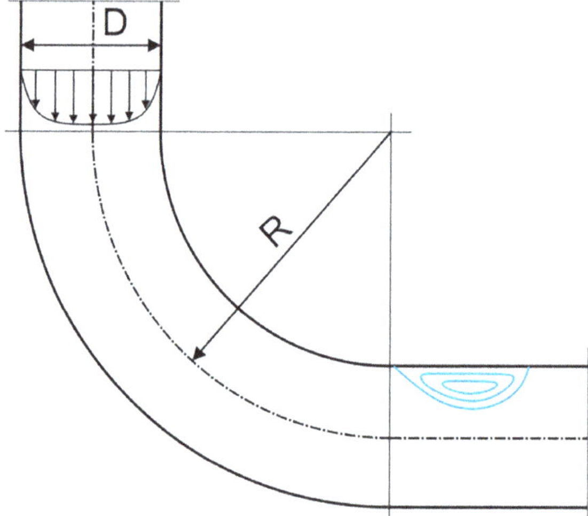

Fig. 5.9 Onset of flow separation at the exit of a pipe elbow

Generally speaking, diffuser angles should be small in order to prevent flow separation. However, an angle of 15° should be adequate for many industrial coating applications.

Another area of concern are elbows in the pipe line system, particularly 90° elbows. If the radius of the elbow is too small for given operating conditions, then a vortex appears just downstream of the elbow, which is undesirable and must be avoided, see sketch in Fig. 5.9. Moreover, the competition between centrifugal and viscous forces may generate a strong secondary flow in the plane normal to the pipe axis. This latter phenomenon is less critical than the vortex at the elbow exit because potential undesirable effects such as mixing or particle migration will cease to exist after the elbow.

Flow visualization experiments revealed that the critical Reynolds number at the onset of the vortex at the elbow exit decreases exponentially with the geometrical ratio of D/R. Specifically, the vortex can be avoided if, for a given value of D/R, the Reynolds number for pipe flow as calculated by Eq. (5.4.10) stays below the critical value. Alternatively, the vortex can be prevented if, for a given Reynolds number and pipe diameter, the radius of the elbow is sufficiently large.

5.5 Duct Flow

Duct flow is relevant for designing the internal geometry of slot dies. In this context, we must differentiate between flow along the axis of a duct, which describes the main flow in the inner distribution cavity, and flow across a duct, which is the main flow in the outer cavity of a dual cavity die. In fact, cross flow is the only flow in the outer cavity if a die is designed perfectly, i.e., if the inner distribution system does not generate any flow rate nonuniformity in the cross-web direction. In industrial applications, however, this ideal situation practically never exists. Therefore, relatively weak axial flow also exists in the outer cavity with the purpose of dampening widthwise flow rate nonuniformities generated by the inner distributions system. More details about die design are discussed in Sect. 8.2.4.

The purpose of the chapter at hand is to analyze characteristic features of duct flow both in the axial and in the cross direction.

5.5.1 Flow Along the Axis of Ducts

5.5.1.1 Pressure Drop

Flow along the axis of a duct is similar to pipe flow, except that the shape of the flow cross-section is anything but circular. The difference between pipe and duct flow is significant because exactly solving duct flow problems requires solving a partial differential equation on the two-dimensional domain of the duct in combination with the Carreau-Yasuda viscosity information.

Analytical equations for calculating the velocity profile of duct shapes that are relevant for die design are difficult to come by. Owen (1954), for example, developed an infinite series for the two-dimensional velocity profile of a gravity-driven flow of a Newtonian liquid in a rectangular channel as follows:

$$v_z(x, y) = \frac{\rho g}{2\mu}\left(a^2 - y^2\right) - \frac{16\rho g a^2}{\mu \pi^3} \sum_{n=1,3,5}^{\infty} \frac{(-1)^{\frac{n-1}{2}} cosh\left(\frac{n\pi x}{2a}\right) cos\left(\frac{n\pi y}{2a}\right)}{n^3 cosh\left(\frac{n\pi b}{2a}\right)} \tag{5.5.1}$$

with 2a = narrow side of rectangle, 2b = wide side of rectangle and g = gravitational acceleration.

More recently, Liu (1983) presented a model for the **viscous** pressure drop-flow rate relationship for power law fluids in ducts of arbitrary cross-section, see Eq. (5.5.2):

$$-\frac{dP}{dz} = \frac{m\dot{V}^{n}}{\lambda^n A^{(3n+1)/2}} \tag{5.5.2}$$

Here, A is the cross-sectional area of the duct, and λ is the viscous shape factor, which depends on the power law index n and the duct geometry but not on the consistency index m, see Sect. 5.5.1.2.

If the volumetric flow rate and the duct cross-section vary along the axis of the duct, which is the case for well-designed die cavities, then the pressure drop along the cavity axis is also determined by inertia effects. Assuming one-dimensional flow, Lee and Liu (1989) and Weinstein and Ruschak (1996) derived the following equation for the **inertial** pressure drop in duct flow:

$$\frac{dP}{dz} = -\frac{2\rho\Gamma}{A^2} V^* \frac{dV^*}{dx} + \frac{2\rho\Gamma}{L_c^5} V^{*2} \frac{dL_c}{dx} \tag{5.5.3}$$

L_c is the characteristic length of the cross-section according to $L_c = \sqrt{A}$, and Γ is the inertia shape factor, which also depends on the power law index and the duct geometry, see Sect. 5.5.1.3. If the flow rate is assumed to decrease linearly along the cavity axis as prescribed by Eq. (6.2.7), and if the cavity cross-section changes along its axis in accordance with Eq. (6.2.8), then the inertial pressure drop can be reformulated as

$$\frac{dP}{dz} = 2\rho\Gamma\left[V^*(x)\right]^2\left[\frac{1}{W[A(x)]^2\left(1+\alpha-\frac{x}{W}\right)} + \frac{1}{L_c^5}\frac{dL_c}{dx}\right] \tag{5.5.4}$$

Furthermore, as explained by Liu (1983), Miller (1972) derived the following expressions for the average wall shear stress $\overline{\tau}$ and the average apparent wall shear rate $\overline{\gamma}_0$:

$$\overline{\tau} = \frac{mV^{*n}}{S\lambda^n A^{(3n-1)/2}} \tag{5.5.5}$$

$$\overline{\gamma}_0 = \frac{V^*}{AS\lambda_N} \tag{5.5.6}$$

S is the wetted perimeter of the duct cross-section and λ_N is the viscous shape factor evaluated for $n = 1.0$. Note that when Eqs. (5.5.5) and (5.5.6) are calculated for Newtonian fluids flowing in a circular duct, then Eqs. (5.4.8) and (5.4.9) are recovered for the wall shear rate and the wall shear stress, respectively. In addition, Liu defined the Reynolds number for duct flow of power law fluids as follows:

$$Re = \frac{\rho V^{*(2-n)}}{mA^{(4-3n)/2}} \tag{5.5.7}$$

Ruschak and Weinstein (2014) took Liu's result for the viscous pressure drop and applied a local power law model as described in Sect. 5.4.2. In particular, they derived the equivalent Newtonian viscosity, i.e., the viscosity giving the same pressure gradient at the same flow rate, to be

$$\mu_e = \frac{m\lambda_N A^{3(1-n)/2}}{\lambda^n V^{*(1-n)}} \tag{5.5.8}$$

where λ_N is the value of λ for $n = 1$. For a given flow rate the unknowns are n, m, μ_e and γ_e with Eqs. (4.4.2), (4.4.4), (5.4.20), and (5.5.5) all evaluated at $\mu = \mu_e$ and $\gamma = \gamma_e$. As for pipe flow, iteration is required and the pressure gradient can then be calculated with Eq. (5.5.2). This approach is most interesting because shear-dependent viscosity data represented by the 5-parameter Carreau-Yasuda equation or even by the 7-parameter Bingham-Carreau-Yasuda model can now be modeled accurately with the 2-parameter power law model, which mathematically is much easier provided that the viscous shape factor is known for a given duct cross-section. A more detailed discussion of the implications of the viscous and inertia pressure drop on die design is presented in Sect. 9.4.5.

5.5.1.2 Viscous Shape Factor

Regarding the viscous shape factor λ for axial duct flow, Liu (1983) used the finite element method for calculating this parameter for several duct cross-sections and as a function of the power law index n. He presented the results in the form of

$$\lambda(n) = \left[d^{1/n}\left(\frac{e}{n} + f\right)\right]^{-1} \tag{5.5.9}$$

Here, d, e and f are fitting parameters, which depend on the shape of the duct. Equation (5.5.9) seems to give reasonable λ-values for $0.1 < n < 2.0$. For circular ducts (pipe flow) the fitting parameters can be calculated analytically and hence exactly, and if $n = 1.0$, $\lambda = 1/8\pi = 0.0397887$. Fitting parameters for calculating λ values of other duct shapes are given in Table 5.1, see also Fig. 5.10.

Table 5.1 Fitting parameters for various duct shapes

Constant	Circle	Semi-circle	Square	Equilateral triangle
d	$2\sqrt{\pi} = 3.5449077$	4.1798	3.7471	4.3821
e	$\sqrt{\pi} = 1.77245385$	1.3344	1.7330	1.0389
f	$3\sqrt{\pi} = 5.3173615$	6.6023	5.8606	6.8785

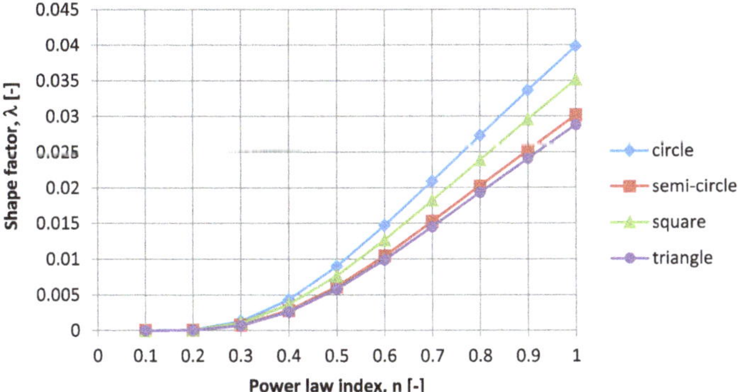

Fig. 5.10 Shape factor as a function of the power law index for various cavity shapes; data taken from Liu (1983)

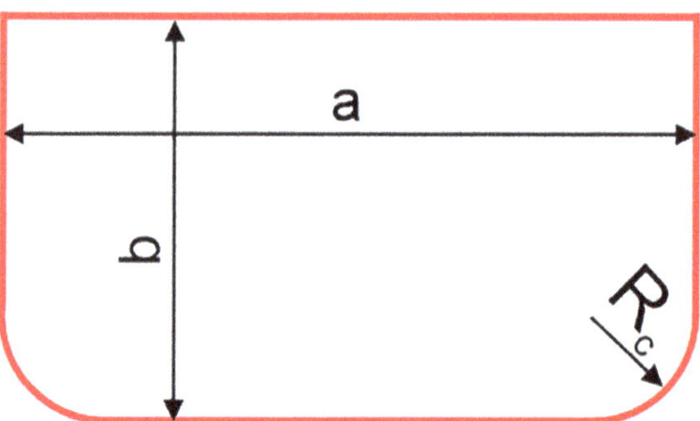

Fig. 5.11 Rectangle

Apart from the semi-circle and perhaps the square, unfortunately, the duct shapes analyzed by Liu are not so relevant for die design. Recently, however, Bebié and Roulier (2020a) used the software program *Mathematica* to solve Eqs. 7, 8, and 12 in Liu's paper and they determined the fitting parameters in Eq. (5.5.9) (Eq. 19 in Liu's paper) in order to calculate the shape factor as a function of the power law index for several duct cross-sections that are more relevant for die design as follows.

We then took these results and, with the help of regression analyses, sought simple functions that were able to express the dependence of the fitting parameters in Eq. (5.5.9) upon the variable geometry of some of the duct cross-sections. In doing so we applied the same procedure as described in Sect. 4.4.5 for determining the concentration dependence of the material parameters in the Carreau-Yasuda equation.

Duct shapes that are used in industry for designing the (inner) distribution cavity of a slot die include the rectangle and the square with two rounded corners, see Figs. 5.11 and 5.13. Here, A is the cross-sectional area, S is the wetted perimeter and L_c is the characteristic length of the duct defined as $L_c{}^2 = A$.

If a rectangular cavity cross-section is used, then it is assumed that the cross-section always starts as a square at the cavity inlet, that the long side a remains constant, and that the short side b decreases from the inlet to the end of the cavity. Consequently, the aspect ratio defined as $v = a/b$ increases from the cavity inlet to its end with $v = 1.0$ at the inlet. See Sect. 6.2.1 for more details on how the size of the cross-section decreases along its axis.

Moreover, rectangular cavity cross-sections with four sharp corners are not used in industrial applications. Instead, and as depicted in Fig. 5.11, the two corners located away from the die slot are rounded with the corner radius R_c to facility milling and polishing during die manufacturing, as well as cleaning during operation.

The shape factor for rectangular cross-sections can be calculated with Eq. (5.5.9) as a function of the power law index and the aspect ratio, whereby the dependence of the fitting parameters d, e and f on the aspect ratio a/b is described by a linear function that was determined by analyzing the results from the finite element calculations. The associated fitting parameters i, j and k as well as the r^2-value, which describes the quality of the fit ($r^2 = 1.0$ would be a perfect fit), are listed in Table 5.2. Note, however, that Bebié and Roulier calculated the shape factor for rectangles with four

Table 5.2 Fitting parameters for shape factor of rectangular cross-section

Fitting Eq.	i_0, j_0, k_0	i_1, j_1, k_1	r^2
$d = i_0 + i_1(a/b)$	3.3840	0.3322	0.9977
$e = j_0 + j_1(a/b)$	0.9919	0.5585	0.9732
$f = k_0 + k_1(a/b)$	4.8882	0.9750	0.9982

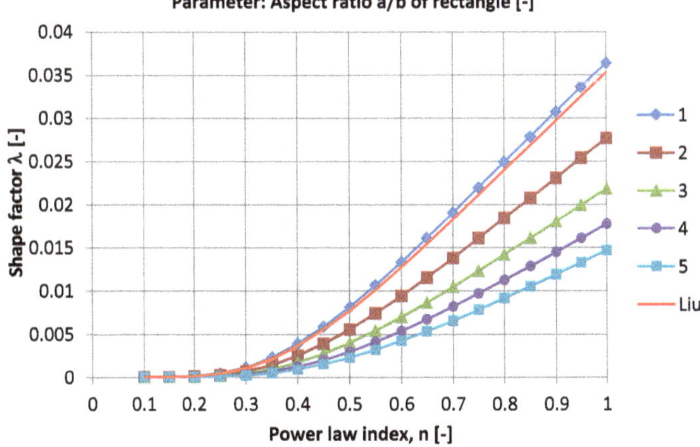

Fig. 5.12 Shape factor of rectangle as a function of the power law index and the aspect ratio

sharp corners because, as will be demonstrated below, the effect of the corner radius on the value of the shape factor is small. In addition, their values calculated with the finite element method over the entire range of the power law index agree to better than 0.4% with values calculated by Liu (1983) or Ruschak and Weinstein (2019).

Equation (5.5.9) for rectangular cross-sections is visualized in Fig. 5.12. a/b = 1 corresponds to a square cross-section, and that shape was also calculated by Liu (1983). As can be seen in Fig. 5.12 our calculations predict slightly higher λ-values than Liu. Specifically, the difference between the λ-values calculated with our fitting parameters and Liu's values is 3.2–5.5% for $0.5 < n < 1.0$.

A square with two rounded corners is an interesting cavity cross-section because it comes close to the circle, which is the best cross-section of all, but which is not interesting for industrial applications owing to its excessive machining efforts (a semi-circle must be machined into each of the adjacent die plates) and hence its excessive manufacturing costs. The corner radius must be smaller or equal to half of the square side. Moreover, the form with R = a/2 is of special interest for die design because then the cross-section consists of half a square and half a circle, see Fig. 5.13.

The shape factor for a square with two rounded corners can be calculated with Eq. (5.5.9) as a function of the power law index and the ratio of corner radius to square side R_c/a, whereby the dependence of the fitting parameters d, e and f on the ratio R_c/a is described by a power law function, which was determined by analyzing

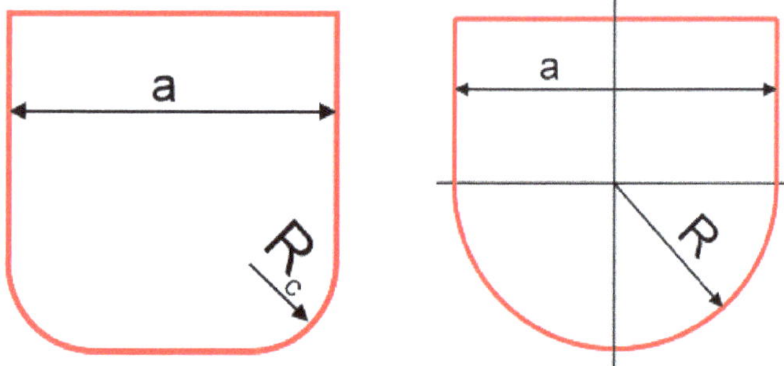

Fig. 5.13 **a** Square with two rounded corners **b** half square plus half circle

the results from the finite element calculations. The associated fitting parameters i, j and k as well as the r^2-value, which describes the quality of the fit, are listed in Table 5.3.

The quality of the regression analysis is not as good as it is for the rectangle. This, however, is not very important because the effect of the corner geometry on the shape factor is only small, see Fig. 5.14. The shape factor increases with increasing corner radius. The λ-curve of a circular cross-section is also shown in this Figure for

Table 5.3 Fitting parameters for shape factor of a square with two rounded corners

Fitting Eq.	i_0, j_0, k_0	i_1, j_1, k_1	r^2
$d = i_0(R_c/a)^{i_1}$	3.6112	−0.0149	0.9729
$e = j_0(R_c/a)^{j_1}$	1.7633	0.00646	0.7292
$f = k_0(R_c/a)^{k_1}$	5.4970	−0.02574	0.9797

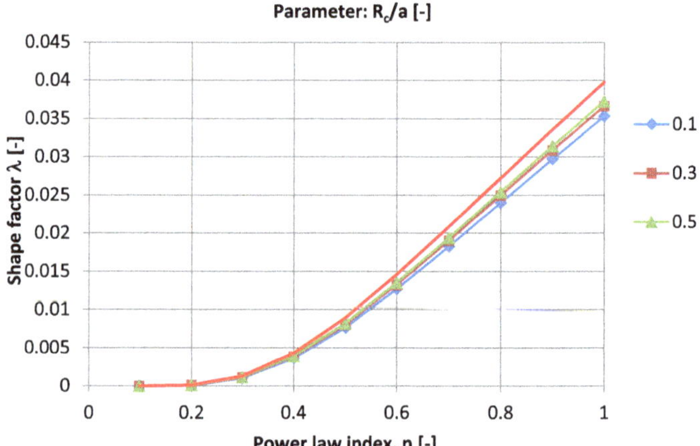

Fig. 5.14 Shape factor of a square with two rounded corners as a function of the power law index and the ratio of R_c/a

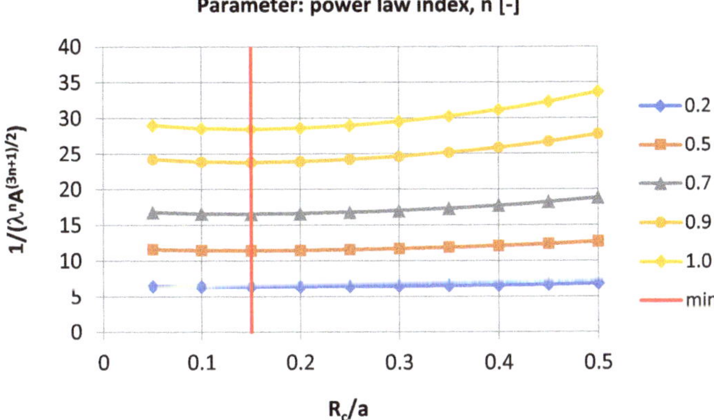

Fig. 5.15 Optimum shape for the square with two rounded corners

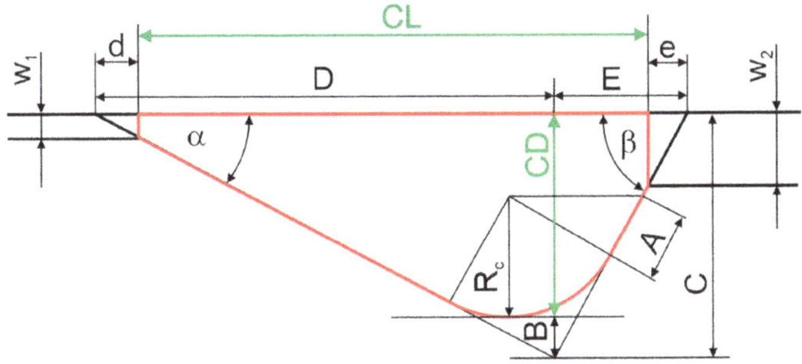

Fig. 5.16 Triangle with rounded 90° corner

comparison with the square for $R_c/a = 0.5$. The λ-values for the circle are higher by up to 10% for $n > 0.35$.

Analyzing this cavity shape reveals that λ increases with increasing R_C/a while the area A decreases with increasing R_C/a. The question therefore arises if there is an optimum shape for this flow cross-section. Using the criterion of minimizing the pressure drop in the cavity, then, according to Eq. (5.5.2), the term $1/(\lambda^n A^{(3n+1)/2})$ should be minimized. Normalizing the cross-sectional area with a^2, this term is plotted as a function of R_C/a and n, see Fig. 5.15. Indeed, an optimum shape exists for $R_C/a = 0.15$, and this is true for any value of n. However, the optimum is very flat, thus indicating that optimizing this cavity cross-section does not bring significant gains in terms of minimizing the pressure drop, particularly if $n < 0.5$.

Duct shapes that are used in industry for designing the outer distribution cavity of a dual-cavity die include the triangle with a rounded 90° corner and the circular segment including the semi-circle, see Figs. 5.16 and 5.18. The rounded corner at the bottom vertex in Fig. 5.16 is necessary from a die manufacturing point of view and, perhaps more importantly, from a contamination and cleaning point of view during coating.

As is explained in Sect. 9.4 on die design, relevant geometrical parameters of these cavity shapes are the cavity length CL and the expansion angle α. It is therefore meaningful to express the cavity depth CD, the area A and the circumference S in terms of these two parameters while also considering the height w_1 and w_2 of the inner and outer die slot, as well as the radius R_c of the rounded 90° corner. Note that the exact area and the circumference are marked in red in the figures.

$$C = \frac{CL + \frac{w_1}{\tan\alpha} + w_2 \tan\alpha}{\frac{1}{\tan\alpha} + \tan\alpha}$$

$$CD = C - R_c[(1 + \tan\alpha)\cos\alpha - 1]$$

$$A = \frac{C^2}{2}\left(\frac{1}{\tan\alpha} + \tan\alpha\right) - \frac{w_1^2}{2\tan\alpha} - \frac{w_2^2 \tan\alpha}{2} - R_c^2\left(1 - \frac{\pi}{4}\right)$$

$$S = C\left(\frac{1}{\tan\alpha} + \tan\alpha + \frac{1}{\cos\alpha} + \frac{1}{\sin\alpha}\right) + w_1\left(1 - \frac{1}{\tan\alpha} - \frac{1}{\sin\alpha}\right) + w_2\left(1 - \tan\alpha - \frac{1}{\cos\alpha}\right) - R_c\left(2 - \frac{\pi}{2}\right)$$

$$R_{c,max} = \frac{C - w_1}{\cos\alpha} \quad 0 < \alpha \leq \pi/4.$$

$$R_{c,max} = \frac{C - w_1}{\sin\alpha} \quad \alpha > \pi/4$$

Ruschak and Weinstein (2018) used the finite element method for computing the viscous shape factor for fife triangular cavity shapes with the following expansion and contraction angles α and β: 30–30, 30–60, 30–45, 45–45, 60–30. The third angle with values of 120, 90, 115, 90, 90 degrees was sharp and not rounded as postulated in Fig. 5.16. The authors argued that this difference is not significant with regard to the viscous shape factor and the pressure drop as long as the duct cross-section is calculated accurately, see Fig. 5.16.

Ruschak and Weinstein resorted to Eq. (5.5.9), which was developed by Liu (1983) for calculating the viscous shape factor as a function of the power law index n. The constants d, e, and f for the fife triangular cavity shapes as determined by Ruschak and Weinstein are listed in Table 5.4.

In addition, Bebié and Roulier (2020a) calculated the viscous shape factor for the 30°–60° triangle with a rounded 90° corner, a form that is often used in industry. Here, the shape factor also depends on the corner radius R_c, which is normalized by the cavity length CL. The value of CL is typically limited due to space restrictions for accommodating the outer cavity in a die bar. Therefore, the varying dimensionless geometrical parameter is $v = R_c/CL$. Note that there is a maximum value for R_c or v, which depends on the expansion angle α and other geometrical parameters as is indicated by the equations below the drawing in Fig. 5.16. $R_{c,max}$ corresponds to the maximum square with side length R_c that can be inserted into the triangle.

The shape factor for a triangle with a rounded 90° corner can be calculated with Eq. (5.5.9) as a function of the power law index n and the geometrical ratio v, whereby the dependence of the fitting parameters d, e, and f on the ratio R_c/CL is described by a linear function. The associated fitting parameters and the r^2-value, which describes the quality of the fit, are listed in Table 5.5.

The quality of the regression analysis is not very good owing to significant data scatter. This, however, is not important because the effect of the corner geometry on the shape factor is very small, i.e., negligible, see Fig. 5.17. Also included in the graph is the curve from Ruschak and Weinstein (2018) for this triangle, but with a sharp corner at the 90° angle, i.e., $R_c = 0$. This comparison confirms the assumption made by Ruschak and Weinstein that the corner radius at the 90° angle has little influence on the shape factor.

As with the triangular cavity shape, the geometry of the circular segment is defined by the expansion angle α and the cavity length CL. Moreover, the cavity depth CD, the area A and the circumference S are expressed in terms of these two parameters while also considering the height w_1 and w_2 of the inner and outer die slot, as well as the radius R of the circle, see Fig. 5.18. Note that the equations for A and S are slight approximations because the areas of the intersection with both slots and the circular segment are calculated as triangles with all three sides being straight instead of one side being curved. Also note that the area and the circumference are marked in red in the figure.

$$R = \frac{CL + \frac{w_1 + w_2}{\tan\alpha}}{2\sin\alpha}$$

$$CD = R(1 - \cos\alpha)$$

$$S \approx 2R(\alpha + \sin\alpha) + (w_1 + w_2)\left(1 - \frac{1}{\tan\alpha} - \frac{1}{\sin\alpha}\right)$$

$$A \approx R^2(\alpha - \sin\alpha\cos\alpha) - \frac{w_1^2 + w_2^2}{2\tan\alpha}$$

Table 5.4 Constants for calculating the viscous shape factor of triangular cavity shapes

	30–30	30–60	30–45	45–45	60–30
d	4.52076	4.08447	4.32751	4.12394	4.24399
e	2.66232	2.94266	2.37790	2.09024	2.27579
f	8.64463	7.31132	7.92910	7.18292	7.62118

Table 5.5 Fitting parameters for the shape factor of a 30°–60° triangle with a rounded 90° corner

Fitting Eq.	a_0, b_0, c_0	a_1, b_1, c_1	r^2
$d = a_0 + a_1(R_c/CL)$	4.2571	0.0399	<0.5
$e = b_0 + b_1(R_c/CL)$	2.1599	0.4493	0.8542
$f = c_0 + c_1(R_c/CL)$	7.5936	0.3299	<0.5

Using the finite element method and the software package *Mathematica*, Bebié and Roulier (2020a) calculated the shape factor for this cavity form as a function of the expansion angle α and the power law index n. Moreover, they subjected the results to a regression analysis based on a power law function in order to determine the dependence upon α of the fitting parameters in Eq. (5.5.9). These parameter values and the r^2-values, which describes the quality of the fit, are listed in Table 5.6.

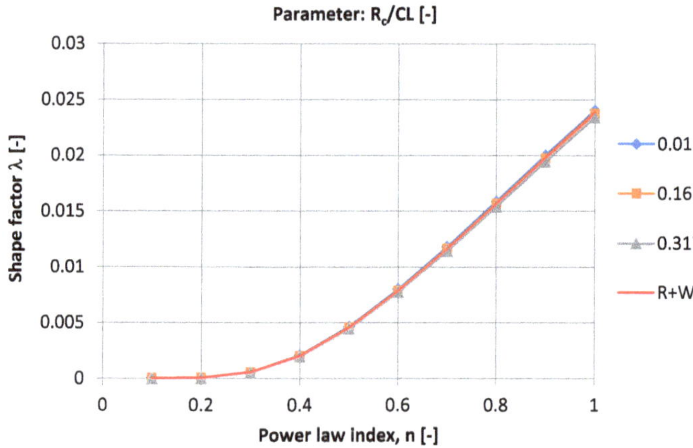

Fig. 5.17 Shape factor for a 30°–60° triangle with rounded 90° corner as a function of the power law index n and the geometrical ratio R_c/CL

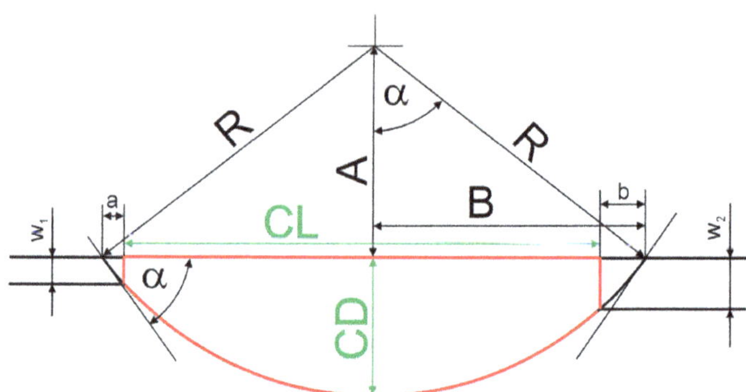

Fig. 5.18 Circular segment

The quality of the regression analysis is very good. The shape factor increases with increasing corner angle α, see Fig. 5.19. The symbols in this graph are the numerical values calculated by Bebié and Roulier, while the lines are generated with Eq. (5.5.9).

When $\alpha = 90°$ then the cavity shape is half a circle and that form was also calculated by Liu (1983), see red line for a comparison in Fig. 5.19. As can be seen, Bebié's data are very close to Liu's values, with a difference of <1% for n > 0.5. The predictions with the regression Eq. (5.5.9) are a bit higher than Liu's values for all n values, with differences of about 2 5.5% for n > 0.5, which is not significant.

5.5.1.3 Inertia Factor

If the cross-section of the inner cavity is not constant along the cavity axis, which is often the case, then the pressure drop in the cavity is also determined by inertia effects (Lee & Liu, 1989). The magnitude of these effects depends on the details of the velocity profile, and these in turn depend on the shape of the duct cross-section and, for power law fluids, on the power law index. Like viscous contributions to the pressure drop can be expressed by the dimensionless shape factor λ, inertia terms can be expressed by a dimensionless inertia factor Γ.

Leonard (1985) defined the inertia factor Γ as follows:

$$\Gamma = \frac{1}{A}\int \frac{u^2}{\hat{u}^2} dA \tag{5.5.10}$$

A is the duct cross-section and $\hat{u}$ is the average velocity. Γ must be calculated numerically, and Lee and Liu (1989) published such data based on the power-law rheological model and for the following duct shapes: circle, semi-circle, tear and semi-tear. The results are shown in Fig. 5.20.

The points in the graph are the numbers from Lee and Liu (1989), and the lines are generated with a regression analysis based on the power law $\Gamma = a_0 n^{a1}$. The regression constants a_0 and a_1 depend on the duct cross-section, and they are listed in Table 5.7. The quality of the fit is good as indicated by high r^2-values.

Note that inertia effects are least felt in "open" duct shapes, i.e., circular cross-section. Using Eq. (5.4.3) in combination with Eq. (5.5.10), the inertia factor for a circular cross-section and for n = 1.0 can easily be calculated by hand to be Γ = 4/3 = 1.333. In addition, Roulier and Bebié (2020b) derived an analytical equation for calculating the inertia factor for power law fluids flowing in a circular duct as

Table 5.6 Fitting parameters for the shape factor of a circular segment

Fitting Eq.	a_0, b_0, c_0	a_1, b_1, c_1	r^2
$d = a_0(\alpha)^{a_1}$	17.6794	−0.3361	0.9930
$e = b_0(\alpha)^{b_1}$	92.3776	−0.8814	0.9917
$f = c_0(\alpha)^{c_1}$	64.1520	−0.5053	0.9992

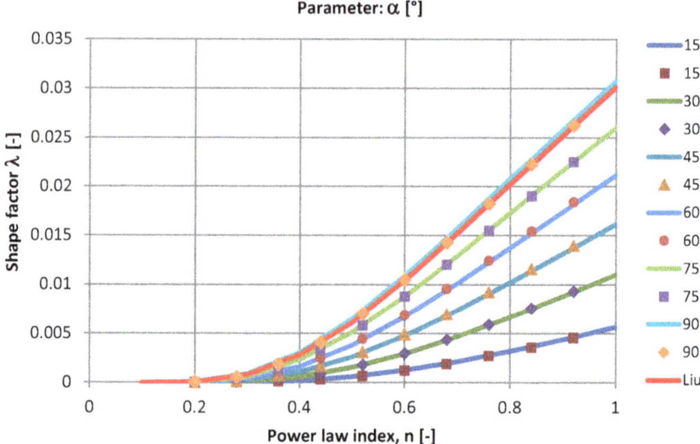

Fig. 5.19 Shape factor for circular segments as a function of the power law index n and the expansion angle α

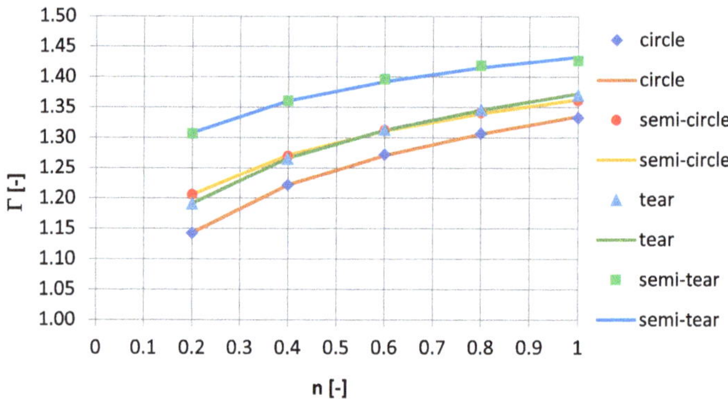

Fig. 5.20 Inertia factor for the circle, semi-circle, tear and semi-tear as a function of the power law index n; data taken from Lee and Liu (1989)

follows:

$$\Gamma = \frac{(1 + 3n)}{(1 + 2n)} \tag{5.5.11}$$

Table 5.7 Regression constants for various duct cross-sections

Duct shape	Circle	Semi-circle	Tear	Semi-tear
a_0 [–]	1.33519	1.36319	1.37253	1.43313
a_1 [–]	0.09636	0.07614	0.08791	0.05658
r^2	0.9997	0.9994	0.9997	0.9931

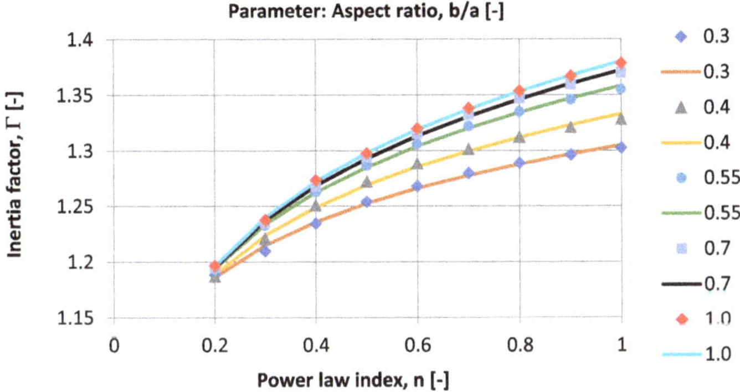

Fig. 5.21 Inertia factor for the rectangle as a function of the power law index n and the aspect ratio b/a

This equation also produces Γ = 4/3 for n = 1.0.

As symmetrical flow cross-sections like the circle or the tear are not used in industrial die design, Roulier and Bebié (2020b) used the finite element method and the software package *Mathematica* for calculating the inertia factor for a rectangle with four sharp corners (Fig. 5.11) and for a square with two rounded corners (Fig. 5.13) as a function of the power law index n and the aspect ratios b/a and R_c/a, respectively. When checking the accuracy of their numerical finite element calculations for the circular cross-section, Roulier and Bebié exactly recovered the Γ-values as predicted by Eq. (5.5.11) for n between 0.3 and 1.0. For n = 0.2, the numerical value deviated by −0.23%. The results for the rectangle are shown in Fig. 5.21.

The symbols in Fig. 5.21 are the numerically calculated values, while the lines through the symbols are generated by the following regression analysis.

$$\Gamma = a_0 n^{a_1} \tag{5.5.12}$$

Here, a_0 and a_1 express the dependence of Γ upon the aspect ratio b/a of the rectangle and both parameters are approximated by a third order polynomial function as follows:

$$a_i = b_0 + b_1\left(\frac{b}{a}\right) + b_2\left(\frac{b}{a}\right)^2 + b_3\left(\frac{b}{a}\right)^3 \tag{5.5.13}$$

The regression constants b_i are listed in Table 5.8. The quality of the fit is good with all r^2-values being >0.99.

Table 5.8 Regression constants for inertia factor of rectangle

	b_0	b_1	b_2	b_3
a_0	1.164070	0.648671	−0.663029	0.230572
a_1	0.005506	0.247396	−0.247655	0.083759

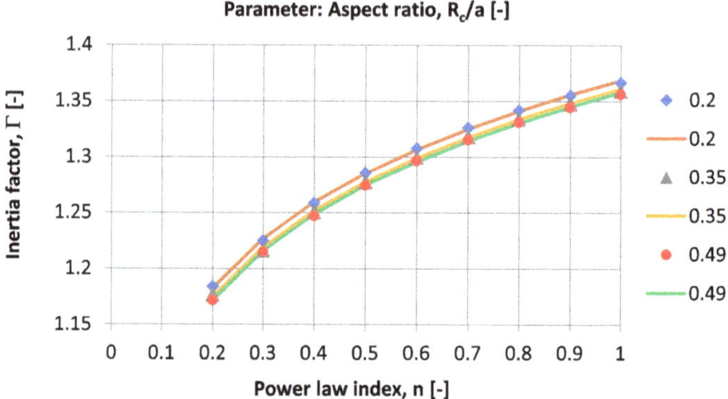

Fig. 5.22 Inertia factor for the square with two round corners as a function of the power law index n and the aspect ratio R_c/a

Table 5.9 Regression constants for inertia factor of square with two round corners

	b_0	b_1	b_2	b_3
a_0	1.388416	−0.127907	0.150559	−0.032543
a_1	0.089810	0.003919	–	–

The results for the square with two round corners are depicted in Fig. 5.22.

Again, the symbols in Fig. 5.22 are the numerically calculated values, while the lines through the symbols are generated by the same power law function shown in Eq. (5.5.12). a_0 and a_1 express the dependence of Γ upon the aspect ratio R_c/a of the duct shape. Here, a_0 is approximated by a 3rd order polynomial function according to Eq. (5.5.13), while a_1 is calculated with the following linear function: $a_1 = b_0 + b_1(R_c/a)$. The corresponding regression constants b_i are shown in Table 5.9. The quality of the fit is equally good with all r^2-values being >0.99. Moreover, it can be seen that changing the corner radius only has a small effect on the inertia factor. However, reducing the cross-sectional area, i.e., if R_c/a approaches 0.5, reduces Γ.

Regarding the axial flow in the outer cavity, inertia effects are generally small if not negligible because the main flow is across and not along the cavity. Nevertheless, Ruschak and Weinstein (2018) used the finite element method for calculating the inertia factor for the fife triangles discussed in Sect. 5.5.1.2. The dependence of the inertia factor upon the power law index was captured by the following equation:

$$\Gamma = \left(c_3 + \frac{1}{n}\right) / \left(c_4 + \frac{1}{n}\right) \tag{5.5.14}$$

The values of the constants c_3 and c_4 for the fife triangles are listed in Table 5.10.

Table 5.10 Constants for calculating the inertia shape factor of triangular cavity shapes

	30–30	30–60	30–45	45–45	60–30
c_3	18.52607	11.38782	12.70275	8.74393	10.96318
c_4	12.14529	7.51606	8.35618	5.76220	7.22850

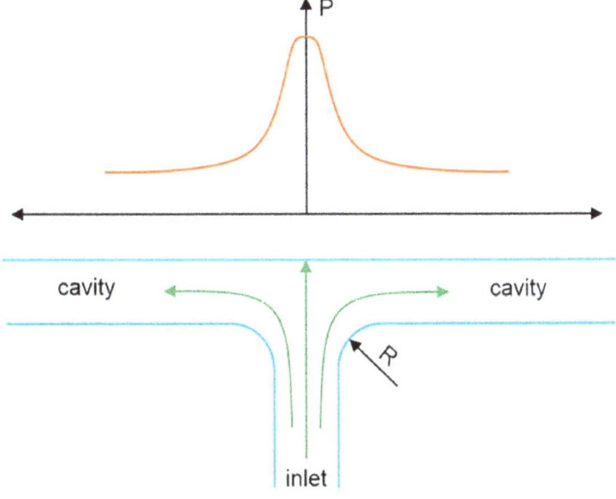

Fig. 5.23 Schematic view of the local pressure peak in the center inlet region of the distribution cavity caused by the stagnation point flow of the incoming fluid

5.5.1.4 Vortices in Axial Duct Flow

Regarding premetered coating processes, duct flow refers to the flow in the distribution cavities of the die. The axial flow in the outer cavity is vortex-free because it is very weak in a well-designed die. The flow in the inner cavity is also without vortices, if the cavity is sufficiently tapered, i.e., if the cavity cross-section decreases along its axis in an attempt to keep the wall shear stress positive to the very end of the cavity. If the inner cavity is of constant cross-section (infinite cavity design), then it is conceivable that a vortex may form toward the end of the cavity. More details about die design and wall shear stress are discussed in Sects. 6.3.1 and 9.4, respectively.

As most industrial dies are center-fed, however, vortices may appear in the vicinity of the T-junction between the inlet duct section and the two arms of the inner distribution cavity. In particular, the flow in the inlet duct generates a stagnation point pressure in the center of the distribution cavity. Figure 5.23 shows a typical but schematic pressure profile in the cavity across the inlet section.

If the magnitude of this pressure peak is sufficiently high and higher than the pressure necessary for driving the axial flow in the distribution cavity, it will locally push more fluid through the die slot, which in turn will ultimately generate a locally thicker film on the coated substrate. To avoid this problem one might be tempted to widen the cross-section of the inlet duct just upstream of the T-junction by increasing the radius R, thus forming a trumpet-like inlet section, see Figs. 5.23 and 5.24. This measure is expected to reduce the velocity of the incoming fluid and hence to reduce the stagnation point pressure in the cavity.

Flow visualization experiments have shown that the velocity of the incoming flow can at best be reduced by increasing R if the Reynolds number of the incoming flow is low enough, see sketch in Fig. 5.24. However, if the Reynolds number is too high, two strong vortices form along each wall of the trumpet shape. Moreover, it looks as if the incoming flow is not affected by the widening cross-section because the arriving fluid continues to jet straight ahead by occupying a cross-section that is not much bigger than the section just upstream of the trumpet, see Fig. 5.24. Obviously, this type of design modification does not help to reduce the stagnation point pressure at the die inlet.

A better approach is to design an inlet diffuser with a much smaller divergence angle in accordance with Abramowitz's inequality (5.4.21). For the flow shown in Fig. 5.24b with a high Reynolds number, the diffuser angle should be on the order of $<5^\circ$ to prevent flow separation. This in turn would require a lot of space in the direction of the incoming flow to suppress an excessive stagnation point pressure, and that space may just not be available in an industrial die.

Another approach to cut and re-distribute the excessive momentum of the incoming flow without generating vortices is to fill the wide inlet section with porous material such as a piece of scrubbing pad (Scotch Brite). This idea was tested experimentally and successfully. However, while this problem solution eliminates the inlet pressure peak and vortices in the die inlet section, it may not be acceptable for industrial coating applications. According to our experience porous materials may act as filter if the coating fluid is a dispersion, and filters will plug up after some time of operation. Also, porous materials used in connection with adhesives cannot be cleaned and hence will also plug up.

Yet another approach for preventing an excessive stagnation point pressure and vortices at the die inlet is to feed the die from one side instead of in the center. This way, perfect inlet geometries can be designed by incorporating a gradual change of the flow cross-section from circle in the supply line to semi-circle or whatever shape the distribution cavity may have. The disadvantage of this concept is that the length of the cavity doubles compared to a center feed, which makes it more difficult to achieve good cross-web uniformities in the outflow of the die. Also, the

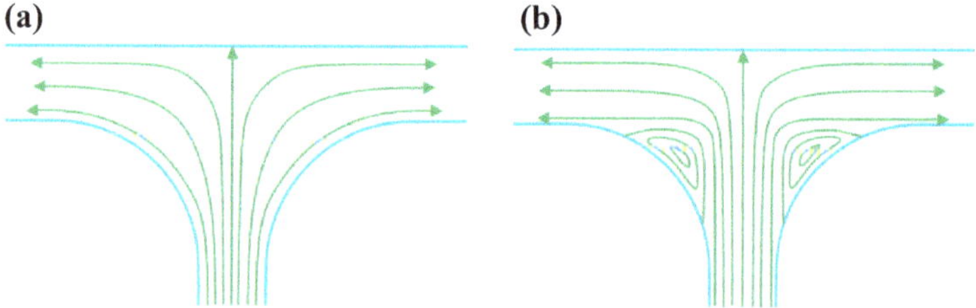

Fig. 5.24 **a** Center feed inlet section of die: without vortices for low Reynolds numbers **b** center feed inlet section of die: with a pair of vortices for high Reynolds numbers

cross-web flow uniformity may be diminished by mechanical effects owing to the unsymmetrical die design in the cross-web direction, i.e., the cavity shape and the hole pattern for the clamping screws.

The best way for preventing flow problems related to the center inlet is to sufficiently reduce the velocity of the incoming fluid stream by widening the flow cross-section upstream of the die, where there is enough space for doing this gently, and by keeping this wider cross-section constant along the stem of the T-junction.

5.5.2 Flow Across Ducts

As mentioned at the beginning of this chapter, flow across ducts only relates to the outer cavity of a dual-cavity die. In particular, all cavity shapes that are of interest in die design consist of a diverging section just downstream of the cavity entrance followed by a converging part toward the exit. In that sense, flow across the outer cavity resembles flow in a slot of varying slot height. However, owing to this complex geometry, analytical equations that describe slot flow are not suitable for describing the cross flow in the outer cavity, except perhaps if the cavity depth CD and the expansion angle α at the cavity entrance are small. In general, therefore, the cross flow of non-Newtonian fluids in the outer cavity must be treated numerically.

Such an approach was taken by Ruschak and Weinstein (2018), who examined both the axial and the cross flow of power law fluids in fife triangular cavity shapes. Specifically, they started by calculating the dominate cross flow called base flow. For shear thinning fluids, this base flow determines a two-dimensional viscosity field, which is relevant for calculating the axial flow. Building on the base flow, the axial flow is then considered as a perturbation of the base flow. Solving all these equations allows the axial flow rate to be computed, which depends on a *viscous shape factor* ζ of the cross flow, which in turn depends on the cavity shape, the Reynolds number and the power law index.

In the complete model of a dual-cavity die, there is coupling between the flows in the inner and outer cavities (Shetty et al., 2012). However, Ruschak and Weinstein (2018) showed that the design and effectiveness of the outer cavity can be address in a simpler but very elegant way by imposing a slightly non-uniform flow exiting the inner slot. This non-uniform flow is modelled as a Fourier series, which can represent any flow variations differing in spatial frequency and phase angle. More specifically, the effectiveness of the outer cavity, i.e., the capability of the outer cavity for damping flow non-uniformities generated by the inner cavity and slot, can be expressed in terms of a *damping factor DF*, which is equal to the difference between the maximum and minimum flow rate/width exiting from the inner slot divided by the flow rate/width in the outer slot. DF is greater than 1, and it depends on the spatial frequency, the shape factor ζ and a dimensionless group called s, which contains the power law index and all geometric parameters defining the outer cavity. More details about the implications of the damping factor DF on die design are discussed in the following section and in Sect. 9.4.

5.5.2.1 Damping Factor of the Outer Cavity

For the case of symmetric flow nonuniformities about the center of the die, i.e., for frown- or smile-shaped cross profiles, Ruschak and Weinstein (2018) defined the damping factor DF as follows:

$$DF = 1 + (j\pi)^2 \frac{\zeta}{s} \tag{5.5.15}$$

Specifically, ς is the shape factor of the cross flow in the outer cavity, and j is the special frequency of the flow nonuniformity. For purposes of designing and optimizing the outer die cavity, j can be set equal to 1 because this comes close to representing long-wave flow rate variations such as frown or smile patterns, which typically are dominant. The damping capacity increases with increasing spatial frequency and decreasing value of the dimensionless geometric group, s, which is defined as

$$s = \frac{{w_2}^{2n+1} W^2}{2^{1+n} n (2 + 1/n)^n L_2 A^{1+n}} \tag{5.5.16}$$

w_2 height of outer slot,
L_2 length of outer slot,
W half of die width for center-fed dies, or die width of end-fed dies,
A cross-sectional area of outer cavity excluding the slots,
n power law index.

Typical values of the damping factor for well-designed dies range from 2 to 20.

5.5.2.2 Viscous Shape Factor

As explained by Ruschak and Weinstein (2018), the shape factor ς for the cross flow over the outer cavity depends on the Reynolds number Re and the power law index n. Normalized with its value at Re = 0, ς is defined as

$$\frac{\zeta}{\zeta_0} = k_4 + \frac{1 - k_4}{1 + k_5 Re_s^{k_6}} \tag{5.5.17}$$

Moreover:

$$\frac{1}{\zeta_0} = k_0 \exp(k_1 n^{k_2}) \tag{5.5.18}$$

$$Re_s = \left[1 + k_3\left(\frac{1}{n} - 1\right)\right] Re \tag{5.5.19}$$

Table 5.11 Parameter values for triangular cavity shapes; data taken from Ruschak and Weinstein (2018)

	30–60	45–45	60–30
k_0	2.68677	2.81201	2.96505
k_1	2.75122	2.61019	2.74614
k_2	0.89663	0.90475	0.89626
k_3	2.96871	2.66878	2.79216
k_4	0.31156	0.31334	0.25265
k_5	0.01418	0.01370	0.001254
k_6	1.82152	1.88633	1.77427

$$Re = \frac{\rho Q_0 (Q_0/A)^{1-n}}{m} \tag{5.5.20}$$

Q_0 is the uniform flow rate/width. Re_s is called a stretched Reynolds number, which relates the shape factor to the Reynolds number. Equations (5.5.17)–(5.5.19) are empirical, and the constants k_0–k_6 depend on the cavity shape. Ruschak and Weinstein (2018) determined these constants for fife triangular cavity shapes by regression from results of the finite element analysis of the cross flow. Here, we reproduce the values for three right-angle triangles with different angles at the cavity entrance and exit, see Table 5.11. The other two shapes examined by the authors are deemed unattractive for die design owing to their excessively long cavity length. All k-values were determined for triangles with a sharp corner at the 90° angle. Such shapes are not used in practice because they are difficult to manufacture and to clean. In practice, therefore, the vertex of such triangles has a rounded corner. The authors argued that this fact can be accounted for approximately by correctly calculating the cavity area while leaving the shape factors unchanged. This argument is supported by the fact that the shape factor for axial flow in a 30°–60° right-angle triangle with a rounded 90° corner does not depend much at all on the corner radius, see Fig. 5.17.

To further investigate the effect of the radius of the vertex corner on the shape factor for the flow across the outer cavity, Bebié (2021) used the software package *Mathematica* for calculating the shape factor ζ_0 of right-angle triangles as a function of the expansion angle α, the ratio of corner radius to cavity length R_c/CL and the power law index n for the limiting case of $Re = 0$, i.e., for the Stokes approximation. Values of other geometric parameters were: $w_1 = 0.25$ mm, $w_2 = 0.4$ mm, $CL = 25$ mm. Note that ζ_0 only depends on the shape but not on the size of the cavity. Hence, R_c/CL is the relevant geometric parameter.

For the case of a sharp 90°-corner, i.e., for $R_c = 0$, Bebié's ζ_0-values were within 1.5% of the values from Ruschak and Weinstein when n was varied in the range of 0.6–1.0. This is a good agreement considering that Ruschak and Weinstein used $w_1 = 0.2$ mm. Furthermore, Bebié found that the ζ_0-values were invariant against exchanging the expansion angle α by the contraction angle β. i.e., the 30–60 triangle produced the same values as the 60–30 shape. This symmetry can be attributed to the Stokes approximation. More importantly, Bebié found that the ζ_0-values do not depend much (<10%) on the radius at the vertex angle, except for large radii in the 45–45 triangle (up to 30%). As an acceptable approximation, therefore, Bebié concluded that ζ_0-values can be calculated with Eq. (5.5.18) from Ruschak and Weinstein, and that the extension to Re > 0 according to Eq. (5.5.17) also applies to triangles with a rounded 90°-corner.

5.5.2.3 Vortices in Flows Across Ducts

The flow across ducts or cavities has been investigated extensively in the scientific literature. The focus of many of these papers is the onset of flow separation, which, typically, was investigated as a function of the cavity shape and the operating conditions. A couple of comprehensive papers relating this topic to die design are from the Group of Ta-Jo Liu in Taiwan, i.e., Lee and Liu (1989) and Lee et al. (1990). They investigated trapezoidal, tear, and semi-circular cavities both experimentally and theoretically. For viscous but shear-thinning fluids, they detected flow separation along the expanding region of all cavities as soon as the Reynolds number exceeded a critical value. They also found that the critical Reynolds number decreased with increasing expansion angle α and with increasing power law index n. For visco-elastic fluids in a trapezoidal cavity, in contrast, they found flow conditions without a vortex in the expanding region for expansion angles of 90° and 45°, but with a vortex on the contracting region. Moreover, the size of that vortex decreased as the contraction angle β was decreased from 90° to 75° to 60°, and it disappeared for $\beta = 45°$. It appears that this fluid anticipated the flow in the outer slot, where the average velocity is much higher than in the contraction region of the outer cavity.

Similar results were obtained when I used a hydrogen bubble technique for visualizing the streamline pattern in a semi-circular outer cavity (Schweizer, 1988, 1997b). A vortex will form on the upstream wall of the cavity, if the Reynolds number of the slot flow entering the outer cavity is too high. Moreover, the size of this vortex increases with increasing Reynolds number. For high Reynolds numbers, the vortex fills out almost the entire cross-section of the cavity.

Furthermore, I determined the onset of the vortex in the outer cavity with a shape of a circular segment. This study was carried out theoretically for Newtonian fluids by using the commercial finite element code FIDAP. The cavity geometry is sketched in Fig. 5.25. The vortex inception depends not only on the Reynolds number but also on the dimensionless geometrical parameters R/w_1, CD/w_1 and w_2/w_1.

R radius of cavity [m],
CD depth of cavity [m],
w_1 height of upstream slot [m],
w_2 height of downstream slot [m].

The angle α is the upstream angle of the cavity, which depends on the cavity geometry as follows:

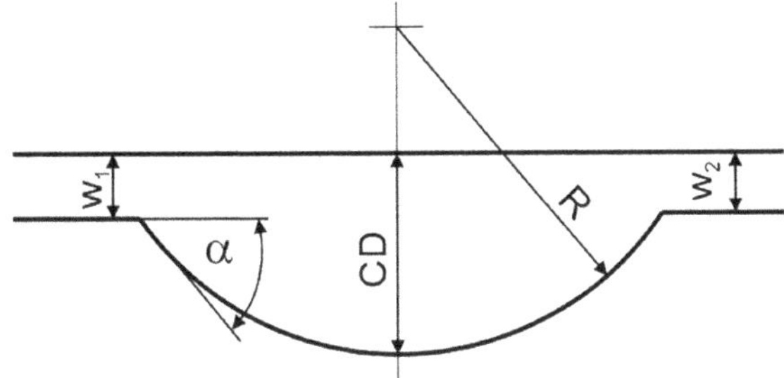

Fig. 5.25 Geometry of outer cavity

$$\alpha = arccos\left(1 - \frac{CD}{R}\right) \tag{5.5.21}$$

19 different geometries were investigated by varying the dimensionless parameters mentioned above over wide ranges, i.e., from semi-circular to very shallow shapes of circular segments.

The purpose of the finite element analysis was, for a given cavity shape, to determine the minimum wall shear stress τ_{min} on the cavity boundary as a function of the flow conditions, which were expressed in terms of the Reynolds number for slot flow, i.e., $Re = \rho Q/\mu$ (Eq. 5.6.13). The onset of flow separation was marked when the minimum wall shear stress along the curved cavity wall changed sign.

At low Re the inertia of the fluid is low enough so that the streamlines are able to turn around the upstream corner and follow the contour of the cavity. As Re is increased beyond a critical value a vortex appears that grows in size as Re further increases. The upstream stagnation point of the vortex is always located near the upstream cavity corner. For low Re the minimum wall shear stress is located at the bottom of the cavity, and that location moves upstream as Re is increased. If flow separation is present, τ_{min} is located near the upstream stagnation point of the vortex, which is found just a short distance below the corner.

Analyzing the computed results, I found that the critical Reynolds number Re_c at the onset of flow separation strongly depends on the upstream angle α of the cavity, and it weakly depends on the height of the upstream slot w_1. In particular, the critical Reynolds number decreases strongly with increasing cavity angle, and it increases slightly with increasing height w_1 of the upstream slot. All other geometrical parameters do not correlate with the critical Reynolds number and hence are insignificant.

Ruschak and Weinstein (2018) used the finite element method for determining the onset of flow separation for viscous but shear thinning fluids (power law) in 5 different triangular outer cavities with sharp corners throughout. Interestingly, they found that, upon increasing the Reynolds number in the 45–45 triangle, a vortex first appeared in the converging region of the cavity, just upstream of the entrance to the outer slot. This is in contrast with the results from Lee et al. (1990), who experimentally detected a vortex in the contraction region of a trapezoidal cavity only for visco-elastic liquids but not for shear-thinning fluids, where flow separation

Table 5.12 Constants for calculating the critical Reynolds number at the onset of flow separation in triangular cavity shapes

	30–30	30–60	30–45	45–45	60–30
k_7	−6.45656	−5.86593	−6.16459	−5.86273	−6.00322
k_8	3.62411	2.74671	3.12531	3.11977	3.59308

only occurred in the expanding region of the cavity. It is also in contrast with the flow across a semi-circular cavity and with the flow across a cavity consisting of a circular segment because in both these cases flow separation occurred in the diverging region of the cavity.

Ruschak and Weinstein defined the Reynolds number for the cavity cross flow as follows:

$$Re = \frac{\rho Q_0 (Q_0/A)^{1-n}}{m} \tag{5.5.22}$$

Q_0 is the uniform flow rate/width, A is the cross-sectional area of the cavity, $\sqrt{A}$ is the characteristic length, $Q_0/\sqrt{A}$ is the characteristic velocity, Q_0/A is the characteristic shear rate, and $m(Q_0/A)^{n-1}$ is the viscosity according to the power law model. The critical Reynolds number at the onset of recirculation can be calculated by the empirical Eq. (5.5.23). The constants k_7 and k_8 were determined by regression, and their values are listed in Table 5.12.

$$ln(Re_c) = k_7(1-n) + k_8 \tag{5.5.23}$$

When focusing on right-angle triangles, Eq. (5.5.23) produces the surprising result that the critical Reynolds number increases with increasing expansion angle α. This trend is opposite from the results in the circular segment, where Re_c decreases with increasing α. The difference may be explained by the fact that Eq. (5.5.23) describes the onset of a vortex in the contracting region of the cavity, while in the circular segment the vortex is located in the expansion region.

Since triangular outer cavities with three sharp corners as investigated by Ruschak and Weinstein (2018) are not common in industrial dies, Bebié (2021) used the software package *Mathematica* for solving the Navier–Stokes equations with the finite element method for power law fluids in triangular cavities with a rounded 90°corner, see Fig. 5.16. Specifically, he analyzed the cross flow in these cavities, and he determined the critical Reynolds number at the onset of recirculation. As was already reported by Ruschak and Weinstein, the exact determination of the critical Reynolds number was hampered by numerical noise. Bebié worked with finite element meshes of about 10,000 elements.

Surprisingly, and in contrast to Ruschak and Weinstein, recirculation in all of Bebié's calculations appeared on the expanding side of the triangle. This was also the case when Bebié analyzed the flow in the 45–45 triangle with a sharp corner at the vertex, exactly as done by Ruschak and Weinstein. Bebié could not explain why

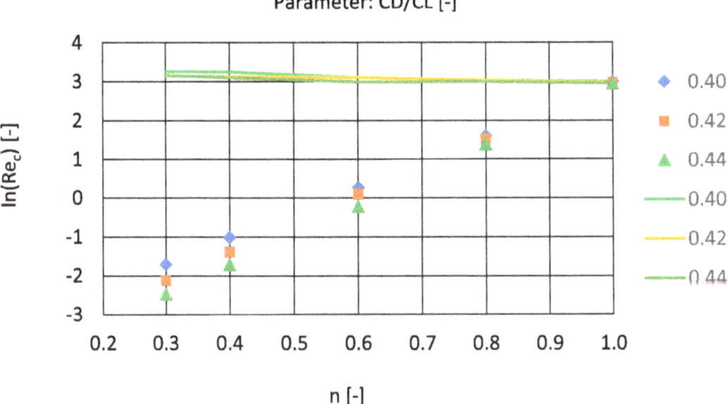

Fig. 5.26 Onset of recirculation in a 45–45 triangle. Symbols: results from Bebié with $\mathrm{Re_c}$ according to Eq. (5.5.22); solid lines: results from Bebié with $\mathrm{Re_c}$ according to Eq. (5.5.24)

different results were obtained when two different people used the same numerical method for solving the same differential equations on the same flow domain and for the same operating conditions.

In any case, Bebié's results are presented in Fig. 5.26 for the 45–45 triangle. The radius $\mathrm{R_c}$ of the 90° corner was held constant at 0.5 cm. The cavity depth CD was varied from 0.7 to 0.9 to 1.2 cm, and the corresponding cavity length CL varied as indicated. The inner and outer slot heights $\mathrm{w_1}$ and $\mathrm{w_2}$ were set to 0.25 and 0.40 mm. When the critical Reynolds number is calculated with Eq. (5.5.22), then the onset of recirculation, i.e., the logarithm of $\mathrm{Re_c}$, depends linearly on the power law index n, and it also depends on the ratio of cavity depth to cavity length CD/CL. In other words, fluids that approach Newtonian flow behavior and cavities that approach the shape of a circular segment postpone the onset of recirculation to higher Reynolds numbers.

Interestingly, however, when the critical Reynolds number was calculated with Eq. (5.5.24), which was taken from Lee et al. (1990), who used the height $\mathrm{w_1}$ of the inner die slot as the critical length instead of $\sqrt{A}$ as Ruschak and Weinstein did, then the critical Reynolds number at the onset of flow separation is largely independent of all other parameters, i.e., it hardly depends on the power law index and on the geometrical cavity parameters, see solid lines in Fig. 5.26. Specifically, the cavity length CL was varied from 1.5 to 3.5 cm, the radius of the vertex corner from 0.5 to 1.7 cm, and the height of the inner slot from 0.025 to 0.04 cm. For all practical purposes, therefore, the inception of flow separation in a 45–45 right-angle triangle with a rounded vertex is described by a constant critical Reynolds number, the value of which, when averaged over 15 different calculations, amounts to $\mathrm{Re_c} = 20.9$ (the average logarithm equals 3.04).

$$Re = \frac{\rho Q_0 \left(Q_0 / w_1^2 \right)^{1-n}}{m} \tag{5.5.24}$$

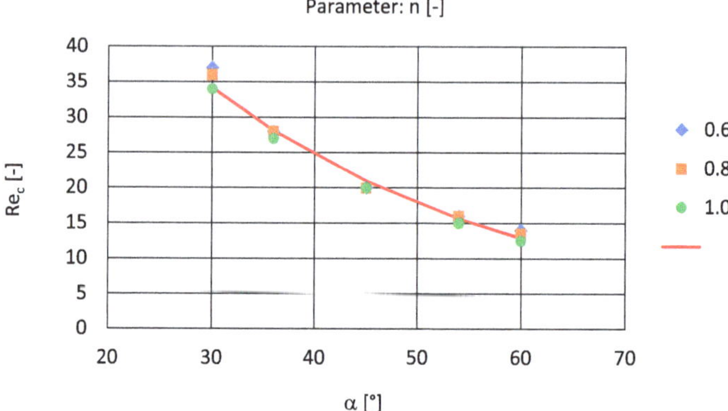

Fig. 5.27 Critical Reynolds number at the onset of flow separation as a function of the expansion angle; symbols are computed values as a function of the power law index; red curve is obtained by regression analysis; CL = 2.5 cm, R_c = 1.1 cm, w_1 = 0.025 cm, CD varied from 0.709 to 0.827

The seeming invariance of the critical Reynolds number indicates that the height of the inner slot, and not the size of the cavity, is crucial for the vortex inception. Moreover, the near independence of the critical Reynolds number from the power law index can be explained by the fact that the viscosity according to the power law, and hence the Reynolds number, strongly depends on the power law index, if the shear rate of the flow is high such as in the inner slot. In contrast, the power law viscosity is almost independent of the power law index if the shear rate of the flow is low such as in the cavity.

Rearranging Eq. (5.5.24) and assuming a constant critical Reynolds number yields the following criterion for the minimum value of the consistency m for preventing the onset of flow separation.

$$m > \frac{\rho Q}{Re_c}\left(\frac{Q}{w_1^2}\right)^{1-n} \tag{5.5.25}$$

The dependence of the critical Reynolds number on the expansion angle α is shown in Fig. 5.27. Again, and as expected intuitively, Re_c decreases with increasing α, which is similar to the result for the flow of Newtonian fluids across circular segments. Moreover, the dependence of Re_c upon α does not change much at all when changing the power law index or other geometric parameters of the cavity. By way of a regression analysis, the critical Reynolds number at the onset of flow separation in right-angle triangles with a rounded vertex can be approximated by

$$Re_c = a_0 e^{a_1 \alpha} \tag{5.5.26}$$

$a_0 = 90.764 \quad a_1 = -0.03250 \quad \alpha = [°] \quad r^2 = 0.985.$

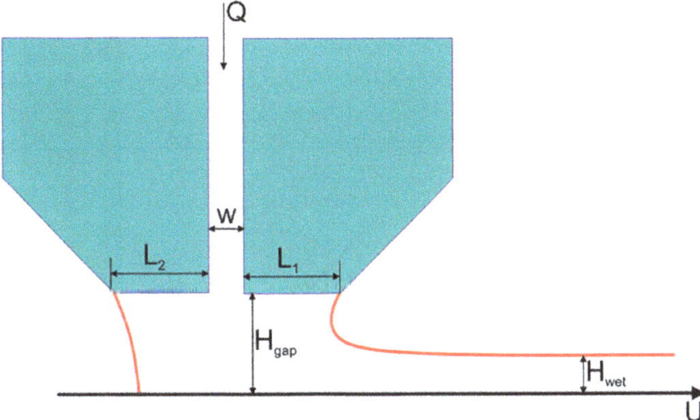

Fig. 5.28 Slot coating process with coating gap formed by the die lips and the moving substrate

In summary, Eq. (5.5.26) contains the essence about the onset of flow separation in right-angle triangles with a rounded vertex. Note, however, that the results above were obtained by theoretical calculations based on the power law constitutive equation. Therefore, predictions of the onset of flow separation are only accurate if the shear rate of both the flow in the inner slot and the flow in the outer cavity are contained in the shear-thinning region of the Carreau-Yasuda viscosity curve.

5.6 Slot Flow

Premetered coating methods are characterized by various types of slot flows. Inside the die, which is common to slot, slide and curtain coating, the narrow slot, which is connected to the distribution chamber, provides the high resistance to flow, which is necessary for a good fluid distribution in the cross-web direction. This type of slot flow is often called a Hagen-Poiseuille flow. It is pressure-driven, and both slot walls are at rest. This slot flow can be modeled by a so called flow rate-pressure drop relationship.

In the slot coating process, the liquid, upon exiting from the die slot, flows through the so called coating gap, which consists of an upstream and a downstream part, see Fig. 5.28. Here, the slot wall formed by the die lips is at rest, but the other wall formed by the substrate surface is moving at web speed U. This type of slot flow is often called a Poiseuille-Couette flow. It is driven by pressure and/or by the motion of the moving substrate. This slot flow can also be modeled by a so called flow rate-pressure drop relationship, but the boundary conditions are different.

In the curtain coating process, a device called suction baffle is placed upstream of the impinging curtain in order to reduce unwanted effects from the air boundary layer dragged along by the moving substrate. An example of such a device is shown in Fig. 5.29. This apparatus is similar in principle to a slot die, but it is operated

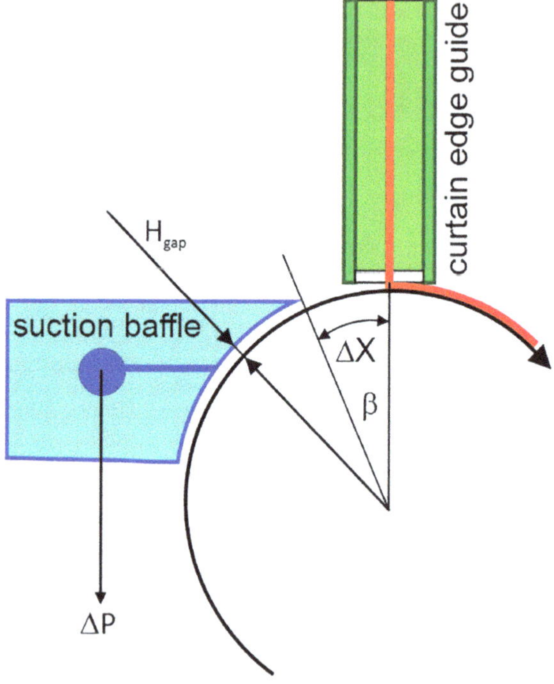

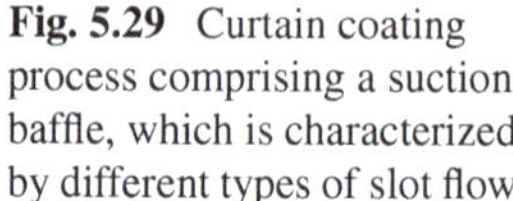
Fig. 5.29 Curtain coating process comprising a suction baffle, which is characterized by different types of slot flow

in reverse, i.e., instead of liquid being pumped through the die and coated onto the substrate, air is being sucked away from the substrate and evacuated through the die.

The purpose of this chapter is to derive analytical formulas, by which these various slot flows can be modeled and their characteristic features can be visualized and quantified.

5.6.1 Constant Viscosity (Newtonian Flow Behavior)

The velocity and shear stress profiles for one-dimensional flow in a straight slot of constant cross-section and with both walls at rest are depicted in Fig. 5.30. The derivation and the solution of the equation of motion for this simple flow can be found, for example, in Bird et al. (1960).

The equations presented here are valid for steady, laminar and rectilinear flows of Newtonian liquids. In addition, gravitational effects are neglected, which is acceptable for slot flows related to premetered coating methods because they refer to liquid flow in the die slot or air flow in the slot of the suction baffle. For these assumptions the one-dimensional equation of motion, i.e., the Navier–Stokes equation, reduces to:

$$\frac{dP}{dz} = \mu \frac{d^2 v_z}{dy^2} \tag{5.6.1}$$

Integrating Eq. (5.6.1) twice with respect to y results in an expression for the velocity profile in the z-direction:

$$v_z(y) = \frac{1}{\mu}\frac{dP}{dz}\frac{y^2}{2} + C_1 y + C_2 \tag{5.6.2}$$

Determining the two integration constants C_1 and C_2 requires two boundary conditions, and they vary from one type of slot flow to the other as explained below.

5.6.1.1 Slot Flow with Both Walls at Rest

The geometrical configuration of this slot flow is shown in Fig. 5.30. The slot length is L and the slot height is w. The pressure at x = 0 is denoted with P_0, and the pressure at x = L is P_L. As is generally accepted in the theory of fluid mechanics, the fluid flowing along a solid wall assumes the velocity of that wall. This leads to the following boundary conditions:

$$\text{BC1:} \quad y = 0: \quad v_z = 0 \tag{5.6.3a}$$

$$\text{BC2:} \quad y = w: \quad V_z = 0 \tag{5.6.3b}$$

The **velocity profile**, which is parabolic in form, can now be expressed in terms of the pressure gradient as follows:

$$v_z(y) = \frac{1}{2\mu}\frac{dP}{dz}\left(y^2 - wy\right) \tag{5.6.4}$$

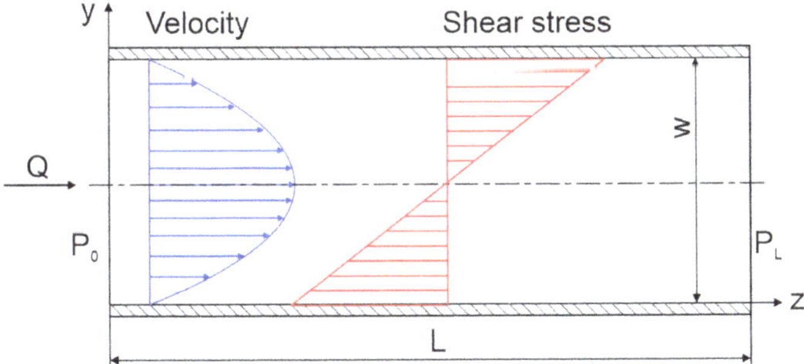

Fig. 5.30 Velocity and shear stress profiles for slot flow

The volumetric flow rate/width of this slot flow is obtained by integrating the velocity profile over the flow cross-section:

$$Q = \int_0^w v_z(y)dy \tag{5.6.5}$$

Performing the integration and solving for the **pressure gradient** yields:

$$\frac{dP}{dz} = \frac{-12\mu Q}{w^3} \tag{5.6.6}$$

Determining the pressure profile in the z-direction requires integrating Eq. (5.6.6) subject to the boundary condition that the pressure must be known either at the slot inlet or at the slot outlet.

$$P(z) = P_0 - \frac{12\mu Qz}{w^3} = P_L + \frac{12\mu Q(L-z)}{w^3} \tag{5.6.7}$$

The pressure changes linearly between P_0 and P_L. Equation (5.6.7) can be used for deriving the so called **flow rate–pressure drop relationship** for slot flow as follows:

$$\Delta P = (P_0 - P_L) = \frac{12\mu QL}{w^3} \tag{5.6.8}$$

Inserting Eq. (5.6.6) into Eq. (5.6.4) yields a **velocity profile** in terms of the volumetric flow rate/width Q, which in turn depends on the operating conditions of the coating process according to Eq. (3.3.4) or (3.3.12).

$$v_z(y) = \frac{6Q}{w^3}\left(wy - y^2\right) \tag{5.6.9}$$

The **average velocity** for the slot flow is obtained by integrating the velocity profile (Eq. 5.6.9) over the slot height and by normalizing the result with the slot height:

$$v_{z,average} = \frac{1}{w}\int_0^w v_z(y)dy = \frac{Q}{w} \tag{5.6.10}$$

Taking the derivative of Eq. (5.6.9) allows us to derive information about the characteristic shear rate of this flow and about the maximum velocity. Note that the derivative of a velocity is equal to a shear rate.

$$\frac{dv_z(y)}{dy} = \frac{6Q}{w^3}(w - 2y) \tag{5.6.11}$$

The maximum velocity is located where $dv_z(y)/dy = 0$, which is at $y = w/2$, i.e., in the middle of the slot. Therefore, the **maximum velocity** for slot flow is

$$v_{z,max} = \frac{3Q}{2w} \tag{5.6.12}$$

Using the average velocity $v_{z,average}$ and the slot height w as characteristic velocity and length, respectively, the Reynolds number for slot flow is

$$Re = \frac{\rho Q}{\mu} \tag{5.6.13}$$

Shear rate information is relevant for determining the local fluid viscosity because most industrial coating fluids are characterized by a shear-dependent viscosity. In that regard, the **wall shear rate** is most important, and it is obtained by evaluating Eq. (5.6.11) at $y = 0$ or $y = w$:

$$\gamma_0 = \frac{\pm 6Q}{w^2} \tag{5.6.14}$$

The characteristic velocity and shear rate properties of this slot flow are best visualized in dimensionless form. Normalizing the velocity with the average velocity and the shear rate with the wall shear rate gives:

$$\frac{v_z(y)}{v_{z,average}} = 6\left[\left(\frac{y}{w}\right) - \left(\frac{y}{w}\right)^2\right] \tag{5.6.15}$$

$$\frac{\gamma(y)}{\gamma_0} = 1 - 2\left(\frac{y}{w}\right) \tag{5.6.16}$$

Equations (5.6.15) and (5.6.16) are graphically displayed in Fig. 5.31.

The shear stress for this type of slot flow is obtained by multiplying the shear rate with the viscosity. Consequently, the wall shear stress is:

$$\tau_0 = \frac{6\mu Q}{w^2} \tag{5.6.17}$$

5.6.1.2 Slot Flow with One Wall at Rest and One Wall Moving

As explained above, this situation typically concerns the coating gap of slot coating or the gap between the suction baffle and the moving substrate. The geometrical configuration of this slot flow is shown in Fig. 5.32. The slot length is L and the slot height is now called gap or coating gap H_{gap}. The pressure at $x = 0$ is denoted P_0, and the pressure at $x = L$ is P_L. The particular form of the velocity profile depends on the boundary conditions, specifically on whether the pressure drop along the slot is positive or negative. As is generally accepted in the theory of fluid mechanics, the fluid flowing along a solid wall assumes the velocity of that wall. This leads to the following boundary conditions:

$$\text{BC1:} \quad y = 0 : u = U \tag{5.6.18a}$$

$$\text{BC2:} \quad y = H_{gap}: \quad u = 0 \tag{5.6.18b}$$

The **velocity profile**, which is parabolic in form, can now be expressed in terms of the pressure gradient as follows:

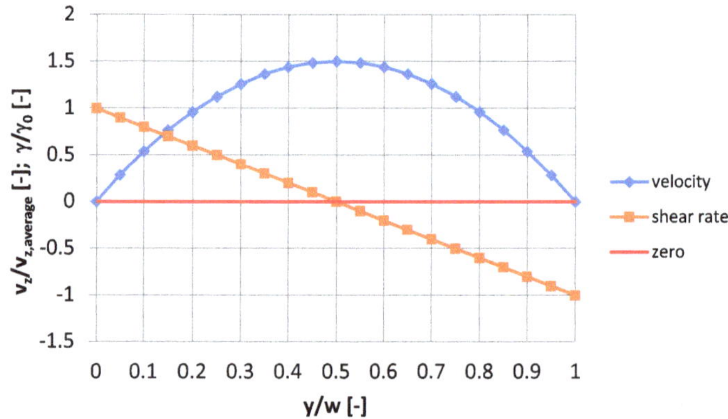

Fig. 5.31 Dimensionless velocity and shear rate profiles for slot flow

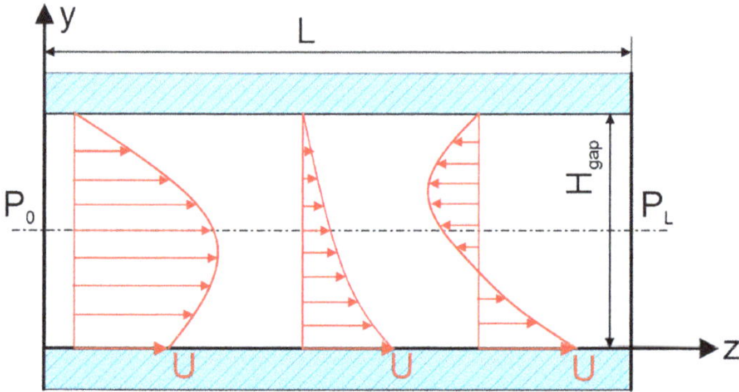

Fig. 5.32 Geometrical configuration of a slot flow with one wall at rest and the other all moving with speed U

$$v_z(y) = \frac{dP}{dz}\frac{y^2}{2\mu} - \left[\frac{dP}{dz}\frac{H_{gap}}{2\mu} + \frac{U}{H_{gap}}\right]y + U \tag{5.6.19}$$

Depending on the specific application, the pressure gradient dP/dz may or may not be known. Therefore, we shall proceed by discussing the downstream and upstream gaps of the slot coater and the suction baffle separately.

5.6.1.3 Downstream Gap of the Slot Coating Process

According to Eq. (5.6.5) the volumetric flow rate/width of this slot flow is obtained by integrating the velocity profile over the flow cross-section. Performing the integration and solving for the pressure gradient yields:

$$\frac{dP}{dz} = \frac{12\mu}{H_{gap}^3}\left[\frac{UH_{gap}}{2} - Q\right] \tag{5.6.20}$$

Determining the pressure profile in the z-direction requires integrating Eq. (5.6.20) subject to the boundary condition that the pressure must be known either at the slot inlet or at the slot outlet.

$$P(z) = P_0 + \frac{12\mu}{H_{gap}^3}\left[\frac{UH_{gap}}{2} - Q\right]z = P_L + \frac{12\mu}{H_{gap}^3}\left[\frac{UH_{gap}}{2} - Q\right](z - L) \tag{5.6.21}$$

The pressure changes linearly between P_0 and P_L. Equation (5.6.21) can be used for deriving the so called **flow rate–pressure drop relationship** for slot flow as follows:

$$\Delta P = (P_L - P_0) = \frac{12\mu L}{H_{gap}^3}\left[\frac{UH_{gap}}{2} - Q\right] \tag{5.6.22}$$

Inserting Eq. (5.6.20) into Eq. (5.6.19) yields a **velocity profile** in terms of the volumetric flow rate/width Q:

$$v_z(y) = \left[3U - \frac{6Q}{H_{gap}}\right]\left(\frac{y}{H_{gap}}\right)^2 - \left[4U - \frac{6Q}{H_{gap}}\right]\left(\frac{y}{H_{gap}}\right) + U \tag{5.6.23}$$

This equation is interesting in connection with premetered coating methods because, owing to the law of mass conservation, the flow rate/width depends on the relevant coating parameters, i.e., on the web speed and the wet film thickness of the coated film, according to Eq. (5.6.24):

$$Q = UH_{wet} \tag{5.6.24}$$

Inserting Eq. (5.6.24) into Eq. (5.6.23) yields another expression for the **velocity profile** in terms of the wet film thickness, which is very interesting and relevant for the downstream gap in slot coating. The dimensionless form of the profile can be written as:

$$\frac{v_z(y)}{U} = \left[3 - \frac{6H_{wet}}{H_{gap}}\right]\left(\frac{y}{H_{gap}}\right)^2 - \left[4 - \frac{6H_{wet}}{H_{gap}}\right]\left(\frac{y}{H_{gap}}\right) + 1 \qquad (5.6.25)$$

In the same way another expression for the **pressure gradient** can be obtained by inserting Eq. (5.6.24) into Eq. (5.6.20):

$$\frac{dP}{dz} = \frac{12\mu U}{H_{gap}^2}\left[\frac{1}{2} - \frac{H_{wet}}{H_{gap}}\right] \qquad (5.6.26)$$

Using Eqs. (5.6.25) and (5.6.26), we now can describe some of the characteristic features of this type of slot flow, i.e., the flow in the downstream gap of a slot coating process.

Figure 5.33 shows the dimensionless velocity profile for various values of H_{wet}/H_{gap}. We can see that the velocity is always equal to the web speed U for y = 0, and it is always equal to 0 for y = H_{gap}. Therefore, the boundary conditions 5.6.18 are satisfied.

If $H_{gap} < 2H_{wet}$, i.e., $H_{wet}/H_{gap} > 0.5$, the pressure drop is negative, i.e., $P_0 > P_L$, and the velocity profile is bowed in the downstream direction. Moreover, the flow is driven by the web motion and also by the pressure gradient, and both effects act in the same downstream direction. If, on the other hand, $H_{gap} = 2H_{wet}$, i.e., $H_{wet}/H_{gap} = 0.5$, then the pressure drop is zero, i.e., $P_0 = P_L$, and the velocity profile is linear. This flow is only driven by the web motion; it is called a pure Couette flow.

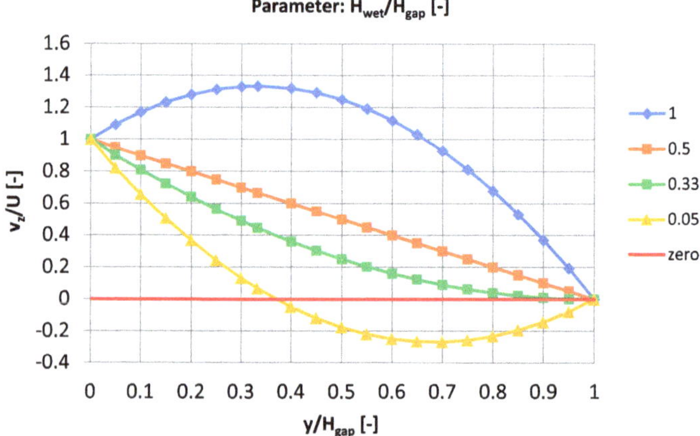

Fig. 5.33 Dimensionless velocity profiles in the downstream coating gap for various values of H_{wet}/H_{gap}

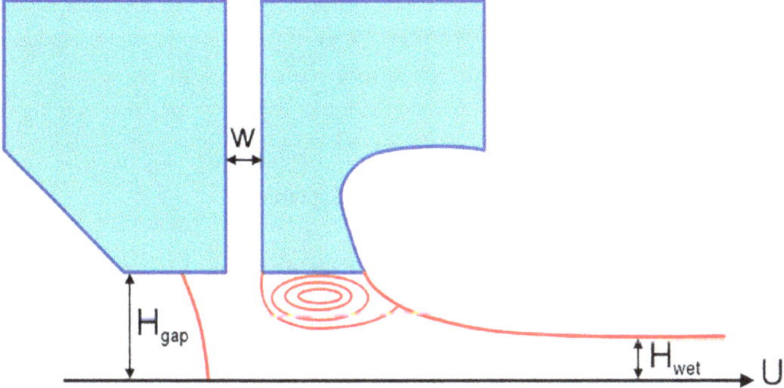

Fig. 5.34 Vortex formation under the downstream die lip in slot coating for $H_{wet}/H_{gap} < 0.333$

If $H_{gap} = 3H_{wet}$, i.e., $H_{wet}/H_{gap} = 0.333$, the pressure drop is positive, i.e., $P_0 < P_L$, and the velocity profile is slightly bowed in the upstream direction. Moreover, the velocity gradient at $y = H_{gap}$ is zero, which marks the onset of back flow or flow separation. This flow is driven by the web motion and the pressure gradient, but the two effects act in opposite directions. Note that positive pressure gradients are not desirable in the downstream gap of the slot coating process because they represent a necessary but not sufficient condition for the onset of ribbing lines. It is therefore safer to operate the slot coating process with negative pressure gradients in the downstream gap.

Finally, if $H_{gap} > 3H_{wet}$, i.e., $H_{wet}/H_{gap} < 0.333$, the pressure drop is positive, i.e., $P_0 < P_L$, and the velocity profile is strongly bowed in the upstream direction, thus marking the presence of backflow, or, as is the case for slot coating, a vortex underneath the downstream lip of the coating gap, see Fig. 5.34. This flow feature is not at all desirable because it tends to degrade the uniformity of the coated film.

As will be explained below, the location across the coating gap, where the velocity turns from positive to negative values, which determines the relative amounts of forward and backward flow in the gap, depends on various process parameters.

The **shear rate** for the flow in the downstream gap of a slot coater can again be derived by taking the derivative of Eq. (5.6.25). The wall shear rates differ in sign and magnitude for $y = 0$ and $y = H_{gap}$, depending on the value of H_{wet}/H_{gap}:

$$\gamma_0|_{y=0} = \frac{2U}{H_{gap}}\left[\frac{3H_{wet}}{H_{gap}} - 2\right] \tag{5.6.27}$$

$$\gamma_0|_{y=Hgap} = \frac{2U}{H_{gap}}\left[1 - \frac{3H_{wet}}{H_{gap}}\right] \tag{5.6.28}$$

The magnitude of the average shear rate across the downstream coating gap equates to be

$$\gamma|_{average} = \frac{U}{H_{gap}} \tag{5.6.29}$$

5.6.1.4 Upstream Gap of the Slot Coating Process

Looking at the upstream gap of the slot coating process, Eq. (5.6.19) can again be used for describing the velocity profile, although in combination with a different boundary condition as explained below.

If the upstream gap is filled with liquid as shown in Fig. 5.35, then the net flow rate in this gap must be zero for steady-state operating conditions because all the liquid entering the gap from the die slot must turn around at the upstream meniscus and leave the gap to flow into the downstream gap. Note that this configuration is not desirable because the flow field contains adjacent streamlines, which move in opposite directions, which in turn causes instabilities and oscillations in the flow.

Nevertheless, calculating the flow rate/width in this gap by integrating Eq. (5.6.19) over the gap height and forcing the flow rate to be zero yields the following expression for the **pressure gradient**:

$$\frac{dP}{dz} = \frac{6\mu U}{H_{gap}^2} \tag{5.6.30}$$

The pressure gradient is always positive, thus indicating that the pressure increases in the main flow direction, i.e., the pressure at the upstream meniscus is lower than the pressure near the die slot.

Inserting Eq. (5.6.30) into Eq. (5.6.19) gives the following expression for the dimensionless **velocity profile** in the upstream gap:

$$\frac{v_z(y)}{U} = 3\left(\frac{y}{H_{gap}}\right)^2 - 4\left(\frac{y}{H_{gap}}\right) + 1 \tag{5.6.31}$$

Note that the velocity profile in the upstream gap only depends on the web speed U and the gap height H_{gap}, but not on the wet film thickness of the coated layer or

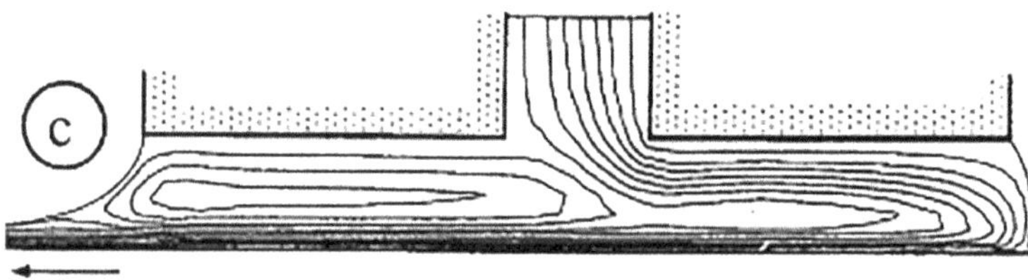

Fig. 5.35 Slot coating process: Upstream gap completely filled with liquid; taken from Durst and Wagner (1997) and reprinted with permission from Springer Nature

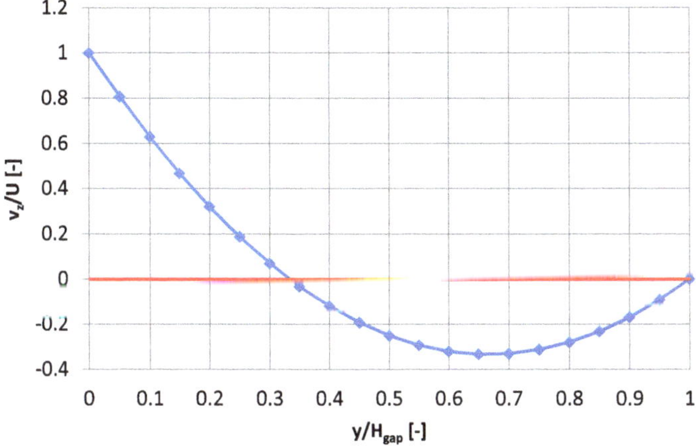

Fig. 5.36 Dimensionless velocity profile in the upstream coating gap

the physical fluid properties. Therefore, there is only one dimensionless form of the profile as shown in Fig. 5.36. The velocity is zero for $y/H_{gap} = 1/3$ and 1.0.

The **shear rate** for the flow in the upstream gap of a slot coater can again be derived by taking the derivative of Eq. (5.6.31). The wall shear rates differ in sign and magnitude for $y = 0$ and $y = H_{gap}$ as follows:

$$\gamma_0|_{y=0} = \frac{-4U}{H_{gap}} \tag{5.6.32}$$

$$\gamma_0|_{y=Hgap} = \frac{2U}{H_{gap}} \tag{5.6.33}$$

The magnitude of the average shear rate in the upstream gap is the same as in the downstream gap, i.e.,

$$\gamma|_{average} = \frac{U}{H_{gap}} \tag{5.6.34}$$

5.6.1.5 Downstream Gap of the Suction Baffle

As mentioned at the beginning of Sect. 5.6, the purpose of the suction baffle is to reduce the air boundary layer, which is carried along by the uncoated substrate. This is accomplished, for example, by operating a slot-die-like apparatus, which is placed closely against the backup roller in order to generate a narrow, concentric gap, through which air is evacuated with the help of a vacuum source as schematically shown in

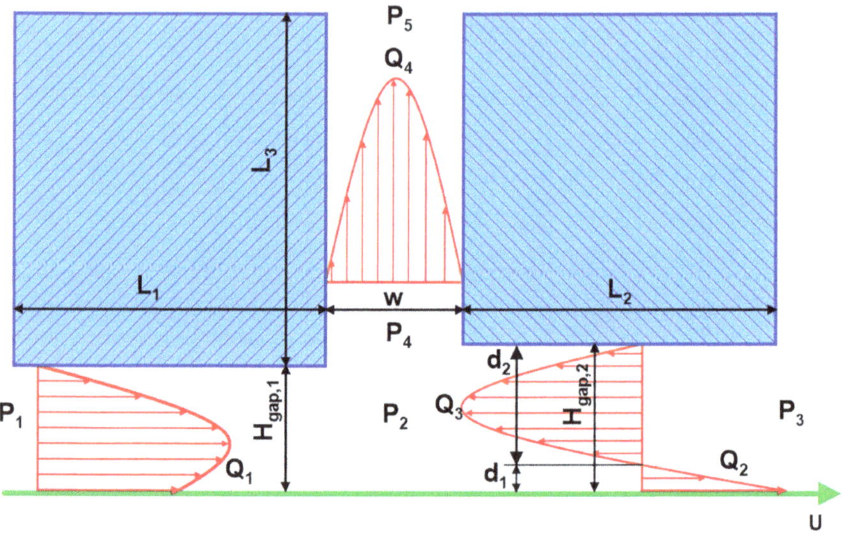

Fig. 5.37 Operating principle of the suction baffle

Fig. 5.29. The geometrical details and the velocity profiles of this configuration are schematically depicted in Fig. 5.37.

Upstream (P_1) and downstream (P_3) of the suction baffle the air pressure is approximately ambient, although P_1 is slightly above ambient due to the stagnation point pressure generated by the air boundary layer impinging onto the upstream side of the baffle body. Normally, this stagnation point pressure is much smaller compared to the pressure drops across the various slots. Applying a vacuum source to the suction slot causes P_2 in the middle of the concentric gap and particularly P_5 at the end of the suction slot to be much below ambient. Looking at the downstream slot (right hand side in Fig. 5.37) we can see that air is flowing against the web motion (Q_3) as long as P_2 is sufficiently lower than P_3, and this is exactly wanted. On the other hand, the moving web will always drag along some amount of air (Q_2), which will eventually impinge onto the backside of the curtain. In contrast to the slot coating process, where the flow in the downstream gap depends on and is controlled by the flow rate or the wet film thickness of the coated layer, the flow here is controlled by the pressure difference $P_3 - P_2$ along the downstream gap, and P_2 is controlled by the vacuum pressure of the suction source P_5.

Pressure P_2 is responsible for driving the air flows in both gaps of the baffle. In addition, there is a difference between P_2 and P_4 owing to a pressure loss in flows around bends. The baffle geometry in Fig. 5.37 is only drawn schematically; a more realistic geometry is depicted in Fig. 5.29. Consequently, flow rate Q_3 has to flow around a bend with an angle of about 45°, while the bend angle for flow rate Q_1 is much higher, i.e., 135°. The pressure drop around a bend can be estimated by multiplying the stagnation pressure generated by the corresponding velocity with a factor ξ, which is representative of the bend geometry (Dubbel, 1994). ξ for a 45° bend is about 0.5, while ξ for a 135° bend is about 1.6.

To visualize and quantify the characteristic features of this flow field, we are interested in knowing the distance d_1, at which the air changes from flowing forward to flowing backward, as well as the flow rates/width Q_3 and particularly Q_2. Specifically, Q_2 is indicative of the performance of the suction baffle because it is a measure for the remaining disturbance, which impinges onto the curtain.

We start our analysis of the slot flow by taking Eq. (5.6.19) and by replacing the differential expression dP/dz by the difference term $\Delta P/\Delta L = (P_3 - P_2)/L_2$ according to the nomenclature in Fig. 5.37. This step is acceptable because we have learned above that the pressure in the gap varies linearly along the gap.

The air velocity is equal to 0 for $y = d_1$ and $y = H_{gap,1}$. Finding an expression for d_1 requires setting $v_z = 0$ in Eq. (5.6.19) and solving a quadratic equation with the coefficients taken from Eq. (5.6.19). The result is as follows:

$$d_1 = \frac{2\mu U L_2}{H_{gap,2}\Delta P} \tag{5.6.35}$$

Consequently, $d_2 = H_{gap} - d_1$. Moreover, d_2 must be positive for backflow to occur. This then allows us to derive the following expression for a minimum value of the pressure drop along the slot that will generate backflow:

$$\Delta P = (P_3 - P_2) > \frac{2\mu U L_2}{H_{gap,2}^2} \tag{5.6.36}$$

Regarding the **flow rates/width** Q_2 and Q_3 the velocity profile (Eq. 5.6.19) must be integrated from 0 to d_1 and from d_1 to $H_{gap,2}$, respectively. The results are as follows:

$$Q_2 = \frac{\mu L_2 U^2}{H_{gap,2}\Delta P} - \frac{2\mu^2 L_2^2 U^3}{3H_{gap,2}^3 \Delta P^2} \tag{5.6.37}$$

$$Q_3 = \frac{2\mu^2 L_2^2 U^3}{3H_{gap,2}^3 \Delta P^2} - \frac{\Delta P H_{gap,2}^3}{12\mu L_2} - \frac{\mu L_2 U^2}{H_{gap,2}\Delta P} + \frac{U H_{gap,2}}{2} \tag{5.6.38}$$

In Fig. 5.38 d_1 and Q_2 are plotted as a function of the pressure difference along the downstream slot for a given set of other parameter values, see the figure caption. Both d_1 and Q_2 decrease very rapidly with increasing pressure difference. It appears that very small values of Q_2, which are desirable, can be reached for $\Delta P > 1000$ Pa. At the same time, the flow in the upper portion of the gap, which moves in the opposite direction, is very strong, i.e., Q_3 is much bigger than Q_2. The corresponding velocity profile is shown in Fig. 5.39.

The location of the maximum backward velocity in the downstream gap is obtained by taking Eq. (5.6.19) and by replacing the differential expression dP/dz by the difference term $\Delta P/\Delta L = (P_3 - P_2)/L_2$ according to the nomenclature in Fig. 5.37,

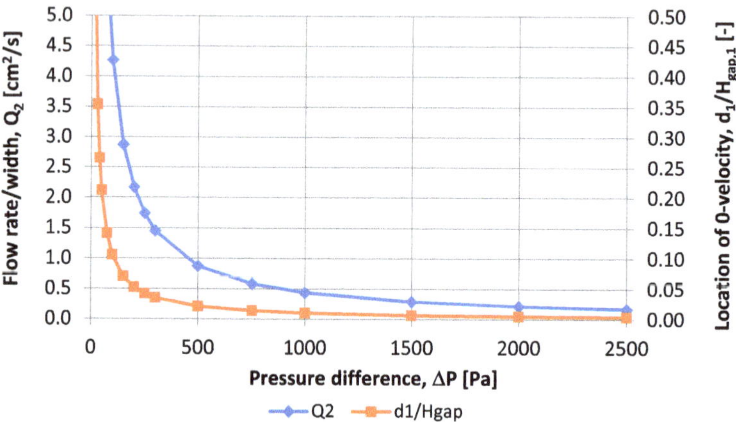

Fig. 5.38 Flow properties in the downstream coating gap. $L_2 = 35$ mm, $H_{gap,2} = 1.0$ mm, $\mu_{air} = 1.819E\text{-}5$ Pas, $U = 500$ m/min

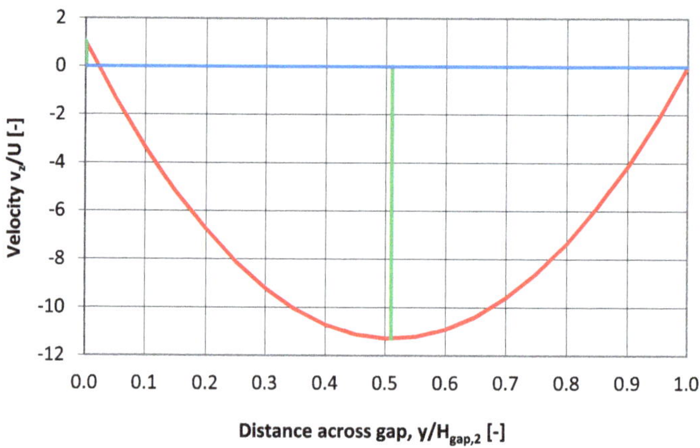

Fig. 5.39 Dimensionless velocity profile in downstream gap; $L_2 = 35$ mm, $H_{gap,2} = 1.0$ mm, $\mu_{air} = 1.819E\text{-}5$ Pas, $U = 500$ m/min, $\Delta P = 500$ Pa

by differentiating this new expression, by forcing the result to be zero, and by solving for y. The result is as follows:

$$y|_{v=max} = \frac{H_{gap,2}}{2} + \frac{U\mu L_2}{H_{gap,2}\Delta P} \tag{5.6.39}$$

Then, the maximum velocity can be calculated according to:

$$v_{z,max} = \frac{U}{2} - \frac{\Delta P H_{gap,2}^2}{8\mu L_2} - \frac{\mu L_2 U^2}{2H_{gap,2}^2 \Delta P} \tag{5.6.40}$$

For the parameter values in Fig. 5.39 the maximum velocity is at $y/H_{gap,2} = 0.511$, and the dimensionless maximum velocity is -11.3, see green line in Fig. 5.39. Using this velocity and the height of the downstream gap as characteristic velocity and length, the Reynolds number for this air flow is 6,145, thus indicating turbulent flow conditions. However, the equations derived above are valid for laminar flows. Consequently, results predicted by these equations are not accurate when turbulent flows are concerned. Nevertheless, the equations still serve well for visualizing the relationships between relevant process parameters.

5.6.1.6 Upstream Gap of the Suction Baffle

The characteristic features of the flow field in the upstream gap of the suction baffle are similar to those in the downstream gap of the slot coater for $H_{wet}/H_{gap} > 0.5$ because both the web motion and the pressure difference along the gap are driving the flow in the same direction. In contrast, however, the **volumetric flow rate/width** must be expressed in terms of the pressure difference and not in terms of the flow rate or the wet film thickness of the coated film. The resulting expression is as follows, with $\Delta P = (P_2 - P_1)$, see Fig. 5.37.

$$Q_1 = \frac{U H_{gap,1}}{2} - \frac{\Delta P H_{gap,1}^3}{12 \mu L_1} \qquad (5.6.41)$$

The flow rate/width Q_1 increases linearly with ΔP, and the flow rate values are of similar magnitude than Q_3 and hence much higher than Q_2. A typical velocity profile in the upstream gap is shown in Fig. 5.40.

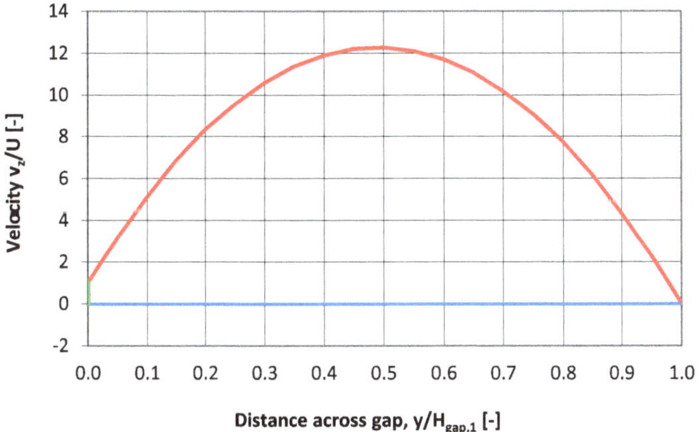

Fig. 5.40 Dimensionless velocity profile in upstream gap; $L_1 = 35$ mm, $H_{gap,1} = 1.0$ mm, $\mu_{air} = 1.819\text{E-}5$ Pas, $U = 500$ m/min, $\Delta P = 500$ Pa

To evaluate the performance of this type of suction baffle, the pressure P_5 must be known because this pressure is indicative of the value that must be generated by the vacuum source. More important than knowing the exact value of P_5 is to know how P_5 depends on the geometrical parameters of the flow configuration depicted in Fig. 5.37, specifically on the height of the various slots. In particular, this knowledge allows the performance of the suction baffle to be optimized in terms of keeping the flow rate/width Q_2 small enough to prevent air entrainment for a given set of curtain coating operating conditions without requiring excessive values for P_5 and Q_4.

Calculations with the equations presented above revealed the following recommendations.

- $H_{gap,1}$ must be kept as small as possible because increasing $H_{gap,1}$ increases the flow rate/width Q_1, which must be evacuated again and hence increases the flow rate/width Q_4 and the required vacuum pressure P_5, all of which is not desirable. On the other hand, $H_{gap,1}$ must be large enough in order for the splice between subsequent substrate rolls to pass the baffle without getting stuck at all and at all times. This minimum gap therefore depends on the quality of the splice, i.e., its thickness, and that in turn depends on the design of the winding equipment of the coating machine, and possibly on the hand work of the operator, for example when applying an adhesive tape across the substrate width.
- The height w of the suction slot of the baffle must be maximized because decreasing w increases the pressure drop along that slot and hence increases the vacuum pressure P_5 that is necessary for removing all the air from the baffle. On the other hand, w must be small enough to guarantee an even suction power across the width of the baffle. This then is an issue of die design, and depends on many parameters such as the length of the suction slot, the cross-sectional size of the distribution cavity, and the effective baffle width, which in turn depends on the coating width and the number of suction ports placed across the baffle width.
- Optimizing the height of the downstream slot $H_{gap,2}$ is more complicated. For small gaps, say <1 mm, the vacuum pressure P_2 must be high (i.e., the absolute pressure value must be low) in order to obtain a small flow rate/width Q_2, which guarantees a good baffle performance. As a result, the average velocity of the air flow Q_1, which is drawn into the baffle through the upstream slot is much higher than the average velocity of the flow Q_3. Therefore, the resulting pressure losses around the respective bends from the concentric gap into the suction slot are in concert with these velocities such that the pressure P_5 is dominated by the air flow in the upstream gap. As the height of the downstream gap $H_{gap,2}$ increases the pressure P_2 that is necessary for obtaining the same desirable flow rate/width Q_2 decreases exponentially. Consequently, the average velocity in the upstream gap as well as the pressure loss around the sharp bend decrease because less air is drawn into the baffle through the upstream gap. As a result, the pressure P_5 also decreases. However, the average velocity of the flow Q_3 increases upon opening the downstream gap and so does the associated pressure loss around the respective bend. For wide enough gaps this effect becomes so strong that the pressure P_5 is eventually dominated by the flow in the downstream gap and it starts to increase

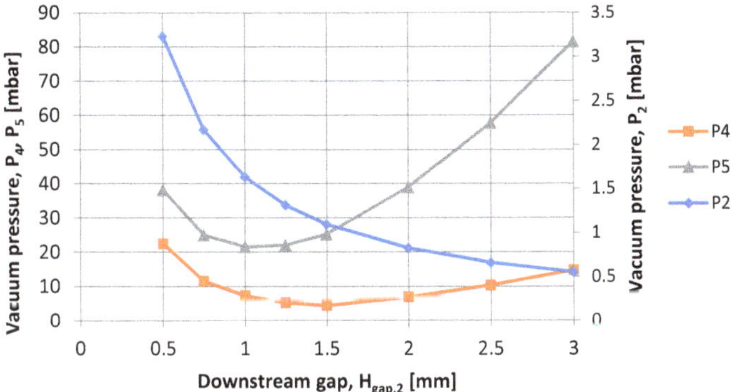

Fig. 5.41 Various vacuum pressures in the suction baffle as a function of the downstream gap $H_{gap,2}$. $Q_2 = 1.0\ cm^2/s$, $U = 500$ m/min, $H_{gap,1} = 1.0$ mm, $w = 0.75$ mm

again. This changing dominance leads to a minimum for the pressure P_5 and hence to an optimum value for the height of the downstream gap $H_{gap,2}$.

An example of this situation is shown in Fig. 5.41. Fortunately, the curve of $P = P_5(H_{gap,2})$ has a flat minimum such that gap values of 0.75–1.5 mm should result in a good baffle performance. In fact, curtain coaters built by Polytype Converting AG with this type of suction baffle and with both the upstream and downstream gaps being set at 1.0 mm worked well for applying pressure sensitive adhesives for coating widths of up to about 2,000 mm and coating speeds of up to about 1,000 m/min. Setting both gaps to the same value also facilitates the manufacturing of the baffle body because the round surfaces in the lower and upper baffle plates now have the same radius, such that they can be machined in one pass with both plates being fastened together.

5.6.2 *Shear Rate Dependent Viscosity (Shear Thinning Flow Behavior)*

5.6.2.1 Slot Flow with Both Walls at Rest

Fully developed flow of power law fluids in a slot with both walls at rest is similar to flow in a pipe. Assuming a coordinate system with its origin in the middle of the slot, i.e., $-w/2 \le y \le w/2$ and neglecting inertia and gravity effects the one-dimensional equation of motion can be written as

$$0 = -\frac{dP}{dz} - \frac{d\tau_{yz}}{dy} \tag{5.6.42}$$

Integrating the pressure over the length L of the slot, thereby assuming that $P_0 > P_L$ and $\Delta P = (P_0 - P_L)$, then integrating the shear stress with respect to y gives

$$\tau_{yz} = \frac{\Delta P}{L} y + c_1 \tag{5.6.43}$$

Implying a symmetrical velocity profile and therefore applying the boundary condition that the shear stress is zero in the middle of the slot, i.e., $\tau_{yz} = 0$ at $y = 0$, results in $c_1 = 0$. Then, using the power law rheological model, the shear stress can be expressed in terms of a velocity gradient as follows:

$$\tau_{yz} = -m \left(\frac{dv_z}{dy} \right)^n \tag{5.6.44}$$

Inserting Eq. (5.6.44) into Eq. (5.6.43), solving the result for dv_z/dy and integrating with respect to y gives:

$$v_z(y) = \left(\frac{-\Delta P}{mL} \right)^{1/n} \left(\frac{n}{n+1} \right) y^{(1/n+1)} + c_2 \tag{5.6.45}$$

Now, applying the boundary condition that the velocity is 0 at the wall, i.e., $v_z = 0$ at $y = w/2$, yields the following expression for the integration constant c_2:

$$c_2 = -\left(\frac{-\Delta P}{mL} \right)^{1/n} \left(\frac{n}{n+1} \right) \left(\frac{w}{2} \right)^{(1/n+1)} \tag{5.6.46}$$

Consequently, the velocity profile can be written as

$$v_z(y) = \left(\frac{\Delta P}{mL} \right)^{1/n} \left(\frac{n}{n+1} \right) \left[\left(\frac{w}{2} \right)^{(1/n+1)} - y^{(1/n+1)} \right] \tag{5.6.47}$$

As always for premetered coating methods the pressure gradient is not known, but the flow rate/width Q is known for a given application, and that parameter is obtained by integrating the velocity profile over the slot height w. The result, i.e., the flow rate–pressure drop relationship, can be found in Bird et al. (1977) or in Ruschak and Weinstein (2014). Slightly rearranging their expression yields

$$\frac{\Delta P}{L} = 2^{(1+n)} \left(2 + \frac{1}{n} \right)^n \frac{m Q^n}{w^{(1+2n)}} \tag{5.6.48}$$

Inserting Eq. (5.6.48) into Eq. (5.6.47) gives the final result for the **velocity profile of a power law fluid in a slot**:

$$v_z(y) = \frac{2^{(1/n+1)}}{w^{(1/n+2)}} \frac{(1+2n)}{(1+n)} Q \left[\left(\frac{w}{2} \right)^{(1/n+1)} - y^{(1/n+1)} \right] \tag{5.6.49}$$

Equation (5.6.49) is only valid for $0 \leq y \leq w/2$. The velocity profile over the entire slot height is obtained by mirroring the profile of the half slot. Using the average velocity for slot flow according to Eq. (5.6.10) yields the following expression for the **dimensionless velocity profile**, which is valid for $0 \leq y/w \leq 0.5$:

$$\frac{v_z(y)}{v_{z,average}} = 2^{(1/n+1)} \frac{(1+2n)}{(1+n)} \left[\left(\frac{1}{2}\right)^{1/n+1} - \left(\frac{y}{w}\right)^{(1/n+1)} \right] \tag{5.6.50}$$

Figure 5.42 shows the half profile as a function of the power law index n.

The shape of the profile only depends on the power law index n but not on the viscosity parameter m nor on the flow rate/width Q. For $n = 1.0$ the parabolic Newtonian profile is recovered with a maximum velocity of 1.5 times the average velocity. For n approaching 0 plug flow is obtained with a uniform velocity equaling the average velocity. All the equations above involving the power law rheological model are not valid for $n < 0$.

Combining Eqs. (5.6.43), (5.6.44), and (5.6.48), the wall shear rate can be derived to be:

$$\gamma_0 = \frac{Q}{w^2}\left(\frac{2}{n} + 4\right) \tag{5.6.51}$$

Consequently, the wall shear stress is:

$$\tau_0 = -m(\gamma_0)^n \tag{5.6.52}$$

As for pipe and duct flow, Ruschak and Weinstein (2014) applied a local power law (LPL) model to slot flow by introducing an equivalent Newtonian viscosity μ_e, defined as that giving the same wall shear stress and pressure gradient as the non-Newtonian liquid:

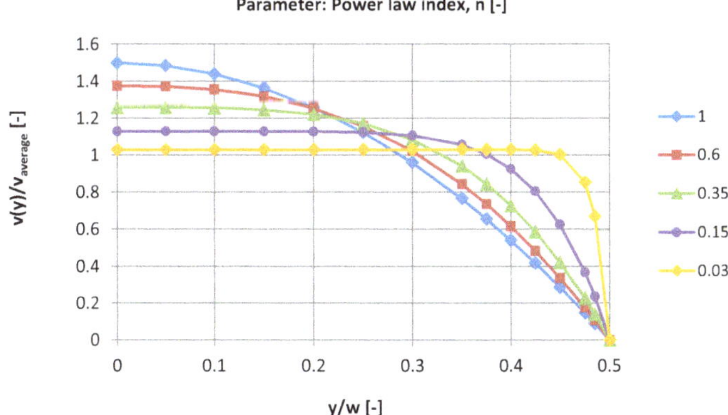

Fig. 5.42 Half of the dimensionless velocity profile for power law fluids in slot flow as a function of the power law index

$$\mu_e = \frac{m}{6} 2^n \left(2 + \frac{1}{n}\right)^n \frac{w^{2(1-n)}}{Q^{1-n}} \tag{5.6.53}$$

Then the procedure for determining m, n, γ_e and μ_e for a given coating application is the same as presented for pipe flow, see Sect. 5.4.2.

5.6.2.2 Slot Flow with One Wall at Rest and One Wall Moving

Slot flow with one still and one moving wall is relevant for slot coating as it describes the flow field of the coating bead. Unfortunately, solving this flow problem for power law fluids is difficult from a mathematical point of view. In particular, a closed analytical expression is not readily available because the solution for the velocity profile depends on the sign of the velocity gradient on the wall. The same is true for other flow parameters such as the flow rate/width or the flow rate–pressure drop relationship. Bird et al. (1977) proposed an analytical approach, and Tsuda (2010) solved the problem analytically and numerically. Here, we follow an analytical approach derived by Bebié (2020).

The flow geometry is sketched in Fig. 5.43. The lower wall at $y = 0$ is at rest while the upper wall at $y = H_{gap}$ is moving with speed U. Two typical velocity profiles of this flow are shown qualitatively. The flow is driven by a constant but negative pressure gradient, i.e., $P_0 > P_L$, and by the motion of one wall, whereby U can be positive or negative relative to the sign of the pressure gradient. Therefore, if the pressure gradient is negative and the wall velocity is positive, then both the pressure and the wall motion drive the flow in the same direction as shown in Fig. 5.43.

The goal is to calculate the velocity profile, and the associated differential equation for power law fluids is:

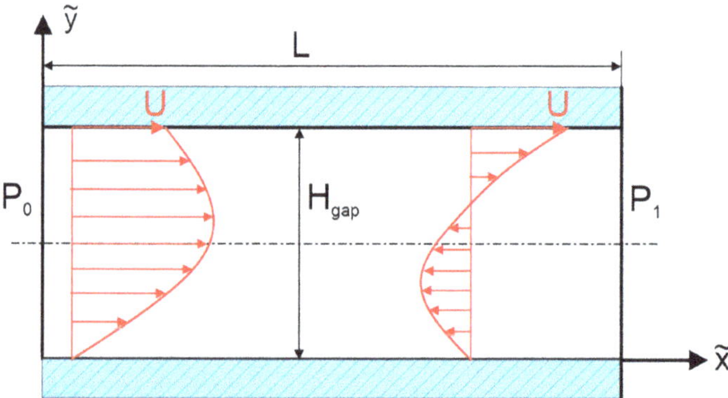

Fig. 5.43 Geometry of slot flow

$$\frac{dP}{d\tilde{x}} = m\frac{d}{d\tilde{y}}\left\{\left[\left(\frac{d\tilde{v}}{d\tilde{y}}\right)^2\right]^{(n-1)/2)}\frac{d\tilde{v}}{d\tilde{y}}\right\} \tag{5.6.54}$$

Note that the treatment of the power in the square bracket depends on the sign of $d\tilde{v}/d\tilde{y}$: $\left(y^2\right)^{(n-1)/2} = y^{(n-1)}\,for\ y > 0$, while $\left(y^2\right)^{(n-1)/2} = (-y)^{(n-1)}\,for\ y < 0$.

Following Liu (1983) for introducing the dimensionless coordinate y and the dimensionless velocities v (liquid) and V (wall or substrate), Eq. (5.6.54) can be made dimensionless:

$$\begin{gathered}\tilde{y} = yH_{gap} \quad 0 \le y \le 1 \\ \tilde{v}(\tilde{y}) = Bv(y), \quad U = BV, \quad B = {H_{gap}}^{(n+1)/n}\left(\frac{-dP}{dx}\frac{1}{m}\right)^{1/n}\end{gathered} \tag{5.6.55}$$

$$\frac{d}{y}\left\{\left[\left(\frac{dv}{dy}\right)^2\right]^{(n-1)/2}\frac{dv}{dy}\right\} = -1 \tag{5.6.56}$$

The boundary conditions are v(0) = 0 and v(1) = V. It is possible to find an analytical solution for the possible velocity profiles by introducing a parametrization. The best and desirable parameter is the wall velocity V. Depending on the wall velocity U, V can assume any positive or negative values. As mentioned above, for V > 0 the motion of the wall supports the action of the pressure gradient.

Unfortunately, however, using V as the parameter is not possible from an algebraic point of view. Instead, other suitable parameters must be sought, which in turn are determined by V. In particular, the search for solutions must consider that the sign of the velocity gradient dv/dy changes somewhere along the profile, for example in the middle of the slot if both walls are at rest. Therefore, the two ranges characterized by a different sign of the velocity gradient must be treated separately.

An obvious parametrization can be accomplished with the velocity gradient at the wall, i.e., with the parameter z as follows:

$$z = \left.\frac{dv}{dy}\right|_{y=0} \tag{5.6.57}$$

The velocity profile is uniquely determined by the boundary conditions v(0) = 0 and dv/dy(0) = z, whereby z can assume any positive or negative value. The wall velocity V associated with z is determined by V = v(1).

Deriving the above showed that the ranges of z > 0 and z < 0 must be treated separately. Therefore, two algebraic expressions resulted for the dimensionless velocity profile.

For z > 0 and V > −n/(n + 1):

$$v(y) = \frac{n}{n+1}\left\{z^{n+1} - \left|z^n - y\right|^{(n+1)/n}\right\} \tag{5.6.58a}$$

For z < 0 and V < −n/(n + 1):

$$v(y) = \frac{n}{n+1}\left\{|z|^{n+1} - \left[|z|^n + y\right]^{(n+1)/n}\right\} \tag{5.6.58b}$$

As can be seen the wall velocity $V_1 = -n/(n+1)$ separates the two ranges. Note the signs for absolute values, which prevent negative arguments from ever being raised to an odd power. Moreover, the two ranges merge smoothly for z = 0.

The wall velocity belonging to z is determined by V = v(1). Unfortunately, an algebraic equation for z = z(V) cannot be derived. Therefore, calculating the dimensionless velocity profile requires an iterative procedure as follows.

1. Set a value for n and calculate V_1.
2. Assume a value for the wall velocity V.
3. Assume a value for z and calculate the velocity profile with the corresponding Eqs. (5.6.58a), (5.6.58b).
4. Change the value of z until the assumed value for V equals the calculated value v(1).

The iteration can easily be programmed in an EXCEL environment by using the goal-seek function. Bebié also solved Eq. (5.6.56) numerically and found good agreement with the analytical Eqs. (5.6.58a), (5.6.58b).

From a coating point of view the procedure above is not useful yet because the flow as modelled so far is driven by a pressure gradient, which is not known for premetered coating methods. What is known, however, is the volumetric flow rate/width Q, which is determined by the wet film thickness of the coated film H_{wet}, and the substrate velocity U according to $Q = UH_{wet}$. As shown above, Q is obtained by integrating the velocity profile over the height of the slot. Here, the corresponding equation takes the following form:

$$Q = \int_0^{H_{gap}} \tilde{v}(\tilde{y})d\tilde{y} = H_{gap}B\int_0^1 v(y)dy = H_{gap}BF(z) \tag{5.6.59}$$

Since the parameter B contains the pressure gradient (see Eq. 5.6.55) the integration of the dimensionless velocity profile yields the flow rate–pressure drop relationship, which then allows the pressure gradient to be eliminated from the equation of the velocity profile. Bebié (2020) showed that the integration of Eq. (5.6.59) can be accomplished analytically, but it must be done for 3 ranges of the parameter z. The reason is that the term $z^n - y$ in Eq. (5.6.58a) changes sign at $y = z^n$ when integrated from 0 < y < 1 when 0 < z < 1. Consequently, the ranges $0 < y < z^n$ and $z^n < y < 1$ must be integrated separately. The resulting equations for the dimensionless flow rate/width F(z) are:

For z > 1 and V > n/(n + 1)

$$F(z) = \frac{n}{n+1}\left\{z^{n+1} + \frac{n}{2n+1}\left[\left(z^n - 1\right)^{(2n+1)/n} - z^{2n+1}\right]\right\} \tag{5.6.60a}$$

For 0 < z < 1 and –n/(n + 1) < V < n/(n + 1)

$$F(z) = \frac{n}{n+1}\left\{z^{n+1} + \frac{n}{2n+1}\left[-\left(1 - z^n\right)^{(2n+1)/n} - z^{2n+1}\right]\right\} \tag{5.6.60b}$$

For z < 0 and V < –n/(n + 1)

$$F(z) = \frac{n}{n+1}\left\{|z|^{n+1} - \frac{n}{2n+1}\left[\left(1 + |z|^n\right)^{(2n+1)/n} - |z|^{2n+1}\right]\right\} \tag{5.6.60c}$$

In order to explain how these dimensionless flow rates can be translated into the real and dimensional world of coating let us look at a slot coating process as sketched in Fig. 5.44. Exemplary values for the geometrical parameters and the operating conditions are listed in the Figure caption.

First, we examine the downstream gap. Based on the law of mass conservation and by comparing the flow in the gap with the flow far downstream of the die, the average velocity in the downstream gap equals $\tilde{v}_{average} = (30/50)U = \left(H_{wet}/H_{gap}\right)U$. Transferring this result to our dimensionless velocities means that the ratio of average velocity to substrate velocity is $v_{average}/V = 30/50 = 0.6$. This ratio determines the value of z because z determines the ratio of $v_{average}/V$. In particular, the average velocity is equal to the dimensionless flow rate F because the dimensionless gap equals 1:

$$\frac{v_{average}}{V} = \frac{F(z)}{V(z)} = \frac{H_{wet}}{H_{gap}} \tag{5.6.61}$$

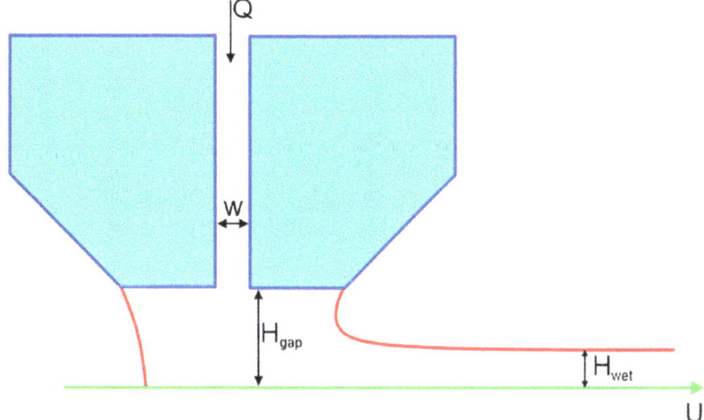

Fig. 5.44 Sketch of slot coating process, U = 50 m/min = 0.833 m/s, H_{wet} = 30 µm, Q = 0.25 cm^2/s, m = 0.33 Pa s^n, n = 0.33, H_{gap} = 50 µm, H_{wet}/H_{gap} = 0.6, w = 150 µm, $L_1 = L_2$ = 1.5 mm

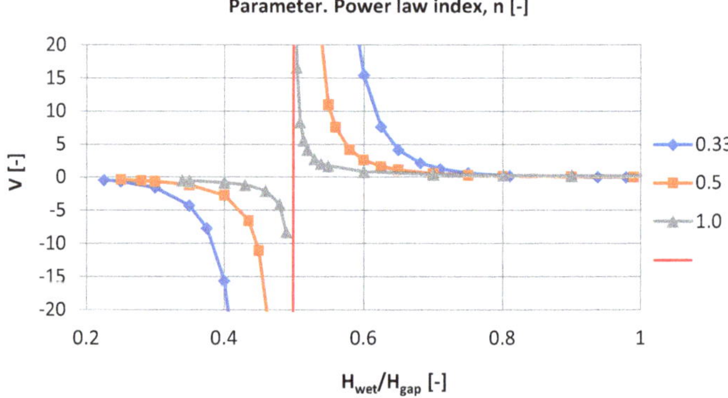

Fig. 5.45 Range of suitable V-values as a function of H_{wet}/H_{gap} and n

With this knowledge the procedure for calculating the velocity profile in the downstream gap of a slot coating process can now be expanded as follows:

1. Set a value for n and calculate V_1.
2. Assume a value for the wall velocity V.
3. Assume a value for z and calculate the velocity profile with the corresponding Eqs. (5.6.58a), (5.6.58b).
4. Iteration 1: Change the value of z until the assumed value for V equals the calculated value v(1).
5. Calculate the dimensionless flow rate F(z) with the corresponding Eqs. (5.6.60a)–(5.6.60c).
6. Calculate the ratio F(z)/V(z).
7. Iteration 2: Change the value of V and execute iteration 1 until the ratio F(z)/V(z) equals the ratio H_{wet}/H_{gap}.

In an EXCEL environment iteration 1 can easily be programmed by using the goal-seek function, while iteration 2 can be carried out by hand with a trial-and-error approach. In support of this procedure, Fig. 5.45 shows the range of suitable V-values as a function of H_{wet}/H_{gap} and n. The pole in this graph, i.e., the vertical red line at $H_{wet}/H_{gap} = 0.5$, indicates the position, at which the pressure gradient dP/dx is 0, i.e., at which the pressure gradient changes sign.

The pressure gradient is negative for $H_{wet}/H_{gap} > 0.5$, and it is positive for $H_{wet}/H_{gap} < 0.5$. Therefore, in agreement with the definition of the coordinate system and the sign of the pressure gradient relative to the coordinate system, the dimensional wall velocity must assume negative values for positive pressure gradients, i.e., for $H_{wet}/H_{gap} < 0$. This in turn assures that the parameter B (see Eq. 5.6.55) is always positive, which is necessary for correctly calculating the dimensional pressure gradient for any value of H_{wet}/H_{gap}.

For the coating example shown in Fig. 5.44, i.e., for $n = 0.33$ and $H_{wet}/H_{gap} = 0.6$, the dimensionless substrate velocity equals $V = 15.62$ and the associated value

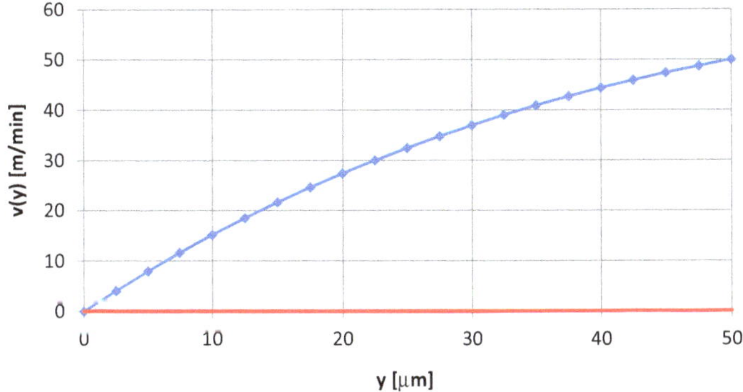

Fig. 5.46 Velocity profile in downstream gap of slot coating for U = 50 m/min, n = 0.33 and $H_{wet}/H_{gap} = 0.6$

for z = 26.33. Using the transformation Eq. (5.6.55), and in particular B = U/V, the corresponding dimensional velocity profile can now be calculated and the result is shown in Fig. 5.46. The flow is driven by the substrate velocity and by a weak pressure gradient. In fact, the pressure gradient is $6.59 \cdot 10^4$ Pa/m and the pressure drop over the downstream coating gap of length 1.5 mm amounts to 98.8 Pa.

For Newtonian flow (n = 1.0) a vortex is present under the downstream die lip if $H_{wet}/H_{gap} < 0.333$, or if $H_{gap}/H_{wet} > 3$, see Fig. 5.34. As vortices are not desirable in coating flows, this fact presents an operating boundary of the slot coating process. The question then arises if and how this boundary can be moved for shear thinning fluids.

For fluids and flow fields that are described with the power law model the inception of the vortex under the downstream die lip, i.e., the onset of back flow, is given if the velocity gradient at the stationary wall is zero, i.e., if z = 0, and this condition is obtained if $V = V_1$.

Figure 5.47 shows that the onset of the vortex shifts to considerably larger coating gaps as the power law index n decreases to low values. This is a welcome and positive result because it opens up the operating window of the slot coating process. The same result was obtained by Tsuda (2010). Backflow under the downstream lip is present if $0 > V > V_1$ and the strongest backflow is obtained when V approaches 0.

Regarding the **flow in the upstream gap** of slot coating this flow field is called a turn-around flow because all the fluid that exits the die slot and initially moves into the upstream gap turns around at the upstream meniscus and eventually is transported downstream by the motion of the substrate. The velocity profile is obtained by forcing the net flow rate in this gap to be zero, which in turn requires changing the calculation procedure listed above as follows:

1. Set a value for n and calculate V_1.
2. Assume a value for the wall velocity V.
3. Assume a value for z and calculate the velocity profile with the corresponding Eqs. (5.6.58a), (5.6.58b).

4. Iteration 1: Change the value of z until the assumed value for V equals the calculated value v(1).
5. Calculate the dimensionless flow rate F(z) with the corresponding Eqs. (5.6.60a)–(5.6.60c).
6. Iteration 2: Change the value of V and execute iteration 1 until F(z) = 0.

The resulting dimensional velocity profiles for the coating example shown in Fig. 5.44 are illustrated in Fig. 5.48 as a function of the power law index n.

Experience shows that V must be adjusted very finely in order to obtain the condition of F = 0. To help in this process the required V-values as a function of the power law index n can be read from Fig. 5.49.

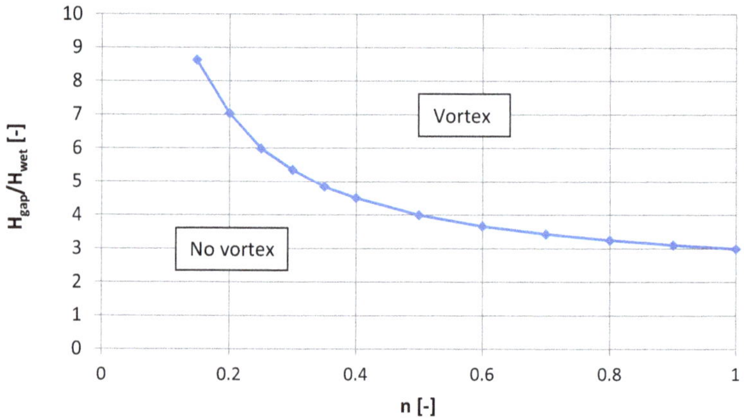

Fig. 5.47 Onset of vortex in downstream coating gap of slot coating

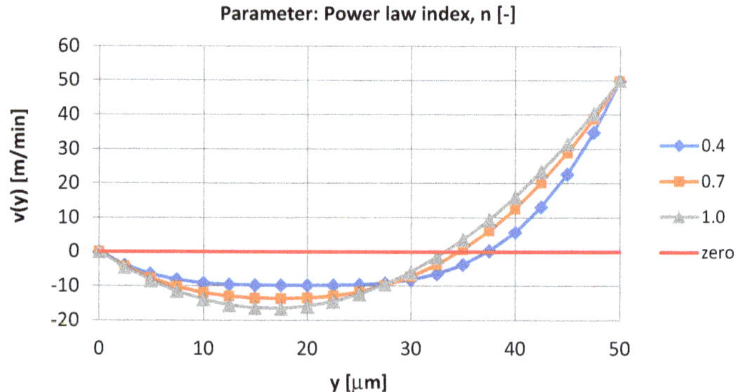

Fig. 5.48 Velocity profiles in the upstream gap of slot coating as a function of the power law index n

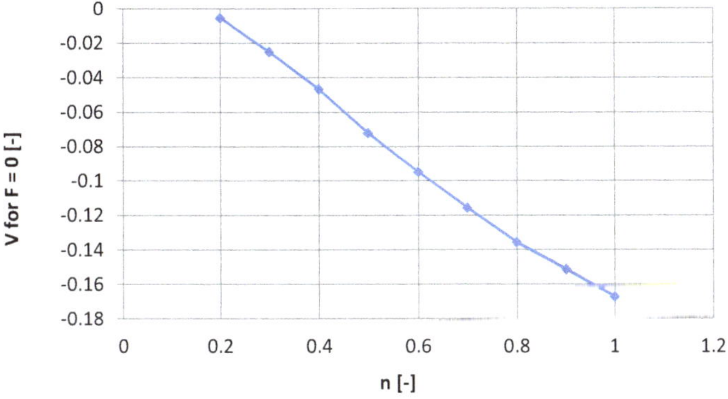

Fig. 5.49 V-values that result in F = 0 as a function of n

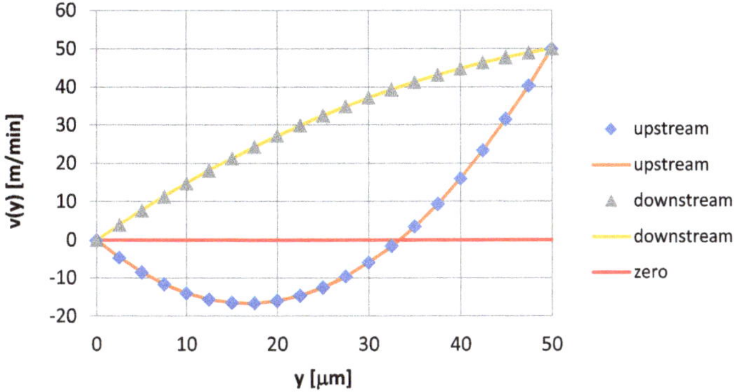

Fig. 5.50 Comparison of the velocity profiles in the upstream and downstream gap of slot coating when calculated with power law and Newtonian flow equations for n = 1.0

Using the power law flow equations presented above for calculating the velocity profiles in the upstream and downstream gap of slot coating for n = 1 and for the operating conditions sketched in Fig. 5.44, and comparing them with the corresponding profiles computed with the Newtonian flow Eqs. (5.6.25) and (5.6.31) resulted in a very good agreement as illustrated in Fig. 5.50. The symbols represent the power law calculations while the lines through the symbols are generated by the Newtonian equations. The deviation between the two profiles is <1% for both profiles. This then suggests that it is safe to use the power law equations for simulating the flow fields in the coating gap of the slot coating process.

As for pipe, duct and slot flow with both walls at rest, it is proposed to solve slot flow problems with one moving wall by adapting the local power law model as described by Ruschak and Weinstein (2014). Since a closed form for an equivalent Newtonian viscosity μ_e, defined as that giving the same wall shear stress and pressure gradient as the non-Newtonian liquid cannot be derived for this flow field, the following calculation procedure is suggested.

1. Calculate the average shear rate in the coating gap according to $\gamma = U/H_{gap}$.
2. Use the Carreau-Yasuda model (Eq. 4.4.1) for calculating the corresponding viscosity.
3. Use Eqs. (4.4.2) and (4.4.4) for calculating m and n according to Ruschak and Weinstein (2014).
4. Proceed with the calculation procedures for flow in the upstream or downstream gap of slot coating as presented above.

5.6.3 *Two-Layer Slot Flow (Newtonian Flow Behavior with Carreau-Yasuda Viscosity)*

Two-layer slot flows are encountered in the simultaneous two-layer slot coating process as described in Sect. 17.16. Here, we focus on the two-layer slot flow inside the die slot as illustrated in Fig. 17.67 where both walls are at rest, and on the two-layer slot flow in the downstream coating gap as depicted in Fig. 17.66 where one wall is at rest and the other moving at web speed U. The former situation has already been solved in a slightly different way than presented here by Bird et al. (1960, Sect. 2.5).

Considering the mathematical difficulties that Bebié (2020) revealed when modeling single layer slot flow for power law fluids (see Sect. 5.6.2.2), particularly for the case of one wall moving at speed U, we purposely keep it simpler here by using Newtonian flow equations but by calculating the local viscosity according to the shear rate-dependent Carreau-Yasuda model. In addition, we only model the case where one wall is moving at speed U. The result can also be applied to the situation where both walls are at rest, simply by setting the wall velocity U equal to zero.

The flow geometry is shown in Fig. 5.51. The total slot height is denoted with w, which, depending on the situation, is equal to the height of the die slot or the

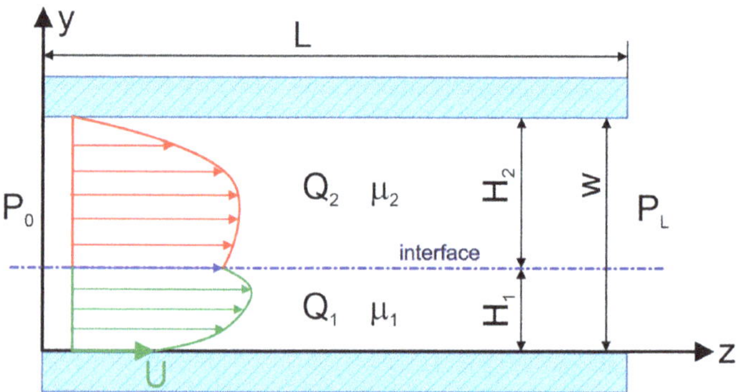

Fig. 5.51 Geometry of two-layer slot flow with one wall moving

height of the downstream coating gap. Each layer is characterized by a thickness H_i, a volumetric flow rate/width Q_i, and a viscosity μ_i. The subscript $i = 1$ refers to the lower layer, while the subscript $i = 2$ applies to the upper layer. The interface between the two layers is located at $y = H_1$. The lower wall of the slot is moving with speed U, which may also assume the value of zero. The slot length is L, and the pressures at $z = 0$ and $z = L$ are denoted with P_0 and P_L, respectively.

Modeling this flow starts with Eq. (5.6.1), which is the differential equation for one-dimensional steady, laminar and rectilinear flow of Newtonian liquids. Applying this equation to each of the layers and integrating twice in each layer results in the following velocity profiles:

$$v_1(y) = \frac{dP}{dz}\frac{1}{2\mu_1}y^2 + C_1 y + C_2 \tag{5.6.62}$$

$$v_2(y) = \frac{dP}{dz}\frac{1}{2\mu_2}y^2 + C_3 y + C_4 \tag{5.6.63}$$

The four integration constants can be determined with the boundary conditions that on both slot walls, the liquid velocity equals the wall velocity and that, at the interface, the momentum transfer is continuous through the interface between the two fluids, and the velocities of both layers are equal as follows:

$$\text{BC1:} \quad y = 0: \quad v_1 = U \tag{5.6.64}$$

$$\text{BC2:} \quad y = H_1: \quad v_1 = v_2 \tag{5.6.65}$$

$$\text{BC3:} \quad y = H_1: \quad \mu_1\frac{dv_1}{dy} = \mu_2\frac{dv_2}{dy} \tag{5.6.66}$$

$$\text{BC4:} \quad y = w: \quad v_2 = 0 \tag{5.6.67}$$

Working out the algebra for determining the integration constants was complicated. Therefore, we introduced the following abbreviations, which helped to simplify and reduce the size of the expressions for the constants.

$$\begin{aligned} &a = \frac{dP}{dz} \quad b = \mu_1/\mu_2 \quad c = H_1^2/\mu_1 \quad d = H_1^2/\mu_2 \\ &e = w^2/\mu_2 \quad f = 1/\mu_1 \quad g = e + c - d \quad h = b - 1 \\ &i = w^3 - H_1^3 \quad j = w^2 - H_1^2 \quad k = w - H_1 \quad \ell = 1/\mu_2 \end{aligned} \tag{5.6.68}$$

Now, the integration constants are given as listed below:

$$C_1 = \frac{ag/2 + U}{H_1(b-1) - bw} \tag{5.6.69}$$

$$C_2 = U \tag{5.6.70}$$

$$C_3 = C_1 b \tag{5.6.71}$$

$$C_4 = \frac{a}{2}(c - d) + C_1 H_1 (1 - b) + U \tag{5.6.72}$$

So far, C_1 and therefore C_3 and C_4 depend on the pressure gradient, which, here, is expressed by the parameter a. For premetered coating methods, this is unfortunate because the pressure gradient is unknown a priori and depends on the flow rate and viscosity of the liquids that are pumped through the slot die. As was shown for single-layer slot flow, the pressure gradient can be eliminated by integrating the velocity profile over the height of the flow and setting the result to be equal to the known flow rate/width of that flow.

For the bottom layer of the two-layer flow, the corresponding requirement is

$$Q_1 = \int_0^{H_1} v_1(y)dy \tag{5.6.73}$$

Carrying out the integration and solving the result for a = dP/dz yields

$$a = \frac{Q_1 - \frac{H_1^2 U}{2(H_1 h - bw)} - U H_1}{\frac{f H_1^3}{6} + \frac{H_1^2 g}{4(H_1 h - bw)}} \tag{5.6.74}$$

Now, the integration constants depend on the thickness of the bottom layer, i.e., the location of the interface, which is still unknown. This problem can easily be solved by integrating the velocity profile of the upper layer and forcing it to be equal to the flow rate/width Q_2 of that layer as follows.

$$Q_2 = \int_{H_1}^{w} v_2(y)dy \tag{5.6.75}$$

$$Q_2 = \frac{ai\ell}{6} + \frac{C_3 j}{2} + C_4 k \tag{5.6.76}$$

Since Eq. (5.6.76) is too complicated to be resolved for H_1, it is easier to assume a value for H_1 and to carry out an iteration, i.e., to vary H_1 until the flow rate Q_2 as calculated by Eq. (5.6.76) is equal to the known flow rate as given by the operating conditions of the coating process. This procedure can conveniently be carried out in an EXCEL environment with the goal-seek function.

The iteration requires an initial value for H_1, which can be quantified by scaling the slot height w with the flow portion taken up by layer 1 according to

$$H_{1,initial} = w\frac{Q_1}{Q_1 + Q_2} \tag{5.6.77}$$

This value can be lowered a bit if $\mu_1 < \mu_2$, and if $U >> 0$. According to our experience, the iteration converges after just a handful of cycles.

The integration constants and the pressure gradient also depend on the viscosities of the two layers. If the viscosities can be described by the Carreau-Yasuda model (Eq. 4.4.1), which is the assumption here, then a shear rate is required for each layer. This parameter is obtained by integrating Eq. (5.6.1) once for each layer. We propose to use the wall shear rate in each layer, and the corresponding results are

$$\left.\frac{dv_1}{dy}\right|_{y=0} = C_1 \tag{5.6.78}$$

$$\left.\frac{dv_2}{dy}\right|_{y=w} = \frac{aw}{2}C_3 \tag{5.6.79}$$

Since both wall shear rates depend on the respective viscosity and the thickness H_1 of layer 1, a second (nested) iteration is required, which again needs initial values for both viscosities. We propose to use the average shear rate (see Eq. 5.6.14) of the total two-layer flow for each layer according to

$$\left.\frac{dv_1}{dy}\right|_{initial} = \left.\frac{dv_2}{dy}\right|_{initial} = \frac{6(Q_1 + Q_2)}{w^2} \tag{5.6.80}$$

To handle both iterations simultaneously, the following calculation procedure has proven helpful in an EXCEL environment.

1. Calculate the initial shear rate (Eq. 5.6.80) and the initial viscosity (Eq. 4.4.1) for each layer.
2. Transfer these initial viscosities into new cells called "old viscosity".
3. Calculate the initial thickness of layer 1 (Eq. 5.6.77) and transfer the result into a new cell that is used for the iteration of H_1.
4. Calculate the pressure gradient (Eq. 5.6.74), all integration constants (Eqs. 5.6.69–5.6.72) and the flow rate Q_2 (Eq. 5.6.76).
5. Use the goal-seek function to vary H_1 until the calculated value of Q_2 is equal to the value given by the coating conditions.
6. For each iteration step, H_1 changes. Therefore, the wall shear rate and the viscosity in each layer changes as well. Calculate the wall shear rates (Eqs. 5.6.78–5.6.79) and the corresponding viscosities (Eq. 4.4.1) in cells called "new viscosity".

7. At the end of the iteration for H_1, the "new viscosity" is different from the "old viscosity". Therefore, transfer the values of the "new viscosity" to the cells called "old viscosity".
8. Repeat the iteration for H_1 by repeating steps 4–7. Now, the initial value for H_1 is equal to the end value of the previous iteration.
9. Repeat step 8 until the new viscosity values become equal to the old values.

According to our experience, step 8 must be repeated just a handful of times. We recommend to write a Macro for steps 5–7, and to copy/paste these commands, say, 10 times to be on the safe side. This procedure is convenient and has worked well for us.

Now that all integration constants are known, the velocity profile of two-layer slot flow can be drawn, and an example for the case of both walls being at rest (U = 0) is shown in Fig. 5.52. The arbitrarily chosen operating conditions, i.e., the flow rate/width and the fife parameters of the Carreau-Yasuda equation are summarized in Table 5.13. The table also contains the calculated wall shear rates and the corresponding viscosities. The slot height was 250 μm. The velocities were made dimensionless with the average velocity of slot flow according to $(Q_1 + Q_2)/w$.

The interface is located at y = 70.4 μm or y/w = 0.28. Reducing the flow rate or the viscosity of layer 1 would move the interface to lower values. Requiring both fluids to be Newtonian (b = 1.0) and of the same low shear viscosity ($\mu_{0,1} = \mu_{0,2}$) recovers the result given by Eq. (5.6.15). Moreover, if $Q_1 = Q_2$ the interface is located in the center of the slot.

Figure 5.53 shows another example for the same operating conditions except that now the wall at y = 0 is moving with the speed U = 100 m/min. The velocity profile represents the flow in the downstream gap of a two-layer slot coater. Owing to the low flow limit (see Sect. 17.5), the process for the operating conditions of Table 5.13 must be operated in the viscous mode, which limits the slot height, i.e., the coating gap, to a maximum value of approximately 1.7 times the total wet thickness of the coated film. Therefore, the slot height w is now set to 133 μm. The resulting

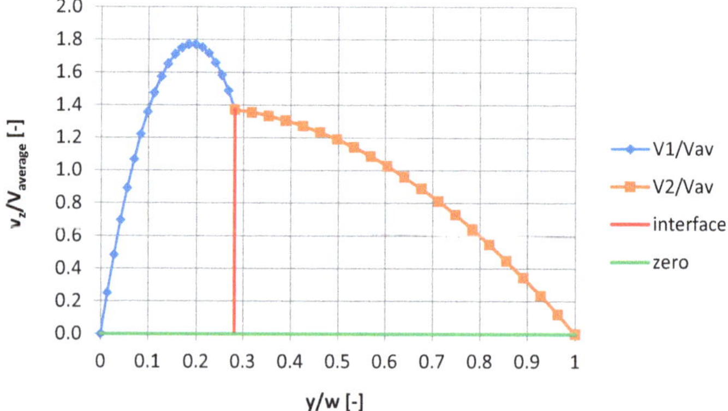

Fig. 5.52 Dimensionless velocity profile of two-layer slot flow with both walls at rest

wall shear rates and the associated viscosities are listed in Table 5.13. The velocities are made dimensionless with the web speed U, and the interface is located at y = 28.9 μm or y/w = 0.22. Increasing the web speed would move the location of the interface to lower values.

The flow of a single layer of Newtonian fluid in the downstream coating gap is characterized by the onset of recirculation along the wall that is at rest if $H_{gap}/H_{wet} > 3$ (see Figs. 5.33 and 5.34). For two-layer flow, the onset of recirculation is postponed to much higher gaps if the viscosity of the upper layer (attached to the wall at rest) is higher than the viscosity of the lower layer, as is the case in the example shown in Fig. 5.53. A simple criterion for the onset of flow separation in two-layer flow of fluids with different rheological properties in the downstream coating gap is not known.

Table 5.13 Operating conditions for example shown in Figs. 5.52 and 5.53

U [m/min]	0		100	
	Layer 1	Layer 2	Layer 1	Layer 2
A_{dry} [g/m^2]	15	25	15	25
H_{wet} [μm]	29.3	48.8	29.3	48.8
Q [cm^2/s]	0.488	0.813	0.488	0.813
μ_0 [mPas]	50	200	50	200
μ_∞ [mPas]	1	1	1	1
γ_c [s^{-1}]	3.0	0.5	3.0	0.5
a [–]	2.0	2.0	2.0	2.0
b [–]	0.6	0.85	0.6	0.85
γ_0 [s^{-1}]	38,709	7,161	32,079	20,907
μ [mPas]	2.11	48.36	2.20	41.33

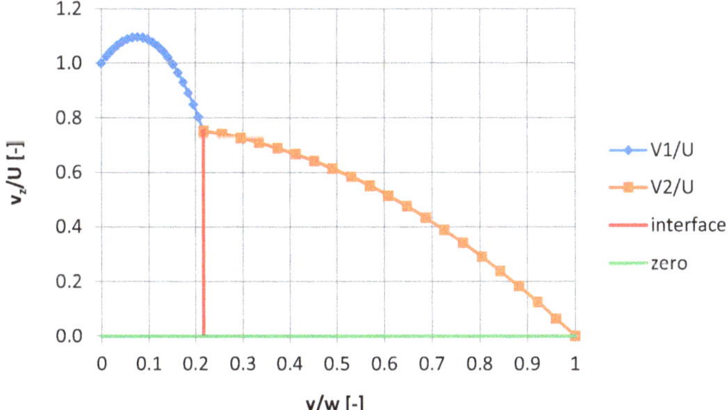

Fig. 5.53 Dimensionless velocity profile of two-layer slot flow with one wall at rest and the other wall moving with speed U

More information about two-layer slot coating can be found in Sect. 17.16.

5.6.4 Vortices in Slot Flow

5.6.4.1 Vortices in the Die Slot

Steady state flow in a slot of constant height is free of vortices. However, vortices may form near the entrance or the exit of slot flow, or if the slot height changes in flow direction. One example, which is discussed in Sect. 5.5.2.3, is the flow across the outer cavity of a dual-cavity die. From a simplified point of view, flow across a cavity may be considered as slot flow with an increasing slot height followed by a decreasing height. This view may be appropriate if the expansion angle and the depth of the cavity are small.

Another potential location for a vortex in the die slot refers to the slot exit of slide dies as illustrated in Fig. 5.54. In particular, if the exit velocity of the slot flow is too high, i.e., if the slot height w is too narrow for a given Reynolds number, then a vortex may form on the die slide just downstream of the exit corner, see details in Sect. 5.7.8. It may be necessary, therefore, to reduce the exit velocity to avoid this vortex, and this can be achieved by locally widening the slot height from w to w_f with a so called flared slot design.

A flared slot exit is nothing but an unsymmetrical diffuser, and the flow field in this geometry is rather similar but not exactly the same as the flow field in the outer cavity having a cross-sectional shape of a triangle or of a circular segment, see Figs. 5.16 and 5.18. As an approximation, therefore, avoiding a vortex in the flared slot exit can be achieved by respecting Eq. (5.5.26). In addition, experience shows that good flow fields are obtained if the flare angle α remains below 15° regardless of the value of the Reynolds number of the slot flow.

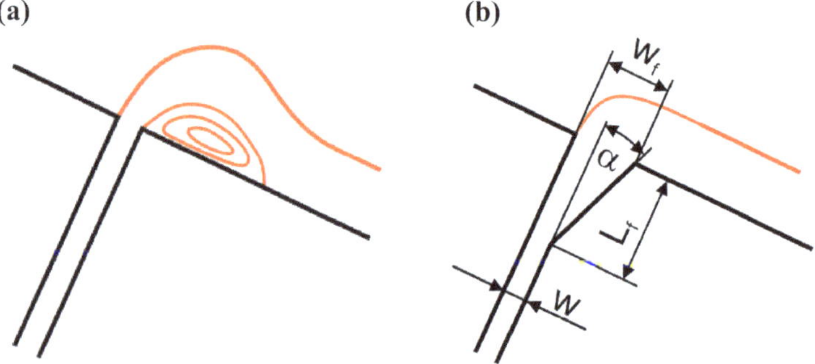

Fig. 5.54 **a** Vortex formation near slot exit owing to excessive liquid jetting **b** prevention of vortex by widening the slot height near exit and thereby reducing jetting

Opening the slot too much also increases the length L_f of the flared slot, and that reduces the overall resistance of the flow in the slot. Consequently, the resulting flow rate uniformity in the cross-web direction will be slightly worse compared to a design with a slot that remains narrow over its entire length. This is another example that requires a compromise, i.e., a design optimization, between a good widthwise flow rate uniformity and the absence of vortices or areas of low wall shear stress.

In slot coating, the flow field of the coating bead is particularly susceptible for vortices. Durst and Wagner (1997) computed this flow theoretically for various geometries and operating conditions, and they showed that both the upstream and downstream gaps can potentially house vortices, particularly when the coating gap is large compared to the wet thickness of the coated film, see Fig. 5.55.

The upstream gap may contain a vortex, or it may be flooded or partially filled with a pure turn-around flow, which is characterized by neighboring streamlines that move in opposite directions. Own flow visualization experiments in the bead of slide coating not only revealed similar flow situations but they also showed that such streamline patterns are unstable and have a tendency to oscillate. This, obviously, must be avoided, and it can be achieved by preventing the upstream gap from being filled with liquid. This in turn can be accomplished by not using, or at least minimizing, vacuum at the upstream meniscus, and by maximizing the coating gap for a given set of operating conditions, see Chap. 17 for more details.

As illustrated in Fig. 5.34, the downstream gap may be filled with a vortex underneath the downstream die lip. Moreover, and as visualized in Fig. 5.47, the inception of this vortex depends on the dimensionless coating gap H_{gap}/H_{wet} and on the power law index n. The vortex can be prevented for Newtonian fluids (n = 1.0) if $H_{gap}/H_{wet} < 3$. This critical gap value increases as the power law index decreases. Preventing this vortex may be difficult or impossible when attempting to coat thin wet film on

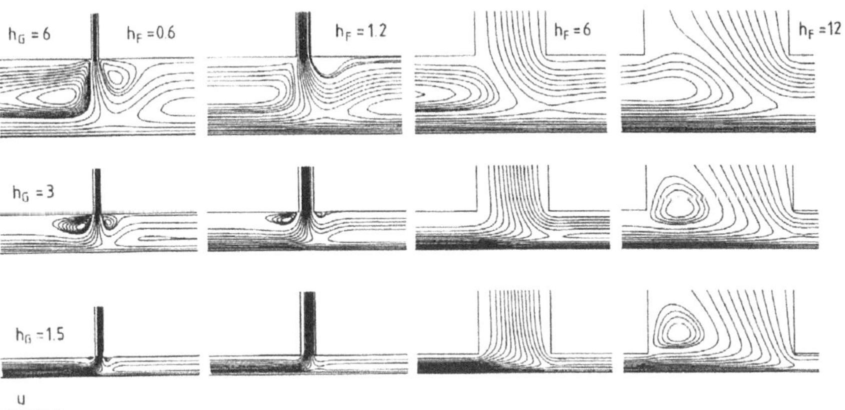

Fig. 5.55 Flow field of the coating bead in slot coating showing various vortices for different geometries and operating conditions; taken from Durst and Wagner (1997) and reprinted with permission from Springer Nature

the order of 10 μm or less. For more details see the topic of the Low Flow Limit in Sect. 17.5.

Another potential vortex is located at the very end of the die slot, just as the liquid is exiting the slot and is being carried away by the motion of the moving substrate, see sketch in the lower right corner in Fig. 5.55. According to Durst and Wagner (1997) this vortex can be avoided if the height of the die slot w is less than about twice the height of the coating gap, i.e.,

$$w < 2H_{gap} \tag{5.6.81}$$

This requirement is difficult to satisfy, particularly if the wet coating gap is small. Assuming $H_{gap} = 75$ μm, then the die slot should be less than 150 μm. However, such a narrow die slot may generate an inacceptable amount of cross profile due to an imperfect mechanical precision. If the precision is ±1 μm, which is a good and low value, then the associated cross profile will be ±2.0% for a die slot of 150 μm and for a Newtonian fluid. Such a large cross profile contribution just from the mechanical slot precision may be inacceptable for products with a high demand on film thickness uniformity. Moreover, this kind of cross profile contribution increases with decreasing power law index. For more details see Sect. 9.4 on Die Design.

Another critical vortex may be located at the exit of the downstream coating gap. As illustrated in Fig. 5.56 this vortex is very likely to be present if the lip angle γ is too large. In this case, and for many operating conditions, the static wetting line is not pinned to the lip corner, but it is located at some undefined position along the lip face. The fluid exiting the coating gap is unable to flow around the lip corner in order to fill the space between the film forming meniscus and the lip face. Consequently, this space is filled with liquid that is separated from the main flow and is rotating, i.e., it is forming a vortex. Experience shows time after time that the coated film is full of lines and streaks as soon as this vortex is present.

Preventing the vortex at the die lip in slot coating is an absolute must. It requires that the static wetting line of the film forming meniscus is pinned to the downstream corner of the die lip.

Fluid-mechanical and thermo-dynamical investigations by Gibbs (1906) showed that the wetting line of a meniscus located on a flat wall can easily be moved by smallest forces, see Fig. 5.57a. However, if the wetting line is located at a sharp

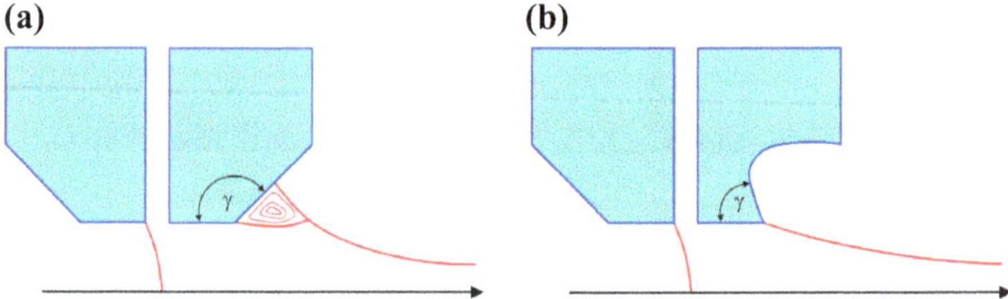

Fig. 5.56 **a** Vortex at the die lip of slot coating **b** prevention of the vortex by smart lip design

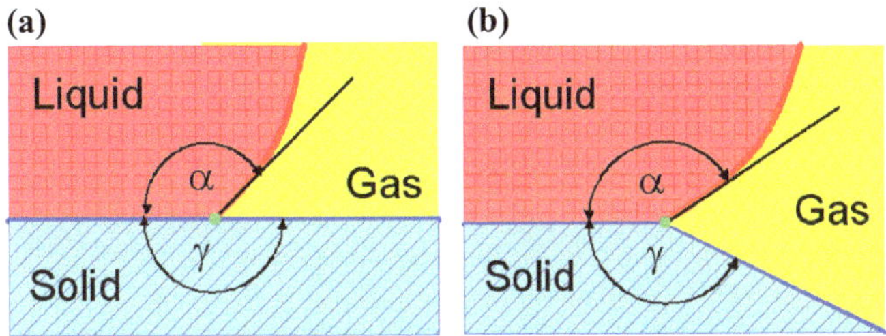

Fig. 5.57 **a** Easy motion of a wetting line located on a flat surface **b** inhibited motion of a wetting line pinned to a sharp corner in the surface

corner in that surface, then that corner is a preferred position for the wetting line, and considerably higher forces are required for moving the wetting line beyond the corner, see Fig. 5.57b. In other words, the presence of a corner in a surface inhibits the motion of the wetting line past the corner.

Gibbs derived an inequality that quantifies conditions, for which the wetting line sticks to the corner as follows (Gibbs inequality).

$$\alpha \leq \theta + (180 - \gamma) \tag{5.6.82}$$

Here, α is the contact angle of the meniscus, and γ is the angle of the corner. Since α is determined by the operating conditions of the liquid flow near the meniscus and the wetting line, the operating window for that flow is maximized by minimizing the corner angle γ. Theoretically speaking, a knife edge would be the best corner form for pinning a wetting line. In an industrial environment, however, knife edges are not acceptable because they are a safety hazard for the operators of coating equipment, and they are easily damaged (bent) through inappropriate manipulations by the operators. In practice, a compromise value for the corner angle has to be found. Figure 5.56a shows a lip corner with a large corner angle, i.e., $\gamma > 90°$, which is often found in industrial slot dies. This lip design is not recommended at all because it promotes the migration of the wetting line along the die face and hence the presence of the vortex. In contrast, if the corner angle is small, i.e., $\gamma < 90°$, then the wetting line stays pinned at the corner for a wider range of operating conditions, see Fig. 5.56b. However, die lips with corner angles of <90° are only recommended if the downstream lip is reasonably long, say if, $L_d > 2$ mm. If the lip length is shorter, then an angle of $\gamma < 90°$ would weaken the mechanical stability of the lip. On the other hand, long lips of $L_d > 2$ mm are not desirable and are rarely needed for industrial coating applications. Therefore, a lip angle of $\gamma = 90°$ has proven to be a good compromise that allows the static wetting line to remain pinned at the lip corner for a wide range of operating conditions. The same arguments apply to the design of the corner of the upstream die lip. However, they are less important because the preferred flow field calls for the upstream coating gap to be free of liquid as much as possible. Consequently, the static wetting line of the upstream meniscus is either pinned to, or it is floating along the lip near the downstream corner of the upstream lip. See Sect. 17.9 on Slot Coating for more details about how to optimize the flow field of the coating bead in slot coating.

5.7 Film Flow

Film flow is relevant for single and multilayer slide and curtain coating, see Fig. 2.1c, f. In slide coating, film flow occurs immediately before coating, and in curtain coating film flow changes to curtain flow before coating. In any case, if film flow is not flawless in any respect, the subsequent flows cannot be perfect either. The following issues will be discussed in detail in this chapter:

- Knowing the thickness of the film on any location along the surface of the slide die is crucial for designing proper edge guides for the film flow.
- Knowing where vortices may occur in the film flow and knowing how to avoid vortices is important for achieving a good uniformity of the film flow.
- Respecting rules for the viscosity and surface tension structuring of multilayer films is necessary for achieving proper wetting of adjacent layers and for avoiding waves in the film.
- Knowing the extent of the boundary layer along both edges of the film flow provides input for designing better curtain edge guides.

5.7.1 *Single-Layer Film Flow for Constant Viscosity (Newtonian Flow Behavior)*

Film flow of Newtonian fluids on an inclined plane as schematically depicted in Fig. 5.58 is well described, for example in Bird et al. (1960). Here, H is the film thickness and β is the inclination angle of the plane relative to the horizontal.

The velocity profile, which is semi-parabolic, is described by

$$v_z = \frac{\rho g sin(\beta) H^2}{\mu}\left[\left(\frac{y}{H}\right) - \frac{1}{2}\left(\frac{y}{H}\right)^2\right] \quad (5.7.1)$$

Integrating the velocity profile over the film thickness gives the volumetric flow rate/width Q, which in turn allows the film thickness to be expressed in terms of the flow rate as follows:

$$H = \left(\frac{3\mu Q}{\rho g sin(\beta)}\right)^{1/3} \quad (5.7.2)$$

The two equations above are often referred to as the Nusselt solution of the two-dimensional laminar film flow problem for Newtonian fluids.

The maximum velocity is found at the free surface of the flowing film. It is important to know this value for calculating the age of the free surface, which in turn is important in determining the dynamic surface tension, see Sect. 8.7.2. Inserting Eq. (5.7.2) into (5.7.1) and solving (5.7.1) for y = H gives the maximum velocity as

$$v_{z,max} = \frac{1}{2}\left(\frac{9\rho g sin(\beta) Q^2}{\mu}\right)^{1/3} \tag{5.7.3}$$

The average film velocity at the end of the (vertical) die slide is important to know because it is equal to the initial curtain velocity. It is found by integrating the velocity profile over the film thickness and dividing it by the film thickness (Bird et al., 1960). Expressing H by Q according to Eq. (5.7.2) yields

$$v_{z,average} = \left(\frac{\rho g sin(\beta) Q^2}{3\mu}\right)^{1/3} \tag{5.7.4}$$

Using the average velocity and the film thickness as characteristic length and characteristic velocity, respectively, the Reynolds number for film flow is

$$Re = \frac{\rho Q}{\mu} \tag{5.7.5}$$

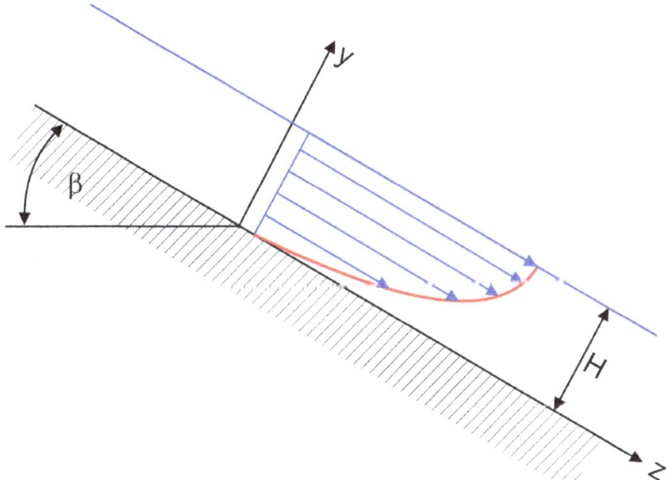

Fig. 5.58 Velocity profile of film flow on an inclined plane

Taking the derivative of Eq. (5.7.1) at y = 0 and expressing the film thickness in terms of the flow rate/width gives the wall shear rate:

$$\gamma_0 = (3Q)^{1/3}\left(\frac{\rho gsin(\beta)}{\mu}\right)^{2/3} \tag{5.7.6}$$

Note that if Eq. (5.7.6) is used in combination with a shear-dependent viscosity as described, for example, by the Carreau-Yasuda model, then determining the viscosity requires iteration because the wall shear rate itself depends on the viscosity. This iteration can easily be implemented by using the "goal-seek" function in an EXCEL environment.

5.7.2 *Single-Layer Film Flow for Shear-Dependent Viscosity (Power Law Flow Behavior)*

As described for pipe and slot flow, and depending on the shear-dependent viscosity, however, Newtonian flow equations are not suitable for characterizing film flows in industrial coating applications. Better suited are power-law flow equations, and they are well described in Bird et al. (1977).

In particular, the thickness of the film H, the wall shear rate γ_0, and the velocity profile $v_z(y)$ are given by the following equations.

Film thickness:

$$H = \left(\frac{m}{\rho gsin(\beta)}\right)^{1/(2n+1)}\left(Q\left(\frac{1}{n}+2\right)\right)^{n/(2n+1)} \tag{5.7.7}$$

Wall shear rate:

$$\gamma_0 = \left(\frac{\rho gsin(\beta)H}{m}\right)^{1/n} \tag{5.7.8}$$

Velocity profile:

$$v_z = \left(\frac{\rho gsin(\beta)H}{m}\right)^{1/n}\frac{H}{(1/n)+1}\left[1-\left(\frac{y}{H}\right)^{(1/n)+1}\right] \tag{5.7.9}$$

In the derivation of Eq. (5.7.9) it is assumed that y = 0 lies in the free surface of the film, and y = H is on the solid surface. The velocity profile according to Eq. (5.7.9) is graphically displayed in Fig. 5.59 as a function of the power law index while all other parameters were kept constant at: $Q = 0.5\ cm^2/s$, $\rho = 1100\ kg/m^3$, $\beta = 30°$, $m = 1.0\ Pas^n$. Note that in this graph the coordinate system has been switched such that y = 0 is in the solid surface of the inclined plane.

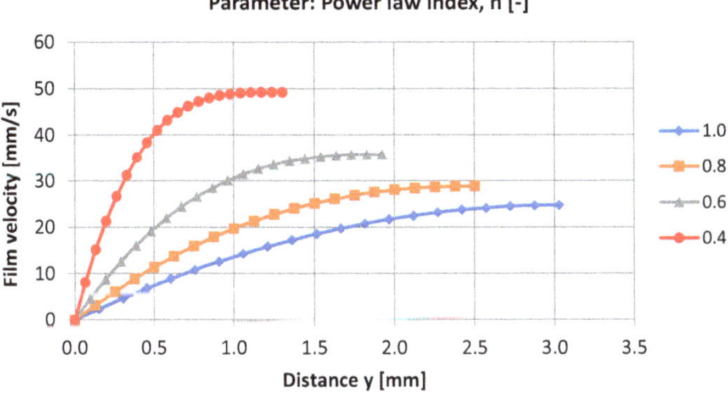

Fig. 5.59 Velocity profile of power law fluid as a function of the power law index

For n = 1.0 (Newtonian fluid) the form of the velocity profile is semi-parabolic as expected, and the profile extends to the free surface, thus revealing a film thickness of 3.03 mm. As n decreases the wall shear rate increases and the viscosity near the wall decreases. Consequently, the film velocity near the wall also increases, which results in a thinner film and the velocity profile is no longer parabolic. The same qualitative behavior was already observed for pipe and slot flow. For n = 0.4 the resulting film thickness is only 1.31 mm.

As explained above, the power law model is often inadequate for solving flow problems if the shear-dependent viscosity can be described by the Carreau-Yasuda model. Therefore, it is proposed to solve film flow problems by adapting the local power law model (LPL) as described by Ruschak and Weinstein (2014) for pipe, duct, and slot flows. Since film flow is not pressure but gravity-driven, an equivalent Newtonian viscosity μ_e, defined as that giving the same wall shear stress and pressure gradient as the non-Newtonian liquid does not exist. Instead, the following procedure is a way forward toward a solution.

5. Assume a wall shear rate.
6. Use the Carreau-Yasuda model (Eq. 4.4.1) for calculating the corresponding viscosity.
7. Use Eqs. (4.4.2) and (4.4.4) for calculating m and n according to Ruschak and Weinstein (2014).
8. Use Eqs. (5.7.7) and (5.7.8) for calculating the resulting film thickness and the wall shear rate.
9. Iterate until the calculated wall shear rate equals the assumed value.

The above procedure can easily be implemented in an EXCEL file by using the “goal-seek” function for the iteration.

Expressing the maximum and average velocities of film flow for power law fluids in terms of the volumetric flow rate/width as was done for Newtonian fluids is too complicated here owing to the complexity of Eqs. (5.7.7) and (5.7.9). It is therefore suggested to evaluate these values in an EXCEL environment. In particular, the

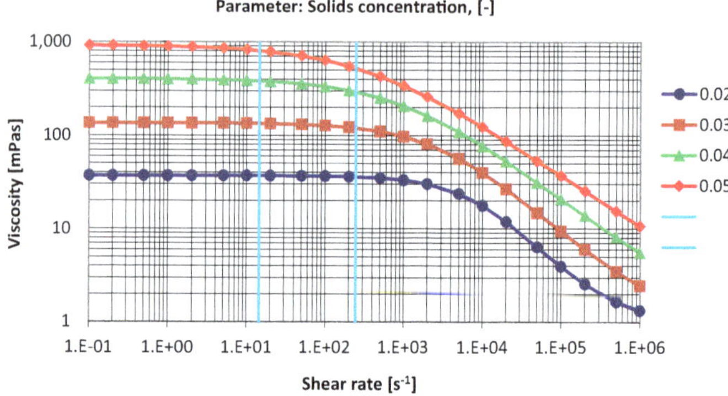

Fig. 5.60 Viscosity curves (Carreau-Yasuda model) of 4 CMC solutions; data taken from Henry (2016) and reprinted with permission from Henry

maximum velocity is obtained by setting $y = 0$ in Eq. (5.7.9), and the average velocity is calculated by performing a numerical integration of the velocity profile over the film thickness. Dividing the film thickness into 50 integration steps is sufficient for obtaining accuracies of <1.5%, which is acceptable for engineering considerations.

Henry et al. (2014) and Henry (2016) experimentally measured the film thickness of several non-Newtonian fluids flowing down an inclined plane. In particular, he measured the thickness of aqueous CMC solutions at 4 different concentrations, and he compared the measured thicknesses with calculated ones according to the Nusselt Eq. (5.7.2), which is valid for Newtonian fluids.

Henry's data are used here for the comparison with predicted film thicknesses according to the local power law model outlined above. The viscosity curves of the four CMC solutions are shown in Fig. 5.60. Henry used the Cross rheological model for approximating the viscosity measurements. Here, the Carreau-Yasuda model is used, and the Yasuda parameter a in Eq. (4.4.1) was adjusted such that the Cross and the Carreau-Yasuda curves became identical. The wall shear rate of the film flow was varied by varying the flow rate/width, see Eq. (5.7.8). The shear rate range covering all the film thickness measurements is marked in Fig. 5.60 by the two vertical blue lines. The lowest shear rates (15–28 s^{-1}) were achieved with the 5% CMC solution, while the highest shear rates (118–245 s^{-1}) are associated with the 2% solution.

It is obvious from Fig. 5.60 that the degree of non-Newtonian flow behavior during the film thickness measurements was modest and only appreciable for the 4 and 5% CMC solutions, while it was insignificant for the 2% solution.

Nevertheless, film thickness data were measured from Fig. 2.9 in Henry (2016) and compared with predictions by the local power law model presented above. In addition, the data were also compared with the Newtonian Nusselt Eq. (5.7.2), see Fig. 5.61.

The predictions with the local power law model are always lower than the predictions with the Nusselt equation, which is reasonable since the shear-dependent viscosity according to the Carreau-Yasuda model is always lower than the low shear

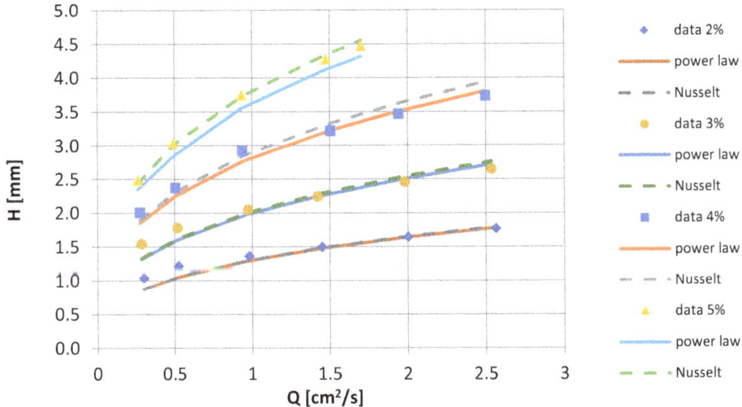

Fig. 5.61 Comparison of measured film thicknesses with the local power law model and the Nusselt model for 4 CMC solutions; film thickness data are taken with permission from Fig. 2.9 in Henry (2016)

viscosity used in the Nusselt equation. The agreement between the power law predictions and the measured film thickness values is in the range of ±5%, which is acceptable for engineering applications, except for low flow rates of about <1 cm^2/s and low solids concentrations of 2 and 3%, where the under-prediction by the power law model is in the range of −10 to −15%. On the other hand, the measured film thickness values for these flow rate and viscosity parameters seem to be suspicious because the Nusselt model also under-predicts the film thickness even though this model is accurate for Newtonian fluids, which is pretty much the case for the 2% CMC fluid. Therefore, the Nusselt model is expected to agree with or at most to slightly over-predict the measured values.

5.7.3 Multilayer Film Flow for Constant Viscosity (Newtonian Flow Behavior)

Regarding premetered coating methods multilayer film flow is encountered on the inclined surface of multilayer slide dies for slide and curtain coating as depicted in Fig. 2.1c, f. Correctly solving multilayer film flow problems for non-Newtonian fluids requires numerical methods for solving a set of differential equations according to the number of different layers, see for example Weinstein (1990) or Henry (2016). This task is challenging, and the necessary tools and skills are often not available to coating engineers working in industrial coating companies. Therefore, an approximate but simpler approach is presented below. The theoretical derivation is based on a 2-layer film flow for Newtonian fluids, for which an exact solution exists, and for which the velocity profile is schematically sketched in Fig. 5.62.

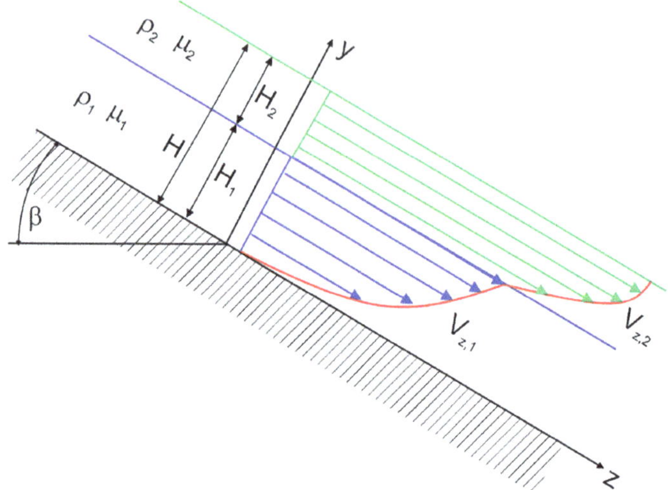

Fig. 5.62 2-layer film flow on an inclined plane

The differential equations for gravity-driven flow of Newtonian fluids in layers 1 and 2 are:

$$\mu_1 \frac{d^2 v_{z,1}}{dy^2} + \rho_1 g_z = 0 \tag{5.7.10}$$

$$\mu_2 \frac{d^2 v_{z,2}}{dy^2} + \rho_2 g_z = 0 \tag{5.7.11}$$

Here, $g_z = g \sin \beta$ and $\mathrm{H} = \mathrm{H}_1 + \mathrm{H}_2$. The boundary conditions are:

$$\mathrm{y} = 0 \quad v_{z,1} = 0 \tag{5.7.12}$$

$$\mathrm{y} = \mathrm{H}: \quad \frac{dv_{Z,2}}{dy} = 0 \tag{5.7.13}$$

$$\mathrm{y} = \mathrm{H}_1: \quad v_{z,1} = v_{z,2} \tag{5.7.14}$$

$$\mathrm{y} = \mathrm{H}_1: \quad \mu_1 \frac{dv_{z,1}}{dy} = \mu_2 \frac{dv_{z,2}}{dy} \tag{5.7.15}$$

Integrating Eqs. (5.7.10) and (5.7.11) with respect to y results in an expression for the velocity gradient in each layer:

$$\frac{dv_{z,1}}{dy} = -\frac{\rho_1 g_z}{\mu_1} y + C_1 \tag{5.7.16}$$

$$\frac{dv_{z,2}}{dy} = -\frac{\rho_2 g_z}{\mu_2} y + C_3 \tag{5.7.17}$$

Integrating Eqs. (5.7.16) and (5.7.17) again gives equations for the velocity profiles in each layer:

$$v_{z,1} = -\frac{\rho_1 g_z}{2\mu_1} y^2 + C_1 y + C_2 \tag{5.7.18}$$

$$v_{z,2} = -\frac{\rho_2 g_z}{2\mu_2} y^2 + C_3 y + C_4 \tag{5.7.19}$$

The 4 constants in the velocity profiles are determined with the help of the 4 boundary conditions. After some algebra the results are:

$$C_1 = \frac{g_z}{\mu_1}[H_1(\rho_1 - \rho_2) + H\rho_2] \tag{5.7.20}$$

$$C_2 = 0 \tag{5.7.21}$$

$$C_3 = \frac{\rho_2 g_z H}{\mu_2} \tag{5.7.22}$$

$$C_4 = g_z\left[\frac{H_1^2}{2}\left(\frac{\rho_1}{\mu_1} - \frac{\rho_2}{\mu_2}\right) + \rho_2 H_1 H_2\left(\frac{1}{\mu_1} - \frac{1}{\mu_2}\right)\right] \tag{5.7.23}$$

The constants of integration depend on H_1 and H_2. These two parameters can be determined by integrating the velocity profile of the first (bottom) layer from 0 to H_1 to obtain the flow rate/width Q_1 of the first layer, and by integrating the velocity profile of the second (top) layer from H_1 to H to obtain the flow rate/width Q_2 of the second layer as follows.

$$Q_1 = \frac{g_z H_1^2}{2\mu_1}[H_1(\rho_1 - \rho_2) + \rho_2 H] - \frac{\rho_1 g_z H_1^3}{6\mu_1} \tag{5.7.24}$$

$$Q_2 = \frac{\rho_2 g_z}{6\mu_2}\left(H_1^3 - H^3\right) + \frac{C_3}{2}\left(H^2 - H_1^2\right) + C_4 H_2 \tag{5.7.25}$$

Rearranging Eq. (5.7.24) provides an expression for H as a function of H_1:

$$H = \frac{Q_1 + \frac{g_z \rho_1 H_1^3}{6\mu_1} - \frac{g_z(\rho_1 - \rho_2)H_1^3}{2\mu_1}}{\frac{g_z \rho_2 H_1^2}{2\mu_1}} \tag{5.7.26}$$

Similarly, rearranging Eq. (5.7.25) provides an expression for $H_2 = H - H_1$ as a function of H_1, C_3 and C_4:

$$H_2 = \frac{Q_2 - \frac{g_z \rho_2}{6\mu_2}\left(H_1^3 - H^3\right) - \frac{C_3}{2}\left(H^2 - H_1^2\right)}{C_4} \tag{5.7.27}$$

Equations (5.7.26) and (5.7.27) cannot be resolved analytically to determine H_1 and H_2, but this system must be solved by iteration. Moreover, the viscosities μ_1 and μ_2 can be taken as shear-dependent according to the Carreau-Yasuda model by calculating the wall shear rate in each layer, i.e., the shear rate at $y = 0$ for the bottom layer, and the shear rate at $y = H_1$ for the top layer in line with Eqs. (5.7.16) and (5.7.17) respectively.

$$\gamma_{0,bottom\ layer} = \frac{g_z}{\mu_1}[H_1(\rho_1 - \rho_2) + H\rho_2] \tag{5.7.28}$$

$$\gamma_{0,top\ layer} = \frac{g_z \rho_2 H_2}{\mu_2} \tag{5.7.29}$$

The following procedure for calculating the shear-dependent μ_1 and μ_2 as well as H_1 and H_2 and hence H can easily be implemented in an EXCEL environment by writing macros for the goal-seek functions required for the iterative procedures for determining μ_1, μ_2. The iterative procedure for H_1 must be carried out by hand as μ_1 and μ_2 must be re-calculated every time H_1 is changed.

1. Assume H_1.
2. Assume μ_1.
3. Use Eq. (5.7.26) to calculate H.
4. Use Eq. (5.7.28) to calculate the wall shear rate in the bottom layer.
5. Use the wall shear rate and the Carreau-Yasuda model for μ_1 to calculate μ_1.
6. Iterate until the assumed and the calculated viscosity values are equal.
7. Assume μ_2.
8. Use Eq. (5.7.22) to calculate C_3.
9. Use Eq. (5.7.23) to calculate C_4.
10. Use Eq. (5.7.27) to calculate H_2.
11. Use Eq. (5.7.29) to calculate the wall shear rate in the top layer.
12. Use the wall shear rate and the Carreau-Yasuda model for μ_2 to calculate μ_2.
13. Iterate until the assumed and the calculated viscosity values are equal.
14. Change H_1 and iterate until $H_{1,assumed} + H_{2,calculated} = H_{calculated}$.
15. Use Eq. (5.7.20) to calculate C_1.
16. Use Eq. (5.7.18) to calculate the velocity profile in the bottom layer.
17. Use Eq. (5.7.19) to calculate the velocity profile in the top layer.

A first guess for the viscosity of the bottom layer μ_1 may be obtained by using the variable power law approach for a single layer film flow as described above, but by taking the sum of the flow rates/width of both layers. Regarding a first guess for the viscosity of the top layer μ_2 the low shear viscosity of the top layer may be good enough. A suitable first guess for the thickness of the bottom layer H_1 is more difficult to calculate and must be found by trial and error because convergence will

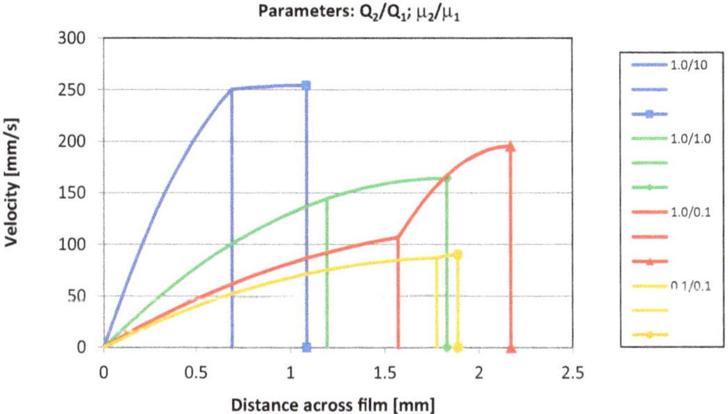

Fig. 5.63 Velocity profiles of 2-layer film flow for different flow rate and viscosity ratios

only be obtained if the first guess is close to the final value. It might be helpful to plot the velocity profile of the 2-layer film flow and to observe how the profiles of the two layers approach each other or move further apart as H_1 is changed. In any case, the thickness of the bottom layer with flow rate Q_1 of a 2-layer film flow is always smaller than the thickness of a 1-layer film with the same flow rate.

Examples of velocity profiles of 2-layer film flows are shown in Fig. 5.63. If the viscosity of the top layer is much higher than the viscosity of the bottom layer for equal flow rates, then the top layer profile is close to plug flow (square symbols). If both layers have the same viscosity, then the velocity profile is semi-parabolic and the 2-layer film behaves like a 1-layer film (diamond symbols). If the viscosity of the top layer is much smaller than the viscosity of the bottom layer for equal flow rates, then the top layer flows much faster than the bottom layer and the velocity profile is characterized by a pronounced inflection point (triangular symbols). If for the same viscosity ratio, however, the top layer flow rate is much smaller than the bottom layer flow rate, then the inflection point of the velocity profile is still present but barely visible, and the top layer is close to plug flow (round symbols).

To check the performance of the 2-layer film flow model presented above, measured film thicknesses for different fluids and flow rates were extracted from Henry (2016) and compared with model predictions. All tests were carried out on a 30° incline relative to the horizontal. Figure 5.64 shows the total film thickness of a 2-layer film consisting of an 80% glycerol bottom layer and a 60% glycerol top layer with both layers containing 0.2% SDS surfactant. These fluids are considered to be Newtonian, and the physical fluid properties are listed in Table 5.14.

The flow rate/width Q_1 of the bottom layer was held constant at 0.31 cm^2/s, and the flow rate/width of the top layer was varied as indicated in the graph. The agreement between measured and predicted values is good with all calculated values being smaller than the measured ones but only by $<1.7\%$ for the tested parameter range.

Figure 5.65 shows thickness data and predictions of a 2-layer film with both layers

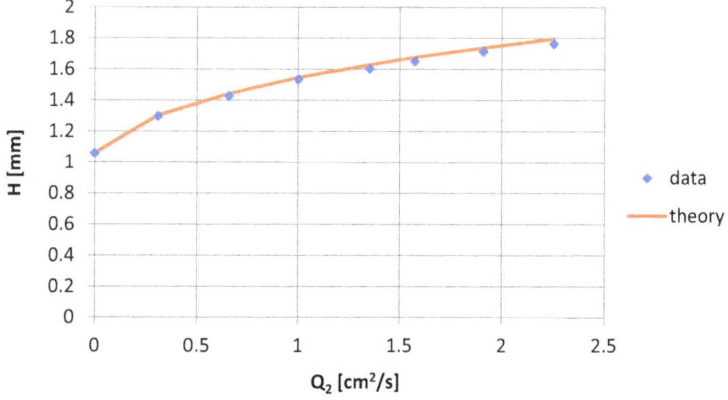

Fig. 5.64 Total film thickness of a 2-layer film as a function of the top layer flow rate; film thickness data are taken with permission from Fig. 2.14a in Henry (2016)

Table 5.14 Physical fluid properties of glycerol

	Solids concentration [–]	Density [kg/m^3]	Viscosity [mPas]	Surface tension [mN/m]
Bottom layer	0.8	1218	77	48.2
Top layer	0.6	1171	16.6	39.9

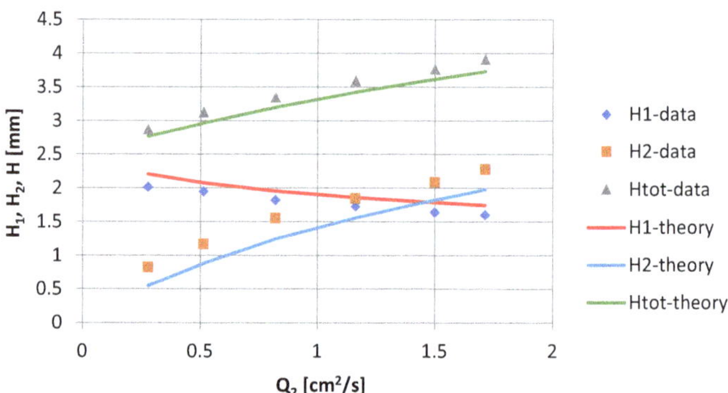

Fig. 5.65 Bottom layer, top layer, and the total thickness of a 2-layer CMC film as a function of the top layer flow rate; film thickness data are taken with permission from Fig. 2.16b in Henry (2016)

consisting of a 4% CMC solution having a shear-dependent viscosity as depicted in Fig. 5.60. The bottom layer flow rate/width was held constant at 0.67 cm^2/s and the top layer flow rate/width was varied. The fluid had a density of 1015 kg/m^3 and a surface tension of 69.9 mN/m. The thickness of the bottom layer was over-predicted while the thickness of the top layer was under-predicted. The total thickness of the 2-layer film was also under-predicted by <5.1% for the tested parameter range.

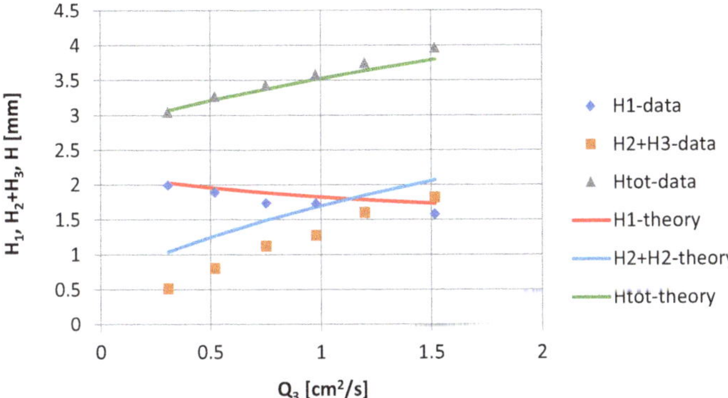

Fig. 5.66 Bottom layer, middle and top layer, and total thickness of a 3-layer CMC film as a function of the top layer flow rate; film thickness data are taken with permission from Fig. 2.18b in Henry (2016)

Finally, Fig. 5.66 shows thickness data and predictions of a 3-layer film with all layers consisting of a 4% CMC solution having a shear-dependent viscosity as depicted in Fig. 5.60. The bottom layer flow rate/width was held constant at 0.67 cm^2/s, the flow rate/width of the middle layer was fixed at 0.33 cm^2/s, and the top layer flow rate/width was varied. Using a 2-layer model for predicting thicknesses of a multilayer film having more than 2 layers, i.e., n layers, requires the model to be modified as follows. The flow rate/width of the top layer is equal to the sum of the flow rates/width of all layers $i = 2 \div n$. All layers $i = 2 \div n$ have the same viscosity, which is the viscosity of the second layer. The corresponding shear rate is calculated with Eq. (5.7.29), where the thickness of the top layer is consistent with the sum of the flow rates/width of layers $i = 2 \div n$. The density of the top layer is an average value of all layers $i = 2 \div n$ according to

$$\widehat{\rho}_2 = \frac{\sum_{i=2}^{n} \rho_i Q_i}{\sum_{i=2}^{n} Q_i} \tag{5.7.30}$$

Steps 1–17 of the procedure outlined above are now carried out with the modifications just discussed, and the so predicted film thicknesses are compared with the measured values in Fig. 5.66.

The thickness of the bottom layer is slightly over-predicted, while the sum of the thickness of layers 2 + 3 is more severely over-predicted. However, predictions of the total thickness of the 3-layer film are quite accurate with a deviation of about −1 to +4.5% over the tested parameter range.

The Reynolds number of a multilayer film flow can still be calculated with Eq. (5.7.5). However, the density is an average value of all layers involved according to Eq. (5.7.30), the flow rate/width is the sum of the flow rates of all layers involved, and the viscosity is the viscosity of the bottom layer taken at the wall shear rate of that layer.

5.7.4 Waves in Film Flow

As sketched in Fig. 5.67 single-layer film flow may be bothered by surface waves, while multilayer films may suffer from surface waves and interface waves. Regarding premetered coating methods, waves may be an issue for single- and multilayer slide and curtain coating when slide dies are used, see Fig. 2.1c, f. Waves of any kind in such film flows must be prevented to avoid deterioration of the uniformity of the coated film.

Driving forces for surface waves are inertia (gravity) and surface tension gradients. Note that the level of surface tension is not as important as the effect of surface tension gradients. Damping forces include viscosity, surface tension (particularly at high levels), and gravity (particularly at small inclination angles). Surface waves have long wave lengths of order 100 mm (Kobayashi, 1992). Moreover, by the time they get stretched through the coating bead, their effect on the nonuniformity of the coated film decreases.

Interface waves are more important than surface waves. They have short wave lengths of the order of 1–10 mm (Kobayashi, 1992), even after the coating process. In addition, the growth rate of the wave amplitude is stronger than for surface waves because the interface of adjacent (miscible) layers is free of tension, and therefore free of a damping force. This is why interface waves can grow appreciably, even on short die slides.

Predicting the consequences of waves in film flow on the uniformity of the coated film is very difficult and conclusive experimental data on wave formation are very sparse. Many models predict neutral stability, i.e., they predict flow conditions, for which waves will exist. However, most of these models do not predict how fast the wave amplitude will grow. A review of wave models in film flow is given, for example, by Smith (1997), Hens and van Abbenyen (1997), and by Henry et al. (2014). A simple model for predicting the onset of surface waves was presented by Benjamin (1957). In particular, he determined a critical flow rate/width Q_c, or a critical Reynolds number Re_c, below which no waves will occur as follows:

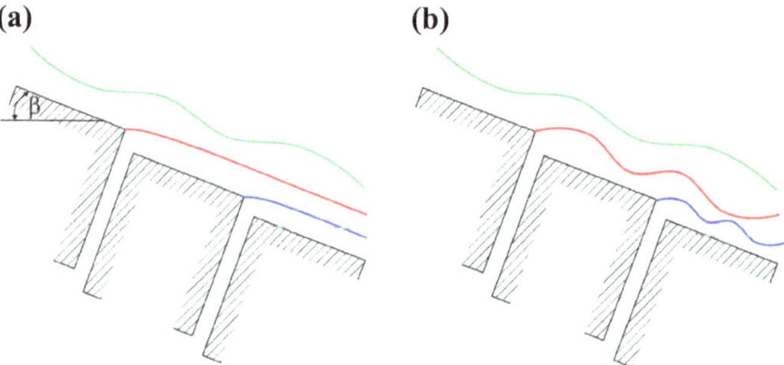

Fig. 5.67 **a** Surface waves in multilayer film flow **b** surface and interface waves in multilayer film flow

$$Re_c = \frac{\rho Q_c}{\mu} = \frac{5}{6} cotan(\beta) \tag{5.7.31}$$

While this model does not include surface tension effects, it says that low viscosity fluids generate surface waves at low flow rates. Also, owing to the cotangent function, films on a horizontal surface are always free of waves, while films on a vertical surface are always susceptible to waves. In other words, film flows of low viscosity fluids require small inclination angles to be free of waves.

Many of the slide dies used in the photographic industry had small inclination angles of <25° in response to the aqueous gelatin-based and relatively low-viscosity photographic emulsions. In the 90ies and thereafter, TSE Troller AG sold multilayer slide dies with an inclination angle of 30° for a great variety of curtain coating applications outside the photographic industry. None of these applications suffered from wave problems, even though the inclination angle turned to vertical for a short distance at the end of the slide, see Fig. 2.1f. Nowadays, several die suppliers offer multilayer slide dies for curtain coating, as they are typically used for slide coating according to Fig. 2.1c. In particular, the slide surface is straight to the very end and has a very steep inclination angle of 75°. Testing such a die with a pressure-sensitive adhesive from BASF revealed no visible waves in the film flow for a flow rate/width of 2.97 cm^2/s and a slide length of about 50 mm. The fluid had a density of 1010 kg/m^3 and a viscosity of 77.3 mPas according to its Carreau parameters and the variable power law approach for these flow conditions. Consequently, the film Reynolds number was 3.88. The critical Reynolds number according to the Benjamin criterion (Eq. 5.7.31) was 0.22, which is vastly different from the experimental observation.

Smith (1997) argued that the critical Reynolds number such as the one calculated by Eq. (5.7.31) does not depend on surface tension because the long wave length of the disturbance, which was used in the underlying mathematical analysis, implies an extremely small curvature of the free surface. However, experience shows that higher surface tensions increase the resistance to wave formation.

Benjamin's analysis did not include the effects of surface tension, nor did it account for the effects of surfactant diffusing from the bulk of the fluid to the film surface. These effects are known to depend on the type and concentration of the surfactant. The model did not account either for the effects of surface tension gradients in the free surface of the film. Ruschak (1987), as well as Schunk and Scriven (1997), addressed these latter issues in interesting publications. In particular, Ruschak studied the formation and damping of waves generated by wind flowing over a thin, horizontal layer of viscous fluid. In contrast, among other systems, Schunk and Scriven investigated wave formation in a falling liquid film.

In a steady film flow on an inclined plane, the free surface of the film is flat and parallel to the plane. Consequently, this flow is free of surface tension gradients, and the system is in chemical equilibrium with the concentrations of all species being uniform throughout. However, when this base flow is disturbed, for example by sinusoidal waves, surface compression and expansion cause small departures from equilibrium. In response to this mechanism, expansion dilutes and compression concentrates the surfactant molecules deposited and adsorbed in the film surface.

Consequently, these differences in surfactant concentration generate surface tension gradients. Schunk points out that these surface tension gradients will either amplify or dampen the disturbance, i.e., the amplitude of the wave, depending on how the gradients are altered by diffusion of surfactant molecules from the bulk to the free surface. Schunk further explains that the damping effect of surfactants depends on the phase lag of surfactant response behind the disturbing wave.

Analyzing such flow systems including the bulk diffusion and surface adsorption of surfactant molecules is complex and difficult. One of the main results derived by Ruschak states that significant surface tension gradients, which oppose flow and hence dampen the amplitude of surface waves, are obtained by maximizing the right-hand side of Eq. (17) in his paper. Interestingly, the term $\Omega = \Gamma_{\infty}^{2}/\left(a\sqrt{D}\right)$ can be extracted from Eq. (17). Maximizing this term will then serve as a criterion for analyzing and ranking different surfactants on their ability to form strong surface tension gradients. Γ_{∞}^{2} is the surface surfactant concentration at saturation, D is the diffusion coefficient, and a is a parameter in the Szyszowski-Langmuir equation of state. All these parameters are physical surfactant properties that can be measured, although only with considerable effort, see Tricot (1997). Note that Ruschak's paper is discussed in more detail in Sect. 5.10.4.

In any case, strong wave damping is obtained with surfactants that are characterized by small diffusion coefficients. Unfortunately, this is in strong conflict with other requirements that surfactants and the resulting surface tension must satisfy. One example relates to curtain coating, where preventing holes in the curtain and assuring high curtain stability requires the (dynamic) surface tension to be low throughout the curtain, i.e., at very short surface ages of the curtain surfaces, and this can only be achieved with surfactants having a high diffusion coefficient. Consequently, optimizing surfactant systems may not be the best strategy for preventing surface waves in film flow. Instead, assuring low enough values for the Reynolds number and using small inclination angles for the die slide as per Eq. (5.7.31) seems a better approach.

The difference between theoretical predictions of the onset of waves in film flow and corresponding experimental observations has been noticed before, particularly in the photographic industry. Hens and Abbenyen (1997) refer to this situation as 'practical stability' if it takes a long distance compared to the length of the die slide for a small wave to grow so much that it deteriorates the uniformity of the coated film. Theoretical models may be good enough to raise a red flag for the coating engineer. However, whether or not predicted waves in film flow are detrimental to the coated film must be determined experimentally. If objectionable surface waves are present, then the best counter measure is increasing the solids concentration of the coating fluid because it not only reduces the flow rate but it also increases the viscosity, both of which tend to render film flows more stable.

Regarding the onset of interface waves in multilayer film flows, the use of theoretical predictions is as questionable as it is for surface waves. Given the practical film stability mentioned above, Kobayashi (1992) combined theoretical calculations with experimental observations from the photographic industry. In particular, he defined a growth length L_e that is required for a perturbation to grow by a factor e. Then,

$L_{e,min}$ is the growth distance of the most amplified wave length. He also defined an amplification factor A_{max}, relating $L_{e,min}$ to the length L_s of the inclined plane. Furthermore, he empirically determined A_{max} to be about 2,000. Kobayashi then calculated the wave behavior, i.e., the surface and the interface mode, of a two-layer film as a function of the viscosity ratio $m = \mu_{top}/\mu_{bottom}$ and found the following results, see also Hens and van Abbenyen (1997). When m is close to unity, then surface waves dominate. However, when $m > 1.4$ or $m < 0.73$, then interface waves dominate, particularly when the top layer viscosity is significantly lower than the bottom layer viscosity. This last situation leads to an inflection point in the velocity profile of the 2-layer film as depicted by the triangular symbols in Fig. 5.63. The presence of an inflection point in the velocity profile is a necessary but not sufficient condition for the existence of interface waves. It is known in the photographic industry that two-layer films can be stable even if the upper layer viscosity is lower than the lower layer viscosity, as long as the upper layer flow rate is small enough. The velocity profile of such a situation is depicted by the round symbols in Fig. 5.63. In addition, this situation is supported and quantified by experimental data generated by researchers at ILFORD Imaging Switzerland GmbH. While their initial data were presented in terms of the wet coat weight, we transformed these values into the equivalent product of liquid density ρ times volumetric flow rate/width Q, which is more meaningful when discussing properties of film flows, see Fig. 5.68.

In particular, high values of the viscosity ratio indicate that the viscosity of the top layer is much lower than the viscosity of the bottom layer. High values of the viscosity ratio require high values of the flow rate ratio for the 2-layer film flow to be stable, i.e., to be without interface waves. This in turn requires the flow rate of the top layer to be much smaller than the flow rate of the bottom layer.

Given this complicated situation regarding the reliable prediction of the onset of interface waves, a cautious design measure is to avoid inflection points in the velocity

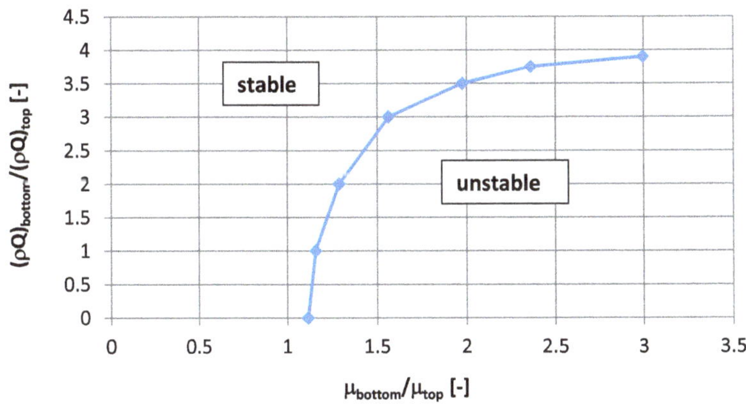

Fig. 5.68 Onset of interface waves in a 2-layer film flow on an inclined plane with inclination angle $\beta = 23°$; data reproduced with permission from ILFORD Imaging Switzerland AG

profile of multilayer films. In turn, this leads to the following rule about the viscosity structuring of multilayer films:

> The viscosity of any layer in a multilayer film must not be lower than the viscosity of the adjacent layer below.

The viscosity of every layer must be determined at the prevailing shear rate in that layer, i.e., at the wall shear rate (bottom layer) or at the interface shear rate toward the layer below. If the above design requirement cannot be implemented, then the flow rate of the low-viscosity layer and the flow rates of additional layers above must be kept as low as possible.

5.7.5 *Standing Waves in Film Flow*

Standing waves in film flow are observed when a film flows into a large pool or against an obstacle. This situation is encountered in slide coating, where the film flows into the coating bead (large pool) before it is taken away by the moving substrate (obstacle) in a direction more or less perpendicular to the direction of the incoming film. As described by Ruschak (1978) the wavy film surface is the result of an interaction between viscous stress gradients and capillary pressure gradients such that the film profile far downstream blends smoothly with the curved surface of the upper meniscus in the coating bead.

High Reynolds numbers are necessary for the amplitude of standing waves to become substantial in slide coating. Figure 5.69a displays a situation with one

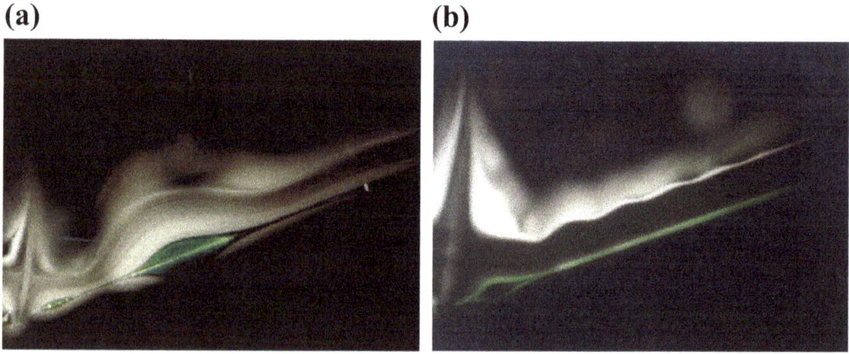

Fig. 5.69 **a** Single standing wave with large amplitude; photo taken from Schweizer (1988) with permission form Cambridge University Press **b** multiple standing waves with small amplitude; photo taken from Schweizer (1988) with permission form Cambridge University Press

pronounced standing wave just upstream of the coating bead for Re = 88.7, Ca = 0.069, and Po = $2.9 \cdot 10^{-9}$.

Here, Po is the property number defined as follows:

$$Po = \frac{g\mu^4}{\rho\sigma^3} \tag{5.7.32}$$

Doubling the Reynolds number produces a film with many standing waves of small amplitude as shown in Fig. 5.69b. Here, Re = 162.6, Ca = 0.196 and Po = $2.9 \cdot 10^{-9}$. The very low values of the property number for both examples imply a low viscosity and a high surface tension. Moreover, the high Reynolds numbers mean a high flow rate/width, i.e., thick coated films and/or high web speeds. In contrast, Fig. 5.76 shows a flow field without any substantial standing wave at all for Re = 16.73, Ca = 0.285, and Po = $9.98 \cdot 10^{-6}$.

Standing waves cannot be prevented for a given set of operating conditions, except by keeping the Reynolds number small enough. However, the effect of standing waves on the uniformity of the coated film is small. Chances for the formation of a vortex underneath the wave crest increase if the wave amplitude is high, see Fig. 5.69a. More information about the onset of this vortex can be found in Sect. 5.7.8.

5.7.6 *Surface Age of Film Flow*

Knowing the age of the free surface of a flowing film is important for single- and multilayer slide and curtain coating, particularly if the fluid is water-based. In this case, the surface tension of the individual layers is adjusted by adding surfactant. Consequently, the resulting surface tension is controlled by the diffusion of surfactant molecules from the bulk of the fluid to the layer surface and so is time-dependent. This issue is specifically important when a multilayer film is established on the die slide before the coating start because each added layer must properly wet the layer below. This in turn can be accomplished by knowing the surface tension of the layer below and by adjusting the surface tension of the layer flowing on top in such a way that the top layer surface tension is a bit lower than the surface tension of the layer below. In curtain coating, it is also important to know the surface tension of the film at the end of the die slide because there the film flow turns into curtain flow, and knowing and controlling the surface tension of the curtain is crucial to curtain stability, i.e., the formation of holes in the curtain.

As depicted in Fig. 2.1c, f, the flow conditions in multilayer slide and curtain coating change on every die plate that produces an additional liquid film, particularly the film thickness and hence the surface velocity of the film flow. As a result, the surface age Δt_s of the film must be calculated for every die plate that generates a new layer according to

$$\Delta t_s = \frac{\Delta L}{v_{z,max}} \tag{5.7.33}$$

ΔL is the length segment of the film, which corresponds to the thickness of the die plate. Mathematical tools for calculating the surface velocity of single and multilayer films are presented in Sects. 5.7.1–5.7.3. For single-layer films, the surface age can be expressed as

$$\Delta t_s = 2\Delta L\left(\frac{\mu}{9\rho g sin(\beta) Q^2}\right)^{1/3} \tag{5.7.34}$$

For curtain coating slide dies that have a round nose at the end of the inclined plane as shown in Fig. 2.1f, the age of the film surface on that curved part can be approximated by dividing the curve into several small angular segments Δβ, provided that the curve has a circular shape of radius R, which is often the case. Then, the inclination angle β changes from one segment to the next, and the arc length of each segment equals ΔL = RΔβ.

All surface age calculations as proposed above can easily be carried out in an EXCEL environment.

5.7.7 Minimum Flow Rate for Film Flow

The issue of a minimum flow rate/width applies to single-layer applications and the upper-most layer of multilayer applications in slide and curtain coating, particularly if the goal is to coat such layers as thinly as possible. Even though there is a large body of scientific literature that addresses the subject of de-wetting, i.e., the break-up of thin liquid layers into rivulets or dry patches, this information is not often useful for industrial coating processes. On the one hand, required physical properties of the liquid/substrate system, such as the contact angle, are not readily available in industrial applications, or they vary from one day to the next because they depend on the degree of cleanliness or contamination of the substrate. On the other hand, the thickness of such films flowing down on the die slide is not constant everywhere. In particular, if the height of the side wall confining the width of the film is higher than the thickness of the film, then the film rises along the wall, i.e., the location of the wetting line is above the level of the uniform film thickness away from the wall (see Sect. 5.9.3 for more details). This phenomenon causes the film to locally thin near the wall, which in turn causes the film to break up before the film located away from the wall. In addition, an industrial coating process cannot be operated at conditions that coincide with the inception of a theoretically or experimentally determined defect. Instead, industrial processes must be operated with a subjectively determined safety margin relative to the onset of a given defect.

As a consequence of this situation, we rely on practical experience by stating that the minimum flow rate/width for film flow can be quantified as follows:

$$Q_{min} \approx 0.1\,\mathrm{cm}^2/\mathrm{s} \tag{5.7.35}$$

Note, however, that the value in expression (5.7.35) is not exact. Small values of that magnitude require that the height of the side wall is very close to the equilibrium film thickness. In particular, care must be taken to avoid local depressions in the film thickness near the side wall. Acceptable minimum flow rates for any given industrial application must be determined experimentally, including a safety margin relative to film break-up.

5.7.8 Vortices in Film Flow

The subject of vortices in film flow refers to film flow on the inclined plane of single- or multilayer slide dies. The photo gallery in Fig. 5.71 illustrates the many possible locations for vortices to occur in this type of flow. The vortices shown here at the end of the die slide refer to the slide coating process. The streamlines in these pictures were visualized by injecting thin filaments of dye and by a hydrogen bubble-producing method adapted to coating flows by Schweizer (1988, 1997b). All vortices are numbered arbitrarily to facilitate their discussion. The numbering scheme is depicted in Fig. 5.70.

Figure 5.71a and b show good flow fields at the upper most and an intermediate slot exit. In particular, the slide surface contains a step at each slot exit with the step height being about equal to the thickness of the film exiting from the slot. In addition, the static wetting line at the upper most slot exit is pinned to a sharp corner,

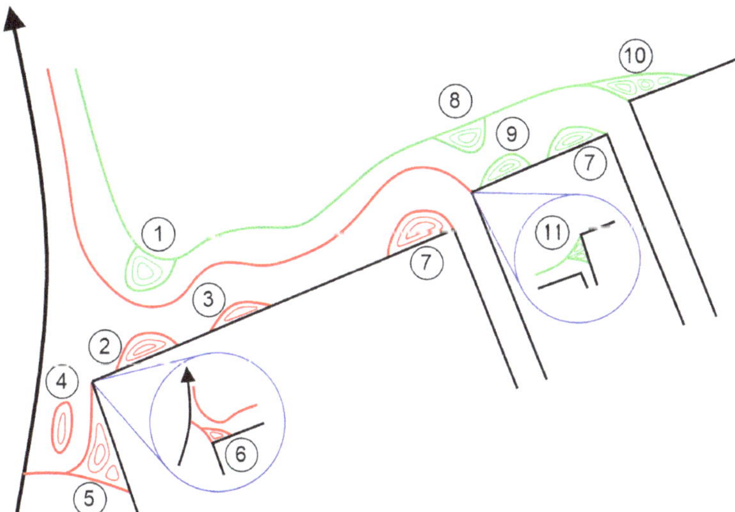

Fig. 5.70 Arbitrary numbering of potential vortices in film flow

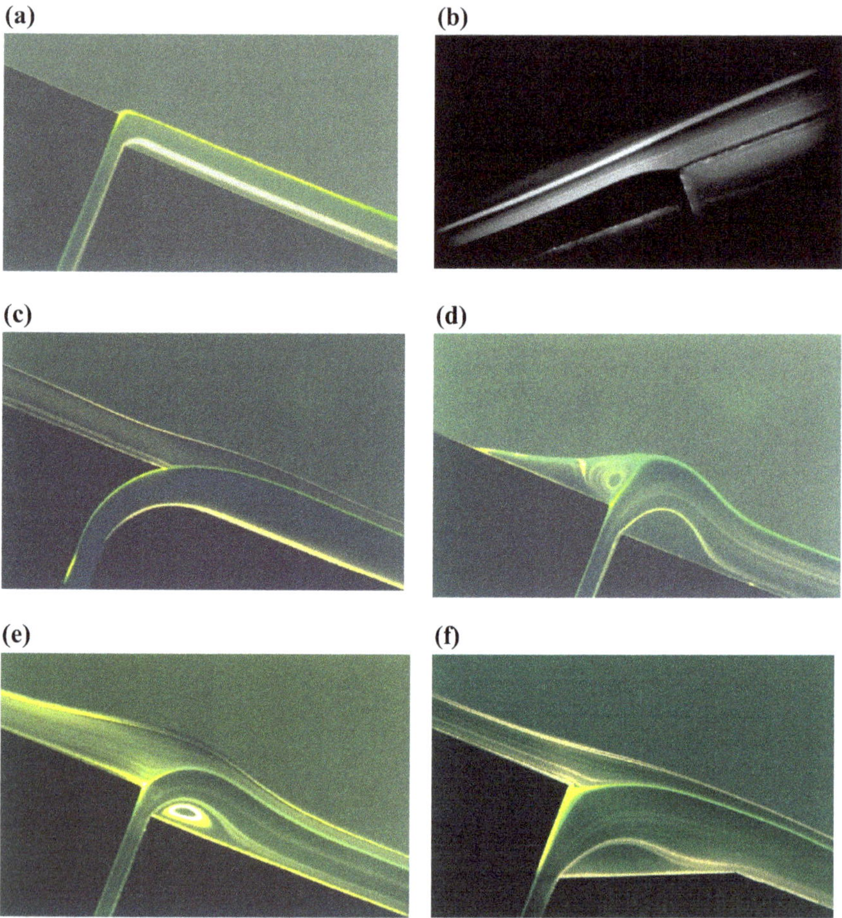

Fig. 5.71 **a** Good flow field at the upper most slot exit; photo taken from Schweizer (1997b) and reproduced with permission from Springer Nature **b** good flow field at an intermediate slot exit; photo taken from Schweizer (1988) and reproduced with permission from Cambridge University Press **c** optimum flow field at an intermediate slot exit; photo taken from Schweizer (1997b) and reproduced with permission from Springer Nature **d** multiple vortices at the upper most slot exit; photo taken from Schweizer (1997b) and reproduced with permission from Springer Nature **e** strong vortex downstream of an intermediate slot exit; photo taken from Schweizer (1997b) and reproduced with permission from Springer Nature **f** strong vortex downstream of an intermediate slot exit even with modified geometry of slot exit; photo taken from Schweizer (1997b) and reproduced with permission from Springer Nature **g** vortex in free surface of upper layer approaching area of merging with lower layer; photo taken from Schweizer (1988) and reproduced with permission from Cambridge University Press **h** vortex in free surface and on slide surface of film approaching the bead of slide coating; photo taken from Schweizer (1988) and reproduced with permission from Cambridge University Press **i** two vortices on slide surface of film with standing waves approaching the bead of slide coating; photo taken from Schweizer (1988) and reproduced with permission from Cambridge University Press **j** vortex in free surface of bead in slide coating; photo taken from Schweizer (1988) and reproduced with permission from Cambridge University Press **k** flow around die lip without vortex in heel flow; photo taken from Schweizer (1988) and reproduced with permission from Cambridge University Press

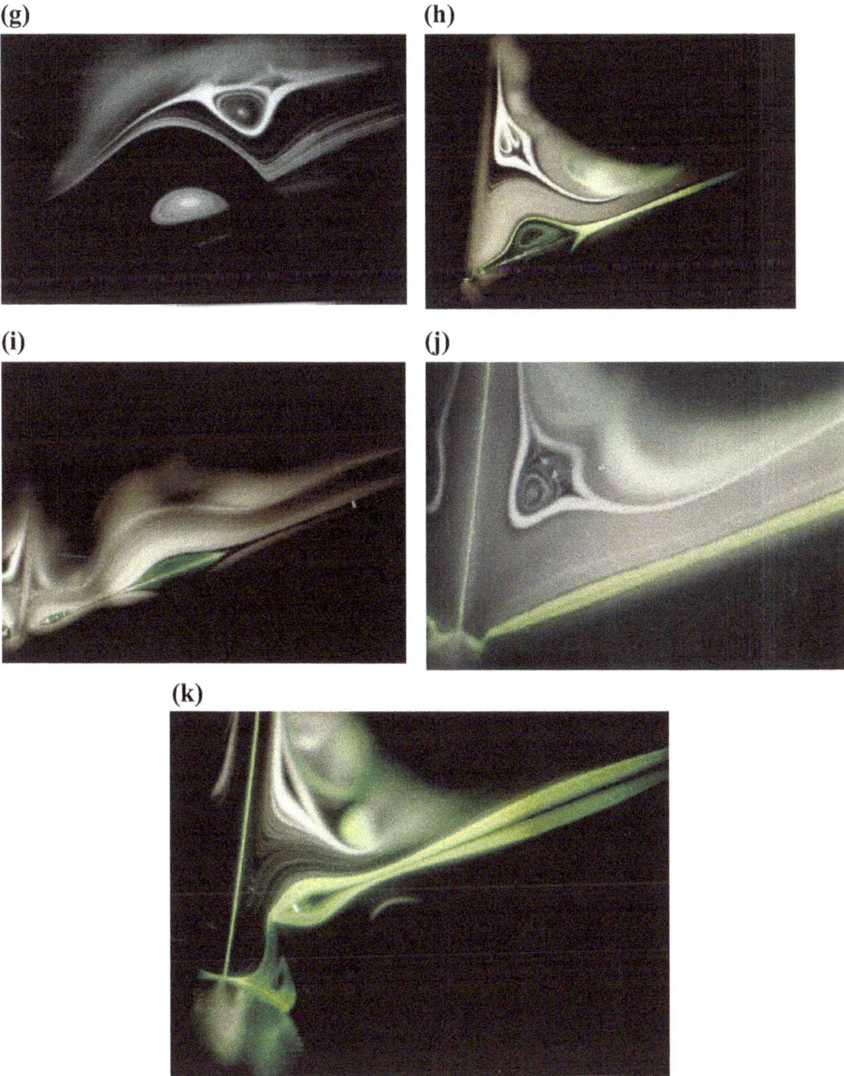

Fig. 5.71 (continued)

and both flow fields are free of vortices, even though the geometry of both slot exits is not optimal owing to the sharp 90°-corner at the transition from slot to film flow. A vortex downstream of the slot exit (vortex 7), therefore, can only be avoided if the velocity (inertia) of the slot flow is low enough to prevent excessive jetting and to allow the film flow to turn around the sharp corner without separating from the slide surface.

Figure 5.71c shows a nearly perfect flow field at the merging point of two adjacent layers. Here, the sharp 90° corner at the slot exit is eliminated by implementing a geometry called a *curved diffuser*. This geometry forces the slot flow to change direction without separating from the flow boundary up to high Reynolds numbers. On the other hand, care must be taken because forced flows at high Reynolds numbers along curved walls, such as flow between concentric cylinders or curved film flow, are susceptible to longitudinal vortices generated by centrifugal forces, see Taylor (1923), or Schweizer and Scriven (1983). Quantitative information about the onset of such vortices in a curved diffuser is not known. A truly perfect flow field would be obtained with the curved diffuser geometry if the step height between the adjacent die plates were a bit higher to match the thickness of the lower layer just like shown in Fig. 5.71b.

Unfortunately, curved diffusers are not found in industrial slide dies because the manufacturing costs of such a geometry are too high. Moreover, neither are multilayer slide dies with steps between each pair of adjacent die plates found in industry, except perhaps for dies that are truly dedicated to only one product, which is rarely the case. If a die is used for coating several products, then achieving an optimum flow field requires adjusting the step height during each product change, which is time-consuming, and delicate work because the die has to be opened frequently. This in turn is not recommended because it increases the chances of damaging the highly precise surfaces of the die plates. Changing the step height is even more complicated if the slide die allows the coating width to be changed because the correct step height would have to be incorporated into the edge guides of the film flow on the die slide, which then would require dedicated edge guides for each product. For those reasons, most industrial multilayer slide dies are free of steps between adjacent die plates.

In Fig. 5.71d the velocity of the slot flow is much too high, thus resulting in a large vortex (vortex 7) just downstream of the slot exit as is indicated but not clearly visualized by only one streamline. More importantly, and even though the slide surface contains a step that is too low compared to the thickness of the film flow, the static wetting line is not pinned to the upstream corner of the slot exit, but is located at some unknown distance upstream of that corner. This results in a separated wedge of fluid that is filled with a series of vortices (vortices 10), the first one of which is nicely visualized by streamlines. The rest of the wedge contains several additional vortices as schematically indicated in Fig. 5.70, the size and speed of rotation of which decreases toward the static wetting line. Experience has shown that the liquid film emerging from the upper most slot, as well as the coated film on the substrate, is susceptible to lines and streaks if vortices 10 are present. This issue is particularly serious if the coating fluid contains low-boiling solvents because the liquid in the wedge next to the wetting line is almost stagnant, promoting evaporation of the solvent and the formation of irregular solid structures. These irregular structures then excessively disturb the film exiting from the slot and flowing toward the coating point.

This problem must be avoided by forcing the static wetting line to be pinned to the upstream corner of the slot exit. For a slide die without steps this can easily be implemented by mounting a bar of metal or hard plastic with a characteristic

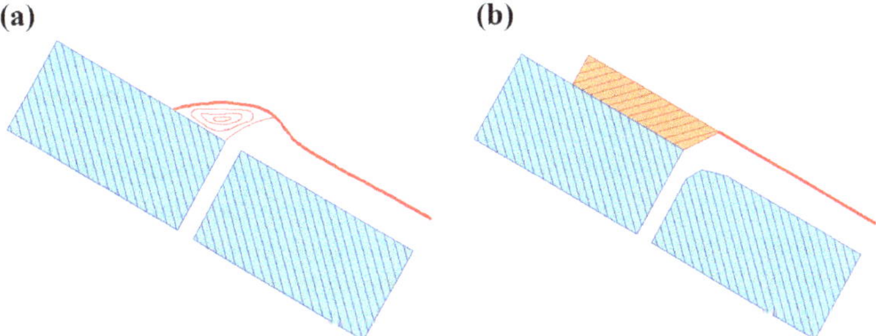

Fig. 5.72 **a** Vortex formation because the static wetting line is not pinned to the upstream corner of the slot exit **b** vortex prevention by mounting a special bar that allows the static wetting line to be pinned to a sharp corner

cross-section onto the upper most die plate as schematically shown in Fig. 5.72. The thickness of the bar should be equal to the thickness of the departing film, and one end of the bar must be beveled to mimic the geometry of the curved diffuser. Fastening the bar to the die plate with double-sided adhesive tape has proven simple and effective. Moreover, changing the bar during product change is easy and allows the bar geometry, particularly the thickness and the length, to be adapted to the thickness of the liquid film and the coating width.

Vortex 7 located just downstream of the downstream corner of the slot exit is better visualized in Fig. 5.71e. Again, the velocity of the flow exiting the slot is too high such that the liquid cannot turn around the 90° corner without separating from the slide surface. The liquid contained inside the vortex is made up of liquid exiting the slot. The presence of this vortex is not at all associated with turbulence. On the contrary, the liquid inside the vortex is driven by the flow surrounding the vortex, and it is rotating slowly as the presence of the slide surface exerts friction to the fluid. The streamline pattern inside a vortex is also nicely shown in Fig. 5.71d. It can be seen that the pattern is made up of ring-like closed lines that are evenly spaced from each other, thus indicating laminar flow conditions. Figure 5.71e also illustrates that the presence of a vortex does not contribute at all to the potential mixing of adjacent fluid layers. The interface between the two layers in this photo is marked with a green dye line, and it does not become wavy or decay, thus indicating laminar conditions and confirming that no liquid is flowing across the interface from one layer to the other or vice versa.

In some dies, the 90° corner is cut off as demonstrated in Fig. 5.71f. The aim of this modification of the slot exit geometry is to avoid the vortex downstream of the slot exit. However, as is evident from this picture, the measure is ineffective, at least for the flow conditions chosen here. The angle measured through the solid corner is 116°, which is not much bigger than 90°. As a result, the cross-section of the vortex may be a bit smaller compared to the 90° corner, but the vortex cannot necessarily be prevented. In addition, cutting the corner away considerably enlarges the cross-section of the flow exiting the slot. This is turn generates an area of low wall shear stress or possibly even generates an additional vortex 11 as visualized by the widening and slowly moving green dye filament located at the end of the upstream slot wall. Areas of insufficient wall shear stress are almost as undesirable as vortices and must be avoided just like vortices. In particular, vortex 7 can be avoided as is explained next.

Flow visualization experiments were carried out with a two-layer film and a slot exit geometry shown in Fig. 5.71e. The height of the slot and the height of the step could be varied. Then the flow rate at the vortex inception was determined for a given fluid, i.e., for a given density and viscosity. The slide angle was 15° and both layers had the same viscosity for all trials. Results from this study can be summarized as follows.

When the critical flow rate/width is integrated into a critical Reynolds number for slot flow according to $Re_c = \rho Q_c/\mu$ (Eq. 5.7.5), then vortex 7 is present if the Reynolds number of a given application is larger than the critical Reynolds number. The flow rate/width of the upper layer has an insignificant effect on the onset of vortex 7. The critical Reynolds number increases with increasing slot height for viscosities of up to about 60 mPas. For higher viscosities, the critical Reynolds number is independent of the slot height. Increasing the step height also results in higher critical Reynolds numbers for any value of the viscosity. Vortex 7 does not exist if the Reynolds number is less than about 3.0, regardless of the values of any of the other parameters.

Figure 5.71g shows vortex 8 in the free surface of the upper layer just before this layer flows on top of the lower layer. This vortex is similar in principle to vortex 1 that may exist in the bead of slide coating as shown in Fig. 5.1, or here again in Fig. 5.71j. Note that in the case of vortex 1 the fluid layer is coated onto a moving substrate while in the case of vortex 8 the fluid layer is "coated" onto another fluid layer. Both times the approaching fluid layer impinges more or less perpendicularly onto the substrate or the other fluid layer, and in both cases, the flow field is characterized by a boundary layer that starts at the corner of the step for layers merging on the die slide or at the dynamic wetting line in the bead, see red line and yellow spot in Fig. 5.73. The length of the boundary layer is determined by a point at which all the fluid flowing down the inclined plane is entrained into the boundary layer, see blue

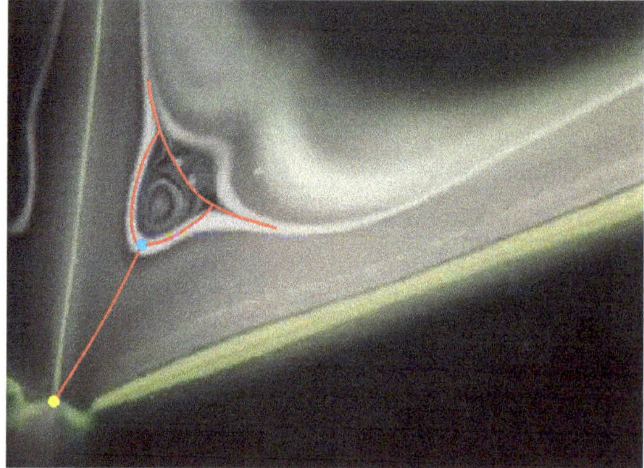

Fig. 5.73 Highlighting the boundary layer and vortex 1 in the bead of slide coating; photo taken from Schweizer (1988) and reproduced with permission from Cambridge University Press

spot in Fig. 5.73. If that point lies below the free film surface, then the free surface streamline branches off into the film flow field to meet the end point of the boundary layer and to form the vortex, see red lines in Fig. 5.73.

As with vortex 7, flow visualization experiments were carried out to determine the onset of vortex 8 in this 2-layer configuration. The experimental set-up and the operating conditions were similar as for vortex 7.

Contrary to vortex 7, the flow rate/width and the physical fluid properties of the top layer are now important regarding the onset of vortex 8. Vortex 8 is only present if the Reynolds number of the top layer (Eq. 5.7.5) is smaller than the critical value, and if the Reynolds number of the bottom layer is large, i.e., >10. This results in a strong jetting effect at the slot exit and forces the top layer to deflect by almost 90° relative to the direction of the incoming film flow as shown in Fig. 5.71g. Reducing the Reynolds number of the bottom layer or the property number of the top layer (Eq. 5.7.32), and increasing either the Reynolds number of the top layer, the slot height, or the step height decreases the chances for vortex 8 to be present.

A special situation, where these measures may not suffice to avoid vortex 8, exists when a very thin top layer of low-viscosity fluid flows onto a thick layer on a slide die without steps between adjacent plates as sketched in Fig. 5.74. This situation may be encountered in industrial coatings when applying a hard protective (scratch-proof) top layer.

A thick layer exiting from the die slot is like a steep and tall mountain, onto which the thin layer cannot climb because the viscous drag exerted by the thick layer onto the low-viscosity layer is insufficient. Consequently, a large puddle (vortex 8) is formed inside which a substantial cross-flow was observed during pilot tests, see Fig. 5.74a. To avoid this undesirable flow field, it is suggested to mount a wedge-shaped bar onto the upstream die plate as depicted in Fig. 5.74b. This bar provides a gradual step in the die slide such that the top layer no longer has to climb up the tall mountain.

Vortex 9 sketched in Fig. 5.70 is similar in principle to vortex 2 near the bead of slide coating as visualized in Fig. 5.71h. Vortex 9 exists only if the Reynolds number of the top layer is higher than around 20 and the Reynolds number of the bottom

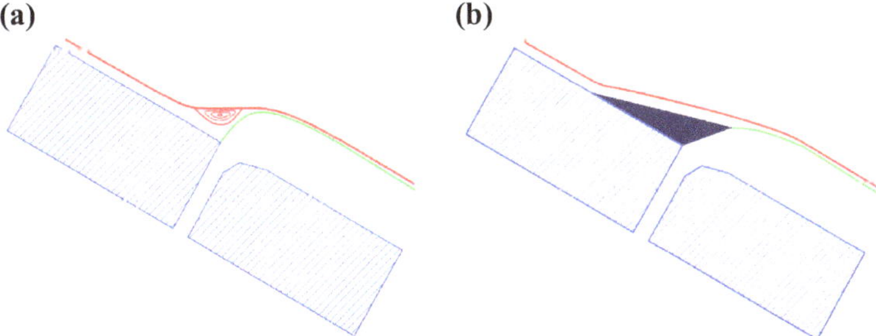

Fig. 5.74 **a** Vortex 8 forming a large puddle when a thin low-viscosity layer flows onto a thick layer **b** preventing vortex 8 by mounting a special wedge-shaped bar

layer is higher than around 30. Both of these Re-ranges are very high and higher than the Reynolds numbers of most industrial coating applications.

In particular, the high Reynolds number of the bottom layer leads to a strong jetting effect upon exiting from the die slot. This in turn causes the thickness of the top layer to locally increase when it approaches the merging area, and the associated pressure increase in flow direction generates the vortex. Increasing w and f or decreasing Re_b and Re_t reduces chances for vortex 9 to be present.

As mentioned above, the physical reasons for vortex 2 at the beginning of the bead in slide coating are the same as for vortex 9, see Fig. 5.71h. In addition to vortices in the coating bead, this picture nicely shows that the impinging flow field of premetered coating processes is a combination of a stagnation point flow and a boundary layer flow. Which of these two flows is dominating depends on the operating conditions and the geometrical parameters of the coating configuration. The approaching liquid film impinges on the substrate more or less perpendicularly, and the streamline pattern then clearly visualizes, and separates the two characteristic flow fields. In particular, fluid elements located only a fraction of a millimeter away from the substrate are still moving opposite to the web direction. Only after these fluid elements are entrained into the boundary layer, whose location is indicated by the sharp bends of the streamlines, do they experience the viscous drag of the substrate, which accelerates all impinging fluid and moves it downstream.

The onset of vortex 2 was determined with flow visualization experiments as reported before. Results can be interpreted by the following dimensionless numbers. Po is the property number according to Eq. (5.7.32). Ca is the capillary number, which is the ratio of viscous to surface tension forces normally defined with the web speed U being the characteristic velocity:

$$Ca = \frac{\mu U}{\sigma} \tag{5.7.36}$$

Ca_{film} is the capillary number of film flow, where the characteristic velocity is equal to the average velocity of film flow according to Eq. (5.7.4).

$$Ca_{fillm} = \left(\frac{\rho g \sin \beta Q^2 \mu^2}{3\sigma^3} \right)^{1/3} \tag{5.7.37}$$

We is the Weber number, which is the ratio of inertia to surface tension forces defined as:

$$We = \frac{\rho Q U}{\sigma} \tag{5.7.38}$$

Vortex 2 is present if the Weber number of a given coating application is smaller than the critical Weber number. Vortex 2 can be avoided by reducing the capillary number of film flow, by reducing the flow rate/width by coating thinner, or by increasing the Weber number by coating faster.

Lowering the viscosity would also reduce the capillary number of film flow. However, if this was accomplished by diluting the coating fluid, then the flow rate/width would increase at the same time, thus negating the purpose of lowering the viscosity. Here, therefore, heating the coating fluid would be a better approach for lowering the viscosity.

Figure 5.71i shows a distinct standing wave in the film just upstream of the bead in slide coating. The local thickening of the film under the wave causes the film velocity to decrease and the pressure to increase, which may lead to the formation of another vortex (vortex 3 in Fig. 5.70) adhering to the slide surface. Experimental data for quantifying the onset of vortex 3 do not exist.

Figure 5.71j shows vortex 1, which adheres to the upper meniscus of the bead in slide coating. The reasons for the existence of this vortex are the same as for vortex 8 as explained above. Once more, flow visualization experiments were carried out to determine the inception of vortex 1. As published by Schweizer (1988) the resulting regression equation takes the form:

$$Ca_c = a_0 Po^{a_1} e^{(a_2 Re Po^{a_3})} \tag{5.7.39}$$

$a_0 = 0.5406$	$a_1 = 0.1441$
$a_2 = 0.8434$	$a_3 = 0.2170$

The fit between experimental data and regression Eq. (5.7.39) is quite good as indicated by an r^2-value of 0.942. Moreover, the experiments revealed that applying a bead vacuum during slide coating has no effect on the inception of vortex 1.

Figure 5.75 is a graphical representation of Eq. (5.7.39). The critical capillary number Ca_c is plotted as a function of the Reynolds number Re and the property number Po. Vortex 1 is present in the flow field if the capillary number of a given

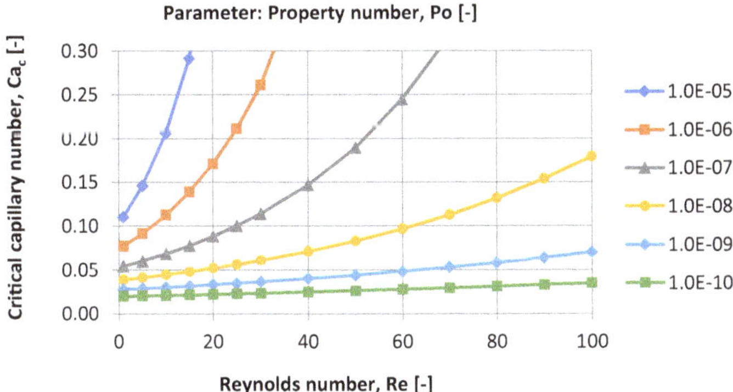

Fig. 5.75 Onset of vortex 1; the vortex is present if Ca < Ca_c; data taken from Schweizer (1988) and reproduced with permission from Cambridge University Press

coating application (Re and Po) is smaller than the critical capillary number. Consequently, vortex 1 can be avoided by coating faster, or by reducing the Reynolds number. The latter can conveniently be achieved by increasing the solids concentration of the coating fluid, which will decrease the flow rate/width and increase the viscosity at the same time. Both these measures act in the same desirable direction.

As sketched in Fig. 5.70, vortex 6 is located at the very end of the inclined die surface, just upstream of the die lip. Vortex 6 is rarely present in slide coating, but when it is, then the dynamic wetting line is located a significant distance downstream from the position of the die lip, but not far enough to generate air entrainment, i.e., the dynamic contact angle is still $<180°$. Vortex 6 can be prevented by taking one or several measures as follows: lowering the web speed, applying bead vacuum, and increasing the momentum of the impinging film. The latter measure can be accomplished by increasing the flow rate/width, by reducing the fluid viscosity, i.e., by diluting the coating fluid, and by letting the film impinge perpendicularly onto the substrate.

Figure 5.71k shows an undesirable flow field in slide coating where the impinging liquid film flows around the die lip. This always causes a separated fluid wedge below the lip, which is filled with a series of vortices 5 because the static wetting line is not pinned to the lip corner, but it is located at an unknown distance below the corner. Vortices 5 are similar to vortices 10, which are located at the other static wetting line. If the flow around the die lip is not too strong, then a pure turn-around flow develops in the heel below the lip, which is free of additional vortices, see Fig. 5.71k. However, if the flow around the lip is too strong, then vortex 4 develops in the turn-around flow, see sketch in Fig. 5.70. Visualizing this flow revealed that the presence of a pronounced turn-around flow, or even of vortex 4, renders the flow field of the heel unstable, and it starts oscillating. Such operating conditions must be avoided in slide coating. This can be accomplished by reducing or turning off the bead vacuum, by reducing the gap between die lip and substrate, or by increasing the solids concentration of the coating fluid. This in turn reduces the momentum of the impinging film (smaller flow rate/width) and increases the viscosity, i.e., the viscous drag exerted by the moving substrate on the fluid. Increasing the web speed will not change the heel flow field significantly because the flow rate/width would increase at the same time, if the wet thickness of the coated film remains unchanged.

In summary, film flow on the die slide offers many opportunities for vortices to form, particularly when associated with slide coating. However, it is possible to operate these flow fields without any vortices at all, and an example of a perfect film and bead flow in slide coating is presented in Fig. 5.76.

Apart from the absence of any vortices, this flow field is characterized by the following: only a very faint standing wave just upstream of the bead, a moderate film thinning at the entrance to the bead, a static wetting line that is pinned to the sharp lip corner, a dynamic wetting line that is located in the projection of the impinging film, a dynamic contact angle that is significantly lower than 180°, and all this without any bead vacuum. Optimum flow fields of the upper part of film flow on the die slide are shown in Fig. 5.71a–c. More arguments for optimizing the slide coating process are discussed in Sect. 18.6.

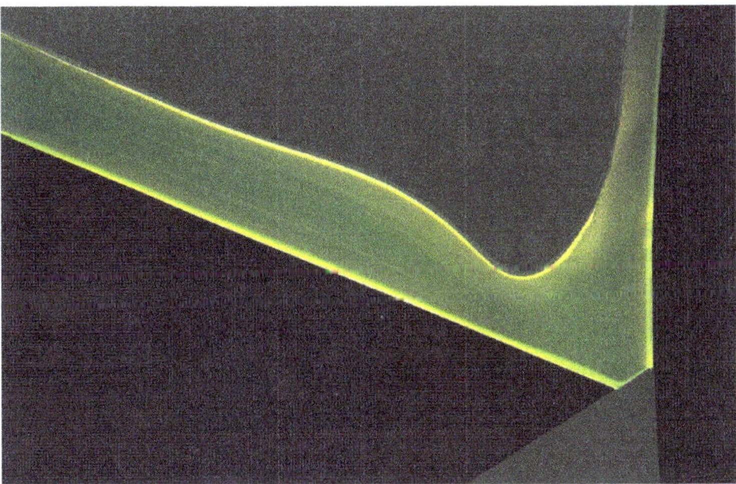

Fig. 5.76 Optimum flow field in slide coating; photo taken from Schweizer (1997b) and reproduced with permission from Springer Nature

5.8 Curtain Flow

In premetered coating methods, liquid curtains are produced either by a slot or a slide die, see Fig. 2.1d, f. The difference in the resulting curtain properties is caused by the initial curtain velocity. If slot dies are used, the initial velocity is equal to the average velocity of slot flow, and if slide dies are used, it is the average film velocity.

Typically, the average slot velocity is higher than the average film velocity, but the curtain velocity below the die lip is higher than the initial velocity. This leads to a significant thinning of the falling liquid film and is particularly so for viscous films on slide dies. The film thinning below the die lip can easily be seen by the eye, see the schematic sketch in Fig. 5.77.

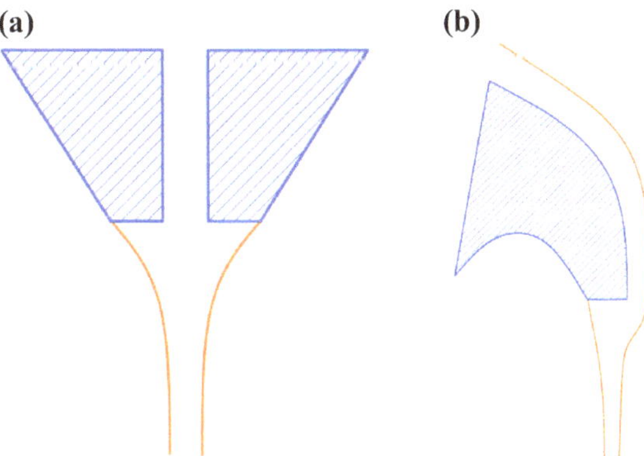

Fig. 5.77 **a** Thinning of initial curtain with slot die **b** thinning of initial curtain with slide die

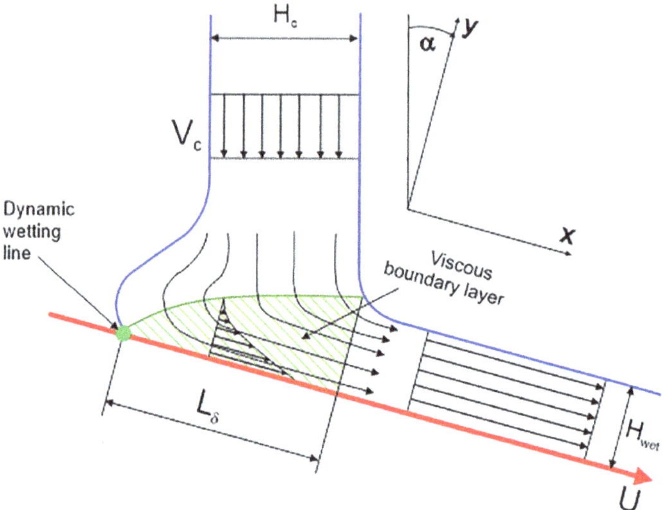

Fig. 5.78 Flow field of the impinging curtain

Just below the die lip, where the slot or film flow changes to curtain flow, there is a short transition region, the length of which is on the order of [mm], and in which shear viscosity is still present. Beyond that transition region, however, the curtain flow is a pure plug flow where shear viscosity is no longer present. The curtain flow is extensional because the curtain velocity increases from top to bottom owing to the gravitational acceleration. Consequently, the curtain thickness decreases from top to bottom, consistent with the law of mass conservation. Due to the extensional character of the curtain flow, the relevant rheological material property is the extensional viscosity, which does not depend on the shear rate but upon the extension rate of the flow.

At the bottom of the curtain, i.e., at the curtain impingement point, the flow turns into a boundary layer flow, in which the drag action of the moving substrate changes the flow direction and accelerates the fluid velocity to web speed, see Fig. 5.78. Observations during air entrainment experiments suggest that the flow inside the boundary layer is again controlled by shear viscosity.

Following is a description of various curtain properties.

5.8.1 Curtain Velocity

Knowing the curtain velocity, and in particular, knowing the impingement velocity of the curtain is important because it influences the coating behavior. Several equations for calculating the curtain velocity have been proposed in the literature. Based on Bernoulli's equation, which applies to inviscid flows and thus to the curtain flow,

Weinstein & Palmer (1997) derived the following expression:

$$V_c = \sqrt{V_{c,0}^2 + 2gz} \tag{5.8.1}$$

Here, z is the vertical coordinate pointing downward and starting at the die lip, and $v_{c,0}$ is the initial curtain velocity according to Eq. (5.7.4) if a slide die is used to form the curtain and the film only contains one layer, or according to Eq. (5.6.10) if a slot die is used. Brown (1961) experimentally measured the velocity of a curtain that was generated by a slot die. Based on his data he modified Eq. (5.8.1) by introducing a distance z_0, which accounted for viscous effects in the transition zone between slot and curtain flow. Empirically, z_0 was defined as

$$z_0 = 2\left(\frac{4\mu}{\rho}\right)^{2/3} g^{-1/3} \tag{5.8.2}$$

Consequently, the curtain velocity was calculated according to

$$V_c = \sqrt{V_{c,0}^2 + 2g(z - z_0)} \tag{5.8.3}$$

Brown's predicted values were only in agreement with his data if z was greater than about three times z_0. Moreover, Kistler (1983) suggested that z_0 was not dependent only upon the viscosity and density of the fluid. More recently, Eggerath (2012) measured the curtain velocity with a laser Doppler anemometer and found that Brown's velocity predictions were higher than his own measurements. Despite these discrepancies, plotting Brown's equation provides valuable insight into the behavior of a falling liquid curtain, see Fig. 5.79.

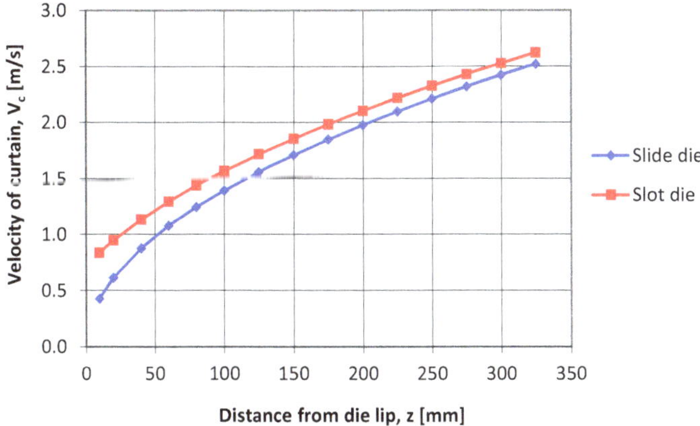

Fig. 5.79 Curtain velocity as a function of the die type

For typical curtain lengths of industrial coating applications, i.e., 100–200 mm, the curtain impingement velocity is about 1.5 m/s = 90/min. This is a low value as curtain coating is considered a high-speed coating method. In most applications, the coating fluid, upon impinging on the substrate, is accelerated to higher velocities.

For a given set of operating conditions and fluid properties, the initial curtain velocity is typically a bit higher if a slot die rather than a slide die is used for generating the curtain. Regardless of the die type, the initial curtain velocity amounts to only about 10% of the impingement velocity.

Using Eq. (5.8.1) and assuming the initial velocity to be 10% of the impingement velocity, then the average velocity of a curtain with length L_c can be approximated to give

$$V_{c,average} \approx \sqrt{gL_c} \tag{5.8.4}$$

5.8.2 *Extension Rate in Curtain Flow*

The extension rate e_c in the curtain can be approximated by neglecting the initial velocity $v_{c,0}$ and by taking the derivative of Eq. (5.8.1) with respect to z as follows, and as depicted in Fig. 5.80:

$$e_c = \frac{dV_c}{dz} = \sqrt{\frac{g}{2z}} \tag{5.8.5}$$

The extension rate is highest at the beginning of the curtain, where the acceleration from film or slot flow to curtain flow is the highest. Moreover, if the extensional

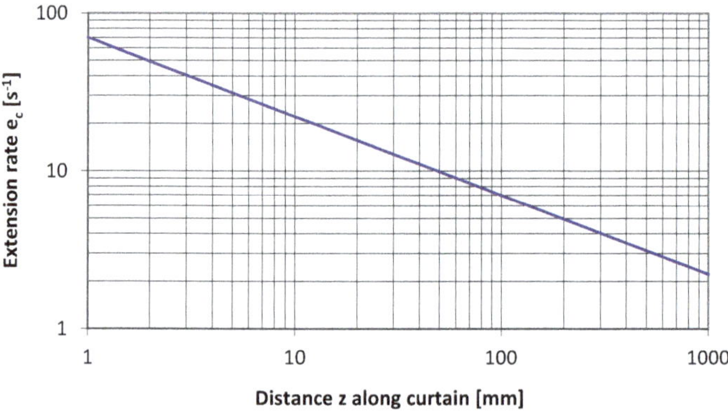

Fig. 5.80 Extension rate for curtain flow

viscosity were to be measured, then it would have to be done for an extension rate range of about 1–100 s^{-1}.

For high-speed applications (e.g., U > 1,000 m/min), where the thickness of the curtain at the impingement point is much larger than the wet film thickness on the substrate, and if the length of the boundary layer is short, high extension rates may also exist just upstream of the viscous boundary layer owing to a local curtain contraction.

5.8.3 *Curtain Fall Time*

Neglecting the initial velocity, Eq. (5.8.1) can be used for estimating the time it takes for liquid to fall through the curtain, i.e., the curtain fall time as follows. L_c is the length of the curtain.

$$t_c = \int_0^{L_c} \frac{dz}{\sqrt{2gz}} = \sqrt{\frac{2L_c}{g}} \tag{5.8.6}$$

If Eq. (5.8.1) is used for calculating the curtain fall time more accurately, then the integral shown in Eq. (5.8.6) must be solved numerically, for example with Gauss quadrature, which can easily be implemented in an EXCEL environment. The result is shown graphically in Fig. 5.81.

For typical industrial curtain lengths in the range of 100–200 mm the curtain fall time is in the range of 100–200 ms. This is important because anything that happens in the curtain, for example, the formation of a hole, takes less than about 200 ms. By the same token, if the curtain properties must be changed, then this change must also happen in less than 200 ms, for example when lowering the surface tension by adding

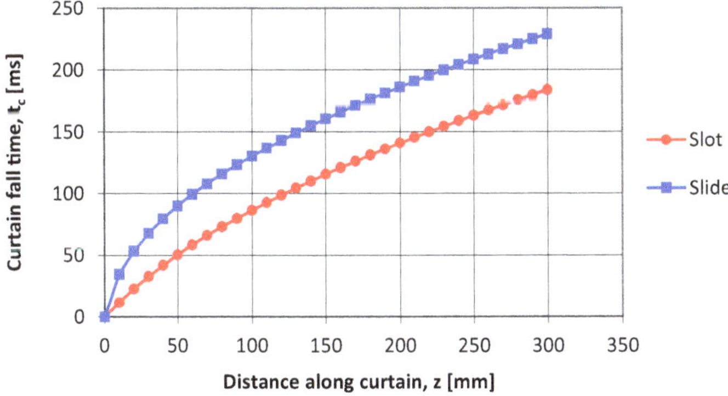

Fig. 5.81 Curtain fall time

a surfactant to prevent the formation of holes. This particular change is challenging because it requires very fast-diffusing surfactant molecules, see Sect. 4.5.

The fall time of a curtain generated by a slot die is typically a bit lower compared to using a slide die, but the order of magnitude remains the same.

5.8.4 Curtain Thickness

In line with the law of mass conservation, the curtain thickness is obtained by dividing the given volumetric flow rate/width by the curtain velocity. Assuming that the initial curtain thickness equals the film thickness at the end of the inclined plane if a slide die is used, or the slot height if a slot die is used, then the curtain rapidly thins out just underneath the die lip owing to the velocity profile discussed above. An example of the thickness profile is shown in Fig. 5.82. At the bottom of industrial curtains, the thickness is typically <200 μm. This small dimension makes it difficult to observe by unaided eye the details of the flow field of the impinging curtain.

5.8.5 Curtain Stability

There is no curtain coating without a stable curtain. In fact, the loss of curtain stability is a process limitation, or one of the boundaries of the operating window of curtain coating (see also Chap. 19 for more detailed information about the operating boundaries of curtain coating). Curtain stability refers to the ability to form and maintain a stable curtain over long operating times on the order of several hours as required by industrial applications. Understanding curtain stability, therefore, is a crucial issue for operating and optimizing industrial curtain coating processes.

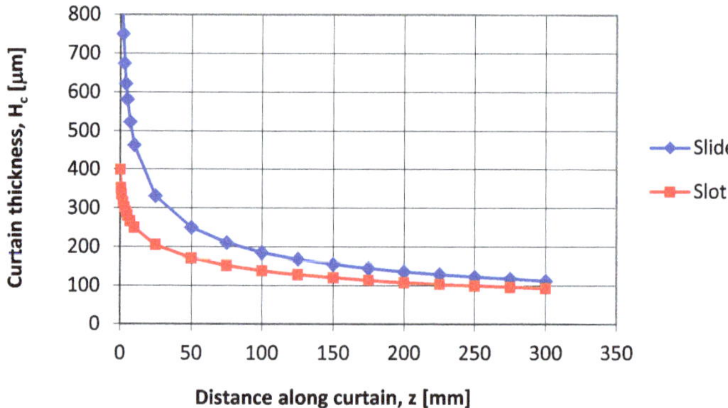

Fig. 5.82 Curtain thickness

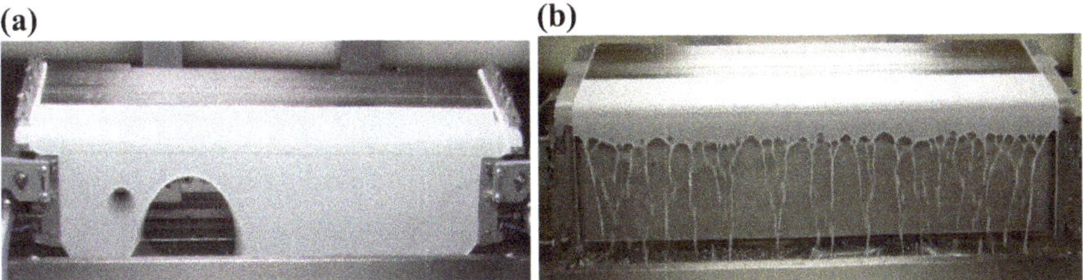

Fig. 5.83 **a** Stable curtain with holes; photo reproduced with permission from Polytype Converting AG **b** unstable curtain; photo reproduced with permission from Polytype Converting AG

The difference between a stable and an unstable curtain is illustrated in Fig. 5.83. The curtain in the left photo is disturbed by a couple of holes, one of them being rather big with a diameter of about 100 mm. Despite this severe defect, the curtain does not disintegrate but remains operational. It is therefore called a stable curtain. The holes are simply transported downstream and out of the curtain. Eventually, the holes are coated onto the substrate where they are being stretched and change their form from more or less round to oval. Such oval defects in the coated film are often referred to as "eggs", see Fig. 5.84.

In contrast, the curtain in the right photo disintegrated in response to a defect. This curtain is called unstable. There no longer appears to be a curtain because the initially coherent curtain changed into many irregular rivulets. Upon closer examination, however, we can see that a little bit of a stable curtain is still present just below the die lip. There seems to be an inception line parallel to the die lip located a couple of centimeters below the lip. Above that line the curtain is stable, but below the line, the curtain disintegrated into rivulets. Experiments revealed that the inception line could be moved up or down in the curtain simply by decreasing or increasing the flow rate.

The situation above is again shown in Fig. 5.85. This time the stable curtain is perfectly homogeneous and coherent, while the disturbed curtain is perfectly unstable with the rivulets being placed a constant distance apart from each other. However, even in this perfectly unstable curtain we can still detect a very short portion of a stable curtain, see Fig. 5.85c. Here, the inception line is located no more than 2 mm below the die lip. This flow configuration is referred to as **short curtain** (Eggerath,

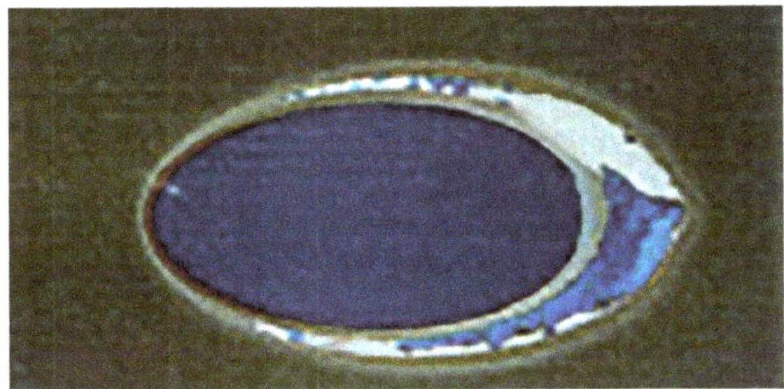

Fig. 5.84 "Egg" defect generated by a hole in the curtain; photo reproduced with permission from ILFORD Imaging Switzerland GmbH

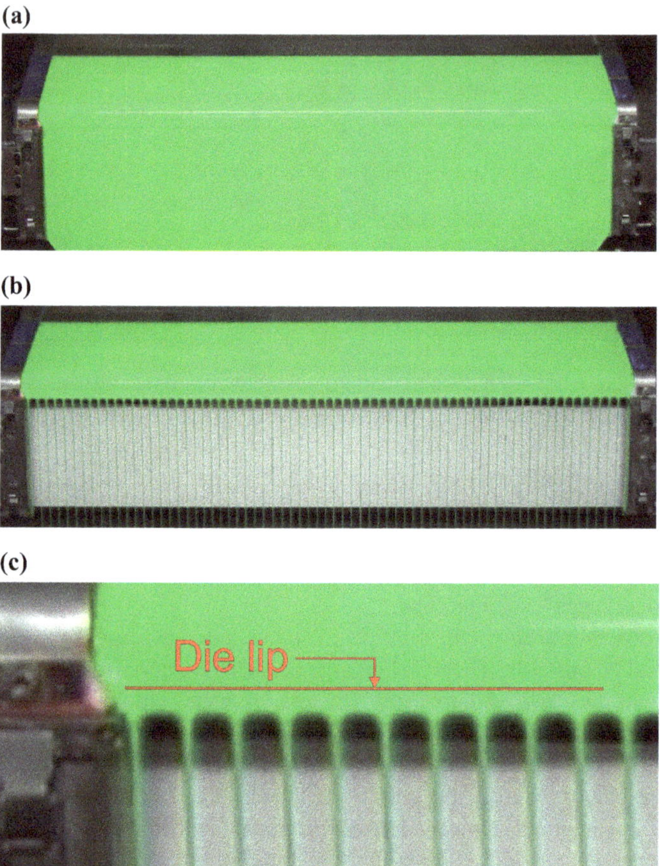

Fig. 5.85 **a** Perfectly coherent and stable curtain; photo reproduced with permission from Polytype Converting AG **b** perfectly unstable curtain; photo reproduced with permission from Polytype Converting AG **c** a very short portion of a stable curtain in an otherwise perfectly unstable curtain; photo reproduced with permission from Polytype Converting AG

2012). By contrast, then, curtains that are stable over the length of the edge guide are called **long curtains**.

Experimental results generated by researchers in the photographic industry suggested that stable curtains can be established if the flow rate/width is above a critical minimum value called low flow limit or minimum flow rate. For aqueous gelatin solutions containing surfactants Greiller (1972), for example, suggested a minimum flow rate/width of about 0.5 cm^2/s.

In parallel, theoretical models were proposed for predicting the loss of curtain stability. Taylor (1959) studied the stability of a water bell by applying a momentum balance on a free edge of the water membrane. Then, Brown (1961) took that idea and applied it to a liquid curtain, see Fig. 5.86. In particular, Taylor determined that

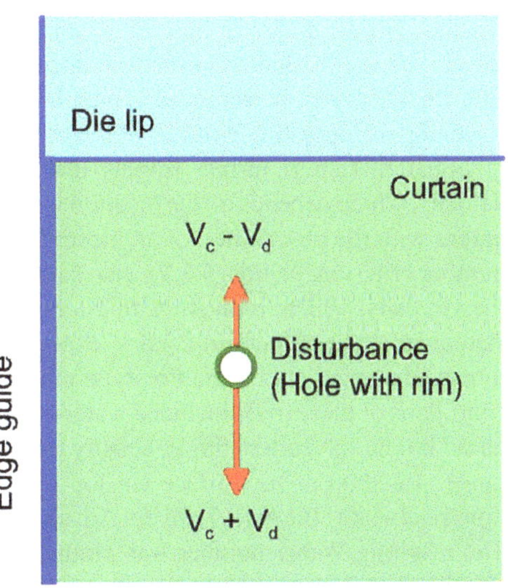

Fig. 5.86 Sketch for analyzing curtain stability

the speed of a disturbance in a thin liquid membrane, i.e., the speed of the rim of the hole in a curtain, equals:

$$v_d = \sqrt{\frac{2\sigma}{\rho H_c}} = \sqrt{\frac{2\sigma V_c}{\rho Q}} \tag{5.8.7}$$

Brown then argued that a curtain is stable if the upward speed of the disturbance is smaller than the downward speed of the curtain. This way, the hole is flushed out of the curtain even if it grows while moving downward, and the curtain is only temporarily disturbed but not broken up completely. Neglecting the initial curtain velocity in Eq. (5.8.1) and relating it to the disturbance velocity then results in the well-known Weber number stability criterion

$$We = \frac{\rho Q V_c}{\sigma} > 2 \tag{5.8.8}$$

The Weber number is the ratio of inertia to surface tension forces. A curtain is said to be stable if We > 2. Solving Eq. (5.8.8) for Q yields

$$Q > \frac{2\sigma}{\rho v_c} \tag{5.8.9}$$

Equation (5.8.9) is a more convenient way for assessing the stability of industrial curtains. It demonstrates that a curtain is stable if the flow rate/width is above a minimum value, which is dependent upon the surface tension. This statement is in agreement with experimental observations made by Greiller. Lowering the surface tension will allow lower flow rates, and so thinner wet films can be coated at a given web speed, and/or lower web speeds are admissible at which a desired wet film thickness can be achieved. Thus, lowering the surface tension is a means for widening the operating window of the curtain coating process.

Equation (5.8.9) further reveals that curtain stability depends on the curtain velocity, which depends on the location within the curtain (coordinate z). This in turn agrees with the observation of an inception line as shown in Fig. 5.83b. The Weber number criterion, or Eq. (5.8.9), also says that curtain stability does not depend on the viscosity, which, intuitively, might be an unexpected conclusion. To check this statement some 40 curtain coating experiment relating to many different applications and carried out on the Polytype pilot machine were evaluated. The operating conditions of these trials included water-based fluids containing surfactants, solvent-based fluids, solventless fluids, density range: 750–1570 kg/m^3, low-shear viscosity range: 46–5600 mPas, surface tension range: 7.4–65.1 mN/m, range of minimum flow rate/width: 0.18–1.17 cm^2/s. All trials were made with a multilayer slide die. The resulting Weber number was plotted as a function of the low shear viscosity, which is representative of the film flow just upstream of the curtain flow, see Fig. 5.87.

While most of the data satisfy the We > 2 criterion, it was possible in a few trials to establish stable curtains for We < 2, particularly if the low-shear viscosity was high, i.e., on the order of 1,000 mPas. Moreover, according to our observations and those as published by Becerra and Carvalho (2011), it is possible to form stable curtains at a much lower flow rate/width than predicted by the Weber number criterion if the fluid has visco-elastic properties. Specifically, they achieved stable curtains at Weber numbers as low as 0.19.

Eggerath (2012) studied the stability of short and long curtains generated by a slot die with Newtonian fluids (aqueous glycerin solutions) of different viscosities. He presented the results in a diagram with the Reynolds number Re on the horizontal axis and the Ohnesorge number Oh on the vertical axis. The latter number is relevant

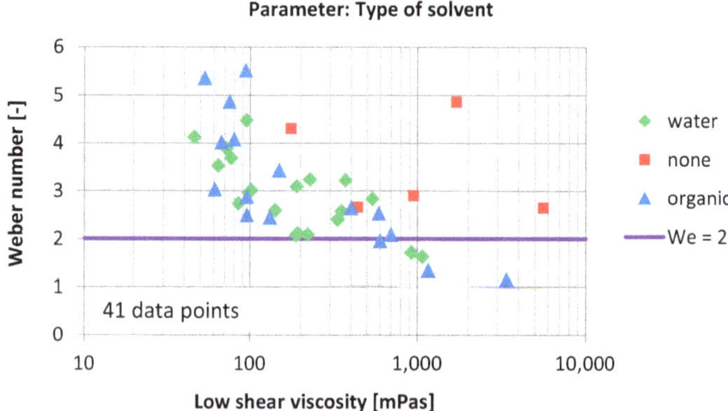

Fig. 5.87 Curtain stability according to the Weber number criterion as a function of the low-shear viscosity; data reproduced with permission from Polytype Converting AG

in the field of jet break-up and drop formation (ink jet printing), which is not far away from the break-up of a thin liquid sheet. The dimensionless Ohnesorge number is defined as the ratio of viscous to inertia and surface tension forces. Eggerath defined Oh and Re as follows:

$$Oh = \frac{\mu}{\sqrt{2\rho\sigma w}} \tag{5.8.10}$$

$$Re = \frac{\rho Q}{\mu} \tag{5.8.11}$$

Here w is the height of the die slot that was used to form the curtains, thus indicating that the constant average velocity of slot flow was used for the derivation of Oh.

In the diagram mentioned above the curtain stability criterion according to We > 2 (Eq. 5.8.8) is represented by a straight line, if both axes are drawn logarithmically. Eggerath tested 12 fluids including water. The viscosity range was from 1 to 1,340 mPas, the surface tension spanned a range from 73.2 to 63 mN/m and the density varied from 1,000 to 1,259 kg/m^3. Curtain stability was determined by starting from a low flow rate value, and by increasing the flow rate until a coherent and stable curtain was established. The evaluation of the data was supported by theoretical calculations to derive an empirical equation that predicts the **stability limit of long curtains** as follows:

$$Oh > \left[\left(\frac{\sqrt{Re}}{1.816}\right)^{3/2} + Re^{3/2}\right]^{-2/3} \tag{5.8.12}$$

Similar experiments were carried out and evaluated for a short curtain. The empirical equation describing the **stability limit of short curtains** is:

$$Oh > \left[\left(\frac{\sqrt{Re}}{0.818}\right)^{3/5} + Re^{3/5}\right]^{-5/3} \tag{5.8.13}$$

The two equations above are graphically displayed in Fig. 5.88 and can be commented as follows.

For long curtains, the minimum flow rate/width depends on the viscosity if Re < 10 and Oh > 0.1 (Regime II). The minimum flow rate/width decreases with increasing viscosity and is lower than the value predicted by the Weber number criterion. In Regime I, curtain stability does not depend on the viscosity, as is also predicted by the Weber number criterion.

For short curtains, the minimum flow rate/width depends on the viscosity over the entire Reynolds number range. Moreover, the minimum flow rate for short curtains is considerably lower than the value for long curtains, thus considerably opening up

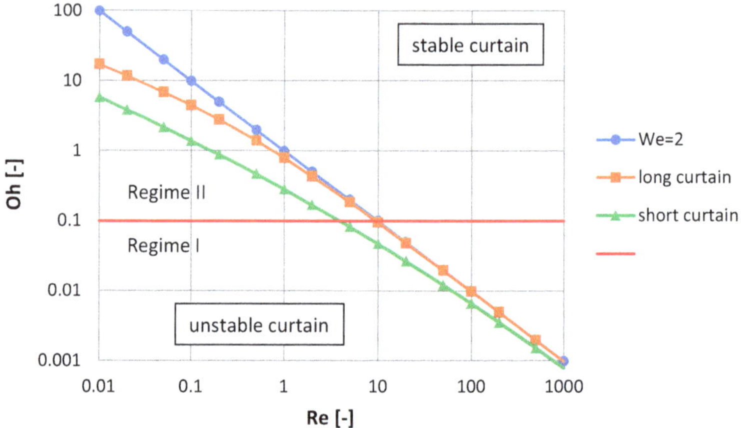

Fig. 5.88 Stability limit of short and long curtains; graph generated with data and permission from Eggerath (2012)

the operability window for curtain stability when compared to the Weber number criterion.

The Reynolds number range for typical curtain coating applications is 0.1–10.

The dimensionless representation Oh = Oh(Re) shown in Fig. 5.88 is not suitable for assessing the curtain stability of industrial curtain coating processes. As mentioned above, a better way is to express curtain stability in terms of a minimum flow rate/width. For that purpose, Eggerath has rearranged Eqs. (5.8.12) and (5.8.13) as follows.

Minimum flow rate/width for a long curtain:

$$Q > \frac{\mu}{\rho}\left[\frac{1}{2}\left(\sqrt{0.167 + \frac{4}{Oh^{3/2}}} - 0.409\right)\right]^{4/3} \tag{5.8.14}$$

Minimum flow rate/width for a short curtain:

$$Q > \frac{\mu}{\rho}\left[\frac{1}{2}\left(\sqrt{1.27 + \frac{4}{Oh^{3/5}}} - 1.13\right)\right]^{10/3} \tag{5.8.15}$$

Figure 5.89 shows the dependence of the minimum flow rate/width upon the viscosity for a long curtain. The parameters for calculating Oh in Eq. (5.8.14) were set to values that promote curtain stability, particularly the surface tension, i.e., $\rho = 1{,}000\ \text{kg/m}^3$, $\sigma = 35$ mN/m and $w = 300\ \mu$m. As already indicated in Fig. 5.88, curtain stability, or the minimum flow rate/width, does not depend on the viscosity, if the viscosity is low ($\mu < 10$ mPas) or the Reynolds number is high ($Re > 10$). For viscosities >10 mPas, the minimum flow rate/width decreases with increasing viscosity. It is disturbing however that, according to Eggerath's data in Fig. 5.89, the

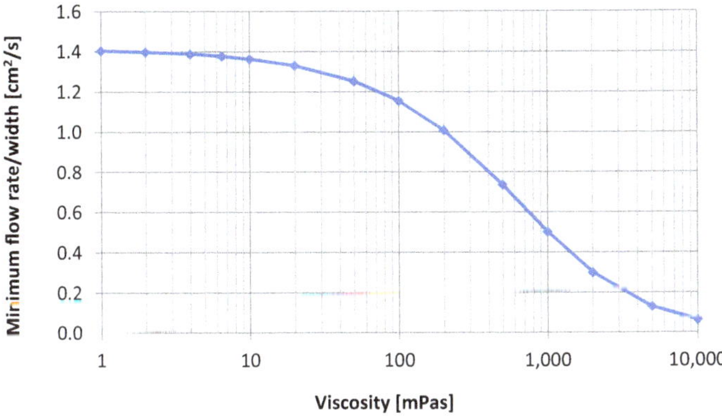

Fig. 5.89 Curtain stability: minimum flow rate/width as a function of viscosity

viscosity must be >200 mPas before the flow rate drops below 1.0 cm^2/s, which is generally considered to be the minimum flow rate/width value for industrial curtain coating applications, see Chap. 19. Moreover, this result does not agree with the data presented in Fig. 5.87. In particular, for surface tensions of about 35 mN/m, minimum flow rate/width values as low as 0.47 cm^2/s could be obtained with aqueous coating fluids containing surfactants and having viscosities of <200 mPas. One explanation for this discrepancy could be the difference in the die used for obtaining the data, i.e., slot versus slide die. Another more plausible explanation is the difference in experimental procedure, i.e., increasing the flow rate until a stable curtain is obtained versus decreasing the flow rate until a stable curtain breaks up. Observations made at Polytype Converting confirm a hysteresis between the two experimental approaches, but the magnitude of the hysteresis is unknown.

In addition, using the Weber number criterion and replacing the position-dependent curtain velocity by the average curtain velocity according to Eq. (5.8.4), then approximate values of the minimum flow rate/width only depend on the surface tension and the curtain length as illustrated in Fig. 5.90. For a surface tension of 35 mN/m the flow rate values according to $We = 2$ are low and, according to Eggerath's data, can only be achieved if the viscosity is on the order of several hundred to several thousand [mPas], which is not in line with the data shown in Fig. 5.87.

Attempting to obtain stable curtains with very low minimum flow rate/width values is often not necessary nor wanted in industrial curtain coating processes for the following reasons:

- As is further explained in Chap. 19, curtains with $Q < 1.0$ cm^2/s often cannot be coated, particularly at higher web speeds, owing to air entrainment.
- A curtain that is stable at a low flow rate/width but that is full of holes is unacceptable in an industrial coating process, see Fig. 5.83a. As is schematically depicted in Fig. 5.91, holes in a curtain, or in a thin film as shown in the figure, are generated by Marangoni flow owing to a local impurity of low surface tension. The surface tension difference between the impurity and the surrounding fluid causes

radial flow away from the impurity, which in turn causes the curtain (or the film) to become locally thin until it breaks up. The strength of this flow increases with increasing local surface tension difference. Consequently, avoiding holes in the curtain requires avoiding, or at least reducing local surface tension differences. In our experience, surface tensions in the curtain of <40 mN/m are helpful in this regard. Remember that the surface age of a falling curtain is between 100 and 200 ms, depending on the curtain length. Therefore, the surface tension must also reach such low values in this time frame. This situation is an issue to be considered for aqueous coating fluids containing surfactants.

- Industrial companies that are interested in introducing curtain coating often first run pilot trials by using a fluid formulation that was optimized for whatever coating method they are using in their production. Many of these fluids contain de-foaming agents to reduce the foam quantity, for example if the fluids are used with a gravure coating method. The purpose of a de-foaming agent is to break up thin liquid lamellas (foam bubbles), but a liquid curtain is just that. Figure 5.83b shows what happens if a fluid containing a de-foaming agent is curtain coated. Premetered coating methods do not generate any foam because the entire fluid delivery system is confined to the exit of the die, and all fluid delivered is immediately coated and

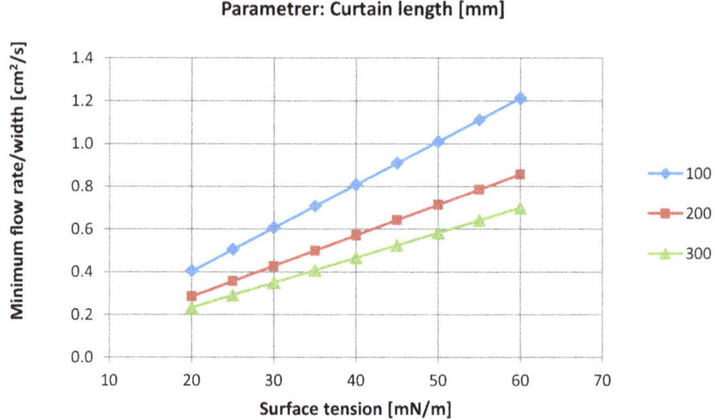

Fig. 5.90 Minimum flow rate/width as a function of surface tension for We = 2

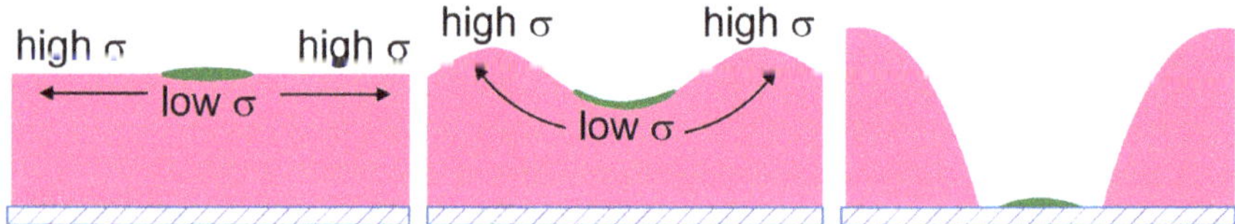

Fig. 5.91 Schematic rendering of Marangoni flow leading to film break-up or formation of holes in curtains

carried away by the substrate. It is obvious, therefore, that de-foaming agents must be banned from fluids intended to be applied by premetered coating methods.

- In high-speed curtain coating applications, such as those found in the paper industry for making thermal paper or in the adhesive industry for applying pressure-sensitive adhesives for making labels, the flow rate/width is so high that the curtain is always stable, regardless of the values of viscosity and surface tension. Coating, for example, PSA with $A_{dry} = 20\ g/m^2$, $C_{w,solid} = 0.65$, $U = 1{,}100$ m/min results in a flow rate/width of 5.53 cm^2/s.
- Sometimes a curtain breaks up not because the flow rate is too low or because contaminations in the fluid generate holes in the curtain as describes above, but because the curtain becomes locally so thin that it will break. One such cause might be the design and the operation of the curtain edge guide, particularly at the bottom of the edge guide where the curtain is thinnest. As sketched in Fig. 5.92 local excessive curtain thinning might be caused by an excessive spreading of the curtain fluid on the edge guide. Avoiding this problem requires a different edge guide design or different operating conditions for the edge guide, or increasing, not decreasing the flow rate in the curtain. This issue is discussed in more detail in Sect. 5.9.4.

Summary

The attractive curtain coating process is characterized by various operating boundaries, one of them being related to the stability of the liquid curtain. Curtain stability is promoted by high volumetric flow rates/width, by low surface tensions, and, to a lesser extent, by high viscosities. It can be concluded that fluids based on organic solvents that inherently have a low surface tension, are preferable over aqueous fluids for achieving stable curtains because the surface tension of aqueous fluids is diffusion-controlled and therefore time-dependent. This may make it difficult if not impossible to achieve the required low surface tensions for low surface ages of <200 ms that are typical of industrial curtain fall times.

However, the curtain coating operating boundary related to curtain stability is often overruled by other operating boundaries related to dynamic wetting and air entrainment. Consequently, industrial curtain coating applications for flow rates/width of

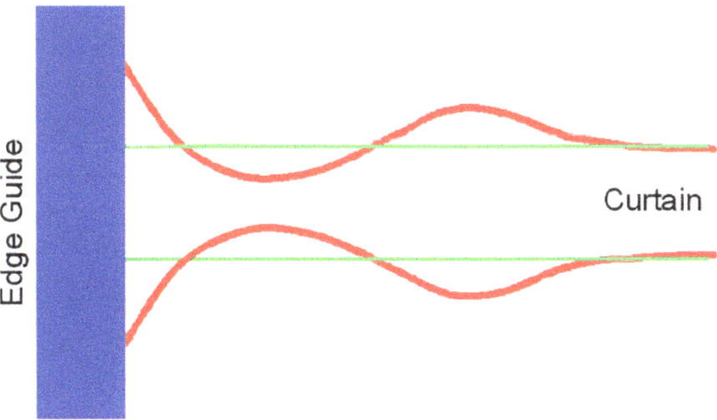

Fig. 5.92 Local thinning of the curtain near the edge guide

<1.0 cm^2/s are rare and limited to low web speeds, thick wet films, and rough or porous substrates.

Curtain stability and hence process reliability is greatly affected by the design of the curtain edge guides. Optimizing curtain stability requires optimization of equipment and product formulation. This in turn requires an intensive dialogue between all parties involved (process engineers, formulation chemists, equipment designers) for implementing curtain coating in an industrial environment.

5.8.6 Curtain Deflection

The phenomenon of liquid curtain deflection is well known and well described, see Kistler (1983) and Kistler and Scriven (1994). It is sometimes referred to as the tea-pot effect, or curtain bend back. In essence, the phenomenon describes the fact that, if a liquid curtain is formed by a liquid film flowing down an inclined plane, the trajectory of the curtain does not always correspond to the ballistic trajectory as suggested by the inclination of that plane. Instead, depending on the volumetric flow rate/width and the physical properties of the liquid, the trajectory is bent backward to some degree, see Fig. 5.93. Sufficiently downstream from the inception of the curtain the trajectory becomes vertical following the orientation of the gravity pull. If the inclination of the film flow plane is vertical, which is the case for most industrial slide

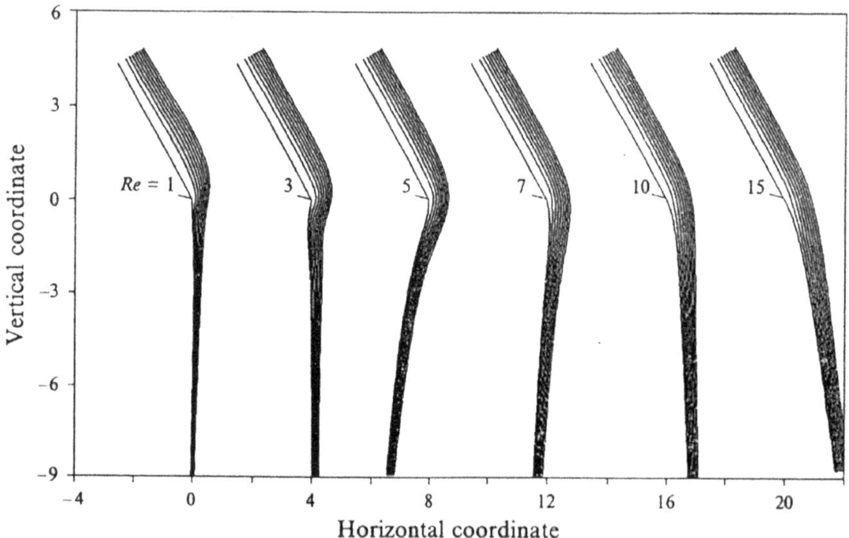

Fig. 5.93 Ballistic and non-ballistic trajectories of liquid curtains; $\beta = 60°$, Po = 1.0; graph taken with permission from Kistler (1983)

Fig. 5.94 **a** No curtain bend-back at low flow rate; photo reproduced with permission from Polytype Converting AG **b** curtain bend-back at high flow rate; photo reproduced with permission from Polytype Converting AG

dies used in curtain coating, the trajectory does not have any forward component. The curtain either falls vertically, or it is deflected backward, see Fig. 5.94.

The physical reason for the curtain deflection is a component of rotation, which is inherent to any film flow on an inclined plane. Owing to the semi-parabolic nature of the velocity profile, the liquid film would like to rotate toward the plane, but the plane provides a counterforce, thus preventing the rotation. However, at the transition point between the film and curtain flows, i.e., at the die lip, the solid plane is suddenly removed, which allows the rotational force to become active in a transition zone just downstream of the die lip. As a consequence, the curtain is bent backward.

Curtain deflection in curtain coating is predominantly observed if a slide die is used for generating the curtain, see Fig. 2.1f. In contrast, no curtain deflection is observed, if a 1-layer inverted slot die is used for making the curtain because the velocity profile in the preceding slot flow is fully parabolic, and hence symmetric, Fig. 2.1d. If a 2-layer inverted slot die is used, some degree of curtain deflection is expected toward one side or the other, depending on the relative flow rates and viscosities of the two liquids, Fig. 2.1e.

The goal of any coating process is to optimize its performance for any given set of specific operating conditions. For curtain coating, this implies controlling the size of the curtain heel at the curtain impingement point, i.e., controlling the location of the dynamic wetting line relative to the location of the curtain, see Fig. 5.78. Process optimization is difficult if the curtain trajectory is not straight and vertical but curved owing to a substantial deflection of the curtain. As a result, the effective curtain impingement angle changes to a generally unknown value, which is not a comfortable position to be in.

Moreover, a strong curtain deflection negatively impacts the acceleration and deceleration phases of fast coating machines. As the curtain deflection depends on the volumetric flow rate/width, and as this flow rate is typically slaved to the coating speed, the effective location of the curtain and thus the impingement point, depends on the coating speed. This may cause problems during the start and stop procedures.

Summarizing the observations and comments presented above, we can state that curtain deflection is an unattractive feature of the curtain coating process, which cannot be prevented if slide dies are used for forming the curtain. Therefore, it is important to understand the extent of the problem, i.e., to be able to quantify the amount of curtain deflection as a function of all relevant process parameters. In addition, it is important to evaluate possibilities for minimizing the undesirable effects of curtain bend back by way of choosing suitable physical fluid properties, or preferably by way of smart equipment design.

General quantification of the tea-pot effect was first published by Kistler and Scriven (1994). They performed numerical calculations of the flow field near the lip of a slide die, and they chose the initial angle of curtain deflection as a means for quantifying the extent of the effect, see Fig. 5.95. Moreover, by choosing a dimensionless representation in terms of the Reynolds and Property numbers, they were able to condense a large amount of data into a single master diagram, see Fig. 5.96. Note that the data points in Fig. 5.96 were extracted from Kistler and Scriven (1994), whereas the bell-shaped (Gauss) curves were obtained by the author of this book by performing an appropriate regression analysis with these data points.

The Reynolds number is calculated according to Eq. (5.8.11), whereas Kistler defined the property number as

$$Po = \frac{\sigma}{\left[(4\mu)^4\left(\frac{g}{\rho}\right)\right]^{1/3}} \tag{5.8.16}$$

The curtain deflection first increases and then decreases as a function of Re. Note that Re corresponds to the flow rate/width for a given fluid with known physical properties. In addition, the peak deflection depends on the physical fluid properties, particularly on the viscosity and the surface tension. Unfortunately, these results are not very useful from a practical point of view because the curtain trajectory and the curtain deflection at the curtain impingement point cannot be extracted. In addition, Kistler did not investigate if and how the curtain deflection depends on the inclination angle of the die lip.

As a consequence, while working at Polytype Converting AG, we decided to embark on an experimental study to obtain quantitative information about the deflection of liquid curtains. Following is a description of the experimental set-up, and a summary of the results obtained from trials carried out in the Polytype Converting pilot facility. All data are reproduced with permission from Polytype Converting AG.

1-layer and 2-layer experiments were carried out by pumping and recirculating various fluids through either a 780 mm wide or 260 mm wide stainless steel multilayer slide die from TSE Troller AG, or a 150 mm wide plexiglass slide die made by Polytype Converting. Various types of curtain edge guides were used for the various dies. Tests showed that the local strong deformation of the curtain near the edge guides during strong bend back was inconsequential for the measurements because the measurements were taken in the middle of the curtain, thus far away from this deformation. Both stainless steel dies were equipped with a lip having a fixed vertical inclination and with a sub-optimal lip tip. In particular, the length of the horizontal land of the lip was 1.0 mm, see Fig. 5.99. As the rear curtain surface did not detach from the near but the far lip corner, i.e., as the static wetting line was always attached to the far corner, this resulted in an initial curtain displacement of also about 1 mm, which was not related to the bend back effect.

As we were interested in finding lip geometries that might reduce or even eliminate the bend back, we built a 150 mm wide slide die made of plexiglass with a cylindrical rotatable lip insert made of aluminum, which allowed us to vary the final inclination of the die lip (angle δ relative to horizontal) from 45° to 150°, see Figs. 5.97 and

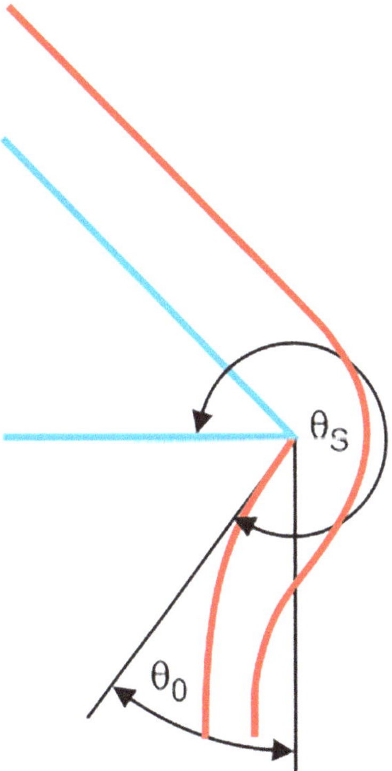

Fig. 5.95 Definition sketch for numerical calculations of the tea-pot effect, adapted from Kistler and Scriven (1994)

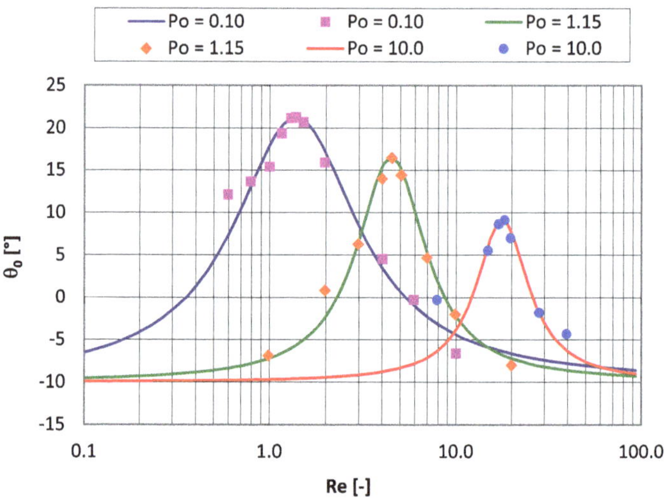

Fig. 5.96 Dimensionless quantification of the tea-pot effect, extracted from Kistler and Scriven (1994); Gauss curve fitting by P. Schweizer

5.98. Note that the inclination angle of the die slide surface was always 30° for all dies used in the trials.

As schematically depicted in Fig. 5.99, a ruler was displaced horizontally relative to a geometrical reference until the tip of the ruler touched the falling liquid curtain. The position of the reference was measured relative to the tip of the die lip. The bend back or the curtain deflection (x-coordinate) was quantified as the difference between the position of the tip of the die lip and the front surface of the curtain. Moreover, the curtain deflection was measured as a function of the vertical distance from the die lip (y-coordinate).

Fig. 5.97 Slide die with cylindrical rotatable lip insert

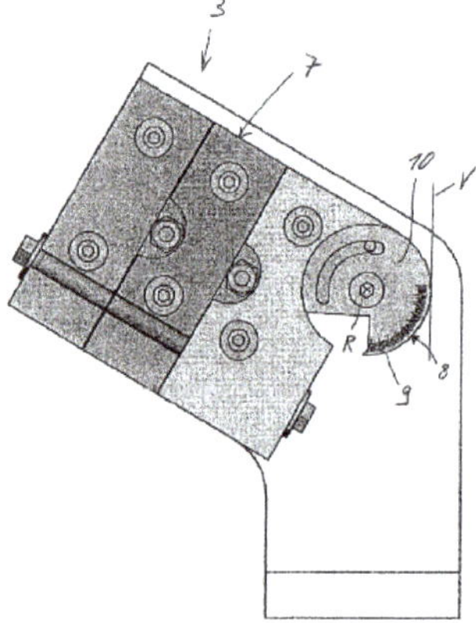

Experiments were carried out with two fluids, i.e., a pressure-sensitive adhesive PSA (Acronal V-215 + additives from BASF) and a polyvinyl alcohol PVA (Mowiol 15–79 from Kuraray). Both fluids were water-based, contained surfactants, and the solids concentration $C_{W,solid}$ was varied to vary the viscosity. The wall shear rate for film flow on an inclined plane for volumetric flow rates/width ranging from 1 to 6 cm^2/s is in the range of 100–300 s^{-1}. Thus, the low shear viscosity was measured with either a 4 mm DIN cup, or a simple Brookfield device.

We measured the deflection of single-layer and 2-layer curtains. For 1-layer curtains, we varied the physical fluid properties (ρ, μ_0, σ), the volumetric flow rate/width Q, and the inclination angle δ of the die lip. Here, the main question was whether or not the curtain deflection could be minimized by choosing a suitable inclination angle of the die lip.

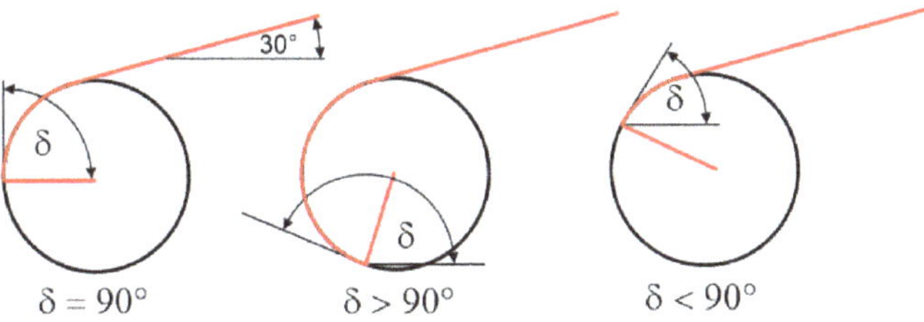

Fig. 5.98 Definition of lip angle δ

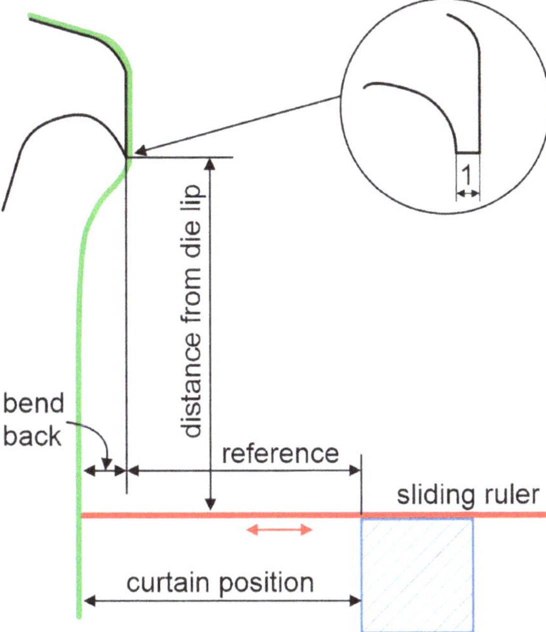

Fig. 5.99 Experimental set-up for quantifying the curtain deflection

For limited experiments with 2-layer curtains, we varied the relative flow rates and surface tensions in an attempt to minimize the curtain deflection.

5.8.6.1 Deflection of 1-Layer Curtain with PSA

Typical trajectories of deflected curtains are shown in Fig. 5.100. The die lip is located at the coordinate [x/y] = [0/0]. The initial curtain displacement, which is caused by the horizontal lip land of 1 mm length, is clearly visible. see Fig. 5.99. There is virtually no curtain deflection for flow rates/width of less than about 3.5 cm^2/s. However, the deflection increases with an increasing flow rate above that value, and the deflection increases also with increasing distance from the die lip, at least for flow rates of >4 cm^2/s. This last observation was surprising. We would have expected the curtain to fall vertically after an initial deflection in a transition zone just underneath the die lip. This was the case for Q = 3.99 cm^2/s, but it was no longer the case for larger Q values, see Fig. 5.100. We could not explain why the curtain did not fall vertically for these high flow rates. Most likely, the explanation can be found in particular rheological properties of this fluid, such as viscoelasticity.

Figure 5.101 shows the curtain deflection as a function of the flow rate for different distances from the die lip. As already seen in Fig. 5.100, there is virtually no deflection

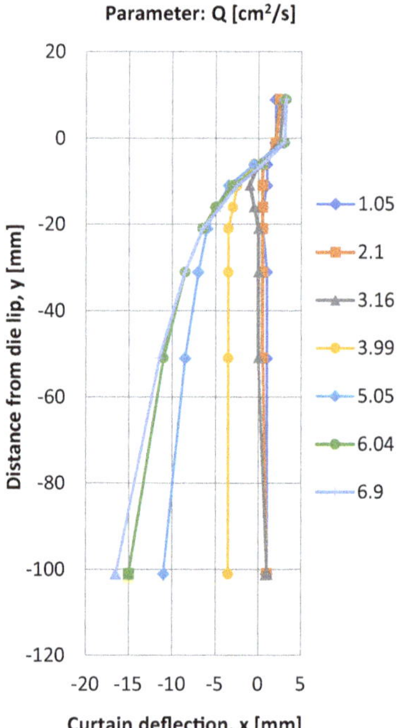

Fig. 5.100 Curtain trajectories of PSA; $\rho = 1015\ kg/m^3$, $\mu_0 = 151$ mPas, $\sigma = 33.1$ mN/m, $\delta = 90°$, Po = 0.30

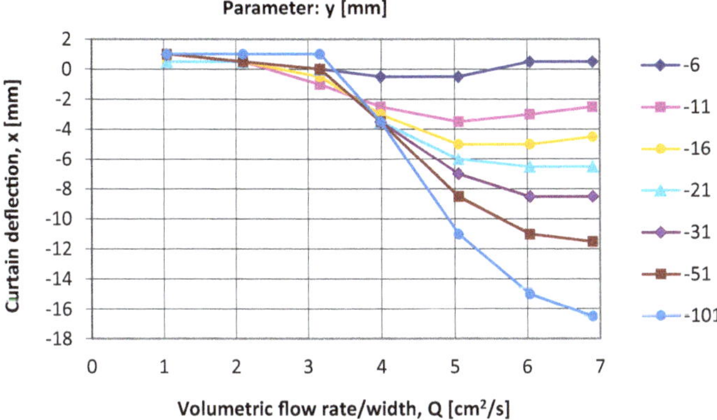

Fig. 5.101 Curtain deflection of PSA as a function of flow rate/width at different distances from the die lip; $\rho = 1015\ \text{kg/m}^3$, $\mu_0 = 151$ mPas, $\sigma = 33.1$ mN/m, $\delta = 90°$, Po $= 0.30$

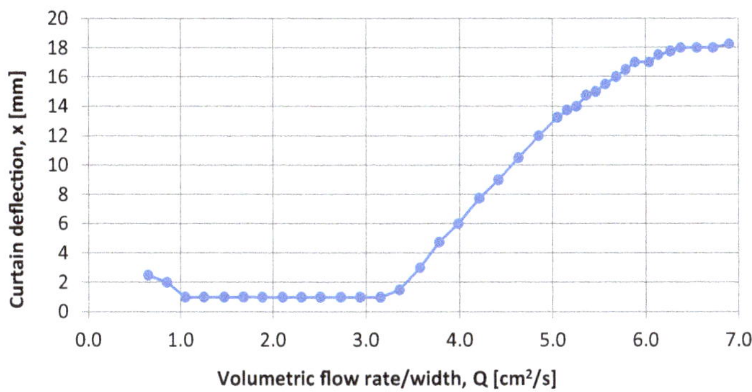

Fig. 5.102 Curtain deflection of PSA at a distance of 101 mm from the die lip, as a function of flow rate/width; $\rho = 1015\ \text{kg/m}^3$, $\mu_0 = 151$ mPas, $\sigma = 33.1$ mN/m, $\delta = 90°$, Po $= 0.30$

for flow rates of $Q < 3.5\ \text{cm}^2/\text{s}$, but the deflection increases sharply for higher flow rates, and it starts to flatten again if Q is higher than about 6 cm^2/s.

Zooming in on the curtain deflection as a function of the flow rate/width at a distance of 101 mm below the die lip (see Fig. 5.102) we see that our results are in qualitative agreement with the findings of Kistler and Scriven (1994) as shown in Fig. 5.96. In particular, the flow rate/width is equivalent to the Reynolds number for a given fluid. Note, however, that this comparison is only qualitative in nature because Kistler computed the initial deflection angle of the curtain at the die lip, while we measured the actual curtain deflection. Also, note that we could not visualize the maximum deflection as was illustrated by Kistler in Fig. 5.96. A flow rate/width of >7.0 cm^2/s for the PSA fluid and the operating conditions under consideration ($A_{dry} = 20\ \text{g/m}^2$, $C_{W,solid} = 0.62$) would require a very high web speed of >1,330 m/min,

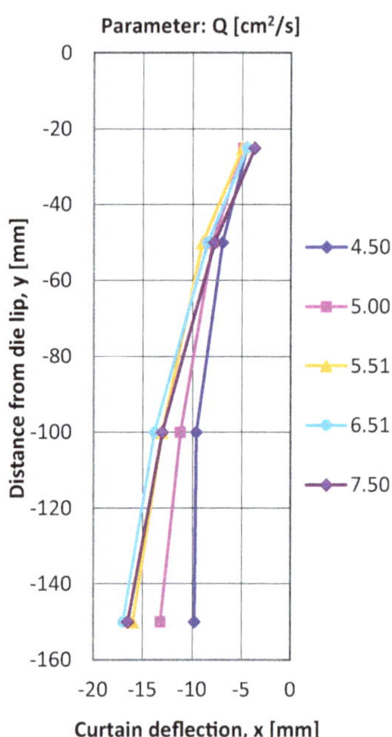

Fig. 5.103 Curtain trajectories of PSA; $\rho = 1020\ \text{kg/m}^3$, $\mu_0 = 520$ mPas, $\sigma = 27.3$ mN/m, $\delta = 90°$, Po $= 0.048$

and to our knowledge, is currently unreached for the production of adhesive labels in the converting industry.

The same qualitative bend-back behavior was observed when the solids concentration of the adhesive was altered to change the low shear viscosity to 520 mPas and 710 mPas, respectively, see Figs. 5.103 and 5.104. For $\mu_0 = 520$ mPas, the curtain deflection was only measured for higher flow rates, as it was negligible for flow rates/width of less than about 3.5 cm^2/s.

Plotting the curtain deflection as a function of the low shear viscosity for $y = 100$ mm and for different flow rates shows a weak decrease of the deflection with increasing viscosity, see Fig. 5.105. This data scatters a bit and so is associated with some uncertainty. The scatter might be because the surface tension also slightly changed upon changing the viscosity. These findings are qualitatively similar to the theoretical calculations by Kistler and Scriven (1994).

Using the special die depicted in Fig. 5.97, we ran many trials with the fluid having a viscosity of 520 mPas. In particular, we varied the inclination angle δ of the die lip in eight steps of 15° from 45° to 150°, all in an attempt to find the optimum geometry that would minimize the curtain deflection for a given set of operating conditions. As the curtain deflection was only significant for higher flow rates, we worked in the limited flow rate range of 4.5–7.5 cm^2/s. Exemplary data are shown in Figs. 5.106 and 5.107 for inclination angles of $\delta = 60°$ and 120°, respectively.

A significant difference in flow behavior is apparent for these two different geometries. For $\delta = 120°$, the behavior was qualitatively similar as seen before, i.e., the curtain deflection increased with increasing flow rate and increasing distance from

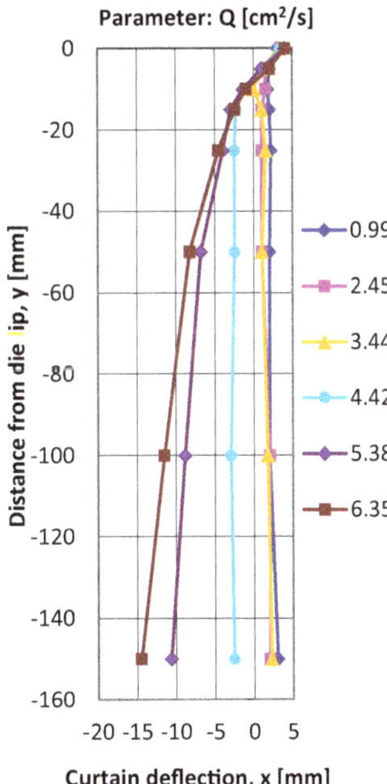

Fig. 5.104 Curtain trajectories of PSA; $\rho = 1022\ \text{kg/m}^3$, $\mu_0 = 710$ mPas, $\sigma = 28.7$ mN/m, $\delta = 90°$, Po $= 0.034$

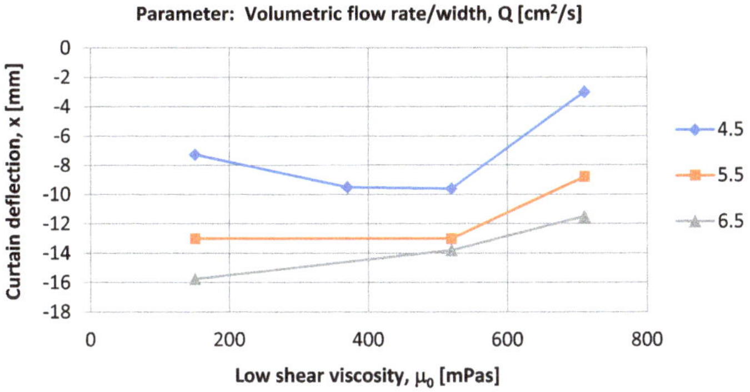

Fig. 5.105 Curtain deflection of PSA as a function of the low shear viscosity and the flow rate/width at a distance of 100 mm from the die lip

the die lip. In contrast for $\delta = 60°$, however, the curtain deflection increased with increasing distance from the die but decreased with increasing flow rate. Note that the deflection could not be measured for $\delta = 60°$ and $Q = 4.5\ \text{cm}^2/\text{s}$ because the deflection was so big that the curtain touched the die body.

The effect of the die lip angle on the curtain deflection is summarized in Fig. 5.108. Here the absolute values of the curtain deflection, i.e., positive numbers, are displayed at a distance of 150 mm below the die lip. Small curtain deflections result for:

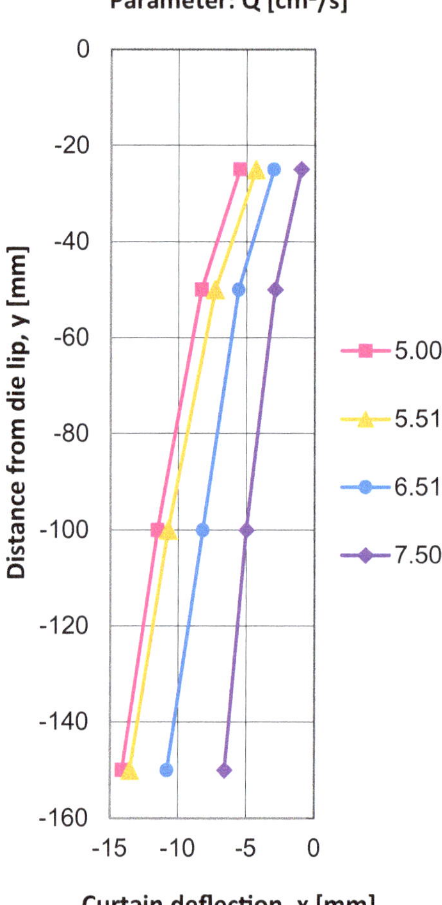

Fig. 5.106 Curtain trajectories of PSA; $\rho = 1020\ \text{kg/m}^3$, $\mu_0 = 520$ mPas, $\sigma = 27.3$ mN/m, $P_o = 0.048$, $\delta = 60°$

- Very high flow rates (e.g., $Q = 7.5\ \text{cm}^2/\text{s}$) and very small lip inclination angles (e.g., $\delta = 45°$).
- Lower flow rates (e.g., $Q = 4.5\ \text{cm}^2/\text{s}$) and very high lip inclination angles (e.g., $\delta = 150°$).
- Small curtain deflections cannot be achieved for any value of the lip inclination angle if the flow rate assumes medium values (e.g., $Q = 5.5\ \text{cm}^2/\text{s}$).
- Note that the terms "high flow rate" and "low flow rate" are relative, as a significant curtain deflection only occurs for flow rates of $Q > 4\ \text{cm}^2/\text{s}$.

These findings are disappointing because there is no single optimum value for the lip inclination angle that results in a small curtain deflection. Instead, the optimum inclination angle, which leads to a small deflection, depends on the flow rate/width. Consequently, process optimization for curtain deflection requires a die with an adjustable lip inclination angle, for example like the die shown in Fig. 5.97. This,

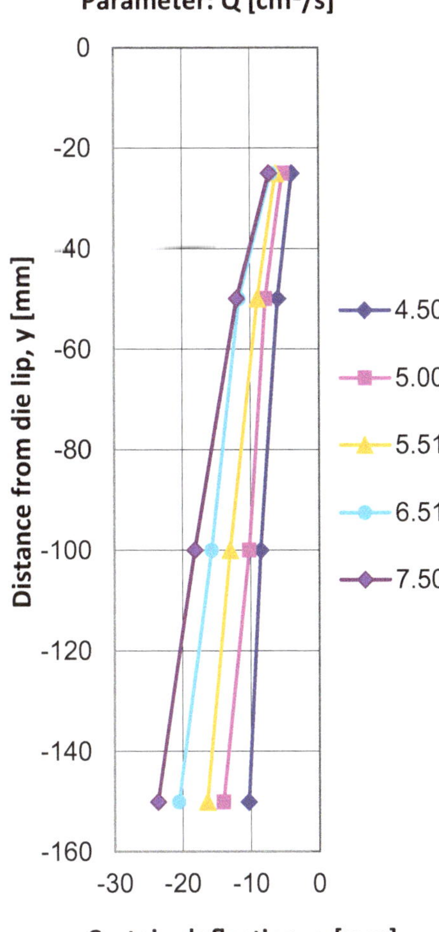

Fig. 5.107 Curtain trajectories of PSA; $\rho = 1020\ \mathrm{kg/m^3}$, $\mu_0 = 520$ mPas, $\sigma = 27.3$ mN/m, Po = 0.048, $\delta = 120°$

however, is not at all an attractive proposition for a production environment because varying the lip inclination angle with the mechanism shown in Fig. 5.97 changes the position of the die lip, which in turn necessitates adjusting the position of the curtain edge guides, and that then necessitates repositioning the entire die body, which is cumbersome. This situation is particularly awkward if, as is the case for the fast-running coating machines, the flow rate changes from low to high values during the acceleration and deceleration of coating starts and stops. Therefore, instead of attempting to suppress excessive curtain bend back with the help of adjustable geometrical parameters, it would be easier to work with surface-type curtain edge guides, which can absorb the curtain deflection without excessively deforming the curtain near the edges. In this scenario, the horizontal die position may have to be adjusted during acceleration and deceleration phases to compensate for the curtain

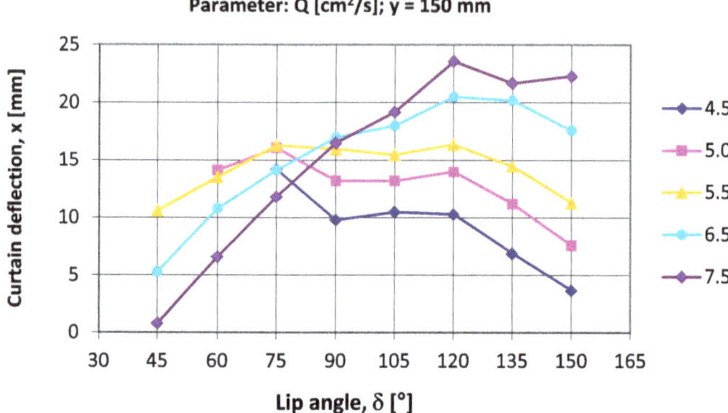

Fig. 5.108 Curtain deflection of PSA as a function of the die lip inclination angle δ and the flow rate/width Q; y = 150 mm, $\mu_0 = 520$ mPas, Po = 0.048

deflection and to keep the curtain impingement point at its optimum position on the coating roll. More information about the design of curtain edge guides is provided in Sect. 5.9.5.

5.8.6.2 Deflection of 1-Layer Curtain with PVA

Additional experiments were carried out with polyvinyl alcohol (PVA, Mowiol 15–79). The shear-dependent viscosity of several aqueous solutions was measured with a Bohlin rheometer. We learned that the critical shear rate was >1,000 s^{-1}, which was the case for solids concentrations of up to about 13%. Moreover, the shear rate of such PVA fluids on a vertical wall (lip of slide die) was <1,000 s^{-1} for flow rates of up to 7.5 cm^2/s and solids concentrations as low as about 4.5%. Therefore, these film flows can be described by the Newtonian low shear viscosity. Since we noticed that the low shear viscosity changed when adding a surfactant, which was necessary for running curtain coating trials, we measured the viscosity of the various PVA solutions with either a 4 mm DIN cup, or a simple Brookfield device.

Typical curtain trajectories and the associated dependence of the curtain deflection upon the volumetric flow rate/width are shown in Figs. 5.109 and 5.110. The curtain width was 149 mm, and the slide angle at the die lip was 90°. Three major differences are apparent relative to the behavior of the aqueous PSA fluids presented above.

- Firstly, curtains made of a fairly viscous and low surface tension PVA solution fall vertically beyond a transition zone just downstream of the die lip, in which the curtain is deflected as a function of the flow rate.
- Secondly, the amount of curtain deflection is much smaller for this PVA fluid than it is for the tested PSA fluids. While the deflection from a vertical die lip was as high as 17 mm for PSA fluids, it was less than 2 mm for this PVA fluid.

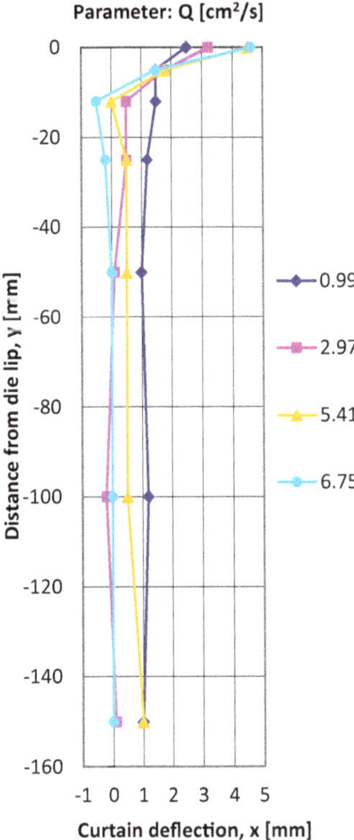

Fig. 5.109 Curtain trajectories of PVA; $\rho = 1018\ \text{kg/m}^3$, $\mu_0 = 1150$ mPas, $\sigma = 20.4$ mN/m, $\beta = 90°$, Po = 0.0125

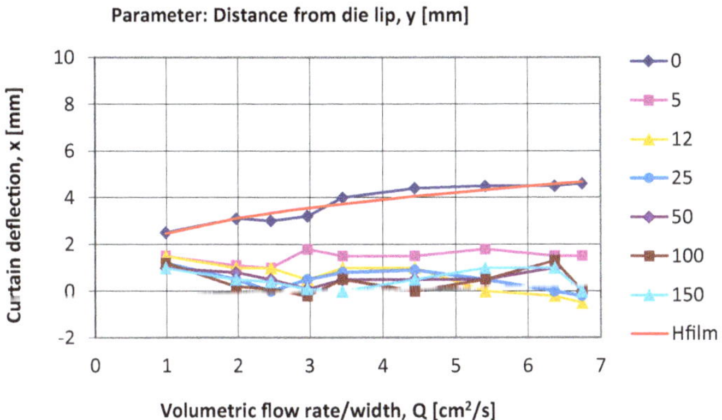

Fig. 5.110 Curtain deflection of PVA as a function of flow rate/width at different distances from the die lip; $\rho = 1018\ \text{kg/m}^3$, $\mu_0 = 1150$ mPas, $\sigma = 20.4$ mN/m, $\beta = 90°$, Po = 0.0125

- Thirdly, the curtain deflection depends much less on the flow rate for this PVA fluid than it does for PSA fluids. As explained above for the PSA fluids, we intuitively expected a vertical trajectory beyond an initial transition zone. However, we did not expect the observed lack of dependence upon the flow rate. As with the form of the trajectory, we could not explain this particular and unexpected behavior. One difference between the two types of fluid was the surface tension, which was much lower for this PVA fluid (20.4 mN/m), than for the PSA fluid (in the range of 27–33 mN/m).

Thus, we decided to evaluate the effect of the surface tension on the curtain deflection by adding different amounts of the surfactant Surfynol PSA 336 to the PVA. Furthermore, we compensated the resulting increase in viscosity by adding water until the viscosity came back to the initial value. All experiments were carried out on a 260 mm wide slide die with a vertical surface of the die lip. The physical properties of the three fluids tested are listed in the caption of Fig. 5.111, which shows curtain trajectories as a function of the flow rate/width.

Now, the results are different from those presented above for the high viscosity, low surface tension PVA fluid, and they are more similar to those of the PSA fluids

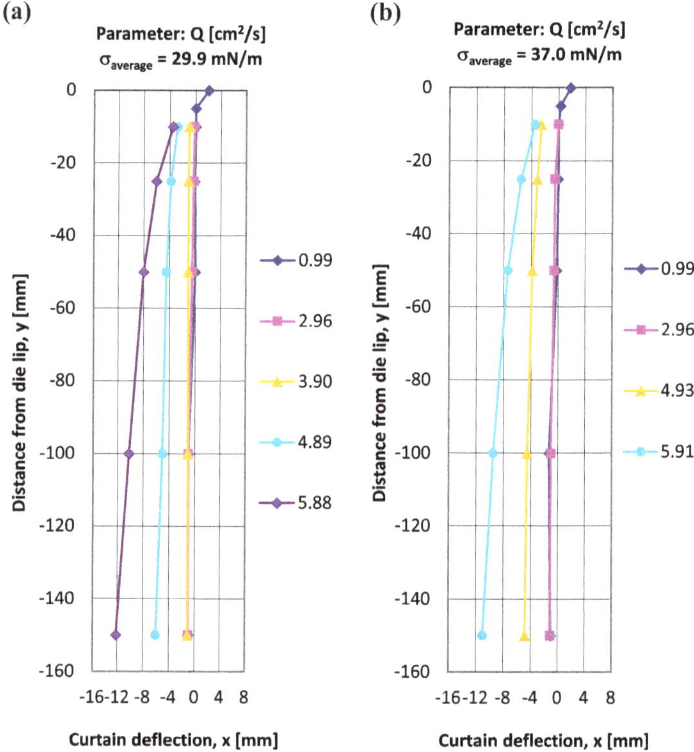

Fig. 5.111 **a** Curtain trajectories of PVA; $\rho = 1011$ kg/m^3, $\mu_0 = 310$ mPas, $\sigma = 29.9$ mN/m, Po $= 0.105$, $\beta = 90°$ **b** curtain trajectories of PVA; $\rho = 1013$ kg/m^3, $\mu_0 = 310$ mPas, $\sigma = 37.0$ mN/m, Po $= 0.130$, $\beta = 90°$ **c** curtain trajectories of PVA; $\rho = 1015$ kg/m^3, $\mu_0 = 310$ mPas, $\sigma = 47.0$ mN/m, Po $= 0.166$, $\beta = 90°$

Fig. 5.111 (continued)

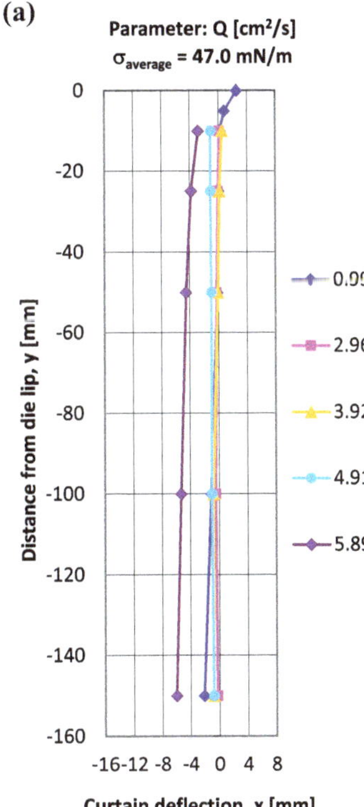

presented above. In particular, not all trajectories remain vertical, but they stay inclined down to distances of >150 mm from the die lip if the flow rate/width is sufficiently large.

Moreover, the curtain deflection depends on the flow rate/width, if the flow rate is sufficiently large. Another view of the same data is shown in Fig. 5.112. The data suggest that

- curtain deflection is negligible for low enough flow rates;
- curtain deflection becomes significant for flow rates above a critical value;
- the critical flow rate value above which curtain deflection is significant depends on the surface tension;
- curtain deflection increases with decreasing surface tension for a given flow rate and a given viscosity.

Comparing these results with the data shown in Figs. 5.109 and 5.110, i.e., very small curtain deflection for $\mu_0 = 1{,}150$ mPas and $\sigma = 20.4$ mN/m, and considering the data shown in Fig. 5.105, i.e., decreasing curtain deflection with increasing viscosity, we might conclude that a viscosity of >1,150 mPas would suppress a significant curtain deflection for a very wide range of flow rates/width, i.e., for $Q < 7$ cm^2/s. This result would also be expected for rather low surface tensions, i.e., 20.4 mN/m,

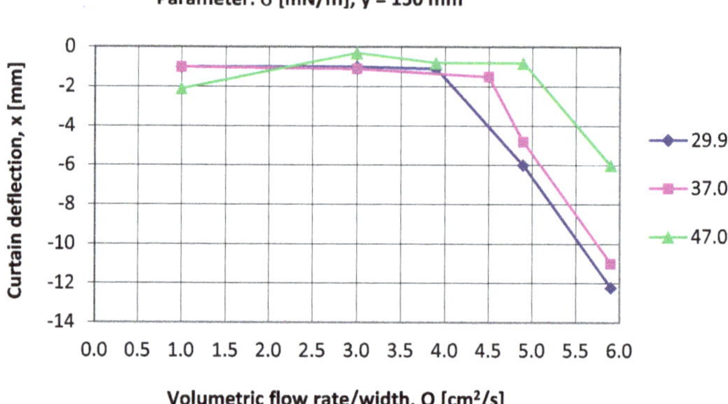

Fig. 5.112 Curtain deflection of PVA as a function of the surface tension and the flow rate/width at a constant distance from the die lip of 150 mm; $\mu_0 = 310$ mPas, $\beta = 90°$

even though we showed for lower viscosities that curtain deflection increases with decreasing surface tension. However, the conclusion proposed above should not be drawn yet because, as already stated above, the available data leave some uncertainty about the effect of viscosity and surface tension on the curtain deflection. Additional experimental data would help to reduce this uncertainty.

Note that for the trials carried out with the 260 mm wide stainless steel die from TSE Troller AG, we used solid body plate-type curtain edge guides without edge fluid that were inclined by 7.6° relative to vertical, see Fig. 4.22. When we repeated a couple of trial points with vertical edge guides, i.e., $\beta = 90°$, $\rho = 1{,}011$ kg/m^3, $\mu_0 = 310$ mPas, $\sigma = 29.9$ mN/m, Q = 4.88 and 5.87 cm^2/s, we obtained exactly the same results with regard to curtain deflection.

5.8.6.3 Deflection of 2-Layer Curtain with PVA

We were interested in evaluating the effect of the relative surface tensions of a 2-layer curtain on the curtain deflection. To that extent, we ran two sets of trials with PVA fluids, and a 260 mm wide stainless steel slide die having a vertical die lip ($\beta = 90°$). In one set, the bottom layer had a high surface tension (PVA without surfactant) and the top layer had a low surface tension (PVA with surfactant Surfynol PSA 336). In the second set, we inverted these two layers. Both layers had a low-shear viscosity of 310 mPas, and the total flow rate/width of both layers was kept constant at 5.5 cm^2/s. The curtain deflection was measured 150 mm below the die lip as a function of the relative flow rates/width of the two layers. The physical fluid properties are given in Table 5.15, and the results of curtain deflection are shown in Fig. 5.113.

The curtain deflection of a 2-layer curtain is small if the surface tension of the top layer is higher than the one of the bottom layer ($\sigma_{bottom}/\sigma_{top} < 1$), and that is so for any value of the ratio of the two flow rates. On the other hand, large curtain deflections

Table 5.15 Physical fluid properties

	σ [mN/m]/ρ [kg/m^3] for trial set 1	σ [mN/m]/ρ [kg/m^3] for trial set 2
Top layer	29.9/1011	48.8/1015
Bottom layer	49.8/1015	28.4/1011
$\sigma_{bottom}/\sigma_{top}$ [–]	1.67	0.58

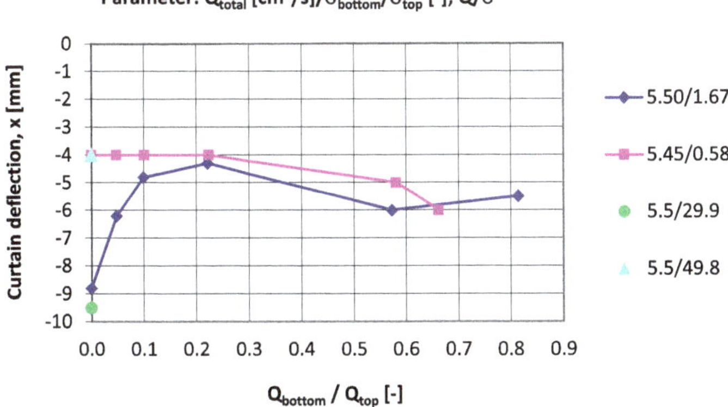

Fig. 5.113 Deflection of a 2-layer PVA curtain as a function of the surface tensions and the flow rates/width of the two layers at a distance of 150 mm from the die lip; $\mu_0 = 310$ mPas, $\beta = 90°$

are present, if the surface tension of the top layer is lower than the one of the bottom layer ($\sigma_{bottom}/\sigma_{top} > 1$), and if the ratio of the bottom layer flow rate to top layer flow rate is small, i.e., <0.2. However, if the ratio of Q_{bottom}/Q_{top} is above about 0.2, the curtain deflection is no longer affected by the relative surface tension values of the two layers, but it is dominated by the surface tension of the top layer alone. Note that the blue triangle and the green circle in Fig. 5.113 are duplicate data points (control measurements) of single-layer curtains having the respective surface tension.

In summary, a favorable bend-back behavior of a 2-layer curtain, i.e., a small curtain deflection, is obtained if the surface tension of the top layer is higher than the one of the bottom layer, and if the flow rate ratio $Q_{bottom}/Q_{top} < 0.2$. Unfortunately, these findings conflict with the requirement that the surface tension of an upper layer should be smaller than the surface tension of a lower layer in a multilayer film or curtain. Satisfying this requirement guarantees proper wetting of adjacent layers and prevents edge withdrawal of the upper layer on the lower layer during film flow on the die slide and during flow after coating on the substrate. Therefore, using the surface tension of the top layer to reduce curtain deflection is usually not an appropriate means in industrial curtain coating applications.

5.8.6.4 Curtain Deflection with Electrostatic Assist

We investigated in principle whether the curtain trajectory can be influenced with the help of an electrostatic field force, i.e., with a force that acts on the liquid curtain like the gravity force but in a different direction, and that can be adjusted in its magnitude. To that extent, we used a 784 mm wide slide die from TSE Troller AG, and we installed an electrostatic bar from Eltex in front of the curtain just a few centimeters below the die lip, see Fig. 5.114.

The bar contained short needles that were spaced some 6 mm apart in the cross-web direction. The electrical circuit was closed by connecting the bar and the curtain (i.e., the die body) to a high voltage power supply. Upon applying an electrical tension of −16 kV resulting in a current of 0.19 mA, the curtain, which initially fell vertically (PVA, Q = 1.58 cm^2/s), was deflected backward as sketched in Fig. 5.114.

This electrostatic bar was normally used for pushing the substrate against the coating roll to reduce the air cushion between the substrate and the roll. As we could not change the polarity of the power supply, we installed another bar with another power supply, which was normally used for removing the electrostatic charge from the substrate surface. The bar was connected to the positive pole of the power supply, and the direction of the electrostatic force was reversed. Thus, we were able to pull the curtain toward the bar. Changing the voltage changed the pull force and thus the curtain trajectory. When we applied −24 kV sparks appeared between the bar and the metallic curtain edge guides, which shortened the electrical circuit.

In summary, we showed that electrostatic field forces are suitable for altering the trajectory of a liquid curtain. A curtain showing excessive bend back can be pulled back into the initial vertical position, although not in a perfect way. Alternatively, an initially vertical curtain can be pushed back into a deflected position. By continuously changing the voltage supply to the electrostatic bar, the changing curtain deflection during the acceleration or deceleration phases of the coating machine could in principle be compensated.

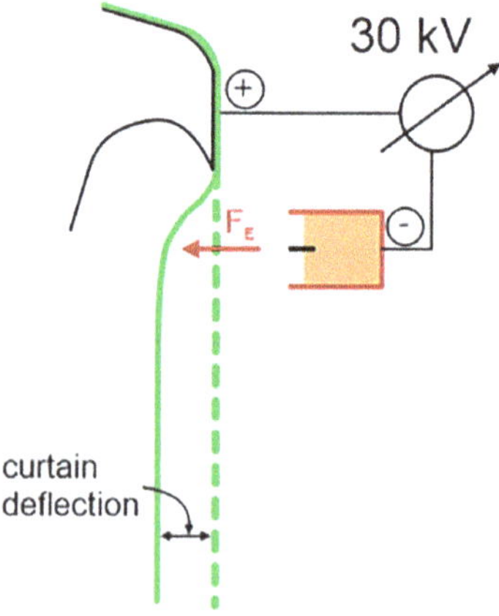

Fig. 5.114 Experimental set-up for studying the effect of an electrostatic field force on the curtain deflection

We believe, however, that the use of electrostatic field forces is not a suitable measure for a production environment because on the one hand, the curtain deflection cannot be prevented, and applying a field force to a deflected curtain will cause undesirable 3-dimensional deformations in the curtain, and on the other hand, short circuits of the high voltage electrical circuit cannot be prevented at all times owing to splashes occurring during the start and stop phases of the curtain coating process.

5.8.6.5 Conclusions on Curtain Deflection

Many experiments were carried out with aqueous PSA and aqueous PVA fluids to characterize and quantify the deflection of liquid curtains that are produced by a slide die. The main results from this study are summarized as follows.

Curtain deflection depends on the volumetric flow rate/width Q, the distance y from the die lip, the inclination angle β of the die lip, the low shear viscosity μ_0, and the surface tension σ.

Curtain deflection can be very significant; values as high as 23 mm have been recorded for PSA with Q = 7.5 cm^2/s and y = 150 mm. Such high flow rate/width values are realistic during the high-speed manufacturing of adhesive labels.

A first critical flow rate/width $Q_{c,deflection}$ exists, that marks the onset of curtain deflection. No curtain deflection is observed below that critical flow rate value, i.e., the curtain drops vertically. For a vertical die lip (β = 90°), the value of $Q_{c,deflection}$ is on the order of 3.5–4.0 cm^2/s for PSA fluids, and 4.0–5.0 cm^2/s for PVA fluids. Moreover, it slightly depends on the low shear viscosity and the surface tension. For $Q > Q_{c,deflection}$, the curtain deflection increases with increasing flow rate, increasing distance from the die lip, decreasing viscosity, and decreasing surface tension. Very small curtain deflections of <2 mm have been observed for any flow rate values (1.0 < Q < 6.75 cm^2/s) for a PVA fluid having a relatively high viscosity (μ_0 = 1,150 mPas) and a relatively low surface tension (σ = 20.4 mN/m).

A second critical flow rate/width $Q_{c,trajectory}$ exists, that describes the shape of the curtain trajectory. For $Q < Q_{c,trajectory}$, the curtain falls vertically from the die lip. However, for $Q > Q_{c,trajectory}$, the curtain trajectory is bent backward, and it remains curved for at least y < 150 mm. For a vertical die lip (β = 90°), the value of $Q_{c,trajectory}$ is on the order of 4.0–4.5 cm^2/s for PSA fluids, and 5.0–5.5 cm^2/s for PVA fluids. Moreover, it slightly depends on the low shear viscosity and the surface tension.

The curtain deflection strongly depends on the inclination angle β of the die lip. Unfortunately, no single value of β exists, for which the curtain deflection becomes small for any flow rate, viscosity and surface tension values. Instead, small curtain deflections result for small inclination angles (β = 45°), if the flow rate/width is relatively large, (Q = 7.5 cm^2/s), or for large inclination angles (β = 150°), if the flow rate/width is relatively small, (Q = 4.5 cm^2/s). Moreover, no small deflection can be achieved for any values of β, if the flow rate/width assumes intermediates values, i.e., Q = 5.5 cm^2/s. This result is disappointing because it does not allow us to suppress excessive curtain deflection by smart equipment design, i.e., by optimizing the inclination angle of the die lip. After all, a die lip with a variable geometry as shown in Fig. 5.97 would be too cumbersome to use in a production environment. Instead, we recommend accepting the curtain deflection, whatever it is, and avoiding excessive curtain deformation in the cross-web direction by using properly designed plate-tape curtain edge guides that can absorb the deflection without deforming the edge zones of the curtain.

The results summarized above apply to single-layer curtains. We also studied the bend back behavior of a 2-layer curtain, and we found that small curtain deflections can be achieved if the surface tension of the top layer is relatively high and is higher than the surface tension of the bottom layer. This is particularly true if the ratio of the bottom layer to top layer flow rate is <0.2. If this flow rate ratio is >0.2, then small deflections can be achieved independently of the surface tension structuring, as long as at least one of the layers has a relatively high surface tension of, say, >45 mN/m.

We also showed that it is possible in principle to alter the curtain trajectory by applying an electrostatic field force to the curtain. This way, a curtain that wants to deflect could be pulled back to its original vertical position. For various reasons, however, we believe that this measure is not suitable for an industrial production environment.

Curtain deflection is a fluid flow phenomenon associated with high flow rates/width in conjunction with slide dies. In curtain coating, high flow rates exist for high-speed applications and/or applications with a high total wet film thickness, i.e., simultaneous multilayer applications. Typical applications, for which curtain deflection must be considered, include:

- Adhesive coatings for label stock manufacturing, i.e., $A_{dry} = 20\ g/m^2$, $C_{w,solid} = 0.6$, and $U > 625$ m/min.
- 1-layer thermal paper coatings at web speeds of $U > 1{,}200$ m/min.
- 2-layer thermal paper coatings at web speeds of $U > 1{,}000$ m/min.
- 3-layer thermal paper coatings at web speeds of $U > 900$ m/min.

5.8.7 Vortices in Curtain Flow

The only location in curtain flow, where a vortex is likely to be present, is just below the die lip, where the film flow changes into curtain flow and detaches from the lip, see schematic diagram in Fig. 5.115.

Vortex formation at the lip of a slide die has been studied extensively by Kistler (1983) and Kistler and Scriven (1994). The formation of this vortex is promoted by the phenomenon of curtain deflection, also known as the teapot effect, see Sect. 5.8.6. Prevention of this vortex is difficult, as the film flow detaching from the die lip tends to push the curtain to the left relative to the position of the vertical die slide as sketched in Fig. 5.115a. Reducing the corner angle of the lip would be helpful but difficult from a manufacturing point of view. Coating the horizontal lip face with a non-wetting material such as Teflon would be helpful too. However, this measure is not durable in practice owing to the loss of non-wetting properties after some time of operation. In contrast, minimizing the size of this vortex by minimizing the length of the lip face L_L has proven doable and acceptable in practice. The minimum lip length depends on the width of the die and the manufacturing capabilities of the die supplier. Lip length values of ≤0.25 mm should be possible and are preferred. In any case, long lip lengths must be avoided to prevent unstable flow situations along the lip face as documented by Kistler and Scriven (1994).

Vortex formation at the lip of slot dies is promoted if the height w of the die slot is too narrow, and if the fluid is prone to the phenomenon of die swell as a result of excessive viscoelastic properties (Macosko, 1994). Again, avoiding this pair of vortices is difficult, but minimizing negative effects from these vortices can be achieved by minimizing the length L_L of the lip faces, and by maximizing the height w of the die slot. The latter is best realized with a slot die equipped with a dual cavity internal design, see Sect. 9.4.2. This design concept allows the inner slot to be as narrow as necessary to achieve an acceptable cross profile and the outer slot to be as wide as possible to minimize the sensitivity to the mechanical slot precision and to design a slot exit flow field with desirable properties such as no die swell.

5.8.8 Neck-in in Curtain Flow

Neck-in in curtain coating is an interfacial phenomenon, which may occur in two different locations and on two different length scales. On the one hand, and as is shown in Fig. 5.116a, a freely falling liquid curtain that is not guided along its sides, will laterally contract or neck-in as a result of surface tension forces (Weinstein and Palmer, 1997). This is an undesirable feature because the width of the curtain decreases with increasing distance from the beginning of the curtain. If projected onto an inboard curtain coating process, the resulting width of the coated film on the substrate would be unknown, and it would depend, among others, on physical fluid properties, i.e., the surface tension, and on the length of the curtain, which again is

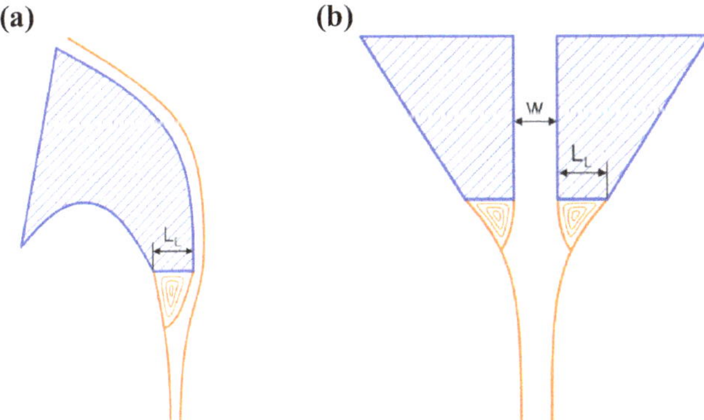

Fig. 5.115 **a** Vortex in the curtain at the lip of the slide die **b** vortices in the curtain at the lip of the slot die

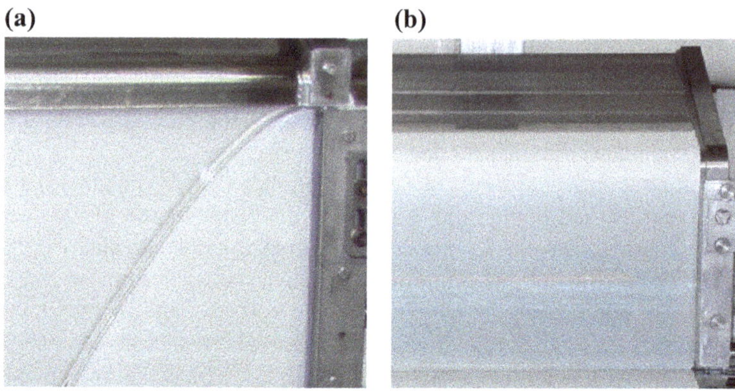

Fig. 5.116 **a** Lateral contraction of an unguided curtain; photo reproduced with permission from Polytype Converting AG **b** guided curtain with constant width along its trajectory; photo reproduced with permission from Polytype Converting AG

undesirable. The situation shown in Fig. 5.116a is referred to as large-scale neck-in. The characteristic length scale is on the order of dm (100 mm).

To overcome these problems for the inboard curtain coating mode, the curtain is usually guided along its trajectory by so-called curtain edge guides, see Fig. 5.116b. Consequently, the nominal width of the curtain is defined by the distance that separates the edge guides.

On the other hand, the bottom of the edge guide is not allowed to touch the substrate surface. Instead, a distance must be guaranteed, which is large enough to let the splice and the loose tail of a flying splice pass without any hindrance whatsoever. According to Polytype Converting's experience, this distance typically is in the range of 0.5–3 mm, i.e., the characteristic length scale is on the order of 1 mm. Consequently, the curtain is not guided over the short vertical distance and neck-in may occur as described above, but this time on a much smaller scale than described above, see Fig. 5.117a. Note that neck-in may occur here irrespective of the geometrical details at the bottom of the edge guide, compare Fig. 5.117a, b.

If significant neck-in occurs, it generates heavily coated edges on the substrate (see Fig. 5.117), which in turn causes highly undesirable winding problems such as barrel rings and telescoping. In the example of Fig. 5.117a, the neck-in causes a reduction of the coating width by about 2 mm per side. In Fig. 5.117b, the neck-in is more severe, thus causing a reduction of the coating width of at least 5 mm. Worse yet, we observed very severe neck-in, which caused a width reduction of some 12 mm per side. It is therefore important to avoid or at least minimize neck-in at the bottom of the curtain edge guides to avoid the heavily coated edges and subsequent winding problems as described above. In the following section, the physical reasons for neck-in are described and modeled theoretically, particularly the neck-in at the bottom of the edge guide. Knowledge gained from these considerations can be used for designing suitable curtain edge guides. This topic is further discussed in more detail in Sect. 5.9.5.

5.8.8.1 Curtain Contraction due to Surface Tension Forces

Neck-in is caused by at least two factors. The most obvious one is the absence of a curtain edge guide, see Fig. 5.116a. The amount of lateral curtain contraction, or the loss of coating width ΔW, depends on the height of the curtain L_c and the initial neck-in angle α at the origin of the neck-in, see Fig. 5.118.

To learn which parameters affect the neck-in angle, and to calculate the value of the neck-in angle as a function of those parameters, we have to resort to the so-called

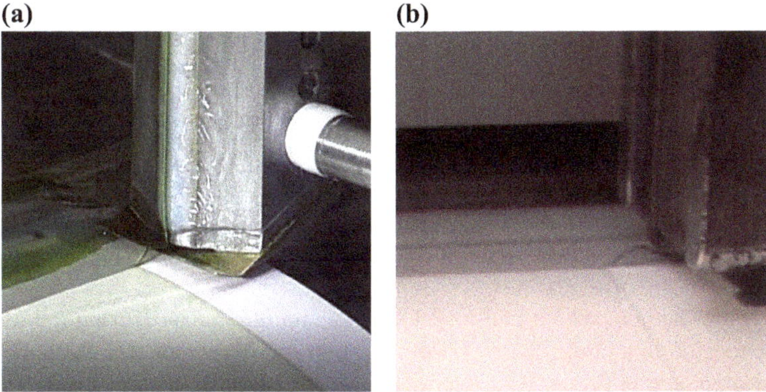

Fig. 5.117 **a** Neck-in at the bottom of a curtain edge guide resulting in a heavily coated edge on the substrate; photo reproduced with permission from Polytype Converting AG **b** neck-in at the bottom of another curtain edge guide resulting in a heavily coated edge on the substrate; photo reproduced with permission from Polytype Converting AG

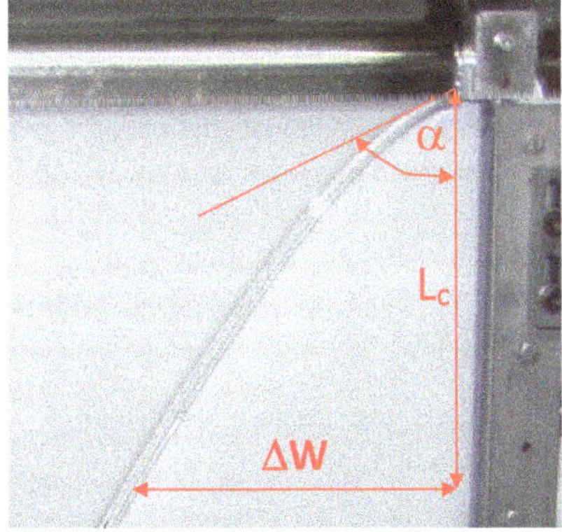

Fig. 5.118 Loss of coating width due to curtain neck-in; photo reproduced with permission from Polytype Converting AG

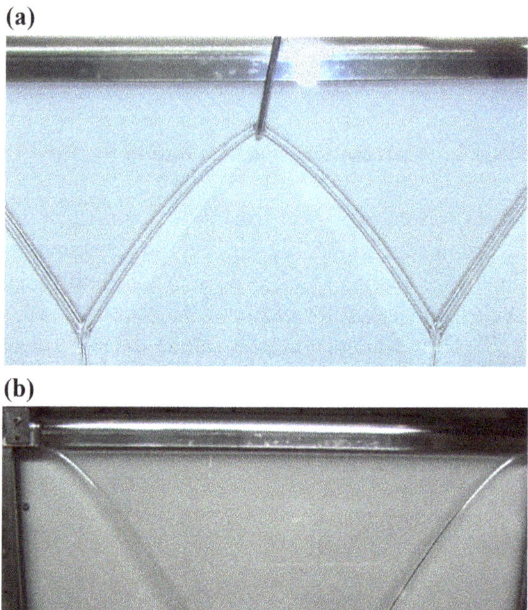

Fig. 5.119 **a** Rim of a broken curtain; photo reproduced with permission from Polytype Converting AG **b** neck-in of an unbound curtain; photo reproduced with permission from Polytype Converting AG

Mach angle method, which is a well-known technique for measuring the dynamic surface tension in a liquid sheet or liquid curtain and is presented in Sect. 4.5.3.

As is depicted in Fig. 4.21, a standing wave called the Mach wave is generated if the surface of the curtain is lightly disturbed by a sharp point source, such as a needle. If the disturbance is weak enough, the curtain will not break, and the shape of the wave front will remain fairly straight. However, if the disturbance is too strong, the curtain will break, see Fig. 5.119a, and the wave front will be bent downward owing to the rim, which gets heavier and heavier as the distance from the point disturbance increases. The shape of the rim of a broken curtain is very similar to the shape of the neck-in of an unguided curtain, compare Fig. 5.119a, b.

According to the theory of Brown (1961), the local dynamic surface tension in the curtain at the point of disturbance can be calculated with Eq. (4.5.1). Rearranging this equation provides an expression for the initial neck-in angle α as follows:

$$\sin \alpha = \sqrt{\frac{2\sigma}{\rho Q V_c}} \tag{5.8.17}$$

Hence, the curtain contraction, or the loss of coating width ΔW can be approximated according to:

$$\Delta W = L_c \tan \alpha \tag{5.8.18}$$

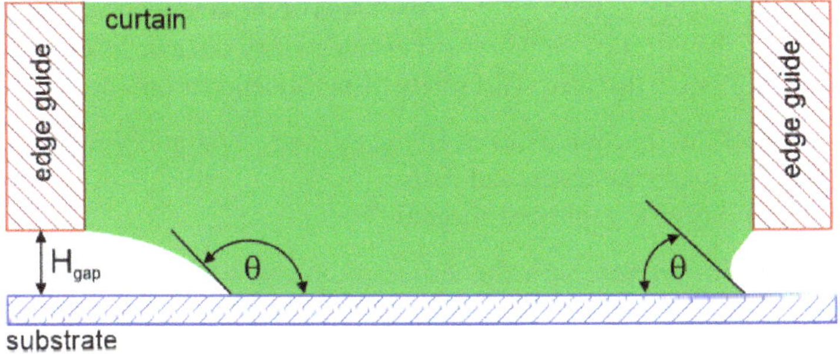

Fig. 5.120 Wetting and spreading between curtain fluid and substrate surface

Obviously, Eq. (5.8.18) is only correct if the wave front, i.e., the form of the rim, is straight. Nevertheless, the qualitative relationship between the 3 parameters in Eq. (5.8.18) is always correct, i.e., the loss of coating width decreases with decreasing curtain height and decreasing Mach angle.

5.8.8.2 Insufficient Wetting Between Substrate and Curtain Liquid

Another reason contributing to the neck-in is improper wetting of the substrate surface by the curtain fluid. Projecting the arguments presented in Sect. 4.5.2 regarding the spreading of a liquid onto the situation of neck-in at the bottom of an impinging curtain, we can say that neck-in is promoted by large contact angles between the curtain liquid and the substrate surface, see left side in Fig. 5.120, and it is not promoted by small contact angles, see right side in Fig. 5.120.

5.8.8.3 Neck-in at the Bottom of the Edge Guide

Regarding neck-in caused by curtain contraction, we have to examine and modify Eq. (5.8.18). In particular, the curtain height L_c must be replaced by H_{gap}, which corresponds to the distance between the bottom of the edge guide and the substrate surface, see Fig. 5.120. Now the lateral extent of neck-in can be approximated by

$$\Delta W = H_{gap} \tan \alpha \tag{5.8.19}$$

Ideally, H_{gap} should be zero, but this cannot be realized for solid edge guides because any splice must be able to pass underneath the edge guide without touching it whatsoever to prevent web breaks. Realistically, therefore, H_{gap} must be minimized subject to considering the splice thickness. For overlap splices, which are worse in terms of thickness than butt splices, but which are more often used in the converting industry, the geometrical situation is presented in Fig. 5.121.

The curtain height is composed of the thickness of the adhesive layer between the two substrates, the thickness of the new substrate, which often is the same as the old substrate, and a safety distance. This safety distance depends on

- the quality of the splice in terms of lateral thickness non-uniformity, which may be caused by blisters, creases, and folds;
- the repeatability of the splice application.

These factors in turn depend on the design and operation of the splicing equipment, on the degree of automation for the splice application, and if manual operations are required for applying the splice on the skills of the operator. Determining the right safety distance is also a matter of risk assessment, i.e., the Plant Manager must decide whether to minimize the number of web breaks at the expense of increased neck-in, or the other way around. On Polytype Converting semi-automatic splicing equipment operated by a skilled crew member, H_{gap} can be set as small as 0.5 mm, and values in the range of 0.5–1.0 mm can be operated with very low risk. If H_{gap} needs to be set to smaller values, say in the range of 0.2–0.5 mm, then a temporary skip-up of the edge guide or the edge guide/die assembly should be considered during the splice passage.

Further referring to Eq. (5.8.17), neck-in can be minimized by minimizing the Mach angle α. This is accomplished by reducing the surface tension and by increasing the momentum of the impinging curtain, which is expressed by the term ρQV_c in Eq. (5.8.17). The curtain momentum is a measure of the force in the flow direction, that the curtain can develop to press onto the rim of the Mach wave, and hence to reduce the Mach angle α. Neglecting the initial velocity, the curtain velocity is proportional to the square root of the curtain length, see Eq. (5.8.1). Moreover, the volumetric flow rate/width Q can be expressed in terms of the dry coat weight, the solids concentration, the density, and the web speed, see Eq. (3.3.4). Consequently, Eq. (5.8.17) can be rearranged to give:

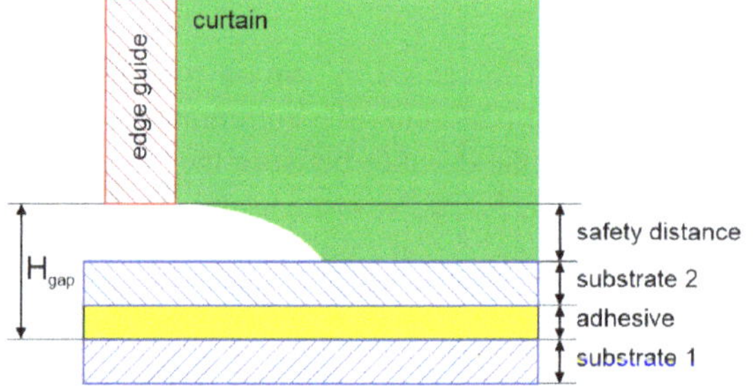

Fig. 5.121 Neck-in at bottom of curtain edge guide; curtain height in relation to the splice thickness

$$\sin\alpha = \sqrt{\frac{2\sigma C_w}{A_{dry} U \sqrt{2gL_c}}} \tag{5.8.20}$$

We can readily see that the neck-in depends on 5 independent coating parameters and that it can be reduced by decreasing the surface tension and the solids concentration, as well as by increasing the dry coat weight, the web speed, and the curtain height. Unfortunately, the effect of changing any one of these 5 parameters upon the Mach angle is dampened by the square root relationship between the parameters. This damping effect is particularly strong for the relationship between the Mach angle and the curtain height.

The surface tension is the only physical fluid property that affects the Mach angle. Low surface tension values keep the neck-in small and thus are preferable. Regarding curtain stability and the associated minimum flow rate/width, we typically request the surface tension of the curtain fluid to be <40 mN/m. However, if neck-in is a severe problem, then the surface tension should be reduced to values of <35 mN/m, and preferably to values of <30 mN/m, if it is possible at all. Note that the surface tension of aqueous fluids containing surfactants is a dynamic property, which depends on the details of the flow field. Therefore, it ideally should be measured directly in the curtain using the Mach angle method as described in Sect. 4.5.3, and at the bottom of the curtain and for the flow rate of the application.

To obtain a better quantitative sense of how much the Mach angle can be reduced by changing one of the 5 process parameters, a base case (n = 1) of typical curtain coating operating conditions was prepared and compared to various scenarios (n = 2–6). One of the parameters at a time was changed by a small but realistic amount such that the Mach angle was reduced, see Table 5.16. Unfortunately, all these angle reductions are modest in quantitative terms, i.e., <−35%. In one scenario (n = 7) all the parameters were changed at the same time by the same amount as was previously done, but now the effect on the Mach angle was significant. However, the effect on other process parameters such as the drying load $M^*_{solvent}$ was also significant and consequently highly undesirable because the solvent quantity to be evaporated increased by a factor of >5.

Table 5.16 Dependence of the Mach angle upon relevant process parameters

n	ρ	σ	C_W	A_{dry}	U	Q	L_c	α	$\Delta\alpha$	$M^*_{solvent}$	$\Delta M^*_{solvent}$
[–]	[kg/m^3]	[mN/m]	[–]	[g/m^2]	[m/min]	[cm^2/s]	[mm]	[°]	%	[kg/h]	%
1	1000	45	0.2	3	400	1.00	100	53.3		288	
2	1000	35	0.2	3	400	1.00	100	45.0	−15.6	288	0
3	1000	45	0.1	3	400	2.00	100	34.5	−35.2	648	125
4	1000	45	0.2	5	400	1.67	100	38.4	−28.0	480	67
5	1000	45	0.2	3	600	1.50	100	40.9	−23.3	432	50
6	1000	45	0.2	3	400	1.00	200	42.4	−20.5	288	0
7	1000	35	0.1	5	600	5.00	200	15.4	−71.1	1620	463

In summary, the Mach angle depends on at least 5 independent process parameters, no single one of which has a significant effect on reducing the Mach angle. Changing several of those parameters at the same time significantly reduces the Mach angle, but it also increases the production capacity owing to the increased coating speed, and it increases the drying load owing to the reduced solids concentration, neither of which are desirable.

The measures presented above for reducing the neck-in were of “global” kind. However, the neck-in at the bottom of the curtain edge guide is a local phenomenon, see Fig. 5.117. Therefore, it is interesting to also examine the local flow field of the curtain along the edge guide to see if other means exist for reducing the neck-in.

Shown as an example in Fig. 5.117b, the curtain liquid flowing right next to the wall of the curtain edge guide is sucked into the evacuation slot at the bottom of the edge guide. Therefore, the curtain locally extends down to that suction slot. A few millimeters away from the edge guide, however, the vertical curtain momentum seems too small because the rim of the Mach wave is bent upward before it is bent back downward again a few more millimeters into the curtain, where the vertical momentum seems again larger. The result is a highly undesirable neck-in with a lateral extension, i.e., with a loss of coating width, of close to 10 mm. If these arguments are true, then the question arises which factors cause the vertical curtain momentum to locally decrease and thereby promote neck-in?

Above, the vertical curtain momentum M_c was defined as follows:

$$M_c = \rho Q V_c \tag{5.8.21}$$

To prevent neck-in, the curtain momentum should be high, and it should not vary locally relative to the mean value, i.e., it should not decrease in the vicinity of the edge guide. The volumetric flow rate/width Q does not only depend on the wet film thickness of the coating and the coating speed as shown in Eq. (3.3.4) but, owing to the law of mass conservation, it also depends on the curtain thickness H_c and the curtain speed V_c as shown in Eq. (5.8.22).

$$Q = H_{wet} U = H_c V_c \tag{5.8.22}$$

Therefore, the curtain momentum can also be expressed as

$$M_c = \rho H_c V_c^2 \tag{5.8.23}$$

Calculating the total derivative of Eq. (5.8.23) and normalizing the expression with M_c visualizes which parameters cause the momentum to vary in the cross-web direction x and in which direction, in case these parameters vary with x by themselves.

$$\frac{dM_c(x)}{M_c} = \frac{d\rho(x)}{\rho} + \frac{dH_c(x)}{H_c} + \frac{2dV_c(x)}{V_c} \tag{5.8.24}$$

Assuming that the coating fluid is homogeneous, then the density does not vary, and hence it does not influence the curtain momentum. However, if the curtain thickness decreases locally (negative value for dH_c in Eq. (5.8.24)), and if in particular, the curtain velocity decreases locally (negative value for dV_c, and amplification factor of 2 owing to the square dependence between M_c and V_c in Eq. (5.8.23)), then the curtain momentum will decrease locally, which in turn may promote neck-in.

Possible thickness variations of the curtain near the edge guide are schematically depicted in Fig. 5.92. Moreover, comparing this figure with Fig. 5.117b it does not seem unrealistic to relate the observed neck-in to the presence of a locally very thin curtain right next to the edge guide surface. Some of these factors causing a non-uniform curtain thickness are related to the presence of viscous boundary layers in slot, film, and curtain flows of premetered coating processes. Another factor is related to the spreading of the curtain fluid on the surface of the edge guide. These topics are further discussed in more detail in Sects. 5.9.4 and 5.9.5.

Regarding the wetting of the substrate surface by the curtain liquid, we can say that a contribution to neck-in can be prevented or at least reduced if the surface energy of the substrate is higher than the surface tension of the curtain liquid. To act on this requirement, we first must know the surface energy of the substrate and the surface tension of the liquid. The surface tension of the curtain liquid can be measured at the curtain impingement point by using the Mach angle method, see Sect. 4.5.3. In contrast, the surface energy of the substrate surface cannot be measured directly, but it can easily be determined with a set of test inks as described in Sect. 4.6.3.

The wettability of a low energy substrate can be improved by treating it with a flame or a corona discharge (atmospheric plasma), which is standard practice in the converting industry for PET and other plastic films. In multi-pass coating applications in the photographic industry, the wettability of the top surface of the previously coated layer pack by the bottom layer of the subsequently coated pack could be improved if the top layer fluid of the previous pack contained a non-ionic surfactant. Alternatively, the wetting of a substrate surface by a liquid can also be improved by lowering the surface tension of the liquid to a value that is below the substrate surface energy. As mentioned above, the surface tension of fluids to be curtain coated should be <35 mN/m, if neck-in is a problem.

5.8.8.4 Modeling Neck-in and Heavy Coated Edges

We attempt to model the size of the heavily coated edge, which is caused by the neck-in at the bottom of the curtain edge guide. The geometrical situation is depicted in Fig. 5.122. We assume that the Mach angle α can be calculated with Eq. (5.8.20). Consequently, the lateral extension ΔW of the neck-in depends on α and on the curtain length H_{gap} according to Eq. (5.8.19).

The uniform flow rate/width Q, which is supplied by the curtain, cannot be deposited onto the substrate over the width ΔW owing to the neck-in. Instead, the flow rate quantity of $Q \cdot \Delta W$ is redistributed into the area of the heavy edge just in-board of the static wetting line. Assuming that this extra fluid is sitting on top

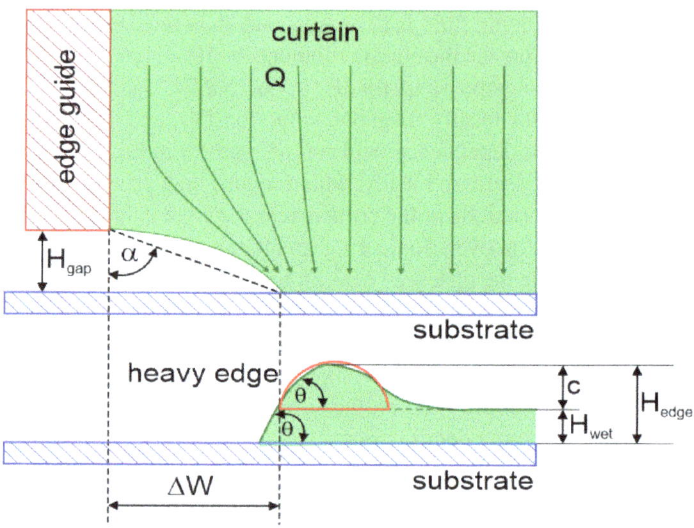

Fig. 5.122 Formation of a heavy coated edge as a result of small-scale neck-in at the bottom of the edge guide

of the already uniformly coated film in the same area, and assuming that the cross-section A of the extra fluid has the shape of a segment of a circle (see red marking in Fig. 5.122), then the flow rate in this circular segment, which is taken away by the moving web equals U · A. Therefore, equating the two flow rates allows us to calculate the size of the circular segment by Eq. (5.8.25).

$$A = \frac{Q\Delta W}{U} \tag{5.8.25}$$

Further assuming that the initial slope of the circular segment is characterized by the contact angle θ (see Fig. 5.122) allows us to calculate all the geometrical details of the circular segment, including the maximum height c, which, in combination with the thickness H_{wet} of the coated wet film, is a measure for the height of the heavy edge H_{edge}, i.e.,

$$H_{edge} = H_{wet} + c \tag{5.8.26}$$

The geometrical details of the circular segment are sketched in Fig. 5.123 and given by the following equations.

$$A = R^2\theta - ab \tag{5.8.27}$$

$$R = \frac{1}{\sin\theta} \tag{5.8.28}$$

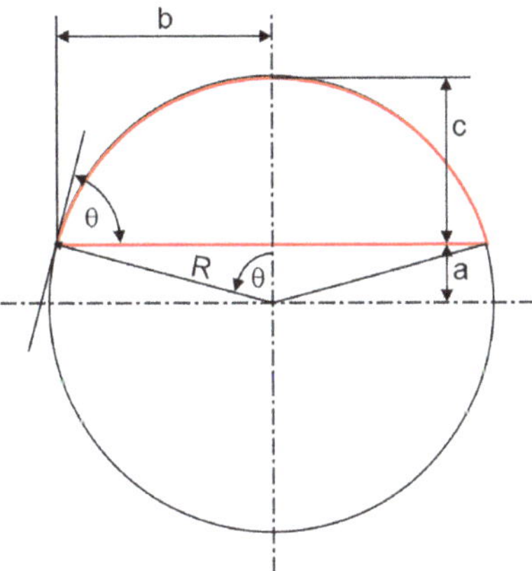

Fig. 5.123 Geometrical parameters of a circular segment

$$a = \frac{1}{\tan\theta} \tag{5.8.29}$$

Inserting Eqs. (5.8.28) and (5.8.29) into Eq. (5.8.27) and solving for b gives:

$$b = \sqrt{\frac{A}{\frac{\theta}{(\sin\theta)^2} - \frac{1}{\tan\theta}}} \tag{5.8.30}$$

$$c = R - a = b\left(\frac{1}{\sin\theta} - \frac{1}{\tan\theta}\right) \tag{5.8.31}$$

Note that the angle θ is measured in units of [rad]. Moreover, even though the circular segment in Fig. 5.123 is drawn for $\theta < \pi/2$, Eqs. (5.8.27)–(5.8.31) are valid for $0 < \theta < \pi$, i.e., for the entire range of complete spreading to complete de-wetting.

All the information needed is now available for calculating the height of the over-thick edge with Eq. (5.8.26). According to this theoretical model, the edge height depends on 8 parameters, namely the liquid density, the surface tension, the solids concentration, the dry coat weight, the coating speed, the curtain height, the gap between the bottom of the edge guide and the substrate surface, and the static contact angle between the curtain fluid and the substrate surface. In addition, H_{wet} and Q are calculated with Eqs. (3.3.3) and (3.3.4), respectively. As the formal relationship between the various parameters is complex and involves various power and trigonometric functions, it is not easy to visualize how each of these parameters affects the edge height. Therefore, we calculated the edge height as a function of each independent parameter, by varying each parameter in a range that is typical for adhesive label applications, see Fig. 5.124. The corresponding base parameters are listed in Table 5.17 and marked with a red square in the associated figures. Moreover, the edge height is normalized by the wet film thickness to render it dimensionless. Hence, a dimensionless edge height of 1.0 signifies no heavy edge, and the edge gets heavier as the dimensionless edge height increases above the value of 1.

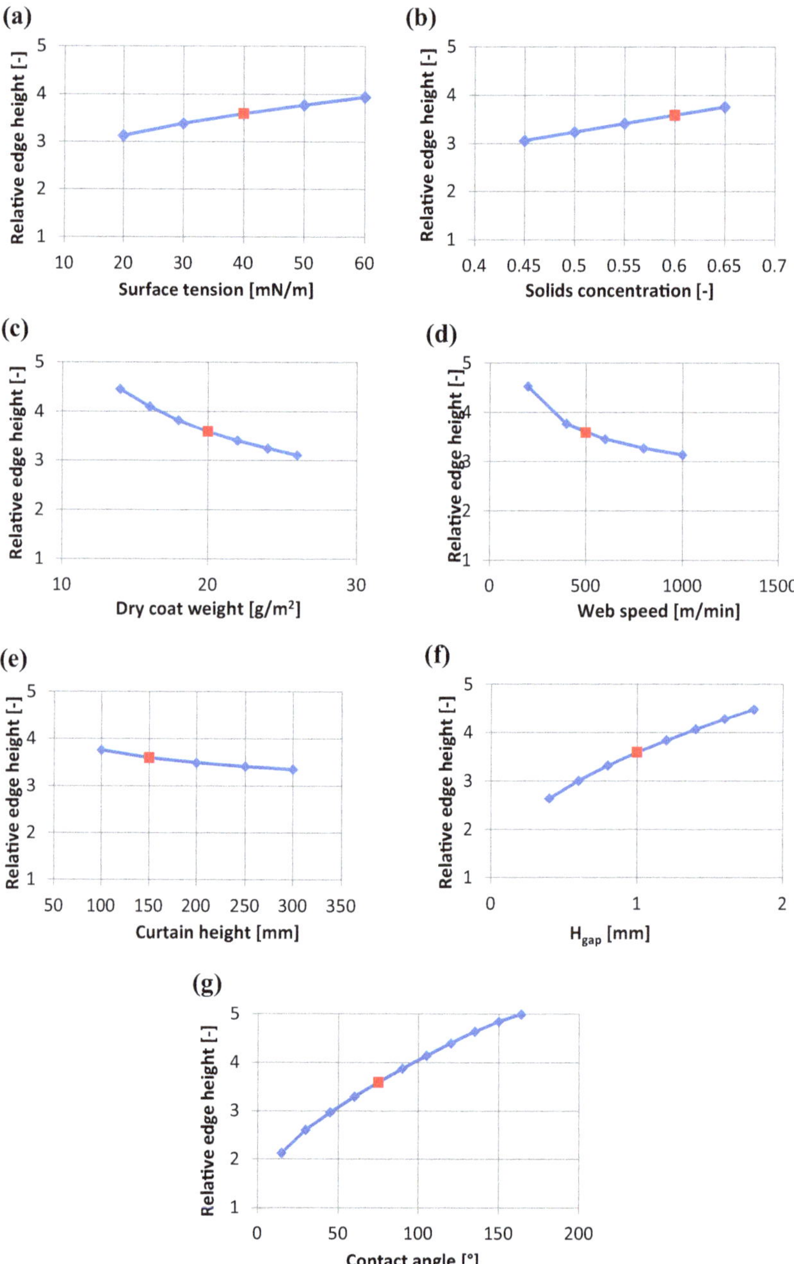

Fig. 5.124 **a** Dimensionless heavy edge as a function of the surface tension **b** dimensionless heavy edge as a function of the solids concentration **c** dimensionless heavy edge as a function of the dry coat weight **d** dimensionless heavy edge as a function of the web speed **e** dimensionless heavy edge as a function of the curtain height **f** dimensionless heavy edge as a function of the gap **g** dimensionless heavy edge as a function of the contact angle

Table 5.17 Base parameter values for typical label application

$\rho = 1020\ kg/m^3$ (constant)	$\sigma = 40\ mN/m$
$C_W = 0.60$	$A_{dry} = 20\ g/m^2$
$H_{wet} = 32.7\ \mu m$	$U = 500\ m/min$
$Q = 2.72\ cm^2/s$	$L_c = 150\ mm$
$H_{gap} = 1\ mm$	$\theta = 75°$

Table 5.18 Modified parameter values for label application

$\rho = 1020\ kg/m^3$ (constant)	$\sigma = 35\ mN/m$
$C_W = 0.50$	$A_{dry} = 25\ g/m^2$
$H_{wet} = 49.0\ \mu m$	$U = 600\ m/min$
$Q = 4.90\ cm^2/s$	$L_c = 200\ mm$
$H_{gap} = 0.5\ mm$	$\theta = 50°$

The size of the heavy edge is most sensitive to changing the contact angle and the gap between the edge guide and the substrate surface, and it hardly responds to changing the curtain height. The influence of the solids concentration, the dry coat weight, and the web speed is not strong either. For the base parameters listed in Table 5.17, the relative edge height is 3.6. Changing all the parameters a bit in the direction that reduces the edge height according to Table 5.18 resulted in a relative edge height value of 1.95, which is by far not perfect yet. Moreover, increasing the coat weight and the web speed and reducing the solids concentration all at the same time interferes with the product formulation, the process productivity and requires additional drying capacity, all of which are not attractive.

In summary, the model predictions of the size of the heavy edge suggest that it is difficult to avoid heavy edges, particularly if the substrate surface is difficult to wet, which is the case for siliconized paper or film in the label industry. These findings agree with our experience from coating trials on the Polytype Converting pilot machine, and with feedback from Polytype Converting's customers. Therefore, it would be helpful to find other ways for eliminating, or at least reducing, heavy edges, for example with smart designs of curtain edge guides. This topic is further discussed in more detail in Sect. 5.9.5.

5.9 Boundary Layer Flow

As the chapter title suggests, boundary layer flow is associated with the boundary that confines a given flow. A boundary layer includes that part of a flow field, in which the presence of the flow boundary is felt, or, in which the resistance to flow is increased owing to the action of viscous forces, which oppose flow. Outside of the boundary layer, the fluid velocity is locally constant. Consequently, the flow inside the boundary layer is characterized by a velocity profile that changes from the

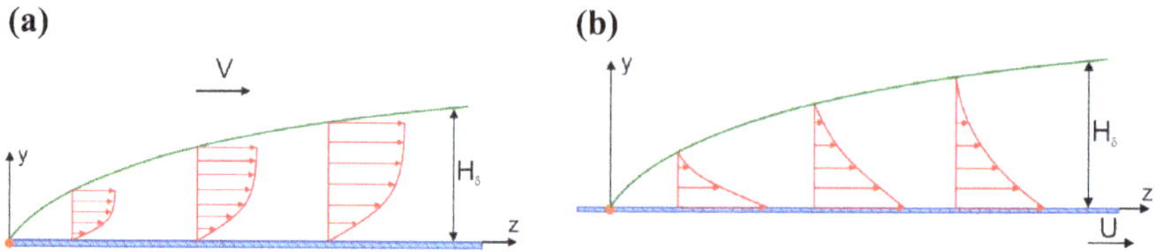

Fig. 5.125 **a** Boundary layer starting at the leading edge of a flat plate in a uniform flow of speed V **b** boundary layer starting at the dynamic wetting line of fluid impinging on a moving substrate of speed U

wall velocity to the velocity outside the boundary layer. The standard reference for boundary layer theory is Schlichting (1965).

Schlichting investigated the boundary layer that develops when a flat and stationary plate is placed into a uniform flow of speed V, see Fig. 5.125a.

The boundary layer starts at the leading edge of the plate and grows in thickness H_δ along the plate according to

$$H_\delta = f\sqrt{\frac{\mu x}{\rho V}} \tag{5.9.1}$$

The factor f in front of the square root is about 5.0. The resulting velocity profile inside the boundary layer was tabulated by Schlichting in a dimensionless form and is graphically depicted by the blue symbols in Fig. 5.126. The parameter f' on the vertical axis is a dimensionless velocity normalized with the speed V of the flow outside the boundary layer. The parameter η on the horizontal axis is a dimensionless y-coordinate defined as $\eta = y\sqrt{V\rho/(\mu x)}$. It turns out that this velocity profile can quite well be approximated by a second-order polynomial function according to

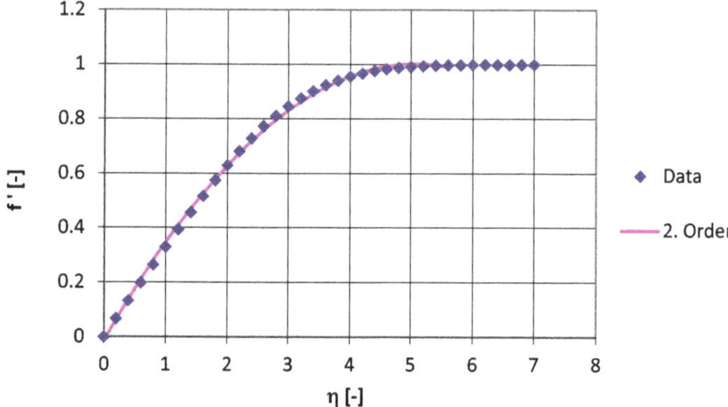

Fig. 5.126 Exact and approximate calculations of the velocity profile inside a boundary layer on a flat plate; data extracted from Schlichting (1965)

$$v(y) = a_0 + a_1 y + a_2 y^2 \tag{5.9.2}$$

The following boundary conditions are considered for determining the constants a_0, a_1 and a_2:

$$y = 0: \quad v(y) = 0 \tag{5.9.3a}$$

$$y = H_\delta: \quad v(y) = V \tag{5.9.3b}$$

$$y = H_\delta: \quad \frac{dv}{dy} = 0 \tag{5.9.3c}$$

Solving this set of equations yields

$$a_0 = 0 \tag{5.9.4}$$

$$a_1 = \frac{2V}{H_\delta} \tag{5.9.5}$$

$$a_2 = \frac{-V}{H_\delta^2} \tag{5.9.6}$$

The wall shear rate for this boundary layer flow can be derived to be

$$\gamma_0 = a_1 = \frac{2V}{H_\delta} \tag{5.9.7}$$

Working with the above approximation for the velocity profile inside the boundary layer requires knowledge of the boundary layer thickness. This point is found where the velocity in the boundary layer is equal to the constant velocity outside the boundary layer. According to Schlichting's data this point is reached for $\eta = 6.0$ where $f' = v(y)/V$ equals 0.999.

Equation (5.9.2) is drawn through Schlichting's data, see the red curve in Fig. 5.126. Note, however, that this approximation is only valid for $\eta < 5$. This limitation is not significant as the velocity at that value has reached 99.16% of the uniform velocity outside of the boundary layer. Therefore, this mathematically simple approximation is good enough for engineering calculations as will be demonstrated below.

It is generally accepted in fluid dynamics that the fluid in contact with a boundary assumes the velocity of that boundary. In premetered coating methods, all solid boundaries that confine the various flow fields discussed above are at rest, i.e., the fluid velocity in contact with those boundaries is zero. The obvious exception is the surface of the substrate to be coated, where the coated fluid in contact with the web assumes web speed U.

Special types of boundary layer flows are Newtonian ($n = 1.0$) pipe (duct), slot, and film flow as discussed in Sects. 5.4–5.7 and depicted in Figs. 5.4, 5.42, and 5.59. In these flows, the presence of the flow boundary is felt throughout the entire flow field, and the velocity profiles are parabolic. For power law fluids, however, the velocity profiles flatten out in the center of the flow field or towards the free surface in film flow as the power law index decreases. In these regions, therefore, the presence of the flow boundary is no longer felt. Note that these flows are usually not called boundary layer flows even though the concept still applies.

For premetered coating processes, the boundary layers of interest are those along the side walls of the various base flows presented above, i.e., the side walls of slot flow inside a die, the side walls of film flow on the inclined surface of a slide die, and curtain edge guides. The velocity profile inside these boundary layers conflicts with the first law of coating technology (see Sect. 3.2) because the velocity there changes in the cross-web direction, which is not desirable but unavoidable.

A special type of boundary layer for all premetered coating methods is associated with liquid impinging on the moving substrate. Examples of these boundary layers are depicted in Fig. 5.74 for slide coating and Fig. 5.78 for curtain coating. This boundary layer is not caused by stationary side walls, but by the moving substrate as it extends uniformly across the entire coating width.

More details of the various boundary layers in coating flows are discussed in the following sections. One focus is on the thickness of the boundary layer because this information quantifies the size of the unwanted velocity profile. Another focus is on reducing the boundary layer thickness. If successful, this would also produce the positive effect of reducing the size of the unwanted velocity profile along the flow boundary. Therefore, a method that achieves this goal and that can be applied to most of the characteristic flow fields of premetered coating processes is discussed next.

5.9.1 Concept for Reducing the Boundary Layer Thickness

Boundary layers are always observed when a viscous liquid flows along a stationary flow boundary. The thickness of the boundary layer increases with increasing viscosity. Since all coating fluids have positive viscosity, eliminating the boundary layer would only be possible, if the flow boundary moved at the same speed as the liquid away from the boundary. This idea was implemented and patented by Schweizer and Troller (1997) in an attempt to eliminate the boundary layer along a vertical curtain edge guide. In particular, they pumped either a low-viscosity fluid, e.g., water or an organic solvent, through a porous plate positioned against the falling curtain, see Fig. 5.127. As a result, a thin low-viscosity film formed on the outside of the plate and adjacent to the curtain. The thickness profile of that plate along the curtain was calculated such that the thickness of the flowing film increased in a way that the resulting surface velocity of the film matched the velocity of the gravity-driven curtain flow at every location along the curtain trajectory.

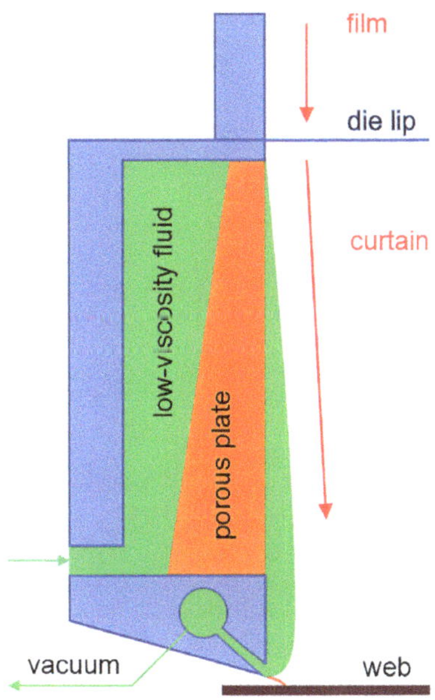

Fig. 5.127 Curtain edge guide without boundary layer in curtain flow

By using this invention, the curtain does not adhere to a stationary solid-body edge guide but to a moving surface provided by the surface of a low-viscosity falling liquid film. In practice, unfortunately, this concept was not viable. On the one hand, porous materials are not suitable for use in industrial coating processes because they are difficult to clean and they tend to plug and hence lose their functionality, particularly when used with adhesives. On the other hand, the flow rate/width of the low-viscosity fluid film required to generate a surface film velocity equal to the curtain velocity was so high that the film formed excessive waves and became unstable. Consequently, these waves disturbed the curtain to an unacceptable level.

Short of the perfect solution, it has long been a practice in the coating industry to reduce the boundary layer thickness by inserting a low-viscosity lubricating fluid, also called edge fluid, between the solid flow boundary and the flow field away from the boundary. Specifically, this practice has been applied when designing curtain edge guides.

While it is difficult to accurately predict the effect of a low-viscosity film on the boundary layer thickness by using simple mathematical formulas for the various characteristic flow fields of premetered coating processes, the effect can be demonstrated and quantified for the simple example of a 2-layer film on a vertical wall.

First, consider a single-layer film of density $\rho = 1000\ \text{kg/m}^3$, viscosity $\mu = 200$ mPas and flow rate/width $Q = 1.0\ \text{cm}^2\text{/s}$. The resulting velocity profile is shown in Fig. 5.128, and the thickness of the boundary layer corresponds to the thickness of the film, i.e., 1.83 mm. Now, consider a 2-layer film with the upper layer consisting of the same fluid as the single-layer film but with $Q_2 = 0.9\ \text{cm}^2\text{/s}$, i.e., 90% of the single-layer film. The lower layer makes up the flow rate difference, i.e., $Q_1 = 0.1$

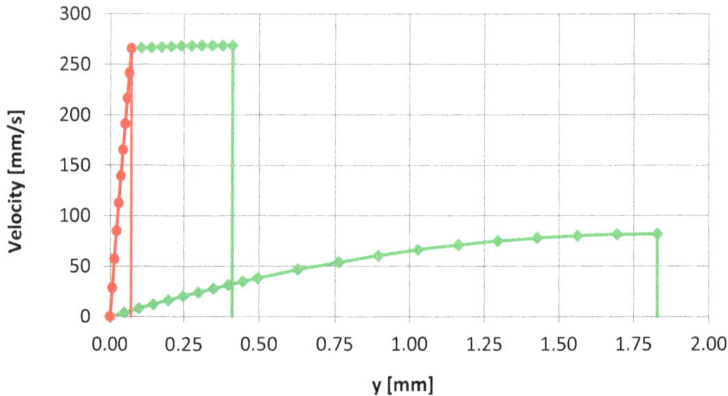

Fig. 5.128 Comparison of velocity profiles of a viscous single-layer film and a 2-layer film with a low-viscosity lubricating layer

cm²/s, and it consists of water, i.e., $\mu_1 = 1$ mPas. The resulting velocity profile is also shown in Fig. 5.128. As can be seen, the velocity of the upper layer is now almost constant, meaning that the effect of the boundary layer is almost entirely confined to the lower layer, and the boundary layer thickness practically decreased from 1.83 mm to about 0.07 mm.

The implementation of the concept of a lubricating layer in other flow fields is further discussed in the following chapters.

5.9.2 Boundary Layer in Slot Flow

The velocity profile across the slot height w (the narrow slot dimension) is parabolic as schematically shown in Fig. 5.129. In contrast, the profile of the average velocity $v_{z.average}$ across the width W of the slot is constant over most of the width except for the two boundary layer regions adjacent to the side walls of the die, see Fig. 5.129.

The average velocity of slot flow is obtained by dividing the volumetric flow rate/width Q by the slot height w, see Eq. (5.6.10). Therefore, if the velocity changes inside the boundary layer, so will also the flow rate/width. In particular, the average velocity inside the boundary layer v_δ can be calculated by integrating Eq. (5.9.2) over the boundary layer thickness and dividing the result by the boundary layer thickness to give

$$v_\delta = \frac{1}{H_\delta}\int_0^{H_\delta} v(y)dy = \frac{1}{H_\delta}\int_0^{H_\delta}\left(a_1 y + a_2 y^2\right)dy = \frac{2}{3}v_{z,average} \tag{5.9.8}$$

Based on the law of mass conservation, therefore, the average flow rate/width inside the boundary layer equals

$$Q_\delta = \frac{2}{3} Q_0 \tag{5.9.9}$$

Inside the boundary layer, only 2/3 of the uniform flow rate outside of the boundary layer can be transported downstream, see the schematic sketch in Fig. 5.130.

The cross-section of a die slot can be considered as a rectangle with a large aspect ratio W/w. Owens (1954) developed an equation for the 3-dimensional velocity profile on a rectangular cross-section in terms of an infinite series. The result is depicted in Fig. 5.131 for a section of the profile near the boundary layer.

Using Owen's equation and calculating the velocity profile along the center line of the slot allows the thickness H_δ of the resulting boundary layer to be estimated as a function of the aspect ratio as shown in Fig. 5.132. Specifically, the boundary layer

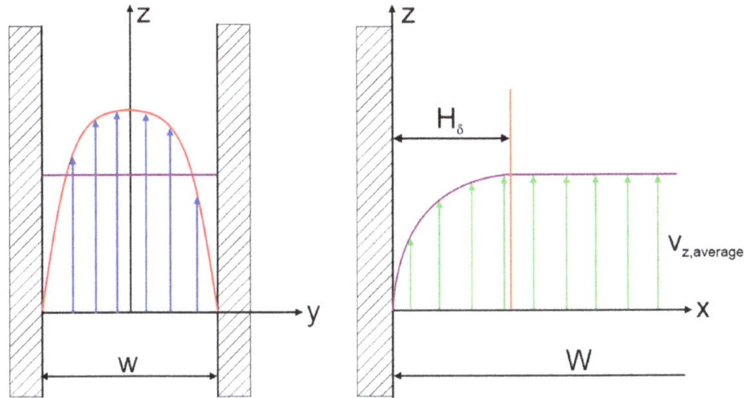

Fig. 5.129 Parabolic velocity profile of slot flow and boundary layer profile of the average velocity in the cross-web direction

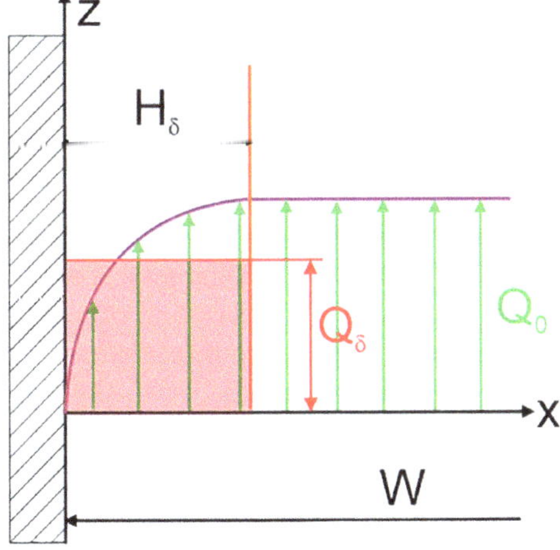

Fig. 5.130 Average flow rate/width inside and outside of the boundary layer

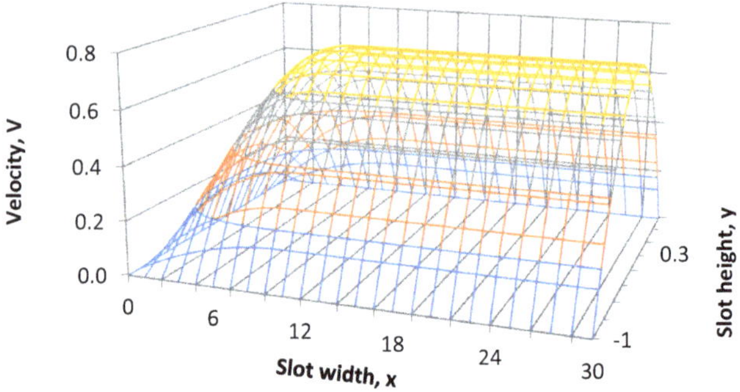

Fig. 5.131 3-dimensional velocity profile of slot flow near side wall

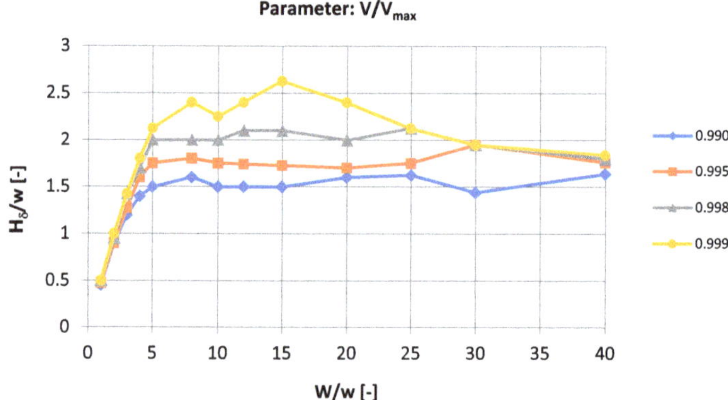

Fig. 5.132 Thickness of boundary layer for slot flow

thickness is obtained when, upon moving away from the side wall, the velocity inside the boundary layer reaches the value of the constant velocity outside the boundary layer.

Even though the calculated results wiggle a bit depending on the value of V/V_{max}, we can conclude that, for W/w > 5, the boundary layer thickness for slot flow approaches a value of about twice the slot height. To a first approximation, furthermore, the boundary layer thickness does not depend on the flow rate or the viscosity (the pressure drop does), but only on the aspect ratio W/w of the slot cross-section. This is a good result considering that the height of die slots is typically in the range of 250–500 μm. Since the width of many industrial dies is in the range of 1,000–2,000 mm the non-uniformity caused by the boundary layer inside the die slot extends to less than about 1 mm on each side, which is inconsequential for most industrial applications. Consequently, the need for implementing means for reducing such small boundary layers in slot flow is small, even though possibilities exist, as is further discussed in the next chapter.

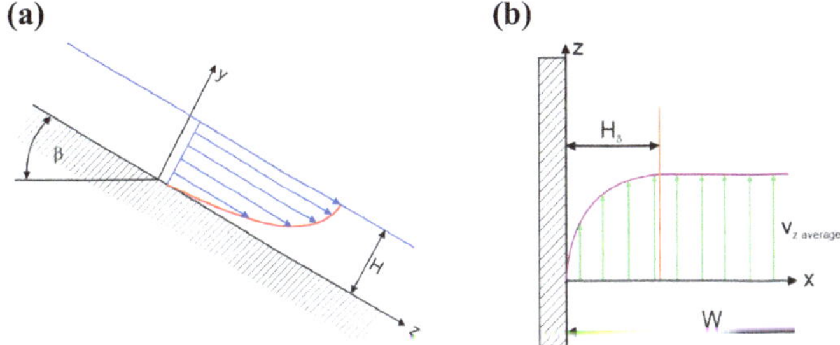

Fig. 5.133 **a** Semi-parabolic velocity profile of film flow **b** boundary layer along the side wall of film flow

5.9.3 Boundary Layer in Film Flow

Film flow on the inclined plane of a slide die is normally bound by two side plates in order to well define the width of the film flow and to prevent any cross flows.

For Newtonian fluids, the velocity profile across the film thickness is semi-parabolic as shown in Fig. 5.133a. In contrast, the velocity profile across the width W of the film flow is constant over most of the width except for the two boundary layer regions adjacent to the side walls of the film flow, see Fig. 5.133b. The form of the boundary layer for slot and film flow is qualitatively similar.

Unlike the boundary layer of a flat plate in a uniform flow or the boundary layer of a liquid jet impinging on a moving substrate, both of which start at a defined point and grow in thickness as the flow continues (see Fig. 5.125), slot and film flow are steady-state flows. Therefore, the boundary layer in slot and film flow does not grow in thickness along the flow but assumes a constant value. In addition, the boundary layer in slot and film flow already existed in one form or another in the proceeding flow, i.e., duct or slot flow, respectively. As a consequence, the thickness of the boundary layer changes in a transition zone between one characteristic flow and the next.

An extensive experimental study was carried out in the pilot facility of Polytype Converting to determine the thickness of the boundary layer in film flow. As illustrated in Fig. 5.134 the boundary layer was visualized for a given set of operating conditions by sprinkling a straight line of chalk powder onto the film surface at the beginning of the film on the slide die. After a short flow distance, the straight powder line changed form in line with the flow in and around the boundary layer. The thickness of the boundary layer was measured off such photographs as shown in Fig. 5.134 at the point where the film speed reached the constant value outside the boundary layer (Fig. 5.134a) or a maximum value as shown in Fig. 5.134b. The significance of a maximum film speed near the boundary layer is further discussed below.

Aqueous polyvinyl alcohol (PVA, Mowiol 15–99 from Kuraray) was used as test fluid. The liquid contained 0.3% of the surfactant Surfynol PSA-336, and the solids concentration was altered to change the low shear viscosity μ_0, which was measured

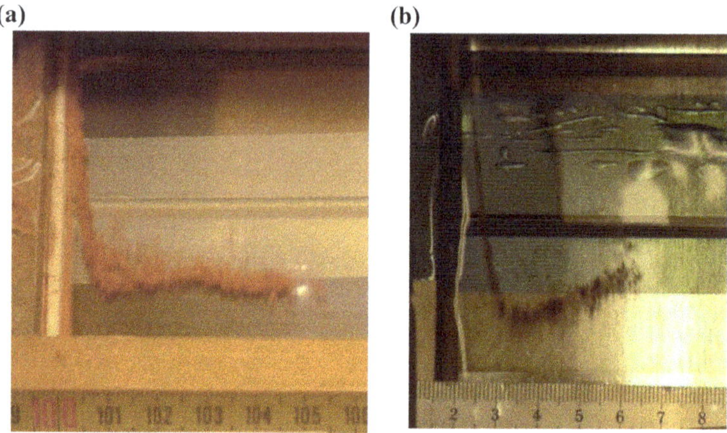

Fig. 5.134 **a** Boundary layer in film flow; $\mu_0 = 317$ mPas, Q = 1.9 cm^2/s; photo reproduced with permission from Polytype Converting AG **b** boundary layer in film flow; μ_0 = 1,628 mPas, Q = 2.6 cm^2/s; photo reproduced with permission from Polytype Converting AG

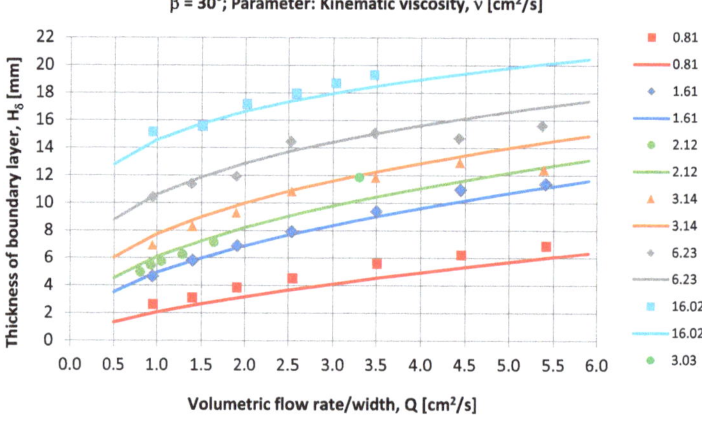

Fig. 5.135 Boundary layer thickness of film flow as a function of the flow rate/width and the kinematic viscosity for a slide angle of 30°

with a Brookfield device. The slide angle β was kept constant at 30°. The viscosity varied from 81 to 1,628 mPas, and consequently, the density varied between 1,005 and 1,016 kg/m^3, thus resulting in a range of the kinematic viscosity ν from 0.81 to 16.02 cm^2/s. The flow rate/width varied from 0.8 to 5.4 cm^2/s.

Results of the study are displayed in Fig. 5.135. The symbols are experimental data, while the curves were obtained from a regression analysis on the data of the following empirical form.

$$H_\delta = a_0 Q^{a_1} \tag{5.9.10}$$

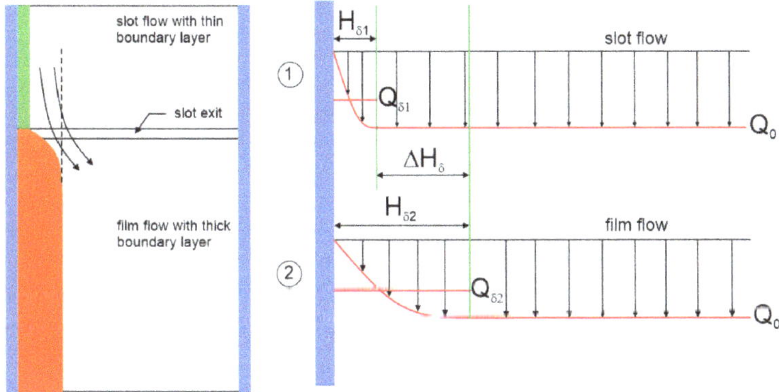

Fig. 5.136 Comparison of boundary layer thickness between slot and film flow

$$a_0 = b_0 + b_1 ln(\nu) \tag{5.9.11}$$

$$a_1 = c_0 \nu^{c_1} \tag{5.9.12}$$

The parameters b_0, b_1, c_0, and c_1 are regression constants, which have values of $b_0 = 2.9367$, $b_1 = 4.1975$, $c_0 = 0.585$, $c_1 = -0.404$.

Figure 5.135 shows that the boundary layer thickness of film flow increases with increasing flow rate/width and increasing low-shear viscosity. Moreover, the boundary layer thickness may be as high as 20 mm, which is much higher than for slot flow. Such high values would be significant in industrial coating processes.

The difference in boundary layer thickness between slot and film flow may have severe consequences on the width-wise thickness uniformity of the film flow. The situation is schematically depicted in Fig. 5.136. In slot flow the boundary layer marked in green (position 1) has a thickness of $H_{\delta 1}$, and the average flow rate/width inside the boundary layer is denoted with $Q_{\delta 1}$, which equals 2/3 of the uniform flow rate/width Q_0 outside the boundary layer, see Eq. (5.9.9). At the slot exit, the slot flow changes to film flow, and the thickness of the boundary layer (marked in orange, position 2) increases to $H_{\delta 2}$. The average flow rate/width inside that layer is denoted with $Q_{\delta 2}$, which also equals 2/3 of Q_0, since Q_0 is the same for both the slot and the film flow.

Now, the total flow rate supplied by the slot flow into the film flow over the boundary layer thickness of the film flow equals

$$V^*_{\delta,1} = H_{\delta,1} Q_{\delta,1} + \left(H_{\delta,2} - H_{\delta,1}\right) Q_0 = Q_0 \left(H_{\delta,2} - \frac{H_{\delta,1}}{3} \right) \tag{5.9.13}$$

In contrast, the total flow rate that can be transported downstream by the film flow inside the boundary layer of that flow is

$$V^*_{\delta,2} = H_{\delta,2} Q_{\delta,2} = \frac{2}{3} H_{\delta,2} Q_0 \tag{5.9.14}$$

The difference between inflow and outflow in the boundary layer of the film flow is positive and equals

$$\Delta V^*_\delta = \frac{Q_0}{3}\left(H_{\delta,2} - H_{\delta,1}\right) \tag{5.9.15}$$

The extra flow rate ΔV^*_δ that is fed into the film flow by the slot flow and that cannot be transported downstream by the film flow generates local cross-flow as indicated by the curved arrows in Fig. 5.136, and it causes the film thickness inside the boundary layer of the film flow to increase, which in turn causes the film velocity to increase. This phenomenon has been confirmed experimentally and is qualitatively illustrated in Fig. 5.134b. Quantification of this effect, however, is difficult because gravity and surface tension forces tend to level the local over-thickness as the film flows downstream. In any case, this situation is unwanted but cannot be avoided. As explained above, the boundary layer thickness $H_{\delta,1}$ in slot flow depends on the slot height but generally is less than 1 mm. On the other hand, the boundary layer thickness $H_{\delta,1}$ in film flow depends on the viscosity and the flow rate/width and can reach values of up to 20 mm. The boundary layer thickness for the flow conditions shown in Fig. 5.134b is just about 18.0 mm. Consequently, the extra flow rate fed into the film flow is proportional to $Q_0(18–1)/3$. This represents a significant increase of 31.5% when compared to the flow rate of $18 \cdot Q_0$ corresponding to a uniform film flow that is not disturbed by any kind of boundary layer effects.

As mentioned above, it is not meaningful to discuss premetered coating flows in terms of the total flow rate pumped into the die. Instead, the relevant flow rate is the flow rate/width Q_0, which is obtained by dividing the total flow rate by the width W of the die slot. By doing so it is assumed that the resulting flow rate/width is constant over the entire width. But now we know that this is not true due to the presence of the boundary layers along both side walls of the flow. Consequently, the effective flow rate/width Q_{eff} outside of the boundary layer is a bit higher than Q_0, see Fig. 5.137.

Q_{eff} can be quantified with Eq. (5.9.16), and the result is shown in Fig. 5.138.

$$\frac{Q_{eff} - Q_0}{Q_0} = \frac{1}{1 - \frac{2}{3}\frac{H_\delta}{W}} - 1 \tag{5.9.16}$$

For most industrial coating applications, i.e., W > 1,000 mm and H_δ < 15 mm, the difference between the uniform and effective flow rate/width is small, i.e., <1%, and can be neglected for engineering calculations.

Film flow in the boundary layer region is often not only disturbed by an uneven velocity profile in the cross-web direction, but also by an uneven film thickness. In particular, the film flow on the inclined surface of a slide die is typically bound by vertical walls on either side of the film. The cross-web film thickness uniformity in the vicinity of the wall is determined by the wall design. Specifically, the film thickness depends on whether the height of the wall is smaller or bigger than the equilibrium film thickness away from the wall, see Fig. 5.139.

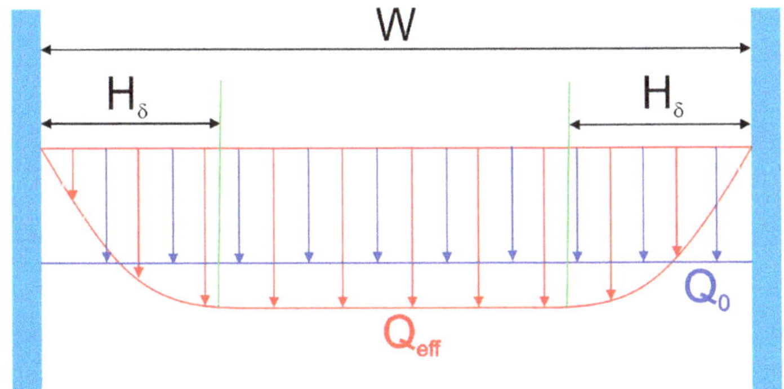

Fig. 5.137 Uniform and effective flow rate/width as a consequence of boundary layers

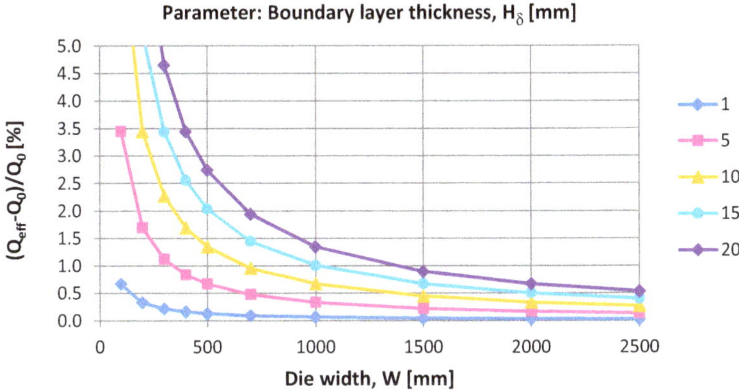

Fig. 5.138 Uniform and effective flow rate/width as a function of the die width and the boundary layer thickness

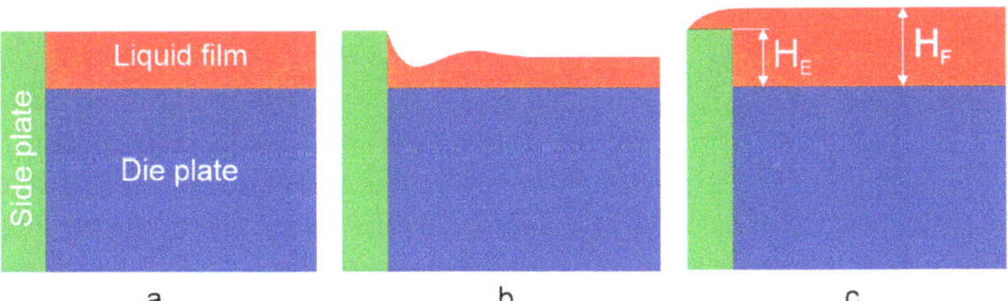

Fig. 5.139 Side wall for bounding the film flow on the inclined die surface. **a** $H_F = H_E$; **b** $H_F < H_E$; **c** $H_F > H_E$

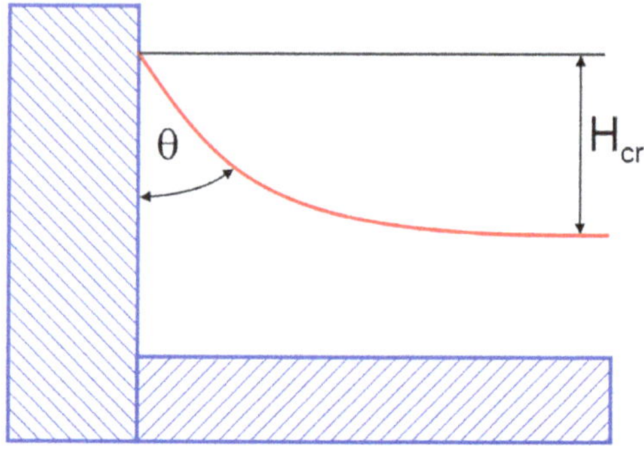

Fig. 5.140 Capillary rise of a liquid film next to a wall

Ideally, the height of the side wall is equal to the film thickness because then and only then is the film thickness even all across the die width. If the wall is higher than the film thickness, then liquid adjacent to the wall will climb up the wall and the static wetting line is located above the equilibrium film thickness. The liquid in the wedge next to the wall is taken from the film next to the wedge, which in turn causes an uneven film thickness as depicted in Fig. 5.139b. Finally, if the wall is lower than the film thickness, the film tends to swap over the wall, which also causes an uneven film thickness in the boundary layer region.

The phenomenon of liquid climbing up a wall is called capillary rise. It was analyzed by Batchelor (1967) for the case of an infinite pool, such that the free surface next to the wedge of rising fluid is not disturbed, see Fig. 5.140.

Batchelor fund that the capillary rise H_{CR} can be calculated by:

$$H_{CR}^2 = \frac{2\sigma}{\rho g}(1 - sin\theta) \tag{5.9.17}$$

The capillary rise, i.e., the position of the static wetting line, not only depends upon the surface tension but also on the static wetting angle θ, i.e., on the wetting behavior between the liquid and the wall surface. The wetting line position, and the surface profile of the adjacent liquid, is not only determined by capillary forces but also by the geometrical details of the surface to be wetted.

As explained above in Sect. 5.6 and based on fluid-mechanical and thermodynamical arguments, Gibbs (1906) showed that the wetting line of a meniscus located on a flat wall can easily be moved by the smallest forces, see Fig. 5.57a. However, if the wetting line is located at a sharp corner in that surface, then that corner is a preferred position for the wetting line, and considerably higher forces are required for moving the wetting line beyond the corner, see Fig. 5.57b. In other words, the presence of a corner in a surface inhibits the motion of the wetting line past the corner. Gibbs' results are quantified in inequality (5.6.82), which defines conditions, for which the wetting line sticks to the corner.

Just like for the downstream corner of a slot die lip used in the slot coating process a compromise value for the corner of the side wall in film flow must be

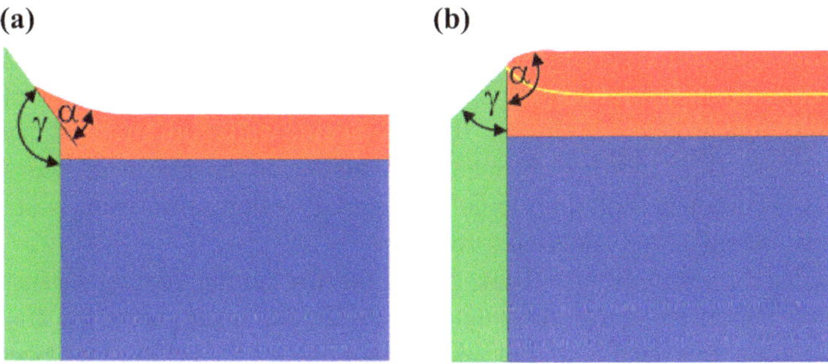

Fig. 5.141 **a** Floating wetting along the side wall beyond a corner with a large angle **b** pinned wetting line at a sharp corner

found. Figure 5.141a shows a side wall with a large corner angle, i.e., $\gamma > 90°$, which, according to several patents, has been used in the coating industry. If the thickness of the liquid film is higher than the position of the corner above the die surface, then it is easily possible for liquid to climb up on the wall beyond the corner, which is not wanted. In contrast, if the corner angle is small, i.e., $\gamma < 90°$, then the wetting line stays pinned at the corner for a wide range of operating conditions, see Fig. 5.141b. In particular, the wetting line will only climb up to the corner if the film thickness is smaller than the corner position above the film floor, see yellow line in the figure. In addition, the wetting line remains pinned to the corner if the film thickness is higher than the corner position.

60° is a good compromise value for the corner angle. Experience shows, however, that a corner angle of 90° is also acceptable for many industrial applications. This larger angle is easier for machining and hence less costly. Corner angles of >90° should be avoided.

As visualized by Eq. (5.9.17) the position of the static wetting line could also be controlled by altering the wetting behavior between the liquid and the side wall. In particular, the wetting line would be located at the height of the free film surface, which is desired, if the contact angle θ were 90°. Many attempts in this regard have previously been made in the coating industry, also by the author of this book. Experience shows, however, that changing the wetting properties of the side wall by coating it with an appropriate material works in principle, but it is not production-worthy. In particular, the altered wetting properties will diminish over time as a result of mechanical scrubbing during cleaning, or as a result of contamination of the wall by residual chemical components from the coating fluid.

Controlling the position of the static wetting line along the side wall of film flow is better accomplished by mechanical rather than chemical means. However, this requirement may be difficult to implement in practice, particularly if simultaneous multilayer applications are considered. The reason is that the film thickness, and thus the height of the side wall, depends on the inclination angle of the slide surface, the local flow rate/width of the layers, as well as on the local densities and viscosities. For non-Newtonian fluids, the viscosity also depends on the local shear rate, which in turn depends on all the other parameters just listed, and which further complicates matters. Moreover, most industrial coating companies want to use a given die for applying several different products. Therefore, they prefer side walls with an adjustable height, if they always want to perfectly match the height of the side wall with the film thickness. Equipment with adjustable geometrical features, however, is difficult to design and manufacture, and therefore costly.

Consider the example of a 3-layer slide die for curtain coating as sketched in Fig. 5.142. Suppose that the slide angle β is 30° in the upper part of the die and 90° (vertical) just above the die lip. Further suppose that all fluids have the same density of 1000 kg/m^3, and that the Newtonian viscosities and flow rates/width are specified as is listed in Table 5.19. In order to optimally design the side wall for the film flow, the film thickness must be known at four different locations, see Fig. 5.142.

Formulas for calculating the thickness of single- and multilayer films have been presented in Sect. 5.7, and the results for the 3-layer film specified above are listed in Table 5.20.

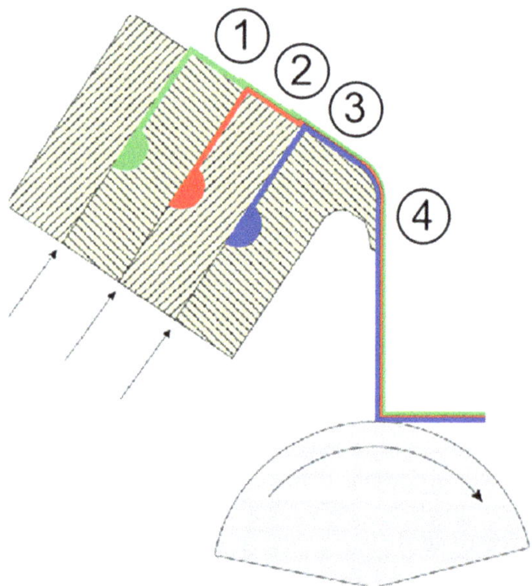

Fig. 5.142 Local film thicknesses of a multilayer film flow

Table 5.19 Viscosities and volumetric flow rates/width of a 3-layer film flow

Layer	Viscosity [mPas]	Volumetric flow rate/width [cm^2/s]
Bottom	75	0.5
Middle	300	1.0
Top	500	0.5

Table 5.20 Thickness of a 1-layer, 2-layer and 3-layer film

Location (See Fig. 5.142)	Film thickness [mm]
1	2.49
2	3.02
3	2.22
4	1.76

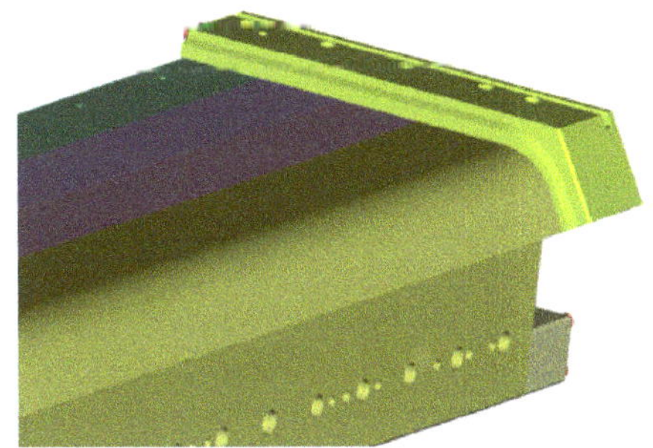

Fig. 5.143 One-piece side wall with changing wall height as per the inclination angle of the die slide; drawing reproduced with permission from TSE Troller AG

In this example, the film thickness first increases in flow direction from one die plate to the next, then decreases again in two steps, which is fairly typical for multi-layer film flow. Now, the challenge is to build a side wall with the same height profile as the film thickness profile. A one-piece solution first offered by TSE Troller AG incorporated a constant wall height for the flow on the 30° slide, and a second, smaller height for the flow on the vertical part of the die lip, see Fig. 5.143. Later, Polytype Converting AG in cooperation with TSE Troller AG developed a two-piece solution, which is also manufactured by TSE Troller. Here, the correct film thickness profile on every die plate and the vertical lip is machined into a suitable type of soft plastic material, for example, Teflon, and that piece of plastic is clamped down onto the slide surface by a rigid stainless steel bracket, see Fig. 5.144. Moreover, the angle γ of the solid corner of the side wall is 90° on every die plate. When coating a different product with the same die, only the soft plastic piece with the correct profile of the wall height must be exchanged. This concept has proven useful and practical in industrial coating processes. Moreover, this concept also works well on TSE Troller dies for applications with changing coating widths.

Note that some die suppliers offer other concepts for minimizing film thickness variations near the side wall of film flow. In particular, they build side walls that are higher than the film thickness, and they build the walls from a material that cannot be wet well by the coating fluid under consideration. As illustrated in Fig. 5.140 and by Eq. (5.9.17), this concept works well if the wetting behavior between the fluid and the wall material is such that the contact angle θ equals 90°. Otherwise, results are less satisfactory because the wetting properties of the wall material would have to be changed every time the chemical composition of the coating fluid changes, and that is difficult if not impossible. For this reason, we prefer the mechanical solution by matching the height profile of the side wall with the thickness profile of the film flow.

As shown in Fig. 5.135 the thickness of the boundary layer in film flow increases with increasing viscosity. Moreover, as shown in Figs. 5.134b and 5.136, thick boundary layers in film flow are detrimental because they prevent the fluid supplied by the slot flow upstream of the film flow from being transported uniformly in the downstream direction, which in turn causes the film to locally thicken near the edge of the boundary layer. This unwanted effect could be reduced if the boundary layer thickness in film flow could be diminished.

The same problem is observed in the boundary layer of curtain flow, see Sect. 5.9.4. There, it has been a common and successful practice for many years to reduce the boundary layer thickness by inserting a thin low-viscosity lubricating layer between the curtain and the stationary, solid curtain edge guide. Various patents to that extent show that this idea has also been tried for reducing the boundary layer thickness in film flow on the inclined die slide. One example is given by Schweizer et al. (2004). Their idea was to build the side wall for film flow from porous material, e.g., stainless steel or glass, through which water or another appropriate organic solvent could be percolated at an adjustable rate to form a lubricating layer between the film and the side wall, see Fig. 5.145. While porous materials serve very well for supplying a fluid stream into another flow without any jetting effects (see Fig. 5.21), they learned that porous materials are not suitable for industrial coating processes because they tend to plug up after a short period of operation and are difficult to clean, particularly when applied with adhesives. In addition, if the height of the porous plate was not equal to the thickness of the adjacent film, then the effect of the lubricating layer was only sub-optimal. Furthermore, changing the geometry of the porous plate with every product change is not production-friendly either.

At Polytype Converting AG we also tested a different approach for feeding a low-viscosity lubricating layer into the side wall of film flow. Specifically, we injected

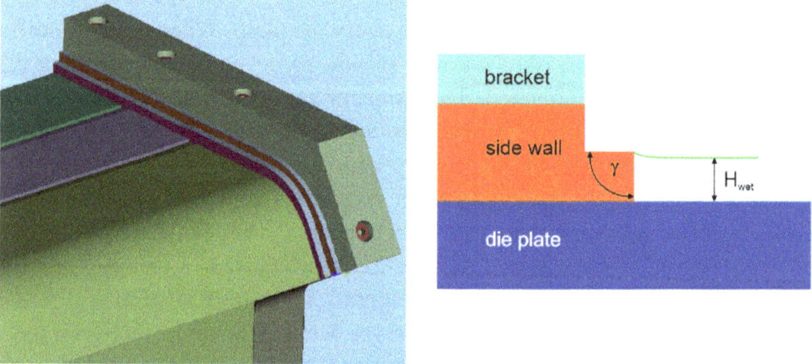

Fig. 5.144 Two-piece side wall with changing wall height on every die plate and according to the inclination angle of the die slide; concept developed by Polytype Converting AG and TSE Troller AG, and manufactured by TSE Troller AG; 3-D drawing reproduced with permission from TSE Troller AG

the lubricating fluid through the side wall of the die into the die slot or the (outer) cavity, see Fig. 5.146.

We observed that the lubricating fluid adhered to the side wall when flowing along and out of the slot, even if some jetting occurred at the injection point. We also saw that the lubricating fluid flowed along the side wall of the film flow down to the end of the inclined die slide. More importantly, we observed that the lubricating fluid that was injected into a given coating fluid (slot) followed that fluid along the entire film and curtain flow, even if that film flowed on top of another film in multilayer applications, see Fig. 5.147. In this experiment, the die and the side wall of the die were made of plexiglass. Furthermore, the side wall of the die extended to form the side wall of the film flow, and it further extended to form the curtain edge guides.

This seemed like a good way to reduce the boundary layer thickness in slot, film, and curtain flow because the lubricating fluid remained exactly where it was supposed to, namely between the side wall and the coating fluid. This result was reproduced for different coating liquids such as aqueous PSA or PVA solutions as well as for a warm (50 °C) micro-porous ink jet formulation.

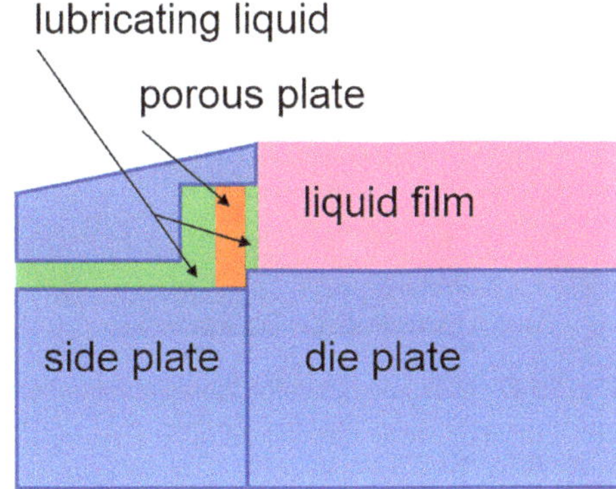

Fig. 5.145 Side wall for film flow made of porous material to inject a lubricating layer between the film and the side wall

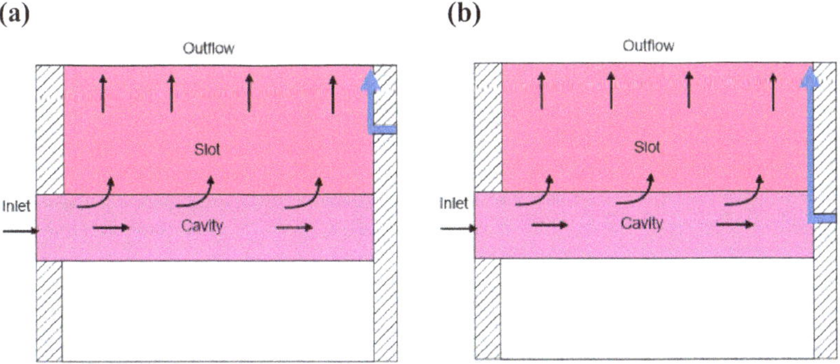

Fig. 5.146 Injection of lubricating fluid into the die slot or the cavity to reduce the thickness of the boundary layer in slot and film flow

Dropping a line of powder onto the film surface for visualizing the boundary layer, i.e., the velocity profile in the boundary layer, with and without lubricating fluid for a given set of flow conditions confirmed the desired effect, at least qualitatively, see Fig. 5.148. A 2-layer PVA film with $\mu = 960$ mPas and a total flow rate/width of $Q = 2.2$ cm^2/s was used in this experiment. Without lubricating fluid, the angle β between the velocity profile of the boundary layer and the side all was approximately 10°, and this angle decreased to around 5° when the lubricating fluid was pumped into both layers.

In Schlichting's considerations and calculations, the boundary layer thickness is dependent on factor f, see Eq. (5.9.1). For the boundary layer generated at the leading edge of a stationary plate placed into a uniform flow, the value of f is roughly 5.0. In principle, altering the value of f could be used to account for and quantify the boundary layer thinning effect when using a low-viscosity lubricating fluid. To that extent, the factor f must also be introduced into Eq. (5.9.10), which calculates the boundary layer thickness of film flow as:

$$H_\delta = f_{film} a_0 Q^{a_1} \tag{5.9.18}$$

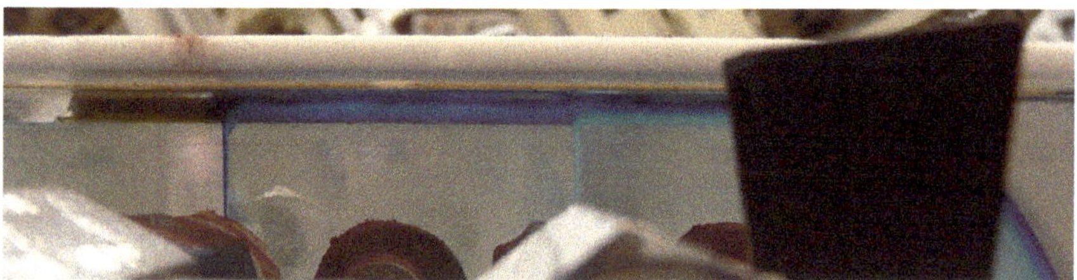

Fig. 5.147 Dark and light blue lubricating fluid following the layer of coating fluid it was injected into; flow rate of edge fluid in each layer: 0.52 L/h; photo reproduced with permission from Polytype Converting AG

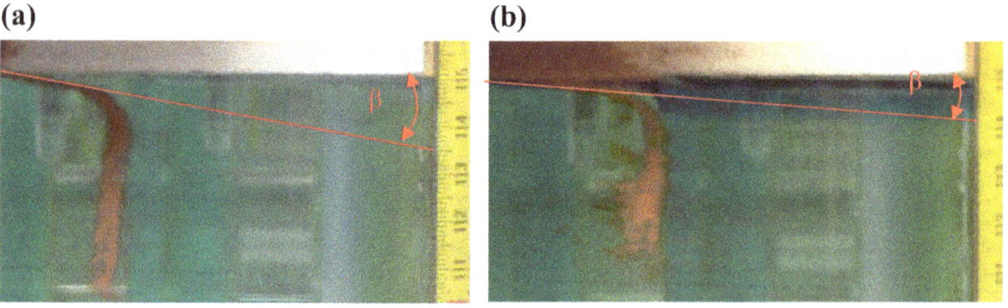

Fig. 5.148 **a** Formation of a boundary layer in 2-layer film flow without lubricating fluid; $\beta = 10°$; photo reproduced with permission from Polytype Converting AG **b** formation of a boundary layer in 2-layer film flow with lubricating fluid in both layers; $\beta = 5°$; photo reproduced with permission from Polytype Converting AG

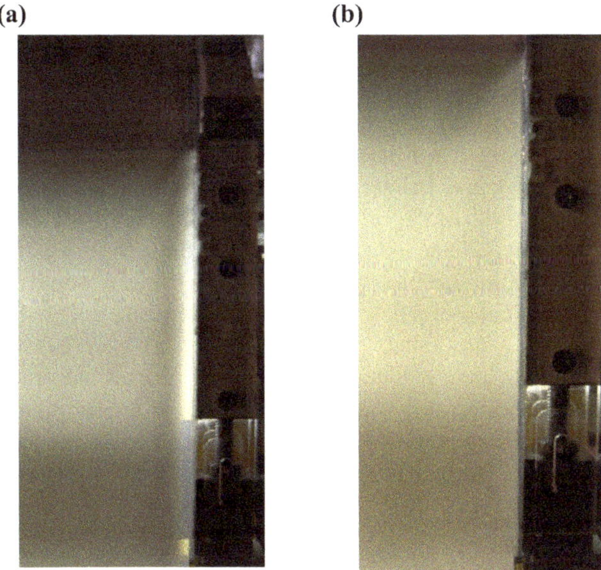

Fig. 5.149 **a** Curtain without lubricating fluid in the film flow: severe thickness variations in the boundary layer of the curtain; photo reproduced with permission from Polytype Converting AG **b** curtain with lubricating fluid in the film flow: uniform thickness in the boundary layer of the curtain; photo reproduced with permission from Polytype Converting AG

The parameters a_0 and a_1 are still calculated with Eqs. (5.9.11) and (5.9.12), respectively. For a side wall without lubricating fluid $f_{film} = 1.0$. Our experiments in the lab and comparisons between different wall designs for film flow and theoretical calculations showed that f_{film} assumes a value of approximately 0.5. Using an edge fluid for film flow, therefore, does not eliminate the boundary layer, but it reduces the boundary layer thickness by about 50%. Additional information about the value of f_{film} is given in Sect. 5.9.5.3 entitled *Inclination of Edge Guide.*

The positive effect of the lubricating fluid on reducing the boundary layer thickness in film flow, and ultimately on improving the thickness uniformity of the subsequent curtain flow, can also be demonstrated when observing a back-lit, opaque curtain, see Fig. 5.149. Here, an aqueous PSA curtain with $\mu = 500$ mPas and $Q = 5.46$ cm^2/s flows along a vertical curtain edge guide. Figure 5.149a shows the situation without lubricating fluid in the film flow, and massive thickness variations are visible in the boundary layer of the curtain. These non-uniformities almost completely disappear upon pumping lubrication fluid in the film flow as shown in Photo 5.149b.

In summary, even though a positive effect has been demonstrated, installing infrastructure for pumping lubricating fluid into the side wall of film flow is complicated and cumbersome to use, particularly when applied to multilayer films. We are not aware that such systems are used in industrial coating processes.

5.9.4 Boundary Layer in Curtain Flow

Curtain coating can be implemented in either the **over-board** or the **in-board** operating mode. In the over-board mode, the freely falling curtain is often not guided along its edges. Consequently, it will laterally contract, or neck-in, as a result of surface tension forces (Weinstein & Palmer, 1997), see Fig. 5.150a. This is an objectionable feature because the width of the curtain decreases with increasing distance from the beginning of the curtain. In an industrial curtain coating process, the resulting width of the coated film on the substrate would be unknown, and it would depend, among others, on physical fluid properties, i.e., the surface tension, and on the length of the curtain, which again is unwanted.

Therefore, to make an over-board curtain coating process industrially viable, the die width must be considerably wider than the width of the substrate, such that the edges of the unsupported curtain fall beyond the substrate edges into a catch pan, from where the excess fluid can be recycled if the curtain consists of only 1 layer. The

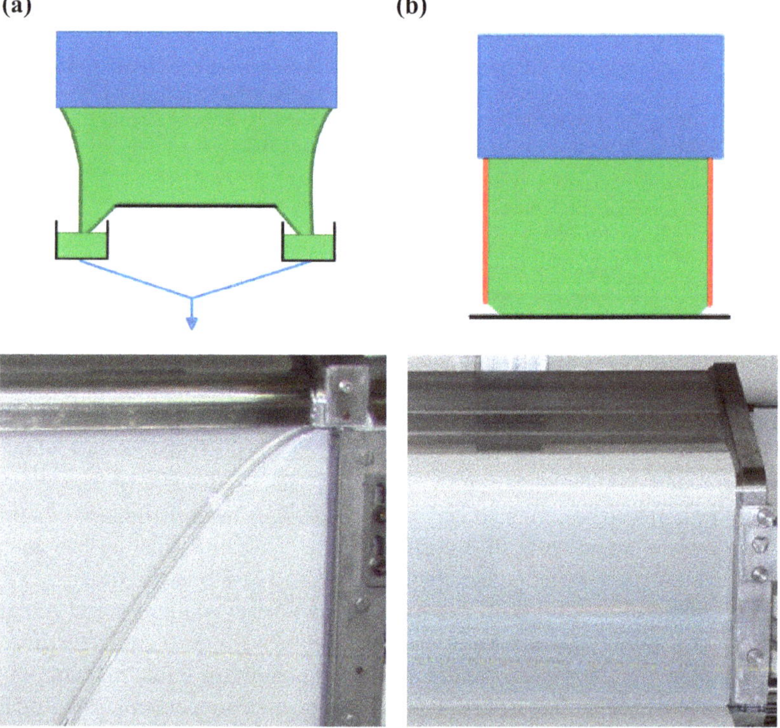

Fig. 5.150 **a** Over-board operating mode: lateral contraction of an unguided curtain; photo reproduced with permission from Polytype Converting AG **b** in-board operating mode: guided curtain with constant width along its trajectory; photo reproduced with permission from Polytype Converting AG

excess die width can be reduced by supporting the curtain with simple edge guides that extend from the die lip to below the position of the substrate. In the over-board mode, the substrate is typically coated over the entire width.

The in-board coating mode does not suffer from the problems described above because the curtain is guided along its trajectory by so-called curtain edge guides, see Fig. 5.150b. Now, the nominal width of the curtain is defined by the distance that separates the edge guides, and the coated substrate is characterized by dry edges. Many industrial curtain coating processes are operated in the in-board mode, specifically those that simultaneously apply multiple layers because the excess fluid generated by the over-board mode would mix in the catch pan and could not be recycled but would be wasted.

Additional information about the in-board and over-board operating modes is provided in Sect. 19.2.5.

In any case, curtain flow along an edge guide is also characterized by a boundary layer, inside which the curtain velocity is lower than the uniform speed outside the boundary layer, see Fig. 5.151.

Even though a boundary layer already exists in the slot or film flow just upstream of the curtain flow, the boundary layer in the curtain more resembles the one on a flat plate in a uniform flow as described by Schlichting (1965). In particular, the thickness of the boundary layer grows from its inception point at the die lip owing to the increasing curtain velocity due to the gravity pull. If the initial curtain velocity $v_{c,0}$ in Eq. (5.8.1) is neglected, and if that curtain velocity is inserted into Eq. (5.9.1), then the thickness of the boundary layer in the falling curtain can be approximated by

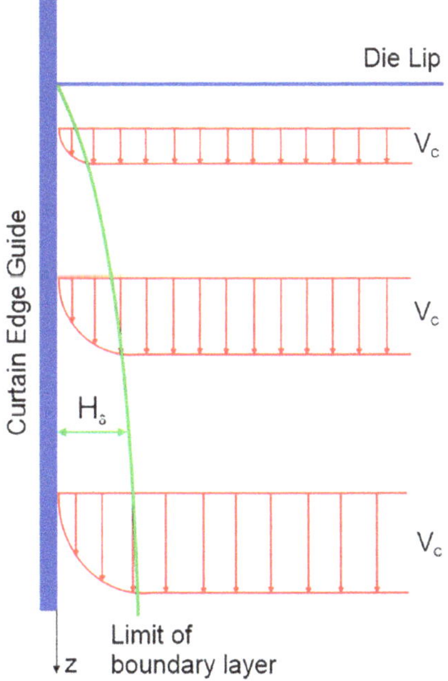

Fig. 5.151 Boundary layer along the edge guide of a falling curtain

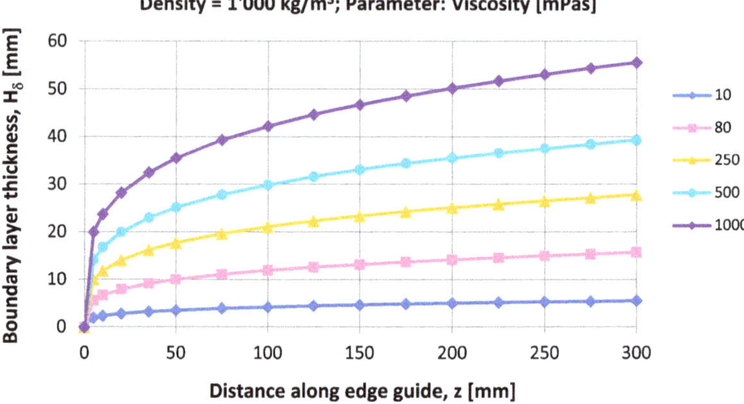

Fig. 5.152 Estimation of boundary layer thickness in curtain flow along a solid body edge guide

$$H_\delta = f_{curtain}\left(\frac{\mu^2 z}{2g\rho^2}\right)^{1/4} \tag{5.9.19}$$

Here, z is the coordinate along the falling curtain, and Eq. (5.9.19) is graphically displayed in Fig. 5.152. For a solid body edge guide, the viscosity in the above equation can be approximated by the viscosity of the flow just upstream of the curtain flow, i.e., slot flow or film flow on an inclined (vertical) surface.

Much like in film flow, the boundary layer in curtain flow can be quite thick, and values of 20 mm or more are not uncommon. This situation is highly undesirable because the curtain impingement velocity in an edge region of significant dimension might be significantly lower than the uniform velocity away from the edge, which in turn might cause local air entrainment.

In an attempt to reduce the boundary layer thickness in curtain flow most industrial curtain edge guides are not built as a solid body. Instead, a thin film of a low-viscosity edge fluid (water of organic solvent) is incorporated in one way or another between the curtain and the edge guide. This film is sometimes called a "lubricating layer", and its effect is schematically depicted in Fig. 5.153. If the edge guide was just a solid body, e.g., a plate of stainless steel, the green velocity profile would be present and the boundary layer thickness would be $H_{\delta 1}$. However, with a thin low-viscosity lubricating layer in place (red velocity profile with film thickness $H_{\delta 2}$) the curtain no longer contacts a solid surface at rest, but a moving surface represented by the surface of the lubricating film and moving at speed v_f. Consequently, the thickness of the boundary layer in the curtain reduces to $H_{\delta 3}$ (blue velocity profile), and the total thickness of the layer with uneven velocity is $H_{\delta 4}$, which is considerably smaller than $H_{\delta 1}$.

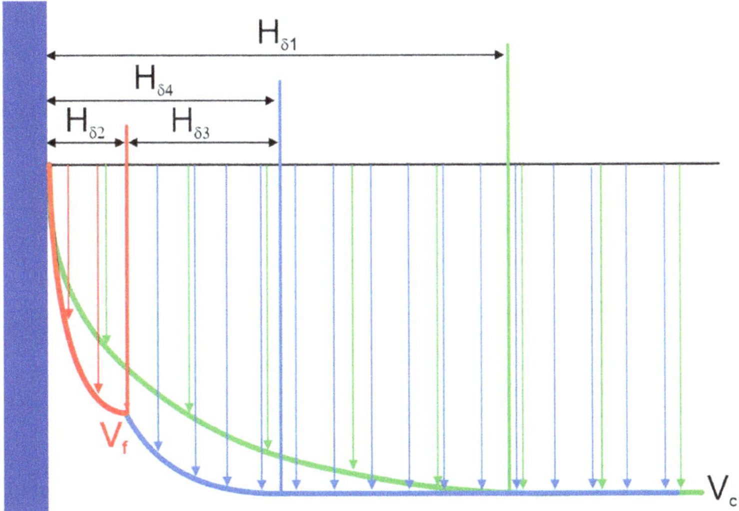

Fig. 5.153 Low-viscosity lubricating layer between curtain and edge guide

Laser Doppler (LDA) measurements of the velocity profile in curtains that were either formed by film flow (Durst et al., 1992) or by slot flow (Katagiri, 1992) showed that the boundary layer is very thin just downstream of the die lip. This is particularly so for curtain emerging from a slot and for low viscosity fluids building a curtain from film flow. The measurements also showed that the boundary layer thickness developed consistent with Eq. (5.9.19), and that the value of the factor $f_{curtain}$ is indeed about 5.0 for solid edge guide surfaces. In addition, Durst's measurements suggested that the factor $f_{curtain}$ reduces to about 2.5, if the curtain adheres to a low-viscosity lubricating film. Unfortunately, Durst did not elaborate as to how exactly this lubricating film was formed. Based on many lab trials, comparisons between different edge guide designs, and theoretical calculations, we estimated the value of factor f to be about 2.2. In summary, and similar to film flow, the LDA measurements confirmed that the boundary layer thickness can be reduced by roughly 50% by using a low-viscosity lubricating film. This idea is implemented in many patented designs for curtain edge guides. More information about edge guide design and the value of the factor $f_{curtain}$ is given in Sect. 5.9.5.2 entitled *Inclination of Edge Guide.*

Regarding the flow rate of the lubricating film, laboratory tests have revealed that suitable values for the flow rate/width are in the range of 0.1–1.0 cm^2/s, and preferred values range from 0.2 to 0.5 cm^2/s. Using water as the edge fluid, the corresponding Reynolds numbers are between 20 and 50. Often, a flow rate/width of 0.35 cm^2/s leads to good results. The optimum value, however, depends on the detailed design of the edge guide, it is product-dependent, and hence it must be determined experimentally by trial-and-error.

Comparing the velocities of the film and curtain flows along the edge guide, it should be noted that apart from a short distance on the very top of the curtain edge

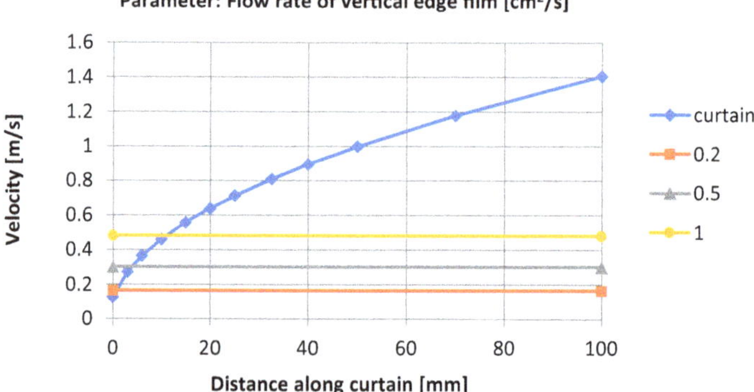

Fig. 5.154 Comparison between curtain and film surface velocity along edge guide

guide, the surface velocity of the film flow is constant and dependent on the viscosity and the flow rate of the fluid, as well as on the inclination angle of the edge guide (often vertical). For water as the lubricating fluid and for the flow rates mentioned in the previous paragraph, the resulting surface velocity of such vertical films is in the range of 0.16–0.48 m/s, see Fig. 5.154. In contrast, the velocity of the curtain increases with increasing distance from the die lip as shown by Eq. (5.8.1). Therefore, the film and curtain velocities do not match except for one point located just a few millimeters downstream from the die lip. Above that point, the film surface moves faster than the curtain, and below that point, the curtain moves faster than the film surface. It can also be noted that the flow rate of the auxiliary fluid would have to be very high if the film surface velocity was to match the curtain velocity at the curtain impingement point. In fact, the required flow rate/width would be 5.0 cm^2/s for a 100 mm long curtain, and $Q = 6.8\ cm^2/s$ for $L_c = 150$ mm. Such film flows would become very wavy and even turbulent, which is not acceptable.

It has been mentioned in Sect. 5.8 that most of the curtain flow is an extensional flow where the concept of shear viscosity does not apply. This, however, is not true for the flow inside the boundary layer along the curtain edge guide, where shear viscosity is present. The question then arises, which is the appropriate viscosity for calculating the boundary layer thickness according to Eq. (5.9.19) in the presence of a lubricating film? The answer is obtained by assuming that the velocity profile of the boundary layer can be approximated with a second-order polynomial function according to Eq. (5.9.2), by moving the origin of the coordinated system to the interface between the curtain and edge fluid, i.e., to $H_{\delta 2}$ in Fig. 5.153, and by changing the boundary condition at that location, i.e., Eq. (5.9.3a), such that the velocity of the boundary layer equals the surface velocity of the lubricating film $v_{film,max}$ according to Eq. (5.7.3). The other two boundary conditions remain unchanged, with the velocity V outside of the boundary layer being equal to the curtain velocity v_c according to Eq. (5.8.1).

Now, the constants a_i in Eq. (5.9.2) change as follows:

$$a_0 = v_{film,max} \tag{5.9.20a}$$

$$a_1 = \frac{-2}{H_\delta}\left(v_{film,max} - v_c\right) \tag{5.9.20b}$$

$$a_2 = \frac{\left(v_{film,max} - v_c\right)}{H_\delta^2} \tag{5.9.20c}$$

Taking the derivative of Eq. (5.9.2) with these new values for a_i yields the following velocity gradient, or the shear rate, inside the boundary layer at the interface with the lubricating film. The corresponding viscosity according to the Carreau-Yasuda equation can now be determined.

$$\gamma_0 = a_1 = \frac{-2}{H_\delta}\left(v_{film,max} - v_c\right) \tag{5.9.21}$$

Consider, however, that the shear rate and viscosity depend on the boundary layer thickness, which in turn also depends on the viscosity. Resolving this dilemma requires iteration by assuming a viscosity value to calculate the boundary layer thickness according to Eq. (5.9.19), calculating the corresponding Carreau-Yasuda viscosity with the help of Eq. (5.9.21), and repeating this procedure until the assumed and calculated viscosity values become equal. Again, the goal-seek function in EXCEL is a suitable tool for accomplishing this task. Remember that the curtain velocity increases continuously between the die lip and the substrate. Therefore, the shear rate inside the boundary layer also increases and the resulting viscosity may decrease, depending on the parameter values of Carreau-Yasuda equation.

Like in film flow, aspects other than a non-uniform velocity profile in cross-web direction contribute to adverse flow effects in the region of the boundary layer in curtain flow. Specifically, excessive spreading of the curtain fluid on a solid or liquid surface of the curtain edge guide results in an uneven curtain thickness profile in the vicinity of the edge guide as sketched in Fig. 5.155a. Of particular interest is the spreading of the curtain fluid on the low-viscosity lubricating fluid, which, quite

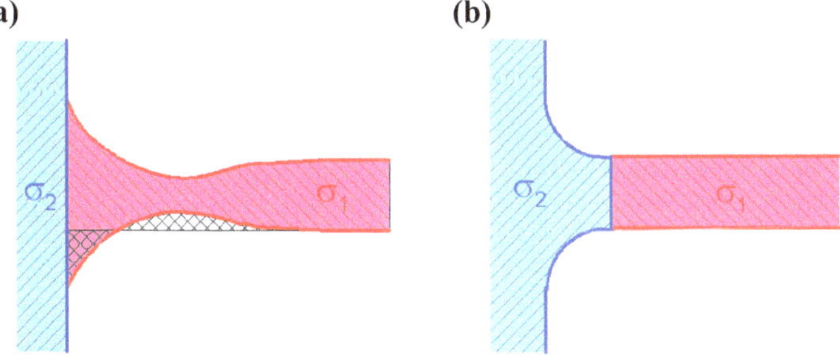

Fig. 5.155 **a** S > 0: spontaneous spreading of liquid 1 (curtain) on liquid 2 (edge fluid) **b** S < 0: retraction of liquid 2 (edge fluid) by liquid 1 (curtain)

often in industrial curtain coating processes, flows down along the curtain edge guide as explained above.

According to the work of Young (1805; see also comprehensive summary by Tricot, 1997) liquid 1 spontaneously spreads on liquid 2 if the spreading coefficient S is positive:

$$S = \sigma_2 - \sigma_1 > 0 \qquad (5.9.22)$$

σ_1 is the surface tension of the curtain fluid and σ_2 is the surface tension of the edge fluid. The wedge of fluid that is drawn from the curtain of uniform thickness by the spreading process is taken from the curtain just outside of the spreading zone, see Fig. 5.155a. This in turn causes the curtain to locally thicken just next to the edge guide and to thin a short distance away from the guide. This effect is not welcome because it weakens the curtain near the edge guide and increases the chances for the curtain to break up and detach from the guide, particularly toward the bottom of the curtain where its uniform thickness is the thinnest.

As illustrated in Fig. 5.155b spontaneous spreading is not possible, if the spreading coefficient is negative, i.e., if $\sigma_2 < \sigma_1$. Indeed, flow is driven in the opposite direction, as the curtain fluid draws edge fluid out of the plane of the edge guide. This effect is not wanted either, but it does not seem to thin out and thus weaken the curtain. An example of this phenomenon is shown in Fig. 5.156. A curtain made of aqueous pressure-sensitive adhesive can draw quite a bit of edge fluid (water in this case) away from the edge guide, particularly toward the bottom of the curtain. The outside edge of the curtain consists of transparent water, but the curtain remains stable and does not detach from the edge guide.

Spreading of the curtain on the edge fluid is not desirable, and spreading can be avoided if the spreading coefficient S = 0. This requires that the surface tension of the edge fluid is equal to the surface tension of the curtain. As explained in Sect. 5.8.5,

Fig. 5.156 Withdrawal of edge fluid by the curtain fluid for S < 0; photo reproduced with permission from Polytype Converting AG

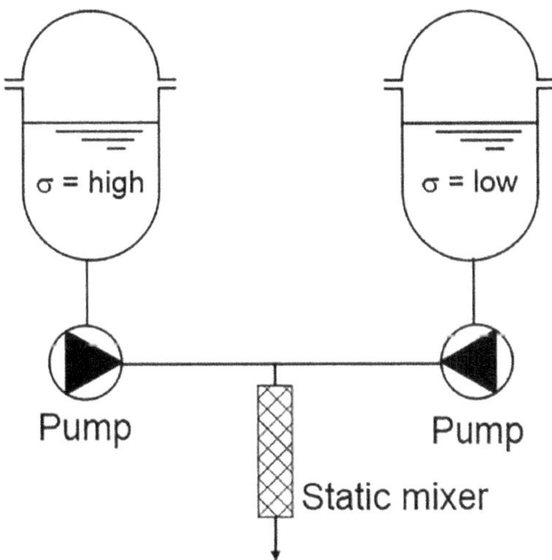

Fig. 5.157 Adjusting the surface tension of the edge fluid by mixing two fluids with different surface tension

acceptable curtain stability in industrial coating processes can be achieved, if the surface tension of the curtain is less than about 40 mN/m. For aqueous coating fluids, this necessitates adding one or several surfactants. Consequently, the surface tension of the coating fluid is typically not constant as it falls along the curtain, so it is difficult to determine the correct surface tension value for the edge fluid. Therefore, using a trial-and-error approach is an appropriate way to determine the best surface tension of the edge fluid.

At Polytype Converting AG, this is accomplished by mixing a fluid with high surface tension, e.g., water with $\sigma = 72$ mN/m, together with a fluid having a low surface tension, preferably $\sigma < 30$ mN/m, as sketched in Fig. 5.157. The latter fluid can be prepared by starting with water and by adding either an organic solvent (solvent concentration in the range of 30–40%, see Fig. 4.17) or a surfactant (surfactant concentration in the range of 0.1–0.5%). For organic coating fluids, i.e., solventless monomers or fluids containing one or several organic solvents, it is more difficult to find an edge fluid that has a low viscosity, i.e., <3 mPas, plus a higher or lower surface tension than the coating liquid, and that is not poisonous or skin-irritating at the same time. This is because the surface tension of many organic solvents is in the range of 20–30 mN/m. In our experience, suitable solvents with a relatively high surface tension include n-methyl-2-pyrrolidone (NMP, $\sigma = 41$ mN/m, $\mu = 1.7$ mPas at 25 °C) or Dowanol PGDA (from Dow, $\sigma = 32.6$ mN/m, $\mu = 2.8$ mPas at 20 °C). Examples of solvents with a relatively low surface tension include iso-propanol ($\sigma = 20.9$ mN/m) or acetone ($\sigma = 22.9$ mN/m). Moreover, the combination of iso-propanol and Dowanol PGDA was suitable for several applications.

Static mixers and small gear pumps from Gather (www.gather-industrie.de) have proven reliable for us for industrial applications. The surface tension of the edge fluid can be adjusted in-line by changing the flow rates of the two fluids. This, however, is not so easy because the total flow rate must always be equal to the optimum value for a good operation of the edge guide, and the flow rate ratio must be found such that the resulting surface tension matches the surface tension of the curtain fluid.

Since edge fluids have low viscosity and low flow rates, the pressure drop across the delivery system is low. Therefore, a simpler gravity-driven delivery system in combination with rotameters (conical glass tube with a floater) works just as well, particularly if the fluid supply vessels can be mounted several meters above the level of the coating station.

The procedure described above is suitable and convenient for finding the best surface tension of the edge fluid for a given curtain fluid. Once that value is known, in-line mixing is no longer required because the correct edge fluid can be prepared right away by adding the correct amount of solvent or surfactant.

In order to test this concept, we prepared two different curtain fluids and two different edge fluids as follows:

- First curtain fluid: PVA + H_2O + ethanol + methylene-blue; $C_{w,PVA} = 0.1035$, H_2O:$C_2H_5O = 70{:}30$, $\rho = 975\ \text{kg/m}^3$, $\mu_0 = 350$ mPas, $\sigma = 32.0$ mN/m.
- Second curtain fluid: PVA + H_2O + surfactant + methylene blue; $C_{w,PVA} = 0.0955$, $C_{surfactant} = 0.002$ (Surfynol PSA 336), $\rho = 1012\ \text{kg/m}^3$, $\mu_0 = 332$ mPas, $\sigma = 31.3$ mN/m.
- First edge fluid: H_2O + ethanol, C_w, ρ, μ, σ = variable.
- Second edge fluid: H_2O + surfactant (Surfynol PSA 336), C_w, ρ, μ, σ = variable.

Four different combinations of curtain and edge fluid were formed, all of which resulted in the same characteristic flow behavior as illustrated in Fig. 5.158 with permission from Polytype Converting AG.

- $S > 0$: Spreading of the curtain fluid on the edge fluid, which causes local thinning of the curtain inside the boundary layer, resulting in a wavy wetting line between the curtain and edge fluid, and causing the curtain to detach from the edge guide if spreading becomes excessive, see Fig. 5.158a.
- $S \approx 0$: Perfect flow behavior of the curtain fluid on the edge fluid, i.e., no spreading, straight wetting line between curtain and edge fluid, stable flow at the suction slot at the bottom of the edge guide, and no local curtain thinning inside the boundary layer, see Fig. 5.158b.
- $S < 0$: Extraction of edge fluid by the curtain fluid, eventually leading the curtain to detach from the edge guide, see Fig. 5.158c, d.

In summary, continuously changing the surface tension of the edge fluid enables the flow field between the curtain and edge fluid to change from bad to optimum and again back to bad flow conditions, affecting local curtain thinning and curtain stability near the edge guide. This situation describes a classic optimization problem that can be solved perfectly with the concept outlined above.

The surface tension of the edge fluid must not only be optimized relative to the surface tension of the curtain fluid, but also in relation to the surface energy of the material of the edge guide body. Many curtain edge guides are made of stainless steel, which has a surface energy of about 55 mN/m. Therefore, if the edge fluid were pure water with a surface tension of 72 mN/m, the edge fluid could not properly wet the edge guide. Instead of forming a uniform and coherent film along the edge guide, the edge fluid would form irregular rivulets, which may or may not be in contact with the curtain fluid. In addition, the flow behavior inside the suction slot at the bottom of the edge guide would deteriorate because the edge fluid would not arrive there as a uniform film, thus increasing the chances for the slot to plug up, particularly when the curtain fluid is an adhesive.

Therefore, proper wetting of the edge guide by the edge fluid is a must, and this requires the surface tension of the edge fluid to be less than about 50 mN/m. However, since the surface tension of the edge fluid should be the same as the surface tension of the curtain fluid, and since the latter should be less than about 40 mN/m to obtain a robust and stable curtain, this requirement will be easy to satisfy.

The edge fluid must not only be optimized to the surface tension, but other chemical and physical properties should also be included in this procedure. This is particularly important to prevent the suction slot at the bottom of the edge guide from plugging prematurely during long industrial coating applications, especially if the coating fluid is an adhesive. The following suggestions may be considered:

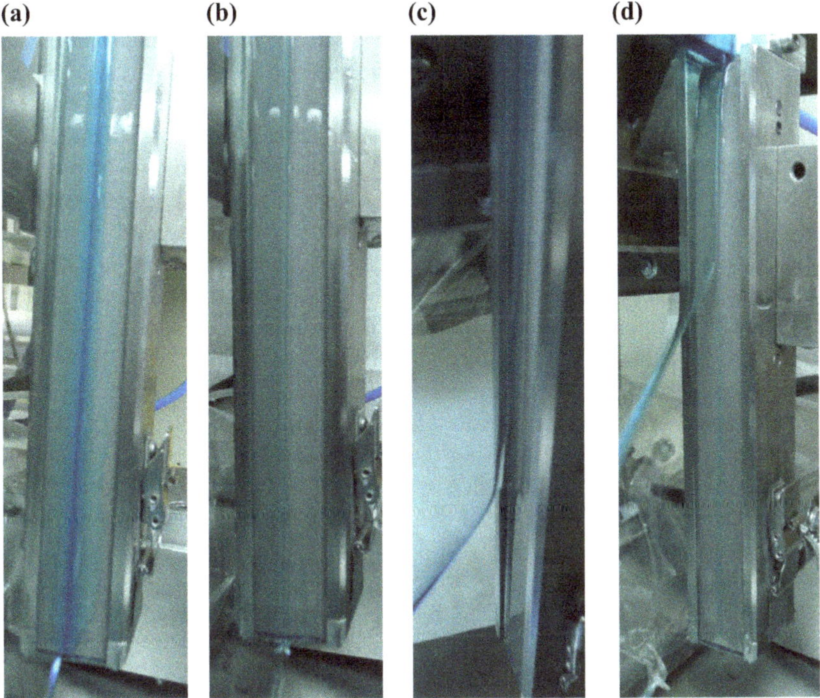

Fig. 5.158 **a** S > 0: spreading of the curtain fluid on the edge fluid **b** S = 0: perfect curtain flow along film of edge fluid **c** S < 0: extraction of edge fluid by curtain fluid **d** S < 0: detachment of curtain from edge fluid

- The pH-value of the edge fluid should match the pH of the curtain fluid to avoid coagulation of the curtain fluid inside the suction slot. The pH of aqueous pressure-sensitive adhesives is in the range of 3.5–6.0.
- The temperature of the edge fluid (and the edge guide body) could be increased to lower the viscosity of the two-phase flow inside the suction slot.
- The edge fluid should be soluble in the curtain fluid. Care must be taken, if the edge fluid is a mixture of water and an organic solvent because not all solvents, such as ethyl-acetate or MEK, can be mixed at will with water. In contrast, ethanol or iso-propanol is miscible with water at any concentration.
- The temperature of the edge guide body could be reduced below the dew point, such that the edge guide starts to sweat. This way, condensation on the edge guide body would form a thin liquid film, which could reduce the chances for the suction slot to plug.

Testing these ideas for several hours in the labs of Polytype Converting produced promising results. In particular, when running a PSA from BASF with an edge fluid consisting of a 16% isopropanol-water mixture having a surface tension of 33 mN/m, the surface of the edge guide was perfectly wetted and the suction slot did not plug or accumulate adhesive for 4 h of operation.

In the previous paragraphs wetting between the curtain edge guide and the edge fluid was accomplished by forcing the spreading coefficient S to become positive, i.e., by adjusting the surface tension of the edge fluid until it was lower than the surface energy of the edge guide material. The same goal could potentially also be achieved by modifying the surface properties of the edge guide. The patents by Güggi and Varli (2001) or by Gugler and Pasquier (2002), for example, describe a mechanical modification, in that the edge guide surface was equipped with many fine longitudinal grooves along the guide, see Fig. 5.159. The purpose of this alteration was to prevent the formation of rivulets in the film of the edge fluid. Our efforts at Polytype Converting AG included sand-blasting the edge guide surface with sand of different particle diameters, also to prevent rivulet formation.

We also focused on rendering the edge guide surface more hydrophilic by coating it with an appropriate chemical material. We cooperated with Buser Oberflaechen-technik AG, a Swiss company specialized in high-quality surface coatings, and with the Department of Materials Science at the Tampere University of Technology in Finland. These specialists used processes such as liquid flame spray coating, powder

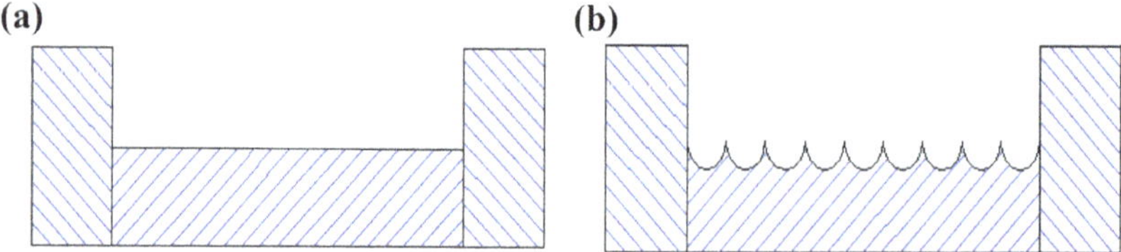

Fig. 5.159 **a** Cross-section of curtain edge guide with a flat surface **b** cross-section of curtain edge guide with a grooved surface, e.g., EP 1 398 084 A1 or WO 03/049870 A1

coating, and sintering for applying different materials including nano-particles of TiO_2, SiO_2, and others. We also tested a hydrophilic tape from TESA. While many of these coatings initially showed promising hydrophilic properties, they all suffered from a loss or at least a degradation of these properties over time and use, i.e., cleaning, and for some of them, the adhesion toward the edge guide material was insufficient.

In summary, optimizing the surface tension of the edge fluid seems a more viable approach for industrial curtain coating processes than mechanically or chemically modifying the surface of the edge guide.

5.9.5 Curtain Edge Guides

Designing suitable curtain edge guides has been a major concern of the coating industry for more than 50 years. One of the main reasons is that the weakest point in curtain coating is located at the bottom of the curtain right next to the edge guide because the curtain is thinnest there. Therefore, any disturbance resulting in further thinning will cause the curtain to break and to detach from the edge guide, which is a catastrophic process failure.

A large number of patents on this topic have been filed over the years, first mainly by photographic companies and later mainly by suppliers of curtain coating equipment for the converting and paper industry. Since most of the photographic companies no longer exist, at least not in the form they had when photographic products were in high demand, it might be worthwhile to check whether or not their patents are still pending. The vast quantity of patents illuminates how designing a good edge guide is a difficult and non-trivial task because the guide must satisfy many process-technological requirements, such as the following:

- Accommodation of a low-viscosity edge fluid to reduce the thickness of the viscous boundary layer along the edge guide, which in turn helps to prevent local air entrainment.
- Absorption of curtain deflection (bend back).
- Prevention of local curtain thinning along the edge guide, which prevents the curtain from detaching from the edge guide.
- Prevention of curtain neck-in at the bottom of the edge guide, which prevents heavily coated edges from being formed, and so avoids telescoping problems during re-winding.
- Accommodation of a variable curtain impingement angle, which allows optimization of the flow field of the impinging curtain, e.g., prevention of air entrainment.
- Good performance for various product formulations, i.e., operating conditions and physical fluid properties.

Most edge guides incorporate both welcome and objectionable design and performance features. The challenge is to find a design with as many good features as possible, and as few as possible negative ones.

Two basic design types are known in the industry, namely line-type and surface-type edge guides. With line-type guides, the curtain is guided by a straight line such as a string, a thin rod, or a narrow groove on a surface. These designs may be appropriate if the curtain exhibits no deflection or bend-back along its trajectory. Moreover, the location of the curtain impingement point is well defined by the position of the guide. With surface-type designs, in contrast, the curtain is not guided by any mechanical means, but it adheres to a moving surface provided by a thin film of lubricating fluid, and it is free to assume any form of trajectory, for example as a result of bend-back. Consequently, the location of the curtain impingement point may not be known precisely. In addition, the consumption of edge fluid is higher than with line-type designs.

Exemplary desirable and undesirable features of both basic design types are discussed in the following paragraphs.

5.9.5.1 Line-Type Edge Guides

An interesting concept of a line guide was patented by Reiter (1992), see Fig. 5.160. The guide is made of two small-diameter wires that run from the die lip to the substrate. The wires are spaced apart by a distance similar to the curtain thickness. Edge fluid is fed between the wires on top and removed again at the bottom. The beauty of this design is that the curtain adheres to a thin membrane of moving edge fluid. This should reduce the friction between the curtain and the edge guide compared to other designs and hence reduce the thickness of the boundary layer. In addition, this guide can be combined with a suction device at the bottom as shown in Fig. 5.161 and patented by Ruschak and Conroy (1994). Here, the positive features are that the edge of the falling curtain is intercepted by a thin blade and vacuumed

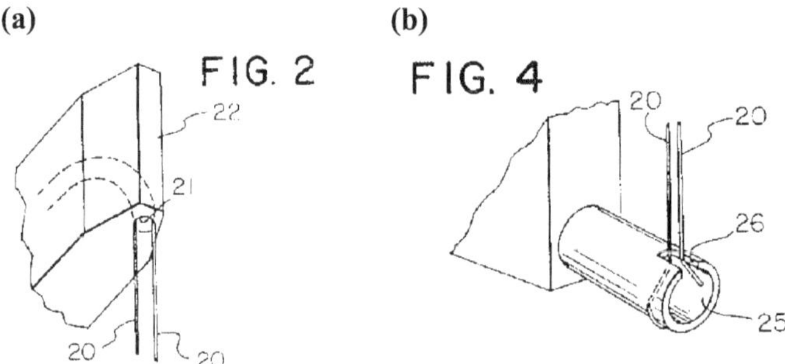

Fig. 5.160 **a** Top of line-type curtain edge guide; US 5,328,726 **b** bottom of line-type curtain edge guide; US 5,328,726

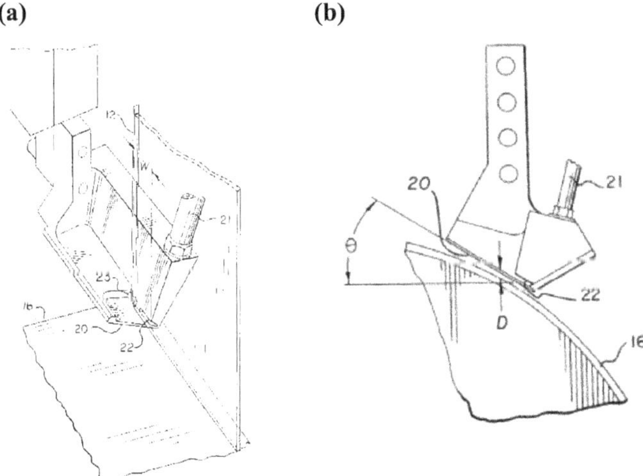

Fig. 5.161 **a** Bottom of line-type curtain edge guide with intercepting blade and vacuum system; US 5,395,660 **b** tilted bottom of line-type curtain edge guide for accommodating variable curtain impingement angle; US 5,395,660

away, thus not introducing any additional friction at that location. Also, the device can easily be tilted and so placed in close proximity to the substrate at any given curtain impingement angle.

Line-type edge guides may be appropriate for single-layer curtains at lower flow rates. As explained in Sect. 5.8.6, curtains with a flow rate of $Q > 3.5\ cm^2/s$ start to deflect appreciably. Consequently, the resulting two-dimensional curtain trajectory is no longer compatible with a straight-line edge guide, and this geometrical mismatch should be avoided. The same arguments may apply to multilayered curtains because, depending on the structuring of viscosities and surface tensions, they tend to bend even at lower total flow rates.

Schweizer et al. (2004) described a line-type edge guide that has self-centering properties for the curtain. The design is similar to the one shown in Fig. 5.127, except that the porous body is not a flat plate but a porous tube that is conical on the inside, or a porous, conical cylinder with one convex surface as sketched in Fig. 5.162. Owing to the flow of edge fluid through the porous material, the outside of the convex surface is entirely covered with a low-viscosity lubricating layer, and the preferred location for the curtain to contact the guide is at the vertex of the convex surface. In an industrial coating environment, the curtain does not always fall perfectly still, but it is in slight motion owing to weak air currents in the factory hall. This motion causes the curtain impingement point to move constantly, which may deteriorate the uniformity of the coated film and/or the performance of the suction slot at the bottom of the edge guide. If the curtain is displaced in machine direction by the amount of Δy, it must expand its width by Δx in the cross-web direction to maintain contact with the edge guide, see Fig. 5.162. However, capillary forces oppose this expansion and push the curtain back to the position of the smallest width, which is at the vertex of the convex surface.

This concept works very well. It is not as stiff as a rod or a pair of wires or a groove, which is good because the flow of a curtain is not stiff either. However, as already mentioned above, porous materials are not viable in industrial coating processes due to their tendency to plug after a short period of operation.

5.9.5.2 Plate-Type Edge Guides

Plate-type edge guides are necessary or at least recommended for operating conditions that result in appreciable curtain deflection, but they also work well for conditions without curtain bend-back. The width of the edge guide should be in line with the expected amount of deflection, see Sect. 5.8.6. Experience has shown that a width of 20 mm is suitable for many applications.

Top of Edge Guide

Plate-type edge guides are also covered with a film of lubricating liquid. The formation of this film requires a small slot die to be located on top of the guide. Moreover, the introduction of the film into the vertical plane of the guide can be accomplished in several ways. Conroy and Ruschak (1992) proposed a curved slot as drawn in Fig. 5.163. Their suggested operating conditions result in a distance L of several millimeters. L is the distance below the die lip, at which the edge fluid comes in contact with the curtain. Consequently, the curtain is not lubricated along these several millimeters, which may cause the boundary layer thickness to grow excessively.

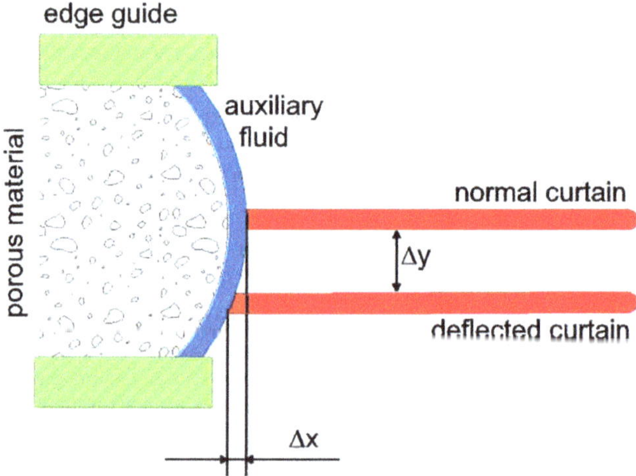

Fig. 5.162 Line-type edge guide with self-centering properties for the curtain

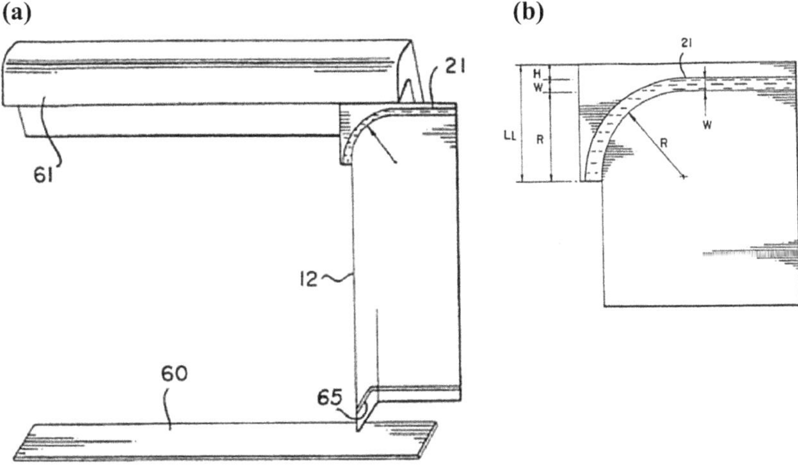

Fig. 5.163 **a** Edge guide with feeding and evacuation system for lubricating fluid: overview from US 5,358,569 **b** Edge guide with feeding system for lubricating fluid: geometrical details

In contrast, Schweizer (2010) shrunk the size of a similar feeding system for the lubricating film such that the distance between die lip and first contact between the curtain and edge fluid decreased to 2.5 mm.

Bottom of Edge Guide

Designing the bottom part of the edge guide is particularly challenging. On the one hand, the edge fluid must be removed with a suction system. Typically, however, not only the edge fluid but also a small amount of curtain fluid is vacuumed away. In any case, these systems must not plug after several hours of operation, even for highly viscous and tacky curtain fluids such as adhesives. As the edge guide cannot extend down to the substrate surface, on the other hand, the curtain must detach from the guide at its bottom, and this separation must be done without generating heavily coated edges. In addition, the coated edges should be well defined and move straight in the machine's direction.

Some designs use protruding elements such as blades or other cutting devices that intersect the curtain several millimeters away (inside) from the edge guide body, see for example Güggi et al. (1995). The purpose of these protruding elements is to cut off the boundary layer of the curtain and any other effects that cause the thickness of the curtain to thin near the guide. The drawback of this idea is that several millimeters of curtain fluid on each side are discarded and, in most cases, wasted, which can be costly, particularly for multilayer applications. Another disadvantage is that the die must twice be built several millimeters wider than the desired coating width, which also increases the investment costs. Figure 5.161 shows a nice example of such a design, which works particularly well for positive curtain impingement angles (see Fig. 5.78), i.e., when the web moves downward at the curtain impingement point

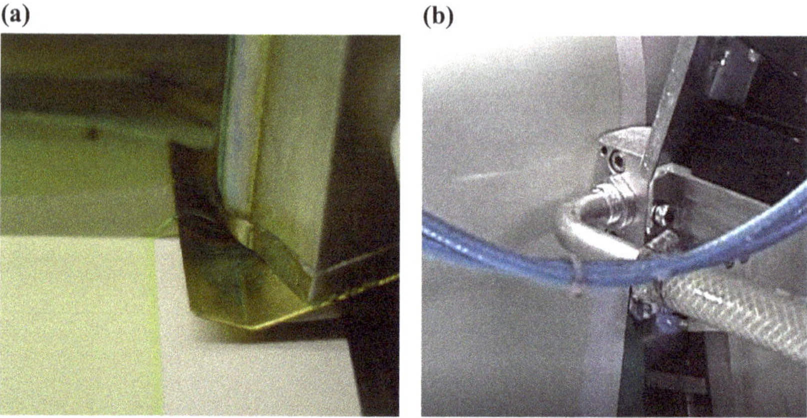

Fig. 5.164 **a** Neck-in at the bottom of a curtain edge guide resulting in a heavily coated edge on the substrate; DE 10 2004 016 923 B4; photo reproduced with permission from Polytype Converting AG **b** slot die-type heavy edge removal device (HERD) from TSE Troller AG; photo reproduced with permission from Polytype Converting AG

because the suction slot is located below the impinging curtain, thus assisting gravity to easily collect and direct the edge fluid to the suction slot for removal.

A second example, which works well when the curtain impinges perpendicularly onto the web, is shown in Fig. 5.164a (Schweizer et al., 2004). The disadvantage of this design is that the protruding blade is bent upwards to facilitate the collection and removal of the edge fluid. Consequently, the cutting edge of the blade is located relatively high above the substrate surface. This causes the detaching curtain to neck-in, which in turn generates excessively heavy-coated edges as visualized by the dark green line on the coated substrate.

As is explained in Sect. 5.8.8 heavy edges caused by neck-in at the bottom of the edge guide, i.e., by the tip of the blade, in this case, cannot be prevented if the point of curtain detachment from the guide is not located very close to the substrate. In industrial coating processes, therefore, this type of edge guide is often used in combination with a so-called heavy edge removal device (HERD). Popular HERD systems are suction devices in the form of round nozzles or narrow slot dies that are placed a short distance downstream of the curtain impingement point. Moreover, they are mounted onto a 2-dimensional positioning system to optimize the cross-web location relative to the location of the heavy edge and the distance to the substrate. They may require a short skip-up during the splice passage, and they must be supplied with ample rinsing fluid to prevent plugging. An example of a patented round design is given by Iwata et al. (1984), who worked for Fuji Photo Film. Figure 5.164b shows an example of a slot die-type HERD manufactured by TSE Troller AG. The coated web is moving from top to bottom, and it is clearly visible that the heave edge upstream of the HERD is removed, but the coating width is also reduced by a small amount. According to our experience, HERD may work well for non-tacky

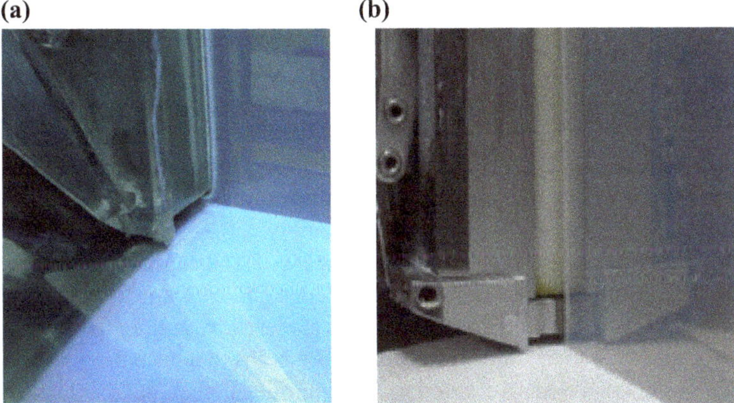

Fig. 5.165 **a** Bottom of curtain edge guide without neck-in and heavy coated edge on the substrate; photo reproduced with permission from Polytype Converting AG **b** bottom of curtain edge guide without neck-in and heavy coated edge on the substrate; photo reproduced with permission from Polytype Converting AG

coating fluids of low viscosity, such as photographic emulsions, but they are not recommended for viscous and tacky fluids such as adhesives.

At Polytype Converting, many development efforts were directed toward designing curtain edge guides without any protruding elements at the bottom. The motivation for achieving this goal was avoiding neck-in, the associated heavy edge and subsequent winding problems. Examples of acceptable flow behavior at the bottom of the curtain edge guide without any neck-in and heavy edges on the substrate are shown in Fig. 5.165.

The pursued design strategy was two-fold:

- To prevent any disturbances along the boundary layer of film and curtain flow, particularly with regard to local thinning of these flows, and in the curtain flow.
- To move the suction slot to the very bottom of the edge guide in an attempt to approach a zero gap between the edge guide and the substrate surface.

Along the way, we confirmed the following facts.

An edge guide with a bottom surface that is parallel to the substrate surface is worse than a version with an inclined bottom surface as illustrated in Fig. 5.166. This is because the chances are higher for trapping debris carried along by the uncoated web or liquid drops that may be generated during the coating start, particularly if the gap between the edge guide bottom and the substrate is very small. If this happens then the coated edge will not be well defined and straight in the machine direction. In contrast, an edge guide with a tilted bottom only makes line contact with the substrate, which is much less critical with regard to generating defects. An edge guide design of this kind is also depicted in Fig. 5.165a.

A suction slot with a variable instead of a fixed geometry at the mouth of the slot is much better for optimizing the performance of the guide regarding the quality of the coated edge. In particular, allowing the distances e, f, and w, as sketched in Fig. 5.167, to vary offers considerable design freedom.

In particular, f must be minimized as much as possible to minimize neck-in at the bottom of the edge guide. In our experience f = 0.2 mm results in a good performance. However, such a small value requires a skip-up during the splice passage to prevent a collision between the guide, the splice, and a subsequent web break. Depending on the design of the edge guide the skip-up must be designed for either the edge guide alone or for the combined die-edge guide assembly.

Comparing the left and right edge guide in Fig. 5.168 shows us that the coating width as defined by the distance between the edge guides is slightly enlarged at the very bottom of the edge guide if e > 0. This is achieved by pulling the static wetting line a short distance to the outside, which in turn helps to reduce the size of a heavy edge that might otherwise be formed as a result of local neck-in.

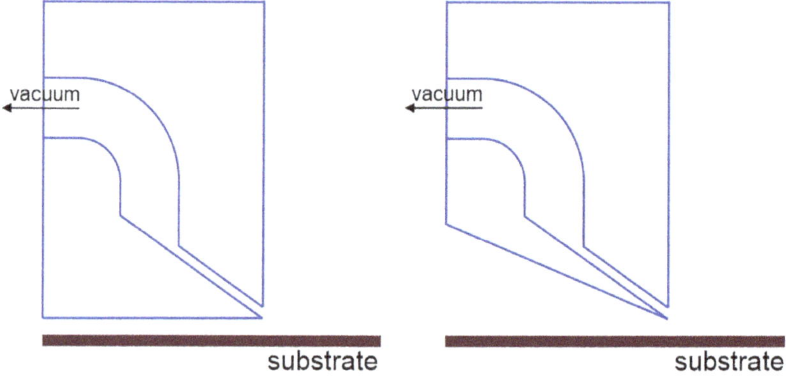

Fig. 5.166 Comparison of edge guides with parallel and inclined bottom surfaces relative to the substrate surface

Fig. 5.167 Suction slot with variable geometry

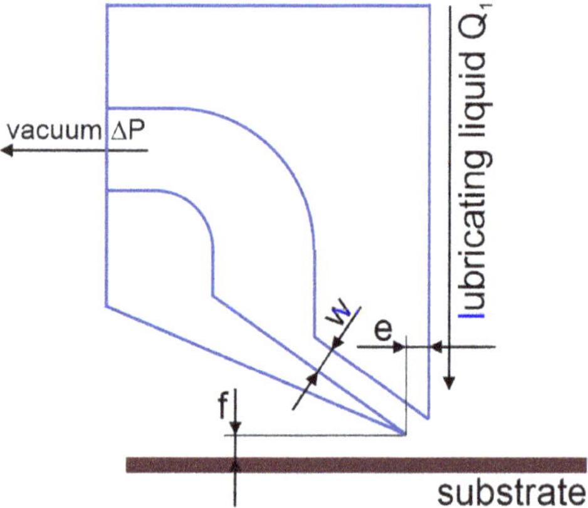

The best performance of such an edge guide design is obtained by optimizing the three geometrical parameters e, f, and w, as well as the two operational parameters Q_1, i.e., the flow rate/width of the edge fluid, and ΔP, i.e., the vacuum pressure of the suction system. As a formal relationship between these fife parameters is not known, their optimum values must be found by an experimental trial-and-error approach. If carried out successfully this procedure results in straight coated edges without any over-thickness, see Fig. 5.165.

If necessary, edge guides must be designed such that they can be positioned very near the substrate surface for curtain impingement angles of $\alpha \neq 0$, see Fig. 5.78. A good example of such a device is shown in Fig. 5.161 because it can be tilted continuously such that it works for a large range of positive impingement angles. Another design concept developed at Polytype Converting AG that works well for negative and positive impingement angles is shown in Fig. 5.169. In particular, the edge guide consists of two pieces that are connected and fixed together, whereby the lower piece can be exchanged and is designed with an inclined bottom surface. The

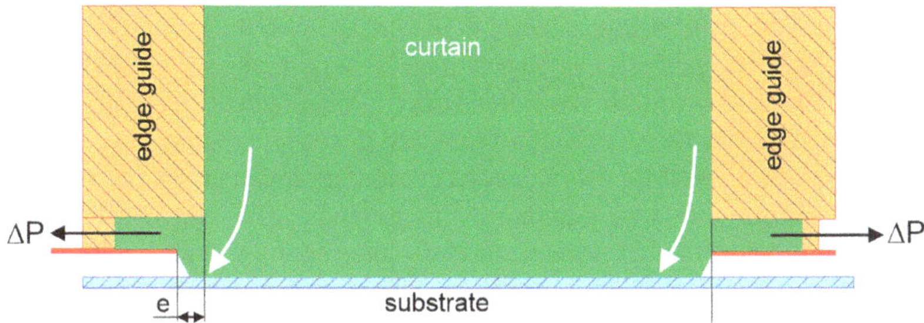

Fig. 5.168 Reduction of a heavy edge by pulling the static wetting line to the outside

Fig. 5.169 Curtain edge guide designed for curtain impingement angles of $\alpha \neq 0$; photo reproduced with permission from Polytype Converting AG

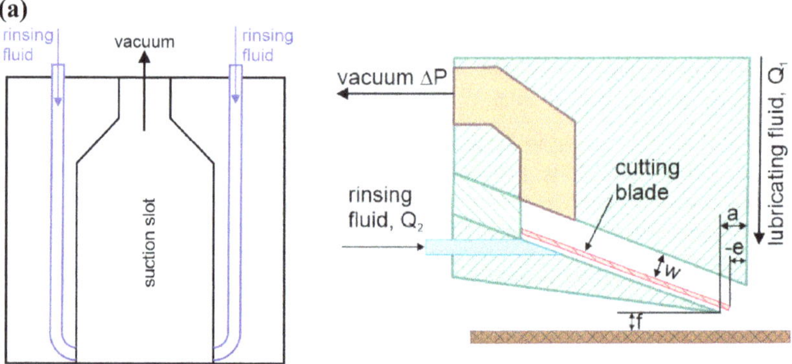

Fig. 5.170 **a** Cross-section of a curtain edge guide with a supply of extra rinsing fluid at the mouth of the suction slot (Holtmann et al., 2000) **b** cross-section of a curtain edge guide with a supply of extra rinsing fluid at the mouth of the suction slot; design by Polytype Converting AG

inclination angle of the bottom surface is in concert with the curtain impingement angle such that the edge guide can be positioned closely to the substrate surface. The performance of this edge guide is not diminished if the mouth of the suction slot is not oriented horizontally. The drawback of this design concept is that an exchangeable lower piece must be built for every desired curtain impingement angle.

Coating viscous and tacky liquids such as aqueous pressure-sensitive adhesives for label applications increases the chances for the suction slot at the bottom of the edge guide to partially plug after a relatively short period of operation. Experience has proven that this problem can be avoided, or at least postponed to much longer operating times, by feeding extra rinsing fluid near the mouth of the suction slot. A good example is sketched in Fig. 5.170a, where the extra fluid is supplied from both sides just inside the suction slot (Holtmann et al., 2000). An alternative design was developed by Polytype Converting AG, see Fig. 5.170b. Here, the rinsing fluid is not fed through two point-sources but through a slot that is located underneath the suction slot and that is as wide as the suction slot.

Ensuring a performant operation of the suction slot for several hours in a production environment requires the use of an appropriate vacuum source for removing the two-phase fluid consisting of air, curtain liquid, lubricating liquid and possibly rinsing fluid, where the latter two are typically the same. A Venturi nozzle driven by compressed air may be adequate for lower-viscosity and non-tacky fluids, but for higher-viscosity and tacky fluids such as PSA a water ring vacuum pump is recommended. This pump is more powerful but also more costly and loud. It may be advisable to install such pumps in a separate room to ensure an acceptable noise level at the coating station. Moreover, for large coating width, e.g., W > 2 m, where the left and right curtain edge guide are far apart, it is also advisable to install a separate vacuum pump for each edge guide.

As a two-phase fluid is evacuated, it is necessary to separate the gas from the liquid to protect the vacuum source from being damaged. This is accomplished by installing a drop separator in the form of a closed vessel between the edge guides and the vacuum pump. In addition, since this vessel will be filled with fluid after some time of operation, it must be equipped with a level sensor, which turns on a pump to periodically empty the vessel when the liquid level is high.

The vacuum pump must be able to generate a pressure of 0.5 bar, or preferably less, as measured in the drop separator vessel. Therefore, the diameter of the hose between the edge guides and the drop separator vessel should be large to keep the pressure drop small, but it should not be too large to keep the wall shear stress high, which prevents the hose from plugging up. In our experience, a hose diameter of 10 mm is adequate for many applications.

A P&ID for the processes of supplying and evacuating edge fluid can now be drawn when considering all the arguments presented in the chapters on boundary layers in film and curtain flow as well as the design of curtain edge guides. The diagram in Fig. 5.171 shows a possible solution that is viable for industrial coating applications. It must be decided on a case-to-case basis whether or not

- it is necessary to supply edge fluid to the film flow on a slide die, and if so how to introduce the edge fluid;
- in-line mixing is needed for adjusting the surface tension of the edge fluid;
- gravity can be used for supplying the edge fluid.

Inclination of Edge Guide

Most of the line-type and surface-type edge guides discussed in the patent literature have a vertical orientation from top to bottom. However, as explained at the beginning

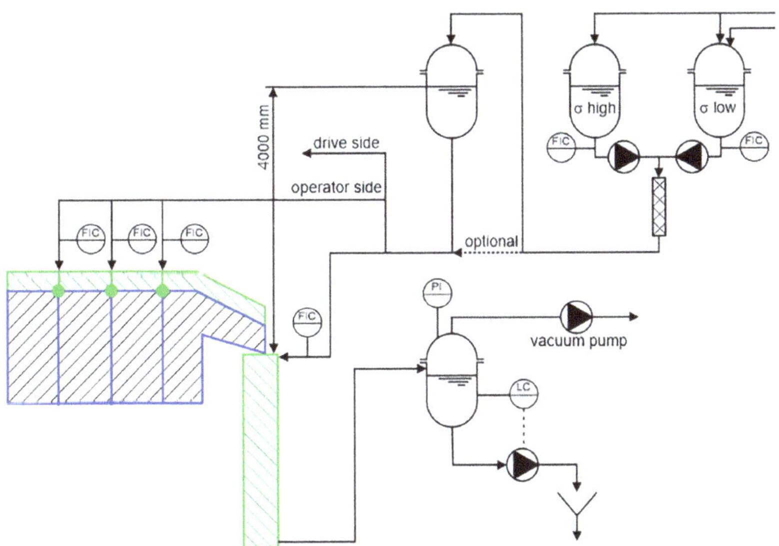

Fig. 5.171 P&ID for supplying and evacuating edge fluid

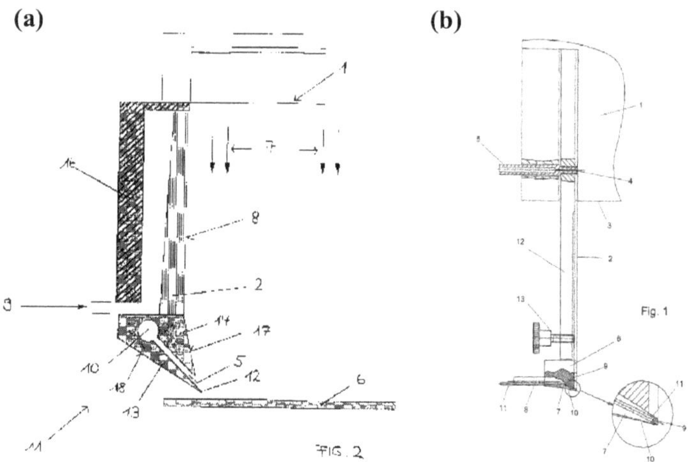

Fig. 5.172 **a** Curtain edge guide with a fixed inclination in the bottom part (Schweizer and Troller. 1997. EP 0 907 103 B1) **b** curtain edge guide with a variable inclination (Metzger et al., 2002. EP 1 377 391 B1)

of Sect. 5.9, the weakest point of a curtain is at the bottom and next to the edge guide because the liquid membrane is the thinnest there. Consequently, some of the patented edge guides incorporate inclined forms where the width of the curtain decreases from top to bottom. The idea is to redistribute curtain fluid toward the thinnest point in the curtain, thereby increasing the local curtain thickness to reduce the chances for curtain break-up.

Two examples of inclined edge guides are shown in Fig. 5.172. Figure 5.172a shows a fixed geometry where the upper ¾ of the edge guides are still vertical, while only the bottom quarter is inclined. In contrast, Fig. 5.172b shows a flexible geometry where the inclination changes gradually from top to bottom with the help of an adjusting screw. Another example is shown in Fig. 4.22 where the edge guide has a constant slope from top to bottom.

While these designs successfully prevent the curtain from breaking away from the edge guide, they tend to generate heavily coated edges because the re-distributed liquid is coated on top of the otherwise uniform film just inside of the edge guide.

At Polytype Converting we conducted a study by placing a straight or bent piece of sheet metal into an opaque curtain, which was back-lit. The shape of the bent piece was similar to the form of the edge guide surface shown in Fig. 5.172a but with a smooth and continuous change of inclination. Moreover, the pieces were positioned into the curtain in every which way, i.e., as the continuation of the sidewall of the film flow on the slide die or somewhere between the edge guides, at various inclination angles relative to vertical, upside down, etc. Each position of the sheet metals disturbed the curtain thickness to some degree, and the thickness variations were nicely visualized by the optical density variations as produced by the back light. From these observations, we then formulated the following arguments for designing

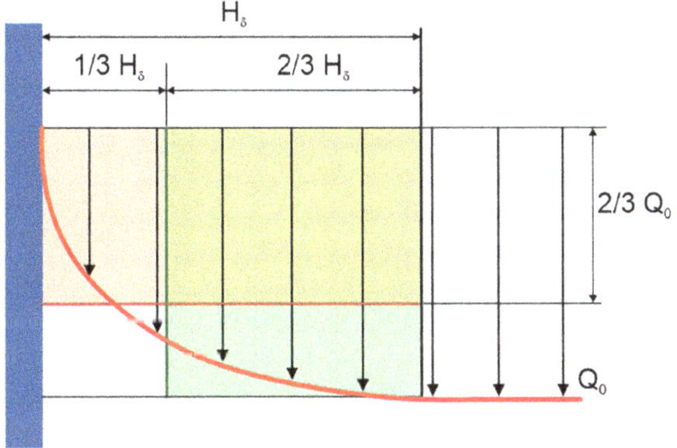

Fig. 5.173 Flow rate/width inside the boundary layer

a surface-type curtain edge guide with a variable inclination from top to bottom (Schweizer, 2010).

Focusing first on the boundary layer in curtain flow, and assuming that the velocity profile of this boundary layer can be approximated by a second-order polynomial function according to Eq. (5.9.2), it was shown above that the average flow rate/width Q_δ inside that boundary layer equals 2/3 of the uniform flow rate/width Q_0 outside of the boundary layer, see Eq. (5.9.9) and Fig. 5.130. If the curtain edge guide was built as a straight, vertical surface, then this lack of flow rate/width equaling 1/3 Q_0 would result in a local reduction of the curtain thickness. However, as illustrated in Fig. 5.173, this thickness reduction can be avoided if the width of the curtain is diminished on each side by ΔW according to

$$\Delta W = -\frac{1}{3} H_\delta \tag{5.9.23}$$

The argument above, i.e., Eq. (5.9.23), may not be perfectly correct from a fluid mechanical point of view. In particular, the boundary layer along the edge guide does not disappear by simply changing the width of the curtain. However, this geometrical measure generates liquid re-distribution right from the very beginning of the curtain since it is related to the boundary layer thickness, and this re-distribution increases the local curtain thickness. Moreover, this re-distribution stays more or less in place since gravitational forces are not available for leveling because the curtain drops vertically, and surface tension forces, which are available for leveling, are weak because the surface tension of the curtain fluid is typically <40 mN/m. The argument above may be called the "break effect" as it is related to the fact that the curtain velocity inside the boundary layer is slowed down relative to the velocity outside of the boundary layer.

In addition, a second argument, which may be called the "dam effect", was already presented in Sect. 5.9.3 and Fig. 5.136, where the transition from slot to film flow was discussed. This argument applies, in the same way, to curtain flow as illustrated in Fig. 5.174.

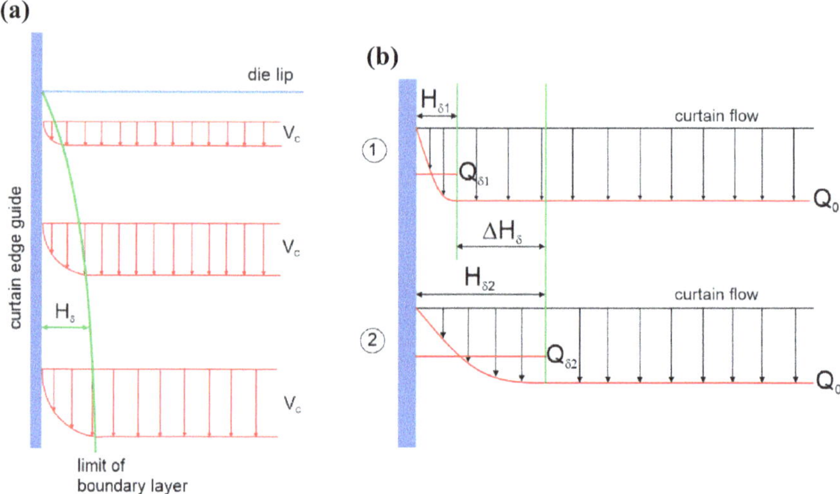

Fig. 5.174 **a** General form of the boundary layer in curtain flow **b** boundary layer in curtain flow at two locations along the curtain

It is assumed that

- The boundary layer in the curtain starts on top of the edge guide, i.e., just below the die lip.
- The boundary layer thickness increases with increasing distance z along the curtain, and it can be calculated with Eq. (5.9.19).
- The velocity profile of the boundary layer can be approximated by a second-order polynomial function according to Eq. (5.9.2).
- The average flow rate/width inside the boundary layer equals 2/3 of the uniform flow rate/width outside the boundary layer.

Now, the total volumetric flow rate V^*_1 at location 1 inside the boundary layer thickness at location 2 equals

$$V_1^* = Q_0\left(H_{\delta 2} - \frac{H_{\delta 1}}{3}\right) \tag{5.9.24}$$

Similarly, the total volumetric flow rate V^*_2 inside the boundary layer at location 2 is calculated to give

$$V_2^* = \frac{2}{3}Q_0 H_{\delta 2} \tag{5.9.25}$$

The difference between these two flow rates is

$$\Delta V^* = \left(V_1^* - V_2^*\right) = \frac{Q_0}{3}(H_{\delta 2} - H_{\delta 1}) \tag{5.9.26}$$

Since the boundary layer thickness at location 2 is larger than at location 1 the right-hand side of Eq. (5.9.26) is always positive, which signifies that the flow rate at location 1, which feeds the flow at location 2, is larger than the flow rate that can be transported downstream at location 2. This in turn generates an accumulation of fluid, which is equivalent to a local thickening of the curtain, which is not desirable. Just with the break effect, it is argued that this curtain thickening can be avoided if the curtain is widened on each side by the amount of

$$\Delta W = \frac{\Delta V^*}{Q_0} = \frac{1}{3}(H_{\delta 2} - H_{\delta 1}) \tag{5.9.27}$$

Now, it is clear that the width on each side of the curtain must be decreased owing to the break effect of the boundary layer on the one hand, and it must be increased due to the dam effect of the boundary layer on the other hand. The sum of these two effects can be quantified as follows:

$$\Delta W = -\frac{1}{3} H_\delta \tag{5.9.28}$$

So far, we assume that the boundary layer in the curtain starts at the beginning of the curtain. For this to be true the flow upstream of the curtain, i.e., the film flow on the inclined die slide would have to be uniform across the flow, such as depicted in Fig. 5.125a. However, we know that this is not the case owing to the boundary layer that is also present in the film flow. Consequently, the average flow rate there is also only 2/3 of the uniform flow rate outside of the boundary layer. Hence, the boundary layer of the curtain is fed with an insufficient flow rate, which results in a local reduction of the initial curtain thickness, which is not wanted. To avoid this effect the width of the film flow should be reduced on each side by ΔW_{film} at the end of the film flow, i.e., on the vertical part of the die lip.

$$\Delta W_{film} = -\frac{1}{3} H_{\delta, film} \tag{5.9.29}$$

While this additional width change is possible but cumbersome in principle from a design point of view, we decided to incorporate the effect into the design of the curtain edge guide. In particular, the boundary layer thickness of film flow is calculated with Eqs. (5.9.18), (5.9.11), and (5.9.12), and the approximate value of f_{film} in Eq. (5.9.18) is 0.5. Even though the change from film to curtain flow is sudden, we believe it is better to distribute the width change associated with the boundary layer of film flow over a transition length L_{trans} defined by the empirical Ansatz that the transition length is proportional to the boundary layer thickness of the film, i.e.,

$$L_{trans} = f_{trans} H_{\delta, film} \tag{5.9.30}$$

and that the required width-change according to Eq. (5.9.29) follows a power law inside the transition length defined as

$$\Delta W_{film} = -\frac{1}{3} H_{\delta, film} \left(\frac{z}{L_{trans}} \right)^k \quad \text{for } 0 < z < L_{\text{trans}} \tag{5.9.31}$$

z is the coordinate along the curtain edge guide. Evaluating the performance of several edge guides designed with the theoretical model presented above revealed that the Ansatz given by Eqs. (5.9.30) and (5.9.31) works well, and good results are obtained for $f_{trans} = 8.0$ and $k = 0.5$.

Now we have a fluid-mechanical model based on boundary layer considerations for calculating and optimizing the curved shape of a surface-type curtain edge guide. In particular, this model accounts for the break and dam effect in the boundary layer of the curtain, and it includes the transition from film to curtain flow, which depends on the boundary layer thickness of the film flow. In addition, the model works for slide and slot dies, which generate different initial curtain velocities. The viscosity of the film flow is calculated with the variable power law method, and the viscosity of the curtain flow is based on the shear rate at the interface between the edge and curtain fluids. All necessary calculations including several iterations can easily be implemented in an EXCEL environment.

As an example, the shape of a right edge guide is calculated for the following conditions: $\mu_0 = 50{,}000$ mPas, $\mu_\infty = 50$ mPas, $\gamma_c = 0.008\ s^{-1}$, $a = 2.0$, $b = 0.27$, $\rho = 1010\ kg/m^3$, $A_{dry} = 18\ g/m^2$, $C_w = 0.59$, $U = 450$ m/min, $Q = 2.27\ cm^2/s$, slide die, slide angle at die lip = 90°, edge fluid = water, $Q_{edge\text{-}fluid} = 0.35\ cm^2/s$, $L_c = 150$ mm. Moreover, 2.5 mm on top of the guide are consumed by the feeding system for the edge fluid, and 10 mm at the bottom of the guide are used for housing the vacuum slot for removing the edge fluid. The shape of the left edge guide is identical but mirrored.

Figure 5.175 shows the result of the calculations. Note that the negative values of ΔW indicate that the curtain width decreases from top to bottom. Also, note that the scales of both axes in the diagram are not identical.

It is obvious that the optimum form of the curtain edge guide depends on the flow rate/width and the rheological properties of the curtain fluid (Fig. 5.176a). Therefore, the edge guide form would have to be changed for every change of these two parameters to preserve the best possible performance of the edge guide. However, at Polytype Converting we also built an edge guide that allowed us to vary the surface inclination continuously at several points along the guide and to change it while the curtain was running (Schweizer, 2010, and Fig. 5.176b). Specifically, the surface of this edge guide was made of a flexible PET film, and it contained 4 screws for laterally moving the position of the PET film. From many tests with this device (viscosity range: 250–510 mPas, flow rate range: 0.94–5.6 cm^2/s) we learned that the theoretical model presented above produces useful results and that the performance of a given edge guide did change a bit upon changing the flow rate or the rheological properties, but the results were still better compared to an edge guide with a straight and vertical

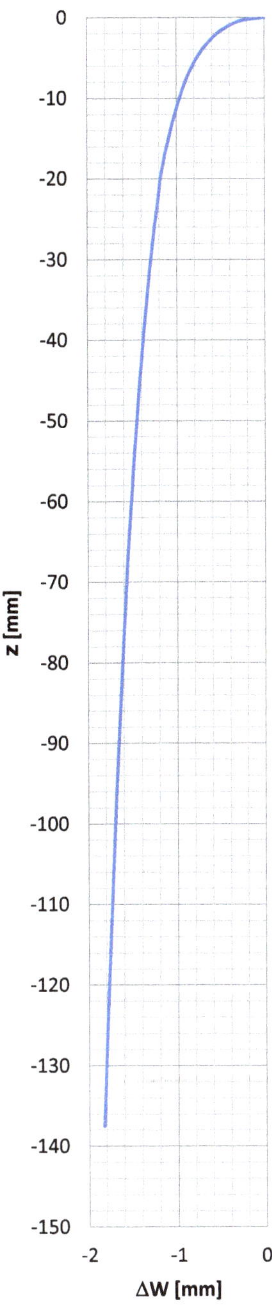

Fig. 5.175 Shape of an optimized curtain edge guide with variable inclination from top to bottom

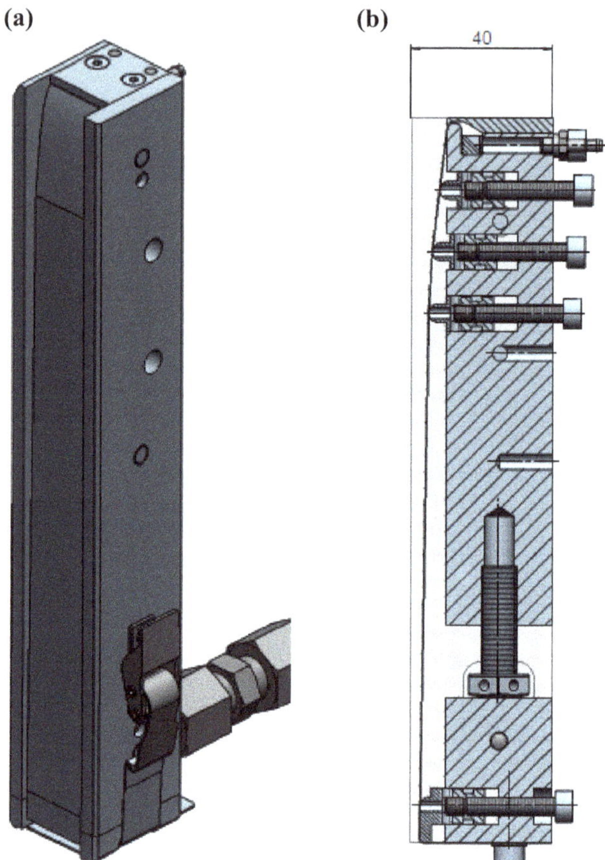

Fig. 5.176 **a** Optimum form of curtain edge guide with a variable inclination for a given set of operating conditions; drawing reproduced with permission from Polytype Converting AG **b** curtain edge guide with a flexible surface (PET film) and a mechanism for continuously changing the surface inclination while the curtain is running; drawing reproduced with permission from Polytype Converting AG

surface. This comparison between an edge guide with a curved surface of variable inclination and a guide with a straight and vertical surface is illustrated impressively in Fig. 5.177. The photo shows the thickness uniformity of a back-lit curtain near the edge guide. While the vertical edge guide produces a curtain with severe thickness variations, i.e., darker and brighter areas just inside of the guide (Fig. 5.177a), the curtain thickness along the curved edge guide is very uniform out to the guide and this for identical flow conditions (Fig. 5.177b).

Furthermore, the tests with the edge guide of adjustable geometry taught us that the parameters f_{film} and $f_{curtain}$ for calculating the boundary layer thickness of film and curtain flow (see Eqs. 5.9.18 and 5.9.19, respectively) are not constant as indicated

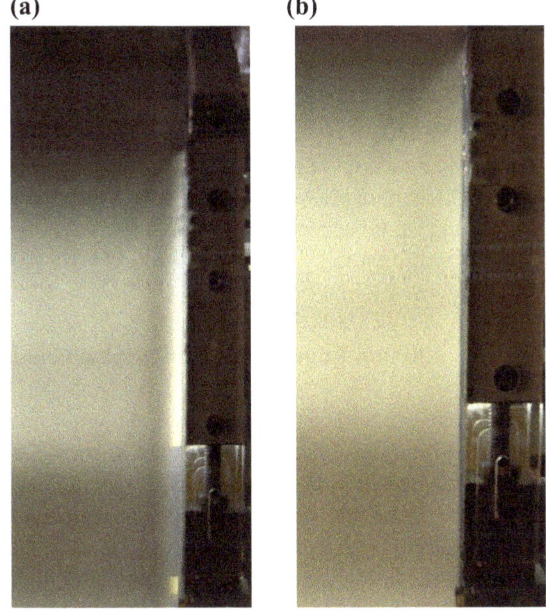

Fig. 5.177 **a** Curtain edge guide with a straight and vertical surface producing severe non-uniformities of the curtain thickness near the edge guide; photo reproduced with permission from Polytype Converting AG **b** curtain edge guide with a variable and optimized inclination producing a curtain with a uniform thickness out to the edge guide; photo reproduced with permission from Polytype Converting AG

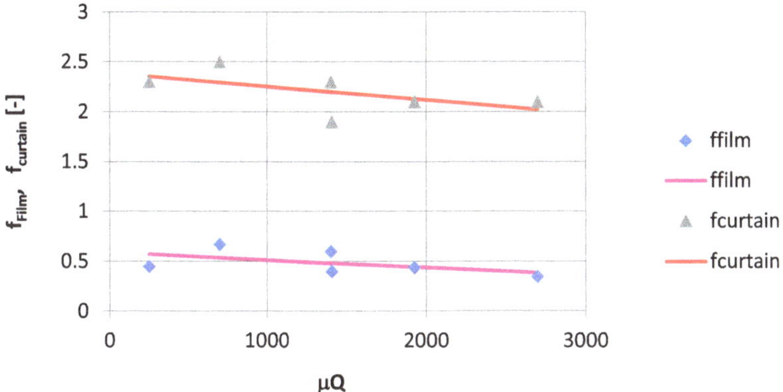

Fig. 5.178 Dependence of the factors f_{film} and $f_{curtain}$ upon the parameter μQ

above, but that they vary slightly with the product of viscosity times flow rate/width, i.e., μQ. The relationships are captured with empirical, linear regression equations and visualized in Fig. 5.178.

For film flow:

$$f_{film} = p_0 + p_1 \mu Q \tag{5.9.32}$$

$p_0 = 0.5921$ and $p_1 = -7.649 \cdot 10^{-5}$.

For curtain flow:

$$f_{curtain} = q_0 + q_1 \mu Q \tag{5.9.33}$$

$q_0 = 2.389$ and $q_1 = -1.352 \cdot 10^{-4}$.

Note that the unit of μ is [mPas] and the unit of Q is [cm^2/s]. Also, the data scatter a bit, which is expressed by relatively low r^2-values of about 0.3. Nevertheless, accounting for this dependency rather than working with constant values for f_{film} and $f_{curtain}$ is worthwhile.

5.9.5.3 Curtain Edge Guide for Variable Coating Width

Many edge-guide designs that allow changing the coating width are known from the patent literature. A very convenient version that has proven viable in industrial coating processes was developed and patented by Ruschak et al. (1999) and by Schweizer and Krebs (2006). The idea is sketched in Fig. 5.179.

The curtain edge guide is attached to a catch pan, which is mounted on a linear guide rail that allows the pan to be moved laterally and allows the curtain or coating width to be changed. The pan movement can be carried out and controlled by a motor or by hand. In particular, the movement can be executed while the curtain is running during production without disturbing the curtain or causing any other problems. A change of coating width can thus be implemented without stopping the coating process.

The point where the catch pan meets the top of the curtain edge guide must be designed and built as a sharp corner, i.e., as a knife-edge in the ideal case. This

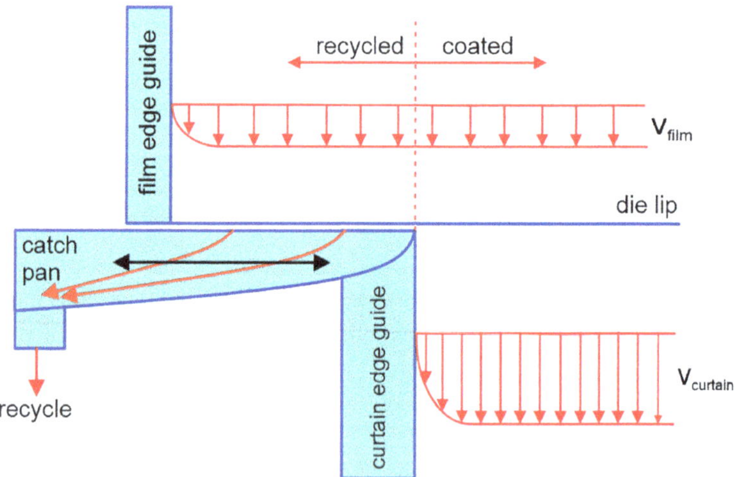

Fig. 5.179 Curtain edge guide for variable coating width

allows fluid arriving from the die lip to be precisely cut into a part of the curtain that is recycled and a part that is coated. Furthermore, this knife edge should not be positioned closely to the die lip but a short distance of about 5 mm below the lip. At this point, the velocity of the curtain is considerably higher than the speed of the arriving film, and therefore the thickness of the curtain is much smaller than the thickness of the film. This way the separation of the coated and recycled fluid streams is clean and without any unwanted side effects.

The recycled fluid can be reused for coating if the curtain consists of a single layer. For multilayer applications, the layers would mix in the catch pan and hence would have to be discarded, which often is excessively expensive, unless the cut-off portion of coating fluids is small and/or the raw material costs of the coating fluids are low.

Recycling coating fluid may generate bubbles in the catch pan. However, usually such bubbles are large in diameter and rise quickly to the free surface where they burst. Micro bubbles or foam are not generated. Therefore, fear of causing coating defects by bubbles in the recycled fluid is not warranted.

Determining the optimum form of the curtain edge guide is different for this configuration when compared to the situation where the curtain edge guide is in line with the film edge guide as explained above. In particular, the flow (film or slot) arriving at the curtain edge guide is free of any boundary layer as depicted in Fig. 5.179. On the other hand, the formation of a boundary layer along the curtain edge guide reduces the curtain velocity inside the boundary layer, which prevents the amount of fluid supplied by the preceding flow to be transported downstream without causing a local accumulation of fluid and hence a thickening of the curtain. This unwanted effect can be prevented by **widening** the curtain in a transition zone just below the die lip. At the same time, the curtain width must be **decreased** to account for the break and dam effect caused by the formation of the boundary layer along the curtain edge guide as described above.

Calculations regarding the transition zone proceed in an analogous way to the previous section, i.e., according to Eq. (5.9.31). However, as the boundary layer of the film flow is not available, the length of the transition zone cannot be made dependent on that boundary layer thickness. Instead, and based on experimental tests, we obtained good results if the transition length was held constant at the arbitrary value of $L_{trans} = 25$ mm. Moreover, since the approaching film flow is free of boundary layers, we arbitrarily choose the boundary layer thickness in the curtain at $z = L_{trans}$ to be responsible for the need to initially widen the curtain. As with the previous case, calculating the boundary layer thickness in the curtain at $z = L_{trans}$ requires iterating by assuming a viscosity, calculating the boundary layer thickness according to Eq. (5.9.1), calculating the shear rate in that boundary layer according to Eq. (5.9.21), calculating the Carreau-Yasuda viscosity for that shear rate, and by repeating this procedure until the assumed and calculated viscosity values are equal.

Now, the initial widening of the curtain in the transition zone can be calculated by modifying Eq. (5.9.31) as follows. Note that the value of the exponent k remains unchanged at $k = 0.5$.

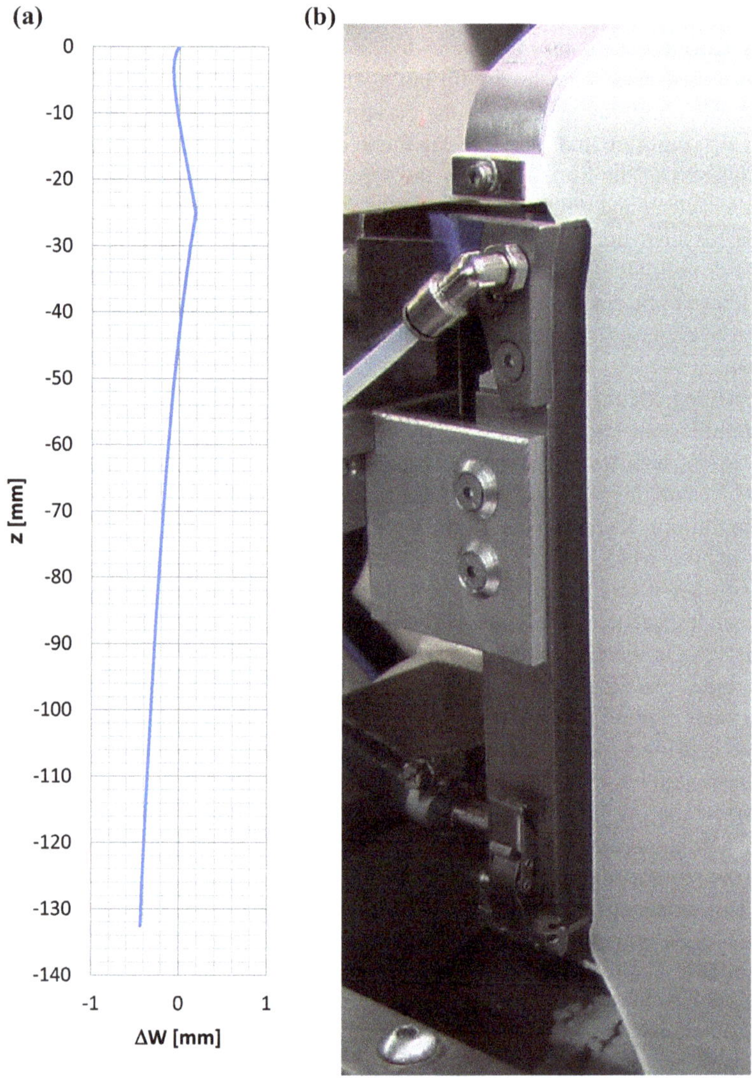

Fig. 5.180 **a** Optimized form of a right edge guide placed far inside of the edge of the preceding film or slot flow **b** optimized form of a left edge guide placed inside of the edge of the preceding film flow for PSA coating; photo reproduced with permission from Polytype Converting AG

$$\Delta W = +\frac{1}{3} H_{\delta,curtain@25mm} \left(\frac{z}{L_{\text{trans}}} \right)^k \text{ for } 0 < \text{z} < \text{L}_{\text{trans}} \quad (5.9.34)$$

Combining Eq. (5.9.34) with Eq. (5.9.28) gives the final shape of the edge guide. Figure 5.180a shows an exemplary result for the same operating conditions as in the previous section, with the difference that 5 mm of the curtain height are consumed by the distance between the die lip and the top of the curtain edge guide, and 5 mm are used for housing the feed slot for the edge fluid. Moreover, an edge guide calculated with this procedure and built for coating a pressure-sensitive adhesive is shown in Fig. 5.180b.

Referring to Fig. 5.179 adverse effects generated by the boundary layer of film flow can be avoided altogether if the die is made wider by at least twice the boundary thickness of film flow because then these effects are separated and discarded by the cutting edge on top of the edge guide for any coating width.

5.9.5.4 Optimization of Curtain Edge Guide Design

Most of the curtain edge guide designs discussed in Sect. 5.9.5 and the many other designs known from the patent literature are characterized by at least one desirable performance criterion. Many of them, however, also include undesirable performance features, and many of the designs were developed by engineers from the photographic industry because it was this industry that initially lifted curtain coating to a higher level. Consequently, many of the edge guide designs were optimized for the product family of photographic films and papers.

In contrast, during my work at Polytype Converting, which is a supplier of coating processes and associated equipment to the converting industry, we made it our goal to develop an edge guide that works well for any kind of application and any type of product family. While the design presented in this chapter may come close to meeting this posted goal, it requires an involved design procedure and an elaborate infrastructure in terms of supplying and removing an auxiliary edge liquid. Therefore, it may be wise and appropriate to spend additional effort to find a simpler edge guide design that works well for new and emerging product families. One such example is a narrow and straight, vertical groove in flat material. The groove is fed with a solvent on top and it does not contain a suction system at the bottom for removing the edge fluid. Instead, owing to the small quantity, the edge fluid can be evaporated during drying.

5.9.6 Boundary Layer in Impinging Flow

All premetered coating methods are characterized by a flow that impinges more or less perpendicularly onto the moving substrate. In slot coating, the impinging flow is slot flow exiting from the die slot. In slide coating, it is the film flow at the end of the inclined die slide, and in curtain coating it is the free-falling curtain. In all processes, the actual coating of the substrate takes place at the point where the impinging flow first comes in contact with the substrate surface. This point is called the dynamic wetting line. In that sense coating is dynamic wetting.

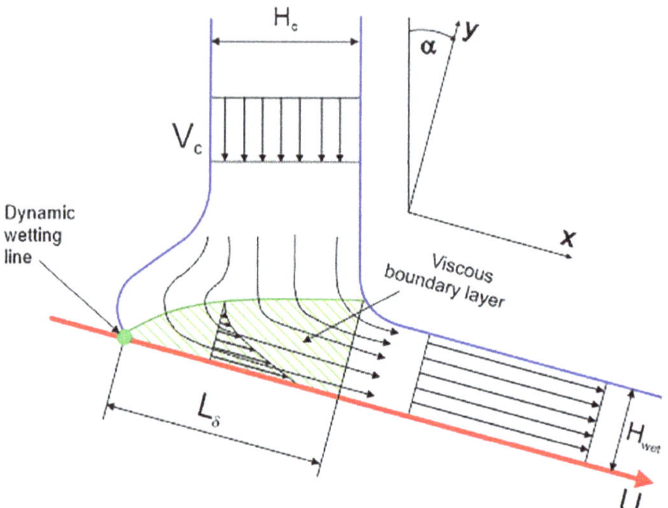

Fig. 5.181 Schematic diagram of an impinging curtain

A streamline pattern of the situation described above is shown in Fig. 5.73 for slide coating, in Fig. 5.35 for slot coating, and it is schematically depicted in Fig. 5.78 for curtain coating. This last figure is shown again here.

Focusing on curtain coating, a boundary layer starts developing at the dynamic wetting line (green spot in Fig. 5.181). The boundary layer grows in thickness along the substrate surface until all the fluid supplied by the impinging curtain is accelerated to web speed and transported downstream in form of plug flow, see the green area in Fig. 5.181. The impinging liquid is affected by the viscous drag action of the moving substrate only inside the boundary layer. Upstream of the boundary layer the arriving fluid has no idea yet that it will momentarily change the flow direction by about 90° and that it will be accelerated to web speed at a high rate. In fact, if the heel of the impinging flow cross-section is well developed, then some of the impinging fluid just upstream of the boundary layer is still moving in a direction opposite to the web motion, see streamlines in Fig. 5.181.

Our own flow visualization pictures and similar photos by Blake et al. (1994) show that the radius of the film building meniscus in curtain coating, i.e., of the transition region between curtain flow and plug flow on the moving web, is very small, i.e., on the order of the wet thickness of the coated film. This means that most of the impinging liquid is accelerated to web speed by the time the boundary layer has reached the front surface of the impinging curtain.

In slide and slot coating the flow field of the impinging fluid is more complicated. While the boundary layer still starts at the dynamic wetting line it is not clear where it ends. In slide coating shown in Fig. 5.73, for example, the boundary layer ends at a vortex located at the free surface of the film building meniscus. However, the size of that vortex and hence the exact location of the end point of the boundary layer are not known. A similar situation may exist in slot coating, where a vortex will form in the downstream coating gap when, depending on the value of the power law index, the coating gap is much larger than the wet film thickness, see Figs. 5.34 and 5.35. In addition, slot and slide coating processes are often operated with a sub-ambient pressure at the wetting meniscus, i.e., with bead vacuum. Consequently, the position of the dynamic wetting line and so the end point of the boundary layer is affected by the pressure difference across the coating bead.

Focusing again on curtain coating, Blake et al. (1994) used boundary layer theory to develop a model for Newtonian fluids that calculates the length of the boundary layer in the curtain impingement zone as sketched in Fig. 5.181. In particular, they argue that the boundary layer is complete when all the fluid supplied by the curtain is entrained in the boundary layer, and that point is reached when the boundary layer meets the front streamline of the curtain. Unlike often seen in slot and slide coating as illustrated in Fig. 5.74, they assume that in curtain coating no vortex is present between the end of the boundary layer and the front surface of the curtain. Theoretical evidence by Kistler (1983) and flow visualization studies by Blake et al. (1994) support this assumption. Consequently, the length of the boundary layer defines the location of the dynamic wetting line relative to the front surface of the curtain. Blake et al. point out that this model is only valid for $\delta << 1$ (definition see below), meaning that the web speed is much faster than the curtain impingement speed. This in turn implies that the thickness of the coated film is much smaller than the thickness of the impinging curtain.

The result of this analysis is an analytical expression for the boundary layer length L_δ normalized with the wet thickness H_{wet} of the coated film as follows:

$$\frac{L_\delta}{H_{wet}} = \frac{3}{20}\frac{(3+2\varphi)}{(1+2\varphi)^2}Re \tag{5.9.35}$$

with $\delta = V_c/U$, $\varphi = \delta \sin\alpha$, and $Re = \rho Q/\mu = \rho U H_{wet}/\mu$. α is the curtain impingement angle on the substrate surface, and α is positive when the substrate moves in a downward direction through the curtain as sketched in Fig. 5.181. The curtain velocity V_c just upstream of the impingement point can be calculated with Eq. (5.8.3).

In addition, Blake et al. (1994) introduced a dimensionless wetting line position ℓ defined as the ratio of boundary layer length L_δ to the curtain thickness projected into the plane of the substrate $H_c/\cos\alpha$:

$$\ell = \frac{L_\delta}{H_c/\cos\alpha} \tag{5.9.36}$$

Inserting Eq. (5.9.35) into (5.9.36) yields the following analytical expression for the relative wetting line position:

$$\ell = \frac{3\delta}{20}\frac{(3+2\varphi)}{(1+2\varphi)^2}\cos\alpha Re \tag{5.9.37}$$

Equation (5.9.37) is an important and useful result. It says that if $\ell = 1$, then the position of the dynamic wetting line coincides with the rear curtain surface, if $0 < \ell < 1$, then the wetting line is located in the projection of the impinging curtain, and if $\ell > 1$, then a heel is formed because the wetting line is located upstream of the impinging curtain.

Blake et al. (1994) extended Eq. (5.9.37) to shear-dependent viscosities by approximating the shear rate in the boundary layer by

$$\gamma = \frac{U - V_c \sin\alpha}{H_{wet}} \tag{5.9.38}$$

The term $[U - V_c \sin\alpha]$ is the velocity difference across the boundary layer, and H_{wet} is an approximate value for the thickness of the boundary layer. Combining the shear rate in the boundary layer with the Carreau-Yasuda Eq. (4.4.1) allows one to calculate an apparent viscosity for the flow inside the boundary layer.

Furthermore, Blake et al. (1994) extended Eq. (5.9.35) to include surface tension effects, which are estimated to become significant if the Weber number We is of order unity or less. The Weber number is the ratio of inertia to surface tension forces defined as $We = \rho QU/\sigma = \rho U^2 H_{wet}/\sigma$, see also Eq. (5.7.38). As a consequence, Eq. (5.9.35) is modified to become:

$$\frac{L_\delta}{H_{wet}} = \frac{3}{20}\frac{(3+2\varphi)}{(1+2\varphi)^2}\left[1 - \frac{30\Sigma}{(1-\varphi)(3+2\varphi)}\right]Re \tag{5.9.39}$$

Based on a momentum balance, the term Σ is estimated to be

$$\Sigma = \frac{(1+2\varphi)}{3}\frac{(1-\sin\alpha)}{We} \tag{5.9.40}$$

Using the finite element method Sünderhauf (1997) calculated the flow field of the impinging curtain for different Reynolds numbers, see Fig. 5.182.

For Re = 2 in this example, the wetting line position is far downstream of the curtain, which looks unrealistic for fluids without strong visco-elastic properties, and most likely would result in massive air entrainment. For Re = 3 the wetting line is located almost under the front surface of the curtain, i.e., $\ell = 0$. For Re = 5 the wetting line is positioned under the rear curtain surface, i.e., $\ell = 1$. For Re > 5 a heel develops, and the size of the heel increases with increasing Re. In fact, a vortex is formed in the heel for Re = 11, which is highly objectionable, and is quite similar to vortex 4 sketched in Fig. 5.70 for the bead of slide coating.

In any case, obviously the flow field of the impinging curtain can be changed by manipulating the length of the boundary layer, and that can be accomplished by changing appropriate process parameters, such as the Reynolds number as demonstrated in Fig. 5.182. The equations above visualize that the boundary layer length depends on 4 dimensionless parameters, namely the impingement angle α, the speed ratio V_C/U, the Reynolds number Re, and the Weber number We. However, in an industrial coating plant, dimensional parameters are preferred over dimensionless ones. The question then arises, which of these dimensional parameters are best suited for changing the length of the boundary layer?

In an industrial coating process, the chemical composition of the coating material and the dry thickness of the coated layer are given and fixed because these two parameters determine the function of the product. Often, the coating speed is also given and fixed because it determines the economic success of the coating process. It may be possible to change the coating speed, but typically not by much and often only towards lower values owing to limitations of the installed drying process. Moreover, changing the web speed changes the shear rate in the boundary layer and hence the apparent viscosity. The wet thickness of the coated layer depends on the solids concentration of the coating fluid. However, decreasing the solids concentration increases the dryer load, i.e., it increases the amount of solvent that must be evaporated, and that in turn may force the web speed to be reduced. It also changes the volumetric flow rate/width and all physical fluid parameters. While the effect of the solids concentration on the density and the surface tension is small, it is significant regarding the viscosity. Equally significant is the effect of the temperature on the viscosity although the preferred range of admissible temperatures is small, say 20–50 °C. Temperature of >50 °C present a safety hazard to the operators, particularly because a die should be handled with bare fingers and not with thick, heat-protecting gloves. Parameter changes as described above also simultaneously affect all dimensionless numbers including the speed ratio, the Reynolds number, and the Weber number, all of which determine the length of the boundary layer. Consequently, it is difficult to see which parameter is most suitable for changing the boundary layer length in the desired direction. By the same token, it is not clear if the streamline plots in Fig. 5.182 give a correct picture of the influence of Re on the flow field of the

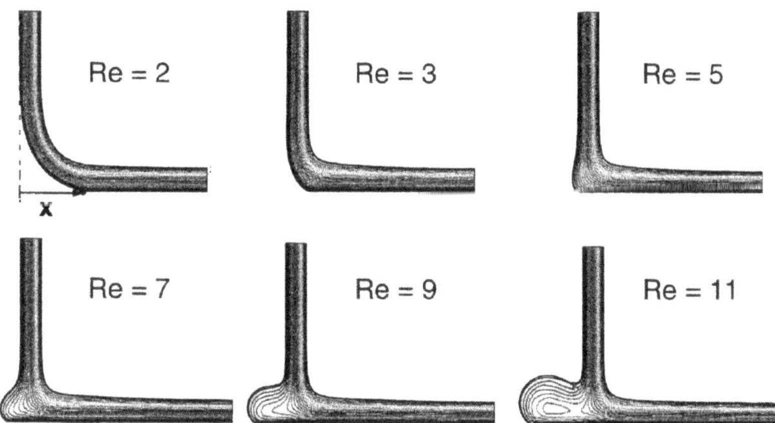

Fig. 5.182 Flow field of an impinging curtain for different Reynolds numbers; streamline plots reproduced with permission from Sünderhauf (1997)

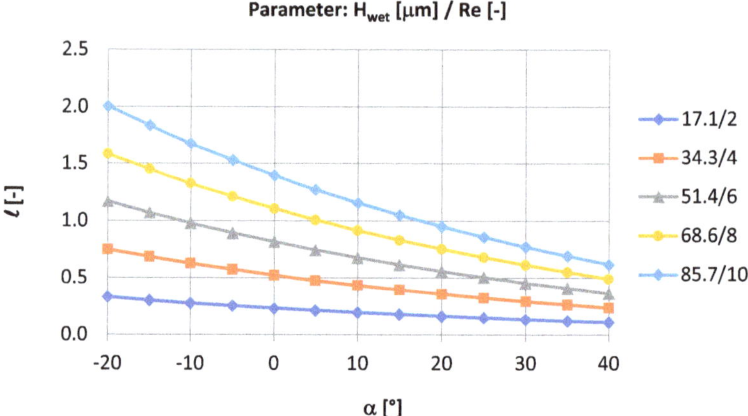

Fig. 5.183 Effect of the curtain impingement angle α on the position of the dimensionless dynamic wetting line

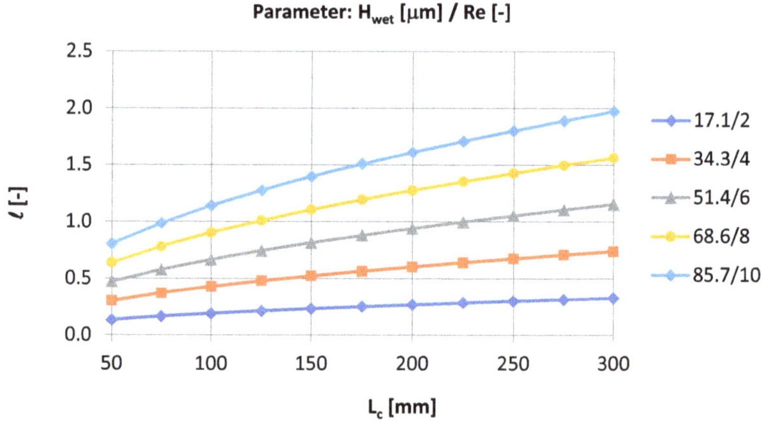

Fig. 5.184 Effect of the curtain length L_c on the position of the dimensionless dynamic wetting line

impinging curtain because it is not known if Sünderhauf accounted for the intricate relationships between the various dimensional parameters.

The way out of this difficult situation is to focus on geometrical process parameters for changing the length of the boundary layer because they are not connected to operational parameters, and they are not or only weakly connected to physical fluid properties, for example, the viscosity via the shear rate according to Eq. (5.9.38). The curtain impingement angle α and the curtain length L_c are available for changing the length of the boundary layer. Figure 5.183 shows how the impingement angle affects the dimensionless wetting line position, and Fig. 5.184 does the same regarding the curtain length. In all calculations the physical fluid properties were kept constant and assumed as follows: $\rho = 1000\ \mathrm{kg/m^3}$, $\mu = 50$ mPas, $\sigma = 35$ mN/m. Moreover, the coating speed was held constant at 350 m/min. When studying the effect of the impingement angle the curtain length was held constant at 150 mm, and when calculating the effect of the curtain length the impingement angle was set to 0°. In all calculations, the curtain impingement speed was approximated by assuming that the initial curtain velocity equals 10% of the gravity contribution. Then the wet film thickness was varied such that the resulting flow rate/width ranged between 1.0 and 5.0 $\mathrm{cm^2/s}$, which is an appropriate range for most industrial curtain coating applications. This procedure resulted in an acceptable range of Reynolds numbers, and it also generated vastly changing Weber numbers (from 16.7 to 83.3), but as explained by Blake et al. the effect of We on the wetting line position is not significant for We >> 1.

Dropping the curtain at negative impingement angles, i.e., onto a substrate that moves in an upward direction, generates longer boundary layers and hence pronounced heels. In contrast, changing the impingement angle to positive values reduces the boundary layer length. These effects increase with increasing Re.

Increasing the curtain length results in longer boundary layers owing to the increased impingement speed, which amplifies the stagnation point character of the flow field, and which forces the upstream half of the impinging curtain to initially flow more in the direction opposite to the direction of the web motion. The effect of the curtain length on the wetting line position is not strong, particularly for low flow rates.

Finally, Fig. 5.185 shows that changing the coating speed has little effect on the wetting line position, particularly if the speed is higher than about 300 m/min. On the one hand, increasing the speed increases the viscous drag exerted on the fluid by the web, which tends to move the wetting line farther downstream. On the other hand, increasing the speed without changing the wet thickness of the coated film also increases the flow rate/width, thereby increasing the stagnation point character

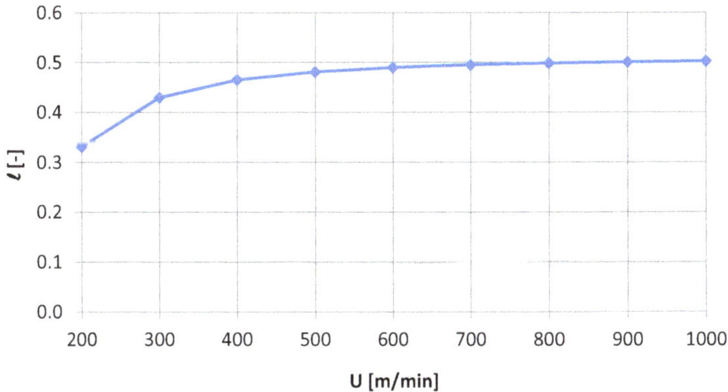

Fig. 5.185 Effect of the coating speed on the position of the dimensionless dynamic wetting line. $\alpha = 0°$, $L_c = 150$ mm, $H_{wet} = 30\ \mu m$, $\rho = 1000\ \mathrm{kg/m^3}$, $\mu = 50$ mPas, $\sigma = 35$ mN/m

of the flow, which in turn tends to move the wetting line in the upstream direction, and therefore tends to negate the effect of the viscous drag.

Trends and effects of other process parameters on the position of the dynamic wetting line are difficult to visualize. It is suggested to implement the equations above in an EXCEL environment and to calculate the wetting line position for any specific coating application. There, it is easy to change any or several of the process parameters to see which of them is best suited for moving the wetting line to the desired position. Further arguments for optimizing the flow field of the impinging curtain are presented and discussed in Sect. 19.4.

5.9.7 Vortices in Impinging Flow

The impinging flow of all premetered coating methods offers a potential location for one or more vortices to form. In all cases, the location is on the upstream side of the impinging stagnation point flow. Vortices are formed if the strength of that part of the stagnation point flow moving opposite to the web motion is much higher than the viscous drag exerted by the substrate surface onto the fluid, thus allowing the turn-around flow to become excessively large.

In slot coating, the vortex is located in the upstream coating gap, and its existence is promoted by a large coating gap relative to the thickness of the coated film and by an excessive bead vacuum. An example of this vortex is indicated in Fig. 5.55.

In slide coating, the vortex is located on the upstream side of the die lip, and an example is sketched in Fig. 5.70. Here too the existence of the vortex is favored by a high bead vacuum.

In curtain coating, the vortex exists when the heel becomes excessively large, see Fig. 5.182. This flow condition is sometimes called *puddling*. A nice example of this vortex was captured by Blake et al. (1994), see Fig. 5.186. In fact, the heel was so big that two vortices formed.

In all premetered coating methods, this type of vortex is also promoted by an excessively low viscosity. Besides lowering or eliminating bead vacuum, therefore, increasing the viscosity by increasing the solids concentration of the coating fluid is

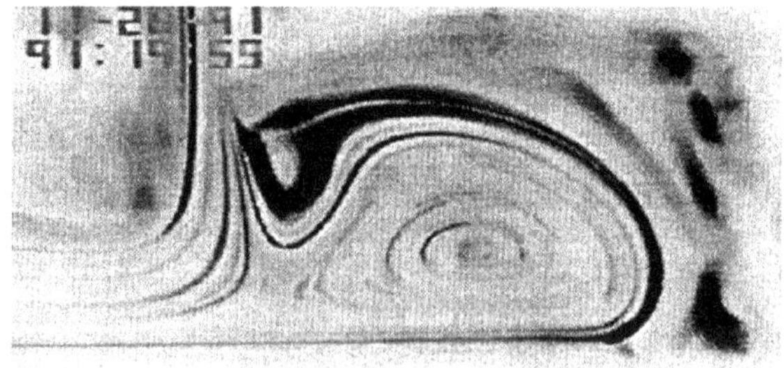

Fig. 5.186 Two vortices in an excessively large heel of an impinging curtain; photo taken from Blake et al. (1994) with permission from AIChE Journal

an effective way of preventing the vortex. This measure also reduces the flow rate in the curtain, which weakens the stagnation point flow and thus reduces the chances for vortex formation.

5.10 Flow After Coating

Flow after coating applies not only to premetered coating methods but to all coating processes. In particular, the coated film remains liquid between the coating point and some distance into the dryer or the radiation curing unit, where, owing to solvent evaporation or polymerization, the fluid viscosity increases continually until the coated layer becomes immobile. The distance between coating and immobilization depends on the size and design of the coating machine, and it may be short, i.e., on the order of 1 m, or it can be considerable, i.e., >10 m. Consequently, the time for flow after coating depends on this distance and the coating speed, and it is typically on the order of a few seconds.

Kheshgi (1997) discussed many phenomena that occur between coating and solidifying. Here, we focus on leveling, flow due to ambient disturbances, and edge withdrawal. Whether or not these phenomena exist in a given coating application, and, if they do exist, how strong they are, depends on many factors, among them the conformity of the coated film as well as the geometry of the web path between coating and drying. These two subjects are addressed next.

5.10.1 Conformity of the Coated Film

Consider coating onto a non-flat substrate such as paper or board. If blade coating or slot coating is used for applying the film then the result is depicted on top of Fig. 5.187. In particular, the surface of the coated film is flat, which is the result of positioning a stationary piece of solid material, i.e., the blade or the downstream lip of the slot die, in close proximity to the uneven substrate and forcing all fluid to flow through this gap while being coated. The downside of this configuration is that the resulting film thickness is uneven in relation to the topography of the substrate, but leveling after coating is not an issue.

In contrast, if the same film is applied to the same substrate by using curtain coating, then the thickness of the coated film is uniform, but the coated film follows the contour of the substrate, thus resulting in a non-flat film surface. This is the reason why curtain coating is called a conformal coating method. The drawback of this configuration is that curtain-coated films on an uneven substrate are subject to leveling between coating and drying, which forces the initially uniform film thickness to become non-uniform.

Other coating methods, such as slide coating or reverse roll coating with steel and rubber-covered rolls produce films with a flatness and thickness uniformity between the two extremes discussed above. Leveling after coating may or may not be substantial.

5.10.2 Web Path Between Coating and Solidifying

The web path between coating and solidifying may be simple and short as shown for a slot coating process in Fig. 5.188a, or it may be complex and long as sketched for curtain coating in Fig. 5.188b.

Whether simple or complex, the web path between coating and drying typically contains several sections of different inclination and hence of different strength of leveling. The web path of the curtain coater shown in Fig. 5.188b even has a section where the coated film moves upside down. Here, the wave amplitude of an undulated film entering this section may grow due to the gravitational pull.

On a fast and wide coating machine involving curtain coating, the level difference between coating and drying may easily be as high as 5 m. Moreover, knowing the drying curve for a given coating application, i.e., the change of solids concentration due to evaporation, and knowing the concentration dependence of the low shear viscosity μ_0 would allow one to estimate the time or distance into the dryer where the viscosity reaches a value of, say, 1,000 mPas. Once $\mu_0 > 1$ Pas, the flow after coating will typically be so weak that quality-related phenomena like leveling will no longer be significant. In any case, the time, during which flow after coating can exist, can be estimated by

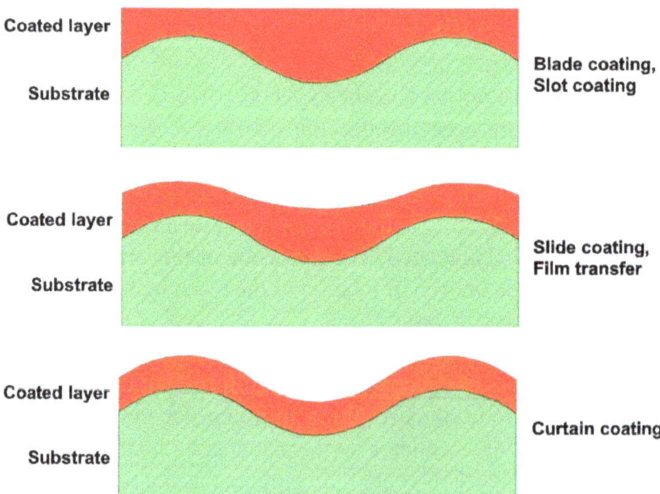

Fig. 5.187 Conformity of the coated film to the topography of the substrate

$$\Delta t = \frac{L_{1000}}{U} \tag{5.10.1}$$

L_{1000} is the web length between the coating point and the point in the dryer, where the low shear viscosity reaches the value of 1,000 mPas.

5.10.3 Leveling

As already alluded to at the beginning of Sect. 5.10, leveling after coating may occur for two reasons. In one situation the uniform thickness of a liquid film coated onto a flat substrate is disturbed after coating by external disturbances such as pressure fluctuations caused by irregular winds between coating and drying, or by coating defects such as ribbing lines, sharp lines and streaks or heavy edges. In the other situation, a liquid film is coated uniformly and conformally onto an uneven substrate, for example with curtain coating, resulting in a film surface that is uneven and more or less parallel to the substrate surface. In any case, locally disturbed film surfaces present just after coating will level to some extent. In the first situation leveling will reduce the initial film thickness non-uniformity, while the non-uniformity will grow in the second case. Obviously, preventing defects such as ribbing lines as well as other external disturbances before and during coating is preferred rather than relying on leveling for reducing or eliminating such defects after coating.

Leveling has been studied for a long time. Orchard (1962), for example, investigated the behavior of a 2-dimensional sine wave disturbance. The free surface of a liquid film disturbed in this way looks like a corrugated sheet as schematically

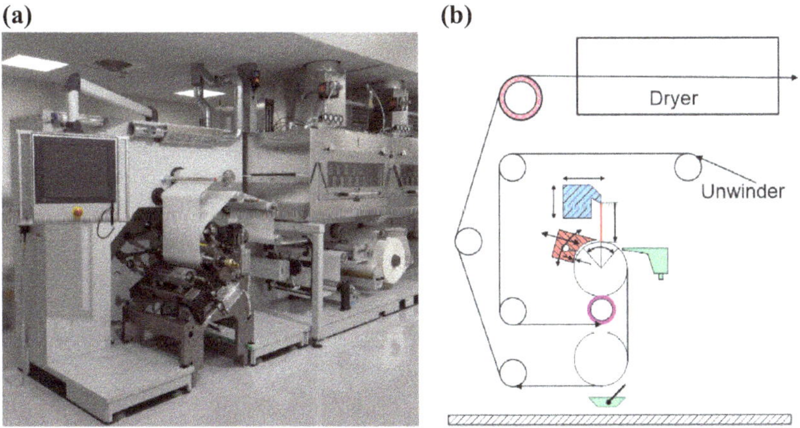

Fig. 5.188 **a** Simple web path for slot coating on pilot machine Techma 3 at Polytype Converting AG; photo reproduced with permission from Polytype Converting AG **b** complex web path for curtain coating

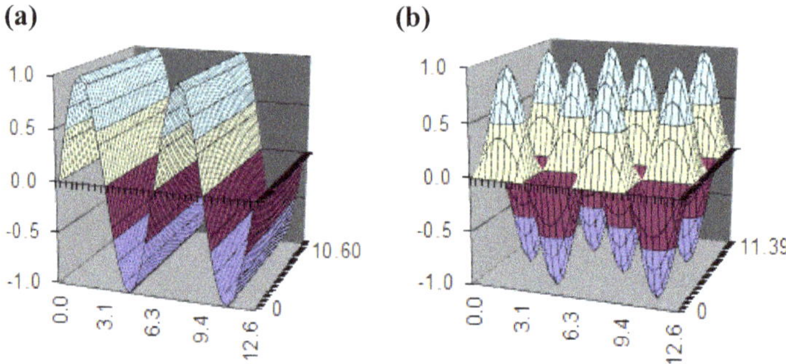

Fig. 5.189 **a** Schematic rendering of a liquid surface disturbed by a 2-dimensional sine wave (corrugated sheet) **b** Schematic rendering of a liquid surface disturbed by a 3-dimensional doubly-sinusoidal wave (egg carton)

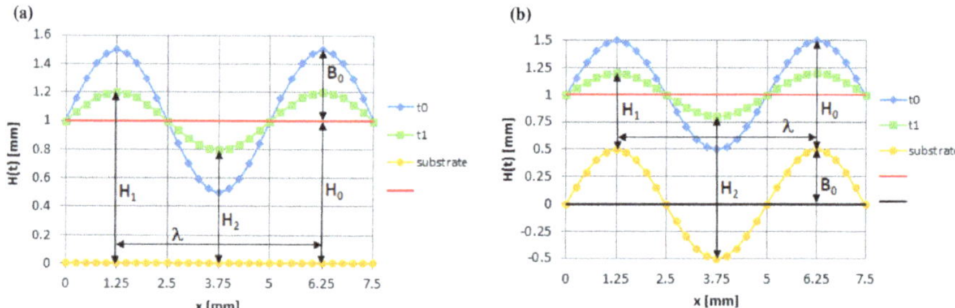

Fig. 5.190 **a** Schematic temporal evolution of a liquid layer being coated with a sinusoidal disturbance onto a flat substrate **b** schematic temporal evolution of a liquid layer being coated uniformly and conformally onto a substrate with a sinusoidal undulation

depicted in Fig. 5.189a, and it might be representative of a film containing a periodic defect such as ribbing lines or a single defect such as a heavy edge. Later, Anshus (1973) extended this work to 3-dimensional doubly-sinusoidal waves, where the disturbed film surface resembles an egg carton as visualized in Fig. 5.189b. This situation may describe a liquid layer being applied uniformly by curtain coating onto an uneven substrate such as paper or board.

The analysis by these authors was carried out for small-amplitude disturbances relative to the uniform thickness of the liquid layer, i.e., for $B_0/H_0 << 1$, see Fig. 5.190.

Small amplitude disturbances decay exponentially until the film surface becomes flat. The underlying flow is driven by surface tension and gravity, while viscosity opposes that flow. In particular, the free surface H(t) of a film containing a doubly sinusoidal disturbance evolves with time as described by Eq. (5.10.2).

$$H(t) = H_0 + B_0 e^{-\alpha t} \tag{5.10.2}$$

H_0 is the uniform thickness of the layer, B_0 is the initial amplitude of the disturbance, α is the leveling rate and t is the time. The physics of the leveling process is contained in the leveling rate α, which is given by

$$\alpha = \frac{\rho k^3 H_{wet}^3}{3\mu(1 + k^2 H_{wet}^2)} \left(\frac{g \cos \beta}{k} + \frac{k\sigma}{\rho} \right) \tag{5.10.3}$$

k is the wave number of a 3-dimensional disturbance according to

$$k = \sqrt{k_x^2 + k_y^2} \tag{5.10.4}$$

with $k_x = 2\pi/\lambda_x$ and $k_y = 2\pi/\lambda_y$, and with λ_x and λ_y being the wave length of the disturbance in the x and y direction, respectively. β is the inclination angle of the film plane relative to the horizontal. Large α-values result in fast leveling, which is particularly promoted by disturbances of short wave length and by thick liquid layers being oriented horizontally. Moreover, high surface tension and low viscosity also result in fast leveling.

A liquid layer with a non-uniform thickness on a flat substrate as depicted in Fig. 5.190a can be considered as a layer with a defect, and the visibility M of this defect can be quantified by relating the local thickness difference to the uniform layer thickness as follows and as sketched in Fig. 5.190a:

$$M(t) = \frac{H_1(t) - H_2(t)}{H_0} = \frac{2B_0}{H_0} e^{-\alpha t} \tag{5.10.5}$$

Here, $H_1(t)$ is equal to H(t) according to Eq. (5.10.2), and the difference between the initial film profile H(0) and H(t) is given by

$$\Delta H(t) = B_0(1 - e^{-\alpha t}) \tag{5.10.6}$$

Furthermore, it is assumed that the amount of liquid flowing away from the top of a wave is equal to the amount arriving in the adjacent trough. Therefore, H_2 can be expressed as

$$H_2(t) = H_0 - B_0 e^{-\alpha t} \tag{5.10.7}$$

Inserting $H_1(t)$ and $H_2(t)$ into Eq. (5.10.5) leads to the expression for M. The visibility M does not need to be zero in an industrial coating process, but it must be small enough to not jeopardize the final uniformity of the coated product. Moreover, the minimum time required for reducing a defect of initial amplitude B_0 to the visibility level M can be calculated by rearranging Eq. (5.10.5) as follows:

$$t_{min} = \frac{ln\left(\frac{M H_0}{2B_0}\right)}{-\alpha} \tag{5.10.8}$$

Equation (5.10.5) can be used to estimate whether a surface deformation of known dimensions present right after the coating point will disappear due to leveling in the available time according to Eq. (5.10.1), i.e., before the layer solidifies during drying or curing. The following example was constructed with arbitrary numbers to visualize the benefit of this procedure.

$\rho = 1000$ kg/m^3, $\mu = 4$ mPas, $\sigma = 35$ mN/m, $H_0 = 70$ μm, $B_0 = 5$ μm, $\lambda_x = 10$ mm, $\lambda_y = 10$ mm, $\beta = 0°$, U = 90 m/min, $L_{1000} = 4$ m. These numbers could be representative of ribbing lines in slide coating. The admissible visibility of the defect is postulated to be M = 0.01, i.e., the residual film thickness non-uniformity relative to the nominal film thickness should not exceed 1%. The various calculations produced the following results: The ratio $B_0/H_0 = 0.071$, which is much smaller than 1, indicates that the coating parameters satisfy the model requirements. The leveling rate $\alpha = 0.842$ s^{-1}. The time required for M to reach 0.01 is 3.2 s, see red point in Fig. 5.191. However, the time available for leveling was only 2.7 s, see green line in Fig. 5.191. The conclusion from this exercise is that leveling is not able to reduce the initial film deformation after coating as requested by the visibility parameter M = 0.01. This discrepancy could be eliminated if the wave length of the disturbance had been 9 instead of 10 mm, if the viscosity was reduced from 4 to 3 mPas, if the surface tension had been increased from 35 to 45 mN/m, or if the coating speed had been lowered from 90 to 75 m/min.

If the leveling equations above are applied to an initially uniform and conformal film applied by curtain coating onto a non-uniform substrate as shown in Fig. 5.190b, then the evolution of M indicates the resulting film thickness non-uniformity due to leveling. In particular, M is now defined as

$$M(t) = \frac{H_2(t) - H_1(t)}{H_0} = \frac{2B_0}{H_0}\left(1 - e^{-\alpha t}\right) \tag{5.10.10}$$

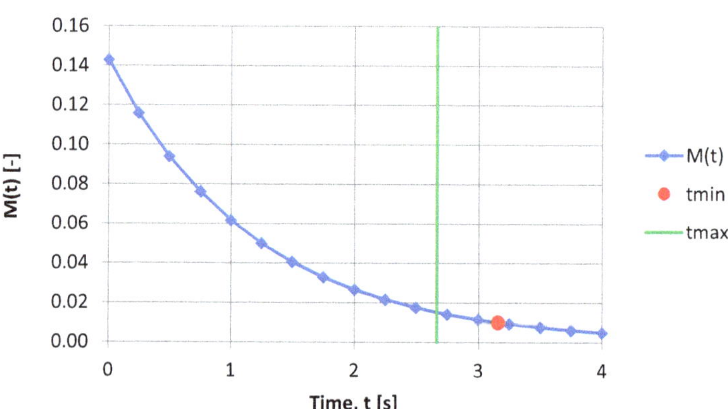

Fig. 5.191 Evolution of the visibility of a film thickness non-uniformity on a flat substrate due to leveling after coating

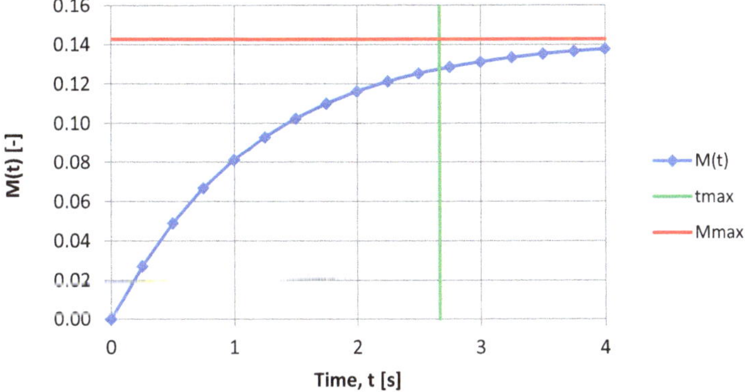

Fig. 5.192 Evolution of the visibility of a film thickness non-uniformity on an uneven substrate due to leveling after coating

Here, $H_1(t)$ and $H_2(t)$ are calculated according to

$$H_1(t) = H_0 + B_0\left(e^{-\alpha t} - 1\right) \tag{5.10.11}$$

$$H_2(t) = H_0 + B_0\left(1 - e^{-\alpha t}\right) \tag{5.10.12}$$

Using the same example as above, the resulting film thickness non-uniformity after the available leveling time of 2.67 s amounts to $M = 0.128$, while the non-uniformity for $t \to \infty$ amounts to 0.143, see Fig. 5.192.

5.10.4 Flow Due to Ambient Disturbances

Often in modeling flows with free surfaces, it is assumed that the gas adjacent to the liquid does not affect the flow inside the fluid. This may be true for a thin viscous layer resting on a horizontal surface, which is stable according to linear stability theory, and it may be justified if a viscous film is flowing down an inclined plane in a quiescent laboratory, but the assumption may not be justified in an industrial coating process, particularly with regard to the flow after coating. On the one hand, fast machines in the converting industry run at web speeds of more than 1,000 m/min, and the maximum speed in the paper industry is about 2,000 m/min. Therefore, the flow in the air boundary layer generated by the web motion between coating and drying can be substantial, and it may well influence the liquid in the coated layer, particularly if the air flow is not perfectly uniform all along the way owing to disturbances from the environment such as operators and vehicles generating wind by moving around in the coating machine or the factory hall, or non-uniform air currents generated by open doors and ventilators of electrical motors, etc. On the other hand, the coated and still liquid layer is exposed to strong air currents upon entering the dryer. In an impingement dryer consisting of slot nozzles, for example, the air velocity at the nozzle exit is typically in the range of 30–40 m/s, and it can be as high as 60 m/s, which is considerable. In addition, dryer nozzles are often made of metal profiles. Consequently, the widthwise uniformity of the air flow exiting these nozzles is not perfect because the mechanical precision of the slots is by far not as high as it is in slot dies for coating processes. It is therefore possible that the thickness uniformity of the coated film is diminished by incessant ambient and non-uniform pressure disturbances that act between the coating and solidifying, and that induce flow in the coated film.

Ambient pressure disturbances may manifest themselves in the final coated product in the form of a defect called diffuse longitudinal streaks or bands. In contrast to sharp lines and streaks, diffuse bands have "soft" edges, and their optical appearance may be faint but is still objectionable in products with high requirements of film thickness uniformity such as photographic films and papers or optical screens.

Ruschak (1987) published a paper addressing the above topic. In that analysis, small deformations in a thin fluid layer caused by ambient pressure disturbances are modeled with lubrication approximation. Effects of viscous, gravitational, and surface tension forces are retained in the model. The model is useful for quantifying the level of disturbances that produce a significant deformation of the layer in the available time, for assessing the wave length of the most dangerous disturbance, for selecting surfactants that are effective in resisting flow, and for discussing the relative importance of several effects that oppose flow. The following is a review of the results of that paper.

The starting point of the analysis, i.e., at $t = 0$, is a horizontal, motionless film of density ρ, constant viscosity μ, surface tension σ, and uniform thickness H_{wet}. The film contains a surfactant of uniform bulk concentration C and an equilibrium surface concentration Γ. For $t > 0$ the film is disturbed by a small ambient, sinusoidal pressure wave of amplitude P_e, wave length λ, wave speed c, and wave number $\alpha = 2\pi/\lambda$.

The driving force for the flow leading to a deformed layer is the pressure disturbance. Opposing this flow are viscous and surface tension forces, as well as gravity in the form of a hydrostatic pressure head that builds up during the deformation. If the liquid contains a surfactant, then additional opposing effects are present owing to surface tension gradients, which are generated by the convection of surfactant in the film surface. However, this opposing effect is only temporary until the gradients of surfactant concentration at the free film surface are eliminated due to the diffusion of surfactant from the bulk to the interface. Thus, a flow system where significant surface tension gradients can be maintained over long times, i.e., until the coated film is immobilized by drying or curing, is least sensitive to ambient pressure disturbances.

The magnitude of the final film deformation is proportional to the following scale factor SF:

$$SF = \frac{P_e}{\left(\rho g + \sigma \alpha^2\right)} \tag{5.10.13}$$

Obviously, high-pressure amplitudes lead to large surface deformations, but a high surface tension, a high gravitational component, i.e., a horizontal instead of an inclined orientation of the film substrate, and a small wave length of the disturbance keep the defect in bounds.

For stationary disturbances, it can thus be concluded that the amplitude of the final film deformation decreases with

- Decreasing amplitude of the pressure disturbance.
- Decreasing wave length of the disturbance.
- Increasing surface tension.
- Increasing density.
- Increasing gravity.

Ruschak's calculations have shown that if the disturbance is equal to the stagnation point pressure resulting from an air jet impinging perpendicularly onto the liquid surface, the minimum air velocity leading to a barely visible coating defect is on the order to 10 mm/s for typical coating conditions (of the photographic industry) and weave lengths of the disturbance between 1 and 100 mm.

Ruschak showed that the amplitude of the surface deformation depends on 3 additional dimensionless numbers, namely:

$$\gamma = \alpha c \tau \tag{5.10.14}$$

$$\chi = \frac{E}{H_{wet}^2\left(\rho g + \sigma \alpha^2\right)} \tag{5.10.15}$$

$$\delta = k\sqrt{D\tau} \tag{5.10.16}$$

Except for the factor 2π in the definition of the wave number, γ is the dimensionless wave speed defined as the ratio of the distance traveled by the ambient pressure wave in the characteristic time τ to its wave length λ. τ is given by $\mu/\left[\alpha^2 H_{wet}^3\left(\rho g + \sigma \alpha^2\right)\right]$, and it is the characteristic time for the deformation of the layer to take place.

χ is called elasticity group, which contains the surface elasticity $E = \beta\Gamma$, and β is the (normally negative) derivative of surface tension with respect to surface surfactant concentration under conditions of equilibrium. As defined here, surface elasticity is strictly an equilibrium property of the surfactant.

The dimensionless group δ, called the diffusion group, is a measure of the importance of diffusion to changes in surfactant concentration at the interface. k is the derivative of bulk surfactant concentration with respect to surface concentration at equilibrium, and D is the diffusion coefficient.

Both the elasticity group and the diffusion group are important in determining whether significant surface tension gradients exist, which in turn retards the surface deformation due to ambient pressure disturbances. The elasticity group depends on the derivative of the surface tension with respect to the surface surfactant concentration. A large derivative favors significant surface tension gradients in the surface. Also important in determining whether significant surface tension gradients exist, however, is the gradient of surfactant concentration along the surface. Such gradients diminish with increasing values of the diffusion group. Thus, important surface tension gradients are favored by χ being large and by δ being small.

When the disturbance is stationary, i.e., $\gamma = 0$, then the amplitude of the surface deformation is solely determined by the scale factor SF given by Eq. (5.10.13) for any value of χ and δ. The film deforms until an equilibrium state is reached, in which the stationary pressure variation is balanced by surface tension and the hydrostatic pressure field in the liquid. The diffusion and elasticity groups only affect the time it takes for the ultimate film deformation to be reached.

The situation is more complicated when the disturbance is not stationary, i.e., when $\gamma \neq 0$, and the effect of increasing wave speed is to decrease the ultimate deformation of the film. For transient disturbances, it thus can be concluded that the amplitude of the final film deformation decreases with

- Decreasing film thickness.
- Increasing wavelength of the disturbance.
- Increasing viscosity.
- Increasing speed of the disturbance.
- Decreasing surface tension.
- Decreasing density.
- Decreasing gravity.

Surface elasticity increases the time it takes for the full deformation of the layer to occur. When χ is small, the surfactant does not affect the deforming flow, regardless of the value of the diffusion group. But when χ is large, the film deformation is reduced unless diffusion effects are even larger.

When diffusion is dominant, significant surface tension gradients do not exist. The final film deformation is approached more rapidly than when diffusion is less important. Diffusion diminishes surface tension gradients that oppose flow.

When χ/δ is small so that diffusion effects are dominant, the elasticity group has no effect on the film deformation.

The values of the dimensionless group χ and δ depend on the bulk concentration of surfactant. To calculate this dependence a surface equation of state is needed, for example, the Szyszkowski-Langmuir model of surfactant adsorption (see also Tricot, 1997), which relates the surfactant concentration in the surface to the bulk concentration under conditions of equilibrium as follows:

$$\Gamma = \Gamma_\infty \left(\frac{C}{a}\right) / \left(1 + \frac{C}{a}\right) \tag{5.10.17}$$

where Γ_∞ is the saturated surface concentration at high bulk concentration, and a is the bulk concentration, at which the surface concentration is half its fully saturated value. The surface tension is related to the bulk concentration by

$$\sigma = \sigma_0 - RT\Gamma_\infty ln\left(1 + \frac{C}{a}\right) \tag{5.10.18}$$

where σ_0 is the value of the surface tension when no surfactant is present and R is the gas constant. From these equations, the surface elasticity can be calculated as

$$E = -RT\Gamma_\infty\left(\frac{C}{a}\right) \tag{5.10.19}$$

Similarly, the rate of change of bulk concentration k with respect to the surface concentration is

$$k = \left(\frac{a}{\Gamma_\infty}\right)\left(1 + \frac{C}{a}\right)^2 \tag{5.10.20}$$

These expressions can be used for evaluating the dimensionless groups χ, δ and χ/δ as follows:

$$\chi = -M\left(\frac{C}{a}\right) \tag{5.10.21}$$

$$M = RT\frac{\Gamma_\infty}{H_{wet}^2(\rho g + \sigma\alpha^2)} \tag{5.10.22}$$

$$\delta = \frac{a}{\Gamma_\infty}\sqrt{D\tau}\left(1 + \frac{C}{a}\right)^2 \tag{5.10.23}$$

$$\frac{\chi}{\delta} = -N\frac{C/a}{(1 + C/a)^2} \tag{5.10.24}$$

$$N = \left(\frac{H_{wet}\alpha^2\sqrt{\tau}}{\mu}\right)\left(\frac{RT\Gamma_\infty^2}{a\sqrt{D}}\right) \tag{5.10.25}$$

Since small surface deformations are achieved for large values of the ratio χ/δ, note that this ratio has a maximum value when the bulk concentration is equal to a. Moreover, it can be concluded that the amplitude of the final film deformation decreases with

- High elasticity, i.e.,
  - thin liquid layers,
  - low gravity,
  - low surface tension.
- Negligible diffusion effects, i.e.,

- low viscosities,
- thick films,
- high gravity,
- high surface tensions,
- large wavelengths,
- small diffusion coefficients.

Also note that there is a value of the surfactant concentration, for which the deformation of the layer is minimized. However, this value varies substantially, and it depends on the values of M and N.

In conclusion, it becomes clear that slowly diffusing surfactants are more efficient than fast diffusing ones in reducing film deformation due to ambient pressure disturbances. Moreover, film deformation decreases with an increasing wavelength of the disturbance and with decreasing viscosity. However, for other bulk parameters, particularly film thickness, surface tension, and gravity, the trends, which lead to a minimum film deformation, are contradictory. This means that the effectiveness of the surfactant cannot be separated from the details of the flow field. In addition, the trends in the bullet list just above conflict with the trends in the two bullet lists farther upstream in this section.

Nevertheless, Ruschak was able to formulate design criteria, which include the effectiveness of surfactants. In particular, he postulated that film deformations due to ambient pressure disturbances decrease with increasing values for the parameters M and N. With respect to the beneficial effects of surfactants, the layer deformation can be reduced most effectively, if

$$M \geq order(1) \tag{5.10.26}$$

$$N \geq order(100) \tag{5.10.27}$$

$$C/a = order(1) \tag{5.10.28}$$

It may also be helpful to minimize the conflicting requirements for surfactant properties by not only using a fast diffusing or a slowly diffusing surfactant but a mixture of both types. In curtain coating, for example, the fast diffusing surfactant will then optimize the behavior of the film flow on the die slide and curtain flow, while the slowly diffusing surfactant will increase the resistance to film deformation between coating and drying.

The content of Ruschak's paper is demanding. Experience has shown that most industrial coating companies are not accustomed to dealing with the various surfactant properties presented here, much less measuring these properties as required by the theoretical considerations and the associated formulas. Tricot (1997) may be able to provide help for these issues as he discussed equipment and procedures for measuring various surfactant properties. For those who are manufacturing coated products with high demands on the film thickness uniformity, it is recommended to spend the necessary time to study and understand Ruschak's paper.

On the other hand, it might be easier and less costly to take a preventive approach by investing in efforts to eliminate, or at least minimize, ambient pressure disturbances in the coating machine, rather than investing efforts into complex product formulations.

An interesting alternative approach is frequently used in the photographic industry. This concept takes advantage of the gelling properties of photographic emulsions, which all use gelatin as their main polymer. Most photographic emulsions have a low solids concentration of order 10%, and such coating fluids gel at temperatures of less than about 30 °C. As a consequence, photographic coating processes are operated at a temperature of about 40 °C. Now, instead of entering a dryer after coating, the coated substrate enters a chill room immediately after coating. This room is equipped, for example, with a roller bed of several meters in length and kept at a temperature of about 5 °C. This temperature shock causes the wet coated film to quickly gel, i.e., to solidify, and hence to become insensitive to ambient pressure disturbances, particularly to those generated in the subsequent drying process. The drawback of this approach is that re-melting of the gelled film must be prevented, which requires rather low drying rates. Moreover, photographic films contain many layers, and they are applied with simultaneous multilayer coating methods, meaning that the wet coated films are thick and contain about 90% water. As a result, dryers for photographic coating machines are very long, some of them several hundred meters, particularly in combination with high coating speeds.

5.10.5 Edge Withdrawal

Withdrawing, i.e., de-wetting is the opposite of spreading, i.e., wetting. According to Eq. (4.6.3), spontaneous withdrawing occurs when the spreading coefficient S is negative. For most industrial coating applications de-wetting is not a problem because most substrates can easily be wetted by the coating fluid. The one well-known exception in the converting industry is the manufacturing of adhesive labels and tapes. These products consist of a silicon-coated release liner, i.e., a silicon-coated paper or film substrate, onto which the adhesive layer is coated before laminating the label or tape material, which also can be a paper or film substrate. The difficulty here is that the adhesive is coated onto a siliconized substrate, which, by requirement, provides a low adhesion toward the adhesive so that the label or tape material can be separated during the final application of the product. Usually, this does not generate any problem because coating, i.e., dynamic wetting, works well and is under control, even though the siliconized substrate surface is not well wettable. However, pseudo-static wetting problems or rather de-wetting problems in the form of spontaneous edge withdrawal occur between coating and solidifying.

Edge withdrawal is a flow after coating. It is driven by capillary effects, i.e., by high surface tensions and high contact angles, and is retarded by high viscosities

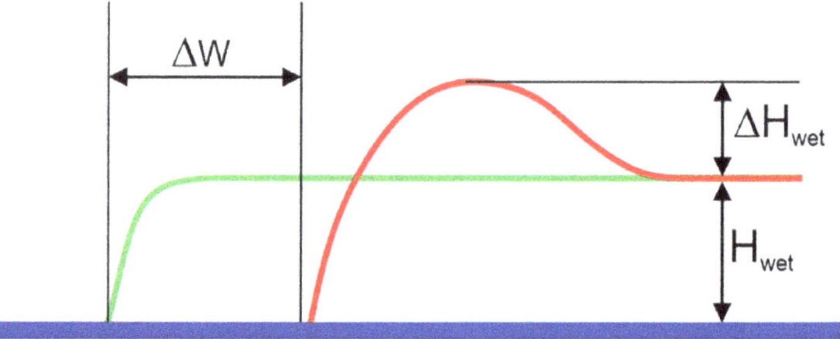

Fig. 5.193 Sketch of an over-thick coated edge as a result of edge withdrawal

as reported by Kheshgi (1997). In our experience, edge withdrawal also increases with increasing thickness of the wet film. The consequence of edge withdrawal is over-thick edges on both sides of the coated substrate, see Fig. 5.193.

One extreme example of edge withdrawal was observed when coating a thick layer of a viscous adhesive (PSA) onto a siliconized substrate for the following operating conditions: $H_{wet} = 383\ \mu m$, $\mu = 23.8$ Pas for a shear rate of $1\ s^{-1}$, $U = 14$ m/min $=$ 0.233 m/s. Slot coating was used for applying the adhesive, and the coated substrate moved slightly upward for a distance of 3 m before entering the dryer. The resulting time between the coating point and dryer entrance was 12.9 s, and measurements revealed that the coating width decreased by a total of 6 mm. In other words, edge withdrawal per coated side amounted to more than $\Delta W = 3$ mm because flow after coating continued for a while inside the dryer.

The consequence of this severe edge withdrawal was massive overly-thick coated edges, which in turn caused winding problems in the form of a defect called "barrel rings", and hence of telescoping rolls. As the thickness of the dried layer was still very high, i.e., 230 μm, it was possible to eliminate the winding problems by mounting angled pairs of narrow squeezing rollers in each edge zone between the dryer exit and the re-winder. These rollers were then able to flatten the overly-thick edges by locally redistributing the dried but still plastic adhesive. Alternatively, winding problems can be avoided by cutting off the overly-thick edges before re-winding.

Considering the problems described just above, why then is the PSA coated onto the siliconized substrate instead of the label stock material? The main reasons are as follows:

- If the label material is a plastic film such as PP, PE or PET, then the air temperature in the adhesive dryer must be kept lower than if the substrate was paper, which in turn increases the dryer length (dryer costs) for a given coating speed.
- If the adhesive was coated onto a paper label, then curl control of the final label material would become more difficult because the required paper moisture might be more difficult to control during the drying process of the adhesive. This issue does not exist if the paper label is simply laminated at room temperature.

- If the adhesive was coated onto a paper label, then some adhesive might penetrate into the porous structure of the paper, thus increasing the adhesive consumption and hence the label costs.

References

Abramowitz, M. (1949). On backflow of a viscous fluid in a diverging channel. *Journal of Mathematical Physics, 28*, 1–21.

Anshus, B. E., (1973). The leveling process in polymer powder painting—a three-dimensional approach. A. Chem. Soc. Div. Org. Coating Plastics Chem. *33*(2): 493–501.

Batchelor, G. K. (1967). *An introduction to fluid dynamics*. Cambridge University Press.

Bebié, H. (2020). *Flow of power law fluids between moving plates*. Personal communication.

Bebié, H. (2021). *Critical Reynolds number for the onset of recirculation in flow of shear-thinning liquids across triangular cavities with a rounded vertex*. Personal communication.

Becerra, M., & Carvalho, M. S. (2011). Stability of viscoelastic liquid curtain. *Chemical Engineering and Processing: Process Intensification, 50*(5–6), 445–449.

Benjamin, T. B. (1957). Wave formation in laminar flow down an inclined plane. *Journal of Fluid Mechanics, 2*, 554–574.

Bird, R. B., Armstrong, R. C., & Hassager, O. (1977). *Dynamics of polymeric liquids: Fluid mechanics*. Wiley.

Bird, R. B., Stewart, W. E., & Lightfoot, E. N. (1960). *Transport phenomena*. Wiley.

Blake, T. D., Clarke, A., & Ruschak, K. J. (1994). Hydrodynamic assist of dynamic wetting. *AIChE Journal, 40*(2), 229–242.

Brown, D. R. (1961). A study of the behavior of a thin sheet of moving liquid. *Journal of Fluid Mechanics, 10*, 297–305.

Christodoulou, K. N., Kistler, S. F., & Schunk, P. R. (1997). Advances in computational methods for free-surface flows. In S. F. Kistler & P. M. Schweizer (Eds.), *Chapter 9 in Liquid Film Coating*. Chapman & Hall.

Conroy, J. E., & Ruschak, K. J. (1992). *Curtain coating method and apparatus*. US Patent 5,358,569.

Dubbel. (1994). *Handbook of mechanical engineering*. In W. Beitz & K.-H. Küttner (Eds.). Springer Nature.

Durst, F., Koo, J.-B., Wagner, H.-G., & Walter, C. (1992). Experimental study of the boundary layer development at the edge guides in curtain coating. In *2nd International Symposium on Coating of Thin Films*. University of Bradford.

Durst, F., & Wagner, H.-G. (1997). Slot coating. In S. F. Kistler & P. M. Schweizer (Eds.), *Chapter 11a in Liquid Film coating*. Chapman & Hall.

Eggerath, D. (2012). *Wissenschaftliche Beiträge zur Vorhangbeschichtung und ihrem industriellen Einsatz*. Ph.D. thesis, University of Erlangen-Nürnberg.

Gibbs, J. W. (1906). *The scientific papers of J.W: Gibbs, Volume I: Thermodynamics* (pp. 314–331). Longmans, Green and Co.

Greiller, J. F. (1972). *Method of making photographic elements*. US Patent 3,632,374

Güggi, M., Pasquier, M., & Schweizer, P. M. (1995). *Process and apparatus for curtain coating a moving substrate*. European Patent 0 740 197 B1.

Güggi, M., & Varli S. (2001). *Method and apparatus for curtain coating*. WO 03/049870 A1.

Gugler, G., & Pasquier, M. (2002). *Verfahren und Vorrichtung zur Vorhangbeschichtung eines bewegten Trägers*. European Patent 1 398 084 A1.

Henry, D. J. D. (2016). *The stability of multi-layer curtain coating*. Ph.D. Thesis, The University of Birmingham.

Henry, D., Uddin, J., Thompson, J., Blyth, M. G., Thoroddsen, S. T., & Marston, J. M. O. (2014). Multilayer film flow down an inclined plane: Experimental investigation. *Experiments in Fluids, 55*, 1859.

Hens, J., & van Abbenyen, W. (1997). Slide coating. In S. F. Kistler & P. M. Schweizer (Eds.), *Chapter 11b in Liquid film coating*. Chapman & Hall.

Holtmann, B., Mena, J. A., Metzger, R., & Frediani, L. (2000). *Method and device for coating a running material web*. WO 01/47643 A1.

Iwata, I., Hosoe, H., & Matsuoka, Y. (1984). Suction nozzle. EP 0139211.

Katagiri, Y. (1992). *Analysis of edge effects in curtain coating*. Paper read at AIChE Spring National Meeting, New Orleans, LU, USA.

Kheshgi, H. S. (1997). The fate of thin liquid films after coating. In S. F. Kistler & P. M. Schweizer (Eds.), *Chapter 6 in Liquid film coating*. Chapman & Hall.

Kistler, S. F. (1983). *The fluid mechanics of curtain coating and related viscous free-surface flows with contact lines*. Ph.D. thesis, University of Minnesota.

Kistler, S. F., & Scriven, L. E. (1994). The-pot effect: Sheet-forming flows with deflection, wetting and hysteresis. *Journal of Fluid Mechanics, 263*, 19–62.

Kobayashi, C. (1992). Stability analysis of film flow on an inclined plane. I. One layer, two layer flow. *Industrial Coating Research, 2*, 65–88.

Lee, K. Y., & Liu, T. J. (1989). Design and analysis of a dual-cavity coat-hanger die. *Polymer Engineering and Science, 29*(15), 1066–1074.

Lee, K.-Y., Wen, S.-H., & Liu, T.-J. (1990). Vortex formation in a dual-cavity coat-hanger die. *Polymer Engineering and Science, 30*(19), 1220–1227.

Leonard, W. K. (1985). Inertia and gravitational effects in extrusion dies for non-Newtonian fluids. *Polymer Engineering and Science, 25*(5), 570–576.

Liu, T. J. (1983). Fully developed flow of power-law fluids in ducts. *Industrial & Engineering Chemistry Fundamentals, 22*, 183.

Macosko, C. W. (1994). *Rheology: Principles, measurements and applications*. VCH Publishers.

Metzger, R., Frediani, L., Holtmann, B., & Hardegger, H. (2002). *Device for coating moving material web*. European Patent 1 377 391 B1.

Metzner, A. B., & Reed, J. C. (1955). *AIChE Journal, 1*, 434.

Miller, C. (1972). *Industrial and Engineering Chemistry Fundamentals, 11*, 524.

Orchard, S. E. (1962). On surface levelling in liquids and gels. Appl. Sci. Res. *11*:451 – 464.

Owen, W. M. (1954). Laminar to turbulent flow in a wide open channel. *Transactions of the American Society of Civil Engineers, 119*, 1174.

Reiter, T. C. (1992). *Curtain coating method and apparatus using dual wire edge guides*. US Patent 5,328,726.

Roulier, A., & Bebié, H. (2020a). *Shape factors of duct cross-sections relevant for die design*. Personal communication.

Roulier, A., & Bebié, H. (2020b). *Inertia factors of duct cross-sections relevant for die design*. Personal communication.

Ruschak, K. J. (1978). Flow of a falling film into a pool. *AIChE Journal, 24*, 705–709.

Ruschak, K. J. (1987). Flow of a thin liquid layer due to ambient disturbances. *AIChE Journal, 33*(5), 801–807.

Ruschak, K. J., & Conroy, J. E. (1994). *Edge removal apparatus for curtain coating*. US Patent 5,395,660.

Ruschak, K. J., Kroon, J. K., & Wakefield, D. A. (1999). *Curtain coating apparatus and method with continuous width adjustment*. European Patent 0 943 961 A2.

Ruschak, K. J., & Weinstein, S. J. (2014). A local power law approximation to a smooth viscosity curve with application to flow in conduits and coating dies. *Polymer Engineering and Science*, 2301–2309.

Ruschak, K. J., & Weinstein, S. J. (2018). Model for the outer cavity of a dual-cavity die with parameters determined by two-dimensional finite element analysis. *AIChE Journal, 64*(2), 708–716.

Ruschak, K. J., & Weinstein, S. J. (2019). Accurate approximate methods for the fully developed flow of shear-thinning fluids in ducts of noncircular cross-section. *Journal of Fluids Engineering, 141.*
Schlichting, H. (1965). *Grenzschicht-Theorie.* Verlag G. Braun.
Schunk, P. R., & Scriven, L. E. (1997). Surfactant effects in coating processes. In S. F. Kistler & P. M. Schweizer (Eds.), *Chapter 11d in Liquid film coating.* Chapman & Hall.
Schweizer, P. M. (1988). Visualization of coating flows. *Journal of Fluid Mechanics, 193*, 285–302.
Schweizer, P. M. (1997a). Control and optimization of coating processes. In S. F. Kistler & P. M. Schweizer (Eds.), *Chapter 15 in Liquid film coating.* Chapman & Hall.
Schweizer, P. M. (1997b). Experimental methods. In S. F. Kistler & P. M. Schweizer (Eds.), *Chapter 7 in Liquid film coating.* Chapman & Hall.
Schweizer, P. M. (2010). *Seitenberandung für Vorhangbeschichtungsverfahren.* European Patent 2 412 446 A1.
Schweizer, P. M., & Krebs, F. (2006). *Vorhangbeschichter mit seitlich verstellbarer Abkantung.* European Patent 1 797 966 A3.
Schweizer, P. M., Krebs, F., & Koestinger, M. (2004). *Vorhangbeschichter und Vorhangbeschichtungsverfahren.* German Patent DE 10 2004 016 923 B4.
Schweizer, P. M., & Scriven, L. E. (1983). Evidence of Görtler-type vortices in curved film flows. *Physics of Fluids, 26*(3), 619–623.
Schweizer, P. M., & Troller, U. (1997). *Process and apparatus for curtain coating a moving substrate.* European Patent 0 907 103 B1.
Shetty, S., Ruschak, K. J., & Weinstein, S. J. (2012). Model for a two-cavity coating die with pressure and temperature deformation. *Polymer Engineering and Science,* 1173–1182.
Smith, M. K. (1997). Asymptotic methods for the mathematical analysis of coating flows. In S. F. Kistler & P. M. Schweizer (Eds.), *Chapter 8 in Liquid film coating.* Chapman & Hall.
Sünderhauf, G. (1997). *Vorhangbeschichtung. Kurzlehrgang über Grundlagen und Verfahren der Beschichtungstechnik.* Lehrstuhl für Strömungsmechanik, Universität Erlangen-Nürnberg.
Taylor, G. I. (1923). Stability of a viscous liquid between two rotating cylinders. *Philosophical Transactions of the Royal Society of London A, 223,* 289.
Taylor, G. I. (1959). The dynamics of thin sheets of fluid: Part III: Disintegration of fluid sheets. *Proceedings of the Royal Society of London A, 253,* 313.
Tricot, Y. M. (1997). Surfactants: Static and dynamic surface tension. In S. F. Kistler & P. M. Schweizer (Eds.), *Chapter 4 in Liquid film coating.* Chapman & Hall.
Tsuda, T. (2010). Coating flows of power-law non-Newtonian fluids in slot coating. *Journal of the Society of Rheology, Japan, 38*, 223–230.
Weinstein, S. J. (1990). Wave propagation in the flow of shear-thinning fluids down an incline. *AIChE Journal, 36*, 1873–1889.
Weinstein, S. J., & Palmer, H. J. (1997). Capillary hydrodynamics and interfacial phenomena. In S. F. Kistler & P. M. Schweizer (Eds.), *Chapter 2 in Liquid film coating.* Chapman & Hall.
Weinstein, S. J., & Ruschak, K. J. (1996). One-dimensional equations governing single-cavity die design. *AIChE Journal, 42*, 2401.
Young, T. (1805). An essay on the cohesion of fluids. *Philosophical Transactions of the Royal Society, 95*, 65–87.

Chapter 6
Wall Shear Stress

Abstract This chapter discusses wall shear stress, which is a fluid mechanical parameter that depends on the viscosity, the flow rate, and the geometry of the flow system. Wall shear stress is also a measure of the self-cleaning capability of a flow system and can increase the productivity of a coating process by reducing the non-productive time related to cleaning and product change-over-time. This chapter introduces time-dependent parameters such as the mean residence time, the residence time spectrum, the transfer function, and the change-over-time. The chapter illustrates how these parameters are controlled by wall shear stress, and so, how the residence time characteristics of a given flow system can be optimized by smart equipment design. Results can be applied to confined flows such as the fluid delivery system and the slot die. Controlling the residence time behavior of a coating process is essential for handling chemically reactive fluids.

6.1 Introduction

The concept of wall shear stress as related to premetered coating processes was described by Schweizer (1997). Wall shear stress is a fluid-mechanical property that depends on the fluid viscosity, the volumetric flow rate, and the characteristic cross-section of the flow domain, i.e., the pipe radius for pipe flow. This constellation is interesting for the coating process designer because the level of wall shear stress can be adjusted at will simply by adjusting the flow cross-section (pipe radius) for any given range of application as specified by the product of viscosity times flow rate. Chapter 5 provides appropriate formulas for calculating the wall shear stress for all the basic flows that are encountered in premetered coating methods.

Wall shear stress does not directly affect the coating process proper, i.e., the process of dynamic wetting, but it influences aspects of the

- Product uniformity, particularly with regard to lines and streaks, diffuse longitudinal bands, and cloudiness if the wall shear stress is too low, which allows vortices to be present in the flow fields of the coating process.
- Process productivity, particularly when affected by the unproductive time of the process such as cleaning the flow system or changing the product, both of which

P. M. Schweizer, *Premetered Coating Methods*, Engineering Materials,
https://doi.org/10.1007/978-3-031-04180-8_6

are related to the residence time behavior of the flow systems, and that in turn is controlled by the level of wall shear stress.

6.2 Cleaning and Contamination

Many scientific studies show (for an overview see Schweizer, 1992) that contamination of flow fields, i.e., the deposition of solid particles or gas bubbles on the boundaries of flow fields, can be prevented if the wall shear stress is above a minimum level everywhere and at all times. By the same token, the cleaning efficiency, i.e., the ability of the flowing liquid to remove contaminations from the flow boundary, also depends on the level of the wall shear stress, and higher wall shear stress means improved cleaning efficiency. In fact, operating a given flow field at a sufficiently high level of wall shear stress prevents contamination in the first place, and makes cleaning unnecessary in the second place, or at least reduces the cleaning time. This feature is interesting because cleaning contributes to the non-productive time of a coating machine.

Schweizer (1997) reported that there is an exponential relationship between the wall shear stress generated by the flow system and the diameter of the smallest particle that can be removed from the flow system. Moreover, experience has shown that the necessary minimum level of wall shear stress depends on the material combination of flow boundary and contaminating matter, e.g., oil droplet on a steel surface, or gas bubble on a plexiglass surface, etc. Here, the word "particle" is used in a generic sense, i.e., it includes any contaminating matter such as droplets, solid particles, gel slugs, bubbles, etc.

Schweizer (1997) therefore postulated that contamination of confined flow systems, i.e., the flow in the fluid delivery system including the distribution cavities and slots inside the die, can be prevented, and the self-cleaning efficiency of a flowing liquid inside such flow systems can be improved, if the wall shear stress remains above a minimum level of order 1 Pa **everywhere** and **at all times**. Here, we prefer to slightly raise the minimum wall shear stress level by requiring

$$\tau_{0,min} = 1 - 10Pa \tag{6.2.1}$$

As is illustrated in Sect. 6.6, several residence time parameters of flow systems including the slot die can be improved if the wall shear stress reached levels of up to 10 Pa.

The term "everywhere" means everywhere between the pressure side of the metering pump and the exit of the die. In the ideal case, this implies the use of a sanitary pipeline system such as those used in the food processing industry, i.e., without cavities and sudden changes of the flow cross-section at the intersection of pipes and valves or any other element present in the delivery system, such as filters, flow meters, pulsation dampeners, sensors, etc. It also implies that diameter changes in the pipeline system are carried out gradually and not abruptly, see Sect. 5.4.4. Experience has shown that it is very difficult if not impossible to buy all the necessary elements of a fluid delivery system with the same diameter.

The term "at all times" means for all formulations that are coated with the same delivery system, i.e., for all values of the product viscosity times flow rate of a given product family. For such cases, $\tau_{0,min} = 1$ Pa applies to the lowest value of μV^* and $\tau_{0,min} = 10$ Pa applies to the highest value of μV^*. The boundaries of the recommended shear stress range, i.e., 1–10 Pa, can obviously normally not be matched exactly by the values of μV^*. Therefore, these shear stress levels serve as guidelines.

Based on explanations given in Chap. 5, it can be concluded that flow fields that respect the above wall shear stress requirement, i.e., positive values above a threshold level, will be free of vortices because the existence of any vortex requires negative wall shear stress values. In addition, it can also be concluded that owing to its low viscosity (see for example Eq. 5.4.9 for Newtonian pipe flow), water is not a good cleaning agent in terms of generating a high wall shear stress necessary for removing contaminants from the flow boundary.

Staying with the example of Newtonian pipe flow and rearranging Eq. 5.4.9 yields an expression for the maximum pipe radius R_0 that will guarantee a contamination-free operation of the fluid delivery system of a coating process:

$$R_0 = \left(\frac{4\mu V^*}{\pi \tau_{0,\min}} \right)^{1/3} \tag{6.2.2}$$

Equation 6.2.2 is graphically shown in Fig. 6.1. The shape of the curves is quite in line with Eq. 6.2.1, i.e., with the recommended range of wall shear stress because,

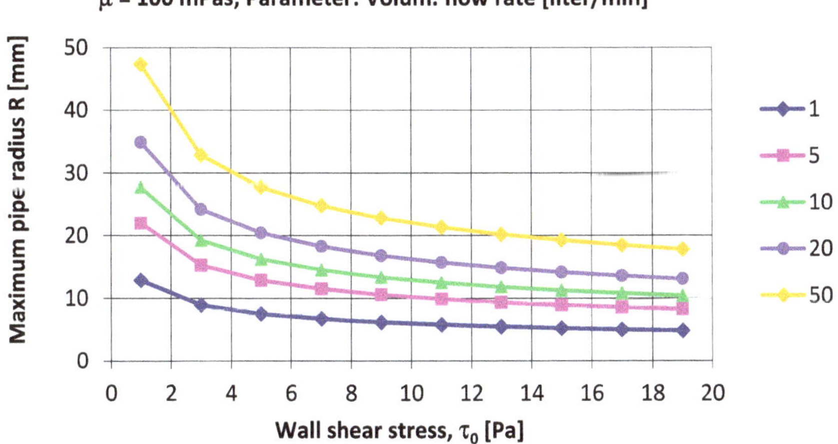

Fig. 6.1 Maximum pipe radius for preventing contamination

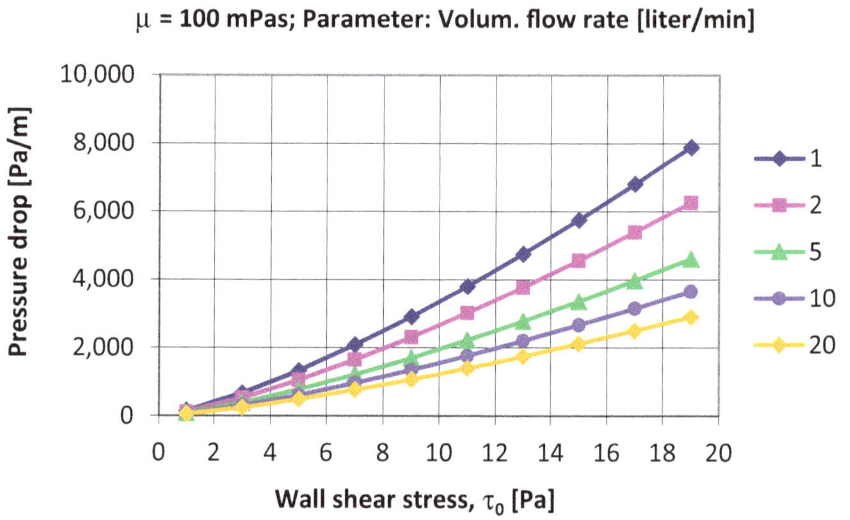

Fig. 6.2 Pressure drop for Newtonian pipe flow as a function of the wall shear stress

depending on the value of the flow rate, the pipe diameter does not change much further if the value of the wall shear stress exceeds the range of 1–10 Pa.

As mentioned above, requiring high wall shear stress requires small flow cross-sections, which in turn results in high pressure drops. Using the example of Newtonian pipe flow and inserting Eq. 6.2.2 into Eq. 5.4.4 yields an expression for the pressure drop as a function of the minimum wall shear stress as follows and as depicted in Fig. 6.2.

$$\frac{\Delta P}{L} = \frac{8\pi^{1/3}\tau_0{}^{4/3}}{4^{4/3}\mu^{1/3}(V*)^{1/3}} \approx 1.845\frac{\tau_0^{4/3}}{(\mu V*)^{1/3}} \tag{6.2.3}$$

In summary, flow systems with a high wall shear stress that do not contaminate or that are easier to clean can be obtained at the price of a higher pressure drop.

6.3 Wall Shear Stress and Die Design

6.3.1 Wall Shear Stress in the Inner Cavity

The main purpose of a die is to distribute liquid in the cross-web direction. Typically, this is accomplished by connecting a distribution channel or cavity with a narrow and long slot. The cavity extends in the cross-web direction from the feed point to the cavity end, and it is characterized by a large flow cross-section (i.e., small resistance

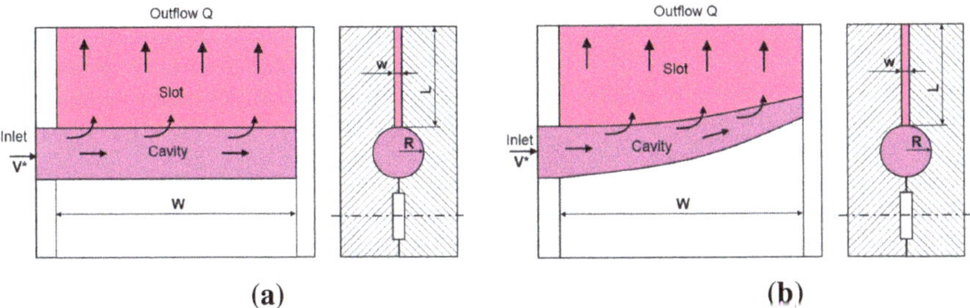

Fig. 6.3 **a** Infinite cavity design. **b** Optimized cavity design

to flow). In contrast, the slot extends in the coating direction from the cavity to the end of the slot, and it is characterized by a small cross-section (i.e., a narrow slot generating a high resistance to flow). This way, liquid will preferably flow in the cross-web direction before flowing out of the die in the direction of the web motion.

The simplest internal die geometry consists of a cylindrical cavity with a constant cross-section, and a narrow slot of constant length. This design concept is called *infinite cavity design*, see Fig. 6.3a. However, besides liquid distribution, a die must also satisfy other performance criteria, an important one being the prevention of contamination, or the ease of self-cleaning. As explained above, the wall shear stress is a crucial parameter in this respect, and high, constant values of wall shear stress are beneficial. Optimized die designs of this kind lead to a banana shape of the distribution cavity and to a slot length that varies across the coating width as depicted in Fig. 6.3b. This design concept is called *coat hanger design*, particularly if the die is center-fed and the internal geometry is symmetrical. Owing to the vastly different characteristic dimensions of the slot and cavity geometries, i.e., small slot height versus large cavity diameter, issues of low wall shear stress typically only apply to the flow in the inner (distribution) cavity, and the flow over the outer cavity, if the die geometry is based on a dual-cavity design.

Regardless of the design concept, the flow rate in the distribution cavity decreases linearly from the initial value at the feed point (cavity entrance) to zero at the cavity end because, at any location in the cross-web direction, the liquid flows out of the cavity into the slot and eventually out of the die. As a consequence, and as shown by Eq. 5.4.9, the wall shear stress also decreases linearly from the initial value to zero, if the cavity has a constant cross-section (infinite cavity design). This is certainly an unwanted performance feature because.

- flow separation may set on and sedimentation is likely to occur near the cavity end if the fluid is a dispersion or suspension;
- the fluid will turn into a solid if it has a yield stress and the shear stress drops below the yield stress level;
- cleaning the die as well as replacing one fluid with another is made difficult, particularly near the cavity end;

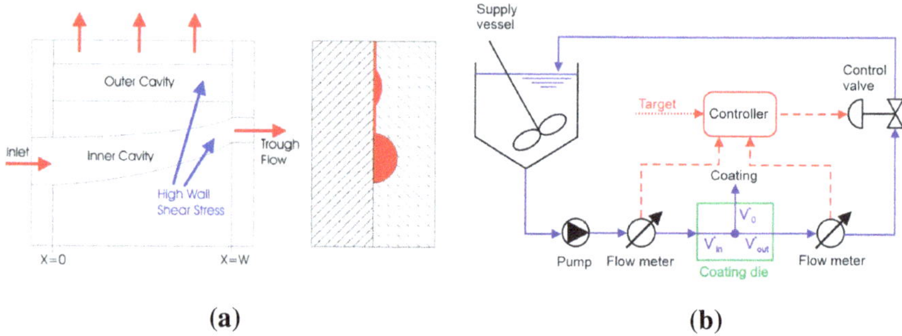

Fig. 6.4 **a** Die with through-flow, preventing areas of low wall shear stress. **b** Controlling both the flow rates entering and exiting the die

- chemically reactive fluids may solidify owing to the excessively broad residence time spectrum caused by the weak flow at the cavity end.

To overcome this deficiency, the distribution cavity must be designed with a cross-section that decreases continually from the entrance to the end, see Fig. 6.3b. For extreme cases that require high wall shear stress to the very end of the cavity such as highly reactive coating fluids, the die may have to be operated in the through-flow mode, where a small amount of fluid pumped into the die does not leave the die through the die slot but through an opening at the end of the cavity see Fig. 6.4a. Consequently, to maintain the feature of a premetered coating process, the flow rate must not only be measured at the die entrance, but also at the die exit, see Fig. 6.4b. Moreover, the difference between these two flow rates, which is the fluid amount leaving through the die slot, must be controlled by a suitable algorithm.

Finally, assuring a high wall shear stress throughout the internal die flow requires the outer cavity to be sufficiently shallow, which reduces the damping capacity of the cavity, and that is not desirable.

For a die with through-flow, the flow rates at the die entrance and exit can be characterized by a so-called **by-pass factor** α, which is defined as follows. The target flow rate V_0^* for achieving the desired coating thickness is based on the law of mass conservation and can be calculated with Eq. 3.3.8, which is repeated here:

$$V_0^* = Q_0 W = H_{wet} U W = \frac{A_{dry} U W}{C_W \rho_{liquid}} \tag{6.3.1}$$

moreover

$$V_{in}^* = V_0^* + V_{out}^* \tag{6.3.2}$$

and

$$\alpha = \frac{V^*_{out}}{V^*_0} \quad (6.3.3)$$

Therefore

$$V^*_{out} = \alpha V^*_0 \quad (6.3.4)$$

and

$$V^*_{in} = (1 + \alpha) V^*_0 \quad (6.3.5)$$

As explained above the flow rate depends on the position x along the cavity axis and decreases linearly from the cavity inlet to the cavity end according to:

$$V^*(x) = a + bx \quad (6.3.6)$$

The constants a and b in Eq. 6.3.6 can be determined with the help of the boundary conditions that at $x = 0$: $V^*(x) = V^*_{in}$ and at $x = W$: $V^* = V^*_{out}$. Therefore, the flow rate along the cavity axis is given by

$$V^*(x) = V^*_0 \left(1 + \alpha - \frac{x}{W}\right) \quad (6.3.7)$$

V^*_0	flow rate exiting the die through the die slot and being coating onto the substrate [m^3/s],
V^*_{in}	flow rate entering the die at the inlet port [m^3/s],
V^*_{out}	flow rate exiting the die at the end of the distribution cavity [m^3/s],
$V^*(x)$	volumetric flow rate [m^3/s],
Q	volumetric flow rate/width [m^2/s],
W	coating width [m],
x	coordinate along distribution cavity [m],
H_{wet}	wet film thickness [m],
U	coating speed [m/s],
A_{dry}	dry coat weight [kg/m^2],
C_W	solids concentration by weight [–],
ρ_{liquid}	liquid density [kg/m^3],
α	by-pass factor, determined by wall shear stress requirement [–].

As explained by Schweizer (1997), the change of the cavity cross-section A along the axis of the distribution cavity is conveniently modeled with a general polynomial form according to Eq. 6.3.8:

$$A(x) = A_0 \left(1 - \frac{x}{\beta W}\right)^{\varphi} \quad (6.3.8)$$

A_0 is the initial cross-section at the cavity entrance, A_e is the end cross-section at $x = W$, W is the width of the die (or the half-width, depending on whether the die is center or side fed), β is a measure for the cross-section at the cavity end, and φ is the taper index, i.e., a measure of how the cross-section changes along the cavity axis, i.e., in the cross-web direction.

The parameter β is defined by considering the boundary condition that at $x = W$: $A(x) = A_e$. Hence:

$$\beta = \frac{1}{1 - \left(\frac{A_e}{A_0}\right)^{1/\varphi}} \tag{6.3.9}$$

From a fluid mechanics point of view, it is meaningful to consider A as an important parameter, with the characteristic length of the cross-section being defined as $h^2 = A$. From a die manufacturing point of view, however, it is more meaningful to work with characteristic cavity dimensions such as diameter, radius, length of the square side, etc. For a circular cavity cross-section, therefore, Eq. 6.3.9 can be re-written as

$$\beta = \frac{1}{1 - \left(\frac{R_e}{R_0}\right)^{2/\varphi}} \tag{6.3.10}$$

R_0 is the initial cavity radius at $x = 0$, and R_e is the radius at the cavity end at $x = W$. The initial cavity radius R_0 can be calculated with Eq. 6.2.2. Regarding the size of the end cross-section, R_e values of <2 mm are possible.

Note that if $R_e = R_0$, i.e., if the cavity cross-section remains constant across the width, then Eq. 6.3.10 results in a division by zero, which is equivalent to β being infinity. This is unacceptable in a computer program. In this case, therefore, β is arbitrarily set to be very large, e.g., $\beta = 10^6$.

Equation 6.3.10 is valid for circular and semi-circular cavity cross-sections as well as for the square shape. In the latter case, the radius R has to be replaced by the square side a. Formulas for the parameter β of other cavity shapes such as the square and the rectangle with two rounded corners are given in Figs. 5.11 and 5.13, respectively, together with equations for other relevant geometrical parameters.

For reasons of simplicity, we stay with the example of Newtonian fluids flowing in a duct of circular cross-section to discuss the properties of the **wall shear stress distribution** in the inner distribution cavity. Looking at Eq. 5.4.9 we now know that V* and R (through Eq. 6.3.8) depend on the coordinate x along the cavity axis. Therefore, inserting Eqs. 6.3.7 and 6.3.8 into Eq. 5.4.9 provides a formula for the wall shear stress distribution τ(x) in the inner cavity. Normalizing this expression with the initial wall shear stress τ_0 taken at $x = 0$, as in Eq. 6.2.2, yields the relative wall shear stress distribution as follows:

$$\frac{\tau(x)}{\tau_0} = \frac{\left(1 + \alpha - \frac{x}{W}\right)}{(1 + \alpha)\left(1 - \frac{x}{\beta W}\right)^{3\varphi/2}} \tag{6.3.11}$$

For shear-thinning fluids according to the power law model, a similar equation for the wall shear stress distribution in the inner cavity can be derived by starting from Eq. 5.5.3, which describes the average wall shear stress in a duct of arbitrary cross-section:

$$\frac{\tau(x)}{\tau_0} = \left(\frac{1+\alpha-\frac{x}{W}}{1+\alpha}\right)^n \frac{1}{\left(1-\frac{x}{\beta W}\right)^{3n\varphi/2}} \tag{6.3.12}$$

Now, the wall shear stress distribution depends on the power law index n. It turns out that working with Eq. 6.3.12 is cumbersome because n is not known a priori, and it changes along the cavity axis. To overcome these difficulties, the local power law model according to Ruschak and Weinstein (2014) must be applied, see also Sect. 5.5.1.1 for more details. As this approach requires iteration at multiple locations along the cavity, it is easier to calculate the accurate wall shear stress distribution only in combination with calculating the pressure drop along the cavity and to work with the simpler Eq. 6.3.11 for finding an initial solution for an attractive wall shear stress distribution. We have experienced that the difference between the two equations (Newtonian and power law flow behavior) is small for most popular cavity shapes, except for the rectangle. However, this small difference justifies the approach proposed here. Moreover, the location along the cavity axis where the wall shear stress falls below the initial value, i.e., the value of the parameter Φ (see Fig. 6.5), is independent of the power law index n.

Note that Eqs. 6.3.11 and 6.3.12 are valid for any shape of the cavity cross-section. As β is determined by R_0, R_e and φ, the relative wall shear stress distribution depends on the three dimensionless parameters R_e/R_0, α and φ. This dependence is depicted in Fig. 6.5 for Newtonian fluids in a circular cavity.

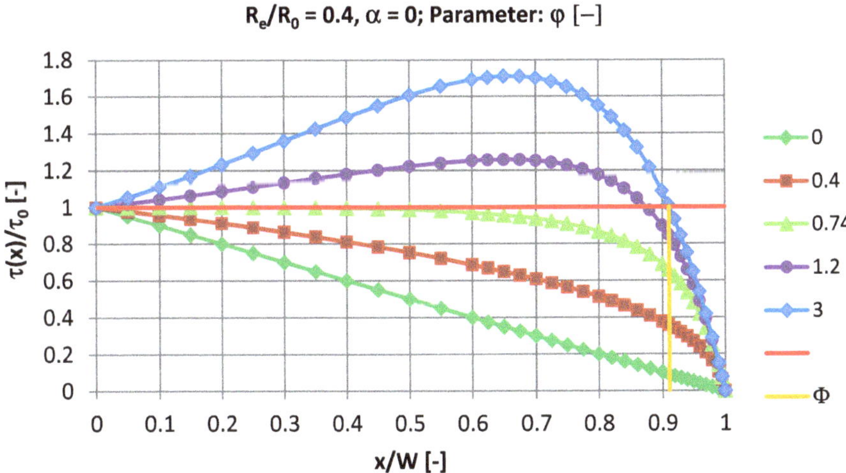

Fig. 6.5 Relative wall shear stress distribution in the distribution cavity

For low φ values, the shear stress decreases below the initial value right from the cavity entrance. For $\varphi = 0$, the cavity cross-section remains constant across the width, and the wall shear stress decreases linearly from the entrance to the end. For $R_e/R_0 = 0.4$ there is a critical φ value of 0.74, above which the shear stress first increases before it decreases below the initial value and further to zero at the cavity end. The location along the cavity axis, at which the shear stress drops below its initial value, is denominated by Φ. In the Fig. 6.5 Φ is about 0.91 for $\varphi = 3.0$.

So far in the discussion above, as well as in the discussion of Schweizer (1997), the taper index φ only assumes constant values. However, we recently discovered that better wall shear stress distributions can be constructed if φ is allowed to vary along the cavity axis according to

$$\varphi = a + b\left(\frac{x}{W}\right)^c \tag{6.3.13}$$

The difference between $\varphi =$ constant and $\varphi =$ variable is visualized in Fig. 6.6. If the parameter b in Eq. 6.3.13 is set to zero, then $\varphi = a$, and if, by trial-and-error, a is set to 0.86, the relative wall shear stress increases right away above its initial value until it reaches a maximum of 1.2. In contrast, if $b > 0$, and if a is set such that the relative wall shear stress does not fall below its initial value over most of the cavity length, and if b and c are varied by trial-and-error until the relative wall shear stress reaches a preferred maximum value (also 1.2 in this example, with $a = 0.7$, $b = 0.71$, $c = 11.0$), then a wall shear stress distribution is obtained that

- remains very constant over most of the cavity length;
- only rises sharply above the initial and to the maximum value over a short distance near the cavity end;

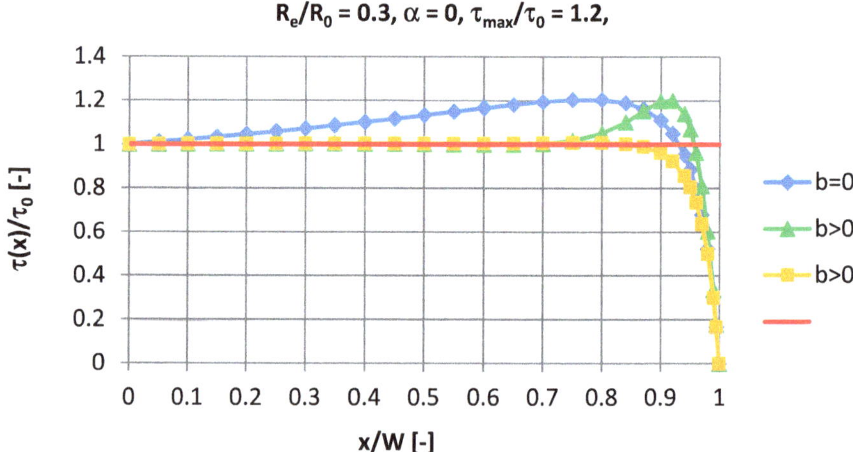

Fig. 6.6 Relative wall shear stress distribution for constant and variable values of the taper index φ

- falls below the initial value at a distance x/W that is larger than in the case of b = 0 for the same maximum wall shear stress value.

In fact, by varying the values of a, b and c, it is possible to obtain wall shear stress distributions that remain above the initial value and that are almost constant up to rather high Φ values. In the example of Fig. 6.6 (yellow curve with square symbols), a = 0.70, b = 0.11, c = 5.30, thus resulting in $\tau_{max}/\tau_0 = 1.008$ and $\Phi = 0.85$. Note, however, that τ_{max} must clearly rise above the initial value to obtain Φ values of >0.9.

In addition, b > 0 results in a lower pressure drop along the cavity than b = 0, which is desirable. In the latter case, the relative wall shear stress increases above its initial value right from the beginning of the cavity, which requires the cavity cross-section to decrease more quickly than if b > 0, and that in turn increases the pressure drop along the cavity.

In our experience, clear rules for determining the best values for the parameters a, b and c do not exist. Therefore, a little bit of trial-and-error is necessary, and we obtained good results by adhering to the following procedure:

- set a such that $\tau(x)/\tau_0$ remains >1 for small values of x/W, say for x/W < 0.5;
- set b such that $\tau(x)/\tau_0$ remains near 1 for larger values of x/W, say for x/W > 0.5;
- set b and c such that Φ reaches the desired value.

The following paragraphs and figures visualize this approach and provide some help in choosing good values for a, b and c. Specifically, preventing the wall shear stress from decreasing right from the cavity entrance can be achieved, if the parameter "a" stays above a critical minimum value, which only depends on R_e/R_0, see Fig. 6.7. As the derivative of Eq. 6.3.11 with respect to x is complex, a_{min} was determined by trial-and-error, see blue symbols in the figure. The so determined a_{min} values can be approximated with the following regression equation that was also found by trial-and-error.

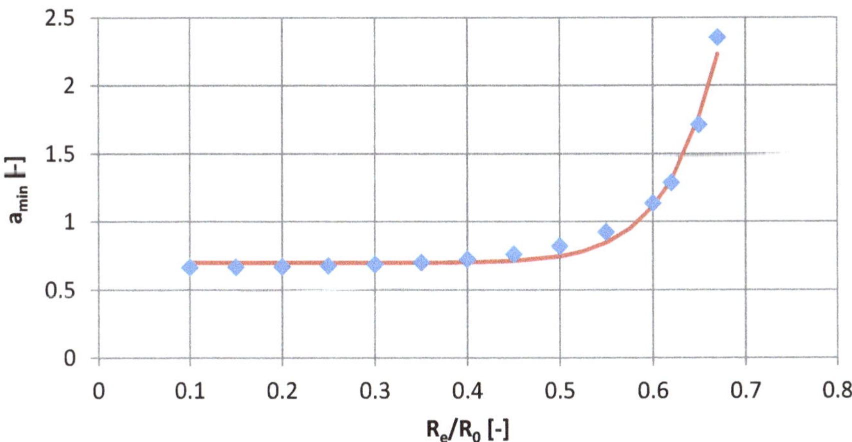

Fig. 6.7 Critical a-value, above which the shear stress remains above its initial value

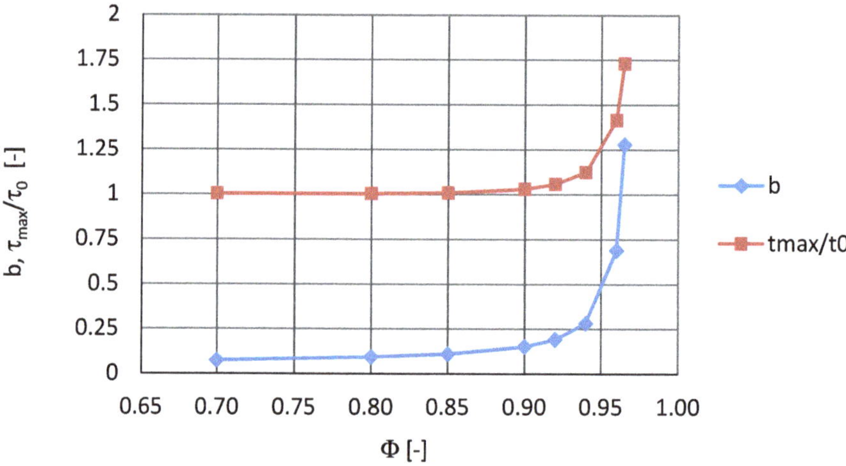

Fig. 6.8 Preferred values of the parameter "b" as a function of Φ; a = 0.70, c = 5.3

$$a_{\min} = a_0 + a_1 \left(\frac{R_e}{R_0} \right)^{a_2} \tag{6.3.14}$$

$a_0 = 0.7$, $a_1 = 180$, $a_2 = 11.9$.

Regarding the parameter "b", a was set to 0.70 and c was set to 5.3. Then, b was varied until the desired value for Φ was obtained. At the same time, the resulting maximum value of $\tau(x)/\tau_0$ was noted and plotted. This data was generated for a cavity cross-section consisting of half a circle and half a square according to Fig. 5.13b. Moreover, the ratio of the characteristic dimension of the cavity, i.e., the square side a in this example, at the cavity end to the cavity beginning was set to $L_{c,e}/L_{c,0} = 0.3$. The so obtained b-values shown in Fig. 6.8 also apply to the following cavity shapes: circle, semi-circle, square, square with two round corners. For $L_{c,e}/L_{c,0} = 0.3$, the maximum attainable Φ-value is about 0.97. If higher Φ-values are desired, then $L_{c,e}/L_{c,0}$ must be further reduced to, say, 0.2. Note that the same Φ-values can be obtained by varying c a bit and by searching for the necessary b-value. However, the difference in the resulting wall shear stress distribution is negligible. Regarding the rectangular cavity shape (see Fig. 5.11), Table 6.1 provides a couple of good values for the parameters a, b, and c that result in high Φ-values.

Figure 6.9 shows that Φ strongly depends on the parameter R_e/R_0. Here, τ_{max}/τ_0 was limited to a relatively low value of 1.2. In generating this figure, φ was varied for a given value of R_e/R_0 until τ_{max}/τ_0 reached the desired maximum value. Clearly,

Table 6.1 Preferred values of the parameters a, b, and c for the rectangular cavity cross-section

a	b	c	R_c [mm]	b_e/b_0 [−]	Φ [−]
0.70	18.0	7.0	2	0.3	0.85
0.66	4.3	5.0	2	0.25	0.90

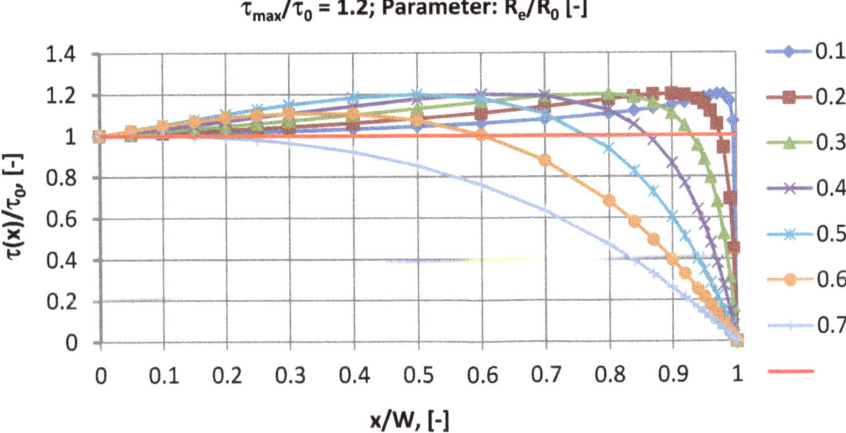

Fig. 6.9 Relative wall shear stress distribution as a function of R_e/R_0 for $\tau_{max}/\tau_0 = 1.2$; $b = 0$

good wall shear stress distributions in the inner cavity require $L_{c,e}/L_{c,0}$ to be < 0.5, preferably < 0.35. However, as will be shown below in Sect. 9.4.8, small $L_{c,e}/L_{c,0}$-values result in a relatively large pressure drop along the cavity, which is not helpful. Therefore, a compromise must be found for acceptable pressure and wall shear stress distributions in the cavity.

At the cavity entrance the wall shear stress assumes the desired minimum value τ_{min}, and the dimensions of the initial cross-section are determined with Eq. 6.2.2. According to the one-dimensional theoretical model presented above, however, the wall shear stress at the cavity end is always zero because the volumetric flow rate decreases linearly to zero. Short of the perfect situation, where the wall shear stress is constant throughout the cavity, an acceptable shear stress distribution is one that remains as constant as possible over most of the cavity length. This requires maximizing Φ, i.e., the point where $\tau(x)$ drops below its initial value, and it requires minimizing the resulting maximum wall shear stress τ_{max}. A nearly constant shear stress is best because the resulting shear rate along the cavity also remains more or less constant, which then decreases the flow sensitivity to non-Newtonian fluid properties, i.e., a power law or Carreau-Yasuda fluid will almost behave like a Newtonian fluid. Moreover, a high Φ value restricts the less-than-perfect die performance to a short section near the cavity end.

There are two known methods to eliminate or at least minimize the problem of the wall shear stress falling below a required minimum value near the cavity end. In one solution, the distribution cavity is terminated a short distance before the end of the die plate as illustrated in Fig. 6.10a. This allows the cavity end to be milled with a 3-dimensional form, which gradually and smoothly diverts the liquid flow from the direction of the cavity flow to the direction of the slot flow. For this design, the height of the die slot is typically determined by inserting a suitable U-shaped shim

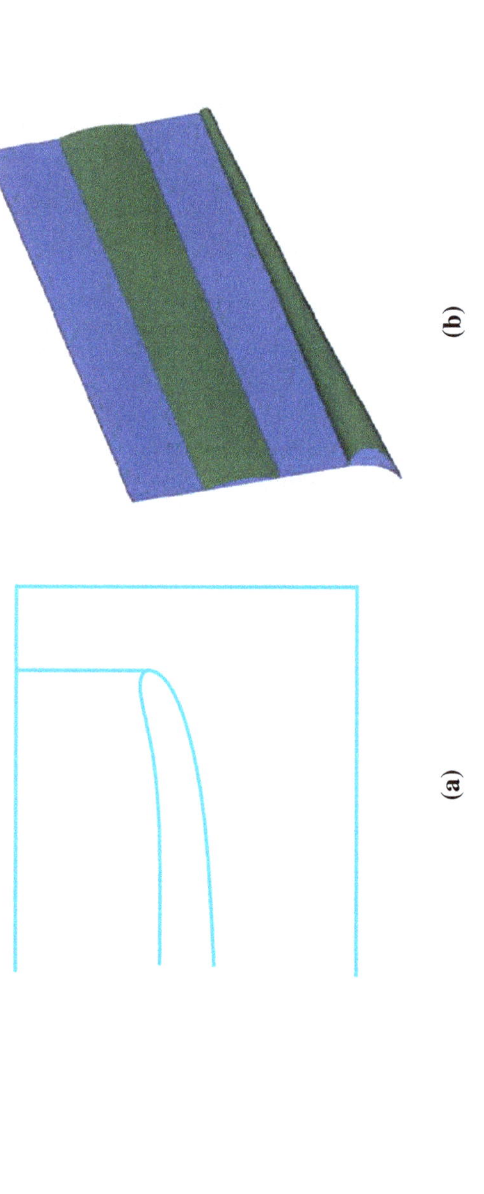

Fig. 6.10 **a** 3-D cavity end by milling the cavity shorter than the length of the die plate. **b** 3-D form attached to the end of the insert for reducing the coating width; drawing reproduced with permission from TSE Troller AG

between adjacent die plates, but the slot height could also be machined into the die plate containing the cavity.

The same effect of flow diversion is obtained in a solution developed by TSE Troller AG. It works well if inserts are used in combination with a cavity that ends at the end of the die plate, and if the form of the insert for the distribution cavity is 3-dimensional, see Fig. 6.10b. Unfortunately, the effect of both solutions on the local level of the wall shear stress cannot be quantified with simple theoretical models, but the shear stress is higher than when the flow dead-ends perpendicularly at the sidewall.

On the other hand, the distribution cavity can be operated with a small amount of **through flow**, i.e., with $\alpha > 0$. From Eq. 6.3.11 we can conclude that the relative wall shear stress value at the cavity end only depends on the by-pass factor α and the ratio of the start-to-end size of the cavity cross-section, i.e., R_e/R_0 for a circular cross-section. Therefore, it is always possible to obtain the same wall shear stress at the cavity end as at beginning of the cavity, i.e., $\tau(x)/\tau_0 = 1.0$ at $x/W = 1.0$, simply by specifying R_e and by searching for the appropriate α value, or vice versa. Figure 6.11 shows such an example of the relative wall shear stress distribution in a circular cavity with through-flow.

R_e/R_0 was set to 0.4 and $\alpha = 0.068$ was found to yield the same wall shear stress at the cavity end as at the cavity beginning. Moreover, $\varphi = 2/3 = 0.667$ was found to always result in a perfectly constant wall shear stress along the entire cavity, if the beginning and end values of the wall shear stress are equal. For $\varphi > 2/3$ the wall shear stress distribution has a maximum between beginning and end. A minimum shear stress value that is below the entrance value results if $\varphi < 2/3$.

An exact expression for the by-pass factor α is obtained for a circular cross-section by solving Eq. 6.3.11 for α, if $x/W = 1$ and $\varphi = 2/3$:

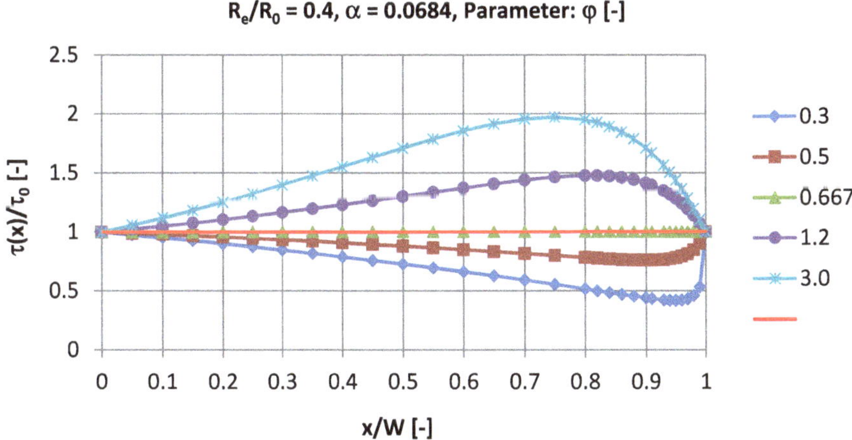

Fig. 6.11 Wall shear stress distribution in a cavity with through flow

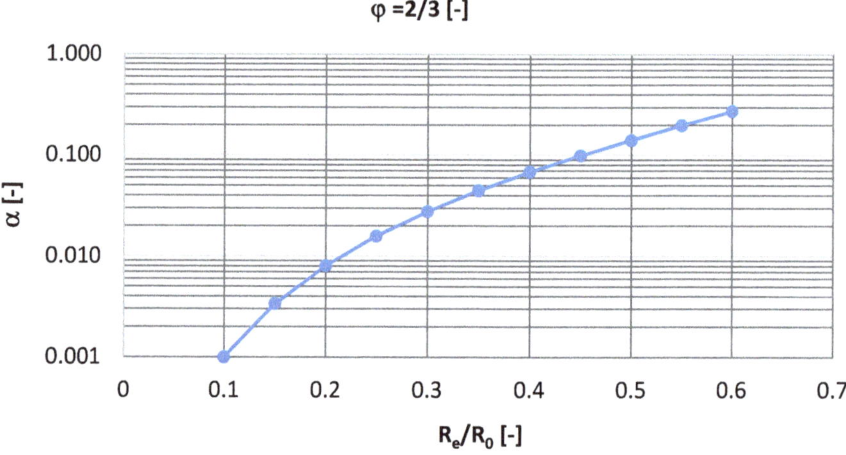

Fig. 6.12 By-pass factor α as a function of R_e/R_0 for $\varphi = 2/3$ resulting in a constant wall shear stress distribution in circular cavities

$$\alpha = \frac{\left(\frac{R_e}{R_0}\right)^3}{1 - \left(\frac{R_e}{R_0}\right)^3} \tag{6.3.15}$$

Equation 6.3.15 is graphically depicted in Fig. 6.12. For non-circular cavity cross-sections, it is more difficult to derive an analytical expression for α. However, the by-pass factor can easily be determined by a trial-and-error approach, i.e., by varying α until $\tau(x)/\tau_0 = 1.0$ at $x/W = 1.0$.

6.3.2 Wall Shear Stress in the Outer Cavity

The concept of **dual-cavity design** is often used for obtaining better cross profile results when many different products with different rheological properties and different operating conditions must be coated with the same die, see Fig. 6.13.

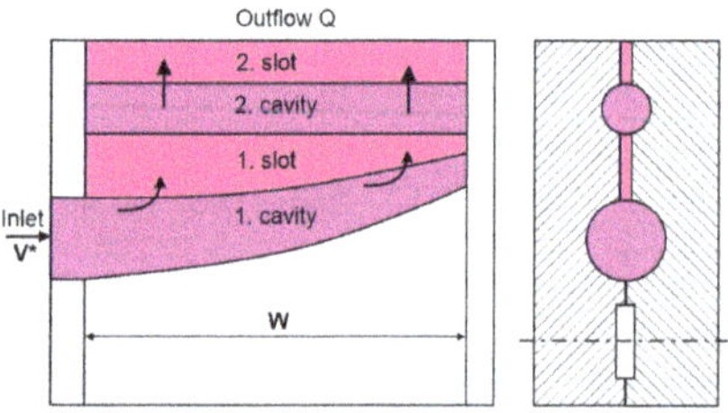

Fig. 6.13 Dual-cavity design concept

In essence, the dual-cavity design incorporates two distribution systems in series, where a distribution system consists of a distribution cavity and an associated slot. The purpose of the outer system is to dampen excessive cross profiles that would be present with only one distribution system. While the flow in the inner cavity is mainly in the direction of the cavity axis, i.e., in the cross-web direction, the flow in the outer cavity is mainly across the cavity and in web direction, i.e., from the inner to the outer slot.

The minimum wall shear stress must be considered when attempting to optimize the geometry of the outer cavity. In particular, the depth of the cavity must be kept small enough to assure a wall shear stress above a wanted minimum level. Note, however, that this design requirement conflicts with obtaining a high damping capacity of the outer cavity relating to cross profiles generated by the die. Selecting the geometry for the outer cavity, therefore, always requires a compromise between good film thickness uniformity in the cross-web direction and good flow behavior with regard to contamination, self-cleaning capability, and favorable residence time properties.

Roulier and Bebié (2020) carried out a numerical study to calculate the minimum wall shear stress for power law fluids flowing across right-angle triangular cavities with a rounded 90°-corner. Specifically, they used the software package *Mathematica* for solving the Stokes approximation of the Navier–Stokes equation on the two-dimensional cavity domain. They argued that neglecting inertia effects is justified because a flow field that generates a relatively high wall shear stress everywhere along its boundary is dominated by viscous forces, which results in low values of the Reynolds number.

As is explained in Sect. 9.4.8, the cavity length CL is a crucial parameter in die design. Therefore, Roulier and Bebié defined a set of fixed base conditions consisting of $CL_0 = 25$ mm, $Q_0 = 1.0\ cm^2/s$, $m_0 = 1.0\ Pa\ s^n$, and they calculated the minimum wall shear stress as a function of the expansion angle α, the power law index n and the radius R_c of the 90°-corner while considering the maximum value of the corner radius, see Fig. 5.16. Radius values were chosen such that the resulting wall shear stress was in the range of 1–10 Pa as suggested in Sect. 6.2. For these conditions, the minimum wall shear stress was always located at the bottom of the cavity, i.e., in the rounded 90°-corner.

Results for the 45–45 triangle are shown in Fig. 6.14. The minimum wall shear stress increases with increasing corner radius owing to the associated decrease in cavity depth. In addition, Newtonian fluids generated higher wall shear stresses than shear-thinning liquids according to the power law. Roulier and Bebié found that the heights w_1 and w_2 of the inner and outer die slots only have a negligible effect on the minimum wall shear stress.

The symbols in the diagram are the finite element calculations, while the lines through the symbols are generated by the following empirical regression equation:

$$\tau_{\min} = e^{a(R_c)+(n-0.4)b(R_c)} \tag{6.3.16}$$

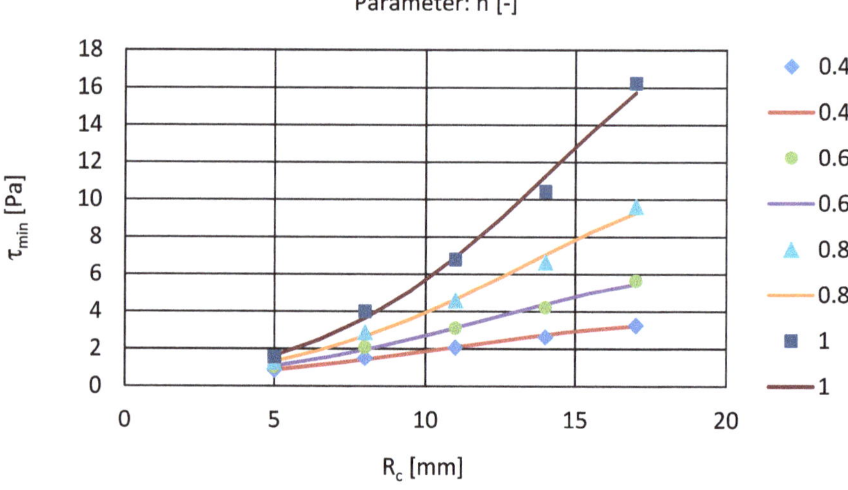

Fig. 6.14 Minimum wall shear stress in a 45–45 triangle; CL = 25 mm, Q = 1.0 cm^2/s, m = 1.0 Pa s^n

Equation 6.3.16 reflects the fact that straight lines are obtained when the logarithm of the minimum wall shear stress is plotted as a function of the power law index n. Moreover, both the functions a(R_c) and b(R_c) are second-order polynomial equations, which describe the dependence of the minimum wall shear stress upon the corner radius R_c.

$$a(R_c) = a_0 + a_1 R_c + a_2 R_c^2 \tag{6.3.17a}$$

$$b(R_c) = b_0 + b_1 R_c + b_2 R_c^2 \tag{6.3.17b}$$

R_c must be taken in units of [mm], and the regression constants are:
$a_0 = -1.21340$; $a_1 = 0.24672$; $a_2 = -0.006267$.
$b_0 = 0.06470$; $b_1 = 0.21645$; $b_2 = -0.003812$.

The maximum difference between numerical and regression calculations is <9%, and for most points, it is <5%.

Similar results are obtained for the 30–60 triangle as depicted in Fig. 6.15. For the same corner radius, the minimum wall shear stress is higher in the 30–60 triangle than in the 45–45 geometry because the associated cavity depth is smaller. The finite element calculations can be approximated with the same regression Eqs. 6.3.16 and 6.3.17, except that the regression constants are now given by:
$a_0 = -1.35933$; $a_1 = 0.36034$; $a_2 = -0.01402$.
$b_0 = -0.25951$; $b_1 = 0.41312$; $b_2 = -0.01653$.

The maximum difference between numerical and regression calculations is <6%, and for most points it is <3%.

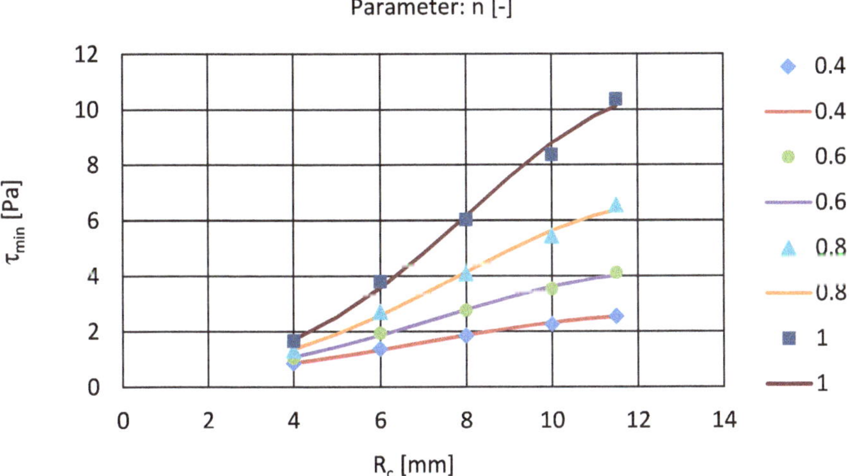

Fig. 6.15 Minimum wall shear stress in a 30–60 triangle; CL = 25 mm, Q = 1.0 cm^2/s, m = 1.0 Pa s^n

Considering the many parameters that affect the minimum wall shear stress in the flow across cavities of right-triangular shape, i.e., geometric parameters (form and shape of the cross-section in terms of α, CL and R_c), physical fluid properties (m, n) and operating conditions (Q), Roulier and Bebié (2020) pointed out the attractiveness of the Stokes approximation for calculating wall shear stresses. In particular, owing to the linearity of the Stokes equation, the following convenient scaling law can be formulated that allows the finite element results presented above to be expanded to arbitrary values of the base parameters CL_0, Q_0 and m_0 introduced earlier.

$$\tau_{\min(form,n)}(Q, m, CL) = \tau_{\min(form,n)}(Q_0, m_0, CL_0)\frac{m}{m_0}\left(\frac{Q}{Q_0}\frac{CL_0^2}{CL^2}\right)^n \tag{6.3.18}$$

The subscript (*form,n*) refers to the shape of the cavity, i.e., to the expansion angle α and the radius R_c of the vertex, and the first term on the right side of the equal sign is calculated with Eqs. 6.3.16 and 6.3.17 while considering the correct regression constants. Now, suppose you want to calculate the minimum wall shear stress for another cavity length CL, and for another radius R_c that results in a desired wall shear stress for a given flow rate/width Q, then the geometric scaling factor is given by $f = CL/CL_0$. Moreover, Eqs. 6.3.17 and 6.3.18 must be modified as follows:

$$\tau_{\min(form,n)}(Q, m, CL) = e^{a(R_c)+(n-0.4)b(R_c)}\frac{m}{m_0}\left(\frac{Q}{Q_0}\right)^n f^{-2n} \tag{6.3.19}$$

$$a(R_c) = a_0 + a_1\frac{R_c}{f} + a_2\left(\frac{R_c}{f}\right)^2 \tag{6.3.20a}$$

$$b(R_c) = b_0 + b_1 \frac{R_c}{f} + b_2 \left(\frac{R_c}{f} \right)^2 \tag{6.3.20b}$$

Consequently, setting a value for R_c results in a corresponding value for τ_{min}. However, iteration on R_c is required, if an exact value of the minimum wall shear stress must result. In addition, the finite element calculations by Roulier and Bebié were carried out for prescribed values of the rheological parameters m and n. More realistically, though, the rheology of industrial coating fluids is not described by the power law model but by the Carreau-Yasuda viscosity curve. Therefore, it is not known à priori how a given liquid will shear-thin as it flows across a given cavity, i.e., the values of m and n are not known à priori. This discrepancy can be eliminated by applying the local power law model as proposed by Ruschak and Weinstein (2014), see also Sect. 5.4.2 for more details. This approach also requires iteration on the Carreau-Yasuda viscosity until the shear stress as calculated by the power law model ($\tau = m\gamma^n$, see Macosko, 1994) equals the shear stress as calculated by Eq. 6.3.19. In EXCEL, both iterations can be carried out with the goal-seek function. As both iterations are coupled with each other but can only be carried out one after the other, they may have to be carried out two or three times until both converge. In any case, this procedure results in values for m and n that realistically describe the flow of a Carreau-Yasuda fluid across a triangular cavity.

The following example visualizes how the radius of the vertex in a 45–45 triangle depends on the flow rate/width Q and the length of the cavity CL, if a specified value of τ_{min} must be achieved, see Fig. 6.16. The Carreau-Yasuda parameters are $\mu_0 = 1.0$ Pas, $\mu_\infty = 0.01$ Pas, $\gamma_c = 0.1\ s^{-1}$, a = 0.7, b = 0.6. The height of the inner slot $w_1 = 0.25$ mm, and the minimum wall shear stress is set to be 2.0 Pa.

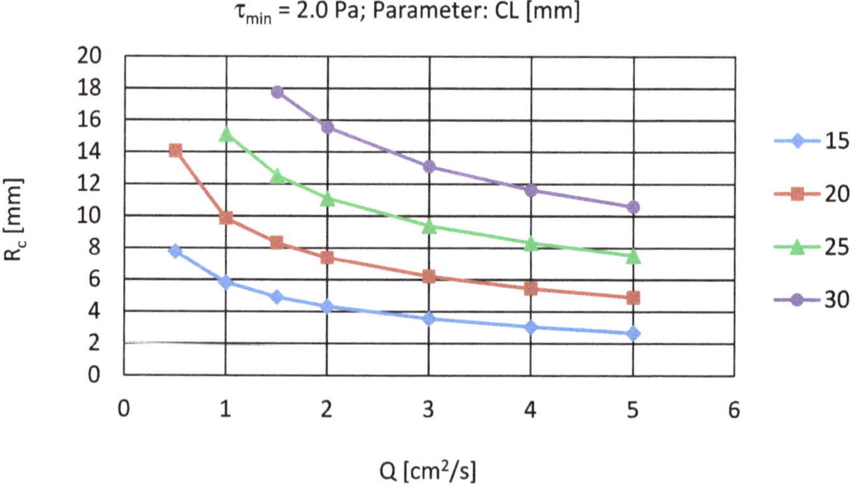

Fig. 6.16 Radius R_c of the vertex in a 45–45 triangle that results in a minimum wall shear stress of 2.0 Pa, as a function of the flow rate/width Q and the cavity length CL

Note that for CL = 25 and 30 mm, a minimum wall shear stress of 2.0 Pa cannot be achieved for the lowest flow rates in this example owing to the limiting maximum radius at the vertex.

As explained above, Roulier and Bebié (2020) determined the minimum wall shear stress with the Stokes approximation. Their motivation was driven by the possibility of scaling the results and by the associated reduction of free parameters that needed to be evaluated. For the parameter range they chose, the resulting wall shear stress levels were always above 1 Pa. For such conditions, the flow field in the cavity was always free of recirculation, and hence the use of the Stokes approximation seemed justified. Nevertheless, Roulier and Bebié compared wall shear stress levels as calculated with the Stokes approximation and with the full Navier–Stokes equations to find out how close to the inception of recirculation the Stokes equations can be used without excessive error. Information about the onset of flow separation was taken from Bebié (2021), see also Sect. 5.5.2.3 for more information. In particular, the Reynolds number Re of the flow across the cavity and the critical Reynolds number Re_c for the vortex inception were calculated with Eq. 5.5.23. They found excessive inaccuracies in the Stokes results when n approached 1.0 but only for small radii at the vertex. Such conditions are not interesting for industrial die design because the resulting wall shear stress is too low. Otherwise, wall shear stresses calculated with the Stokes approximation deviate by less than about 5% from the accurate values if $Re < 0.8Re_c$ and $n \leq 0.8$.

In summary, achieving a high minimum wall shear stress in the outer cavity requires small cavity dimensions, which in turn reduces the damping capacity of the cavity. Therefore, finding the best shape and size of the outer cavity requires a compromise by weighting the benefits of a high wall shear stress against those of a high damping factor. More information about this issue is given in Sect. 9.4 on die design.

6.4 Effect of Variable Coating Width on Wall Shear Stress Distribution

A die with a variable coating width offers great operational flexibility at low cost. The die width can be changed in different ways, see Chap. 10. In a preferred manner offered by TSE Troller AG, the coating width is changed by appropriately plugging the distribution cavities and the metering slots in the die. This in turn is accomplished by inserting so-called deckles into the cavities and slots from the outside. However, this can only be done, if the cross-section of the distribution cavity is constant from the maximum down to the minimum coating width. As visualized in Fig. 6.5, this is an unfortunate situation because the wall shear stress decreases linearly in a cavity having a constant cross-section. Consequently, the "loss of wall shear stress" in a die with variable coating width ($W_{min}/W_{max} < 1$), and therefore also the loss of self-cleaning capability as well as the increase in change-over-time (see Sect. 6.5) can be

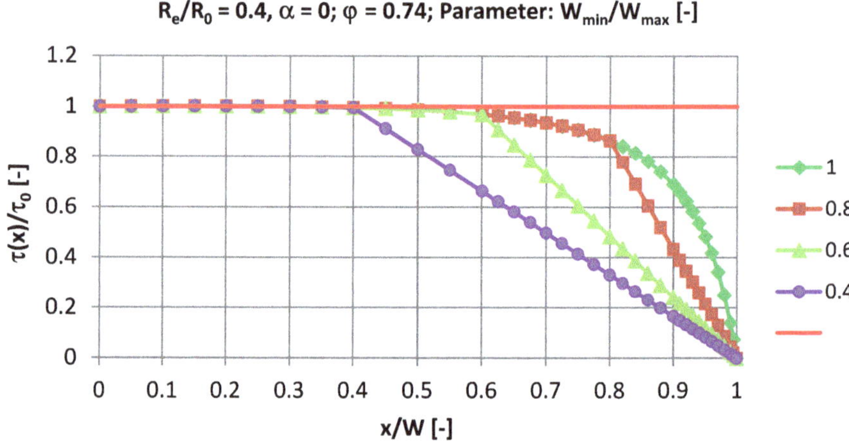

Fig. 6.17 Sub-optimal relative wall shear stress distribution in a die with variable coating width

considerable compared to a die with a fixed coating width and an optimized cavity shape ($W_{min}/W_{max} = 1$), see Fig. 6.17.

One possibility of overcoming the problem of losing wall shear stress at the cavity end in a die with variable coating width was developed by TSE Troller AG and is shown in Fig. 6.18. Compared to the deckle form shown in Fig. 6.10b, the deckle used here for varying the width only plugs the outer slot, the outer cavity, and part of the inner slot. This way, a part of both the inner slot and the inner cavity is filled with stagnant fluid during the coating operation. However, the cross-section of the inner cavity still decreases from the beginning to the end, which allows the wall shear stress to stay high and more or less constant up to W_{min}, although not quite as high as it would be for $W = W_{max}$ because the flow rate for $W < W_{max}$ is lower than the initial design flow rate V^*_0. Stagnant fluid areas inside a die cannot be tolerated for reactive fluids. Therefore, during the cleaning or changing of the fluid, stagnant fluid can be easily and effectively purged by opening the bleeding valve located in the side plate of the die.

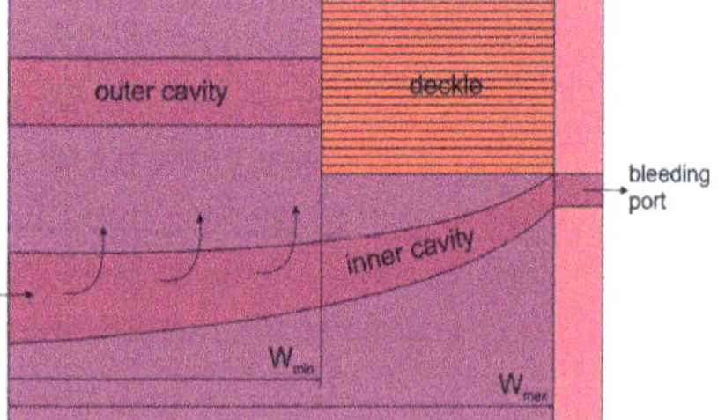

Fig. 6.18 Modified deckle form for varying the coating width; design concept from TSE Troller AG

6.5 Wall Shear Stress and Fluids with a Yield Stress

Materials that are characterized by a rheological property called yield stress are not liquid but solid if the shear stress of the flow is lower than the yield stress of the material. Yield stress can be detected with a rotational rheometer because the viscosity increases rapidly if the shear rate is lowered below a critical low value. Yield stress may also be observed by the eye, for example, if the free surface of a liquid in an unstirred vessel is not flat but disturbed in irregular ways. Materials with yield stress are difficult to handle in coating processes.

As seen in the previous Sect. 6.4, a potential problem area is the flow inside the die because the wall shear stress may fall below the yield stress if the internal die geometry is ill-designed. One example is shown in Fig. 6.19. The fluid was a suspension containing small iron particles for manufacturing video and audio tapes. The die had a dual-cavity design with a tapered inner cavity. However, as the die was used on a pilot machine, the taper was only minimal in order to generate acceptable cross profiles for many different coating applications. Consequently, the wall shear stress dropped rapidly and considerably toward the cavity end, which caused the coating fluid to solidify in a substantial part of the inner cavity. This solid material, in turn, changed the geometry of the inner cavity, which then resulted in an unacceptable cross profile.

Another example is shown in Fig. 6.20. Here, the coating material was a suspension used for manufacturing electrodes for Li-ion batteries. The internal die design was based on a concept developed by FMP Technology (fmp-technology.com). In particular, liquid distribution in cross-web direction is accomplished in part by fans

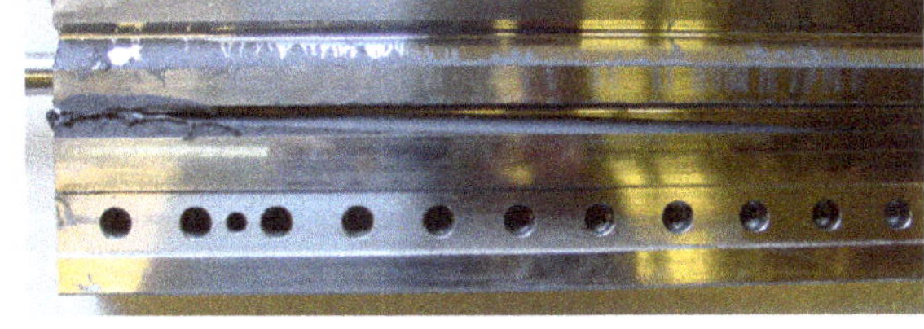

Fig. 6.19 Partly plugged inner cavity owing to solidification of a yield stress material in combination with insufficient wall shear stress; *photo* reproduced with permission from Polytype Converting AG

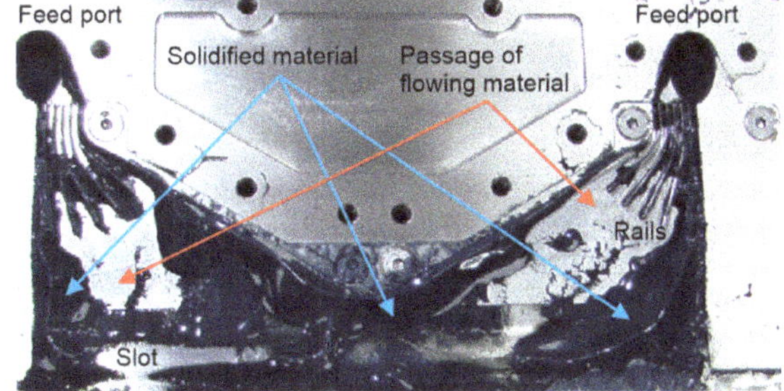

Fig. 6.20 Massively plugged inner die space owing to solidification of a yield stress material in combination with insufficient wall shear stress; *photo* reproduced with permission from Polytype Converting AG

of rigid rails that force the liquid to flow side-wise from a feed port towards the die slot. In order for the final liquid distribution to be acceptable, a relatively large and deep space is required between the end of the rails and the slot, which facilitates sufficient flow in the cross-web direction. The function of this space is similar to but stronger than the function of the outer cavity of the die shown in Fig. 6.19.

The flow conditions in Fig. 6.20 were such that the wall shear stress in the space between the rails and the slot dropped below the yield stress. Consequently, the coating material solidified over almost the entire die width. However, as the material was continuously pumped into the die, a couple of channels formed at random locations, where the shear stress became high enough, and so the material was able to flow towards the slot and exit the die.

Avoiding these solidification problems for the dual-cavity design concept would require a stronger taper for the inner cavity as discussed in Sect. 6.3. This would be easy to implement, provided that the level of yield stress is known. Avoiding these problems with the FMP design concept might be more difficult because the space between the rails and the slot would have to be made shallower, but this may in conflict with the need for providing sufficient cross flow to obtain acceptable cross profiles.

6.6 Wall Shear Stress and Residence Time

Most coating machines are designed to handle several different fluids. Therefore, the subject of contamination and cleaning of the flow systems, i.e., a fluid delivery system including the die, is important because these operations take time and reduce the productivity of the coating process. Even if the fluid delivery system does not contaminate and, therefore, does not need to be cleaned, simply replacing an old fluid with a new fluid takes time as well. This so-called change-over-time δ depends on several relevant factors. Some of them are time-related, and another one is the wall shear stress, which, as is demonstrated above, can be controlled through smart equipment design. Understanding these factors allows us to minimize the change-over-time for any given situation. We now introduce several time-related parameters, namely the mean residence time, the residence time spectrum, and the transfer function, see also Grassmann (1970) and Schweizer (1992). While the following equations apply to laminar flow in pipes, the underlying physical concepts also apply to the flow inside the die, i.e., to cavity and slot flow.

6.6.1 *Mean Residence Time*

The mean or average residence time $\overline{t_r}$ is equal to the time required by the flow to fill up the volume V of the flow system, e.g., the pipeline of the delivery system. For pipe flow:

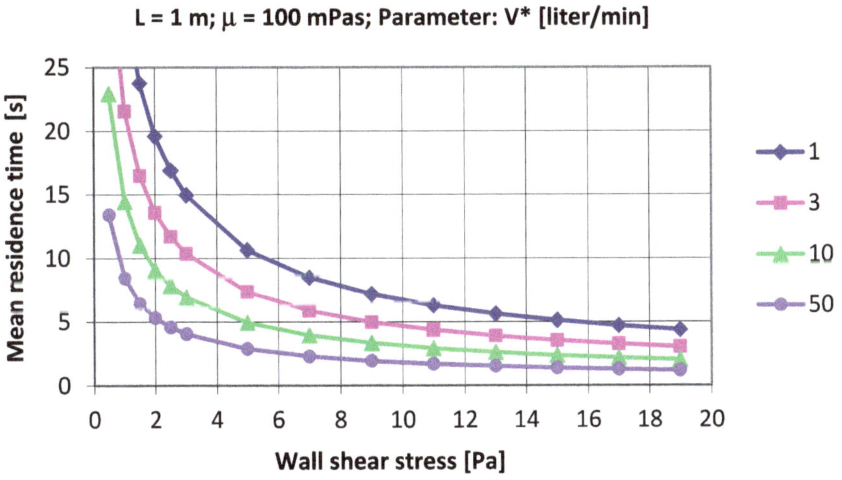

Fig. 6.21 Mean residence time for pipe flow as a function of wall shear stress

$$\overline{t_r} = \frac{\pi R^2 L}{V^*} = \frac{L}{\overline{v}} \tag{6.6.1}$$

R is the pipe radius, L is the pipe length, V^* is the volumetric flow rate and $\overline{v}$ is the average velocity of pipe flow. Short mean residence times can be achieved by choosing small dimensions of the flow system, in particular small pipe radii, and high flow rates. Similarly, the mean residence time of other flow systems such as slot or cavity flow can easily be calculated by dividing the volume of the flow system by the flow rate. Note that the mean residence time does not depend on the rheological fluid properties.

The mean residence time of Newtonian pipe flow can be expressed in terms of the wall shear stress by replacing R in Eq. 6.6.1 with Eq. 6.2.2.

$$\overline{t_r} = L\left(\frac{\pi}{V^*}\right)^{1/3}\left(\frac{4\mu}{\tau_0}\right)^{2/3} \tag{6.6.2}$$

The result is graphically shown in Fig. 6.21. Increasing the wall shear stress decreases the mean residence time. Moreover, and depending on the flow rate, increasing the wall shear stress beyond about 10 Pa does not reduce the residence time by much anymore.

6.6.2 Mean Residence Time Distribution of a Slot Die

The time it takes for a fluid element to flow from the cavity entrance at x = 0 to a given point x along the cavity and from there through the die slot, and as the case may be, across the outer cavity and the outer slot (see Fig. 6.11) is equal to the sum of the times it takes to flow through each of these sub-domains. A general but accurate expression for calculating the flow time between two arbitrary points is given by

$$t_r = \int_{x_1}^{x_2} \frac{dx}{v} \tag{6.6.3}$$

where v is the velocity along the corresponding streamline. For calculating the average or mean residence time it seems appropriate to replace v in Eq. 6.6.3 with the average velocity of the fluid in a given sub-domain. For pipe flow, this is equal to the volumetric flow rate divided by the flow cross-section according to Eq. 5.4.6, and for slot flow, the average velocity equals the volumetric flow rate/width divided by the slot height according to Sect. 5.6.10. For a single cavity die, therefore, and considering that the flow rate and the flow cross-section vary along the cavity axis according to Eqs. 6.3.7 and 6.3.8, the average time $\overline{t_r}(x)$ required by a fluid element to flow from the cavity inlet to the slot exit at any arbitrary position x equals

$$\overline{t_r}(x) = \frac{A_0}{V_0^*} \int_0^x \frac{\left(1 - \frac{x}{\beta W}\right)^{\varphi}}{\left(1 + \alpha - \frac{x}{W}\right)} dx + \frac{L(x)w}{Q_0} \tag{6.6.4}$$

The integral in the above equation must be evaluated numerically for a large number of steps, i.e., >1,000, for the result to become independent of the step size. For die designs where the banana shape of the inner cavity is not pronounced, the length L of the die slot is nearly constant across the die width, and the mean residence time distribution is mostly determined by the flow in the inner cavity. The flow through the slot merely adds a nearly constant time at any width-wise location. An example of the mean residence time distribution normalized with the mean residence time at the cavity entrance is shown in Fig. 6.22 for the following parameter values: circular cavity cross-section, $R_0 = 6$ mm, $R_e = 2$ mm, $R_e/R_0 = 0.333$, $\beta = 1.038$, $\varphi = 0.6667$, W = 500 mm, L = 40 mm, w = 300 μm, $V^*_0 = 5$ L/min, $Q_0 = 1.667$ cm^2/s, $\overline{t_r}(slot) = 0.072s$.

In this example, it takes only 0.072 s for the fluid to flow from the cavity entrance straight up to the slot exit, whereas the time required to first flow through 98% of the cavity before flowing up through the slot is about 8.3 times longer, i.e., 0.6 s. The sharp increase of the mean residence time at the very end of the cavity is an effect of the theoretical model because the flow impinges perpendicularly onto the sidewall of the cavity. This effect can be avoided if the geometry at the cavity end is modified with a rounded insert as shown in Fig. 6.9.

Flow visualization experiments were carried out to investigate the residence time behavior of slot dies. These experiments revealed the following qualitative results.

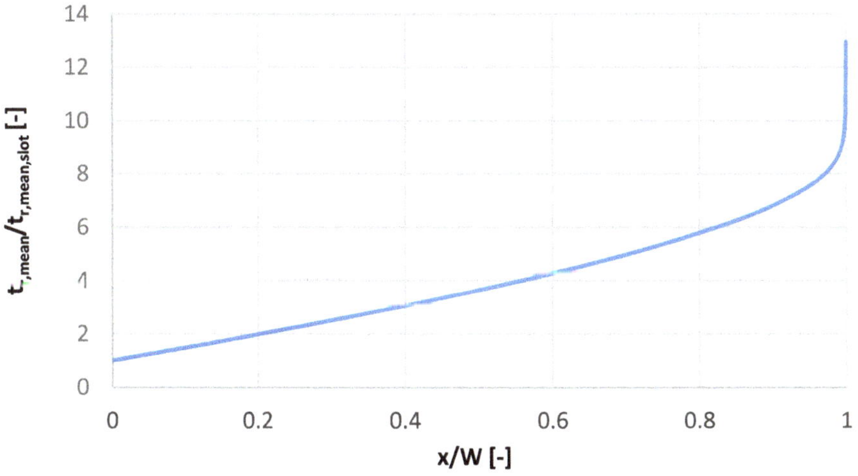

Fig. 6.22 Mean residence time distribution of a single-cavity slot die

- Streamlines located near the wall in the delivery line (slowly moving fluid) end up in the end portions of the die slot (with respect to the cross-coordinate x), whereas the fast-flowing liquid from the center of the delivery line flows through the center portion of the die slot. This observation is true for center- or end-fed dies, and the situation for the former case is sketched in Fig. 6.23.
- The residence time characteristics of a die strongly depend on the residence time characteristics and the hydrodynamic features of the flow domain upstream of the die (i.e., the fluid delivery system), such as pipe bends, valves, measurement devices, and also the length of the pipeline. This fact makes exact theoretical predictions of residence time parameters even more difficult because it requires a considerable extension of the 3-dimensional computational domain.
- The mean residence time distribution of flow through a slot die is difficult to calculate with simple flow models. Accurate theoretical calculations require solving

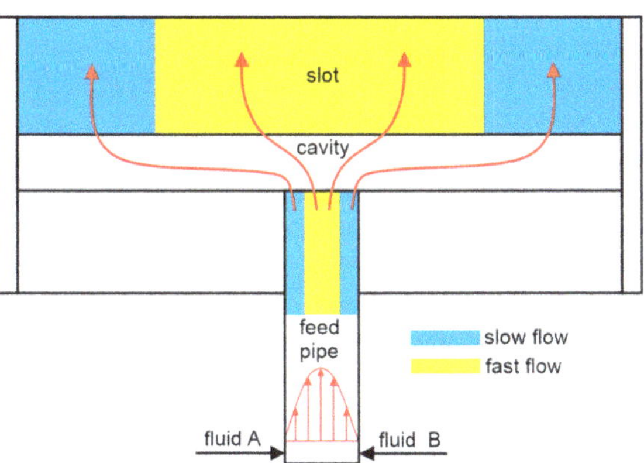

Fig. 6.23 Qualitative effect of 3-dimensional fluid flow on residence time distribution of a slot die

the 3-dimensional flow equations for the geometrically complex domain of the fluid delivery system and the internal die geometry.

6.6.3 Residence Time Spectrum

The residence time spectrum $E(t_r)$ is a normalized distribution function (histogram) indicating the fraction of the fluid stream that leaves the flow system (vessel or pipeline) between t and t + Δt. It is a measure for the distribution of ages of all elements of the fluid stream leaving the flow system, and thus it is also a measure of the distribution of residence times t_r of the fluid within the flow system. For Newtonian pipe flow, $E(t_r)$ can be calculated as follows:

$$E(t_r) = \frac{\overline{t_r}^2}{2t_r^3} = \frac{1}{t_r^3}\frac{\pi^2 L^2 R^4}{2V*^2} \tag{6.6.5}$$

with

$$t_{r,\min} = \frac{\overline{t_r}}{2} = \frac{\pi R^2 L}{2V*} \tag{6.6.6}$$

The existence of a minimum residence time $t_{r,min}$ for pipe flow is based on the fact that fluid in the center of the pipe flows faster than fluid adjacent to the pipe wall. For Newtonian pipe flow, the maximum velocity is twice as high as the average velocity, hence the factor 2 in Eq. 6.6.6, see Sect. 5.4.1. For power law fluids, the maximum velocity decreases with decreasing power law index n until the maximum velocity approaches the average velocity as the power law index approaches 0. Consequently, the factor 2 in Eq. 6.6.6 must be adjusted accordingly by setting $r = 0$ in Eq. 5.4.14 and by calculating the maximum velocity as a function of n. As is evident, the factor decreases from 2 to 1 as n decreases from 1 to 0 according to the term (1 + 3n)/(1 + n).

The residence time spectrum $E(t_r)$ for laminar pipe flow of a Newtonian fluid as calculated by Eq. 6.6.5 is graphically shown in Fig. 6.24. Narrow residence time spectra, which are characterized by a short tail of the curve, can be achieved by choosing small dimensions of the flow system, i.e., small pipe radii, high flow rates, and by avoiding dead flow areas, i.e., areas of flow separation (vortices), which in turn requires high wall shear stress.

6.6.4 Transfer Function

The transfer function $J(t_r)$ describes the fraction of the fluid stream that has a residence time of less than t_r. For pipe flow, $J(t_r)$ can be calculated as follows by considering

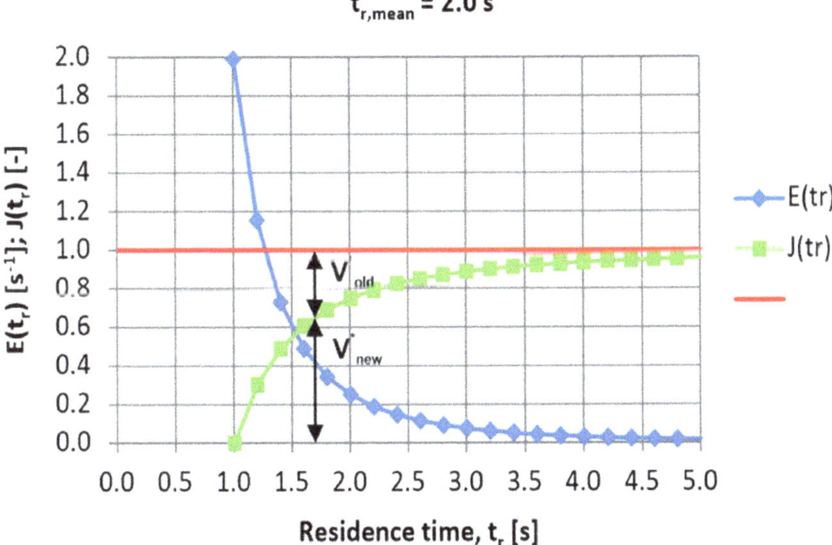

Fig. 6.24 Residence time characteristics for pipe flow

the boundary condition that $J(t_r) = 1$ for $t_r \rightarrow \infty$. Note that Eq. 6.6.6 applies here as well.

$$J(t_r) = \int E(t_r)dt_r = 1 - \frac{1}{4}\left(\frac{\overline{t_r}}{t_r}\right)^2 \qquad (6.6.7)$$

The characteristic residence time features for laminar pipe flow (Eqs. 6.6.5 and 6.6.7) are graphically depicted in Fig. 6.24 for the following operating conditions: $V^* = 6$ L/min, L = 1 m, R = 8 mm, $\overline{t_r} = 2$ s.

Note that all old fluid is replaced by new fluid when the value of the transfer function $J(t_r)$ approaches 1.0.

6.6.5 Change-Over Time

The transfer function can be used to define the change-over-time δ, i.e., the time required by the flow to replace a certain amount of old fluid by new fluid. For pipe flow, δ can be calculated by solving Eq. 6.6.7 for t_r as follows:

$$\delta = \frac{\overline{t_r}}{2}\frac{1}{\sqrt{1 - J(t_r)}} = \frac{\pi R^2 L}{2V^*}\frac{1}{\sqrt{1 - J(t_r)}} \qquad (6.6.8)$$

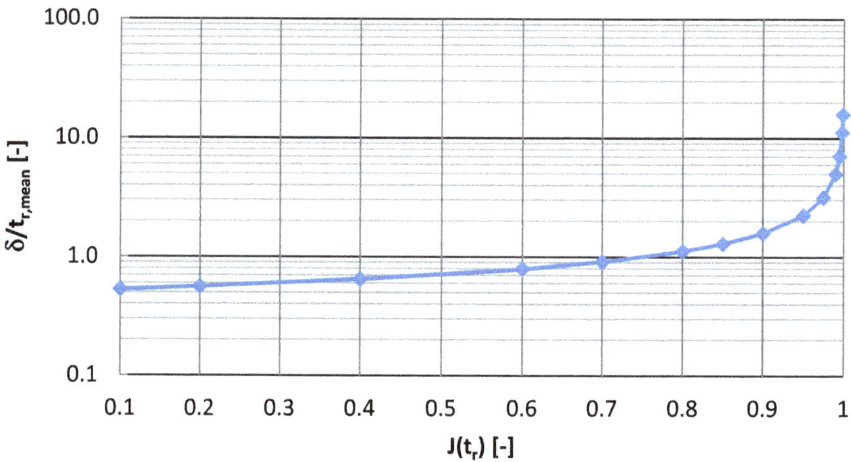

Fig. 6.25 Normalized change-over-time for pipe flow as a function of the transfer function

Again, small flow cross-sections (small pipe radii) and high flow rates will result in short change-over-times. A complete fluid change-over is reached when $J(t_r)$ assumes a value of very close to 1. Note that $J(t_r) = 1.0$ is not admissible because it would result in a division by 0. Figure 6.25 shows a plot of the change-over time normalized with the mean residence time as a function of the transfer function. About 75% of the fluid is replaced from the flow system (pipeline) after a time equal to the mean residence time. However, it takes about 16 mean residence times for replacing 99.9% of the old fluid from the flow system.

Using Eq. 6.2.2, the change-over-time can be expressed in terms of the wall shear stress. The result for pipe flow is given by Eq. 6.6.9, which is graphically presented in Fig. 6.26.

$$\delta = \frac{L}{2\sqrt{1 - J(t_r)}} \left(\frac{\pi}{V^*}\right)^{1/3} \left(\frac{4\mu}{\tau_0}\right)^{2/3} \tag{6.6.9}$$

Again, increasing the wall shear stress shortens the change-over-time. However, wall shear stress values beyond about 10 Pa do not reduce the change-over-time by much anymore. Note that the change-over-time is on the order of minutes, while the mean residence time is on the order of seconds, see Fig. 6.21.

6.7 Wall Shear Stress and Reactive Fluids

The concepts of minimum wall shear stress and residence time are of great importance when coating chemically reactive fluids. The issue is illustrated by the example of a customer of Polytype Converting, who carried out preliminary coating trials with a

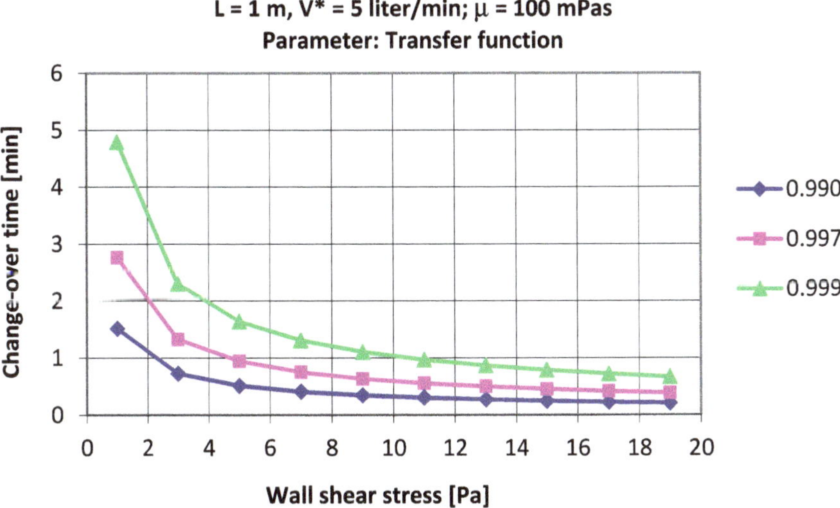

Fig. 6.26 Change-over-time for pipe flow as a function of wall shear stress

very reactive fluid by using a 250 mm wide center-fed slot die that was equipped with an internal geometry similar to the infinite cavity design, see Fig. 6.3a. Unfortunately, the material started to solidify after only a few minutes of operation. For us, this observation was not surprising because the ill-designed distribution cavity violated the requirement of a minimum wall shear stress by far. Consequently, the residence time spectrum was too wide and the mean residence time was too long.

In cooperation with TSE Troller AG, we convinced the customer to trust in the various concepts presented in this chapter, and TSE Troller came up with a cavity design that incorporated the wall shear stress concept to the utmost possible in concert with their machining capabilities. Specifically, the die was now of production size with a width of about 1,500 mm, and the ratio of end to entrance cavity cross-section A_e/A_0 was 0.16. The performance of this die was much more satisfactory because the worry-free coating of this highly reactive fluid was now possible for several hours.

The improved die performance can be explained by looking at the mean residence time for the flow in a circular cavity of length W, where the volumetric flow rate V^* decreases linearly along the cavity axis according to Eq. 6.3.7 and the cavity cross-section A changes according to Eq. 6.3.8 with β defined by Eq. 6.3.10. In accordance with Eqs. 6.6.3 and 6.6.4 the mean residence time can be calculated with

$$\overline{t_r} = \int_0^W \frac{1}{v(x)} dx = \int_0^W \frac{A(x)}{V^*(x)} dx = \frac{A_0}{V_0^*} \int_0^W \frac{\left(1 - \frac{x}{\beta W}\right)^{\varphi}}{\left(1 - \frac{x}{W}\right)} dx \tag{6.7.1}$$

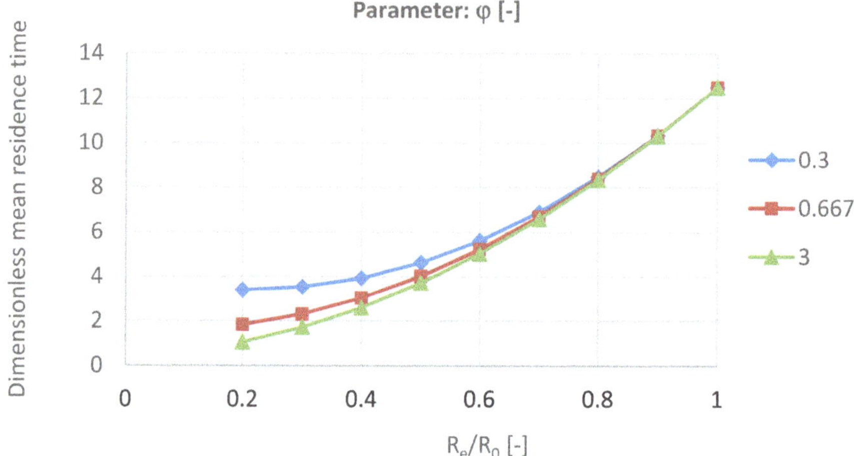

Fig. 6.27 Dimensionless mean residence time of flow in a circular die cavity

Performing the integration numerically by using 1,000 steps and making the result dimensionless with the term WD_0/V^*_0 produces the following graphical result when plotted as a function of the geometrical parameters R_e/R_0 and φ (Fig. 6.27).

For $R_e/R_0 = 1$, which corresponds to the infinite cavity design, the dimensionless mean residence time is about 12.5. However, when R_e/R_0 is reduced to 0.2, which indicates a strongly tapered cavity form, the mean residence time drops by a significant factor of at least 7. This happens provided that the parameter φ, which is a measure for how the cavity cross-section changes along the cavity axis, is above the minimum value according to Eq. 6.3.12. When $\varphi > \varphi_{min}$, the wall shear stress does not drop below its initial value until towards the end of the cavity, and the value of φ has little effect on the mean residence time.

References

Bebié, H. (2021). Critical Reynolds number for the onset of recirculation in flow of shear-thinning liquids across triangular cavities with a rounded vertex. *Personal communication.*

Grassmann, P. (1970). *Physikalische Grundlagen der Verfahrenstechnik*. Sauerländer AG.

Macosko, C. W. (1994). *Rheology: Principles, measurements and applications*. VCH Publishers.

Roulier, A., & Bebié, H. (2020). Wall shear stress in flow across triangular cavities. *Personal communication.*

Ruschak, K. J., & Weinstein, S. J. (2014). A local power law approximation to a smooth viscosity curve with application to flow in conduits and coating dies. *Polymer Engineering and Science* 2301–2309.

Schweizer, P. M. (1992). Fluid handling and preparation. In E. Cohen & E. Gutoff (Eds.), *Chapter 2 in modern coating and drying technology*. VCH Publishers Inc.

Schweizer, P. M. (1997). Control and optimization of coating processes. In S. F. Kistler & P. M. Schweizer (Eds.), *Chapter 15 in liquid film coating*. Chapman & Hall.

Chapter 7
Dynamic Wetting and Hydrodynamic Assist

Abstract The concept of the hydrodynamic assist in dynamic wetting was first presented by Blake, Clarke and Ruschak from Kodak. Hydrodynamic assist is a property common to all premetered coating methods. Specifically, the momentum of the approaching liquid is exploited to postpone the onset of air entrainment to higher web speeds by placing the location of the dynamic wetting line into the projection of the approaching liquid stream. This chapter discusses why premetered coating processes are superior to self-metered methods with regard to dynamic wetting. It also explains that exploring hydrodynamic assist requires process optimization to ensure that the dynamic wetting line is indeed located inside the impinging liquid stream and not too far upstream or downstream.

7.1 Introduction

The physical process acting at the dynamic wetting line, i.e., at the point where the liquid to be coated touches the substrate for the first time, is called dynamic wetting. This process is present in every coating process, but the location of the dynamic wetting line relative to the point on the substrate where the coated film is taken away by the moving web depends on the specific coating method, compare Figs. 7.1 and 7.3.

Today, intensive research in dynamic wetting is still going on because the underlying physical concepts are not yet fully understood. A comprehensive review of the subject was given by Blake and Ruschak (1997).

In industrial coating processes, dynamic wetting depends on the capillary number Ca and the Weber number We, i.e., on the relative magnitude of capillary, viscous, and inertia forces. The values of Ca and We can vary from small to large, and they depend on the specific coating application. Wetting failure is equivalent to the onset of substantial air entrainment; it represents one of the operating boundaries, at least for premetered coating methods. It is generally accepted by the wetting experts that wetting failure occurs when the dynamic wetting angle approaches 180°. Note that the dynamic wetting angle is not a material property such as the static wetting angle

P. M. Schweizer, *Premetered Coating Methods*, Engineering Materials,
https://doi.org/10.1007/978-3-031-04180-8_7

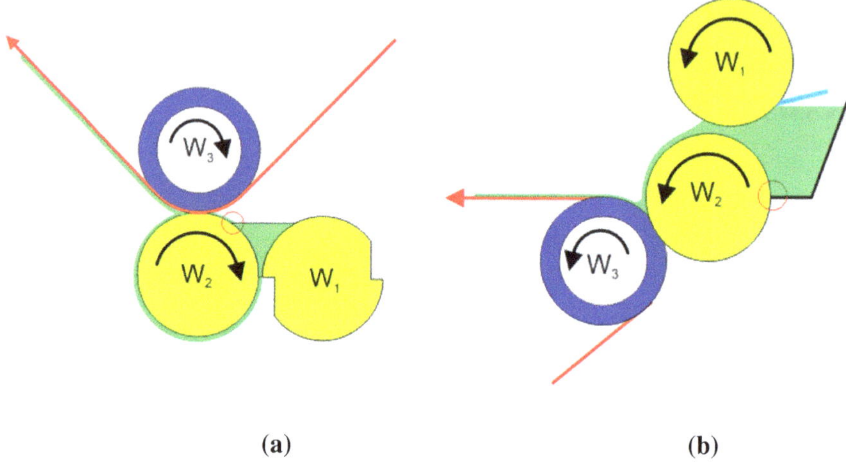

Fig. 7.1 **a** Comma coater with nip feed. **b** Reverse roll coating with vertical nip feed

presented in Sect. 4.6 because it depends on several process parameters such as the web speed and the viscosity of the wetting fluid.

Blake and Ruschak (1979) reported that, upon increasing the substrate speed, the dynamic wetting line, which extends straight across the substrate for lower speeds, becomes unsteady and decays into straight-line segments of triangular shape just before the dynamic wetting angle reaches 180°. Consequently, the wetting line across the substrate has the appearance of a sawtooth line. The triangular segments may be small in size, i.e., just a few millimeters, or they may extend several centimeters in the direction of the substrate motion. An example of a large saw tooth wetting line is shown in Fig. 11.2.

For operating conditions resulting in a large dynamic contact angle, an air film forms between the liquid and substrate in every triangular segment, and air is entrained into the liquid, often in form of a string of air bubbles originating at the vertex of every triangle. Many experiments have been carried out to determine the web speed at the onset of air entrainment, mostly for the configuration of a substrate plunging into a quiescent pool. A review of these results appeared in Blake and Ruschak (1997). While no experiment produced the same result as was obtained in any other experiment, they all showed that the maximum speed of wetting increased with decreasing viscosity. This is an important result, not the least for premetered coating methods. Should air entrainment ever be an issue in an industrial coating application, and if every other available means for preventing air entrainment failed, then lowering the viscosity will eliminate air entrainment. As explained in Sect. 12.2, this measure can easily be implemented for slide and curtain coating by resorting to the concept of the so-called carrier layer.

Unfortunately, theoretical models for predicting the dynamic contact angle as a function of the operating conditions of a given coating method do not exist. In particular, models that are simple enough to be used in an industrial coating environment are not known. Attempts of modeling dynamic wetting of a coating method by solving the Navier–Stokes equations are possible but they face the problem of a force singularity at the dynamic wetting line, see Blake and Ruschak (1997). The origin of the singularity is a discontinuity in velocity at the dynamic wetting line. Blake and Ruschak explained that, in the frame of reference, in which the wetting line is stationary, "the limiting velocity as the wetting line is approached along the meniscus is zero, but because of the no-slip boundary condition, the limiting velocity as the wetting line is approached along the solid is the velocity of the solid". Numerical experts have developed ways for handling the singularity, for example by relaxing the no-slip boundary condition at the solid substrate, i.e., by introducing a so-called slip coefficient, see Christodoulou et al. (1997). While this approach produces a solution of the Navier–Stokes system, the result cannot by accurate because it depends on an assumption needed for quantifying the slip coefficient. Kistler (1983) showed that the force singularity, i.e., the need for a contact angle boundary condition, vanishes from the hydrodynamic equations for the limiting case of infinite capillary number. However, such operating conditions are generally not present in industrial coating processes, except perhaps for high-speed curtain coating.

Solving the Navier–Stokes equations for a flow field with dynamic wetting yields a macroscopic results, and we have seen that predictions solely based on first principles are not possible. Therefore, to further advance the understanding of dynamic wetting, additional models based on a microscopic or molecular scale are needed. Various such models have been proposed in the past, and one that has persisted and generated promising results is called *molecular-kinetic theory* (MKT). An early review of this concept was given by Blake and Ruschak (1997). Accordingly, the dynamic contact angle depends on the substrate velocity, and for simple cases, this dependence is due to the disturbance of adsorption equilibria at the wetting line. MKT produced an expression relating the wetting line speed (the web speed) to the dynamic contact angle. Unfortunately, however, this expression contains parameters that are not readily available in industrial coating applications. A more recent review of the MKT can be found in Duvivier et al. (2013).

According to our experience, the lack of a complete understanding of dynamic wetting is not crucial for successfully operating industrial coating processes. The reason for this view is the concept of hydrodynamic assist to dynamic wetting, which was first introduced by Blake Clarke and Ruschak (1994), and which is described in the following chapter.

7.2 The Concept of Hydrodynamic Assist

Hydrodynamic assist is an important property of all premetered coating methods that gives these methods a decisive advantage over all other coating processes. In particular, dynamic wetting in premetered coating methods, i.e., the actual coating of the substrate, is characterized by a hydrodynamic assist in the form of momentum

Table 7.1 Comparison of typical impingement velocities for curtain, slide, and slot coating

	Curtain	Slide	Slot
Impingement velocity [m/s]	1.72	0.092	0.40
v/v_{curtin} [%]	100	5	23

carried by the approaching liquid, which facilitates the prevention of air entrainment at the dynamic wetting line up to high web speeds. As explained in Sect. 5.9.6, the liquid in slot coating approaches the substrate in form of slot flow; in slide coating, it is in the form of film flow on the inclined die surface, and it is in the form of the impinging curtain in curtain coating.

According to Blake et al. (1994), the momentum M of the approaching liquid can be quantified by

$$M = \rho Q v \cos\alpha \tag{7.1.1}$$

Here, ρ is the liquid density, Q is the volumetric flow rate/width, v is the impingement velocity, and α is the impingement angle of the incoming liquid. For slot coating, v is the average velocity of slot flow according to Eq. 5.6.10. For slide coating, v is the average velocity of film flow according to Eq. 5.7.4, and for curtain coating, v is the curtain impingement velocity according to Eq. 5.8.1. In Table 7.1 the impingement velocities of the three premetered coating methods are compared for the following arbitrary but typical operating conditions. The initial velocity for the curtain velocity is taken as the average film velocity on a vertical surface.

$L_c = 150$ mm, $\beta = 30°$ (slide), $\beta = 90°$ (curtain), $w = 300$ μm, $Q = 1.2$ cm^2/s, $\rho = 1000$ kg/m^3, $\mu = 30$ mPas.

Curtain coating has by far the highest hydrodynamic assist not only of the premetered but of all coating methods.

By comparison, Fig. 7.1 illustrates the dynamic wetting situation for nip-fed or pan-fed roll and blade coating methods.

Figure 7.1a shows a nip-fed comma coater where dynamic wetting is similar to a tape plunging into a pool. This is the worst case of all since the hydrodynamic assist is just about zero, except perhaps for a few millimeters of static liquid height because the plunging surface draws the dynamic wetting line a short distance below the surface of the liquid pool. Figure 7.1b shows a reverse 3-roll coater with a vertical nip feed. This situation has more dynamic wetting assistance than the horizontal nip feed. On the one hand, the bottom of the pan acts as a shield to block the air boundary layer on the uncoated roller surface. On the other hand, wetting of the roller surface W_2 is assisted by a few centimeters of static liquid height in keeping with the depth of the vertical nip feed. However, all these means for assisting dynamic wetting are only static in nature and hence smaller than the hydrodynamic assistance found in premetered coating methods, particularly when compare to curtain coating.

In gravure coating, with pressurized chambers the chamber pressure facilitates dynamic wetting. However, initial wetting failure is observed when the gravure pattern consists of small, individual cells because the returning empty cells carry quite a bit of air into the chamber when they pass under the leading blade. This type of air entrainment can be avoided by using a dual-chamber design in combination with a helical groove pattern. As illustrated in Fig. 7.2, the high pressure in the upstream chamber forces the coating liquid in the grooves to flow upstream against the motion of the cylinder and past the leading blade of the chamber.

Experience shows that it is very difficult to operate roll, blade, and gravure coating methods without excessive air entrainment, particularly if the viscosity is not very low. Consequently, many small air bubbles accumulate in the liquid in the pan or nip feed after only a short period of operation. A nice example of this undesirable behavior is shown in Fig. 5.2, where the initially clear silicone fluid turned into a micro-foam.

Focusing again on premetered coating methods, hydrodynamic assist is only available in the projection of the impinging fluid stream, i.e., if the dynamic wetting line is located in the projection of the impinging liquid. If the dynamic wetting line is located too far upstream or downstream relative to the location of the impinging stream, then air entrainment may occur at low speeds and/or vortices may form. The situation is depicted in Fig. 7.3.

The upper row of photos and sketches shows an optimized flow field of the impinging fluid for all premetered coating methods. Optimized refers to the fact that the dynamic wetting line is located in the projection of the incoming fluid, meaning that the momentum of the approaching fluid can be used to facilitate dynamic wetting without air entrainment. In the lower row of photos and in the streamline pattern, the wetting line for all processes is located too far upstream, i.e., outside of the projection of the incoming fluid stream, and hence the momentum is not available for the prevention of air entrainment at higher web speeds. In addition, the flow field near

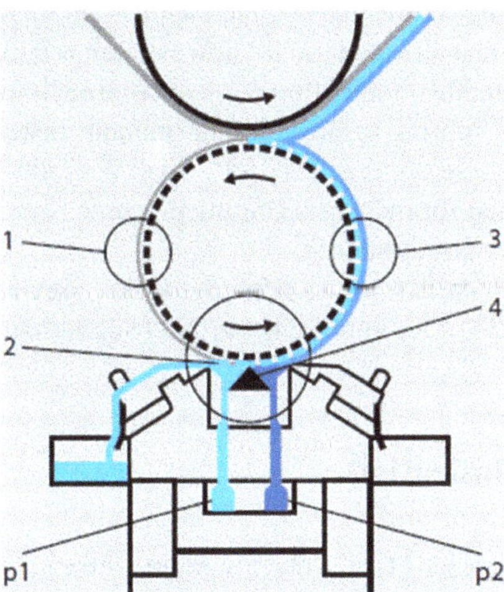

Fig. 7.2 AGS-2500 dual chamber gravure coating system from Polytype Converting AG (www.polytype-converting.com)

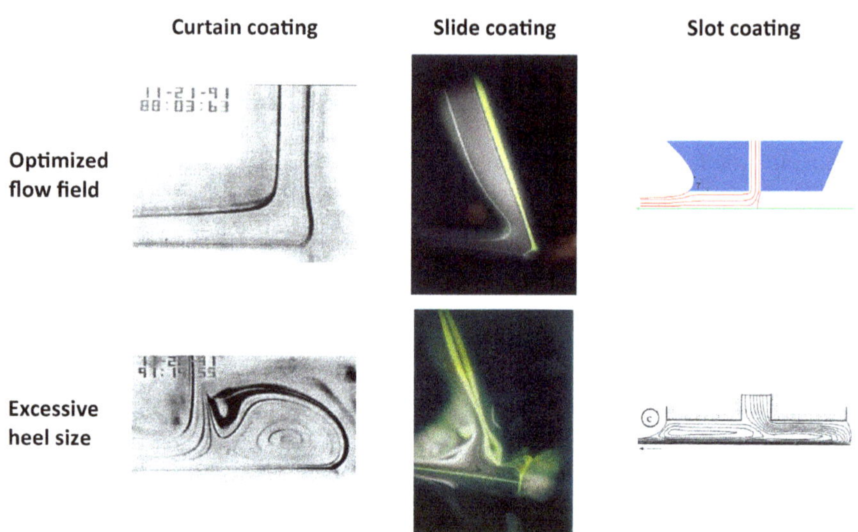

Fig. 7.3 Impinging fluid in curtain, slide, and slot coating; the curtain coating photos are taken from Blake et al. (1994) with permission from AIChE Journal; the slide coating photos are taken from Schweizer (1988) with permission from Cambridge University Press; the bottom streamline pattern in slot coating is taken from Durst and Wagner (1997) with permission from Springer Nature

the wetting line is characterized by a vortex, which is highly undesirable. Such flow fields must be avoided.

Note that both photos of slide coating are purposely rotated by 90° to emphasize the similarity of the impinging flow fields of curtain, slide, and slot coating.

Objectionable flow fields are also obtained with all premetered coating methods if the dynamic wetting line is located too far downstream relative to the position of the incoming stream. Again, the momentum of the incoming fluid stream cannot, or only partly, be explored for fighting against air entrainment. Instead of forming vortices, the dynamic contact angle approaches 180°. Such operating conditions are highly undesirable and not good for industrial coating processes because small disturbances may trigger massive air entrainment.

Photographs of such situations are difficult to take. Figure 7.4 shows a computed but unrealistic flow field of an impinging curtain (Sünderhauf, 1997).

7.3 Process Optimization

As discussed by Schweizer (1997) the impinging flow of all premetered coating methods presents a classic opportunity for process optimization. Specifically, optimum flow conditions are obtained when the dynamic wetting line is located in the projection of the incoming fluid stream. In contrast, flow conditions where

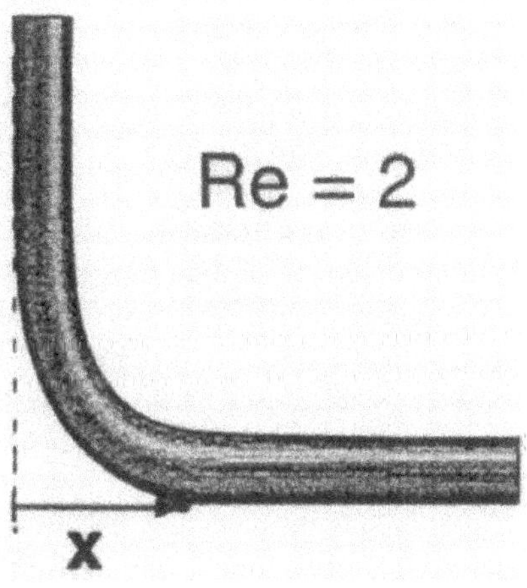

Fig. 7.4 Impinging curtain with the dynamic wetting line located too far downstream; from Sünderhauf (1997) with permission

the dynamic wetting line is located too far upstream or downstream relative to the location of the incoming fluid stream must be avoided. This kind of optimization can easily be carried out for curtain coating because Blake et al. (1994) developed a theoretical model, by which the dimensionless location of the dynamic wetting line ℓ can be calculated for a given set of operating conditions, see Eq. 5.9.37 in Sect. 5.9.6. Regarding the optimization criterion, Blake et al. (1994) resorted to maximizing the pressure that is exerted on the boundary layer by the impinging fluid stream, and they argued that best conditions are obtained when the dimensionless wetting line position assumes a value of approximately 2.0. In contrast, Schweizer (1997) used arguments of process robustness for claiming that the best wetting line position should coincide with the steepest slope of the upstream branch of the pressure distribution in the impinging curtain. This is the case when the dimensionless wetting line position assumes a value of about 1.5. In any case, Blake et al. (1994) proposed that best conditions are obtained when the dynamic wetting line is located in the vicinity of the rear curtain surface, i.e., at $\ell = 1.0$, and our experience confirms that good coating results are obtained for $0.5 < \ell < 2.0$, which is in agreement with the opinion of Blake et al. Moreover, our experience also indicates that Eq. 5.9.37 is a useful tool for optimizing industrial coating processes. While working at ILFORD, for example, we successfully redesigned the coating machine based on calculations with this tool in order to allow positive curtain impingement angles, and we adjusted the viscosity of many product formulations such that the wetting line position came to lay in the preferred range. An improved product uniformity and a reduction of product waste resulted from these actions. In addition, air entrainment was hardly ever an issue during 15 years of running numerous curtain coating trials on the Polytype Converting pilot machine for applications across the entire spectrum of the converting industry because we used Eq. 5.9.37 for determining the best process geometry prior to the trials, and this for speeds of up to 1,000 m/min.

Should air entrainment be observed even though the dynamic wetting line is located inside the preferred range, then Schweizer (1997) argued that the momentum of the incoming fluid stream must be increased. From this point of view, operating premetered coating processes with an impingement angle of $\alpha = 0°$ is optimal, i.e., when the incoming fluid stream impinges perpendicularly onto the substrate surface. However, this may not always be possible because, as illustrated in Fig. 5.9.59, the impingement angle also affects the location of the dynamic wetting line. Therefore, a compromise must be sought and other means to affect the process performance must be considered. In curtain coating, for example, the impingement momentum can conveniently be increased according to Eq. 7.1.1 by increasing the curtain impingement velocity, which in turn results from increasing the length of the curtain. Note, however, that this is not a strong measure because the impingement velocity depends on the square root of the curtain length. Moreover, changing the curtain length requires a new set of curtain edge guides.

Optimizing the bead flow field of slide and slot coating with regard to the position of the dynamic wetting line is more difficult than for curtain coating because a similar theoretical model does not exist. The main reason is the lack of a suitable geometric reference position, such as the front surface of the curtain. While the formation of an excessive heel in curtain coating can be visible to the naked eye, the bead flow fields in slot and slide coating are not visibly accessible owing to the fact that these processes are typically operated with a vacuum box. Also, increasing the momentum of the incoming fluid stream with only geometrical means is more difficult than for curtain coating. In slot coating, it could be achieved by reducing the height of the die slot, but perhaps only at the expense of an increased cross profile. In slide coating, increasing the slide angle would increase the impingement momentum, but care must be taken to avoid waves in the film flow.

Optimizing slot and slide coating processes is mostly based on a trial-and-error approach with bead vacuum being the only convenient and formulation-independent parameter. Therefore, it is suggested to operate these processes without bead vacuum, if air entrainment is not observed. If air entrainment is obvious, then the bead vacuum should be increased to a value that is slightly higher than the value at which air entrainment disappears. If air entrainment cannot be avoided this way, then the viscosity has to be lowered, either by diluting or heating the coating liquid and/or by applying the concept of the carrier layer. If, on the other hand, the presence of an excessive heel or vortex is suspected in slide coating even without any bead vacuum, then the fluid viscosity should be increased and the gap between the die lip and the substrate surface should be minimized. If the same suspicion exists in slot coating, then maximizing the coating gap and possibly increasing the viscosity should result in the desired improvement.

References

Blake, T. D., Clarke, A., & Ruschak, K. J. (1994). Hydrodynamic assist of dynamic wetting. *AIChE Journal, 40*(2), 229–242.

Blake, T. D., & Ruschak, K. J. (1979). A maximum speed of wetting. *Nature, 282*, 489–491.

Blake, T. D., & Ruschak, K. J. (1997). Wetting: Static and dynamic wetting lines. In S. F. Kistler & P. M. Schweizer (Eds.), Chapter 3 in *Liquid film coating*. Chapman & Hall.

Christodoulou, K. N., Kistler, S. F., & Schunk, P. R. (1997). Advances in computational methods for free-surface flows. In S. F. Kistler & P. M. Schweizer (Eds.), Chapter 9 in *Liquid film coating*. London: Chapman & Hall.

Durst, F., & Wagner, H.-G. (1997). Slot coating. In S. F. Kistler & P. M. Schweizer (Eds.), Chapter 11a in *Liquid film coating*. Chapman & Hall.

Duvivier, D., Blake, T. D., & De Coninck, J. (2013). Toward a predictive theory of wetting dynamics. *Langmuir, 29*(32), 10132–10140.

Kistler, S. F. (1983). *The fluid mechanics of curtain coating and related viscous free surface flows with contact lines*. Ph.D. thesis. University of Minnesota, Minneapolis.

Schweizer, P. M. (1988). Visualization of coating flows. *Journal of Fluid Mechanics, 193*, 285–302.

Schweizer, P. M. (1997). Control and optimization of coating processes. In: S. F. Kistler and P. M. Schweizer (Eds.), Chapter 15 in *Liquid film coating*. Chapman & Hall.

Sünderhauf, G. (1997). Vorhangbeschichtung. Kurzlehrgang über Grundlagen und Verfahren der Beschichtungstechnik. Lehrstuhl für Strömungsmechanik, Universität Erlangen-Nürnberg.

Part II
General Properties of Premetered Coating Methods

Abstract

Part II of the book addresses various properties and requirements that are common to all premetered coating methods. It opens with a description of the fluid delivery system and it continues by discussing mechanisms that affect the thickness uniformity of the coated film both in machine and cross-web direction. The latter issue is devoted to designing the inner geometry of slot dies.

Next, concepts for changing the coating width and handling splices as they pass through the coating process are illustrated. A large section of Part II deals with the attractive property of all premetered coating methods to simultaneously apply several different fluid layers. Specifically, the issue of interlayer mixing is addressed and guidelines for designing single- and multilayer films are offered. Part II ends by proposing a theoretical economic model that is helpful for making investment decisions. Specifically, the model visualizes how the unit price of coated materials can be affected by technical features and the operating conditions of the coating process.

Chapter 8
Preparing, Conditioning, and Delivering Coating Fluids

Abstract Elaborate fluid conditioning and delivery systems are an integral part of all premetered coating methods. Achieving a high uniformity of the coated film is impossible if the fluid is not uniform and homogeneous in terms of the solids concentration and temperature. This chapter illuminates the various processes and the associated equipment that might be included in a well-designed fluid conditioning and delivery system. Specifically, a layout of an ideal delivery system is proposed. Also, properties and requirements of the following processes and equipment are discussed: degassing, pumping, measuring flow rate, filtering, controlling temperature, mixing, changing coating fluids, damping pulsations, recycling fluids, cleaning-in-place, designing vessels and pipe-line systems.

8.1 Introduction

Systems for preparing, conditioning, and delivering coating fluids are paramount for any coating method, and they are particularly important for premetered coating processes because the best coating method cannot produce a uniform film if the fluid to be coated does not meet homogeneity and uniformity requirements concerning temperature and concentration of solid and liquid components.

Fluid delivery systems for premetered coating methods have the following purposes:

- **Fluid preparation**, encompasses dissolving chemical components and mixing different liquid components. Some coating companies prepare their fluids entirely from scratch, others buy major chemical components from suitable suppliers and add special chemicals to generate proprietary formulations, and still others buy the entire coating liquid from a supplier.
- **Fluid conditioning**, includes filtering and degassing the liquid as well as setting and controlling the desired temperature to obtain the desired homogeneity with regard to solids concentration and temperature of the fluid. Filtering and degassing may be carried out by the external supplier of the coating fluid. Also, degassing may not be necessary in the coating plant, if the time between liquid preparation and coating is long enough for spontaneous degassing to take place, and if no air

P. M. Schweizer, *Premetered Coating Methods*, Engineering Materials,
https://doi.org/10.1007/978-3-031-04180-8_8

is introduced during the transport of the liquid from the supplier to the coating plant. An example of such a situation is aqueous pressure sensitive adhesive (PAS), which sometimes is coated directly from the tank truck that delivers the adhesive. In-line filtering, however, is still recommended for such applications.

- **Fluid delivering and premetering**, occurs before coating (hence premetering) and comprises the processes of pumping and measuring the exact amount of fluid that is required for obtaining the desired wet thickness of the coated film for a given web speed and coating width. As the necessary fluid quantity depends on the web speed, the pump speed is often enslaved to the web speed according to a suitable control algorithm guaranteeing a constant and correct film thickness during acceleration and deceleration of the coating machine at the start and stop of the coating process.
- **Facilitating the transition from batch to continuous processes**, i.e., from fluid preparation as well as the filtering and degassing processes to coating, drying, and winding.

Schweizer (1992) provided an in-depth discussion of the various processes that make up fluid delivery systems. Following is a description of what we believe represents an ideal fluid conditioning and delivery system for premetered coating processes. The layout is based on the system that was implemented and successfully used by ILFORD Imaging AG in Switzerland. Whether or not all components of the ideal delivery system must be implemented for a particular situation depends on a variety of factors such as.

- Product performance requirements.
- Operational requirements.
- Degree of automation.
- Availability of funds for investments.
- Existing equipment and installations.
- Building layout.

A delivery system consists of several "elements" such as vessels, pipes, pumps, filters, flow meters, etc. Some of the elements are necessary, others are optional. Each element can be implemented in various ways, resulting in different levels of process performance and several levels of investment. In designing a delivery system, it is not only important to look at its elements, but also at how the variety of elements are positioned relative to each other (floor and elevation plan). The layout and the elements of an ideal fluid delivery system are schematically shown in Fig. 8.1. The pros and cons of each element as well as tips for filling and operating the delivery system are discussed in the next section.

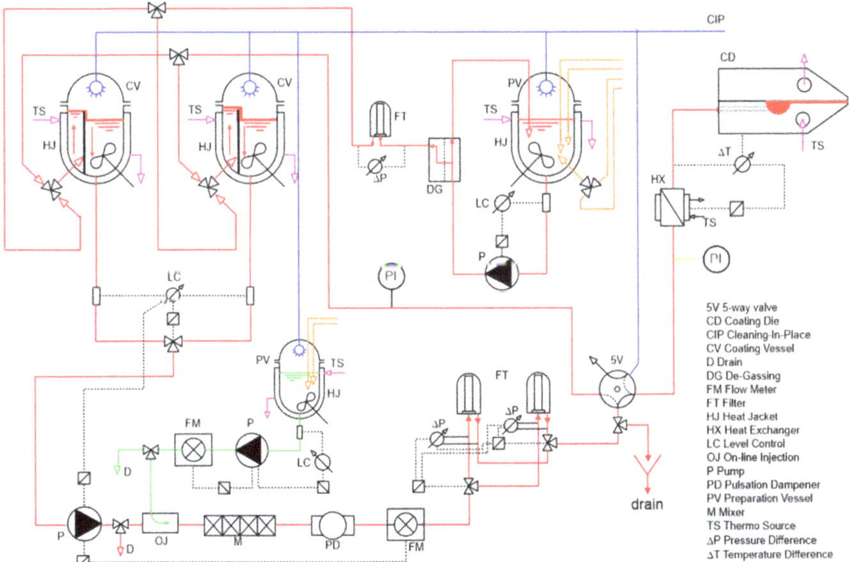

Fig. 8.1 Schematic diagram of an ideal fluid delivery system for premetered coating method

8.2 Elements of the Fluid Conditioning and Delivery System

8.2.1 Preparation Vessel

Preparation or mixing vessels are containers in which all the chemicals (solids, solvents from storage tanks, etc.) are added and mixed according to the product formulation specifications (recipe). Sometimes, ready-made coating solutions are brought in form an external manufacturer and delivered in some sort of container (vat, barrel, supply vessel, tank truck, etc.).

Adding and mixing chemical components should be accomplished without entraining air. If air is entrained during these process steps, then it must be removed later on with great effort. Air entrainment can be prevented in several ways, for example by.

- Using tall and slender vessels instead of short and wide ones.
- Using closed vessels that are operated under vacuum.
- Using external high-shear mixing units.
- Mounting stirrers off-center and possibly from below through the bottom of the vessel.
- Adjusting the stirrer speed to the liquid level in the vessel and the fluid viscosity.
- Placing the exit of liquid feed tubes and funnels for solids near the bottom of the vessel and/or near the vessel wall (fanned pipe endings).

- Mounting feed pipes for liquid additives into the bottom of the vessel (preferred option).

Each vessel must have a bottom outlet valve with a liquid level control positioned just downstream of the outlet valve to detect when the vessel is empty.

For coating applications at temperatures above ambient, the mixing vessel must be equipped with a double jacket for heating purposes.

In most cases, fluid preparation is a discontinuous batch operation. To generate a continuous coating process, the finished coating fluid must be transferred from the preparation vessel into one of two coating vessels, or it must be temporarily stored in a large-volume tank.

8.2.2 *Coating Vessel*

If no intermediate storage tanks are used, then two coating vessels are needed in parallel, one from which fluid is supplied to the coating applicator, and another one, which is cleaned and refilled while the first vessel is being emptied.

Stirrers should be mounted from below through the bottom of the vessel. The stirrer speed must be adjusted to the liquid level in the vessel and to the fluid viscosity to prevent air entrainment. At low liquid levels, the stirrer must be turned off to prevent air entrainment because it will no longer be fully immersed in the fluid.

Tall and slender vessels are better than short and wide ones. The liquids to be transferred should enter the vessel through bottom inlet ports to prevent splashing and air entrainment if the liquids are fed from the top of the vessel and dropped onto the free surface of the liquid already in the vessel. Closed vessels prevent contamination of the liquid by dirt from the environment. Closed vessels also allow the use of spray nozzles as used in cleaning-in-place (CIP) systems.

For coating applications at temperatures above ambient, the coating vessel must be equipped with a double jacket for heating purposes.

Each vessel must have a bottom outlet valve with a liquid level control positioned just downstream of the outlet valve to detect when the vessel is empty.

8.2.3 *Degassing*

Practical experience with slide and curtain coating has shown that all gas bubbles present in the coating fluid will pass through the coating process and end up destructively on the substrate. This is in contrast with roll and blade coating methods where it was observed that bubbles in the coating fluid may disappear during coating. The explanation is based on the difference between the pressure in the nip of a roll coater or under the blade tip of a blade coater when compared to the impingement pressure of slide and curtain coating. Theoretical estimations for typical operating conditions of the nip and blade pressure reveal that they are 100–10,000 times higher than the impingement pressure of slide and curtain coating. These higher pressures can force air bubbles to dissolve in the coating fluid as they pass through the nip or under the blade. The behavior of slot coating is comparable to blade coating owing to the relatively high pressure in the coating bead between the die lip and the substrate. In summary, therefore, degassing requirements for premetered coating processes are higher than for other coating methods.

Degassing is a residence time problem. In general, the time required for proper degassing is different from the time required to prepare a new batch of coating fluid in the preparation vessel. Moreover, the flow rate required for operating degassing equipment near optimum conditions is typically different from the flow rate required by the premetered coating process (which is determined by the wet film thickness, the web speed, and the coating width). For these reasons, degassing is best accomplished offline and de-coupled from the coating process. A convenient location for the degassing process is the transfer line between the preparation and coating vessel because degassed fluid can be fed into the coating vessel, and fluid rich in gas can be recycled into the preparation vessel, thereby preventing any loss of coating fluid, see Fig. 8.1.

As explained by Schweizer (1992), the term "degassing" refers to two subsequent processes, namely *fluid degassing* and *bubble removal*. Degassing means lowering the concentration of the gas that is dissolved in the liquid phase. Therefore, degassing requires the liquid to be exposed to conditions that promote the dissolution of air. As a result, degassing may temporarily increase the number of bubbles in the coating fluid. Omitting the degassing process increases the chances that dissolved air comes out of the solution while the fluid is pumped from the supply vessel into the coater, for example as a result of an excessive pressure drop along the delivery line.

After degassing, bubbles purposely generated this way and bubbles already present in the liquid, for example as generated by an improper mixing process, must all be removed in a second process step. Additional sources of air in the coating fluid include bubbles generated by leaky pipe fittings in the delivery system (if the system is operated at sub-ambient pressure) and air that gets trapped during the filling process of the delivery system. How the delivery system can be operated at above-ambient pressure, and how it can be filled without trapping air is discussed later in this chapter.

Based on Raoult's law for ideal or diluted multicomponent systems the solubility of a gas in a liquid can be expressed as

$$x_i = \frac{y_i P}{P_i} \tag{8.1}$$

x_i molar concentration of component i in liquid phase [kmol/kmol],
y_i molar concentration of component i in gas phase [kmol/kmol],
P total pressure [Pa],
P_i (equilibrium) vapor pressure [Pa].

The vapor pressure is a material property, which increases with increasing temperature. Therefore, it is evident from Eq. 8.1 that a particular gas component can be removed from a liquid by reducing its molar concentration in the liquid phase, which in turn is facilitated by.

- Lowering the total pressure, i.e., applying vacuum.
- Increasing the vapor pressure by increasing the liquid temperature.

For water, for example, the vapor pressure increases by a factor of 5.35 when the temperature is increased from an ambient 20–50 °C.

Bubble removal is controlled by the bubble rise velocity. For spherical bubbles, and as long as the bubble Reynolds number is <2, the rise velocity is described by Stoke's law according to (Bird et al., 1960):

$$v_b = \frac{d_b^2 g(\rho_l - \rho_g)}{18\mu} \tag{8.2}$$

v_b bubble rise velocity [m/s],
d_b bubble diameter [m],
g acceleration of gravity [m/s^2],
ρ_g gas density [kg/m^3],
ρ_l liquid density [kg/m^3],
μ liquid viscosity [Pas].

The bubble rise time t_b, which is the decisive parameter, is obtained by dividing the thickness of the liquid layer H, through which the bubble must rise, by the bubble rise velocity. Neglecting the gas density owing to $\rho_l >> \rho_g$ yields

$$t_b \approx \frac{18H\mu}{\rho_l g d_b^2} \tag{8.3}$$

Moreover, the isentropic ideal gas law states that the diameter of a gas bubble increases with decreasing pressure according to

$$d_b = d_{b,0}\left(\frac{P_0}{P}\right)^{1/3\beta} \tag{8.4}$$

P_0 ambient pressure [Pa],
P absolute vacuum pressure [Pa],
d_b bubble diameter at vacuum pressure P [m],
$d_{b,0}$ bubble diameter at ambient pressure P_0 [m],
β isentropic exponent, $\beta = 1.4$ for air [−].

A dimensionless form of Eq. 8.4 is visualized in Fig. 8.2. It becomes obvious that significant bubble enlargement factors can only be obtained for absolute pressures of <0.2 bar.

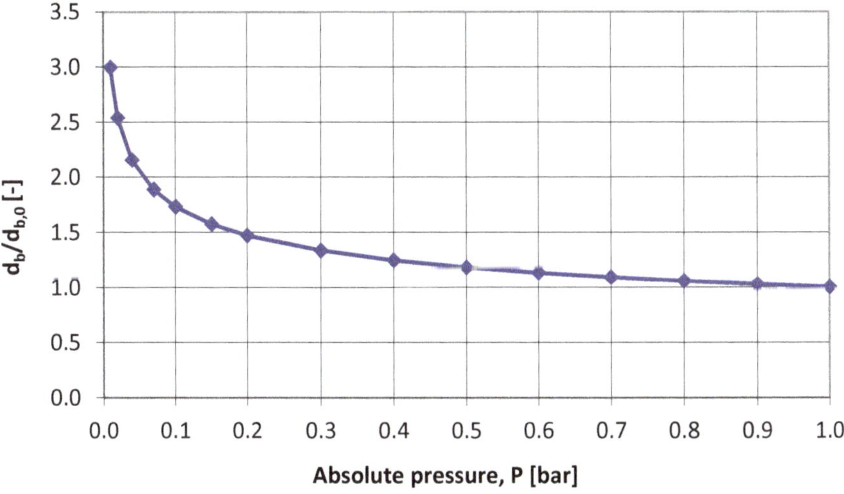

Fig. 8.2 Enlargement of an air bubble as a function of the absolute pressure P

Furthermore, Eq. 8.3 visualizes that it is ineffective to degas liquids in large vessels owing to the large height of the liquid level. Hence degassing times become excessively long, particularly if the viscosity is high and the required bubble diameter is small. For $\mu = 75$ mPas, H = 1 m and $d_b = 100\ \mu$m, for example, the degassing time amounts to about 230 min, i.e., almost 4 h. If the admissible maximum bubble diameter were reduced to 75 μm the degassing time would increase to 6.8 h.

Therefore, spreading the bubble-containing liquid into a thin layer is an obvious means for improving the efficiency of the degassing process. The best orientation of the liquid film is perpendicular to the orientation of the field force. The field force drives the bubbles to move across the liquid layer to its free surface where they collapse. If the field force is gravity, the film should be horizontal, which is unacceptable because such a film cannot flow. Using film flow on an inclined plane is a better de-bubbling solution. The time it takes a fluid element to flow along an incline of length L can be estimated for Newtonian fluids by dividing the slide length by the average film velocity according to Eq. 5.7.4. Then, the minimum diameter of the bubble that can be removed this way is obtained by equating the time of film flow on the slide to the bubble rise time through the film thickness according to Eq. 8.3 and by solving for the bubble diameter as follows:

$$d_b \geq \sqrt{\frac{18\mu Q}{\rho g L\cos\beta}} \tag{8.5}$$

As depicted in Fig. 8.3 the layer thickness H_{bubble} for the bubbles to rise equals $H_{film}/\cos\beta$.

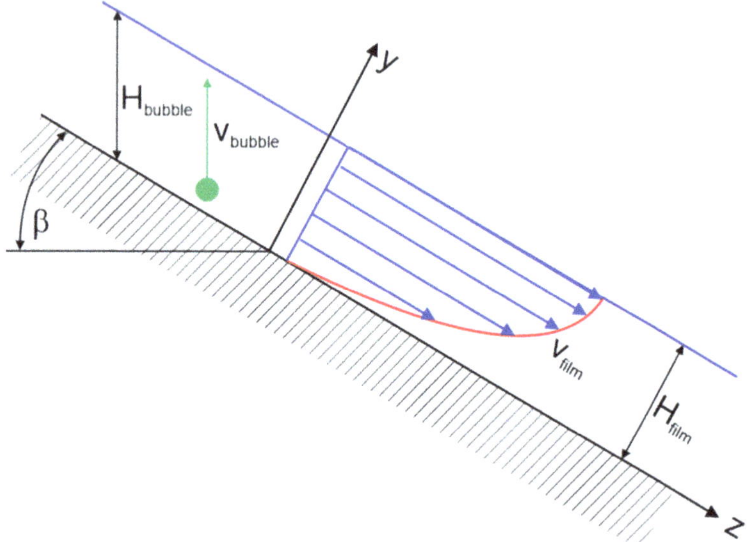

Fig. 8.3 Sketch of bubble removal in film flow

For the example of $\rho = 1000$ kg/m^3, $\mu = 75$ mPas and Q = 1.5 cm^2/s Fig. 8.4 shows that bubbles of <100 μm diameter cannot be removed with this process if the slide length should not be excessively long, i.e., >2 m. Also, a slide angle of $\beta = 30°$ seems to be an optimum value because smaller angles cannot reduce the

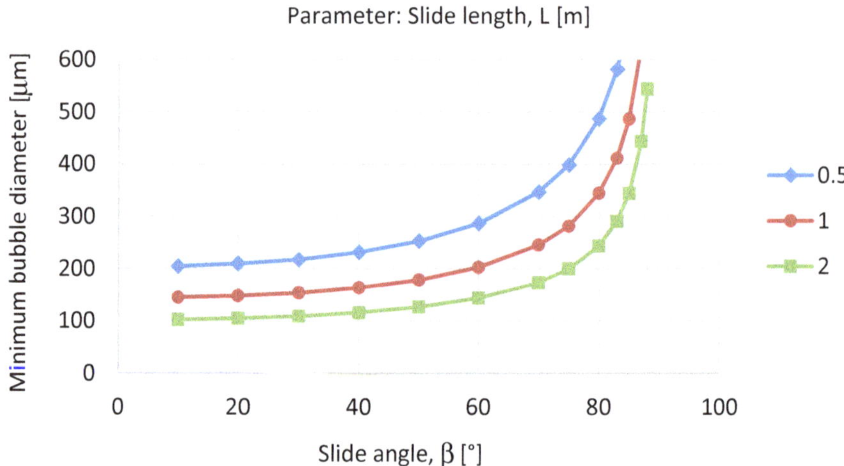

Fig. 8.4 Degassing with film flow: minimum bubble diameter as a function of slide angle and slide length

bubble diameter by much, and larger angles increase the minimum bubble diameter regardless of the length of the inclined plane.

A big performance increase regarding bubble removal is obtained by replacing the gravitational force with a centrifugal force. Specifically, the gravitational acceleration should be replaced by the centrifugal acceleration as follows:

$$g \to r\omega^2 \tag{8.6}$$

Here, r is the radius of rotation and ω is the angular velocity measured in units of [rad/s]. ω depends on the number of revolutions n expressed in units of [rpm], i.e., revolutions per minute, according to

$$\omega = \frac{2\pi n}{60} \tag{8.7}$$

The ratio of centrifugal to gravitational acceleration is plotted in Fig. 8.5 as a function of the rotational speed and the radius of rotation. As is demonstrated, the field force can easily be increased by one to two orders of magnitude with reasonable values of n and r. Therefore, degassing processes and equipment that are based on centrifugal rather than gravitational acceleration are much more effective and should be preferred.

Degassing equipment with a rotating film flow on an inverted cone (degassing centrifuge) was first described by Schubert et al. (1972). Its design features are illustrated in Fig. 8.6, and its operating principle is explained in Fig. 8.7. The liquid is pumped to the bottom of an inverted cone from where the fluid flows up along the inside wall of the cone. The driving force for this flow is the component of

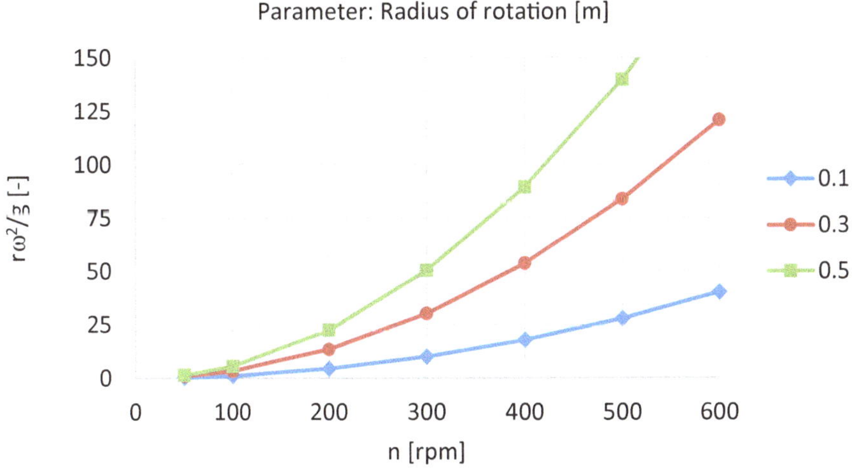

Fig. 8.5 Comparison of centrifugal and gravitational acceleration as a function of rotational speed and radius of rotation

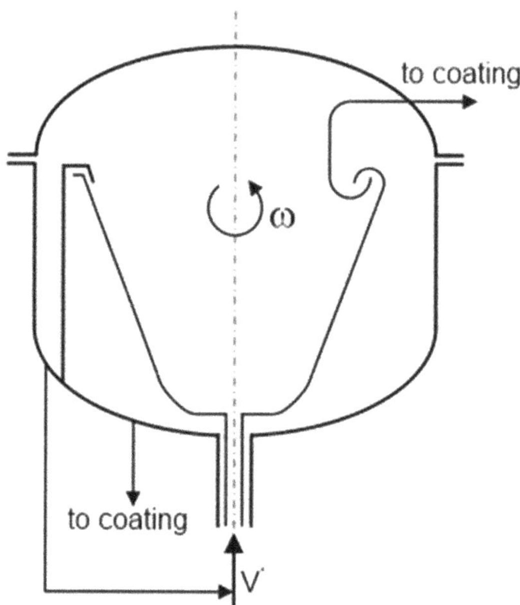

Fig. 8.6 Sketch of design features of degassing centrifuge

the centrifugal acceleration that is parallel to the cone wall, i.e., r(z) $\omega^2 \sin\alpha$ with α being the cone angle. As r(z) increases with increasing z, so does the centrifugal acceleration, but the volumetric flow rate/width of the film flow and the resulting film thickness as well as the average velocity decrease. In reality, gravity opposes the upward flow a bit, but as shown in Fig. 8.5 this effect can be neglected if the centrifugal acceleration is much larger than the gravitational acceleration. For Newtonian fluids the average film velocity and the film thickness can be estimated with the following equations:

$$\overline{v}_{film} = \left(\frac{\rho \omega^2 V^{*2} \cos\alpha}{12 \pi^2 \mu z} \right)^{1/3} \tag{8.8}$$

$$H_{film} = \left(\frac{3 \mu V^*}{2 \pi \rho \omega^2 z^2 (\tan\alpha)^2 \sin a} \right)^{1/3} \tag{8.9}$$

Thc time it takes for a fluid element to flow from the bottom to the top of the inclined cone surface can be calculated with Eq. 6.6.3, but the integration over z is best carried out numerically. While the liquid is flowing upwards, bubbles contained in the liquid are exposed to the centrifugal acceleration, which acts in a direction perpendicular to the cone axis. Consequently, the bubbles move horizontally through the liquid film from the cone surface to the free film surface where they collapse or accumulate in form of a foam layer. The maximum distance of the bubble path

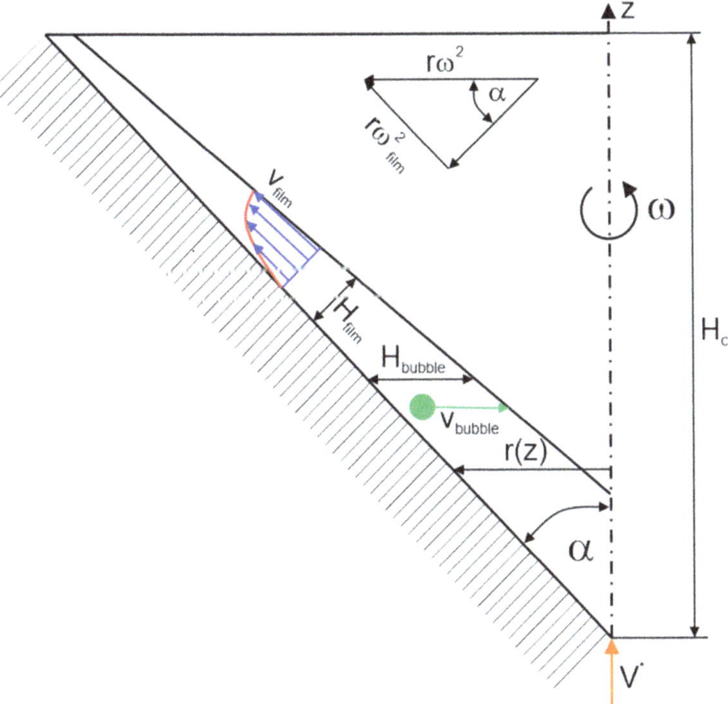

Fig. 8.7 Sketch of the operating principle of degassing centrifuge

changes with the coordinate z and can be calculated according to

$$H_{bubble} = \frac{H_{film}}{\cos\alpha} \tag{8.10}$$

At the upper end of the cone surface, the bubble-free liquid can be separated from the foam layer with a skimming blade as sketched on the left side in Fig. 8.6. Alternatively, if the degassing apparatus is properly designed and all of the liquid is free of bubbles and without a foam layer, the liquid can be collected in a pool and removed from it by inserting a pipe into the pool as shown on the right side in Fig. 8.6. The driving force for emptying the pool is the high static pressure generated by the centrifugal acceleration.

Neglecting the gas density, the bubble velocity through the liquid film can be estimated with Stoke's law (Eq. 8.2) by replacing the gravitational with the centrifugal acceleration as follows.

$$v_{bubble} = \frac{d_b^2 \rho z \omega^2 \tan\alpha}{18\mu} \tag{8.11}$$

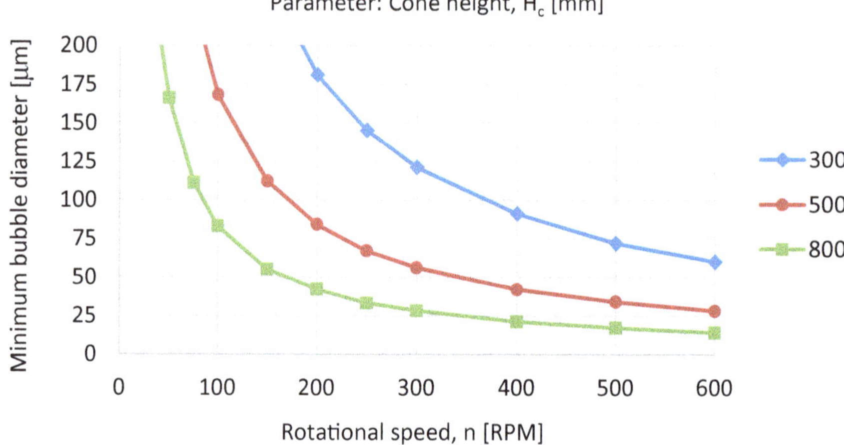

Fig. 8.8 Minimum bubble diameter that can be removed with a degassing centrifuge

The maximum time it takes a bubble to move through the liquid film can also be calculated with Eq. 6.6.2. The integration must be carried out along the horizontal bubble path from 0 to H_{bubble}, and it also is easily carried out numerically.

The performance of this degassing device in terms of the smallest bubble that can be removed at the top of the film flow along the cone can be determined by setting desired operating conditions, by assuming a bubble diameter to carry out all the necessary calculations, and by iterating on the bubble diameter until the time for the bubble to flow through the film on top of the cone equals the time it takes the film to flow from the bottom to the top of the cone. The result of this procedure is shown in Fig. 8.8, where the minimum bubble diameter is plotted as a function of the rotational speed n and the height H_c of the cone for the following operating conditions: $\alpha = 30°$, $V^* = 13.5$ L/min, $\rho = 1000$ kg/m^3, $\mu = 75$ mPas.

It is evident that film flow exposed to centrifugal acceleration can remove much smaller bubbles compared to gravity-driven film flow, see Fig. 8.4. In addition, heating the fluid to lower its viscosity and applying vacuum to the enclosing vessel to enlarge the bubble diameter further reduces the diameter of the smallest removable bubble, such that it should be possible to remove bubbles as small as 10 μm with reasonable equipment size and for a wide range of operating conditions. In our view, this process and its associated apparatus is the only one that satisfies all desirable requirements for degassing as follows:

- Distribution of the fluid to be degassed into a thin layer (film flow) to minimize the path of bubbles from the bulk of the liquid to its free surface, where bubbles either collapse or burst spontaneously.
- Application of vacuum to facility dissolution of dissolved air and to increase the bubble diameter, which increases the bubble rise velocity.

- Increase of the field force that provides buoyancy (centrifugal acceleration instead of gravity).
- Perpendicular orientation of the field force relative to the direction of the film flow.
- Increase of the fluid temperature, which facilitates the degassing process, thereby decreasing the viscosity, which in turn increases the bubble rise velocity.
- The residence time of the fluid inside the degassing equipment must be longer than the time it takes for the smallest bubble to be removed to rise to the free liquid surface.
- The dimensions of degassing equipment, as well as its operating parameters, must be optimized for any specific application.

Furthermore, as the operating principle of the degassing centrifuge is solely based on film flow, the modeling and optimizing capabilities are better than for alternative degassing processes. To our knowledge, unfortunately, a degassing centrifuge as depicted in Fig. 8.6 is not available commercially. However, a similar version of this degassing method was later patented by Metso Paper for applications in the paper industry. They also patented a multi-stage version with a higher degassing capacity (Kesti & Forsten, 2005).

Other degassing principles include the approach from FrymaKoruma, which now belongs to ProXES (www.proxes.com) as illustrated in Fig. 8.9. The liquid is pumped onto a spinning disc located in a vessel that is under a vacuum. On the disc, the liquid

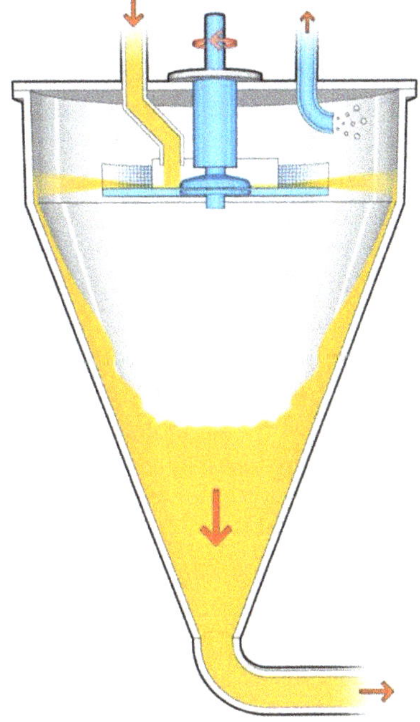

Fig. 8.9 Visualization of the operating principle of the FrymaKoruma degassing apparatus; Image courtesy of FrymaKoruma AG, Theodorshofweg 6, 4310 Rheinfelden, Switzerland; Copyright © 2022; All rights reserved

forms a thin film, which flows to the rim of the disc. There, with the help of a sieve, small droplets are formed that fly away from the disc and impinge with high speed onto the vessel wall. The so degassed fluid then collects at the bottom of the vessel and is pumped back to ambient pressure.

In this apparatus, the centrifugal acceleration does not act across the thickness of the film but along the plane of the film. Therefore, the major degassing step occurs when the droplets impinge on the vessel wall. At Polytype Converting we had successful experiences with this process for curtain coating with several different liquids, although only on a pilot scale. For some fluids, however, the maximum acceptable flow rate for achieving good degassing results was very low and much lower than the flow rate required by the curtain coating process. Using a degassing apparatus with a higher capacity would have overcome this problem.

A multi-stage version of this degassing process is offered by Voith Paper (www.Voith.com). The company Netzsch, on the other hand, developed a degassing process where the liquid is fed onto and distributed over a rotating disk, the plane of which is oriented vertically, i.e., the axis of rotation is oriented horizontally. The effectiveness of this approach must be lower compared to the methods described above because neither the gravitational nor the centrifugal accelerations are available for moving the air bubbles in a direction perpendicular to the direction of the film flow on the disk.

A different degassing principle, which is sometimes used for treating coating fluids, was developed and introduced some 20 years ago by Membrana, now part of the 3M Company Animated illustrations of the operating principle can be found on their web site (www.3M.com/Liqui-Cel). Accordingly, the basic unit cell is made of a small hollow fiber, the wall of which consists of a membrane that is permeable to gases but not to liquids. The liquid to be degassed flows outside of the small hollow fiber, and then a vacuum is applied at the inside of the tube. The resulting concentration gradient causes gases dissolved in the fluid to migrate through the membrane where they are transported away. This operating principle is implemented in an apparatus that consists of an outer housing with four connecting ports, two liquid and two gas ports. Inside, many of the small membrane hollow fibers are wrapped around a center tube. A baffle for the liquid phase is created in the middle of the unit. This maximizes the surface area and improves the gas transfer efficiency. The liquid enters the shell-side port and travels into the distribution tube. Liquid is forced radially over the fibers on each side of the baffle and exits through the collection tube and the second shell-side port. The hollow fibers are open from one end of the unit to the other. Vacuum is pulled from both gas side ports of the unit.

The degassing performance of this unit cannot be predicted for a given application. Consequently, trials must be performed for every fluid to be degassed to determine the appropriate size and number of the units. In addition, each fluid should have its own dedicated degassing unit consisting of a housing and a membrane cartridge. This approach prevents cross-contamination between subsequent fluids in a production environment. The degassing units of those fluids needed for a particular coating run (i.e., single-layer or simultaneous multilayer applications) are installed into their respective fluid delivery systems, see Fig. 8.1. The degassing units from the previous production run are placed onto an off-line rack for cleaning and storage purposes. Cleaning after each use is important to extend the lifetime of these units. Very long production runs with the same fluid may require intermediate cleaning of the membrane cartridge.

Degassing is a complicated process; its performance depends on many parameters such as flow rate, rheological fluid properties, surface tension, the gas content of the incoming fluid, etc. The degassing performance for a given degassing principle and a given application cannot be predicted accurately based on physical principles but must be determined experimentally. If an industrial coating company is interested in a degassing process, therefore, it is more meaningful to establish a good relationship with the supplier of this process, rather than buying the process from the supplier of the coating machine because the knowledge of and the experience with degassing of the coating machine supplier is often insufficient. To help in this interaction the operator of the coating process should know the largest bubble diameter that can be tolerated in the final coated product without causing any degradation of product performance.

If degassing of the coating fluid is necessary at all for a given application, then scientific studies (Mair & Sangl, 2004) and Polytype Converting's observations during pilot coating and degassing experiments showed that premetered coating methods require a vacuum degassing process, and that hydro-cyclone degassing units and other simpler devices are not sufficient. Moreover, chemical degassing by using de-foaming agents is not helpful and not needed for premetered coating methods because these processes, including their fluid delivery system, do not generate any foam owing to the absence of recirculation of excess fluid. De-foaming agents are particularly detrimental for curtain coating because a liquid curtain is nothing but a thin fluid lamella that will be destroyed by such chemicals if its thickness is sufficiently small.

8.2.4 Pump

Gear pumps are very well suited for premetered coating methods due to their low-amplitude flow rate fluctuations, which is related to the relatively high number of teeth per revolution, and hence the high number of small-volume delivery chambers per revolution. Gear pumps were used successfully for many years in the photographic industry, which had very high requirements on the uniformity of the delivered flow rate. However, due to the high shear forces between adjacent teeth during teeth engagement and between the gear wheels and the pump housing, simple gear pumps are not recommended for shear sensitive liquids such as latex-containing formulations because they tend to cease up after a short time of operation. Instead, gear pumps with two driven wheels or preferably progressing cavity pumps are more suitable. The latter pump type is very versatile for many coating fluids including water- or solvent-based and solventless fluids as well as solutions, emulsions, and suspensions. In the pilot facility of Polytype Converting, our experience with progressing cavity pumps

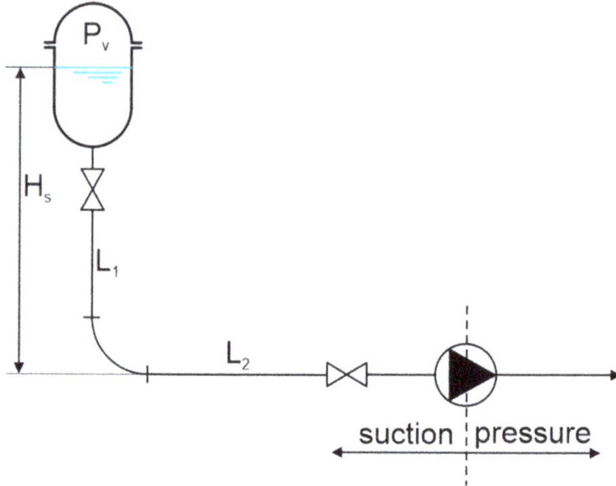

Fig. 8.10 Minimum static pressure as a function of the pressure drop on the suction side of the pipeline

was positive. Beinlich (www.beinlich-pumps.com) is a supplier of such pumps with a low flow rate range, which are suitable for slow and/or narrow coating machines. For highly loaded suspension such as slurries, a multi-diaphragm pulseless pump may be preferable over rotor—stator designs because they are less likely to suffer from abrasion problems caused by the solid particle in the fluid. Suppliers of such pumps include Hydra-Cell metering pumps from Wanner Engineering (www.hydra-cell.com) and smoothflow pumps from Tacmina Corporation (www.tacmina.com).

It is imperative that the fluid pressure on the suction side of the pump does not drop below the vapor pressure of the fluid to prevent gas generation due to flashing or cavitation. This can be achieved by respecting the NPSH value (**N**et **P**ositive **S**uction **H**ead). NPSH has the dimension of a length, which defines a static pressure head that must be larger than the difference between the system pressure on the suction side of the pump and the vapor pressure of the coating fluid at operating temperature. Better yet, the pressure on the suction side of the pump should not drop below atmospheric pressure to prevent air entrainment through leaky fittings and defective pump seals, etc. This can be ensured by respecting inequality 8.12. The formula requires the sum of the static pressure generated by the height difference H_s between the liquid in the supply vessel and the pump and the super-ambient pressure P_v imposed on the liquid in the supply vessel, if applicable, to be equal to or higher than the sum of the various pressure drop contributions ΔP_i generated by the flow in the pipeline system.

$$P_v + \rho g H_s \geq \sum_i \Delta P_i \tag{8.12}$$

As sketched in Fig. 8.10, the pressure drop along the suction side of the pipeline is composed of the pressure drop along straight pipes of length L_i according to Eq. 5.4.13, and of pressure drops across valves, elbows, and other components of the delivery system. Pressure drop values for such components can be obtained from equipment manufacturers or they can be found in engineering handbooks such as Dubbel (1994).

To ensure that the static height difference H_s between the liquid in the supply vessel and the pump is large enough and consistent with inequality 8.12, and to ensure that the pipeline on the pressure side of the pump does not contain horizontal sections, or worse, sections with a downward flow, the pump is best installed one floor level below the coating station. If this is not possible, then the pump must at least be placed on a short pedestal located on the floor of the coating room and not onto a cart or table for the convenience of the operator. In addition, if such undesirable pipe sections cannot be avoided, for example if an obstacle must be bypassed, then a bleeding valve must be installed at the highest local point of the pipe line just upstream of the downflow as indicated in Fig. 8.11.

According to our experience, imposing a super-ambient pressure P_v to the free surface of the liquid in the supply vessel should be avoided. Equation 8.1 teaches that the molar concentration of a component i in the liquid phase is proportional to the molar concentration of that component i in the gas phase and to the total pressure in the flow system, and it is inversely proportional to the vapor pressure. This explains why a degassing effect is obtained when the liquid in the vessel is exposed to a sub-ambient pressure, i.e., to vacuum, because air that is dissolved in the liquid is forced to come out of solution. The opposite effect is achieved if the liquid is exposed to a super-ambient pressure because now, more air is able to dissolve in the liquid. Unfortunately, this air is not visible. However, when the system

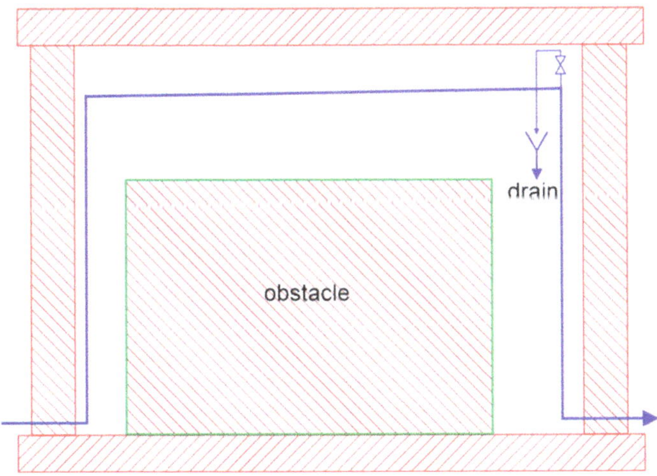

Fig. 8.11 Installation of a bleeding valve just upstream of a pipe section with downflow

pressure drops suddenly by a significant amount, then the dissolved air comes out of solution and forms bubbles. Locations of a sudden expansion may be found at the pump outlet or behind valves. This effect can also be observed when opening a beer or champaign bottle.

Once during a coating trial on a pilot machine, we applied vacuum to the supply vessel in an attempt to remove air bubbles that were entrained by inappropriate mixing during the preparation of the coating fluid. Then, we imposed a super-ambient pressure in the vessel to help pumping the fluid and to satisfy the NPSH requirement. The liquid was a clear polymer solution and the hoses of the delivery line between the vessel and the pump as well as between the pump and the slot die were also clear. So, we could clearly see that the liquid between the vessel and the pump inlet was free of any air bubbles. However, upon leaving the pump, the fluid was full of bubbles.

Concluding from this observation, the NPSH requirement must not be satisfied by applying a super-ambient pressure to the liquid in the supply vessel, but by providing enough static height between the vessel outlet and the pump inlet, and by minimizing the pressure drop in the feed line between the vessel and the pump inlet by using a sufficiently large diameter of the hose or pipe.

Ideally, the speed of the pump motor is controlled by a signal from the flow meter. If a flow meter is not available, then, in principle, fluid metering can be achieved by the metering pump alone, i.e., by calibrating the pump. Based on our experience, however, we do not recommend this approach because flow rate values obtained in this way are often associated with uncertainty, which negates one of the advantages of premetered coating methods, see Sect. 3.3.5. In particular, the pump calibration depends on the total pressure drop downstream of the pump. Therefore, the pump calibration must be repeated as soon as the fluid rheology changes, or any of the geometric parameters of the feed line such as the length or diameter of the pipe, or the thickness of the shim that determines the height of the die slot. In addition, the calibration must be carried out with the die in place. This may be acceptable if the width of the die is small, say 200 mm. However, if the die is 2 m wide, then this task would become cumbersome if not impossible to carry out. Finally, for slot coating, the total pressure drop also includes a contribution from the flow through the downstream coating gap. Calibrating the pump with a running slot coating process is not possible. In conclusion, therefore, using a flow meter for measuring and controlling the flow rate of premetered coating methods is the preferred approach.

If the flow rate range to be covered is large, then two or more pumps, possibly of different capacities, need to be installed. A pump that is operated near the lower limit of its operating range will turn too slowly and will generate flow rate fluctuations that translate into film thickness variations, which might be objectionable in the coated film.

Special care must be given to the selection of the pump seal, i.e., single- or double-acting mechanical seal, with or without barrier fluid. Using a barrier fluid is important when pumping suspensions. Mechanical seals can be avoided altogether by using a pump with a magnetic coupling. However, such couplings are characterized by a low maximum torque that can be transmitted. Hence, such pumps are limited to

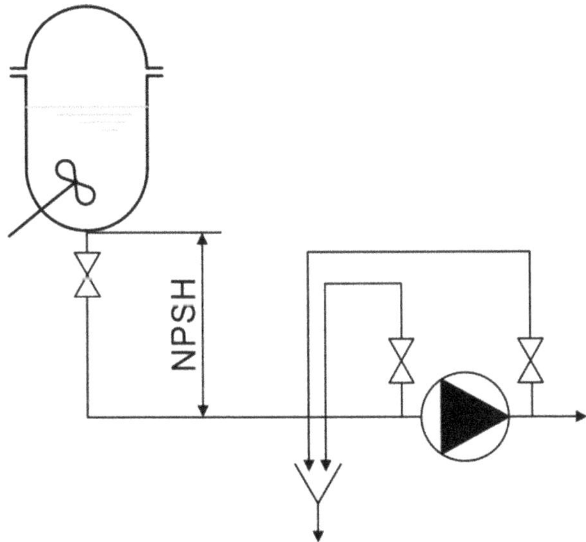

Fig. 8.12 Bleeding valves for properly filling the metering pump

applications with a relatively low pressure drop, i.e., applications with a low flow rate and/or a low viscosity.

Special care must also be given to the selection of the coupling between the pump and the motor/gear box assembly, i.e., magnetic coupling, mechanical coupling with a rigid or flexible shaft, or a universal joint. Shear-sensitive fluids have been observed to cease-up universal joints.

As sketched in Fig. 8.12, bleeding valves must be installed immediately upstream and downstream of the pump to facilitate proper filling and degassing of the pump and the pipe lines upstream and downstream of the pump, particularly on the suction side of the pump. Bleeding valves are typically small in size and have a small pipe diameter. Their purpose is to be temporarily opened to let air escape that has collected during filling of the suction side, as well as in domes and other areas of flow separation inside the delivery line during coating. The direction of the pipe flow through the bleeding valve must be upwards to facilitate the purging of the air, which is driven by gravity force.

It is recommended to connect a short piece of transparent hose to the downstream end of the bleeding valve, and to terminate this hose in a waste bucket or drain. This allows visual observation of the flow exiting from the bleeding valve. Typically, pure air flow comes first, followed by a mixture of air and coating fluid. The valve can be closed as soon as the pure liquid flow becomes visible.

Before starting the pump motor, the delivery system must be set to the recirculation mode, and the suction side of the pump, i.e., the pipe line between the supply vessel and the pump inlet, must be filled with coating fluid according to the following procedure:

1. Open the bleeding valve on the upstream side of the pump.
2. Open the outlet valve of the supply vessel.
3. Assure that the static pressure between the vessel outlet and the pump inlet is large enough to generate fluid flow. This will be the case as long as the NPSH requirement is satisfied.
4. Observe the flow in the transparent hose downstream of the bleeding valve and close the valve after pure liquid flow becomes visible.
5. Turn on the pump.
6. Temporarily open the bleeding valve on the downstream side of the pump to let air escape that might have been trapped in the piping hardware just downstream of the pump.

This procedure allows the pump to purge most components of the delivery system. Moreover, it allows the pressure to slowly build up and to increase the flow rate to the specified value. When all air is purged and the pump is running steadily at the correct flow rate, then the operating mode of the delivery system can be changed from recirculation to coating. Following this procedure results in clean starts for all premetered coating methods.

8.2.5 Flow Meter

Non-obstructive flow meters are the device of choice. Coriolis-type flow meters, which measure a mass flow rate, are most accurate (±0.1%) but are also the most expensive and need more space for installation than flow meters based on the principle of magnetic induction. In contrast, magnetic flow meters measure a volumetric flow rate. They are adequate for aqueous applications. Their accuracy is about ±0.5%; they are 2–3 times less expensive than Coriolis flow meters and they need less space for installation. However, fluids must be electrically conductive (>3 μS).

If the flow rate range to be covered is large, then two or more flow meters, possibly of different capacities, need to be installed to guarantee the required accuracy.

If the flow rate is very small, for example for applications on a slow and narrow machine such as a pilot machine, then it may be difficult to find an appropriate flow meter. Bronkhorst (www.bronkhorst.com) offers Coriolis-type devices for very low flow rates. However, in the Pilot Center of Polytype Converting, we experienced problems with high pressure drops across the flow meter, particularly when working with highly viscous fluids and when using a pressurized vessel for pumping the fluid. The high pressure drops were caused by the rather small diameter of the measuring tube inside the device.

8.2.6 Filter

The main filter should be placed off-line between the preparation and coating vessel and downstream of the degassing step. A second filter, sometimes called a police filter (one, or two in parallel), must be mounted downstream of the metering pump and of the online injection device, and upstream of the 5-way return valve. In this configuration, it is possible to properly replace a contaminated filter cartridge or bag of a parallel filter assembly during the coating operation without upsetting the flow rate. In-line filters are of particular importance if reactive chemicals such as hardeners are injected online (see Fig. 8.1) because unwanted precipitation may occur at the injection point.

Filters with disposable cartridges or bags are the filters of choice. The filter housing must be of polished stainless steel to reduce contamination of the inside of the housing and to facilitate cleaning.

The filter housing must be equipped with bleeding valves (at least on the inlet sides of the filter) to ensure proper evacuation of the air inside the housing, particularly during the initial filling of the filter housing. Just as with the metering pump, the exit pipe of the bleeding valve must be mounted in an upward direction, and the valve must be followed by a transparent hose that terminates in a drain or waste bucket, see Fig. 8.12. After turning on the metering pump, the bleeding valve of the filter must temporarily be opened until clean and air-free coating liquid becomes visible in the transparent hose.

The filter housing must be equipped with a drain valve mounted in the bottom of the housing.

The pressure drop across the filter must be monitored to prevent filter breakthrough. If a coating campaign takes longer than the time for maximum pressure build-up, then two filters have to be installed in parallel with automatic switching that is controlled by the maximum filter pressure.

New and unused filters are dry; hence their pores are filled with air. Experience has shown that it is often difficult to remove the air from new filters by simply pumping coating fluid through the filter. Therefore, it is recommended to first soak the new filters in the main solvent of the coating fluid (i.e., water or organic solvent) before mounting the filter into its housing. Soaking a new filter in a low-viscosity liquid will help to purge air bubbles from the inside of the filter medium.

If parallel filters are installed then they both must be filled during the filling of the delivery system according to the steps outlined above. This way the second filter is ready for use during a production run. If filters must be replaced more than once during long runs, then the valve feeding liquid to the fresh cartridge must be opened slowly, such that the feedback control loop between the flow meter and the pump can properly increase the pump speed. This way, the housing of the fresh filter can be filled slowly (the bleeding valve must be open) while the old filter is still active and is feeding liquid to the coating station.

8.2.7 Temperature Control

Control of the fluid temperature is necessary to guarantee isothermal operating conditions of the die. Isothermal operating conditions require the temperature difference between the die body and the coating liquid to be zero, or at least smaller than a maximum admissible value. Isothermal operating conditions prevent cross profiles generated by a viscosity change in the cross-web direction as a result of a temperature change of the coating liquid as it flows through the distribution chamber of the die.

If the die is operated at room temperature, then fluid temperature control may be omitted, if all fluids are also kept at room temperature at all times. Slow changes of the room temperature, e.g., seasonal changes between summer and winter, are acceptable as long as both the die body and the coating liquids follow these changes in parallel.

A simple way of controlling the fluid temperature is to use jacketed preparation and coating vessels in conjunction with insulated pipes of the delivery system. However, the jacket temperature must be slightly higher than the desired coating temperature at the die to account for heat losses during fluid flow through the delivery system. The required temperature compensation in the vessel depends on the flow rate, thermal conductivity, etc., and must be determined for each application by trial and error.

An alternative (preferred) way is to install a heat exchanger in the delivery system, see Fig. 8.1. A double-pipe heat exchanger augmented with a static mixer for enhanced heat transfer is a good option. Plate-type heat exchangers of the sort that are used in the food and dairy industry are also suitable because they are of sanitary design. A disadvantage of heat exchangers is the added fluid hold-up of the delivery system. In-line heat exchangers must be operated with a feedback control loop allowing for heating and cooling.

8.2.8 In-Line Mixing

Coating of chemically reacting fluids, for example, an aqueous polymer containing a hardener, is a principal problem, particularly if all reactive components are already present in the fluid supply vessel, and if the components are reacting fast. For such situations, the fluid viscosity changes (increases) over time, and a changing viscosity negatively affects the performance of the die and the coating process. The issue is particularly cumbersome, if the characteristic time, which leads to a significant viscosity change as a result of the reaction, is shorter than the time it takes for emptying the supply vessel.

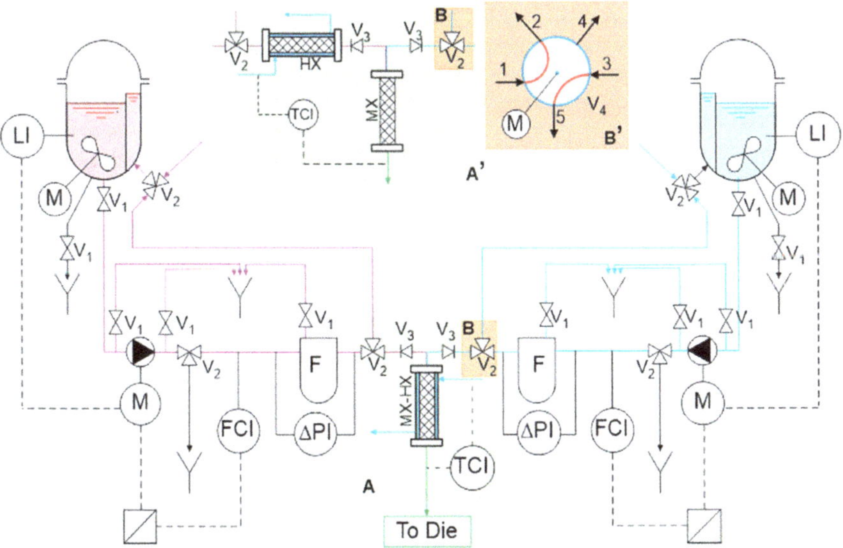

Fig. 8.13 P&ID of in-line fluid mixing system for premetered coating methods

Legend:

M motor;
F filter;
MX mixer;
HX heat exchanger;
V_1 ball valve, manually operated;
V_2 3-way valve, manually operated;
V_3 one-way valve;
V_4 5-way valve (motorized).

For premetered coating methods, problems related to chemically reacting fluids can be avoided, or at least greatly reduced because 100% of the fluid, which is pumped from the supply vessel to the coating station, is immediately coated onto the substrate. The key to avoiding such problems is to not mix the reactive component(s) in the supply vessel, but to inject them into the supply line of the main component, and to thoroughly in-line mix them just downstream of the injection point. This way, the time during which the chemical reaction can affect the viscosity, can be reduced to a few seconds or a few minutes at most. Moreover, if a viscosity change occurs between the injection and coating points, it will be constant over time and so more manageable because fluid supply systems are operated under steady-state conditions.

Following is a more detailed description than sketched in Fig. 8.1 of a fluid supply system, which shows suitable hardware components necessary for implementing an in-line mixing system for chemically reactive components.

Figure 8.13 shows a P&ID of two identical fluid supply systems suitable for premetered coating methods. Let us assume that the red fluid depicted on the left side in the diagram is an aqueous polymer such as gelatin and that the blue fluid on the right is a fast-acting hardener. Each branch consists of.

- a metering pump and a flow meter with a feedback loop for controlling the flow rate;
- a filter, possibly in parallel execution, with a differential pressure gauge for monitoring the filtration performance;
- several bleeding valves for purging air from the system;
- a drain valve at the lowest point of the pipe line;
- a fluid recirculation loop.

The two branches are brought together in the center of the diagram, and the relevant hardware components are located in an area shaded in green and marked with A.

Combining the two fluids takes place in a T-junction. Both pipe lines leading up to the T-junction are equipped with a one-way valve V_3, thus preventing the opposing fluid from entering upstream into the other pipe line. Immediately downstream from the exit of the T-junction follows a static mixer, which is embedded in a double pipe heat exchanger. This device serves two purposes, namely physically mixing the two fluid components, and removing heat generated by the chemical reaction.

The physical mixer length depends on the mixing ratio, i.e., on the flow rates of the two fluid streams, and the required mixing homogeneity. The thermal mixer length depends on the heat, which must be removed from the reacting fluid mixture. The longer length determines the size of the mixer. Suppliers of static mixers, such as Sulzer (www.sulzer.com), can help with the correct sizing of the equipment.

While the system presented above is fairly simple but workable, it may suffer from two drawbacks described next. These drawbacks may be overcome by installing the following optional equipment.

The first option involves replacing the 3-way valve with a 5-way valve. The 3-way valve design comprises a short piece of pipe between the T-junction and the one-way valve of the reactive component, which cannot be purged when the coating campaign is stopped. This may allow the reactive components to react for a long time, which, in the worst case, may result in the solidification of the coating fluids inside the pipeline system. Replacing the 3-way valve V_2 (see orange area marked with B in the diagram) with a 5-way valve V_4 (see orange insert marked B') will avoid this problem. Now, the reactive fluid enters the valve at port #3 and it leaves at port #5. The return port #4 is idle at this time. In addition, a neutral cleaning fluid, e.g., water or a suitable solvent enters the valve at port #1 and leaves at port #2. For this valve position, the cleaning agent is not involved in the process. It is either being recycled or waiting to be used in case port #2 is dead-ended. Upon turning the rotating body of the valve counter-clockwise by one position, the reactive fluid is now recycled through port #4, while the cleaning agent leaves the valve at port

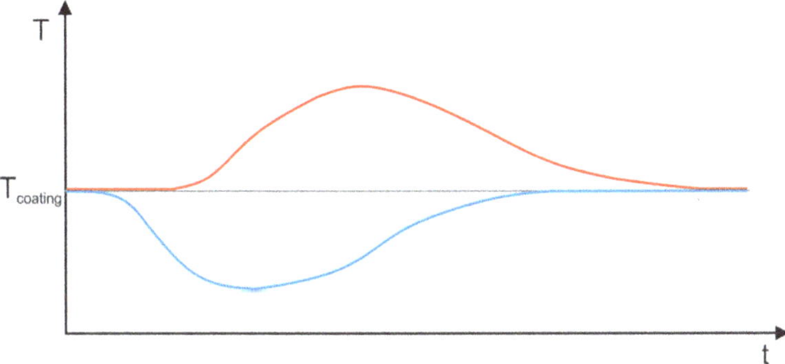

Fig. 8.14 Temperature profiles depicting pre-cooling or post-cooling of reacting fluids

#5, thereby cleaning the pipeline all the way to the T-junction, which then stops the chemical reaction. The 5-way valve may be motorized or operated manually.

The second option involves pre-cooling instead of post-cooling. Using a combined mixer-heat exchanger as shown in the main diagram in Fig. 8.13 (see area shaded in green and marked with A) may result in a temporary temperature increase of the reacting fluids, see red curve in Fig. 8.14. This temperature increase will be limited and eventually eliminated by the cold energy provided by the heat exchanger. The necessary cooling energy is related to the temperature of the reacted fluid via a temperature control feedback loop. However, sensitive fluids may not be able to support any kind of temperature increase. For such situations, the heater-mixer equipment can be replaced by individual heaters and mixers (see green insert marked with A'). Now, the heat exchanger is mounted upstream of the T-junction, such that it will cool down only the main flow prior to the chemical reaction, see blue curve in Fig. 8.14. The heat exchanger may be of sanitary plate design, or it may be of double-pipe design involving a static mixer for increased heat transfer. Downstream of the T-junction we still recommend using a static mixer for the physical mixing of the two fluids. However, this device can now be of simple design, even involving a plastic mixer, which can be disposed of after use. The supply of cooling energy is still controlled by measuring the temperature of the reacting fluids.

8.2.9 In-Line Fluid Change

When operating a premetered coating method, the question arises whether or not it is possible to change the coating fluid without stopping the coating machine. If possible, this would have a positive effect on the process productivity because unproductive downtime could be greatly reduced.

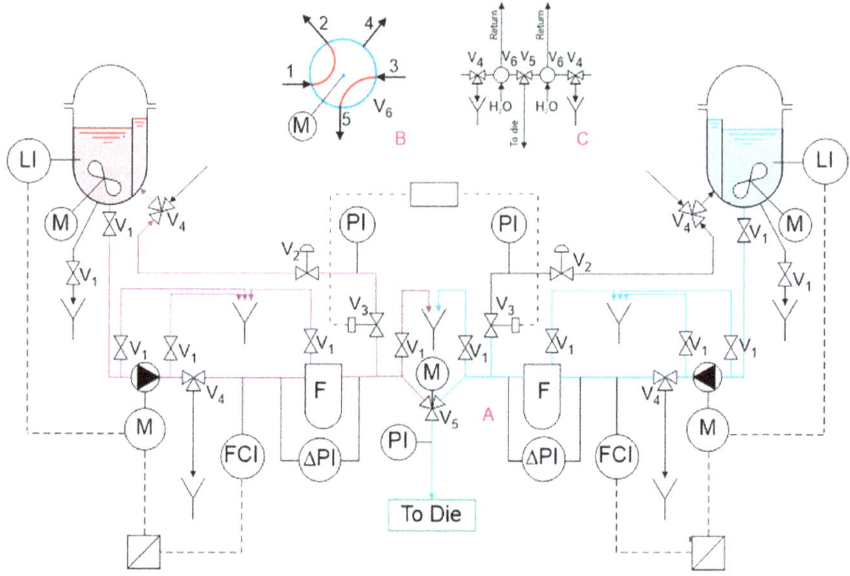

Fig. 8.15 P&ID of in-line fluid change-over system for premetered coating methods

Legend:

M motor;
F filter;
V_1 ball valve, manually operated;
V_2 needle valve, manually operated, for adjusting the pressure in the return line;
V_3 magnetic valve, for directing the fluid to the die or back to the vessel;
V_4 3-way valve, manually operated;
V_5 3-way valve, motorized;
V_6 5-way valve, motorized.

With in-line fluid change, the cleaning of the delivery system is avoided as the old fluid is directly pushed out by the new fluid. While neither the pumps nor the coating machine is stopped during this change process, thus insinuating no downtime, a short period still exists, in which both fluids are present in the delivery system. Consequently, a mixture of fluids is coated onto the substrate, and the resulting product cannot be sold. The ratio of old fluid to new fluid continually changes during this period. As discussed in Sect. 6.6.5 the change-over-time, i.e., the time it takes the new fluid to completely replace the old fluid in the delivery system, can be calculated for pure pipe flows.

The P&ID in Fig. 8.15 shows several variations of how an in-line fluid change system can be implemented. Moreover, experimental results have been reported, which verify the practical implementation of one of the design options. More specifically, it is shown that in-line fluid change systems can also be implemented successfully for the curtain coating process.

The valves V_1 upstream and downstream of the pumps, on top of the filter housing, and upstream of the motorized 3-way valve serve to purge the air from the pipes. It is particularly important that the pipes upstream of the junction point of the two fluids are free of air at all times. Therefore, different elevations for V_1 and V_5 at this location may be necessary (see green area labeled A in P&ID).

An in-line fluid change requires that both pumps run at the correct speed in keeping with the desired flow rate before carrying out the change-over. This is accomplished by directing one fluid from the supply vessel to the die (blue fluid on the right side in P&ID), and by circulating the other fluid with the return pipe starting just upstream of the 3-way valve at the junction point (purple fluid on the left side in P&ID).

The fluid flowing to the die generates a certain pressure in the pipe between the die and the junction point. This pressure must be measured and displayed. For the fluid change-over to work well, the pressure in the return pipe of the other fluid must be equal to the pressure upstream of the die. As the flow rate and the viscosity in the feed pipe going into the die are generally different from the corresponding values in the return pipe of the other fluid, the pressure in the return pipe can be adjusted and made equal to the pressure in the feed pipe by opening or closing the needle valve in the return loop.

The fluid change-over is accomplished by simultaneously turning the 3-way valve V_5 and the two magnetic valves V_3, just upstream of the junction point. Good results can only be achieved if all these valve motions are carried out very quickly. This explains why the 3-way valve V_5 must be driven by a high torque motor and the ball valves V_3 by electrical coils. Note in particular that pneumatic drives for these valves may not be suitable because they may not be rigid and hence fast enough.

As an alternative, all the equipment located inside the green area labeled A in Fig. 8.15, i.e., the two bleeding valves V_1, the two magnetic valves V_3 and the motorized 3-way valve V_5, could be replaced by one motorized 5-way valve V_6 as shown in the inserted green area labeled B in the top center-left of the P&ID. Here, the old fluid directed to the die enters the 5-way valve at position #3 and leaves it at position #5. In contrast, the new fluid waiting to be switched enters the valve at position #1 and leaves it at position #2, which is the beginning of the return pipe. Port #4 is the beginning of the return pipe of the old fluid; it is idle at this time. Now, upon turning the valve body counter-clock wise by one position, the old fluid is still entering at position #3 but leaving at position #4, then leading back to the supply vessel. On the other hand, the new fluid still enters at position #1 but now leaves at position #5, which goes to the die, and the return port #2 is idle. Again, the valve rotation has to be carried out quickly, which requires a torque motor. This configuration seems simpler than the one presented earlier because only one element has to be moved, and there is absolutely no issue related to degassing all parts of both pipe lines.

If it is considered risky to change the coating fluid without washing the die, then a configuration as shown in the inserted green area labeled C in the top center-right of the P&ID can be implemented. Here, the fluid return loop starts at a 5-way (or possibly a 4-way) valve V_6, which allows the coating fluid to be directed either to the die or back to the supply vessel, and which also allows cleaning fluid (e.g., water) to enter at this valve point to wash the die while the coating fluid is re-circulated. The old and new fluids are exchanged at the subsequent motorized 3-way valve V_5. While the new fluid is being coated, the entire re-circulation loop of the old fluid can be cleaned by entering cleaning fluid through the 3-way valve V_4 at the bottom of the supply vessel. The manually operated 3-way valve V_4 upstream of the 5-way valve V_6 in the green insert labeled C allows the recirculation loop to be drained after the cleaning cycle. Obviously, both supply systems are equipped identically.

The main configuration shown in Fig. 8.15 was tested experimentally in the Pilot Center of Polytype Converting. In particular, we were interested to see if an in-line fluid change-over could be implemented in conjunction with the curtain coating process. The main question was whether or not the curtain would break up during the switching of the valves. Note that we did not have suitable magnetic valves V_3 or a suitable motorized 3-way valve V_5. Instead, we used manually operated 2-way and 3-way ball valves, and we needed two people for turning the 3 valves at the same time. A third person was observing the behavior of the curtain during the valve motion.

If the liquid supply to the curtain is temporarily starved or even interrupted during the valve motion, then the curtain will thin out and becomes unstable. This is manifested by a break-away of the curtain from one or both edge guides. If adhesion from both edge guides is lost, then the curtain may completely disintegrate up to the die lip. However, as fluid starvation is only temporary, the unstable curtain will re-establish itself momentarily without the need for any interaction from the operator. Nevertheless, if adhesion between the curtain and its edge guides is lost, this will cause a big mess on the coated substrate. On the Polytype curtain coater such a mess was prevented from entering and contaminating the dryer by temporarily engaging a full-width scraper blade installed downstream of the curtain impingement point, see Fig. 19.2.

Starvation and detachment of the curtain may occur if the return valve of the new fluid is still open when the 3-way valve is switched because this prevents the new fluid from being entirely directed to the die. In contrast, if the return valve of the new fluid is closed before the 3-way valve is switched, then the pump of the new fluid is temporarily working against a closed valve, which causes the pressure to increase. Upon switching the 3-way valve this over-pressure will cause extra liquid to be delivered into the curtain, which in turn causes an over-thickness of the coated film on the substrate. However, the curtain does not become unstable or break up at all. The latter situation is definitely preferable over the former one. Even more preferable is optimizing the switching times of all 3 valves involved. We have seen that it is possible to prevent any kind of upset in the curtain, i.e., starvation or over-thickness, by properly fine-tuning the switching times. This was a bit difficult when carried out by hand, but it is easily accomplished by a suitable computer algorithm. No further action is required downstream of the curtain impingement point for such an optimized configuration. However, it may still be reassuring for the operator to temporarily engage the full-width scraper blade.

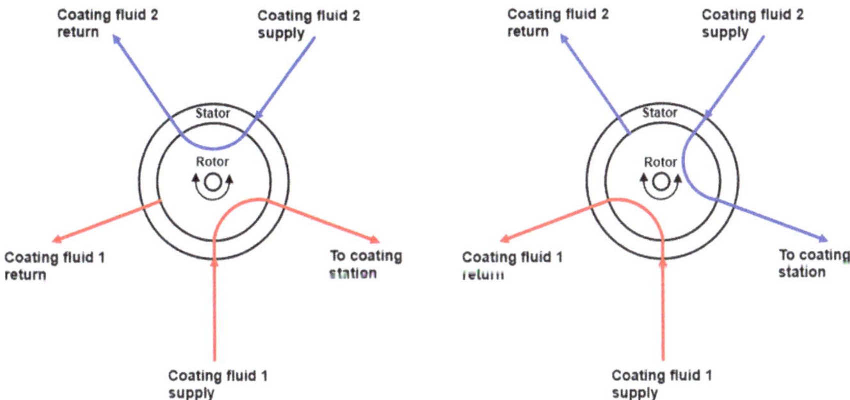

Fig. 8.16 Operating principle of a 5-way valve

Comparing the two hardware options presented in Fig. 8.15 (see green areas with 3-way valve and 5-way valve) we must emphasize that we did not experimentally test the version with the 5-way valve because such a device was not available at the time.

Working with a 3-way ball valve (see green areas with 3-way valve and 5-way valve in Fig. 8.15), we observed that, as soon as the ball, which contained an elbow-shaped flow duct, was rotated a bit to close the flow cross-section of the incoming old fluid, a little bit of flow cross-section of the new fluid on the other inlet port was opened. Consequently, fluid was always flowing through the valve during the turning of the valve ball, either 100% old fluid, or a changing mixture of old and new fluid, or 100% new fluid. This feature is very advantageous for curtain stability because it can prevent the starvation of liquid in the curtain if properly operated as explained above. This continuous liquid flow cannot be accomplished with 5-way valves, at least not with the version depicted in Figs. 8.15 and 8.16.

Here, the liquid flow through the valve is temporarily interrupted while the rotating body containing the elbow-shaped liquid ducts is moved from one position to the other. It is not clear whether or not this leads to an unstable curtain, even if the valve rotation is carried out at a high speed and so in a short time.

In summary, we demonstrated that in-line fluid change is possible for all premetered coating methods. However, geometrical details of selected 3- and 5-way valves and perhaps further trials must decide as to which hardware option is most suitable. One might think that the problem of temporary fluid starvation is less important for slide and slot coating because the issue of curtain stability does not exist. However, we do not believe so because fluid starvation may cause the coating bead between the die lip and the substrate surface to break up, and that in turn could cause serious quality problems in terms of lip contamination and subsequent formation of lines and streaks.

8.2.10 Pulsation Dampener

Turning pump rotors or switching valves at the outlet of supply vessels or filter housings, for example, generate pulsations in the flow rate. Proper pump design (number of teeth in gear wheels, rotational speed of pump) will reduce pulsations. Remaining pulsations, if objectionable, can be eliminated by pulse dampeners, designed either as vessels with an included air pocket or as a piece of hose with an elastic wall. Elastic hoses are preferred over vessels because of their smaller fluid hold-up (shorter residence time and narrower residence time spectrum) and better flow features (higher wall shear stress, easier to clean, etc.) If vessels are used as dampeners, their size and the size of the air pocket must be properly designed and compliant with the desired damping rate, see Schweizer (1992).

In our experience, PVC hoses have an adequate damping capacity as no additional pulse damping is required when used in combination with gear or progressing cavity pumps. This damping capacity is reduced if PVC hoses are fortified with a textile net to withstand higher fluid pressures. Hoses that are lined with a solvent-resistant material have practically no damping capacity, and require additional means for damping pressure pulses.

FMP Technology (www.fmp-technology.com) has developed and patented a pulsation dampener that consists of an inner hose made of an elastic material and an outer rigid pipe. The space between the two pipes can be pressurized such that the gas pressure is higher than the liquid pressure inside the elastic hose. Pulsation damping is obtained via a deformation of the elastic material by the pressure pulses in the liquid (FMP Technology, 2011).

8.2.11 Pipe Line

Fluid delivery lines can be constructed from pipes or hoses. To prevent contamination and to facilitate filling as well as cleaning in place, wall shear stress levels of higher than 1–10 Pa must be obtained for all operating conditions, see Chap. 6. Consequently, lines of different diameters should be installed, if the product of volumetric flow rate times effective viscosity varies greatly from one application to the next. Moreover, sharp corners in the pipeline must be avoided. To facilitate the spontaneous evacuation of air during the filling stage and coating operations, the pipeline downstream of the pump should always have an upward orientation in its flow direction. If downhill sections cannot be avoided, a bleeding valve should be installed on the highest point upstream of the downhill section to facilitate the purging of air, see Fig. 8.11. Drain valves must be installed at the lowest point of the downhill sections and also at the lowest points of siphons. Horizontal pipe sections must be avoided as they do not drain completely.

If hoses are used for delivering solvent-based coating fluids, then the inside of the hose must be covered with a solvent-resistant material such as Teflon. Consequently, such hoses are stiffer than PVC hoses that are suitable for aqueous fluids. They require larger bending radii, and they cannot dampen pulsations.

If the coating temperature is much above ambient, then all pipe sections should be insulated to prevent heat loss between the coating fluid and the environment. The inside of pipes can be polished or Teflon-coated to prevent contamination and to facilitate cleaning.

8.2.12 T-Junction

Most delivery systems contain several T-junctions. With regard to air pockets, T-junctions with one pipe section being dead-ended and oriented upward (no through-flow of coating fluid) are of particular concern, see Fig. 8.17. Such T-junctions are used, for example, for mounting sensors such as pressure gauges. Upon filling the delivery system, dead-ended pipe sections will always be filled with a big air pocket that is difficult to remove by the flowing coating fluid, see Fig. 8.17a. In contrast, dead-ended pipe sections oriented downward are filled with fluid during filling of the delivery system, but these sections cannot be drained during fluid change or cleaning, and so must also be avoided, see Fig. 8.17b. To avoid problems with air pockets during coating, T-junctions that house sensors must be oriented upward and must be equipped with a bleeding valve as shown in Fig. 8.17c.

If the T-junction consists of a 3-way valve, for example for taking fluid samples or for draining the delivery system, then a through-flow in all 3 pipe sections can be established. However, as described above for the bleeding valves for pumps and filters, the third section of the T-junction should point upward to facilitate proper and complete air purging. Should this not be possible, then the connecting hose should first be bent upwards to form a siphon before ending in a bucket or drain, see Fig. 8.17c.

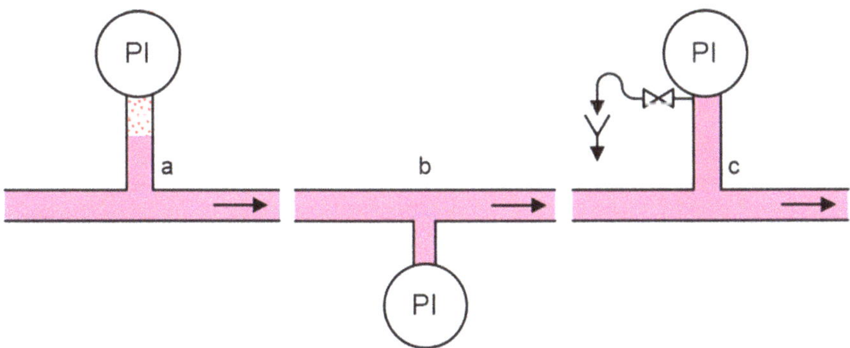

Fig. 8.17 T-junctions with dead-ended pipe sections; **a**, **b** incorrect mounting, **c** correct mounting

8.2.13 Valves and Fittings

All valves and fittings such as elbows, T-junctions, etc. must be of sanitary design, i.e., they must be free of domes, gaps, steps, sudden changes in flow cross-section, etc., where air and fluid can be trapped and where the wall shear stress is too low. Expansions in flow cross-section must be gradual to prevent flow separation. In particular, the admissible divergence angle must be related to the flow conditions, i.e., the Reynolds number, see Sect. 5.4.4.

The requirement of a sanitary design may be dropped for applications where cross-contamination between subsequent coating campaigns is not a problem, i.e., where all coating fluids are chemically compatible, such as different grades of aqueous pressure-sensitive adhesives.

8.2.14 5-Way Valve and Recirculation Loop

Excessive fluid waste during the filling stage of the delivery system can be prevented and the product change-over time can be shortened by installing a re-circulation loop. Such a re-circulation loop also allows one to save coating fluid after emergency stops as well as during start-up and stop procedures of a coating campaign.

The return valve of the recirculation loop is often best designed as a 5-way valve, see Fig. 8.16. Depending on the valve position, the coating fluid is either allowed to pass to the coating station while the cleaning/rinsing fluid is re-circulated, or the coating fluid is recirculated while the cleaning/rinsing fluid is passed through the downstream section of the delivery system and on to the coating station.

The return valve in the delivery system should be located downstream of the pumps (main stream and injected streams), the filters, and the pulsation dampener, but upstream of all other elements. This way, filter housings, static mixers, and other large containers such as pulsation dampeners can be pre-wetted with coating liquid and purged of air before starting the coating process.

The return section of the recirculation loop should enter the coating vessel through a valve at the bottom of the vessel. Inside the vessel, a small section around the inlet of the return fluid should be shielded off by sheet metal (over-flow weir), such that the returning fluid must first flow upward to the top of the weir. This configuration prevents the liquid from short-circuiting between the inlet and the outlet, and it allows for degassing and defoaming of the fluid during the filling stage of the delivery system. If the return line enters the vessel through a valve in the cover on top, then the fluid splashes down onto the free surface of the liquid in the vessel, which generates air entrainment. Moreover, the end of the return line has a downward orientation, which is difficult to purge of air.

The length of the delivery line downstream of the return valve should be minimized.

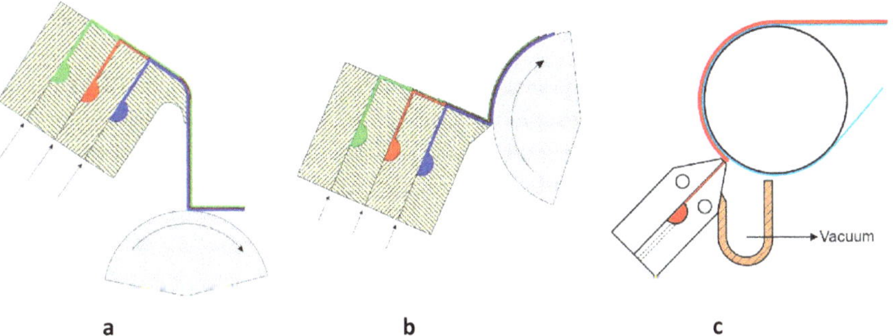

Fig. 8.18 **a** Slide die for curtain coating, **b** Slide die for slide coating, **c** Slot die for slot coating

8.2.15 Die

The die slot(s) must be oriented with the slot exit pointing upward to facilitate the purging of air bubbles that are transported into the die during the filling process or coating. This is automatically given with slide dies for slide and curtain coating, Fig. 9.18a, b. For slot coating, the die must be positioned between 7 and 8 o'clock (or 4 and 5 o'clock) on the backup roll, Fig. 8.18c. The often seen 3 or 9 o'clock positions are not very good and not recommended from an air bubble purging point of view.

Inverted 1-layer or 2-layer slot dies used for curtain coating are more difficult to purge of air bubbles because the opening of the die slot always points downward and therefore in the wrong direction, Fig. 2.1d, e. For such situations, it is recommended to mount bleeding valves into both side plates of the die, with the location of the valve coinciding with the highest point of the (inner) distribution cavity. These valves must temporarily be opened during the filling of the die. It may even be necessary to periodically open the valves for a short period during coating.

As a reminder, problems related to air purging during filling and operating the die can be avoided altogether by designing the cavities such that the wall shear stress remains above a minimum value of 1–10 Pa for all intended applications. See Sect. 6.3 for more details on this topic.

8.2.16 Cleaning-In-Place (CIP)

The inside of fluid delivery systems is generally not accessible for mechanical cleaning devices such as brushes, etc. (exception: cleaning pigs). To prevent replacing pipes and disassembling pumps and other elements for cleaning purposes to prevent cross-contamination between subsequent coating runs, cleaning-in-place (CIP) is recommended. CIP systems should include all vessels and the entire delivery system including the die. CIP systems may encompass elaborate computer-controlled cleaning cycles involving detergents, solvents, acids, bases, etc. that are optimized for a particular cleaning task.

In summary, selecting appropriate equipment components for building a fluid delivery system is best accomplished by cooperating with suitable equipment suppliers. Their experience is invaluable for optimizing each component concerning the range of operating conditions under consideration.

References

Bird, R. B., Stewart, W. E., & Lightfoot, E. N. (1960). *Transport phenomena*. John Wiley & Sons.

Dubbel. (1994). Handbook of mechanical engineering. In: W. Beitz & K.-H. Küttner (Eds.), Springer Nature.

FMP Technology GmbH. (2011). Vorrichtung zur Dämpfung von Pulsationen. German patent DE 202011002772 (U1).

Kesti, E., & Forsten, E. (2005). Method and apparatus for degassing coating material. WO 2007074211 A1.

Mair, B., & Sangl, R. (2004). Use of deaerator chemicals and mechanical deaerators for the degassing of coating colors. PTS Technical Report.

Schubert, H., Rumpf, H., Leschonski, K., & Hofmann, F. (1972). Entgasen hochviskoser Flüssigkeiten im Fliehkraftfeld. *Chemie-Ing. Tech., 44*(8), 497.

Schweizer, P. M. (1992). Fluid handling and preparation. In E. Cohen & E. Gutoff (Eds.), *Chapter 2 in Modern coating and drying technology*. New York: VCH Publishers Inc.

Chapter 9
Film Thickness and Film Thickness Uniformity

Abstract One of the attractive features of premetered coating methods is the fact that the nominal thickness of the coated film is pre-determined by the law of mass conservation. For a constant coating width, in particular, the thickness only depends on the flow rate/width and the coating speed. Interesting is the question about the uniformity of the nominal film thickness, both in the machine and cross-web direction. This chapter shows that the film thickness uniformity in the machine direction is controlled by the design of the pump as well as the uniformity of the motors that drive the pump and pull the substrate through the coating machine. Achieving a good film thickness uniformity in the cross-web direction is more complicated because it depends on the design and performance of the slot die. Therefore, a large portion of this chapter is devoted to die design. Specifically, theoretical models for designing the die-internal geometry are presented, the performance of such dies is analyzed, and ways for optimizing die design are proposed.

9.1 Introduction

This chapter analyses the film thickness and the film thickness uniformity of coated films generated by all of the three premetered coating methods. In particular, various parameters are identified, which influence the film thickness uniformity in both the machine (length profile) and cross-web (cross profile) direction. All premetered coating methods use dies for distributing the coating fluid in the cross-web direction. Therefore, identifying and examining those parameters, which affect the film thickness uniformity, will tell us how to best design, fabricate and operate dies such that the resulting film thickness uniformity meets a specified requirement.

9.2 Nominal Film Thickness

For all premetered coating processes, including single-layer applications and all individual layers of simultaneous multilayer applications, the nominal wet film thickness

P. M. Schweizer, *Premetered Coating Methods*, Engineering Materials,
https://doi.org/10.1007/978-3-031-04180-8_9

in the machine direction is determined by the physical law of mass conservation. As already discussed at the very beginning of this book, i.e., Eq. 2.1, which is repeated here, the wet film thickness is calculated according to:

$$H_{wet} = \frac{Q}{U} = \frac{V^*}{WU} \tag{9.1}$$

H_{wet} wet film thickness [m],
Q volumetric flow rate/width [m^2/s],
V* volumetric flow rate [m^3/s],
U web speed [m/s],
W coating width (die width) [m].

From a coating process point of view, the wet film thickness is the relevant thickness parameter. However, from a product performance point of view, it is the dry thickness or the dry coat weight, which is relevant. The dry coat weight also obeys the law of mass conservation. Since it is related to the wet film thickness through the solids concentration and the fluid density, it is calculated by Eq. 9.2:

$$A_{dry} = \frac{\rho_{liquid} C_{w,solid} V^*}{WU} \tag{9.2}$$

A_{dry} dry coat weight [kg/m^2],
ρ_{liquid} density of liquid [kg/m^3],
$C_{W,solid}$ solids concentration by weight [–].

9.3 Film Thickness Uniformity in Machine Direction

The term ***length profile*** is synonymous to film thickness nonuniformity in the web or machine direction. The absence of a length profile indicates a film with a perfectly uniform thickness.

Assuming an ideal situation, in which the density, the solids concentration, and the coating width are constant for a given application, then Eqs. 9.1 and 9.2 show that the wet film thickness and the dry coat weight only depend on the volumetric flow rate and the web speed. Taking the derivative of these two equations yields:

$$\frac{dH_{wet}}{H_{wet}} = \frac{dA_{dry}}{A_{dry}} = \frac{dV^*}{V^*} - \frac{dU}{U} \tag{9.3}$$

dH_{wet} variation of wet film thickness [m],
dA_{dry} variation of dry coat weight [kg/m^2],
dV* variation of volumetric flow rate [m^3/s],
dU variation of web speed [m/s].

dH_{wet}/H_{wet} is a measure for the film thickness nonuniformity. If dH_{wet} is zero, then the film thickness is perfectly uniform. Similarly, the other expressions in Eq. 9.3 are measures for the coat weight nonuniformity, the flow rate nonuniformity, and the web speed nonuniformity, respectively.

Note that the product of $\rho_{liquid}V^*$ in Eq. 9.2 equals the mass flow rate M^*. Therefore, Eq. 9.3 remains valid, if the term dV^*/V^* is replaced by the term dM^*/M^*.

Equation 9.3 shows a deterministic relationship between the film thickness or the coat weight and the volumetric flow rate and the web speed. In practice, however, variations of the flow rate and the web speed do not occur in a deterministic (smooth, predictable) way, but rather in a stochastic (random, unpredictable) way. Therefore, the dependence of the film thickness nonuniformity upon the flow rate and web speed nonuniformities is more accurately expressed by Eq. 9.4:

$$\frac{dH_{wet}}{H_{wet}} = \frac{dA_{dry}}{A_{dry}} = \pm\sqrt{\left(\frac{dV^*}{V^*}\right)^2 + \left(\frac{dU}{U}\right)^2} \tag{9.4}$$

For all premetered coating methods, the uniformity of the film thickness or the coat weight only depends on the uniformity of the flow rate (= pump rate) and on the uniformity of the web speed. In particular, the film thickness does not depend on the physical fluid properties such as rheological properties, nor on the geometry of the coating process, such as the gap between adjacent rollers. This is a difference and a major advantage when compared to other coating processes such as roll and blade coating. Moreover, since variations of flow rate and web speed can easily be kept very small by applying standard process control algorithms (web speed variations on Polytype Converting coating machines, for example, are typically $<\pm 0.1\%$), film thickness variations of $<\pm 1\%$ can easily be achieved with all premetered coating methods. This is true as long as not only the pump speed is kept constant, but also as long as a pump type is used that is capable of delivering fluid at a constant rate, such as a gear pump.

For non-ideal situations, in which the liquid density and the solids concentration cannot be kept constant, the coat weight variations can be estimated by using Eq. 9.5:

$$\frac{dA_{dry}}{A_{dry}} = \pm\sqrt{\left(\frac{d\rho_{liquid}}{\rho_{liquid}}\right)^2 + \left(\frac{dC_{w,solid}}{C_{w,solid}}\right)^2 + \left(\frac{dV^*}{V^*}\right)^2 + \left(\frac{dU}{U}\right)^2} \tag{9.5}$$

Here, $d\rho_{liquid}/\rho_{liquid}$ is a measure for the variation of the liquid density, $dC_{W,solid}/C_{W,solid}$ is a measure for the variation of the solids concentration, dV^*/V^* is a measure for the variation of the volumetric flow rate, for example when caused by a pump type that is characterized by excessive pulsations, and dU/U is a measure for the variations of the web speed. Note that the term dW/W, which is a measure for the variation of the coating width, is not included in Eq. 9.5 because dW is usually zero for all premetered coating methods. Equation 9.5 also visualizes the importance of preparing coating fluids that are very uniform in terms of solids concentration and temperature because $d\rho_{liquid}$ and $dC_{w,solid}$ are only non-zero, if the fluid shows variations in these two parameters.

9.4 Die Design (Film Thickness Uniformity in Cross-Web Direction)

The term ***cross profile*** is synonymous with film thickness nonuniformity in the cross-web direction. The absence of a cross profile indicates a film with a perfectly uniform thickness.

The subject of film thickness uniformity in cross-web direction is much more complicated than in machine direction because it is determined by the performance of the die. Specifically, it is influenced by many more independent parameters, notably the flow rate, the density, and several rheological parameters, as well as some 15 geometric parameters that define the internal geometry of the die. The purpose of the current chapter, therefore, is to explain the underlying physical concept for achieving a good cross-web fluid distribution while at the same time obtaining desired wall shear stress and residence time properties.

9.4.1 Concepts for Liquid Distribution

In many of the selfmetered coating methods including pan- and nip-fed roll and knife coating, the distribution of the coating fluid in the cross-web direction is driven by gravity, see schematic diagram in Fig. 9.1.

Moreover, the flow fields in a pan or a nip contain free surfaces, therefore surface tension may play a role in liquid distribution. However, both surface tension and gravitational forces are often weak when compared to opposing viscous forces. To

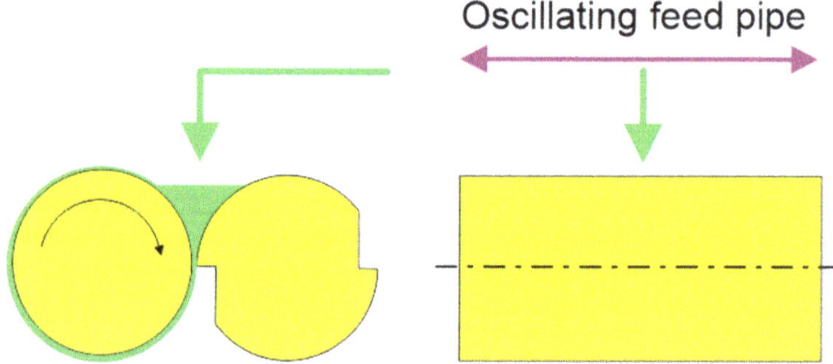

Fig. 9.1 Schematic rendering of liquid distribution in the nip of a comma coater

overcome this deficiency the feed pipe is sometimes oscillating across the coating width as depicted in the figure.

In contrast, premetered coating methods use slot dies for distributing the fluid in the cross-web direction. This process is driven by pressure and inertia forces, which are typically much stronger than gravitational and surface tension forces. As illustrated in Fig. 9.2, this distribution concept combines a flow domain with low resistance, i.e., a cavity with a large cross-section and so with a small pressure drop, together with a domain with high resistance, i.e., a narrow slot with a high pressure drop. Because flowing liquid always takes the path of least resistance, it prefers to first flow along the cavity, i.e., across the coating width, before flowing in the machine direction by exiting through the slot. In addition, the flow field inside the die is confined by solid walls, therefore allowing other properties of the slot die such as wall shear stress and residence time to be optimized by smart geometrical forms of the flow boundaries, see Sect. 6.3.

This fluid distribution system can easily be modeled by forcing the pressure drop along the cavity to be much smaller than the pressure drop across the slot. Assuming a Newtonian fluid and a circular cavity of constant cross-section, then the pressure drop in the cavity is given by Eq. 5.4.2, and the pressure drop across the slot is given by Eq. 5.6.8. Further recognizing that the fluid viscosity in the cavity is different from the one in the slot owing to the different characteristic dimensions and hence to the different characteristic shear rates as was proposed by Weinstein and Ruschak (2004), the performance of such a die can be quantified as follows:

$$\frac{\Delta P_{cavity}}{\Delta P_{slot}} \propto \frac{\mu_{cavity}}{\mu_{slot}} \frac{w^3 W^2}{L A^2} = \varepsilon \ll 1 \tag{9.6}$$

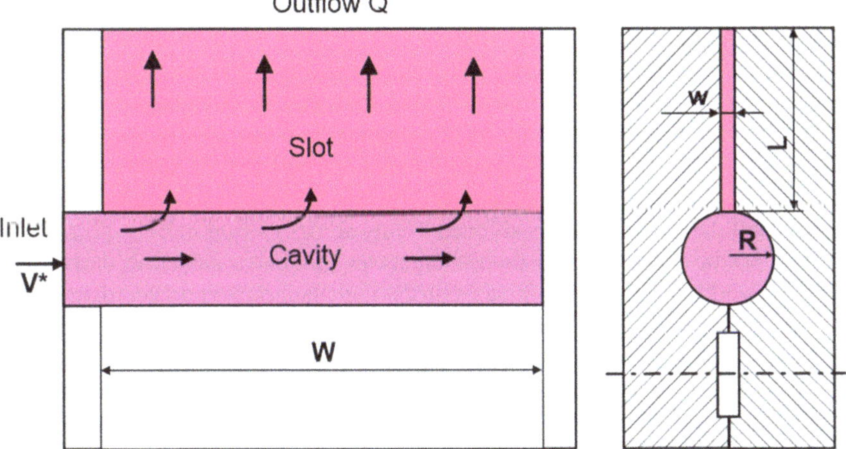

Fig. 9.2 Pressure-driven distribution system in a slot die

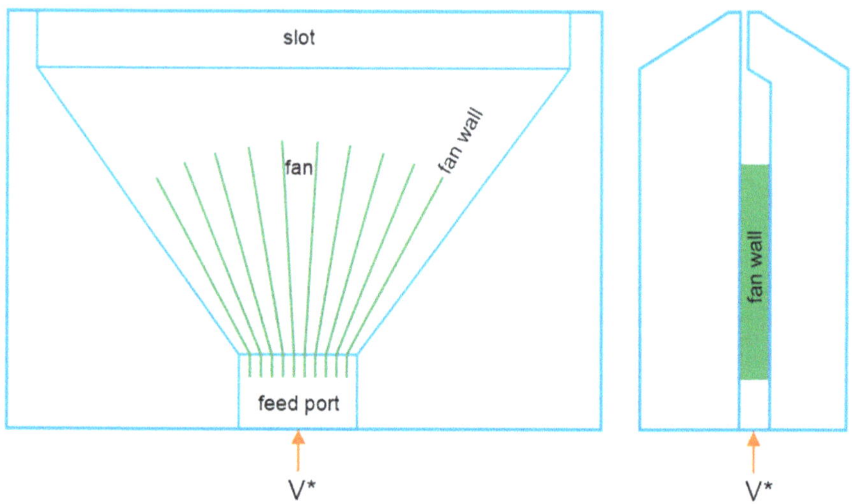

Fig. 9.3 Alternative fluid distribution system in slot die from FMP Technology

Here, A is the characteristic cross-sectional area and W is the length (coating width) of the cavity, and w and L are the height and length of the slot. With this simple view, a good fluid distribution is obtained by forcing ε to be small. i.e., by a long and narrow slot, and by a large and short cavity. This concept is widely used in the converting industry for designing slot dies. It is often referred to as the classical design concept.

In contrast, Prof. F. Durst from the University of Erlangen in Germany founded the company FMP Technology in 2006. Subsequently, he developed and patented an alternative distribution system, the concept of which is shown in Fig. 9.3 (Durst & Bülent, 2008). Specifically, the distribution system consists of a 250 mm wide element that contains a plurality of diverging distribution channels that are arranged in a fan-like fashion. The channels are formed by thin walls that are oriented perpendicularly relative to the surface of the distribution element and that have a height of about 10 mm. The walls begin in the feed port and they end far short of a narrow slot located at the outflow of the distribution element. The feed port has a rectangular flow cross-section and the fan walls are positioned evenly across the width of the port. As the velocity profile of the incoming fluid stream in the feed port is parabolic, the resulting flow rate/width entering the various distribution channels varies accordingly. This undesirable effect could be compensated by varying the width of each channel at the exit of the feed port, but this was not so for the die we had at Polytype Converting. In any case, the velocity profile at the exit of the fan, and hence the corresponding profile of the flow rate/width, is highly non-uniform owing to the boundary condition that the fluid adjacent to each fan wall is at rest. Therefore, the large space between the end of the fan and the narrow slot has the function of an outer or secondary cavity as depicted in Fig. 9.7 for the conventional die design. This space is necessary for allowing sufficient cross flow until the velocity profile becomes uniform before the fluid exits through the narrow slot.

The advantage of this approach is that liquid distribution in the cross-web direction is forced to occur mechanically by the walls of the fan structure. Consequently, according to the patent description, the fluid pressure remains more or less constant across the width, even if the flow rate or the fluid viscosity change from one application to the next. This led FMP Technology to create the advertisement slogan One For All, i.e., one die for all applications.

The disadvantage of this design approach is that the concept of a minimum wall shear stress as discussed in Sect. 6.3 cannot be implemented and explored for a good purpose because the channel width between the feed port and the narrow slot is relatively large and must remain constant throughout the distribution element. Moreover, several distribution elements must be arranged next to each other if the coating width is larger than 250 mm, which is the width of a single distribution element. This in turn requires the single fluid stream generated by the metering pump to be divided into several feed streams that must have the same flow rate. Finally, as explained below, the "one for all" design feature is not a unique trait of the FMP approach because it can also be implemented by the conventional die design as explained above, although with the same disadvantage that the minimum wall shear stress concept cannot be implemented. Furthermore, a "one for all" die is not an important design feature for industrial coating companies because most of them only coat one product family, such as adhesive labels, photographic films or paper, or battery electrodes, where the formulations for the various product variations are similar enough such that the fluid rheology and the flow rate do not vary much. Such small variations can easily be handled by the conventional die design including the concept of a minimum wall shear stress. If a coating company produces two vastly different product families such as thin water-like fluid films applied at high web speeds and thick honey-like fluid coated at low speeds, then it is very likely that these two families are manufactured on different coating machines with different dies that can be optimized for their respective range of operating conditions.

9.4.2 *Film Thickness Nonuniformity*

As mentioned in Sect. 9.4.1, one common property of all premetered coating methods is the fact that they use dies for the purpose of liquid distribution in the cross-web direction. It is well known that the cross profile of a coated film is a complex issue because it is determined by a series of subsequent processes, namely

- Liquid distribution (flow inside the die).
- Film flow on an inclined plane (for slide dies only).
- Flow in the curtain (curtain coating only).
- Coating (application of the liquid film onto the moving web).
- Leveling (flow after coating).

- Drying or curing.

Here, we are only concerned with what happens inside the die, i.e., we are focusing on the cross profile at the exit of the die slot. At that location, the cross profile is composed of two classes of contributions, namely **deterministic** and **stochastic** contributions, see also Sect. 15.3.3.2 in Kistler and Schweizer (1997). As schematically shown in Fig. 9.4, the cross profile is represented as a band characterized by a slope, which is indicative of all deterministic contributions, and a band width, which is indicative of all stochastic contributions. Any measured cross profile data will always be contained inside the band. Moreover, the difference between the highest and lowest point of the band is denoted with ε_{max}, which is characteristic of all cross profile contributions. ε_{max} is expressed as a single number, e.g., 4.24%, or in the form of ±2.12%. In the same way, the difference between the highest and lowest point of the slope of the band, i.e., the red curve in the figure, is denoted with β_1, which is characteristic of all deterministic contributions. It too is expressed as a single number, e.g., 1.68%, or in the form of ±0.84%.

Referring to Fig. 9.4, x is the coordinate in the cross-web direction and W is the width of the cavity. $x = 0$ coincides with the feed port of the distribution cavity in the die, and $x = W$ is at the end of the cavity. Therefore, Fig. 9.4 shows the cross profile of a side-fed die, and W corresponds to the coating width. A center-fed die would have a symmetrical cross profile with the image in the figure being mirrored to the other side of the feed port. Moreover, W corresponds to half of the coating width.

Q is the volumetric flow rate/width exiting the die slot, and dQ(x) indicates a small variation of Q. Hence, dQ(x)/Q is a measure for the cross profile at the exit of the die slot. Note that for all premetered coating methods and in accordance with

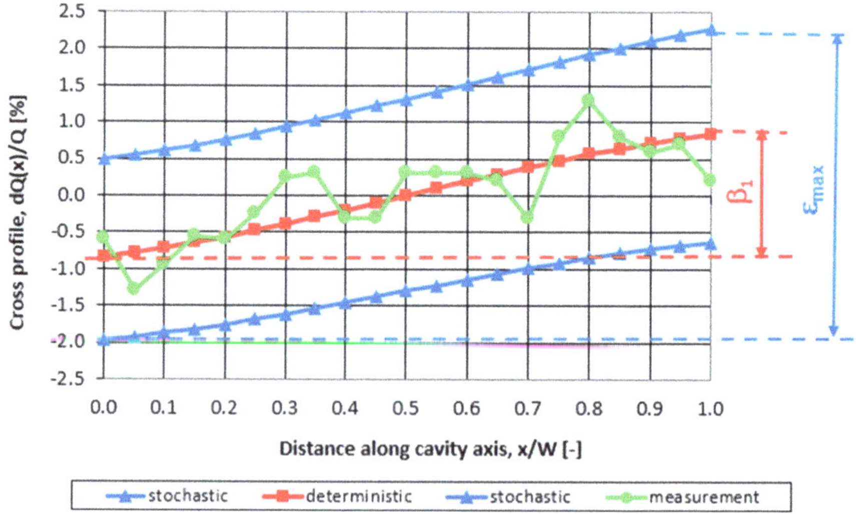

Fig. 9.4 Schematic representation of a cross profile

Eq. 9.1, Q is related to the wet film thickness and the web speed according to

$$Q = U H_{wet} \tag{9.7}$$

On the other hand, Q is also determined by the flow filed in the die slot, and slot flow of a Newtonian fluid is described by the so-called flow rate–pressure drop relationship, see Sect. 5.6, Fig. 9.5 and Eq. 9.8.

$$Q = \frac{\Delta P w^3}{12 \mu L} \tag{9.8}$$

ΔP pressure drop across die slot [Pa],
L length of die slot [m],
w height of die slot [m],
μ Newtonian fluid viscosity [Pas].

For a perfect die, Q is constant and does not depend on the coordinate x. For such a situation, the red line in Fig. 9.4 is flat and horizontal, and deterministic contributions to the cross profile are zero. On the other hand, if Q is not constant across the coating width, then the variation of Q must be caused by variations of one or several of those parameters upon which Q is dependent, see Eq. 9.8. An analytical expression describing this relationship is obtained by normalizing the total derivative of Eq. 9.8 as follows.

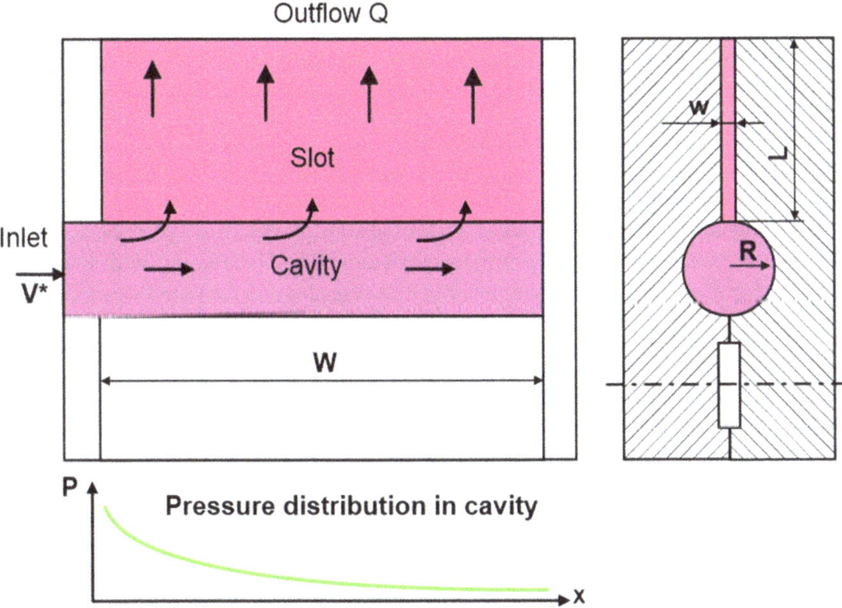

Fig. 9.5 Schematic diagram of "infinite cavity" die geometry

$$\frac{dQ(x)}{Q} = \frac{d\Delta P(x)}{\Delta P} + \frac{3dw(x)}{w} - \frac{dL(x)}{L} - \frac{d\mu(x)}{\mu} \tag{9.9}$$

Note that all terms on the right-hand side of Eq. 9.9 can be deterministic or stochastic in nature. **Deterministic** contributions give raise to smooth and gradual (long wave length) changes of the film thickness. Often, terms like "smile" or "frown" (for center fed dies) or "wedge" (for side fed dies as shown in Fig. 9.4) are used to characterize deterministic contributions to the cross profile. Specifically, positive β_1-values result in a smile profile, while negative β_1-values produce a frown pattern.

Factors that contribute to the deterministic cross profile include:

- Die internal fluid flow.
- Degree of isothermal operating conditions.
- Mechanical deflection of die plates due to internal fluid pressure or due to expansion and contraction resulting from non-isothermal operating conditions.

Typically, deterministic contributions can be visualized and quantified with the help of theoretical models. In the scope of this book, deterministic contributions refer to contributions from fluid flow effects, i.e., the pressure drop in the inner cavity.

Stochastic contributions, on the other hand, give raise to erratic and random (short wave length) changes of the film thickness. Since the exact location of such film thickness variations is random and cannot be predicted, they must be assessed and treated statistically and in an average sense. Factors that contribute to the stochastic cross profile include:

- Local variation of the slot height (mechanical precision).
- Temporal variation of physical fluid properties (process control).
- Temporal variation of the volumetric flow rate (process control, pump design).

Let us first focus on deterministic contributions to the cross profile. Referring to Fig. 9.5 and Eq. 9.9, $\Delta P(x)$ is the pressure drop across the die slot. Since the pressure at the slot exit is always and everywhere equal to the ambient pressure, which is constant for a given application, ΔP is really a measure for the lateral pressure distribution in the cavity. As shown at the bottom of Fig. 9.5, the pressure in the cavity is normally not constant but changes (most often decreases) with increasing distance from the cavity entrance, just like the pressure drop in pipe flow. Increasing pressure may be observed if the cavity cross-section changes along its axis and if the flow is characterized by a high Reynolds number. In any case, therefore, the term $d\Delta P/\Delta P$ in Eq. 9.9 is normally not zero, which is unfortunate because it is one of the origins of cross profile. However, if the pressure distribution in the cavity can be quantified, for example with the help of a suitable theoretical fluid flow model, then a perfect die can still be designed by searching for a lateral distribution of the slot length $L(x)$ such that the sum of $d\Delta P(x)/\Delta P - dL(x)/L$ for any value of x equals zero. This philosophy is applied in sophisticated die design procedures (see example in Fig. 9.6), and it is often referred to as "coat hanger design", which is in contrast to the "infinite cavity design" shown in Fig. 9.5.

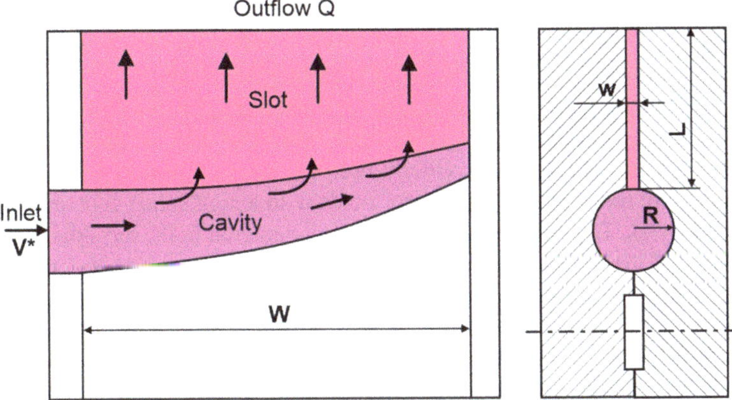

Fig. 9.6 Schematic diagram of "coat hanger" die geometry

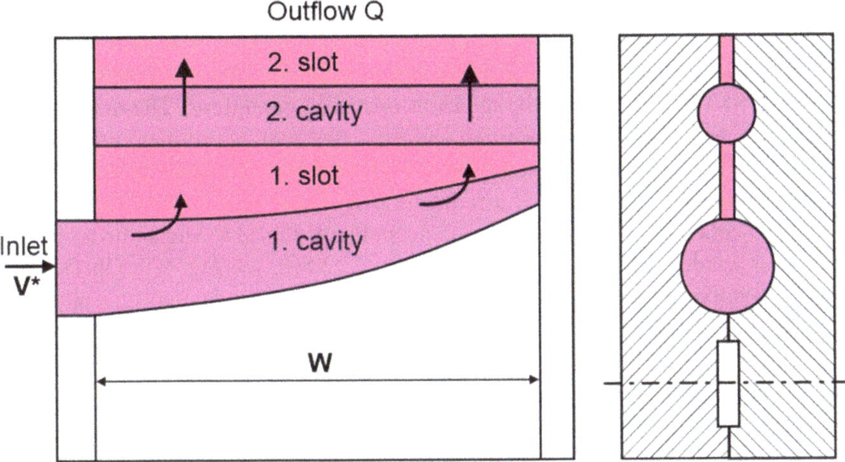

Fig. 9.7 Schematic diagram of a "dual cavity coat hanger" die geometry

However, precisely calculating the pressure drop along the distribution cavity is a challenging task because it depends on the following parameters:

- Geometrical details of the cavity, i.e., shape, cross-sectional size, and length.
- Volumetric flow rate/width.
- Physical fluid properties, i.e., density and rheological parameters.

With appropriate simulation tools and optimization procedures, perfect dies can be designed by determining the die internal geometry for one flow rate/width and one set of physical fluid properties such that the cross profile becomes zero. If such a die is then used for a different flow rate/width, or for a fluid with different physical properties, a deterministic cross profile will result because a different optimum geometry would be needed for this set of operating conditions.

Adequate rheological fluid data require the measurement of the viscosity at coating temperature over the range of relevant shear rates. Die internal flows (cavity, slot) are characterized by parabolic velocity profiles. For that reason, low shear rates, i.e., $<1\ s^{-1}$, are relevant, particularly for fluids with a high solids content that might have yield stress. The highest shear rates are encountered in the die slots. For power law fluids the magnitude of the maximum shear rate γ_{max} can be calculated from Eq. 5.6.51:

$$\gamma_{max} = \frac{Q}{w^2}\left(\frac{2}{n} + 4\right) \tag{9.10}$$

Here, n is the power law index, w is the height of the die slot, and the volumetric flow rate/width, Q, is related to the web speed and the wet film thickness according to Eq. 9.7. Typically, the maximum shear rate is in the range of 10^3–$10^5\ s^{-1}$.

In summary, a pre-defined range of operating conditions (flow rate/width, fluid properties) in combination with a die having a fixed geometry will generate a deterministic cross profile for all but one set of operating conditions. The magnitude of these cross profiles can be quantified and predicted with appropriate tools. For an optimized die, the internal geometry is chosen such that the cross profile does not exceed a specified value ε_{max} for a defined range of operating conditions.

The so-called "dual cavity design" concept is often used to widen the acceptable operating window of a given die, e.g., $\varepsilon_{max} < \pm 2.5\%$, see Fig. 9.7. On the other hand, a non-optimized die (standard design, infinite cavity design) typically has a "forgiving" internal geometry that produces inferior but still acceptable cross profiles for a wide range of operating conditions. Here, the terms "acceptable" and "wide" are not well specified. However, according to our experience, standard dies are suitable for less demanding applications, i.e., if $\varepsilon_{max} > \pm 3$–$5\%$. Excellent cross profiles of $\varepsilon_{max} < \pm 1\%$ can be achieved for dedicated applications, i.e., for applications in which the range of all relevant parameters is narrow.

Now, if the pressure drop and slot length terms in Eq. 9.9 cancel each other perfectly at any location x, or if they are at least very similar in magnitude, then it becomes important to also avoid lateral variations of the slot height (i.e., $dw = 0$) and the viscosity (i.e., $d\mu = 0$), see Eq. 9.9.

Looking at the viscosity, it is well known that this physical fluid property strongly depends on the solids concentration and the temperature of the liquid, i.e., $\mu = \mu(C_{W,solid}, T)$. The effect of a lateral viscosity variation as a result of a lateral concentration and/or temperature variation upon the cross profile can be expressed and quantified as follows:

$$\frac{dQ(x)}{Q} = -\frac{1}{\mu}\frac{\partial \mu}{\partial T}\frac{\partial T}{\partial x}dx + -\frac{1}{\mu}\frac{\partial \mu}{\partial C_{w,solid}}\frac{\partial C_{w,solid}}{\partial x}dx \tag{9.11}$$

Here, the terms $\partial\mu/\partial T$ and $\partial\mu/\partial C_{W,solid}$ are measures for the viscosity variation as a function of the temperature and the solids concentration, respectively. Both are physical material properties that must be determined experimentally and that are typically not zero. Moreover, the terms $\partial T/\partial X$ and $\partial C_{W,solid}/\partial X$ are measures for the lateral variation of the temperature and the solids concentration of the fluid. If either of these parameters is not zero, then some amount of cross profile will be generated as a result of a corresponding lateral viscosity variation as in Eq. 9.11. Therefore, we can easily recognize the importance of operating the die isothermally, meaning that the temperature of the fluid entering the die must be equal to the temperature of the die body. If this is not the case, then the die body acts as a heat exchanger, and the fluid changes its temperature as it flows in the distribution cavity from the entrance to the opposite end. The same argument applies to the solids concentration of the fluid, which must be very uniform when the fluid enters the die. Therefore, proper mixing is important, particularly if the fluid contains solid particles with a density that is higher than the one of the solvent, or if the fluid is chemically reactive.

The necessary degree of isothermal operating conditions, i.e., the maximum admissible difference between the die and fluid temperatures, depends on

- the sensitivity of the viscosity upon the temperature, which is a material property and which must be measured for any given fluid;
- the magnitude of ε_{max};
- which portion of ε_{max} will be "used up" by rheological and flow rate effects, as well as by effects of die plate deflection, see below.

Isothermal operating conditions are easiest implemented if fluid preparation, fluid delivery and fluid application, i.e., coating, take place at room temperature because no special measures must be taken. If the coating is carried out at a different temperature, however, then the die body must be tempered at coating temperature by suitable means such as feeding tempered water through long holes in each die plate, and a heat exchanger must be installed just upstream of the die entrance to adjust the fluid temperature to the coating temperature with the help of a suitable control algorithm as depicted in Fig. 8.1.

While we recommend operating the die isothermally, a non-isothermal operating mode can in principle be used to compensate for a deterministic cross profile contribution generated by fluid flow effects in the distribution cavity. If, for example, the die produces a frown-shaped cross profile, i.e., more liquid is exiting the die slot in the center than at both ends, then this profile can be eliminated by increasing the die temperature to a value above the temperature of the fluid. This will cause the fluid temperature to increase from the die entrance to the cavity end, and so it will decrease the viscosity accordingly, thus reducing the flow resistance through the die slot at both ends of the die, and hence increasing the local flow rate/width accordingly.

The fluid pressure inside the die varies in cross-web and machine direction; it is highest at the die inlet and lowest along the slot exit. The average pressure level inside the die is proportional to the volumetric flow rate and the fluid viscosity. In principle, fluid pressure tends to separate adjacent die plates, which in turn affects the height of the slots. Since the internal pressure distribution is three-dimensional, so is the mechanical plate deflection and the resulting deformation of the die slots, which directly translates into a contribution to the deterministic cross profile.

Die plate deflection is minimized by a high modulus of elasticity of the plate material and by thick die plates. In our experience, internal fluid pressures for typical slide die applications are low to moderate, i.e., <10 bar, often <5 bar, so that die deflection and the associated cross profile remain insignificant if the plate thickness is sufficient. Specifically, a plate thickness of 30–50 mm is sufficient for fluid pressures <5 bar. Slot die applications, where fluid pressures may be moderate to high, i.e., >10 bar, may require thicker die plates.

Shetty et al. (2012) coupled flow (truncated power law model) and deformation (classical beam theory) analyses to estimate slot deformations resulting from internal fluid pressure and non-isothermal operating conditions. Based on comparisons with two-dimensional and three-dimensional finite element calculations they found that the beam model was suitable for this purpose. While unwanted effects from non-isothermal operating conditions can be avoided by imposing isothermal operating conditions, they also found that slot deformations caused by internal fluid pressure strongly depend on whether the flow in the inner and outer cavity as well as in the inner and outer slot are Newtonian or power law. Moreover, they suggested that the plate thickness should be determined such that the flow rate non-uniformity generated by plate deformation related to fluid pressure should not exceed the contributions from the mechanical slot precision. More details about modeling die deformation are presented in Sect. 9.4.6.

Turning now to slot height variations generated by the manufacturing process of the die. The die slot is formed by placing two appropriately machined plates together. Hence, the slot uniformity (mechanical precision) depends on the flatness and straightness of the two plates, and that is determined by the manufacturing steps the plates are subjected to, i.e., milling, grinding, lapping, etc.

In accordance with Eq. 9.9 for fixed-lip dies having two distribution systems in series (dual cavity design), the effect of the mechanical slot non-uniformity on the cross profile can mathematically be modeled as follows:

$$\frac{dQ}{Q} = \frac{3}{DF}\frac{dw_1}{w_1} + 3\frac{dw_2}{w_2} \tag{9.12}$$

w_1 height of inner slot [m],
w_2 height of outer slot [m],
dw_1 variation of inner slot height [m],
dw_2 variation of inner slot height [m],
DF damping factor of outer cavity [–].

Damping factors of well-designed dies are typically in the range of 2–20. Hence, what happens in the inner slot is much less important compared to what happens in the outer slot. To take advantage of this situation, the inner slot of well-designed slide dies is typically narrow, which helps to keep deterministic cross profile contributions low, and the outer slot is wide, which reduces the sensitivity to the mechanical precision. This approach may not be possible for slot dies used with the slot coating process, as the outer slot height is often tied to the wet film thickness, and should not be chosen freely. Consequently, it may be difficult if not impossible to coat very thin films with an excellent cross profile because thin films require a narrow outer die slot, which in turn is sensitive to mechanical precision.

While Eq. 9.9 serves well conceptually for visualizing and explaining the parameters that affect the cross profile generated by a slot die, it cannot accurately be applied to industrial coating applications because it is derived for Newtonian fluids. However, a more realistic result is obtained by using the flow rate–pressure drop relationship for slot flow of shear-thinning fluids based on the power law model, i.e., Eq. 5.6.48. Solving this equation for Q and calculating the normalized total derivative yields

$$\frac{dQ(x)}{Q} = \frac{1}{n}\frac{d\Delta P(x)}{\Delta P} + \left(\frac{1}{n} + 2\right)\frac{dw(x)}{w} - \frac{1}{n}\frac{dL(x)}{L} - \frac{1}{n}\frac{d\mu(x)}{\mu} \tag{9.13}$$

Now, all factors contributing to the width-wise flow rate non-uniformity depend on the power law index n, and the effect of each factor is amplified for $n < 1$, which is typical for most industrial coating fluids. While the optimization concept of forcing the effects of a varying pressure drop across the slot and a varying slot length to cancel each other at any location x can still be applied, it is now even more important to avoid any width-wise variation of the viscosity and to minimize variations of the slot height, i.e., to use a die having a high mechanical precision. The need for a high mechanical precisions is particularly important for dies with a fixed slot height that is either machined into one of the die plates or that is set by inserting a (replaceable) precision shim between the two die plates. In contrast, the need for mechanical slot precision is less important for dies with an adjustable slot height. Here, the slot height can be adjusted while coating using screws or thermal expansion bolts that are placed a few centimeters from each other across the entire die width until the online measured cross profile satisfies the uniformity requirements.

Note that Eq. 9.12 describes a deterministic cross profile contribution from the mechanical slot precision. This view may be appropriate when looking at an existing die where the mechanical slot height non-uniformity in the cross-web direction can be inferred by measuring the surface profiles of both die plates in the areas of the die slots. When designing a new die, however, this information is not (yet) available, so that the mechanical slot height variation dw must be assumed in a stochastic sense. Therefore, the cross profile contribution from the mechanical slot precision is better quantified by considering Eq. 9.13 and by re-writing Eq. 9.12 for power law fluids as follows:

$$\frac{dQ}{Q} = \pm\sqrt{\left[\frac{1}{DF}\left(\frac{1}{n_1} + 2\right)\frac{dw_1}{w_1}\right]^2 + \left[\left(\frac{1}{n_2} + 2\right)\frac{dw_2}{w_2}\right]^2} \tag{9.14}$$

Here, n_1 and n_2 are the power law indices of the flows in the inner and outer slots, respectively. Typically, n_1 and n_2 are not equal owing to the preferred design approach, which postulates that the outer slot height w_2 is larger than the inner one w_1 to reduce the sensitivity of the cross profile to the mechanical slot precision. It is suggested that both n-values are determined with the local power law model proposed by Ruschak and Weinstein (2014) as described in Sects. 4.4.4 and 5.6.2. Depending on the values of the various parameters in Eq. 9.14, particularly for n values in the range of about 0.5–1.0, stochastic cross profile contributions from the mechanical slot precision are lower than deterministic ones. The opposite is true if n is less than about 0.5, and the difference in either direction can be as large as 30%.

Dies with a fixed slot height that is machined into one of the die plates can be manufactured more precisely than dies that use shims for setting the slot height because it is easier to achieve a very flat surface on a thick metal plate than on a thin metal shim. However, a very high mechanical precision, i.e., $dw < \pm 1\ \mu m$, cannot be achieved with machines only, even if the best grinding and lapping machines are used. Instead, hand lapping is required, and that requires an elaborate and highly precise measurement system that is able to visualize and quantify the topography of the die plate over the entire plate width. Knowing the topography of the die plate, a skilled mechanic then uses special lapping paper to reduce by hand the height of the mountain tops across the plate. It is obvious that such a procedure, which may require several iterations, is labor-intensive and hence costly. When purchasing a die, therefore, it is important to not only consider the investment costs but also the operating costs. As illustrated in Sect. 15.5 it may be very worthwhile to purchase a more expensive die that produces a better cross profile due to having a higher mechanical precision as in Eq. 9.14. Specifically, the higher precision allows the coat weight to be reduced by a small amount, which in turn reduces raw material costs.

Fixed-lip dies are recommended for all premetered coating applications at operating temperatures of, say, <60–80 °C. As such dies have no mechanisms for adjusting the cross profile during coating, they are easier to use and the coating process runs more steadily and stably. However, it is important to use a well-designed and optimized die, which already beforehand guarantees an acceptable cross profile for a specified range of applications. Since the physical principles of die design are well understood, at least for viscous fluids, this requirement can easily be satisfied. Consequently, dies with an adjustable slot height are the device of choice for high-temperature applications such as hot melt or polymer extrusion, or applications with highly visco-elastic fluids. On the one hand, it is uncertain whether or not the mechanical precision, which must be machined into the die plates at room temperature, can be maintained when the die is heated up to temperatures of >100 °C owing to uncontrolled mechanical deformations as a result of thermal expansion. On the other hand, modeling the flow of visco-elastic fluids is much more difficult and less understood than flow modeling of shear-thinning, viscous fluids. Hence, an à priori guarantee of an acceptable cross profile for a specified range of operating conditions cannot be given with great certainty, and using an online mechanism for adjusting the die performance after the die has been designed and manufactured is an acceptable means for eliminating this disadvantage.

For premetered coating methods, the need for routinely measuring the thickness of the coated film is smaller than for self-metered methods because, as the term "premetered" implies, metering (= measuring) efforts focus on measuring and controlling relevant parameters (the web speed and the flow rate) before (= upstream of) the coating process. Appropriate flow meters are discussed in Sect. 8.2 and by Schweizer (1992). The web speed is typically measured indirectly by using a rotational encoder that picks up the rotational speed of a suitable roll in the coating machine. Here, no slip between the roll surface and the web must be guaranteed.

On industrial coating machines, the flow rate is often slaved to the web speed by an algorithm that is depicted in Fig. 9.8. In particular, the control unit is fed with the nominal flow rate, which depends on the wet film thickness, the web speed, and the coating width according to Eq. 9.1. If the measured flow rate does not correspond to the nominal value, then the control unit increases or decreases the speed of the pump motor until the flow rate discrepancy disappears.

Sometimes, it is appropriate, at least temporarily, to measure the thickness of the coated film, even if the coating method is of the premetered type. This is particularly true when premetered coating methods are newly introduced in industrial coating plants and are associated with uncertainty and doubt by the new user. For such cases, any of the well-known and commercially available non-contacting measurement

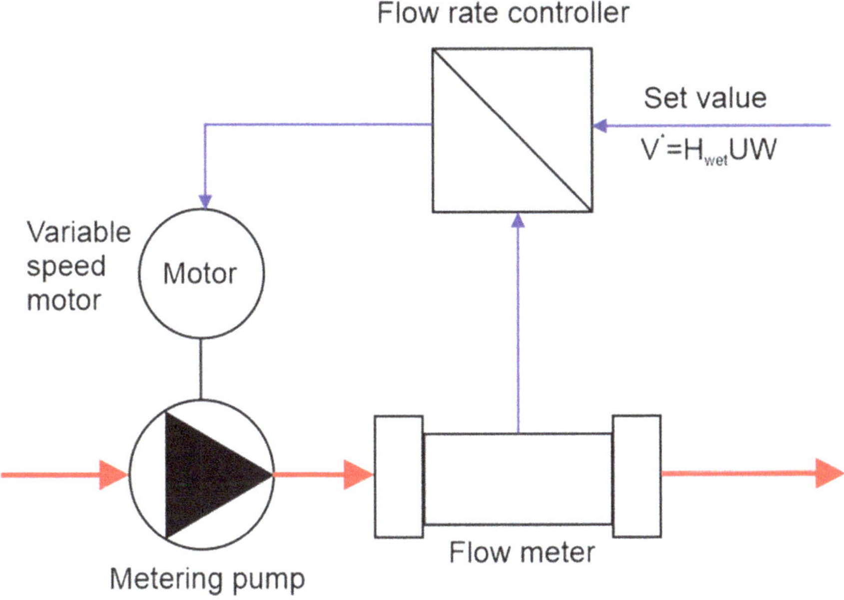

Fig. 9.8 Control algorithm for slaving the flow rate to the web speed

methods are suitable. Specifically, a point source that emits radiation of a particular frequency range in the electromagnetic spectrum is used in combination with a detector that measures the loss of energy due to selective absorption by the coated film. Depending on the frequency or wave length of the emitted energy, different measurement methods are distinguished. Infrared absorption, microwave absorption, and x-ray fluorescence belong to this class of measurement techniques. For an overview of these methods see for example Schweizer (1997a).

9.4.3 Recent Literature on Die Design

Theoretical modeling of die design has been done extensively in the past, and a comprehensive summary was written by Secor (1997). Since then, additional papers on this topic have appeared from various authors, most notably from the group of Ta-Jo Liu at the National Tsing Hua University in Hsinchu, Taiwan, and from the group around Ken Ruschak and Steve Weinstein at the Rochester Institute of Technology in Rochester, New York.

Durst et al. (1997) investigated the performance of a dual-cavity side-fed die by comparing cross profile predictions from an analytical model and from three-dimensional finite element calculations with experimental measurements. The latter aspect in particular was not often included in previous publications. Here, experimental cross profiles were generated in cooperation with TSE Troller Schweizer Engineering, who provided the slot die, and ILFORD AG in Switzerland, who provided a pilot coating machine and their facility for measuring cross profiles. Using the power law model, the authors concluded that the difference between numerical predictions and experimental data was less than 1%, and predictions from the analytical model were only slightly worse, provided that the Reynolds number was small, i.e., <5. Nevertheless, the authors pointed out the attractiveness of analytical models because the computational time was on the order of seconds, while it was on the order of hours for the 3-D model, at least for the computer power at that time.

Yu et al. (1997) carried out a similar study by comparing cross profile predictions from four different one-dimensional models with three-dimensional finite element calculations. 1-D modeling means that the flow equations are averaged over the cross-section of the cavity. The authors argued that owing to the lack of experimental cross profile data, the validity of 1-D modeling can best be examined by comparing the predictions with 3-D calculations. The 1-D models included

- The truncated power law approach by Yuan (1995), which compared the average wall shear rate along the cavity with the critical shear rate according to the Carreau rheological model to determine whether the flow in the cavity is Newtonian or shear-thinning.
- The analytical switching approach by Weinstein and Ruschak (1996), which is based on a linearization procedure to derive a switching criterion to predict whether or not the flow in the cavity is Newtonian.

- The Newtonian cavity flow approach, which assumes a constant (Newtonian) viscosity for the cavity flow regardless of the actual shear rate.
- The standard power law approach of Lee and Liu (1989), which assumes power law behavior in both the cavity and the slot, irrespective of shear rate.

They further assumed that slot flow is always shear-thinning according to the power law model, while the flow in the cavity is either Newtonian or shear-thinning. For the 3-D calculations the viscosity was modeled with the Carreau equation. They concluded that cross profile predictions by all the tested 1-D models are reasonably accurate, provided that the characteristic shear rate in the cavity meets the assumption of the model. If this is not the case, then the models that can handle both shear-thinning and Newtonian flows are preferred. However, if inertial effects become significant, i.e., if the Reynolds number at the cavity entrance is higher than about 25, then none of the 1-D models produced accurate cross profile predictions.

Pan et al. (1999) used a 1-D model to study the die performance for chemically reactive fluids. Specifically, the viscosity expression included the effects of chemical reaction and heat transfer. They found that it is much more difficult to generate good cross profiles with reacting fluids than with non-reacting liquids. They also found that the slot height is the most sensitive parameter for improving the cross profile. However, reducing the slot height caused the pressure drop across the die to increase excessively for the highly viscous electronic packaging materials that were the subject of their study. Consequently, they developed an algorithm based on a fourth-order polynomial shape function, with which the form of a choker bar could be optimized to obtain acceptable cross profiles.

Shetty et al. (2012) investigated the behavior of a dual-cavity die subject to bar deformation generated by pressure and temperature effects. Die bars may deform due to non-uniform fluid pressure across the die width, and due to expansion and contraction associated with non-isothermal operating conditions. Bar deformation was modeled with classical beam theory, thereby accounting for the varying bar thickness due to the cavities. Flow and deformation analyses were coupled, whereby the flow in the cavities was assumed to be one-dimensional. Furthermore, the equations were linearized about a perfect flow distribution. Inertia effects were neglected, and the fluid rheology was treated with the truncated power law model.

The authors re-emphasized findings from an earlier paper (Ruschak & Weinstein, 1997), in which they showed that the flow in the outer cavity is fundamentally different from the flow in the inner cavity because the primary flow direction is across the outer cavity and not along its axis. For a shear-thinning fluid, therefore, the cross-flow determines the viscosity that is relevant for the axial flow. Moreover, if inertial effects are significant, the momentum of the cross-flow opposes flow along the cavity axis. These observations were not accounted for in dual-cavity models published previously, and hence these models were deficient.

The purpose of the outer cavity is to reduce flow non-uniformities generated by the inner distribution system, i.e., by the inner cavity and the inner slot, and to reduce the sensitivity of the die to variations in operating conditions, i.e., flow rate and physical fluid properties. In a comprehensive model of a dual-cavity die, the flows in the inner and outer cavities are coupled, which makes the analysis difficult. During the design and optimization of a new die, however, there is a need for a simpler but useful tool that can quantify the effectiveness of the outer cavity. To that extent, the authors defined a damping factor, which is based on imposing a slightly nonuniform flow exiting the inner slot, and that is represented by a Fourier series.

By comparing the results from the beam theory with two-dimensional finite element calculations, this paper confirmed the simplifying assumption that the slot heights deform linearly along the length of the slots. Further comparisons between the beam model and two- and three-dimensional finite element calculations show that results from the beam model are adequate approximations to 2-D finite element calculations, which in turn agree well with 3-D calculations away from the center of the die. Therefore, results generated by the beam model are useful when the design goal is to limit slot deformations to fabrication tolerances, which was the aim of this analysis. On the other hand, the beam model significantly underestimated slot deformations resulting from temperature variations in the die bar. Hence, this result emphasizes the need for tight temperature control of the die body and for assuring isothermal operating conditions, which require the temperature of the fluid entering the die to be equal to the temperature of the die body.

In 2014, Ruschak and Weinstein published an important paper when they introduced the local power law (LPL) approximation to noncircular ducts such as cavity cross-sections of slot dies, see also Sects. 4.4.4 and 5.5 for more details. As illustrated in Sect. 4.4.4, the viscosity behavior of most coating fluids can well be represented by the Carreau-Yasuda (CY) model (Eq. 4.4.1), which connects regions of constant viscosity at low and high shear rates with a region of shear-dependent (shear-thinning) viscosity at intermediate shear rates. The transition between these regions is continuous and smooth. While predictions with Newtonian flow equations are only accurate if the viscosity is constant, and while results with the classical power law model are only acceptable if the shear rate of the flow coincides with the shear rate of the shear-thinning region of the CY model, the truncated power law model can handle all three shear rate regions with a continuous but not smooth transition. Consequently, predictions in the transition regions are inaccurate, particularly if the transition region is broad, i.e., if the value of the parameter a in the Carreau-Yasuda equation is small, i.e., <2. In contrast, all deficiencies of these models are eliminated by using the local power law model, and flow predictions are more accurate over the entire shear rate range of a given viscosity curve. Moreover, the flows can be simulated with the classical power law model, which is modest in computational load.

The local power law model is constructed as a tangent to the viscosity curve viewed in a double-logarithmic coordinate system at an appropriate viscosity. For duct and slot flows, Ruschak and Weinstein proposed an equivalent Newtonian viscosity, which is the Newtonian viscosity giving the same pressure gradient as the local power law model for the same flow rate. The authors also provided recipe-like instructions for the implementation of the LPL model on a personal computer, which can easily be accomplished in an EXCEL environment.

In summary, the local power law model is a versatile, accurate and easy-to-use tool for calculating viscous pressure drops in slot and duct flows of coating dies.

Ruschak and Weinstein (2018) also developed a model for the outer cavity of a dual-cavity die. They investigated five different triangular shapes, and they started by calculating the two-dimensional flow of power law fluids across the cavity up to a critical Reynolds number, which marked the onset of recirculation. Knowing the details of the cross-flow provided information for calculating the axial flow, which, in reality, is coupled to the flow in the inner cavity. However, as they did in their 2012 paper with Shetty, they imposed a slightly nonuniform flow exiting the inner slot and represented it as a Fourier series, which is able to represent any kind of flow variation. This approach allowed them to assess and quantify the performance of the outer cavity. Specifically, the effectiveness of the outer cavity can be measured by the difference between the maximum and minimum values of the flow rate in the inner slot divided by the flow rate in the outer slot. This ratio is called damping factor DF, which has values of >1. Increasing DF-values indicate increasing damping capacity of the outer cavity, and so improved flow uniformity at the exit of the die.

In addition, finite element calculations by Ruschak and Weinstein confirmed an assumption often made in past publications that the pressure across the outer cavity is nearly constant. However, they found that the pressure is slightly higher than that given by a fully developed flow in the outer slot. They attributed this difference to losses incurred by the flow approaching the outer slot. In any case, this difference is not critical when designing a new die because it might be viewed as a virtual increase of the outer slot length, which in turn improves the effectiveness of the outer cavity.

Modeling the outflow from the inner slot by a Fourier series further showed that if that flow is asymmetric relative to the center of the die, then more flow is distributed to one side of the die, and it becomes more difficult for the outer cavity to even out this maldistribution. Therefore, the die should be symmetrical in all aspects, and, in particular, the feed port of the inner cavity should be located in the center of the die.

Ruschak and Weinstein also showed that the damping factor decreases significantly with decreasing power law index, and they attributed this behavior to the high shear thinning in the outer slot, which in turn lowers the pressure drop in this flow. This result was true for all five triangular cavity shapes they investigated. Furthermore, they found that the 45–45 triangle has the highest damping capacity because it has the largest area for a specified cavity length. In contrast, the damping capacities of the 30–60 and the 60–30 triangles are quite similar. Concluding from these findings, the dual-cavity design concept is most effective for Newtonian fluids, i.e., for high values of the power law index, although significant damping of flow nonuniformities exiting from the inner slot can still be achieved for lower n-values.

When inertial effects are important, then the best cavity shape may be the one that best resists the onset of flow separation. The authors found that shear thinning promotes recirculation. Furthermore, comparing the five triangular cavity shapes they investigated with regard to damping capacity and resistance to recirculation, they suggested that the 45–45 triangle is an attractive choice for the best cavity shape.

Finally, Ruschak and Weinstein derived an optimization criterion regarding the partition of the outer working length between the outer cavity length CL_2 and the outer slot length L_2, see Fig. 9.9.

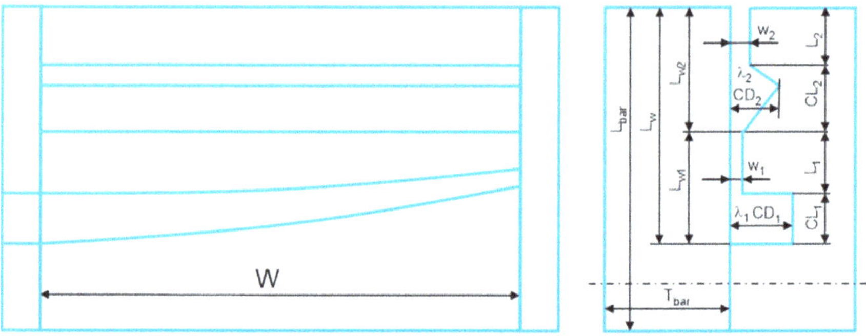

Fig. 9.9 Definition of geometrical parameters for internal and external die geometry

In their 2019 paper, Ruschak and Weinstein used full numerical solutions to compare the accuracy of two approximate but accurate and computationally efficient one-dimensional models that accommodate a general viscosity curve, for example the Carreau-Yasuda model that calculates the viscous pressure drop in non-circular ducts. The first model is the local power law (LPL) approximation to a smooth viscosity curve, see Ruschak and Weinstein (2014). In this first publication, they established the feasibility of this model to flows in non-circular ducts, but they did not establish the accuracy of the method. The second method is an improved calibration of the method of Kozicki et al. (1966, KCT), which is based on the classical Rabinowitsch-Mooney equation. In addition, The KCT method arbitrarily uses the hydraulic diameter, and the model parameters are calculated solely from Newtonian duct flow, which causes inaccuracies. Therefore, Ruschak and Weinstein replaced the hydraulic diameter with an equivalent diameter. Furthermore, both methods require one-time numerical calculations for the power law viscosity model to quantify the so-called shape factor, which is necessary for calculating the pressure drop. Shape factors were calculated and accuracy checks were carried out for ten different duct shapes, including the square, the rectangle with four different aspect ratios, the semicircle, and the 30° right angle triangle, all of which are potentially interesting for die design.

The authors concluded that both the LPL and the KCT methods produce quite accurate results with pressure gradient errors of less than 1%. The KCT method consistently resulted in slightly smaller errors than LPL. Nevertheless, the choice between the two methods may be determined by computational issues, and Ruschak and Weinstein prefer the LPL approach because it is a bit easier to implement, particularly when the volumetric flow rate is specified and the pressure gradient must be calculated, which is always the case in applications for premetered coating methods.

Considering all the published information, designing a good die now becomes a manageable task with quite affordable computational efforts that can be implemented in an EXCEL environment. A recipe-like procedure for carrying out this task is proposed later in Sect. 9.4.7.

A coating company in need of a new die can now choose between designing the internal die geometry on their own or completely relying on a die supplier for designing and manufacturing the equipment. In the latter case, understanding the fluid-mechanical concepts as outlined in this book is invaluable for selecting the best supplier. Equally important is evaluating and comparing the capabilities of different suppliers for achieving the desired mechanical slot precision because that aspect may become decisive if the internal die geometry is well designed. Even though dies with higher mechanical precision are more expensive than less precise devices, it may still be wise not to buy the cheaper die. This is because higher initial investment costs may quickly be recovered by lower material costs over the lifetime of the die. This is due to the enhanced cross profiles produced by the more precise die, and therefore to the smaller film thickness that can be coated for the final product to perform as required.

9.4.4 Die Geometry

As illustrated in Fig. 9.9, the internal and external geometry of a dual-cavity die is determined by at least 15 geometrical parameters. W is the length of the cavity, which corresponds to the die or coating width, if the die is side-fed, or it corresponds to half of the die or coating width if the die is center-fed. Usually, only half of the die is investigated when using one-dimensional models for assessing internal flow properties of center-fed dies, thereby assuming symmetry about the inlet location and neglecting potential inlet effects associated with the stagnation point flow at the entrance to the inner cavity. L_w is the working length, i.e., the portion of the die bar that is used for housing the inner (L_{w1}) and outer (L_{w2}) distribution systems. λ is indicative of the shape of the cavity cross-section, i.e., it defines whether the shape is circular, rectangular, triangular, or of any other form. The other parameters are self-explanatory.

Designing and optimizing the geometry of a die requires assigning the best possible value to each of the 15 geometrical parameters for a given set of operating conditions. This is an important task, which is best accomplished by theoretically modeling various aspects that affect the performance of the die.

9.4.5 Modeling Die Performance

The primary purpose of a die is liquid distribution in cross-web direction. A well-designed die performs well if the resulting cross profiles for a specified range of applications lie within a specified band width that is quantified by the parameter ε_{max}, see Fig. 9.4. However, a good die incorporates additional quality-related performance criteria. Specifically, a good die is characterized by a wall shear stress that is above a minimum value everywhere inside the die and for all intended applications. High wall shear stress avoids recirculation and results in preventing contamination of the die-internal flow field, which in turn avoids the need to clean the die during a product change. High wall shear stress also results in a short mean residence time and a narrow residence time spectrum. Consequently, fluid change-over times are short and reactive fluids can be coated without problems. Fortunately, both the pressure drop across the die and the wall shear stress inside the die are fluid-mechanical parameters that can be modeled and optimized by optimizing the geometry of the die-internal flow domains for a given set of applications.

From a fluid-mechanical point of view, therefore, three-dimensional flow modeling would be optimal because three-dimensional flow effects such as

- inlet effects;
- the transitional flow between the inner cavity and the inner slot;
- end effects in the inner cavity;
- the coupling of the flow in the inner and outer cavities

could be simulated accurately. However, as pointed out by Ruschak and Weinstein (2019), computational efforts for modeling the entire die-internal flow field are enormous, not the least owing to the significant disparity of relevant geometrical parameters. In particular, characteristic dimensions of relevant geometrical parameters are of the following order:

- slot height: 10^{-4} m,
- slot length: 10^{-2} m,
- cavity diameter: 10^{-2} m,
- cavity length: 10^{+0} m.

Moreover, comparing 3-D modeling with two different one-dimensional models (KCT and LPL, see Sect. 5.4.2), Ruschak and Weinstein concluded that 1-D calculations of the pressure drop in the inner cavity, i.e., in ducts of arbitrary cross-section and for power law fluids, are rather accurate, i.e., within 1% of the 3-D calculations. In addition, three-dimensional inlet effects in center-fed dies can be minimized by minimizing the stagnation point character of the inlet flow field, i.e., by making the diameter of the inlet pipe, i.e., the stem of the T just upstream of the inner cavities, as big as the characteristic dimension of the cavity cross-section, see Sect. 5.5.3. Inlet effects in side-fed dies can be avoided altogether. In particular, using electro-erosion technology, a short adapter piece can be inserted into the delivery line just upstream of the cavity entrance, inside which the flow cross-section changes very gradually and smoothly from circular in the feed pipe to whatever shape of the inner cavity cross-section. End effects in the inner cavity can be minimized by imposing a local three-dimensional design to the cavity end rather than letting the flow impinge head-on and perpendicularly onto the side wall of the die. An example of such a 3-D design was developed by TSE Troller AG and is shown in Fig. 6.10b. Finally, as argued in Sect. 5.9.2, the thickness of the boundary layer in slot flow is small, i.e., less than twice the slot height, which, typically, amounts to <1 mm.

For all these reasons, therefore, our one-dimensional model of choice for **modeling the flow in the inner cavity** is the local power law (LPL) model, not only for its ease of implementation in an EXCEL environment, but also for its ease to optimize the die geometry for a set of different applications, which very often is a requirement when designing a new die. In such situations, it is more important that the chosen die geometry works well for all specified operating conditions rather than that the design works perfectly well for just one condition and not so well for all the others. In that sense potential inaccuracies of the 1-D model can be considered as an additional application such that they lose their importance amidst the non-perfect cross profiles of all but one application considered. Complete three-dimensional modeling may be appropriate if a new die is dedicated to just one application, which, however, is rare in the converting industry.

More specifically, equations for calculating the pressure drop in terms of viscous and inertia effects for shear-thinning fluids in ducts of arbitrary cross-sections are presented in Sect. 5.5. The same chapter contains an equation for calculating the desired minimum wall shear stress. Better yet, rearranging Eq. 5.5.5 allows the size of the initial cavity cross-section to be calculated such that the resulting wall shear stress assumes the desired value for a specified set of operating conditions. Finally, Sect. 6.3.1 provides information for calculating the wanted wall shear stress distribution along the inner cavity.

The flow along the inner cavity combined with the flow through the inner slot constitutes a fluid distribution system. This system can be optimized for one specified base set of operating conditions by calculating the slot length at every location x along the cavity such that the pressure drop across the slot equals the cavity pressure at that location according to

$$L_1(x) = \frac{P_{cavity}(x) w_1{}^{(1+2n_1)}}{2^{(1+n_1)}\left(2 + \frac{1}{n_1}\right)^{n_1} m_1 Q^{n_1}} \tag{9.15}$$

For all other operating conditions, this so designed die produces widthwise flow nonuniformities dQ(x), the magnitude of which is called β_0, and it is defined as

$$\beta_0 = \frac{Q(x = W) - Q(x = 0)}{Q_0} \tag{9.16}$$

For a single-cavity die, β_0 is a measure for the final deterministic contribution to the cross profile. However, β_0 can be further reduced by adding a second distribution system consisting of an outer cavity and an outer slot.

Regarding the **outer cavity**, the main flow is not along the cavity axis as it is in the inner cavity, but it is across the cavity. For calculating relevant flow properties, in particular the damping capacity of the outer cavity, we follow the approach of Ruschak and Weinstein (2018). An expression for the damping factor DF is given by Eq. 5.5.14 in Sect. 5.5.2. Now, the final widthwise flow rate nonuniformity is called β_1, and it is calculated according to

$$\beta_1 = \frac{\beta_0}{DF} \tag{9.17}$$

For sizing the cross-section of the outer cavity such that the minimum wall shear stress assumes a preferred value for a specified set of operating conditions, we follow the work of Roulier and Bebié (2020), who investigated the flow across right-angle triangular cavity shapes with a rounded vertex. Relevant equations are summarized in Sect. 6.3.2.

The theoretical models for investigating the flow in the inner and outer cavity presented here are used in Sect. 9.4.8 for analyzing the performance of slot dies as a function of relevant operating conditions, rheological fluid properties and geometric parameters. This work culminates in Sect. 9.4.9, where the design of slot dies, i.e., the inner die geometry is optimized for a specified range of operating conditions and physical fluid properties.

9.4.6 Bar Deflection

In this section, a simple attempt is made to account for the widthwise flow nonuniformity due to mechanical deflection of the die bars, which is generated by the internal fluid pressure. Bar deflection results in widening the slot height along the axis of the slot. This would not be a problem if the fluid pressure was uniform across the width of the die, which it is not. Figure 9.10 shows the uneven three-dimensional pressure

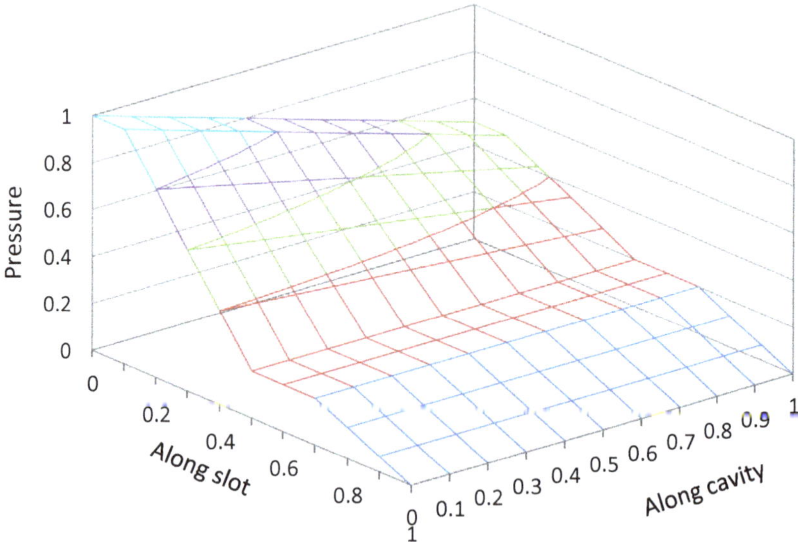

Fig. 9.10 Schematic three-dimensional pressure field inside a slot die

field inside the die in a qualitative and dimensionless form. The pressure is highest at the inlet of the die; it decreases continuously across the die width, and in a step-wise fashion along the slot. At the slot exit, the pressure is uniform at the ambient level.

The effect of bar deflection is expressed by Eq. 9.12 or, in a more general form, by the second term on the right side of Eq. 9.13. This term is amplified by a factor of at least 3, which points out the importance of knowing the widthwise variation of the slot height dw(x) and keeping it as small as needed. Bar deflection adds a deterministic contribution to flow nonuniformity, which is in contrast to the stochastic contribution stemming from the mechanical precision of the slot height. Shetty et al. (2012) suggested that these two contributions should be of the same magnitude, i.e., dw(x) should not exceed the order of 1 μm.

As shown and explained by Shetty et al. (2012), modeling bar deflection is a complex problem involving three-dimensional mechanical stress calculations of die bars, which are exposed to static bolt loads, dynamic loads due to internal fluid flow, and thermal stresses caused by the presence of a non-isothermal environment. The issue is already difficult for a simple slot die but it is considerably more complex when considering the stack of adjacent bars of a multilayer slide die. Here, the analysis is purposely kept simple and on an analytical level to visualize those relevant parameters that contribute to flow nonuniformity when bar deflection is driven by fluid flow. Consequently, the following assumptions apply, see also Figs. 9.11 and 9.12:

- Only dynamic effects of fluid flow on bar deflection are considered.
- Bar deflection is assumed to result from bending only; effects of shear forces are neglected.

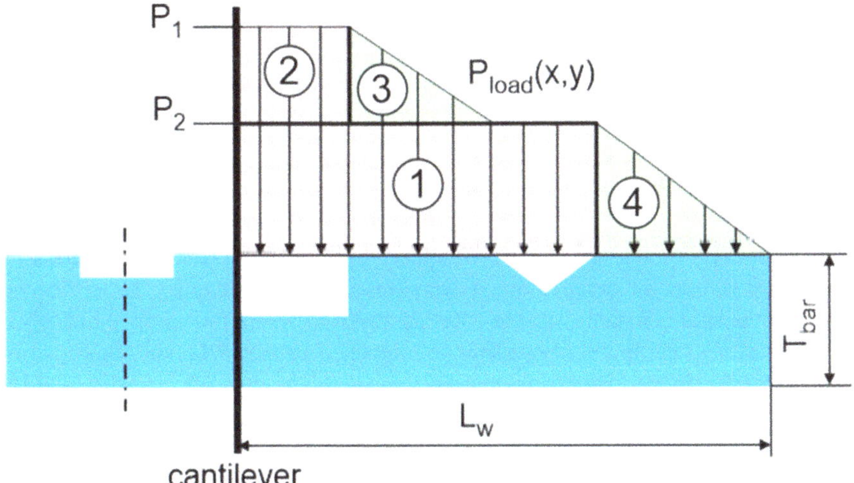

Fig. 9.11 Nonuniform pressure load on die bar

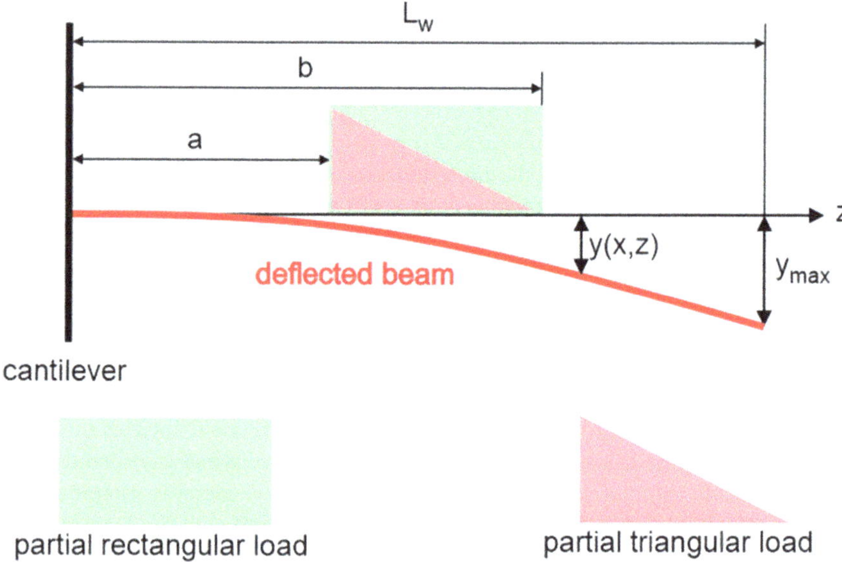

Fig. 9.12 Deflection of cantilevered die bar

- The die bars are treated as cantilevered beams, meaning that no portion of the clamping section of the bars enters the computations.
- The three-dimensional problem of bar deflection is reduced to a two-dimensional computation of the widthwise variation of the slot height dw(x).
- The die bars are exposed to piece-wise constant and piece-wise linearly varying sub- loads corresponding to the fluid pressure distribution in the inner and outer cavity and inner and outer slot, respectively.
- Only the bar containing the cavities is assumed to deform, while the adjacent bar remains undeformed.

The deflection of a cantilevered bar increases with increasing z, see Fig. 9.12. The maximum deflection occurs at the end of the bar at $z = L_W$. The load of the bar is proportional to the fluid pressure at any particular position along the slot, and its distribution in the z-direction is sketched in Fig. 9.11. In particular, the load is constant at P_1 over the inner cavity, it decreases linearly in the inner slot to the level P_2, which is held constant over the outer cavity from where it decreases linearly in the outer slot to the ambient pressure. As depicted in Fig. 9.11, the total load on the bar can be divided into triangular and rectangular sub-loads, and the total bar deflection is equal to the sum of the deflections attributed to those individual loads. Since the pressure in the inner cavity, and often also the length of the inner cavity and the length of the inner slot, vary along the x-axis, so does the load on the bar and hence the bar deflection y(x, z).

Using two-dimensional finite element analysis, Shetty et al. (2012) confirmed the assumption that the slot heights deform linearly along the slot, i.e., the deflection line y(z) at any fixed location x is very close to a straight line. Therefore, the problem of calculating the deflection $y = y(z)$ reduces to the simpler task of determining the maximum deflection y_{max} at the end of the bar where $z = L_w$. Equations for y_{max} can be found in engineering handbooks such as Dubble (1994), and they are:

Partial rectangular load:

$$y_{max} = \frac{P_{load}}{24EI}\left(4\left(a^2 + ab + b^2\right)L_w - a^3 - ab^2 - a^2 b - b^3\right) \tag{9.18}$$

Partial triangular load:

$$y_{max} = \frac{P_{load}}{60EI}\left(\left(15a^2 + 10ab + 5b^2\right)L_w - 4a^3 - 2ab^2 - 3a^2 b - b^3\right) \tag{9.19}$$

The geometrical parameters a, b and L_w are defined in Fig. 9.12, where L_w is the working length, i.e., the length of the cantilevered bar, which is composed of the length CL_1 of the inner cavity, the length L_1 of the inner slot, the length CL_2 of the outer cavity and the length L_2 of the outer slot, see Fig. 9.9. P_{load} is the load on the bar in units of a force, E is the modulus of elasticity, and I is the moment of inertia. The latter parameter is approximated by Eq. 9.20, which accounts for the fact that the strength of the die bar is weakened by the intruding outer, and especially the inner cavity.

$$I = \frac{W(T_{bar} - CD_1)^3}{12} \tag{9.20}$$

Here, W is the length of the die bar in the x-direction but excluding the thickness of the side plates (see Fig. 9.9), T_{bar} is the bar thickness, and CD_1 is the depth of the inner cavity.

Since the load is measured in units of force, i.e., [N], its value is obtained by multiplying the fluid pressure over the area it is acting. Therefore, the formulas for the 4 sub-loads shown in Fig. 9.11 can be written as follows.

$$\begin{aligned} P_{load,1} &= P_2 W(b - a) \\ a = 0, &\quad b = CL_1 + L_1 + CL_2 \end{aligned} \tag{9.21}$$

$$\begin{aligned} P_{load,2} &= (P_1 - P_2)W(b - a) \\ a = 0, &\quad b = CL_1 \end{aligned} \tag{9.22}$$

$$\begin{aligned} P_{load,3} &= (P_1 - P_2)W\frac{(b - a)}{2} \\ a = CL_1, &\quad b = CL_1 + L_1 \end{aligned} \tag{9.23}$$

$$\begin{aligned} P_{load,4} &= P_2 W\frac{(b - a)}{2} \\ a = CL_1 + L_1 + CL_2, &\quad b = CL_1 + L_1 + CL_2 + L_2 \end{aligned} \tag{9.24}$$

For power law fluids the pressures P_1 and P_2 in the inner and outer cavity are calculated based on Eq. 5.5.48 in combination with the local power law approach as follows:

$$P_1 = \Delta P_1 + \Delta P_2;\ P_2 = \Delta P_2 \tag{9.25}$$

$$\Delta P_1 = L_1 2^{1+n_1}\left(2 + \frac{1}{n_1}\right)^{n_1} \frac{m_1 Q^{n_1}}{w_1{}^{1+2n_1}} \tag{9.26}$$

$$\Delta P_2 = L_2 2^{1+n_2}\left(2 + \frac{1}{n_2}\right)^{n_2} \frac{m_2 Q^{n_2}}{w_2^{1+2n_2}} \tag{9.27}$$

Remember that P_1, CL_1 and L_1 are usually a function of the coordinate x across the coating width.

It seems likely that the maximum bar deflection according to Eqs. 9.18 and 9.19 is overpredicted because the moment of inertia (Eq. 9.20) is based on the depth of the inner cavity only. Although the intrusion of the inner cavity into the die bar weakens its bending stiffness, it does not seem to be correct if the moment of inertia is calculated for a bar thickness of (T_{bar}–CD_1) because the bar material between the two cavities increases the stiffness to some extent. While this effect cannot be accounted for with the simple theory of deflection of cantilevered beams, results for determining the minimum bar thickness that keeps flowrate nonuniformities generated by bar deflection below the desired level will be on the safe side.

So far, the focus was on calculating the nonuniform bar deflection in the cross-web direction if the bar is loaded by fluid pressure. However, we are not really interested in bar deflection but rather in the width-wise change of the slot height dw(x) as a result of bar deflection. In particular, the goal is to calculate deterministic contributions to flow rate nonuniformities according to the second term on the right-hand side of Eq. 9.13. In reality, dw is not only a function of the coordinate x across the die but also of the coordinate z along the slot. In addition, the inner and outer slots typically vary in height for a dual cavity geometry. So, the question arises as to which position along the z-axis is most representative for evaluating the effect of a varying slot height on the flow rate nonuniformity. Is it in the inner or outer slot, is it at the slot exit, or at some internal location in the outer slot? The answer is not clear, but since the situation in the outer slot is more crucial to the width-wise flow nonuniformity than the one in the inner slot due to the damping capacity of the outer cavity, it is suggested to determine the width-wise variation in slot height dw(x) at the mid-point of the outer slot. Assuming linear behavior of the bar deflection with respect to z, the maximum bar deflection y_{max} as calculated by Eqs. 9.18 and 9.19 must therefore be multiplied by a factor F, which is less than 1, and which is defined as

$$F = 1 - \frac{L_2}{2L_w} \tag{9.28}$$

The maximum contribution to flow nonuniformity from bar deflection is obtained by comparing bar deflection at the die inlet, where it is maximal, and at the end of the cavity, where it is minimal. Calling this parameter $\varepsilon_{deflection}$ and referring to Fig. 9.11, the second term in Eq. 9.13 can now be calculated by

$$\epsilon_{deflection} = \left(\frac{1}{n_2} + 2\right)\frac{1}{w_2}\left[\left(F\sum_{i=1}^{4} y_{max,i}\right)\bigg|_{x=0} - \left(F\sum_{i=1}^{4} y_{max,i}\right)\bigg|_{x=W}\right] \quad (9.29)$$

As visualized by Eq. 9.20, flow nonuniformity due to bar deflection is controlled by the bar thickness T_{bar}. Therefore, Eq. 9.29 can be used to determine the minimum bar thickness that keeps deterministic flow rate nonuniformities from bar deflection below a specified maximum value. Corresponding exemplary calculations are presented below in Sect. 9.4.8 entitled *Analyzing die performance.*

So far, die deformation was only considered for a simple die consisting of just one slot. However, in simultaneous multilayer coating processes slide dies with multiple slots are used, and the issue of bar deflection due to fluid pressure gets more complicated. Consider a stack of die elements as sketched in Fig. 9.13.

Slot #2 is formed by bars B and C, both of which deflect not only due to the pressure P_2 in that slot, but also due to the pressures P_1 and P_3 in the adjacent slots, and to a lesser extent, due to pressures in slots still farther away. If all slots have identical cavity geometries, then the direction of deflection of bar C (positive or

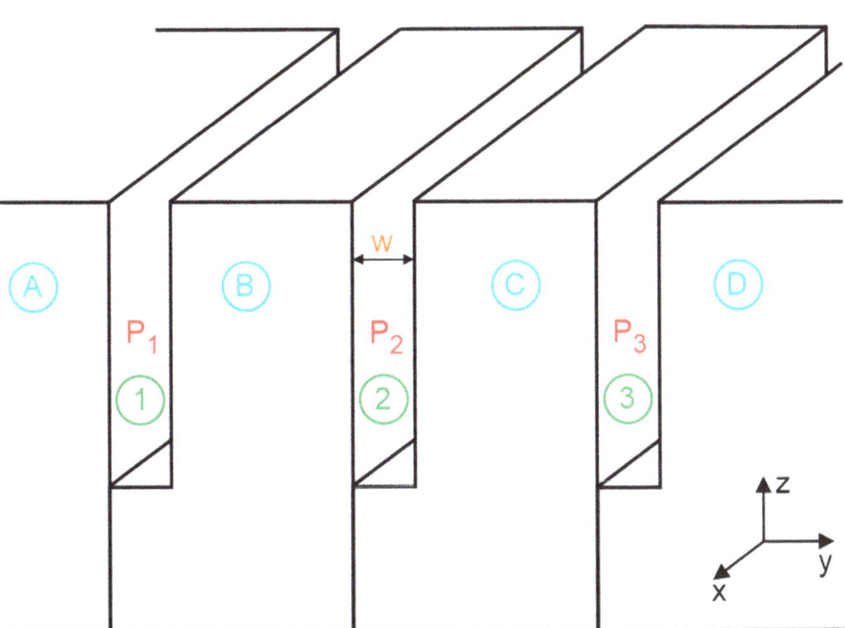

Fig. 9.13 Stack of cantilevered bars of a multilayer slide die

negative y-direction) depends on the relative magnitude of the pressures P_2 and P_3. The same is true for bar B, except that pressures P_1 and P_2 are relevant. In other words, the net load on a bar inside a multilayer slide die is the difference between the loads (pressures) on either side of the bar.

If both bars B and C deflected by different but constant amounts at any location x across the die, then the slot height w would be different from its original nominal value, but it would be constant in the x-direction for a given z-position, and the effect on the width-wise flow nonuniformity would be negligible.

Since the deflection of bars B and C depends on the relative magnitude of the pressures P_1, P_2 and P_3, which are rarely, if ever, identical in a realistic environment, variation of the slot height w in the x-direction must be expected according to one of the four following scenarios:

1	$P_2 > P_1$ and $P_2 < P_3$	Both bars bend toward the negative y-direction
2	$P_2 > P_1$ and $P_2 > P_3$	The bars bend away from each other
3	$P_2 < P_1$ and $P_2 > P_3$	Both bars bend toward the positive y-direction
4	$P_2 < P_1$ and $P_2 < P_3$	The bars bend toward each other

Now, it is obvious that the width-wise variation of the slot height dw(x) is the result of uneven deflections of two adjacent die bars, and that the net pressure acting in that slot is equal to the difference of the resulting pressure loads on each of the adjacent bars according to

$$P_{net} = (P_2 - P_1) - (P_3 - P_2) = 2P_2 - P_1 - P_3 \tag{9.30}$$

If P_{net} is negative, then adjacent bars bend toward each other. Moreover, since $dw(x/W = 0)$ is always greater than $dw(x/W = 1)$, the slot is narrower at the cavity inlet than at its end, and the resulting flow pattern is in the form of a smile or a positive wedge. In contrast, if P_{net} is positive, the bars bend away from each other, and the slot opens. It does more so at the cavity inlet than at the end, and the resulting flow pattern is in the form of a frown or a negative wedge.

The arguments above apply to any particular slot of a multilayer slot die. If P_2 corresponds to the pressure in the bottom or top slot, then P_1 or P_3 is equal to the ambient pressure, which is assumed to be zero within the scope of this model.

Calculating bar bending effects on the flow rate nonuniformity of a multilayer slide die is cumbersome. Therefore, a more pragmatic approach is to focus on the slot with the highest pressure drop and dimension the bar thickness such that $\varepsilon_{deflection}$ (Eq. 9.29) does not exceed a small value of, say, 0.5%.

The following qualitative recommendations improve the mechanical stability of a stack of die bars concerning nonuniform thermal and mechanical loads:

- Temper both the fluid flow and the clamping section of each bar by adding water holes to both sections.

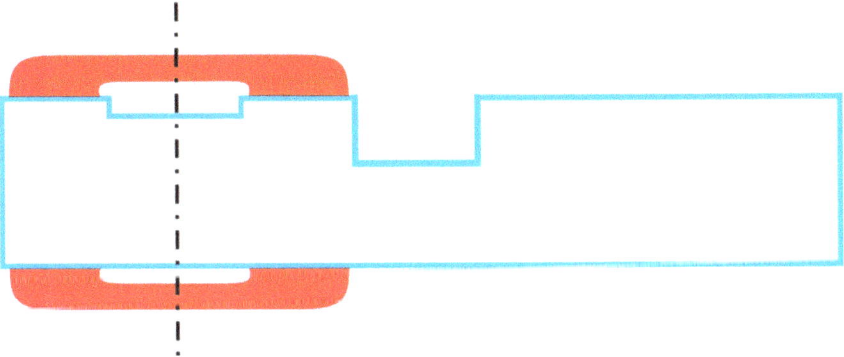

Fig. 9.14 Whiffle plates to distribute the bolt force to the clamping land areas

- Keep all die bars at the constant coating temperature at all times, i.e., during coating, cleaning, assembly/disassembly, storage, transport between coating machines, etc. This is particularly important if the coating temperature is above ambient.
- If a change in the thermal state of the die is necessary, it has to be done slowly.
- Operate the die isothermally at all times, i.e., the temperature of all coating fluids must be equal to the temperature of the die bars.
- If the clamping section is lightly grooved at the location of the bolt holes as shown in Fig. 9.14, then use whiffle plates at both ends of the bar stack to distribute the bolt force to the clamping land areas only.
- The length of the clamping section should be at least as long as the length of the fluid flow section.
- The die bar should be at least 5 cm longer on each side than the width of the cavities to minimize end effects in slot height variation caused by a local lack of symmetry (Poisson effect in response to thermal or mechanical loads). If the cavities end at side plates, then the extra bar length should be implemented in the clamping section.

9.4.7 *Die Design Procedure*

With all the information presented in Chaps. 5 and 6 or published in the scientific literature, it is now possible to define recipe-like instructions for designing and optimizing the geometry of a new die, or as the case may be, for taking the geometry of an existing die and calculating the cross profile and other performance-related properties such as the level of wall shear stress for a given set of operating conditions. The design steps outlined below are based on some 40 years of experience, first by using and designing dies for the photographic industry, and later at TSE Troller Schweizer Engineering and Polytype Converting for providing dies to industrial coating companies. These steps have proven to be viable and useful.

1. Specify several sets of operating conditions, i.e., several different products, preferably from the same product family, that must be coated with the same new die. Each set of operating conditions must include the volumetric flow rate, the density, and rheological information such as the 5 parameters of the Carreau-Yasuda model (Eq. 4.4.1). The volumetric flow rate, in turn, depends on the dry coat weight, the web speed, the coating width, the density, and the solids concentration (Eq. 3.6.8). It also depends on the by-pass factor α, which may be non-zero if one of the coating fluids is chemically reactive (Eq. 6.3.7). Keep in mind, however, that die manufacturers will not be pleased if you specify 63 applications or some such high number. Usually, 5–10 applications should suffice.
2. Out of the specified applications choose one (the design base), for which the die geometry will be optimized. This product, when coated at the specified operating conditions, will be characterized by a cross profile that is only determined by the mechanical slot precision. All other products will show a small cross profile with a smile or frown form, depending on how the various operating parameters deviate (larger or smaller) from the corresponding parameters of the design base. The choice of the design base may fall on the product with the highest annual production volume, or the product with the most demanding uniformity requirement. The choice may also fall on the product containing the most expensive raw materials, or on any other product that is selected for any other good reason in a given situation.
3. Specify a value for ε_{max}, for example 4 or $\pm 2\%$, i.e., specify a band width that must not be exceeded by any of the cross profiles produced by the specified applications, see Fig. 9.1. Remember, though, that specifying a wide range of operating conditions, where the various process parameters differ greatly from each other, and allowing only a small value for ε_{max} at the same time may not be acceptable for physical reasons because no die geometry may exist that can satisfy both requirements.
4. Specify the number of cavities and the geometrical parameters of the die, including the die width, the location of the feed port, the working length, the partition of the working length for the inner and outer distribution system (L_{w1} and L_{w2}), the height of the inner and outer slot, the shapes of the inner and outer cavity, see Fig. 9.7.
5. Specify the level of wall shear stress for the inner and outer cavities, which in turn specifies the initial size of the inner cavity and therefore the initial length of the inner slot, as well as the depth of the outer cavity. This process requires judgement from the die designer. On the one hand, specifying a high wall shear stress makes it more difficult to achieve good cross profiles for all of the specified applications. On the other hand, and as explained in Sect. 6.2, the recommended range for the wall shear stress is 1–10 Pa. For the inner cavity, therefore, it is suggested to assign a value of 5 Pa to the design base by applying the local power law approach to Miller's equation for the wall shear stress and the wall shear rate (Eqs. 5.5.5, 5.5.6 and 5.5.7). Then, the characteristic length of the cavity shape must be adjusted by hand, or better yet by iteration, until the

desired level of wall shear stress is achieved. For this geometry, the resulting wall shear stress must be calculated next for all other applications. If all stress levels fall within the recommended range, then all is fine. However, if most of the stress levels are below the level of the design base or are even below the lower bound of the recommended range, then the level of the design base can be increased a bit in an attempt to bring all stress levels into the desired range. The same procedure can be applied in the opposite direction if the stress levels of the other applications are higher than the level of the base design. The same procedure can also be applied for determining the size (depth) of the outer cavity. Here, Eqs. 6.3.19 and 6.3.20 must be considered. If this approach does not lead to an acceptable result because the various stress levels are always spread too far apart from each other, then the die designer must decide on whether to relax the wall shear stress requirements or to eliminate some of the applications from this design project. Also, the wall shear stress level in the inner and outer cavities for the design base must not be equal, as long as both values are above the minimum level of 1 Pa.

6. Specify the wall shear stress distribution in the inner cavity by specifying the ratio of the characteristic cavity dimension at the end and the entrance of the cavity (Eq. 6.3.9), and by setting the desired value for the parameter Φ, which defines the location along the cavity where the wall shear stress falls below its initial value, see Sect. 6.3. This process requires assigning values to the parameters a, b, and c, which determine the value of the taper index φ (Eq. 6.3.13). φ in turn determines the taper function of the inner cavity cross-section (Eq. 6.3.11). This process requires a bit of experience as no clear rules are known for choosing the best values for the parameters a, b and c, but helpful guidelines are discussed in Sect. 6.3.1.
7. Calculate the maximum pressure inside the die, which is located at the entrance of the inner cavity, by calculating the pressure drops across the inner and outer slot. This involves applying the local power law model to Eq. 5.6.48 for the pressure drop and using Eq. 5.6.51 for the shear rate and Eq. 5.6.53 for the equivalent viscosity.
8. Calculate the pressure drop along the inner cavity with Eq. 5.5.4 for accounting for inertial contributions, and by applying the local power law model to Eq. 5.5.2 for the viscous contributions. The latter task requires iteration, which, in EXCEL, can conveniently be accomplished with the "goal-seek" function. Note, however, that the iteration must be carried out at every calculating step along the cavity because the values of many relevant parameters change continuously. To avoid excessive work by initiating each iteration by hand, it is suggested to modify the underlying macro routine for the goal-seek function. Specifically, the three commands of the routine should be copied as many times as there are calculating steps, and the cell numbers should then be adjusted by hand. As the pressure profile along the cavity axis typically is free of steep gradients, some 50 calculating steps should suffice, perhaps with a slightly finer mesh toward the cavity end. Calculating the pressure drop requires knowledge

about the viscous and inertia shape factors, and that information is provided in Sects. 5.5.1 and 5.5.2 for various cavity shapes.

9. The information gained from calculating the pressure profile in the inner cavity now allows the correct wall shear stress distribution to be calculated as a function of the power law index, which varies slightly along the cavity axis, particularly toward the very end of the cavity, see Eq. 6.3.12.
10. For a dual-cavity design, calculate the length of the inner slot by solving Eq. 5.6.48 for L_I such that the pressure in the inner cavity minus the pressure drop across the outer slot equals the pressure drop across the inner slot according to

$$L_1(x) = \frac{\left(P_{cavity}(x) - \Delta P_2\right) w_1{}^{(1+2n_1)}}{2^{(1+n_1)}\left(2 + \frac{1}{n_1}\right)^{n_1} m_1 Q^{n_1}} \tag{9.31}$$

This die geometry generates a flat cross profile in terms of deterministic contributions for the base operating conditions as long as the die is operated isothermally, and as long as mechanical slot deformations from internal fluid pressure are negligible.

11. Calculate the damping factor DF of the outer cavity by using Eq. 5.5.14.
12. Use Eq. 9.14 for calculating the band width of the cross profile due to stochastic contributions related to the mechanical precision of the die slots.
13. Save the so obtained die geometry, which was optimized for the base operating conditions, in a new file and do not change it anymore.
14. Calculate the resulting cross profiles for the other specified operating conditions.
15. Check if all the cross profiles are within the initially specified value for ε_{max}. If not, decide whether to relax the wall shear stress requirement, whether to drop one or the other application from the initially specified range, or whether to change specific geometric parameters such as the height of the inner and/or outer slot.

9.4.8 *Analyzing Die Performance*

Using the procedure described in the previous chapter, we wrote a die design program in EXCEL. This program serves well not only for designing an optimized die for a given set of operating conditions, but also for visualizing the performance of that die when the operating conditions are changed, and that is the purpose of this chapter.

The program is contained in a single file, but it has a structure by using different sheets for carrying out the various tasks, such as defining different sets of operating conditions, specifying geometrical parameters, calculating the initial size of the inner cavity that results in a desired wall shear stress, specifying a desired wall shear stress distribution in the inner cavity, calculating the maximum pressure at the die inlet, specifying the size of the outer cavity that results in a desired wall shear stress and calculating the resulting damping factor, and calculating the pressure drop in the inner cavity as well as the length profile of the inner slot that results in a perfect flow distribution for the base operating conditions.

The comfort of working with the program was enhanced by writing macros for carrying out the various iterations with the "goal seek" function. Moreover, using the "lookup" function allows the program to work well for various cavity shapes and various sets of operating conditions.

To demonstrate the die performance, we constructed 13 exemplary sets of operating conditions as listed in Table 9.1. Specifically, we varied the volumetric flow rate/width Q_0 and all 5 parameters of the Carreau-Yasuda model. In contrast, the through-flow parameter α, the density ρ were kept constant.

Set number 1 was used as the base conditions for optimizing the die geometry. Moreover, the various geometric parameters and wall shear stress requirements were chosen as follows:

- Maximum coating width: $W_{coating} = 1500$ mm.
- Number of cavities: 2.
- Feed location: center.
- Length of cavity: $W = 750$ mm.
- Shape of inner cavity: half square + half-circle, i.e., square with $R_c = a/2$.
- Initial size of inner cavity such that the initial wall shear stress $\tau_0 = 5$ Pa.
- Distance, at which the wall shear stress in the inner cavity drops below the initial value: $\Phi = 0.85$.
- Ratio of end-to-initial cross-section of the inner cavity: $L_{c,e}/L_{c,0} = 0.3$.

Table 9.1 List of thirteen sets of operating conditions

Number	V^*_0	α	Q_0	ρ_{liquid}	μ_0	μ_∞	γ_c	a	b
	[dm^3/min]	[–]	[cm^2/s]	[kg/m^3]	[mPas]	[mPas]	[s^{-1}]	[–]	[–]
1	4.5	0	1	1000	1000	10	0.1	0.7	0.6
2	2.25	0	0.5	1000	1000	10	0.1	0.7	0.6
3	13.5	0	3	1000	1000	10	0.1	0.7	0.6
4	4.5	0	1	1000	1000	10	0.1	0.7	0.3
5	4.5	0	1	1000	1000	10	0.1	0.7	0.9
6	4.5	0	1	1000	100	10	0.1	0.7	0.6
7	4.5	0	1	1000	5000	10	0.1	0.7	0.6
8	4.5	0	1	1000	1000	10	1	0.7	0.6
9	4.5	0	1	1000	1000	10	10	0.7	0.6
10	4.5	0	1	1000	1000	10	0.01	0.7	0.6
11	4.5	0	1	1000	1000	1	0.1	0.7	0.6
12	4.5	0	1	1000	1000	100	0.1	0.7	0.6
13	4.5	0	1	1000	1000	10	0.1	2.0	0.6

- Height of die plate: $L_{bar} = 120$ mm.
- Thickness of die plate: $T_{bar} = 40$ mm.
- Total working length: $L_w = 80$ mm.
- Ratio of inner to outer working length $v = L_{w1}/L_{w2} = 1.0$.
- Length of inner working length: $L_{w1} = 40$ mm.
- Length of outer working length: $L_{w2} = 40$ mm.
- Height of inner slot: $w_1 = 250$ μm.
- Height of outer slot: $w_2 = 400$ μm.
- Shape of outer cavity: 45–45 right-angle triangle with a rounded vertex.
- Length of outer cavity: CL = 30 mm.
- Radius of vertex in outer cavity such that the resulting minimum wall shear stress $\tau_{min} = 1.5$ Pa.
- Mechanical nonuniformity of inner slot: $dw_1 = \pm 1$ μm.
- Mechanical nonuniformity of outer slot: $dw_2 = \pm 1$ μm.
- Treatment of mechanical nonuniformity: stochastic.

The resulting, optimized die geometry for the base conditions is shown in Fig. 9.15.

The widthwise change in the length of the inner die slot is very minimal. In fact, the length changes from 18.88 mm at the cavity inlet to 18.50 mm at the cavity end. In contrast, the size of the inner cavity cross-section changes strongly across the width. Figure 9.16 shows the profile of the square side *a* of the cavity form. This information can be used for programming the milling machine that carves the cavity during manufacturing.

The initial square side of 21.117 mm was determined such that the initial wall shear stress equals 5.0 Pa. The taper index φ for the inner cavity cross-section varied across the width according to the constants a = 0.70, b = 0.11, c = 5.30. In addition,

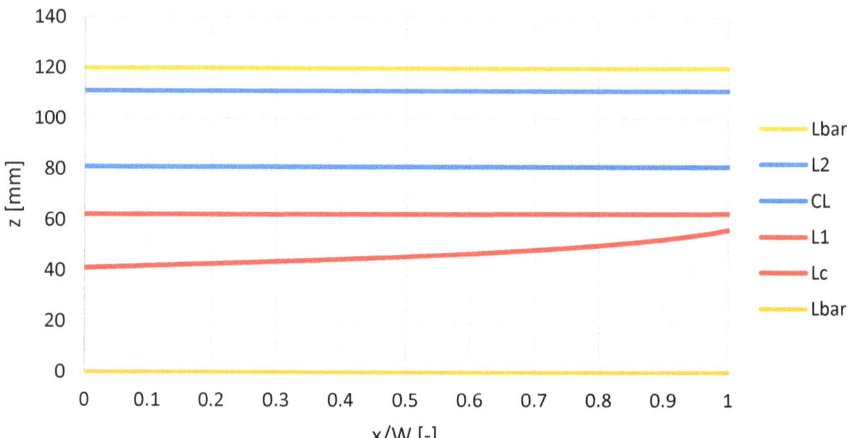

Fig. 9.15 Optimized cavity geometry; the die bar is marked in yellow, the inner cavity in red, and the outer cavity in blue

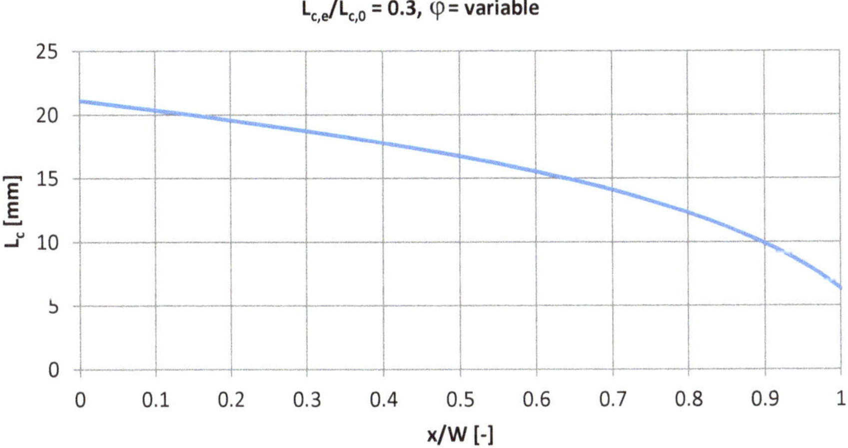

Fig. 9.16 Widthwise profile of the square side a of the inner cavity shape

the radius of the vertex of the outer cavity was determined to be 17.374 mm, thus resulting in a minimum wall shear stress of 1.5 Pa.

Figure 9.17 visualizes the normalized wall shear stress distribution in the inner cavity. The shear stress remains nicely constant until it drops below its initial value at x/W = Φ = 0.85 as requested. Note that this Φ-value does not change, even if the operating conditions and rheological fluid properties are changed later during the operation of the die.

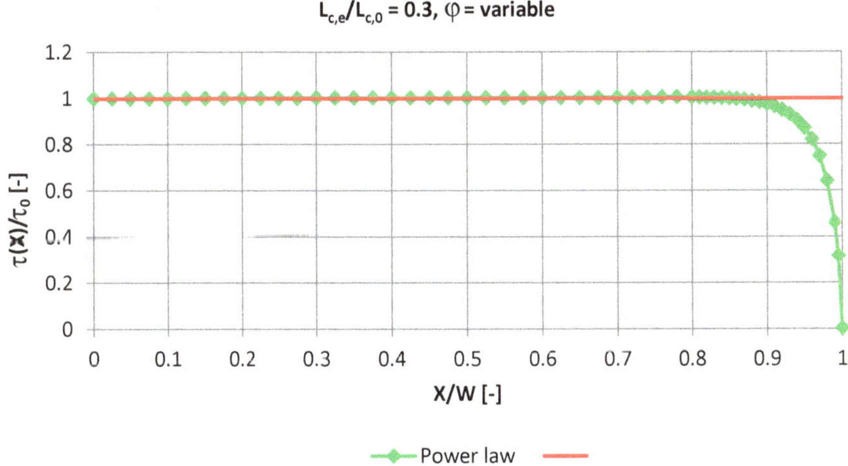

Fig. 9.17 Normalized wall shear stress distribution in the inner cavity

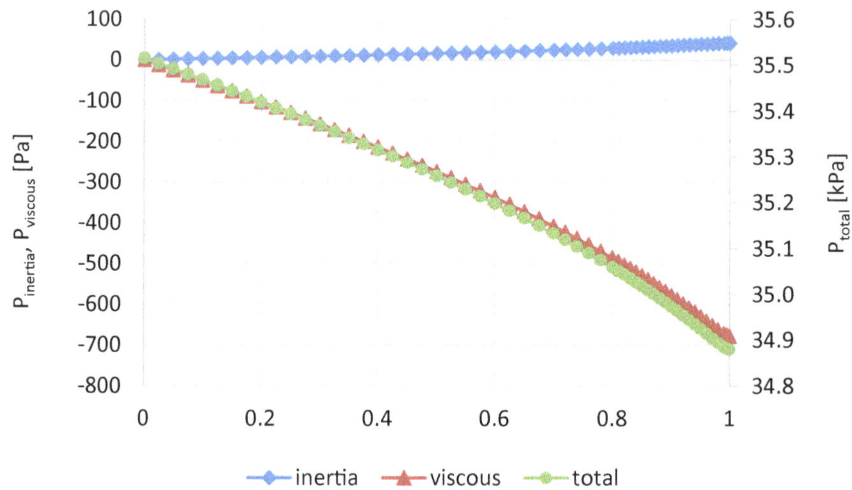

Fig. 9.18 Viscous, inertia (primary vertical axis) and total pressure drop (secondary vertical axis) along the inner cavity

For the base conditions, the maximum fluid pressure at the cavity inlet was calculated to be 35.5 kPa, and the total pressure drop along the inner cavity amounted to 635 Pa. As depicted in Fig. 9.18, this total pressure is composed of a viscous component, which decreases along the cavity, and of an inertia component, which slightly increases. For the base conditions, the inertia pressure drop is much smaller in magnitude than the viscous contribution, but the former contribution increases with increasing Reynolds number.

The Reynolds number as calculated with Eq. 5.5.7 amounted to 15.51 at the entrance to the inner cavity. This parameter decreases to zero in a non-linear fashion along the cavity, all in accordance with the volumetric flow rate, which decreases linearly to zero at the end of the cavity.

Results of the die performance when running the operating conditions 2–13 through this optimized die are summarized in Table 9.2. Specifically, the inner cavity was designed for an initial wall shear stress of $\tau_0 = 5.0$ Pa, and it ranges from 0.88 to 26.3 Pa for the other operating conditions. While this spread is wide, the lowest value is closed to the lower bound of the recommended range of 1–10 Pa, so overall, this result is quite acceptable. Similarly, the outer cavity was designed for a minimum wall shear stress of $\tau_{min} = 1.5$ Pa, and the other conditions resulted in wall shear stresses of 0.21–7.22 Pa. Here, the lowest value is rather low, but at least not negative. Nevertheless, it might be appropriate to require a minimum wall shear stress of 2.0 Pa and to check if the resulting increase of the flow nonuniformities for conditions 2–13 is still acceptable.

The initial value of the power law index n_l in the inner cavity as calculated with the local power law model is close to the specified Carreau-Yasuda parameter b of the 13 operating conditions, but it varies in response to the variation of the other

Table 9.2 Performance data of 13 different operating conditions

#	τ_0	m_1	n_1	τ_{min}	m_2	n_2	P_{max}	ΔP_{cavity}	Re_0	β_0	β_1	DF	ε_{max}
	[Pa]	[Pas^n]	[–]	[Pa]	[Pas^n]	[–]	[kPa]	[Pa]	[–]	[–]	[–]	[–]	[–]
1	5.00	0.37	0.644	1.50	0.375	0.639	35.5	635	15.5	0	0	2.32	2.01
2	3.21	0.375	0.639	0.96	0.372	0.644	20.9	424	6.0	−0.39	−0.16	2.37	2.19
3	10.21	0.35	0.657	3.03	0.372	0.642	85.8	1,005	70.1	0.88	0.39	2.28	2.37
4	1.26	0.122	0.566	0.47	0.165	0.464	16.7	130	53.6	1.24	0.93	1.33	3.30
5	26.30	0.789	0.903	5.39	0.784	0.906	542.8	3,482	4.7	1.5	0.15	9.96	1.72
6	0.88	0.036	0.815	0.21	0.041	0.771	17.8	75	119.7	1.67	0.41	4.08	2.05
7	23.31	1.907	0.615	7.22	1.873	0.621	114.0	3,117	3.2	−1.77	−0.82	2.16	3.02
8	11.43	0.834	0.647	3.32	0.762	0.681	64.9	1,525	6.8	−0.98	−0.35	2.80	2.34
9	23.94	1.319	0.728	5.93	1.038	0.814	137.9	3,194	3.8	−1.04	−0.18	5.70	2.03
10	2.29	0.148	0.683	0.66	0.159	0.656	23.8	267	36.4	0.91	0.37	2.48	2.26
11	4.62	0.383	0.611	1.44	0.375	0.619	21.2	584	15.8	−1.91	−0.89	2.14	3.14
12	8.79	0.358	0.815	2.12	0.405	0.771	177.8	1,140	12.0	1.41	0.32	4.46	1.94
13	5.04	0.382	0.637	1.54	0.398	0.623	35.5	641	15.2	−0.03	−0.01	2.17	2.07

parameters specified in Table 9.1. This indicates that the local power law model is better to account for the effects of other parameters than the classical power law model. Moreover, as a consequence of requiring a constant wall shear stress for most of the cavity length (85% in this example), the power law index n_1 as well as the consistency m_1 also remain constant. Indeed, these two parameters do not vary by more than 1% over at least 92% of the cavity length for the 13 sets of specified operating conditions. Similar results can be deduced for n_2 in the outer cavity. Here, widthwise variations of the rheological parameters are insignificant because the flow conditions hardly vary along the cavity for well-designed dies.

Damping factors range from 1.33 to 9.96. Higher numbers could be achieved by relaxing the requirement for a minimum wall shear stress, which, for this example, is not recommended as explained above. The overall die performance with regard to the cross profile is measured by the parameter ε_{max} (see Fig. 9.4), which ranges from 1.72 to 3.30%. Figure 9.19 shows two exemplary cross profiles for conditions 3 and 7. Condition 3 produces a smile profile, whereby contributions from fluid flow effects amount to $\beta_1 = 0.39\%$, and the overall profile width accounting for fluid flow and mechanical slot nonuniformity effects is $\varepsilon_{max} = 2.37\%$. In contrast, condition 7 produces a frown profile, whereby contributions from fluid flow effects amount to $\beta_1 = -0.82\%$ and the overall profile width is $\varepsilon_{max} = 3.02\%$. For both profiles, the contributions from fluid flow effects are smaller than those associated with the mechanical precision of the die slots. However, if, for this exemplary project, an overall profile width of $\varepsilon_{max} = 3.0\%$ is acceptable, then both these profile forms are also acceptable because they meet the overall requirement.

Regarding the form of the cross profile, frown shapes result if the initial flow rate in the inner cavity and the high shear viscosity are lower, and the low shear viscosity, the critical shear rate as well as the Yasuda parameter *a* are higher than the

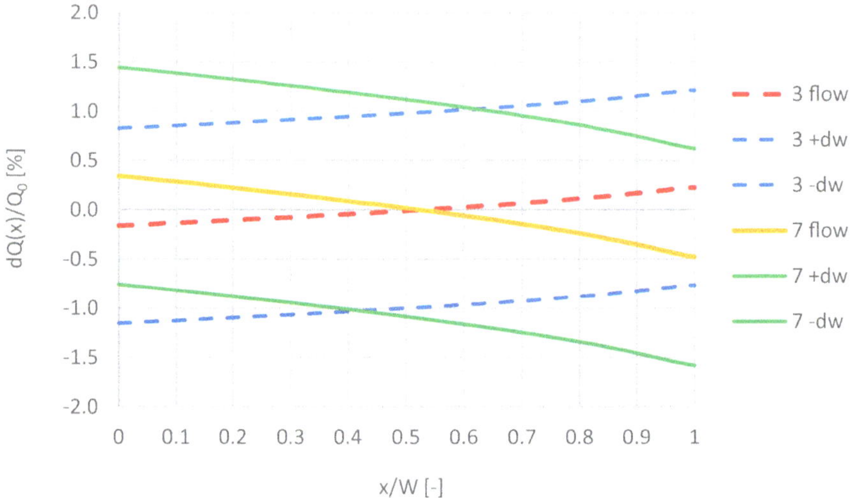

Fig. 9.19 Cross profile examples for operating conditions 3 and 7

respective base values. In contrast, smile profiles are generated if these parameters are on the other side of the base values. However, increasing or decreasing the Carreau parameter b relative to the base value does not change the profile form. Remember that parameter b is a measure for the shear-thinning behavior of the fluid.

Figure 9.20 is an attempt to meaningfully visualize the quantitative effects of the volumetric flow rate/width and the 5 Carreau-Yasuda parameters on the total flow

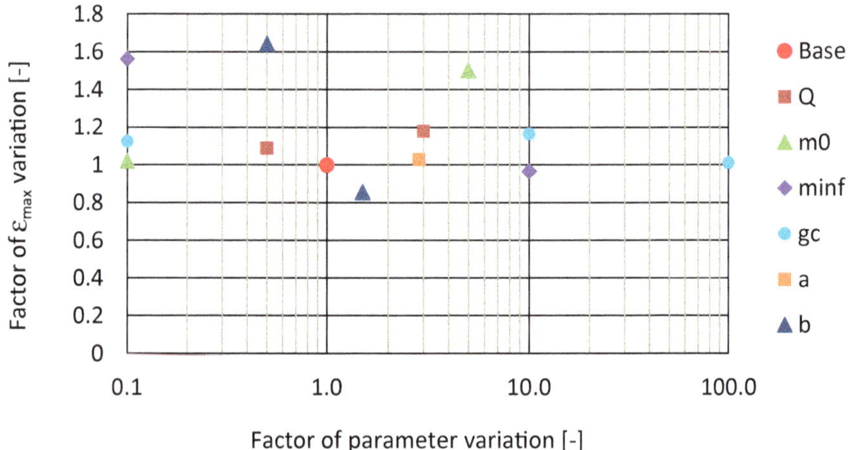

Fig. 9.20 Total flow rate nonuniformity as a function of variation of the 6 independent operating parameters; Q = volumetric flow rate/width, μ_0 = low shear viscosity, μ_{inf} = high shear viscosity, γ_c = critical shear rate, a = Yasuda parameter, b = Carreau parameter related to shear thinning

rate nonuniformity. The x-axis shows the factor, by which each of the 6 independent parameters was varied relative to its base value, and the y-axis shows the factor, by which the resulting ε_{max} value varied relative to its base value. The base conditions are marked with a red circle. If, for example, the parameter b was varied by a factor of 1.5, i.e., from 0.6 to 0.9, then ε_{max} decreased by a factor of 0.86, i.e., from 2.01 to 1.72%. However, when the critical shear rate was varied by a factor of 100, ε_{max} increased by a negligible factor of 1.01.

Nine of the twelve parameter variations are closely located to the base value, which is desirable. Specifically, they result in less than about 20% variation of the total flow rate nonuniformity. However, only two of these points, i.e., when the high shear viscosity and the parameter b are higher than the respective base value, result in a lower total flow nonuniformity than the base case. In contrast, three of the operating points are located far away from the base case, i.e., when the low shear viscosity is much higher or the high shear viscosity and the parameter b is (much) lower than the base values. Such situations are detrimental and should be avoided because they result in ε_{max} values that are about 40–50% higher than the base case. Changing the Yasuda parameter *a* has no significant effect on the flow rate nonuniformity.

The **trade-off between obtaining a high flow rate uniformity and a high initial wall shear stress** is illustrated in Fig. 9.21 for the **inner cavity**.

Designing a new die for a lower or a higher initial wall shear stress in the inner cavity results in a lower or higher pressure drop along the inner cavity, but this effect is compensated by the associated length profile of the inner slot that gives a perfect flow uniformity for the base conditions. While the total cross profile as expressed by ε_{max} increases with increasing wall shear stress, the differences between operating conditions are relatively small. Specifically, when increasing the wall shear stress

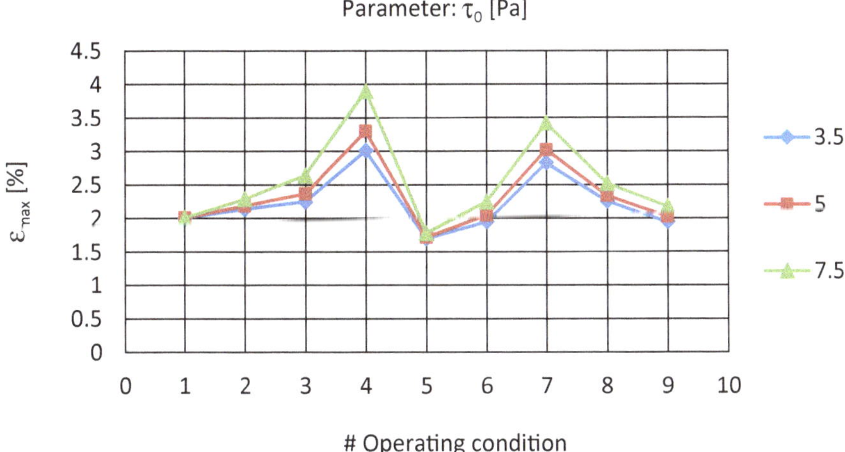

Fig. 9.21 Total flow rate nonuniformity as a function of the initial wall shear stress in the inner cavity for the operating conditions 1–9

by 1.5 Pa from 3.5 to 5.0 Pa, the increase in ε_{max} varies from 1.2 to 8.8% with an average increase of 4.0%, and when increasing the wall shear stress by 2.5 Pa from 5.0 to 7.5 Pa, the increase in ε_{max} varies from 3.4 to 15.4% with an average increase of 7.6%. In addition, increasing the wall shear stress from 3.5 to 5.0 Pa increases the maximum pressure at the die inlet by an average of 20.0%, and a shear stress increase from 5.0 to 7.5 Pa results in an average pressure increase of 15.9%. In other words, increasing the wall shear stress in the inner cavity increases the operating costs of the coating process because the pump has to work harder to overcome the higher pressure drop across the die.

Using the base operating conditions in Table 9.1 for calculating the pressure drop in the inner cavity as a function of Φ and $L_{c,e}/L_{c,0}$ revealed that ΔP_{cavity} increases with decreasing ratio $L_{c,e}/L_{c,0}$, and it increases only slightly with increasing Φ, at least up to a "higher value" that depends on $L_{c,e}/L_{c,0}$, see Fig. 9.22. Beyond this "higher" Φ-value, the pressure drop increases sharply until a maximum Φ-value is reached that also depends on $L_{c,e}/L_{c,0}$. In particular, a practical maximum Φ-value that prevents excessive pressure drops in the inner cavity is about 0.85 for $L_{c,e}/L_{c,0} = 0.4$, 0.90 for $L_{c,e}/L_{c,0} = 0.3$, and 0.95 for $L_{c,e}/L_{c,0} = 0.2$. These findings are important because they confirm that high values for Φ and low values for $L_{c,e}/L_{c,0}$, which are necessary for obtaining a desirable wall shear stress behavior in the inner cavity, can be used without interfering much with the resulting cross profile. Therefore, the combinations of $L_{c,e}/L_{c,0} = 0.3$ with $\Phi = 0.85 - 0.90$ for non-reactive fluids, and $L_{c,e}/L_{c,0} = 0.2$ with $\Phi = 0.90 - 0.95$ for reactive fluids will result in a pleasing wall shear stress behavior. As already mentioned above, the main price that must be paid for this result is the higher pressure drop across the die.

Increasing the minimum wall shear stress in the **outer cavity** by increasing the radius of the vertex, i.e., by reducing the depth of the cavity, always decreases the

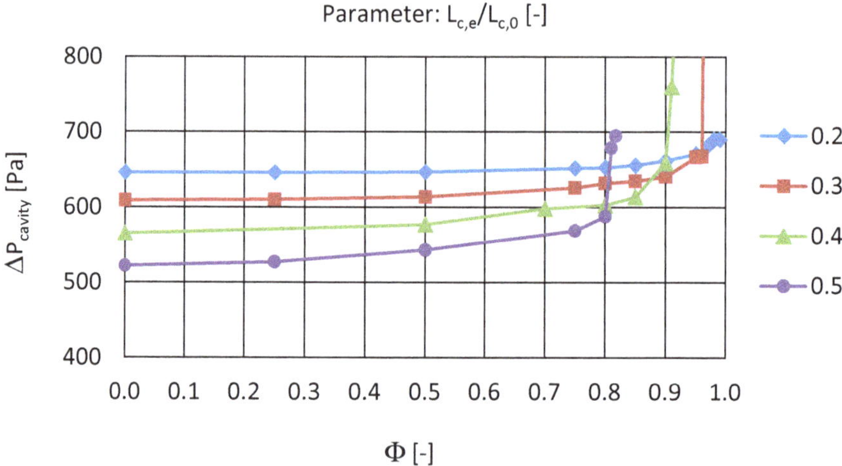

Fig. 9.22 Pressure drop in the inner cavity as a function of Φ and $L_{c,e}/L_{c,0}$

damping factor. The magnitude of this effect depends on different parameters, and it is strongest by

- Increasing the cavity length CL, see Fig. 9.23.
- Decreasing or increasing the expansion angle of the triangle relative to $\alpha = 45°$; the latter angle always results in the highest damping capacity.
- Reducing the volumetric flow rate/width.
- Reducing the low shear viscosity, and in particular, by reducing the power law index, i.e., the parameter b in the Carreau-Yasuda equation, see Fig. 9.24.

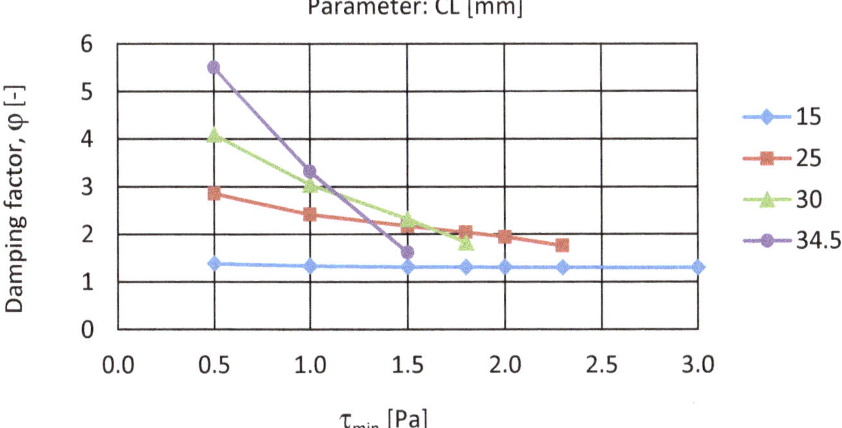

Fig. 9.23 Reduction of the damping capacity of the outer cavity by increasing the minimum wall shear stress as a function of the cavity length CL

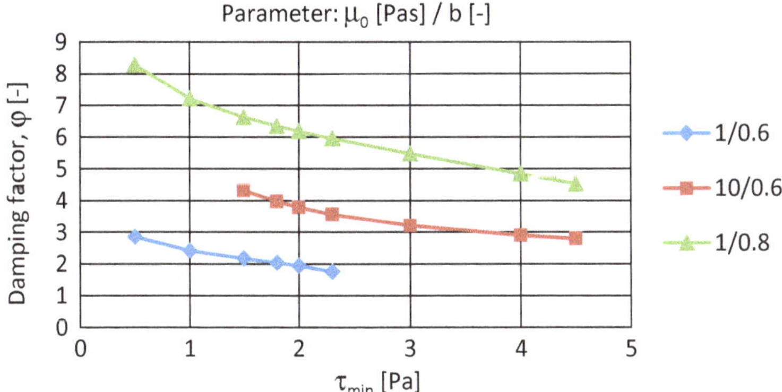

Fig. 9.24 Reduction of the damping capacity of the outer cavity by increasing the minimum wall shear stress as a function of the low shear viscosity μ_0 and the power law index, i.e., the parameter b in the Carreau-Yasuda equation

9.4.8.1 Maximum Versus Minimum Coating Width

Every new die must always be designed for the maximum intended coating width. In order to assess the effect of a reduced coating width on the wall shear stress distribution in the inner cavity and on the widthwise flow rate uniformity, the cross-section of the initial cavity at the reduced width must be determined as a first step. This cross-section becomes the end cross-section of the new cavity, and it determines the new value of the ratio $L_{c,e}/L_{c,0}$. The three constants a, b, and c of the taper index φ remain unchanged, but the parameter β changes as per the new value for $L_{c,e}$. Together with the new, reduced coating width (cavity length), the resulting initial wall shear stress and the wall shear stress distribution as well as the pressure drop in the inner cavity can now be calculated. The initial wall shear stress will be lower because the initial flow rate will be lower consistent with the reduced coating width while the initial cavity cross-section remains unchanged.

To quantify the effects of a width reduction, we take the die that was designed for set 1 of the operating conditions in Table 9.1, and we reduce the coating width from 1,500 to 1,000 mm. The initial wall shear stress in the inner cavity drops from 5.0 to 3.85 Pa, i.e., it is only 77% of the value associated with the maximum coating width. Moreover, as illustrated in Fig. 9.25, the wall shear stress drops below the initial value right from the beginning of the cavity, which is not attractive. The deficit of wall shear stress in the shorter cavity is considerable, particularly in the second half of the cavity.

Regarding the flow rate uniformity, the difference between the two cavities is much smaller. On the one hand, the nonuniformity generated by flow in the inner cavity is, when averaged over the 9 sets of operating conditions, about 46% higher

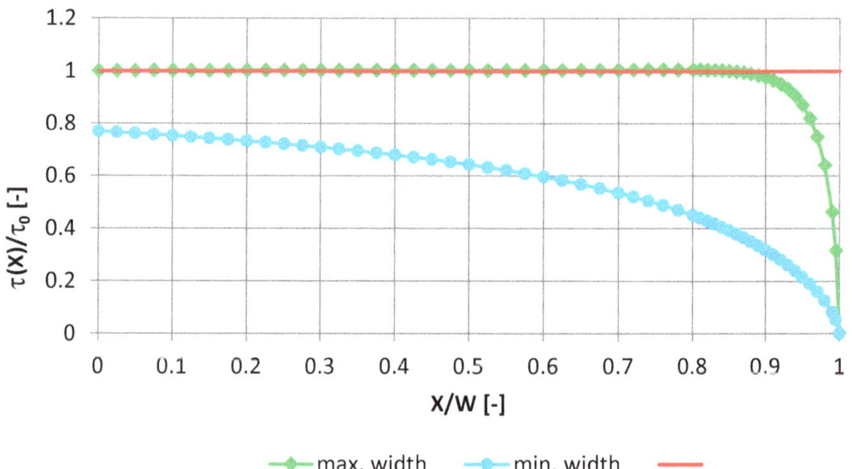

Fig. 9.25 Comparison of wall shear stress distribution in an inner cavity for maximum and minimum coating width

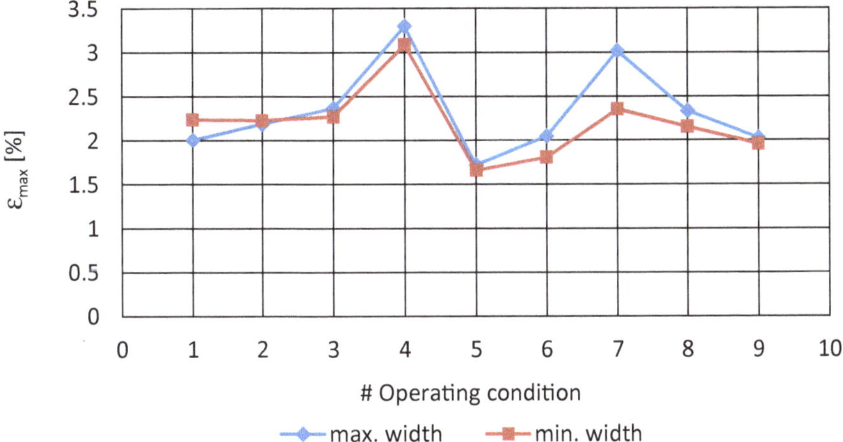

Fig. 9.26 Comparison of the total flow nonuniformity ε_{max} between a 1,500 mm and a 1,000 mm wide die for nine different operating conditions

in the shorter cavity than in the longer. On the other hand, the damping factor of the narrower die is on average about 43% higher than the factor of the wider die. As illustrated in Fig. 9.26, therefore, these two effects cancel each other in such a way that the total flow nonuniformity as measured by ε_{max} does not vary much over the nine sets of operating conditions. In addition, all profiles from the narrower die are of the smile type.

This result is surprising because changing the coating width does not change the flow in the outer cavity, and hence the damping capacity of the outer cavity could be expected to remain unchanged. However, the damping factor DF (Eq. 5.5.14) is inversely proportional to the geometric parameter s (Eq. 5.5.15), which in turn is proportional to the square of the cavity length. This then explains why the damping factor increases upon reducing the coating width.

9.4.8.2 Single Versus Dual Cavity Design

According to Eq. 9.6, increasing the length of the die slot improves the cross profile for a specified range of applications. In practice, however, this concept cannot be implemented at will because longer die slots increase the size and weight of the die, which makes die handling more cumbersome, and space limitations on the coating station set bounds on the die size. Furthermore, die manufacturers prefer to work with a standard size for the die plates, at least for a specified range of coating widths, e.g., 1,000–2,000 mm, to keep raw material costs low and logistics aspects manageable. Using the dual cavity design concept (Fig. 9.7) is a way to achieve good cross profiles without excessively long die slots. Therefore, it is interesting to compare the two design approaches regarding the resulting flow rate non-uniformities for the operating conditions 1–9 defined in Table 9.1.

In particular, the dual cavity die is the one in our example that was optimized for operating condition 1. The single cavity die was also optimized for the same set of operating conditions. In addition, the total working length $L_w = 80$ mm and the initial wall shear stress $\tau_0 = 5$ Pa were kept the same as for the dual cavity version. An attractive feature of the dual cavity design is the fact that the inner slot can be made narrower to obtain a good cross profile, while the outer slot can be made wider to reduce the sensitivity to the mechanical imprecision of the die slot. As this feature is not available for the single cavity design, the performance of the dual cavity die can be compared to single cavity dies having different nominal slot heights. Results are shown in Fig. 9.27.

Focusing first on the single cavity dies, calculations reveal that increasing the nominal slot height increases β_1, i.e., the contributions from fluid flow, but it reduces the effect of the mechanical slot precision on the total flow nonuniformity. Specifically, for operating conditions 1, 2, 3 and 7, 8, 9, mechanical slot precision dominates effects from fluid flow, and the total flow nonuniformity decreases with increasing slot height. In contrast, for conditions 4, 5, 6, fluid flow effects dominate the mechanical precision, if the slot height is >300 μm, and the total nonuniformity starts increasing again with increasing slot height. For conditions 4 and 5, parameter b in the Carreau-Yasuda equation differs significantly from the base value, and for conditions 6 (and also 7) it is the low shear viscosity that is considerably different from the base value.

Looking at the dual cavity design, most of the ε_{max} values are lower than those of the single cavity design, thus indicating that a dual cavity die performs better than a single-cavity die for a given working length and a prescribed wall shear stress in the inner cavity. However, conditions 4 and 7 are exceptions to these findings because

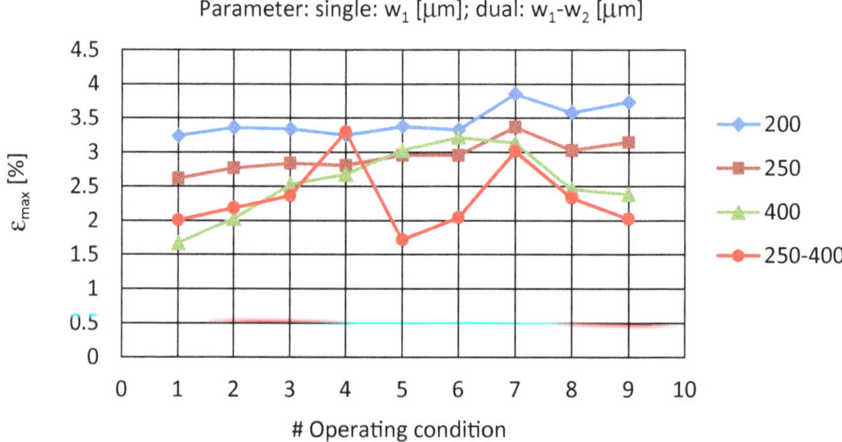

Fig. 9.27 Comparison of total flow nonuniformity ε_{max} for single cavity dies with different slot height and dual-cavity die for operating conditions 1–9

fluid flow contributions are much higher than for all other conditions, and because, at the same time, the damping factor is lower, thus resulting in high ε_{max} values. For condition 4, parameter b is much lower and for condition 7 the low shear viscosity is much higher than the respective base values.

In summary, dual cavity designs tend to produce better total flow uniformities than single cavity dies. Exceptions are related to large differences of rheological parameters relative to the base values, specifically the low shear viscosity and the parameter b, which is indicative of the shear-thinning behavior of the fluid. Moreover, the slot height for single cavity dies should not be too narrow nor too wide, with the best values being in the range of 300–400 μm.

9.4.8.3 Center Versus Side Feed

Using a side-feed instead of a center-feed is equivalent to doubling the die width. Comparing the two feed locations addresses the question of how the die width has to be scaled if the cross profile and the wall shear stress in the inner and outer cavities should not change upon increasing the die width. Qualitatively speaking and relative to a center-fed die, the initial flow rate in the inner cavity of an end-fed die is twice as high and the pressure drop along the cavity is higher. Therefore, the working length L_{w1} of the inner system must be increased to increase the length of the inner slot L_1 in such a way that the flow rate nonuniformity β_0 generated by the inner slot does not change by much. In addition, doubling the die width reduces the damping factor of the outer cavity as is explained in the Sect. 9.4.8.1 above. On the other hand, assuming that the inner and outer working lengths are equal, as is done in our example, then the outer working length must be increased as well, which in turn increases the damping factor. For reasons of simplicity, we kept the length of the outer cavity constant at 30 mm, as was done for the center-fed die. To obtain a quantitative insight into this matter, we took as a baseline the die that was optimized for operating conditions 1, for a center feed, and a coating width of 1,500 mm. We then designed and optimized a side-fed die for the same operating conditions and coating width, i.e., by doubling the cavity length, while maintaining the wall shear stress requirements in the inner (5 Pa) and outer (1.5 Pa) cavity. Next, we varied the working length L_w of the side-fed die to find that value that resulted in the same cross profiles for various operating conditions. Results of such calculations are shown in Fig. 9.28.

If only the cavity length is increased (doubled) without changing any other geometrical parameters, then the flow rate nonuniformity increases significantly for all operating conditions when compared to the center-fed die, which is marked by a red line in the figure. However, ε_{max} decreases steadily if the total working length L_w of the side-fed die is increased. For the geometric parameters of the example, a working length of about 130 mm results in approximately the same flow rate nonuniformity as for the center-fed die for all operating conditions. However, increasing the working length increases the length of both the inner and outer die slot, which in turn increases the maximum pressure inside the die. Specifically, increasing the working length from 80 to 120 mm increases the maximum pressure by a factor of about 1.3,

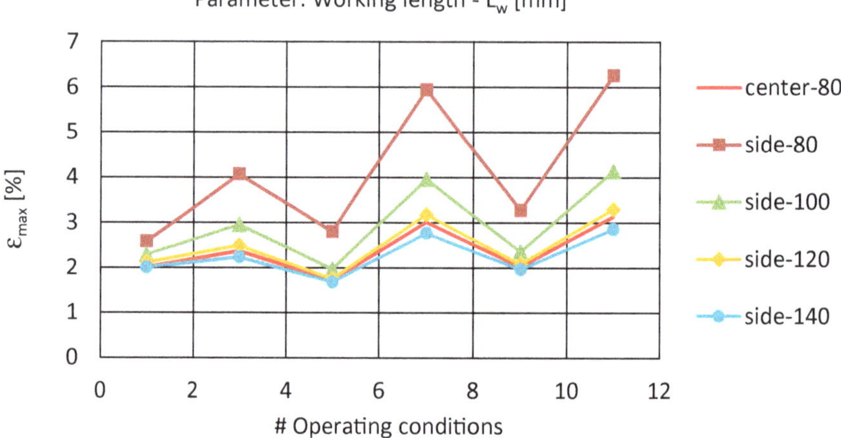

Fig. 9.28 Comparison of ε_{max} between a center-fed die and side-fed die of different working lengths for different operating conditions

i.e., from 85.6 to 197.3 kPa for operating condition 7, which is significant. Moreover, the increased pressure and the increased working length in side-fed dies increases bar deflection, which must be counter-acted by thicker die bars, and that, together with the longer working length, increases the weight of the die and decreases its ease of handling.

Given all these consequences, a center-fed die is preferred over a side-feed design, provided that inlet effects associated with a center inlet port can be suppressed, see Sect. 5.5.1.4. Ruschak and Weinstein (2018) also favor center-fed over side-fed dies. They argue that a die should be symmetric regarding its center in all respects to maximize the damping capacity of the outer cavity.

9.4.8.4 Bar Deflection

As described in Sect. 9.4.6, bar deflection is driven by the fluid pressure inside the die, and it varies not only along the cavity axis but also along the inner and outer die slots. The highest bar deflection is observed at the exit of the outer slot at the location of the cavity inlet. As a compromise, furthermore, we suggest observing bar deflection not at the exit of the die slot, but in the middle of the outer slot length. Figure 9.29 shows an example of the widthwise deformation at the midpoint of the outer slot. Specifically, Δw_2, which is the difference of the outer slot opening between the beginning and the end of the cavity, is plotted as a function of the distance along the cavity. The profile is valid for a 1,500 mm wide side-fed, dual-cavity die with a working length of 80 mm, a bar thickness of 40 mm, and for operating conditions 1 as presented in the previous Section. The slot deformation is generated by a maximum pressure of 26.5 kPa. Shetty et al. (2012) suggests ensuring that the effect of bar

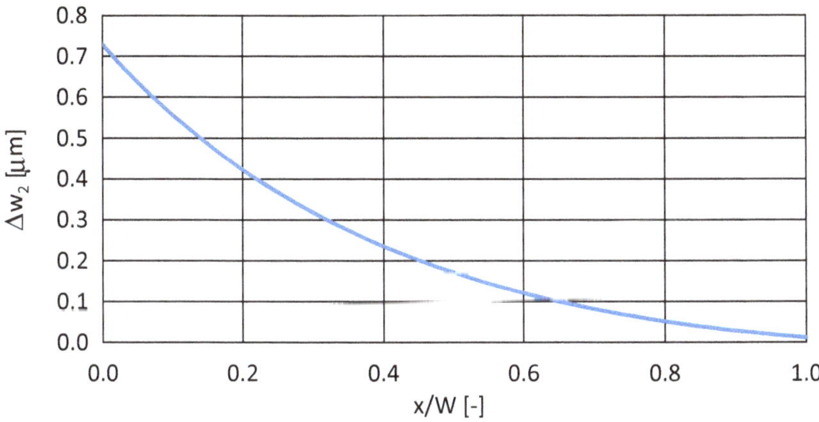

Fig. 9.29 Example of the deformation of the outer die slot w_2 as a result of die-internal fluid pressure

deformation, i.e., slot deformation, should not exceed the effect of mechanical slot precision, which, in our example, is specified as $dw = \pm 1\ \mu m$. Therefore, a maximum slot opening of 0.72 μm as shown in Fig. 9.29 should be acceptable.

Figure 9.30 shows how bar deformation (Δw_2) decreases upon increasing the bar thickness, and how it increases upon increasing the working length, which in turn increases the maximum pressure in the die. The data with the solid lines were calculated for operating condition 1, which resulted in maximum pressures of 26.5, 44.0, and 61.6 kPa for the working length of 80, 100, and 120 mm. In contrast, the

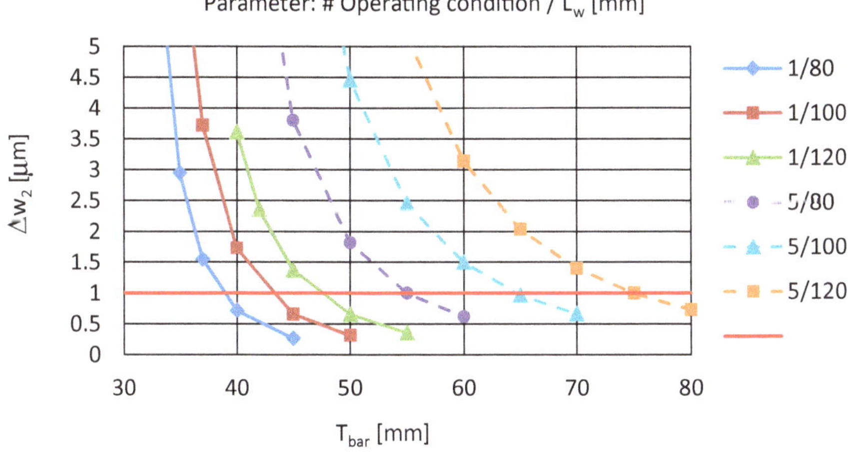

Fig. 9.30 Bar deformation driven by maximum fluid pressure, which increases with increasing working length, as a function of the bar thickness

data with the dashed lines were calculated for operating condition 5, which resulted in much higher maximum pressures of 401.6, 673.1, and 944.5 kPa for the same working length of 80, 100, and 120 mm. As can be seen, bar deflection can be kept at a low value of 1 μm, if the maximum pressure in the die is kept low, say <1 bar, even if the working length is as high as 120 mm. However, much thicker bars in the range of 55–75 mm are required, if the same low bar deformation must be achieved for higher fluid pressures in the range of 5–10 bar. Remember, though, that calculated numbers of bar deformation presented here are only approximate, as they were generated with a simple, analytical model based on the deformation of a cantilevered beam. Shetty et al. (2012) also used beam theory for calculating bar deformation. Their comparison with 2-D and 3-D finite element calculations revealed that beam theory underestimates deformation by several percent. However, they argued that results generated with the beam model are useful and adequate during die design if the goal is to limit slot deformation to the level of fabrication tolerances.

9.4.9 *Optimization of Die Design*

In the example featured in the previous chapter, most of the parameters determining the internal and external die geometry were chosen arbitrarily. This approach, however, is not an appropriate way for designing a good die, much less for designing the best possible die for a given set of operating conditions. The purpose of this chapter, therefore, is to define and explain criteria for optimizing die design from a fluid-mechanical and mechanical point of view. In particular, the goal is to find quality-related criteria for specifying each of the 15 geometric parameters introduced in Sect. 9.4.4 that determine the die geometry, see also Fig. 9.9. Additional information about optimizing die design can be found in Schweizer (1997b).

9.4.9.1 Coating Width

Finding the right coating width is a non-trivial task because it depends on many different parameters. As explained in the *Introduction* to this book, the average size of coating equipment sold by Polytype Converting AG to the converting industry during the years 2008–2016 was 1.525 m. This is quite a bit shorter than typical coating widths in the paper industry, where modern paper machines are up to 10 m wide, or even a bit wider.

As is explained in Sect. 9.4.8, shorter dies are much easier to handle than wider ones. While criteria for properly scaling the die width without losing desirable properties related to cross profile and wall shear stress are known and explained in the previous chapter, increasing the coating width results in higher and thicker die bars, and hence in heavier dies that are more cumbersome to handle. With wider dies, it is also more difficult to achieve mechanically uniform die slots and uniform coating gaps for slide and particularly slot coating, where coating gaps are <100 μm for many applications.

The right coating width may be dictated by the productivity arguments of the coating machine. As described in Chap. 15, the annual volume of coated product generated by one machine is obtained by dividing the annual production time by the specific machine utilization SMU (Eq. 15.1). This ratio is proportional to the product of coating speed times coating width. In other words, doubling the coating speed or doubling the coating width gives the same increase in machine productivity. Increasing the coating width requires a wider building and a wider machine. Moreover, as explained in the previous chapter, increasing the width increases the cross-section and thus the weight of the die, if it has the same performance features with regard to cross profile and wall shear stress as a narrow die. This in turn decreases the handling comfort concerning the mounting and dismounting of the die from the coating station, opening and re-assembling the die for inspection and cleaning, and it increases the non-productive time of the coating process. In addition, it may not be possible to increase the width at will, if the new width is dictated by the format of the coated product. On the other hand, increasing the coating speed requires a longer dryer, which necessitates a longer or a taller building if the dryer can be folded. Furthermore, a speed increase may be limited by a critical value that marks the onset of wetting failure (air entrainment). Based on our experience, a trend to machines with a width of much beyond 2 m is not in sight, except for the functional layers of thermal paper, but that sector can be considered a part of the paper rather than the converting industry. The main reason for staying with narrower machines is problems with web handling. Specifically, substrates like plastic film or paper that are produced on very wide machines and cut to smaller widths for converting purposes are not straight and hence cause increasing problems with increasing width when pulled through a converting machine. Consequently, the trend in the converting industry is towards faster machines with a maximum width of about 2 m. An example is the manufacturing of adhesive labels, where some machines run at speeds of >1,000 m/min.

9.4.9.2 Feed Location

As was demonstrated in the previous chapter regarding the overall performance related to the cross profile and the wall shear stress behavior of a die, and as recommended by Ruschak and Weinstein (2018) as regards the effectiveness of the outer cavity, a center-fed die is highly preferred over a side-fed version. Care must be taken, however, to suppress unwanted inlet effects related to the stagnation point flow at the cavity inlet. Using a dual-cavity design concept is helpful in this regard owing to the damping capacity of the outer cavity. Assuring low inertia of the arriving flow may also be necessary by sufficiently widening the diameter of the delivery line and the inlet section of the die.

9.4.9.3 Number of Cavities

As explained in the previous chapter, dies containing two cavities tend to produce smaller total flow nonuniformities than single-cavity dies for equal wall shear stress requirements, equal working lengths, and an equal range of operating conditions. The main reason for this difference is the possibility for the dual-cavity design to make the inner slot narrow for obtaining a good flow distribution, and to make the outer slot wide for reducing the sensitivity of the die to the mechanical imprecision of the slots. The main exception to these findings is operating conditions with a strong shear thinning behavior, i.e., with a low value of the parameter b in the Carreau-Yasuda equation because they result in a low damping capacity of the outer cavity. This result was also found by Ruschak and Weinstein (2018). In conclusion, therefore, the dual-cavity design is generally preferred over the single-cavity design.

9.4.9.4 Total Working Length

Determining the best working length is also an important task. On the one hand, increasing the working length improves the flow uniformity as shown in Fig. 9.28. In fact, even single-cavity dies would perform very well for a wide range of operating conditions, if only the working length was long enough. On the other hand, increasing the working length increases bar deformation, which must be suppressed by increasing the bar thickness, and this, in turn, once again increases the weight of the die thereby decreasing its handling comfort.

The goal is to find the minimum working length that results in acceptable cross profiles for a specified maximum coating width and range of operating conditions. Increasing the spread of the operating conditions and increasing the coating width requires a longer working length. Working lengths in the range of 80 ± 20 mm should result in good cross profiles for many applications and center-fed dies having a width of $\leq$2,000 mm. At the end of the day, however, the choice of working lengths may be limited by the availability of bar dimensions at the die manufacturer.

9.4.9.5 Ratio of Inner to Outer Working Length

The effect of varying the ratio v of the inner to the outer working length on the total flow rate nonuniformity is illustrated in Fig. 9.31. For all calculations, the total working length was assumed to be 80 mm, and the wall shear stress in the inner and outer cavity was specified to be 5.0 and 1.5 Pa, respectively. Note that operating condition 14 is equal to condition 7, except that the low shear viscosity μ_0 has a value of 2,000 instead of 5,000 mPas. Interestingly, the curves for all but the base operating conditions have a minimum. The location of the minimum depends on the details of the operating conditions and it ranges from about 0.5–1.0. Moreover, the minimum is fairly flat, which is fortunate for the following reasons.

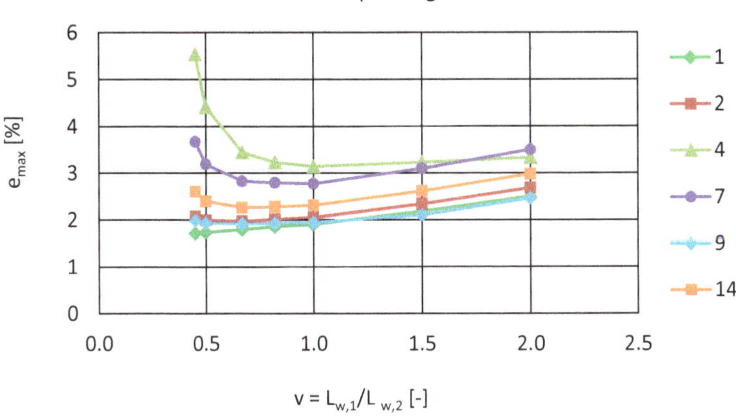

Fig. 9.31 Effect of the ratio v of the inner to outer working length on the total flow rate nonuniformity as a function of different operating conditions

If $v > 1$ and increasing, then $L_{w1} > L_{w2}$ and L_2, i.e., the length of the outer slot, decreases to rather small values. In contrast, if $v < 0.5$ and decreasing, then $L_{w1} < L_{w2}$ and L_1, i.e., the length of the inner slot, also decreases to low values of just a few millimeters. In the experience of die manufacturers, excessively short lengths of the inner and outer die slots cause problems during manufacturing. Specifically, the processes of grinding, lapping, and hand lapping cannot be carried out with the necessary precision. To avoid such problems, die manufacturers request both slots to have a minimum length of 10 mm.

In addition, if $v < 1$ and decreasing, then $L_{w2} > L_{w1}$, and L_2, as well as CL_2, the length of the outer cavity, increase. As a consequence, it was not possible to achieve a minimum wall shear stress of 1.5 Pa in the outer cavity for $v < 0.67$. The outer cavity was a 45–45 right angle triangle in this example.

In summary, small and large v-values must be avoided. Fortunately, best values range from about 0.8–1.0, and, owing to the flat minimum, they result in a minimum flow rate nonuniformity for a wide range of operating conditions.

9.4.9.6 Shape of Inner Cavity

Slot dies are built with many different cross-sectional shapes of the inner cavity. Several of these shapes that are deemed interesting are discussed in Sect. 5.5.1.2. But why are so many cavity shapes found in industrial dies? Is there not one shape that outperforms every other shape, thereby making all other shapes obsolete? To answer these questions, we carried out a comparative study. Specifically, we selected the following shapes: circular, semi-circular, rectangular with two rounded corners, square with two rounded corners, and half-circle + half-square. The corner radii for the square and rectangle were set to 2 mm. We used the operating conditions listed

Table 9.3 Performance comparison of inner cavity shapes

Shape	ΔP_{cavity} [Pa]	Rank	ε_{max} [%]	Rank	Cumul. rank
Half s + Half c	3117	1	2.78	2	3
Circle	3793	3	2.66	1	4
Rectangle	3738	2	2.94	4	6
semi-circle	4246	5	2.88	3	8
Square	4237	4	3.06	5	9

in Table 9.1 and the geometric parameters specified just after that table in Sect. 9.4.8 to optimize the die geometry for different shapes of the inner cavity. We wanted all shapes to produce the same initial wall shear stress of 5 Pa, and they needed to have the same wall shear stress distribution with $\Phi = 0.85$. We then compared and ranked all cavity shapes with regard to the total pressure drop along the inner cavity ΔP_{cavity} and the total flow rate nonuniformity ε_{max}. During this exercise, we learned that the relative ranking of the various shapes did not change upon changing the operating conditions. Therefore, we are limiting our reporting to operating condition #7, which resulted in a relatively wide spread in the two parameters of interest. Results are summarized in Table 9.3.

Regarding the pressure drop in the inner cavity, the shape of half a circle + half a square gives the lowest value, followed by the rectangle and the circle. This result is due to the cross-sectional area, which is largest for this shape and an initial wall shear stress of 5 Pa. Note that using Eq. 5.5.2 for calculating the viscous pressure gradient and comparing the shape factor and the cavity size of the various shapes does not give the same ranking because the gradient is not constant along the cavity. Regarding the total flow rate nonuniformity, the circle is on top, followed by the half circle + half square, and the semi-circle. Adding the two rankings, then the winner is half-circle + half-square, closely followed by the circle and the rectangle in third place. However, since the circle is a symmetric form, thus requiring machining of adjacent die plates, which is more expensive and hence not attractive for manufacturing, the shape of half-circle + half-square is the clear winner of this competition and the preferred shape for the inner cavity. This result is further supported by the fact that the effort to machine a tapered cavity does not change much at all between the various cavity shapes.

9.4.9.7 Initial and End Size of Inner Cavity

The initial size of the inner cavity is determined by the requirement of the desired wall shear stress value at the cavity inlet according to Eq. 5.5.5. Recommended wall shear stress values are in the range of 1–10 Pa. Therefore, 5 Pa is a good value for the base operating conditions, for which the die geometry is optimized because it leaves ample room above and below this value for the wall shear stress of other operating conditions that must be coated with the same die. The wall shear stress value of the base conditions may have to be adjusted up or down a bit if the stress levels generated by the other operating conditions are all much below or above the base value.

Similarly, the end cross-section of the inner cavity is determined by the wanted form of the wall shear stress distribution along the cavity according to Eq. 6.3.12. The best distribution is the one where the stress stays constant at its initial level throughout the cavity. This, however, is only possible with throughflow, which is not often encountered in the converting industry, except perhaps for highly reactive fluids. Without throughflow, and keeping with the one-dimensional theoretical model, the flow rate in the inner cavity drops to zero at the cavity end, and so does the wall shear stress. Therefore, the second-best distribution is the one where the wall shear stress stays constant at the initial level as far as possible to a value $x/W = \Phi$ along the cavity, and then, without first increasing, drops below the initial level and approaches zero. A good example of such a wall shear stress distribution is shown in Fig. 9.17, where $\Phi = 0.85$. An almost flat stress distribution along most of the cavity requires the taper index φ to vary along the cavity according to Eq. 6.3.13. In addition, high Φ-values can only be achieved if the ratio of end-to-initial cavity cross-section is small. On the other hand, excessively high Φ-values must be avoided because they result in excessively high pressure drops along the inner cavity. In our experience, good compromises are obtained for the following parameter combinations: $L_{c,e}/L_{c,0} = 0.3$ with $\Phi = 0.85$–0.90 for non-reactive fluids, and $L_{c,e}/L_{c,0} = 0.2$ with $\Phi = 0.90$–0.95 for reactive fluids. As mentioned in the previous chapter, the main price that must be paid for good wall shear stress behavior is a higher pressure drop across the die, which requires the pump to work harder.

9.4.9.8 Taper of Inner Cavity

The taper of the inner cavity, i.e., the way the cavity changes its cross-sectional size along its axis, is also determined by the desired wall shear stress distribution. Specifically, the taper depends on the taper index φ, which varies along the cavity axis according to Eq. 6.3.13. Thus, the constant b in Eq. 6.3.13 must be >0. Best values for the constants a, b and c depend on the cavity shape, and they must be found through trial-and-error. Recommendations for this process are provided in Sect. 6.3.1.

The values for the constants a, b and c must be chosen such that the wall shear stress does not exceed its initial value early along the cavity because a stress increase is equivalent with a decrease of the cavity cross-section, and this, in turn, increases the pressure drop along the cavity, which is not desirable. Moreover, a wall shear stress increase above its initial value does occur if high Φ-values are required for the rectangular cross-section. Again, the values for a, b and c must be chosen such that the stress distribution looks more like the green curve with $b > 0$ in Fig. 6.6 than the blue curve where $b = 0$.

9.4.9.9 Length of the Inner Slot

Once the total working length L_w and the ratio $v = L_{w1}/L_{w2}$ are specified, then the inner working length is given by

$$L_{w1} = \frac{L_w}{1 + \frac{1}{v}} \tag{9.32}$$

In addition, once the inner working length and the initial size of the inner cavity are specified, the initial length of the inner slot is given by

$$L_1 = L_{w1} - CL_1 \tag{9.33}$$

CL_1 is the length of the inner cavity; its initial value depends on the shape and size of the inner cavity. For circular and semi-circular cavities, CL_1 is equal to the diameter of the circle, and for square, rectangular, and half-square + half-circle cavities, it is equal to the side of the square. Along the cavity axis, the size and hence the length of the cavity reduces according to Eq. 6.3.8, and the length of the inner slot reduces according to Eq. 9.31 during design optimization for the base set of operating conditions.

9.4.9.10 Shape of Outer Cavity

As with the inner cavity, many shapes for the outer cavity are used in the industry, including semi-circles, circular segments, half-tears, and triangles with a rounded vertex. Ruschak and Weinstein (2018) investigated five different triangular cavities with sharp vertexes. They concluded that the 45–45 right-angle triangle is the best form because it has the highest damping capacity of all triangular shapes owing to its highest value of the constant k_9, which relates the size to the length of the cavity according to

$$k_9 = \frac{A_2}{CL_2^2} \tag{9.34}$$

In other words, the 45–45 right-angle triangle has the largest area for a given cavity length. This shape has also a fair resistance to flow separation, i.e., to the onset of vortex formation, which must be avoided.

In addition, we strongly recommend designing the outer cavity such that the minimum wall shear stress assumes positive values, thereby implying that the flow is free of vortices. This can be achieved by rounding the vertex corner of the triangle. Moreover, the level of wall shear stress can be adjusted by changing the radius of the vertex corner according to Eq. 6.3.19. However, we have learned that achieving wall shear stress levels in the range of 1–10 Pa, as we suggest for the inner cavity, is more difficult in the outer cavity because the vertex radius cannot be changed at will owing to its limited maximum value, see Fig. 5.5.7. Therefore, a more realistic range for the minimum wall shear stress in the outer cavity is 1–5 Pa when optimizing the design for the base operating conditions.

Flow separation incurs if the Reynolds number of the flow across the cavity exceeds a critical value. As visualized by Eq. 5.5.26 and Fig. 5.5.18, this critical value increases with decreasing the expansion angle α of the triangle. Therefore, if any Reynolds number of the intended applications exceeds the critical value of the 45–45 triangle, then the shape of the outer cavity has to be changed. Currently, however, the only alternative is the 30–60 right-angle triangle because relevant flow parameters are only available for this cavity form. The price for this change is a reduced damping capacity of the outer cavity.

Semi-circular outer cavities are less attractive than triangular shapes because the initial expansion angle is 90°, thus resulting in flow separation at lower Reynolds numbers. Cavities having the form of a circular segment are also less attractive owing to the value of the parameter k_9 (Eq. 9.34), which is smaller than for the 45–45 right-angle triangle. However, if the vertex radius of the triangle reaches its maximum value to obtain a high enough minimum wall shear stress, then the triangular form changes into the form of a circular segment.

9.4.9.11 Depth of Outer Cavity

The depth CD of right-angle triangular cavities depends on the radius of the vertex (see Fig. 5.5.7), and the radius in turn is determined by the requirement of a minimum wall shear stress. In the inner cavity, a small cavity depth results in a high wall shear stress, but also in a high pressure drop along the cavity, thus negatively affecting the flow rate uniformity for a specified set of operating conditions. Likewise, in the outer cavity, a small cavity depth results in a high wall shear stress but also in a reduced damping factor. In both cases, the best compromise between a high wall shear stress and a high flow rate uniformity must be found by weighing the two quality-related parameters, and by relaxing the requirement of one of them in favor of the other. The two parameters are also affected by the spread of the operating conditions. For large spreads, it might be difficult if not impossible to find an acceptable compromise. The best way forward in such a situation could be to split the operating range into two parts and to dedicate an optimized die to each of the parts, thereby facilitating an acceptable compromise for both the flow rate uniformity and the wall shear stress.

9.4.9.12 Ratio of Outer Cavity Length to Outer Slot Length

The question of how to best partition the outer working length between the length of the slot and the length of the cavity was answered by Ruschak and Weinstein (2018) as follows. According to Eqs. 5.5.14 and 5.5.15, the damping factor DF of the outer cavity is indirectly proportional to the geometric group *s*, and s is indirectly proportional to $L_2A_2^{(1+n)}$. Therefore, the maximum damping capacity of the outer cavity is obtained by maximizing $L_2A_2^{(1+n)}$, with L_2 and A_2 being the slot length and the cross-sectional area of the outer cavity. Moreover, $A_2 = k_9CL_2^2$ (Eq. 9.34), and $L_{w2} = L_2 + CL_2$. The outer working length L_{w2} has already been optimized and its value is known as explained above in this chapter.

Considering that $L_2 = L_{w2} - CL_2$, the function $f(CL_2)$ that needs to be maximized is:

$$f(CL_2) = (L_{w2} - CL_2)k_9^{1+n}CL_2^{2+2n} \tag{9.35}$$

Taking the derivative of $f(CL_2)$ and setting it to be zero results in the ratio of CL_2/L_{w2}, which subsequently leads to the ratio of L_2/L_{w2} as follows:

$$\frac{L_2}{L_{w2}} = 1 - \frac{CL_2}{L_{w2}} = \frac{1}{3+2n} \tag{9.36}$$

Now, L_2 and CL_2 can be expressed in terms of the working length and the power law index:

$$L_2 = \frac{L_{w2}}{3+2n} \tag{9.37}$$

$$CL_2 = 2L_{w2}\frac{1+n}{3+2n} \tag{9.38}$$

Ruschak and Weinstein point out that both L_2 and CL_2 are not very sensitive to n, with $L_2/L_{w2} = 0.2$ for $n = 1.0$ and 0.29 for $n = 0.2$. These numbers indicate that the outer slot takes up less than 1/3 of the outer working length. Consequently, chances are high that the outer slot length may assume values that are less than 10 mm. This, however, would conflict with the requirement of die manufacturers that the slot length should be >10 mm to facilitate the manufacturing steps of grinding, lapping, and hand lapping. We believe that this mechanical constraint should override the fluid mechanical optimization criterion, such that Eq. 9.37 is only applied if the resulting value for L_2 is >10 mm.

9.4.9.13 Height of Inner and Outer Slot

As explained in Sect. 9.4.2, one of the advantages of a dual cavity design is that the inner slot can be made narrow to minimize cross profile contributions from fluid flow effects, and the outer slot can be made wide to reduce the sensitivity to the mechanical imprecision of the die plates that form the slots. Moreover, analyzing the behavior of single cavity dies reveals that, generally speaking, increasing the slot height increases the fluid flow contributions to the cross profile but reduces the sensitivity to the mechanical slot precision, see Sect. 9.4.8 and Fig. 9.27. Unfortunately, other criteria for optimizing the slot heights of a dual cavity die are not known.

To visualize the effect of changing the inner and outer slot heights, we took the dual cavity die that was optimized for operating condition 1 in the example in Sect. 9.4.8, and we calculated the total flow rate nonuniformity ε_{max} for operating conditions 4 and 7, if the inner slot height was set to 200, 250 and 300 μm, and, for each of these inner slot heights, the outer slot height was fixed at 300, 400 and 500 μm. Results are shown in the following figures.

Specifically, Figs. 9.32, 9.33 and 9.34 compare the effect of slot height variations as a function of the mechanical slot precision (dw = ±0.5, 1.0 and 2.0 μm) for operating condition 7. Figure 9.35 compares the effect of slot height variation for operating conditions 4 and 7 for dw = 1 μm. The calculations reveal that increasing the inner slot height w_1.

- increases β_0, i.e., the flow nonuniformity exiting from the inner slot;
- slightly increases the damping factor DF;
- slightly decreases the effect from the mechanical slot precision dw.

such that ε_{max} increases for dw = 0.5 and 1.0 μm, and also for dw = 2.0 μm, but only if w_2 = 300 μm. If w_2 = 400 μm, then μ_{max} is hardly affected, and if w_2 = 500 μm, then ε_{max} decreases. The effect of increasing w_2 depends on the value of dw. If dw = ±0.5 μm, then ε_{max} increases with increasing w_2. If dw = ±1.0 μm, then w_2 = 400 μm gives less total nonuniformity than w_2 = 300 μm, and w_2 = 300 μm gives less total nonuniformity than w_2 = 500 μm. This is also the case for different operating conditions. In contrast, if dw = ± 2.0 μm, then w_2 = 400 μm is better than w_2 = 500 μm, and that is better than w_2 = 300 μm. Generally speaking, the effects of mechanical slot precision dominate contributions from fluid flow to the flow rate nonuniformity, at least for the conditions of this example. Exceptions relate to situations where dw is very small, i.e., 0.5 μm and where w_1 is wide, i.e., >300 μm, and w_2 is also wide, i.e., >400 μm. The lowest, or very close to the lowest

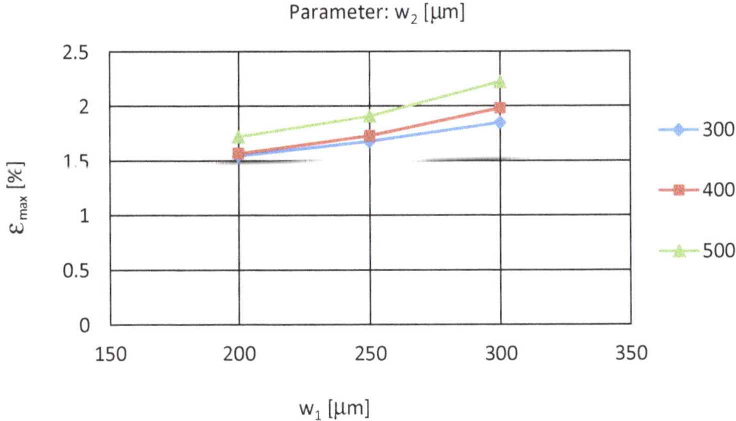

Fig. 9.32 Effect of varying the inner and outer slot height on the total flow rate nonuniformity; dw = ±0.5 μm, operating condition 7

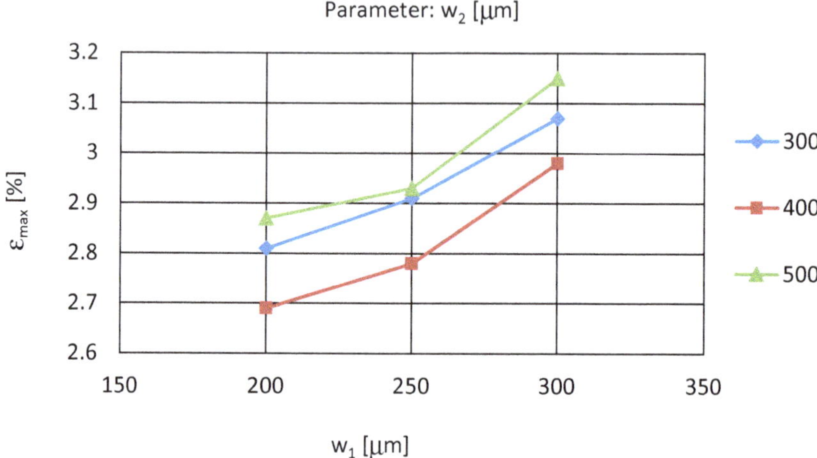

Fig. 9.33 Effect of varying the inner and outer slot height on the total flow rate nonuniformity; dw = ±1.0 µm, operating condition 7

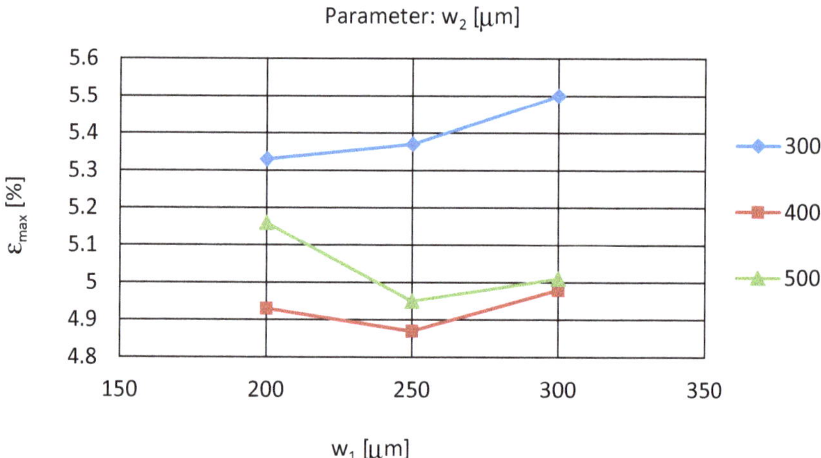

Fig. 9.34 Effect of varying the inner and outer slot height on the total flow rate nonuniformity; dw = ±2.0 µm, operating condition 7

value, of ε_{max} is obtained if $w_1 = 200$ µm and $w_2 = 400$ µm. This is true for the range of mechanical slot precision and operating conditions tested in this example. Therefore, the combination of $w_1/w_2 = 200/400$ µm is an interesting suggestion. If these values are not optimum, they at least serve as a good starting point for optimizing the geometry of a new die.

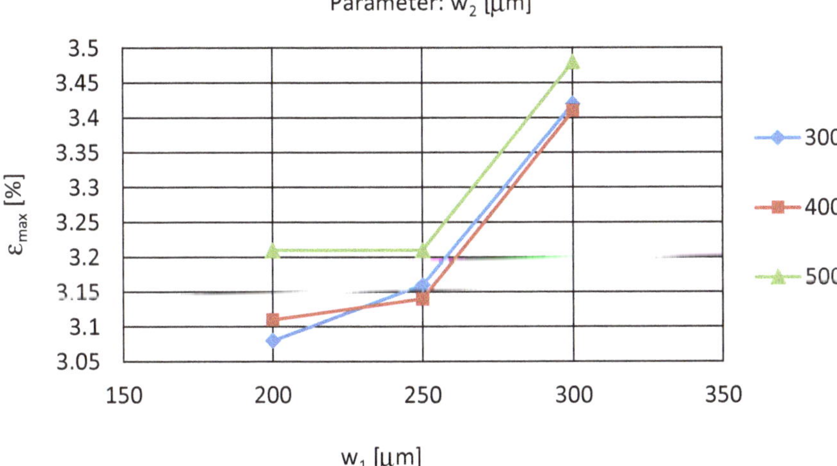

Fig. 9.35 Effect of varying the inner and outer slot height on the total flow rate nonuniformity for operating conditions 4 and 7; dw = ±1.0 μm

9.4.9.14 Height of Die Bar

The total height of the die bar includes the fluid flow section, i.e., the working length L_W for housing the inner and outer distribution systems, as well as the clamping section. Optimizing the design of the clamping section is a complex mechanical challenge because it must address issues such as

- what is the best number, diameter, and widthwise pattern of the bolts?
- what is the best bolt torque, and what is the best sequence for applying the torque in the cross-web direction?
- should all bolts prevent or allow slip or stick, or should one bolt assure stick while the other bolts should allow slip, for example, if the die is operated above room temperature?
- how should the die be fixed to the support with regard to stick/slip?

The overriding goal is to design the clamping section such that it does not lead to excessive deformations of the die slots. Optimizing the design of the clamping section exceeds the knowledge of the author of this text. Therefore, it is recommended to rely on the expertise of the die manufacturers. Nevertheless, having worked with dies from different suppliers, we can report that the ratio of fluid flow section to total bar height is often in the range of 0.45–0.60.

9.4.9.15 Thickness of Die Bar

As explained in Sects. 9.4.6 and 9.4.8, the bar thickness must be determined in response to the maximum fluid pressure inside the die and the details of the cavity design. Generally speaking, the bar thickness should be as small as possible and as large as necessary to limit the widthwise slot deformation generated by bar deflection in response to internal fluid pressure to the level of the mechanical slot precision.

Equation 9.29 may be helpful to estimate the effect of bar deformation on the flow rate nonuniformity as a function of the bar thickness. In addition, relying on the experience of the die manufacturer is recommended. Often, however, die manufacturers do not optimize the bar thickness, but they work with a couple of standard thicknesses, which have proven reliable for different ranges of the maximum fluid pressure.

9.4.9.16 Mechanical Slot Precision

For good dies, the mechanical slot precision, i.e., the variation of the flatness profile of the die plates, typically is in the range of $\pm 0.5 - \pm 2.0$ μm. Achieving such high precision, or even a better one for special applications requires hand-lapping and an elaborate measurement system for visualizing the topography of the die plates. Knowing the topography of the die plate points the lapping specialist to the "mountains" in the plate surface that must be removed. Therefore, improving the mechanical slot precision increases the manufacturing costs of the die.

As explained in Sect. 15.5, the optimum slot tolerance should be determined based on economic considerations. Specifically, improving the slot tolerance often reduces the cross profile, and that, in turn, allows the nominal wet film thickness to be reduced, although by only a small amount. Nevertheless, premetered coating methods allow the flow rate to be adjusted by very small amounts. Therefore, if the annual coated surface and the specific raw material costs of a given product are sufficiently high, then the annual savings in manufacturing costs may easily outweigh the higher investment costs in a more precise slot die.

9.4.9.17 Optimization Procedure

Die design optimization can be carried out in two ways, either

- by specifying a value of the working length L_W and by finding those values of all other geometric parameters that minimize the total flow rate nonuniformity ε_{max} for a given set of operating conditions or
- by specifying a maximum value of ε_{max} that should not be exceeded by any of the operating conditions, and by finding those values of all other geometric parameters that minimize the total working length L_W.

So far, we have specified optimization criteria for all parameters that define the internal die geometry, except for the height of the inner and outer die slot, for which such criteria are not known. Moreover, visualizing the effect of both slot heights on the flow rate uniformity does not reveal clear trends, see Figs. 9.32, 9.33, 9.34 and 9.35. Also, finding the best slot height values for a set of several operating conditions as was done in the example described above is very labor-intensive. What is needed is a computer program that carries out an automated procedure in the form of a full-factorial experimental design plan to find the best values for the slot heights, or the best values for any of the geometric parameters. When working in the photographic industry, such a program was available. It was a Fortran code consisting of many nested do-loops that calculated the cross profile for every specified operating condition while every geometric parameter, i.e., one at a time, was varied over several steps. This procedure generated several thousand solutions in a short time, many of which resulted in a cross profile that was better than the specified maximum value for ε_{max}. Of those solutions, we then selected the result that was least sensitive to ever-present fluctuations of the flow rate and the rheological parameters.

A procedure as described above is currently not available in the EXCEL-based die design program used for the examples presented in Sect. 9.4.8. Shetty et al. (2012) also developed an EXCEL-based die design program. They suggested working with the add-in function SOLVER, an optimization routine, which can enforce constraints. It is not clear, however, if SOLVER would work properly in an environment as presented in this chapter because implementing the local power law model for calculating the pressure drop along the inner cavity requires iteration at every intermediate step along the cavity.

References

Dubbel. (1994). Handbook of mechanical engineering. In: W. Beitz & K.-H. Küttner (Eds.), Springer Nature.

Durst, F., & Bülent, U. (2008). Beschichtungswerkzeug zum Auftragen eines Flüssigkeitsfilmes auf ein Substrat. German patent DE 102008041423.

Durst, F., Sünderhauf, G., Troller, U., & Schweizer, P. M. (1997). Flow in a two-cavity die: Comprehensive study between analytical model, finite element analysis and experiment. In *Proceedings of the 2nd European Coating Symposium*. Université Louis Pasteur, Strasbourg, France.

Kistler, S. F., & Schweizer, P. M. (1997). *Liquid film coating*. Chapman & Hall.

Kozicki, W., Chou, C. H., & Tiu, C. (1966). Non-Newtonian flow in ducts of arbitrary cross-sectional shape. *Chemical Engineering Science, 21*(8), 665–679.

Lee, K. Y., & Liu, T.-J. (1989). Design and analysis of a dual-cavity coat hanger die. *Polymer Engineering and Science, 29*, 1066.

Pan, J.-P., Liu, T.-J., & Wu, P.-Y. (1999). Theoretical analysis on extrusion die flow of electronic packaging materials. *AIChE Journal, 45*(29), 424–431.

Roulier, A., & Bebié, H. (2020). Wall shear stress in flow across triangular cavities (personal communication).

Ruschak, K. J., & Weinstein, S. J. (2019). Accurate approximate methods for the fully developed flow of shear-thinning fluids in ducts of noncircular cross-section. *Journal of Fluids Engineering, 141*(11).

Ruschak, K. J., & Weinstein, S. J. (1997). Modeling the secondary cavity of two-cavity dies. *Polymer Engineering & Science, 37*(12), 1970–1976.
Ruschak, K. J., & Weinstein, S. J. (2014). A local power law approximation to a smooth viscosity curve with application to flow in conduits and coating dies. *Polymer Engineering and Science, 54*, 2301–2309.
Ruschak, K. J., & Weinstein, S. J. (2018). Model for the outer cavity of a dual-cavity die with parameters determined by two-dimensional finite-element analysis. *AIChE Journal, 64*(2), 708–716.
Schweizer, P. M. (1992). Fluid handling and preparation. In E. Cohen & E. Gutoff (Eds.), *Chapter 2 in Modern coating and drying technology*. VCH Publishers Inc.
Schweizer, P. M. (1997a). Experimental methods. In S. F. Kistler & P. M. Schweizer (Eds.), *Chapter 7 in Liquid film coating*. Chapman & Hall.
Schweizer, P. M. (1997b). Control and optimization of coating processes. In S. F. Kistler & P. M. Schweizer (Eds.), *Chapter 15 in Liquid film coating*. Chapman & Hall.
Secor, R. B. (1997). Analysis and design of internal coating die cavities. In S. F. Kistler & P. M. Schweizer (Eds.), *Chapter 10 in Liquid film coating*. Chapman & Hall.
Shetty, S., Ruschak, K. J., & Weinstein, S. J. (2012). Model for a two-cavity coating die with pressure and temperature deformation. *Polymer Engineering and Science, 53*, 1173–1182.
Weinstein, S. J., & Ruschak, K. J. (1996). One-dimensional equations governing single-cavity die design. *AIChE Journal, 42*, 2401.
Weinstein, S. J., & Ruschak, K. J. (2004). Coating flows. *Annual Review of Fluid Mechanics, 36*, 29–53.
Yu, Y.-W., Wu, P.-Y., & Liu, T.-J. (1997). Validity of one-dimensional equations governing extrusion die flow. *AIChE Journal, 43*(11), 3117–3120.
Yuan, S. L. (1995). A flow model for non-Newtonian liquids inside a slot die. *Polymer Engineering and Science, 35*, 577.

Chapter 10
Concepts for Varying the Coating Width

Abstract The choice between variable or constant coating width and the choice between variable or constant web width bears significant consequences for the product quality and process productivity, and hence to the manufacturing costs of coated products on the one hand, and with regard to the investment costs for coating equipment on the other hand. The process of selecting the best number and size of different coating and web widths in an attempt to minimize both the investment and operating costs is a classic optimization problem. In this chapter, various positive and negative aspects of the options for changing the width are considered and evaluated.

10.1 Introduction

Dies with a constant coating width are much simpler in design and therefore easier and faster to assemble and disassemble because they consist of fewer and simpler parts. Typically, deckles, inserts, or shims in combination with additional peripheral equipment are used to vary the coating width. Some die manufacturers offer systems where the deckle position can be varied continuously and adjusted manually or with the help of a motorized positioning system. With other deckle systems, the side plates of the die, or even the die plates themselves must be disassembled before the deckles can be inserted.

Some deckle systems require that the cross-sections of the distribution chambers must be constant, at least near the ends of the chambers and over the range of the variable coating width. Commonly applied process performance criteria require

- a tight seal between the deckles and the die;
- no leakage of coating fluid in the range of the deckled coating width;
- no interference between the deckles and the coating process.

In general, symmetrical deckling on both sides is easier than unsymmetrical deckling on only one side because in the latter case, either the die has to be mounted off-center, or the web has to be threaded off-center.

P. M. Schweizer, *Premetered Coating Methods*, Engineering Materials,
https://doi.org/10.1007/978-3-031-04180-8_10

In the following sections, we focus on dies with fixed lips (as opposed to dies with deformable lips). One of the manufacturers of such dies is TSE Troller AG, which uses deckles or shims for varying the coating width. Deckles can be inserted without disassembling the die plates, but the side plates of the die and perhaps other peripheral parts must first be removed. On the other hand, shims can only be replaced, if the die plates are completely disassembled. Shims are typically U-shaped and come in different thicknesses so that the height of the die slot can also be varied if so desired. Moreover, the coating width can only be varied in steps because a separate shim is needed for each individual coating width. The same is true if individual deckles are used for each coating width. However, the coating width is continuously adjustable if long deckles are used, the position of which can be adjusted at will. Movable deckles cannot satisfy the requirement of a tight seal between the deckles and the die. Consequently, additional measures must be taken to prevent liquid from leaking out of the die slot in the deckled areas. For reasons of mechanical stability, the maximum deckle length is limited; it depends on the cross-sectional size of the cavities and the width of the die.

10.2 Width Change for Slide Dies

Figure 10.1 illustrates a manually operated deckle system of a multilayer slide die for curtain coating from TSE Troller AG. Figure 10.1a shows the slide edge guide, which is attached to the adjacent slide plate, and both elements are pressed against the die slide by levers attached to the backside of the uppermost die plate. Moreover, the

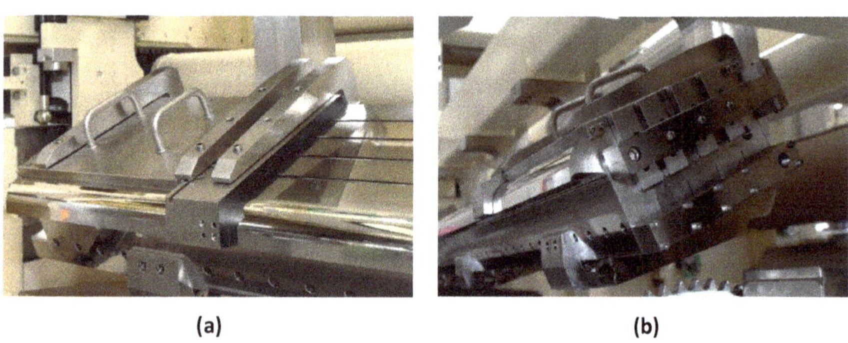

Fig. 10.1 **a** Deckle system of a multilayer slide die for curtain coating: the slide edge guides and the slide plate are pressed onto the slide surface by levers attached to the backplate of the die; photo reproduced with permission from TSE Troller AG. **b** Deckle system of a multilayer slide die for curtain coating: the deckles are attached to deckle plates, which are pressed against the side of the die plates by a rotatable lever for quick release of the deckles; photo reproduced with permission from TSE Troller AG

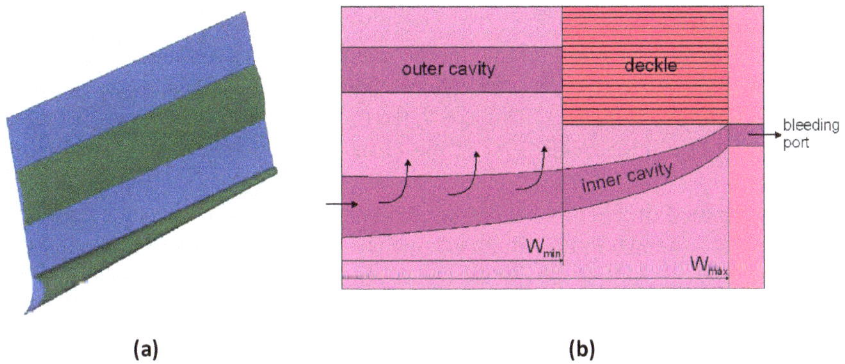

Fig. 10.2 **a** Example of a deckle for plugging both cavities of a dual cavity die; drawing reproduced with permission from TSE Troller AG. **b** Example of a deckle plugging only the slots and the outer cavity of a dual cavity die; design concept by TSE Troller AG

underside of both elements contains soft and compliant sealing material that prevents coating fluid from leaking out of the die slots. The slide plate is as wide as the deckles are long. When changing the coating width, i.e., the length of the deckles, then the width of the slide plate must be changed as well. Figure 10.1b shows a rotatable lever that presses the deckle plates, to which the deckles are attached, against the side of the die plates. Opening the lever allows quick access to the deckle plates, which must be pulled carefully by hand to remove the deckles. Changing the coating width involves stopping the coating process, cleaning and emptying the die, removing the deckles and other equipment parts, inserting other deckles with a different length, re-assembling other hardware, filling the die, and resuming coating. The operators of the pilot machines at Polytype Converting AG managed to carry out such a width change of a 3-layer curtain die in about 45 min.

Deckles can either plug both cavities, or only the outer cavity of a dual cavity die. In the former case, deckles may look as visualized in Fig. 10.2a. Deckles of this kind are mechanically very delicate because the parts that plug the die slots are very thin. Moreover, the inside part of the deckle that plugs the inner cavity may be designed with a three-dimensional rounded form to facilitate the turn-around flow from the cavity into the slot. This design feature also prevents the wall shear stress from vanishing at the end of the cavity, which is preferable.

The deckle shown in Fig. 10.2a has a three-dimensional form for the inner cavity, which is perfect because it matches the cavity taper all the way to the cavity end, and that taper guarantees high wall shear stress throughout the cavity. The drawback of this design is that such deckles cannot be mounted by sliding them into the cavities from the sides as described above. Instead, the die must be disassembled completely, which increases the downtime and hence decreases the process productivity. Deckles that can be installed from the outside without opening the die require a constant cross-section over the entire length of the deckle, and the inner cavity must also be of the same constant cross-section for the distance between the maximum and minimum coating widths. As discussed in Sect. 6.4, the drawback of this approach is that the wall shear stress in the outer parts of the inner cavity drops to zero, which is not wanted.

Using a partial deckle that only plugs most of the inner slot, the outer cavity, and the outer slot as shown in Fig. 10.2b is a way of avoiding these disadvantages. This design requires a bleeding valve at the end of the inner cavity, which must be opened during the filling and cleaning of the die. Moreover, this design can only be used with fluids that are not chemically reactive because the end of the inner cavity and part of the inner slot underneath the deckle are both filled with stagnant fluid while coating and that fluid portion would solidify if reactive.

TSE Troller AG also offers a motorized and computer-controlled deckle system, by which the coating width can be changed continuously and much more quickly than a manual system, see Fig. 10.3. In fact, TSE Troller claims that a complete width change involving all the process steps described above can be executed in about 10 min, which is attractive in terms of process productivity.

Fig. 10.3 Motorized and computer-controlled deckle system for continuously changing the coating width; photo reproduced with permission from TSE Troller AG

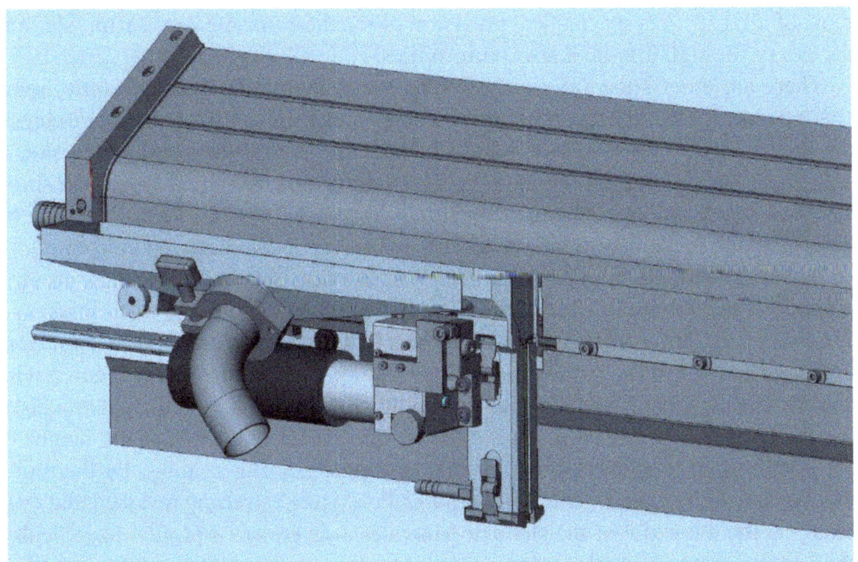

Fig. 10.4 Convenient and fast system for continuously changing the coating width in single-layer curtain coating applications; drawing reproduced with permission from Polytype Converting AG

The drawback of this approach is the need for a constant cross-section of the inner cavity over the length of the deckles, and the associated loss of high wall shear stress as discussed in Sect. 6.4.

As already explained in Sect. 5.9.5 and visualized in Fig. 5.9.55, a very convenient and fast way for changing the coating width in single-layer curtain coating applications exists by combining the curtain edge guide with a catch pan, and by mounting this equipment assembly onto a linear rail that allows a variable placement of this assembly relative to the die lip, see Fig. 10.4. The die is always operated at full width, and the coating width is defined by the lateral position of the knife-edge on top of the curtain edge guide. The portion of the liquid supplied by the die that flows into the catch pan can be recycled. This concept only works for single-layer coatings, such as PSA because the removed portion of a multilayer curtain would mix in the catch pan, and so would have to be discarded.

10.3 Width Change for Slot Dies

Slot dies with a fixed geometry cannot be equipped with a deckle system as described above for slide dies because such deckles cannot achieve a tight seal in the blocked portion of the metering slot. However, liquid leakage from the slot must be avoided during slot coating. Ideally, therefore, a different die is needed for each individual coating width. The advantages of this approach are short change-over times, and most of all, excellent mechanical precision of the die slots, e.g., <±1 μm, which is necessary for high film thickness uniformities.

There are other known design solutions for changing the coating width. Some coating companies insert a shim into the outer metering slot in an attempt to block the flow to obtain the preferred width. If the shim is a bit thicker than the slot, it will seal properly, but the die must be disassembled to insert and fasten the shim. This will result in longer change-over times. Moreover, the die slot will slightly be deformed due to the over-thick shim, leading to a reduced cross-web uniformity. If, on the other hand, the shim is a bit thinner than the slot, then the shim can be inserted without disassembling the die (difference between slot height and shim thickness, $\Delta w \approx 25$ μm), but it will not seal. To prevent leakage, and to prevent the shim from being pushed out of the slot by the internal fluid pressure, the shim is secured with an adhesive tape that is applied over the die lip as sketched in Fig. 10.5a. Obviously, the thickness of the adhesive tape must be quite a bit thinner than the height of the coating gap to prevent mechanical interference during coating. Furthermore, flow visualization experiments carried out at Polytype Converting revealed that only plugging the outer slot or the slot exit generates a severe cross profile. Specifically, fluid in the inner cavity flows through to the cavity end and then up the inner slot and into the outer cavity. Because the passage through the outer slot is blocked by the shim, and owing to the low flow resistance in the outer cavity, the fluid then easily flows inwards and away from the cavity end to the end of the shim. However, a collision between two flow streams occurs at that location because the correct flow rate also arrives from the inner slot. Hence, all the returning flow in the outer cavity must be added to the correct flow, which is completely unacceptable. In summary, therefore, this approach cannot work and is not recommended for any application.

To overcome the seal problem with deckles in slot dies, TSE Troller AG has developed an alternative approach involving replaceable lip inserts in combination with

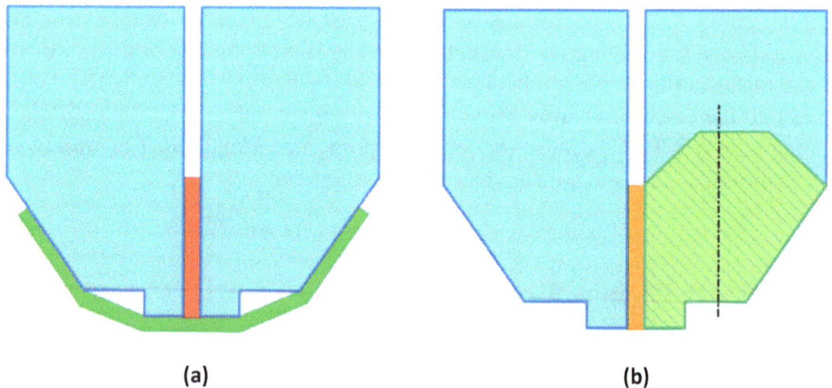

Fig. 10.5 **a** Example of width change with shim in the outer slot and adhesive tape. **b** Example of width change with replaceable lip containing sealing element, in combination with deckles for plugging both cavities and the inner slot; design concept by TSE Troller AG

standard deckles as described in Sect. 10.2. As shown in Fig. 10.5b, the replaceable lip is mounted into a prism and fastened to the die plate with a set of screws. One of the surfaces of the lip insert essentially makes up one of the sides of the outer metering slot. Where the slot is to be sealed, the lip is equipped with a specially designed compressible seal material, which, upon inserting and fastening the lip, achieves a tight seal without mechanically deforming the slot.

The disadvantage of this approach is a longer change-over time if screws are used to fasten the lip. If hydraulic bolts are used, the change-over time can be reduced, but the hydraulic system adds complexity and cost to the coating equipment. Another disadvantage is the lower mechanical precision of the outer metering slot, which translates directly into increased cross-web nonuniformity according to Eq. 9.14. With replaceable lips, therefore, it is not possible to achieve high film thickness uniformities, particularly if the coated films are thin.

Coating width variation can also be achieved with a slot die of fixed geometry, in which the slot heights and the coating width are defined by a U-shaped metal shim, see Fig. 10.6.

With this approach, the mechanical slot precision is not good either: $\Delta w \approx \pm 3\,\mu m$ if the die plates are of good mechanical precision and if the shim is lapped. In addition, the inner and outer metering slots must be of the same height. Moreover, the die must be disassembled to replace the shim. This not only leads to high change-over times, but the frequent opening of the die increases the risk for mechanical damage of sharp edges and lapped surfaces.

Based on the arguments above, the biggest drawback of slot dies with variable coating width is their inability to achieve coated films with an excellent cross-web uniformity, due to the inferior mechanical precision of the slot heights compared to dies with a fixed coating width.

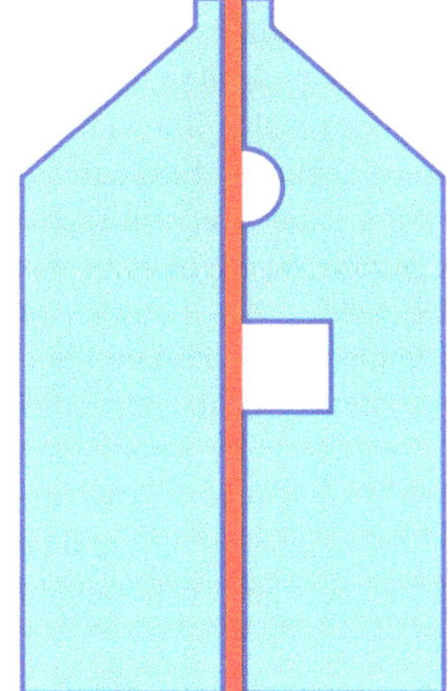

Fig. 10.6 Coating width variation with U-shaped shim

Choosing the best system for changing the coating width may also involve economic considerations. In particular, changing the coating width requires the deckle system of the die to be adjusted, and therefore the coating process must be interrupted. Consequently, a frequent change in coating width will considerably reduce the productivity of the coating process. Moreover, increasing the number of different web widths reduces the lot sizes per coating campaign because in most cases, the coating process must also be interrupted to change the web width (repositioning of web steering sensors, etc.). Quantification of these effects can be accomplished with the help of an economic model, which was introduced by Schweizer (1997), and which is also presented in Chap. 15. Specifically, process productivity is calculated in terms of the specific machine utilization SMU and the substrate yield Y_S, both of which relate the productivity to all relevant process parameters, including the coating width, the loss of substrate yield due to uncoated edges as well as start and stop losses, and the gross lot size, which reflects the number of web widths. The parameter SMU is measured in units of time it takes to produce a unit surface of coated film. The substrate yield Y_S measures the ratio of usable coated web surface to uncoated gross web surface.

High productivity is synonymous with low SMU and high Y_S values. However, a frequent change of web and/or coating width will lead to high SMU and low Y_S values because it will not only increase the non-productive change-over time, but it will also decrease the gross lot size since the annual volume of a given product will then be manufactured in many small lots instead of a few big ones.

Furthermore, the unit price of substrate stock will increase for the reason that the annual base surface of a given product will have to be made up of many small lots of rolls of different width instead of a few big lots of rolls of just a few different widths.

Based on the arguments presented above, it is apparent that width variation is much more easily accomplished with slide dies than with slot dies. It is therefore important to reflect on an appropriate operating strategy for slot dies.

One possible strategy could focus on securing a continuous production of coated products by having backups of all relevant die parts but not for all coating widths. The idea is that if a width-relevant part becomes damaged, then production is kept up by coating a different width while the damaged part is being repaired.

For a slot die with replaceable lips, for example, two die plates containing the cavities, one die plate containing the lip prism, and one lip insert per coating width are needed. On the other hand, for high-precision dies with fixed geometry, one complete die is needed per coating width. Finally, two complete dies and one shim per coating width are needed for dies that use shims to set the coating width.

We used the economic model as proposed by Schweizer (1997) for comparing the investment costs between the different operating strategies for slot dies as a function of the number of different coating widths. The model also accounted for associated differences in product quality and consequences in operating costs resulting from different effects on productivity as described by the economic model. Results showed that the option with shims is likely not the most attractive option when viewed in a balanced way. Moreover, the option with dedicated dies per coating width seems attractive, if the number of different widths does not exceed 3.

In summary, coating width variations are much more easily accomplished with slide dies than with slot dies due to reasons of sealing and liquid leakage. The major disadvantages of a slot die equipped with a variable coating width include longer change-over times, increased risk of mechanical damage during die manipulations, and, most of all, lower mechanical slot precision, which is synonymous with inferior film thickness uniformity in the cross-web direction. We therefore strongly recommend using slide or curtain coating over slot coating whenever possible, particularly if the number of different coating widths is high. Moreover, increasing the number of coating widths increases the product unit costs, which is why the number of different coating widths should be kept small.

Reference

Schweizer, P. M. (1997). Minimization of product unit costs. Example on p. 761 in Chapter 15 entitled Control and optimization of coating processes In S. F. Kistler & P. M. Schweizer (Eds.), *Liquid film coating*. Chapman & Hall.

Chapter 11
Splice Passage and Coatability

Abstract Slot and slide coating are characterized by narrow gaps between the die lip and the substrate surface that extend across the full coating width. In contrast, curtain coating does not have narrow gaps across the width, but they exist between the bottom of the curtain edge guides and the substrate at the edges of the coated film, and upstream of the impinging curtain between the suction baffle and the substrate. These gaps must be designed and operated in such a way that the splice between subsequent substrate rolls can pass through the gaps without hindrance. This chapter describes the type of splices that are typically encountered in the converting industry. In addition, operating instructions are proposed that minimize the harm inflicted by the splice to the uniformity of the coated film during the splice passage.

11.1 Introduction

In all premetered coating methods, the liquid distribution in cross-web direction is accomplished with the help of dies. In slot and slide coating, the die is positioned very closely to the web, thus generating narrow coating gaps between the die lip and the web. These gaps extend across the entire coating width. More precisely, coating gaps in slot coating processes are extremely small, i.e., on the order of the wet film thickness, typically 10–100 μm. In slide coating, gaps are still small but not as narrow as in slot coating, typically 100–250 μm, particularly for multilayer applications. In contrast, the distance between the die lip and the substrate is much larger in curtain coating compared to slot and slide coating, i.e., on the order of 100 mm (typical range: 50–300 mm).

However, even though the distance between die lip and web is large in curtain coating, narrow gaps still exist between the web and peripheral equipment. Specifically, narrow gaps are related to the baffle for removing the air boundary layer, and, if the coating width is narrower than the web width, to the curtain edge guides. Regarding the curtain edge guides, the bottom of this device must be positioned very closely above the web surface to prevent heavy edges due to contraction of the unguided curtain as a result of surface tension forces. The closer the edge guides are placed to the web, the better is the quality of the coated edges, and the lower are

P. M. Schweizer, *Premetered Coating Methods*, Engineering Materials,
https://doi.org/10.1007/978-3-031-04180-8_11

the chances for contamination of the dryer and sticky edges at the re-winder. Gaps associated with curtain edge guides typically range from 0.2 to 1.0 mm.

Similarly, if the baffle for removing the air boundary layer on the uncoated web is not placed closely to the web, excessive amounts of air can pass underneath the baffle and impinge on the back side of the curtain, thereby disturbing the curtain and potentially generating excessive coating defects, which may even lead to massive air entrainment. Typical gaps between the baffle and the web are in the range of 0.5–1.0 mm.

11.2 Types of Splices

The converting industry distinguishes between butt and overlap splices. Butt splices as sketched in Fig. 11.1a are often thinner and of better quality than overlap splices, but their application requires stopping the coating machine with the help of an accumulator. Moreover, to minimize the down-time and to maximize the splice quality, i.e., avoiding wrinkles and blisters, the tapes should be applied by an automized tape applicator system. In some cases, both sides of the web ends must be taped to sustain the required web tension.

In contrast, overlap splices can be applied on the fly, and they come in different forms, see Fig. 11.1b, c. The total splice thickness, or the extra thickness of the splice compared to a single layer of substrate, depends on the splice design and the thicknesses of the substrate and the adhesive tape being used. Special adhesive tapes as shown in Fig. 11.1c have a thickness of about 100 μm. Double-sided tapes as shown in Fig. 11.1b are about 50–60 μm thick, and single-sided tapes as shown in Fig. 11.1a have a thickness in the range of 30–50 μm.

11.3 Splice Passage

The passage of splices through the coating bead presents a severe disturbance to the local flow field. Indeed, the total thickness of the splice may be higher than the coating gap for many applications. Trying to pass such splices through the coating gap would inevitably lead to web breaks, which is not acceptable. Clearly, thinner splices are preferred because they present less of a disturbance to the coating bead.

Whether or not a given splice can be pulled through the coating bead without withdrawing the die or other equipment such as curtain edge guides or the vacuum chamber depends on the relative magnitude of the coating gap and extra splice thickness. We have noted that applications with wet film thicknesses of well above 100 μm may lend themselves to pulling the splice through the coating bead. However, care must be taken with regard to the stability of the splice, if the substrate is paper. Very thick wet films may be loaded with a lot of solvent, and this solvent may penetrate into the paper, thus causing the paper to swell laterally between the coating and

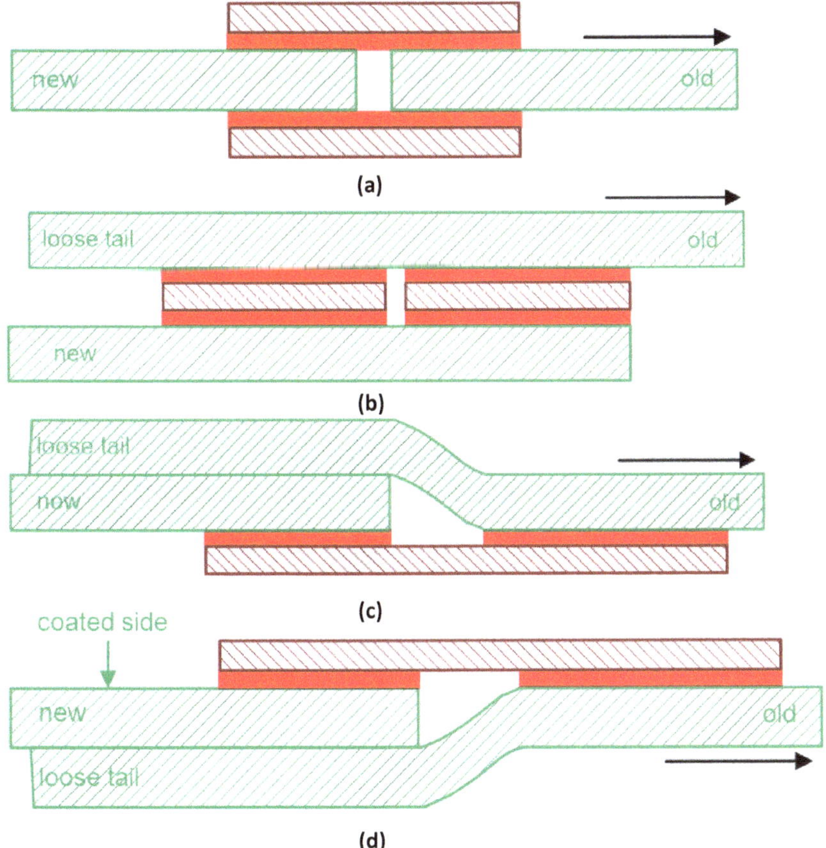

Fig. 11.1 **a** Butt splice. **b** Overlap splice. **c** Overlap splice. **d** Inverted overlap splice

drying points. At the splice, a second layer of uncoated substrate is taped to the coated substrate. This layer of the uncoated web will not swell. However, as it is attached to the swelling paper, a substantial tension across the splice will develop, which may cause the splice to break. This effect may be intensified if it takes place at elevated temperatures in the dryer.

If the wet film thickness and the associated coating gap are too small for the splice to be pulled through the coating gap, then the coating equipment must temporarily be withdrawn by a short distance, and the fluid supply to the die may have to be re-circulated. This procedure is called *skip out*. Normally, the skip out is computer-controlled, and the timing is adjusted and optimized to minimize associated material losses. The draw-back of a skip out movement is that the coating process is often interrupted for a short time. Consequently, the process must be re-started after the splice has passed through. Our experience has shown that the propensity for the formation of lines and streaks increases after every coating re-start. In the worst case, the interruption of the coating process must be prolonged, such that the operator can clean the die lip or other equipment, which facilitates the coating re-start.

Focusing on slot coating, which has the narrowest gaps of all premetered processes, the best way for avoiding problems related to temporarily stopping and re-starting the coating process is to install a fluid delivery system of the kind that is used for *intermittent coating* (see Sect. 17.14), where it is necessary to obtain close to perfect leading and trailing edges for every coated patch without generating defects in the coated film.

11.4 Splice Coatability

The term "splice coatability" addresses issues related to the passage of a splice through the coating point. "Good splice coatability" implies that the passing splice does not significantly disturb the coating process. Hence, no action is required to temporarily widen narrow gaps, for example by withdrawing the die and/or peripheral equipment. Moreover, the quality of the coated film is acceptable all the way up to the splice and immediately after the splice. Good splice coatability without any loss of product typically requires perfect butt splices as shown in Fig. 11.1a.

On the other hand, "poor splice coatability" is often associated with overlap splices, see Fig. 11.1b, c. They are characterized by loose and sometimes long tails, possibly on both web ends, which represent large disturbances to the coating process, and which may physically damage the coating equipment. Overlap splices are often used in high-speed coating because they can be applied without stopping the web. For slot and slide coating, however, the die and peripheral equipment should temporarily be withdrawn during splice passage to widen the narrow gaps and thereby prevent mechanical damage to the coating equipment.

In curtain coating, any type of splice can easily pass through the curtain, but the curtain edge guides must temporarily be skipped-up as already mentioned in Sect. 5.9.5. Normally, skip-up or skip-out motions are computer-controlled, and the timing for opening and closing the gap can easily be adjusted such that the time of out-of-perfect coating conditions is minimized.

To avoid any material losses caused by defects generated during the splice passage in curtain coating requires either butt splices or overlap splices with the loose tail tucked under the beginning of the new substrate as illustrated in Fig. 11.1d, and it requires optimization of the flow field of the impinging curtain as explained in Sect. 7.2. The importance of the latter aspect is visualized by the following example that happened many years ago at ILFORD in Switzerland during multilayer curtain coating of photographic products. In particular, we observed that temporary air entrainment occurred every time the butt splice passed through the curtain. A sample of this defect is shown in Fig. 11.2. The web moved in the downward direction, and the saw tooth wetting line that is typical for air entrainment is clearly visible. However, air entrainment disappeared again spontaneously after a short while because the disturbance had long moved downstream. The length of the saw teeth varied from

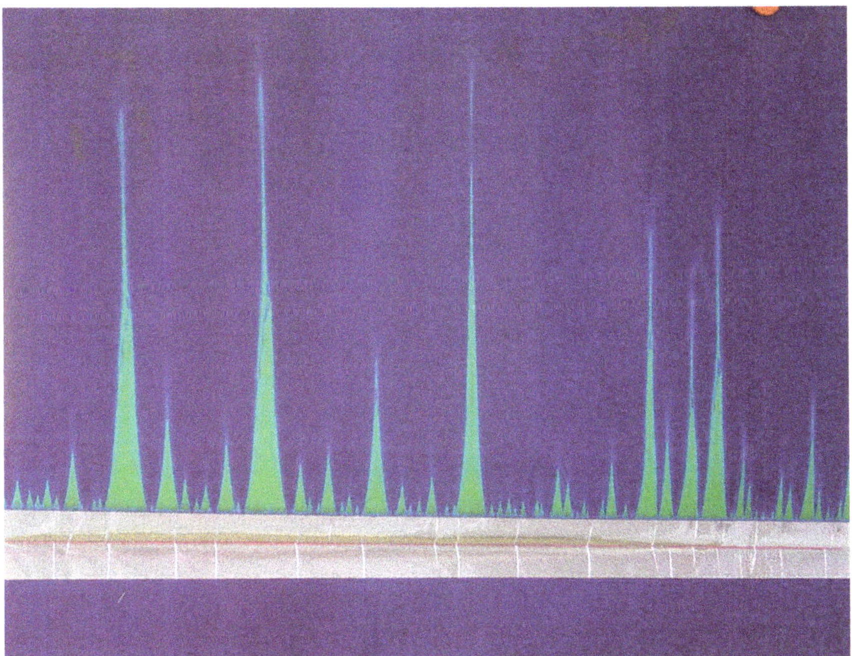

Fig. 11.2 Temporary air entrainment during splice passage through a multilayer curtain of photographic products; *photo* reproduced with permission from ILFORD Imaging Switzerland GmbH

a few centimeters to a few meters, depending on the coating speed, the total film thickness, and the rheology of the individual layers in the curtain. The splice was of the butt type. Interestingly, nevertheless, this small disturbance was able to generate significant material losses splice after splice.

Upon examining the coating conditions with the help of Eq. 5.9.37, we realized that the dynamic wetting line was located much too far behind the impinging curtain. Therefore, the hydrodynamic assist of the impinging curtain could not be utilized, which explained why a weak disturbance could trigger massive air entrainment, though only temporarily. At that time, the web departed from the coating roll in a horizontal direction toward the chill section and the dryer, thus allowing only negative curtain application angles. Note, though, the range of admissible application angles was larger compared to a situation where the web departed from the roll in a slight upward direction, see Fig. 11.3a, b. Then, we decided to modify the web path downstream of the coating point by installing a suction arc such that the substrate was able to leave the coating roller in a downward direction as shown in Fig. 11.3c. This design modification allowed us to move the curtain impingement point on the roll from $-9°$ to $+5°$. As per Eq. 5.9.37, this design change also reduced the size of the curtain heel considerably, such that the hydrodynamic assist could now be fully

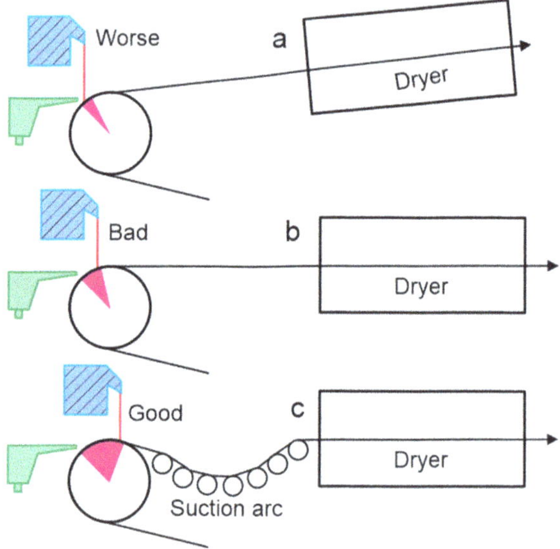

Fig. 11.3 Different web paths through the curtain coating station resulting in different ranges of admissible curtain impingement angles as visualized by the purple wedge in the coating roll

exploited and air entrainment disappeared for all products. In fact, only 1 mm after the splice the coating uniformity was perfect again, and the product was sellable.

In summary, the issue of splice quality in connection with premetered coating methods must be carefully evaluated. Butt splices, which present the smallest disturbance to the coating process, generate extra investment costs due to the accumulator and the tape applicator. The length of the accumulator is related to the maximum intended web speed and the degree of sophistication of the tape applicator. Typically, butt splices are only used for low speeds of up to a couple of hundred [m/min] to avoid excessively large accumulators. On the other hand, overlap splices generate extra investment costs due to positioning systems that allow the temporary withdrawal of various coating equipment during the splice passage. Overlap splices are necessary for high-speed applications. While this author was working there, Polytype Converting AG successfully installed several splicing units for curtain coating of pressure-sensitive adhesives at speeds of >1,000 m/min.

Chapter 12
Simultaneous Multilayer Coating Capability

Abstract The ability of simultaneously applying several different liquid layers is an attractive feature of premetered coating methods. This chapter describes several ideas that improve the functionality of the coated product and/or increase the productivity of the coating process by designing multi-layered films.

12.1 Introduction

One of the big advantages of premetered coating methods over other processes is their ability to coat several layers at the same time. The emphasis is on ***simultaneously*** coating several layers because, as explained in detail in Sect. 15.4, the economic advantages of this approach are enormous. Consider manufacturing a 5-layer product with a roll or blade coating process. The substrate would have to be sent five times through the coating and drying machine. The associated five coating starts and stops incur material losses of the substrate and coating fluids, and chances for producing coating defects such as lines and streaks at every start increase. Moreover, layers coated early on have to pass through the dryer several times, which may further degrade the performance of the functional layers. Alternatively, five coating and drying machines would have to be built in series, which would drive the space requirements as well as the investment and operating costs to unbearable levels.

In contrast, using the slide or curtain coating process, these five layers can be coated in one pass on only one coating and drying machine. Granted, the dryer would be longer than the one of a single-layer coating machine, but not five times longer because losses at the beginning of the drying process to heat the substate and the coated film to the drying temperature occur only once and not five times. Also, it is very likely that layers 2–5 may be prepared with a higher solids concentration than the first layer, thus reducing the drying load because these layers don't affect the performance of the coating process as much as the bottom layer does.

Another attractive feature of simultaneously applying at least two layers is the ability of coating ultra-thin layers with a dry thickness of <1 μm, and even down into the nanometer range. This is possible as long as the sum of the thicknesses of

P. M. Schweizer, *Premetered Coating Methods*, Engineering Materials,
https://doi.org/10.1007/978-3-031-04180-8_12

all the layers in the layer pack is high enough not to interfere with the operating boundaries of the chosen process, see Schweizer (2013).

Eastman Kodak was the first company to be granted significant patents for multilayer coating methods, i.e., Russel (1956) for the slide coating process, and Hughes (1970) for curtain coating. Since that time, extensive research activities at universities and industrial laboratories around the world together with practical experience in many coating companies (particularly from the photographic industry) have resulted in a high level of fundamental understanding of these coating processes.

Photographic color *paper*, for example, consists of 6–8 layers, and this product is typically manufactured in one pass using curtain coating at speeds of several 100 m/min. Photographic color *films* contain a dozen layers or more. Typically, they are produced in 3 passes of 3–5 layers each using slide or curtain coating at speeds also of several 100 m/min, but not quite as high as those for color paper.

Nowadays, coating activities in the photographic industry have been reduced considerably, but new product families have emerged that take advantage of simultaneous multilayer coating capabilities, such as photo-quality ink jet papers, optical films, high-grade thermal papers, pressure-sensitive adhesives for labels (Herma, 2007), packaging materials on paper or board substrates, electrodes for Li-ion batteries (Diehm et al., 2020) and OLEDs (de Vries, 2018).

12.2 Examples of Multilayer Products

Figure 12.1 shows an example of a multi-layered product. It is a cross-section of a 10-layer photographic film manufactured by ILFORD Imaging Switzerland GmbH. The substrate is Melinex, a voided polyester, and the total dry thickness of all coated layers is 23 μm. As shown, the thickness of each layer is very uniform, and all interfaces between adjacent layers are straight and uniform.

Following are a more examples that visualize the almost endless possibilities for designing multilayered products to improve the function of the final product and/or the productivity of the coating process.

Figure 12.2 shows an adhesive label, where the PSA, which typically is coated as one layer, is now designed as a 3-layer structure. Traditionally, the total thickness of the adhesive layer is relatively thick, i.e., about 20 μm, to allow the cutting tool that defines the size and form of the label to penetrate through the label material, plus a bit into the adhesive layer, but not down to the siliconized substrate. However, the bulk of this thick layer (marked in blue) does not need adhesive properties. Therefore, material costs could be reduced by using a thin adhesive for layer 1 that has adhesion properties optimized for the material onto which the label is applied, i.e., a glass bottle or a paper envelope, and by using another thin adhesive for layer 2 with adhesion properties optimized for the label material. The thick bulk layer, however, is made of a simpler and cheaper polymer that has no adhesion properties.

Figure 12.3 illustrates the concept of the so-called **carrier layer**, the purpose of which is to improve the productivity of the coating process, see for example Dittman

Fig. 12.1 Cross-section of the photographic film ILFOCHROME CPA.1 K; photo reproduced with permission from ILFORD Imaging Switzerland GmbH

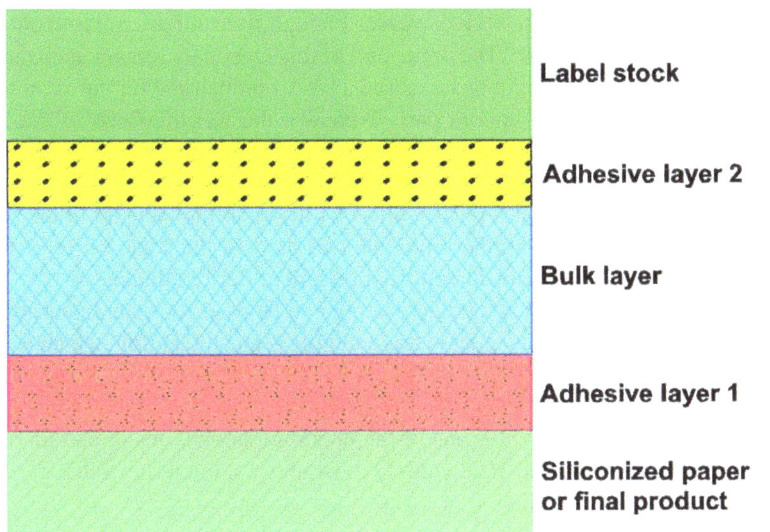

Fig. 12.2 Example of a 3-layered adhesive structure for labels

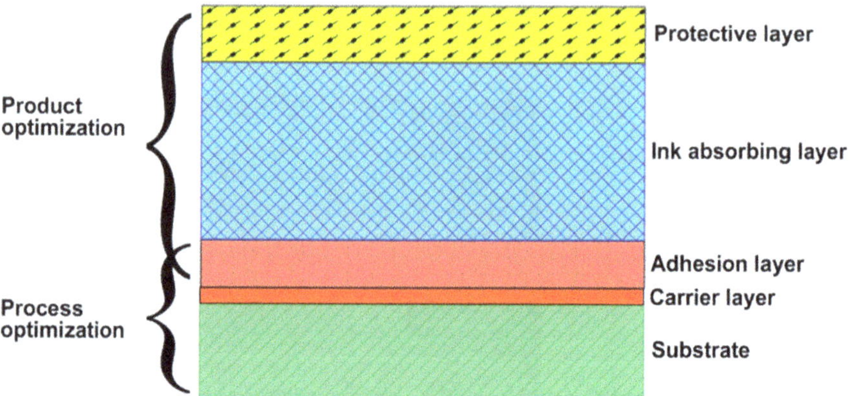

Fig. 12.3 Exploring multilayer coating technology by implementing the carrier layer concept, which improves the process productivity

and Rozzi (1977) or Koepke et al. (1983). In this example, the product is a 3-layer ink jet paper. Suppose that the adhesion layer at the bottom of the layer pack has a high viscosity. This may limit the maximum coating speed to excessively low values to prevent air entrainment at the dynamic wetting line. This limitation could be avoided by splitting the adhesion layer into two layers that are of identical chemical composition, but the bottom part of this layer, i.e., the carrier layer, is diluted so much that the resulting viscosity decreases enough to avoid air entrainment at the desired higher coating speed. The upper part of this layer may remain unchanged, or the solids concentration might be increased a bit to compensate for the extra solvent that is introduced into the bottom part. Typically, the wet thickness of the carrier layer is small, i.e., just a few microns. Best values for the thickness and the reduced viscosity must be found by experiment. Instead of 3 layers, this product now has 4 layers, but this complication is readily offset by the resulting increase of the process productivity. After drying, furthermore, the two parts of the adhesion layer reunite to one layer of uniform chemical composition and of desired thickness.

Figure 12.4 visualizes how curtain coating can be used for manufacturing ultra-thin functional layers. The product in this example is an optical film, which is coated onto a clear plastic substrate such as PET film. The dry thickness of such films is in the range of 10–100 nm.

As is explained in Chap. 19, one of the operating boundaries of curtain coating requires the volumetric flow rate/width Q to be above a minimum value of about 1 cm^2/s to facilitate the formation of a stable curtain and to prevent air entrainment at low speeds. Now, if the goal is to produce an ultra-thin dry film of just a few nanometers, then the thickness of the corresponding wet film cannot exceed a few micrometers. Following Eq. 2.1 this would require the coating speed to be at least several 100 m/min, which might be much too high for coating companies in this industry. This problem can be overcome by adding an inert bulk layer, which has no product function, but that helps to increase the sum of all layer flow rates/width

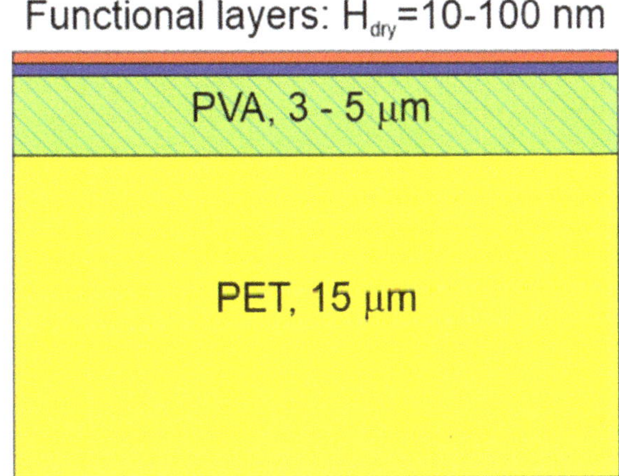

Fig. 12.4 Example of using an additional bulk layer to allow ultra-thin functional layers to be applied with curtain coating

to about 1 cm^2/s. Moreover, if this layer consisted of just a simple polymer, e.g., polyvinyl alcohol (PVA), water, and perhaps a small quantity of surfactant, it would not deteriorate the product function because its effect is as if the PET substrate was not just 15 μm but perhaps 18 μm thick as indicated in the figure, see Schweizer (2013).

This concept was tested on the pilot machine of Polytype Converting AG as follows. The bulk layer was an aqueous PVA solution with a solids concentration of 0.1045 and a dry thickness of 3.5 μm. The single functional layer was also aqueous PVA with a concentration of 0.0285 and doped with a blue dye. Its flow rate was adjusted such that the sum of the bulk and functional layers added up to 0.8 cm^2/s. When this 2-layer pack was curtain coated at a speed of 135 m/min, then the dry thickness of the functional layer was 46 nm. Had we doubled the speed to 270 m/min, which is not an excessively high value, the dry thickness would have decreased to 23 nm, a truly ultra-thin film. Unfortunately, we could not do that because we ran out of fluid.

We also saw that checking the optical uniformity of this very thin film was difficult, even though the film was dyed for a better appearance because the film was barely visible by the naked eye. This is of no concern, though, when using a premetered coating method because the film thickness in the machine direction is always correct and very uniform, and the thickness in cross-web direction is of similar quality as long as the internal die geometry is optimized for this type of application.

Another way of exploring the advantages of simultaneous multilayer coating is depicted in Fig. 12.5. The example refers to curtain coating a single layer of an aqueous pressure-sensitive adhesive for making labels when the adhesive machine is built in line with a proceeding machine for siliconizing the substrate. It is known in this industry that high coating speeds can be achieved, i.e., up to 1′000 m/min or even higher, if the adhesive is coated onto a paper substrate. However, if the substrate is a plastic film such as PET for clear-on-clear applications, then the maximum speed must be reduced to about half that value for the following reasons. On the one hand,

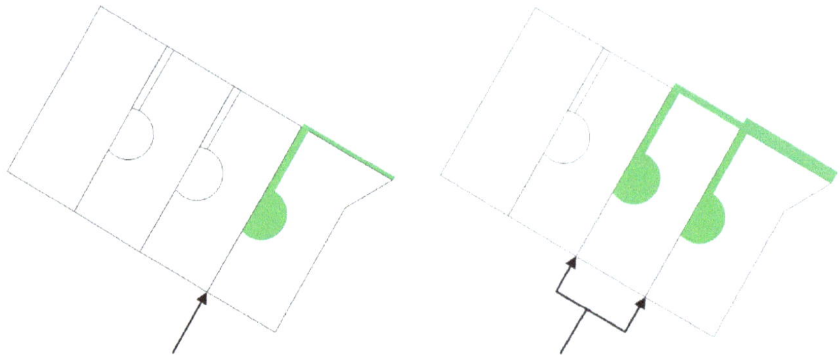

Fig. 12.5 Exploring multilayer coating technology by feeding low flow rate applications through one die slot, and by splitting high flow rates and feeding them through two slots to achieve very uniform cross profiles for a wide range of flow rates

the speed limitation is imposed by the given length of the silicone dryer because PET is thermally less stable and cannot be heated as high as paper. On the other hand, the residual moisture in the adhesive layer of clear-on-clear products must be very low, say <1–2%, to obtain a clear film, while the residual moisture in paper labels can be as high as 5–6%. Therefore, excellent cross profiles in the adhesive layers can be achieved for both low and high flow rates that differ by a factor of about 2 by using a multilayer slide die, and by feeding the low flow rate through just the bottom slot while splitting the high flow rate into equal parts and feeding them through two die slots. After passing through the die the two films on the die slide reunite again to a single layer of uniform concentration and of correct flow rate.

A few years ago, Polytype Converting AG was able to convince a customer of this concept. Soon thereafter, this customer realized that he now had a multilayer die, with which he could also coat truly multilayered products. He did not waste much time before he started developing such materials. We think this is the perfect attitude when involved with premetered coating methods because it takes full advantage of the many attractive features these processes offer for improving product properties and process productivity.

References

De Vries, I. (2018). Advances in organic & printed electronics processing. In *Proceedings of the 3rd Thin Film Technology Forum*. Karlsruhe Institute of Technology. Karlsruhe, Germany.

Diehm, R., Kumberg, J., Dörrer, C., Müller, M., Bauer, W., Scharfer, P., & Schabel, W. (2020). In situ investigations of simultaneous two-layer slot die coating of component-graded anodes for improved high-energy Li-ion batteries. *Energy Technology, 8*, 1901251.

Dittman, D. A., & Rozzi, F. A. (1977). Method of multilayer coating. US Patent 4,001,024.

Herma. (2007). Converter of the month. Coating & Converting, March/April.

Hughes, D. J. (1970). US Patent 3,508,947.

Koepke, G., Frenken, H., Bussmann, H., & Brawatzki, K. (1983). Verfahren zur Mehrfachbeschichtung von bewegten Bahnen. European Patent EP 001 0074 A2.
Russel, T. A. (1956). US Patent 2,761,417.
Schweizer, P. M. (2013). Coating of ultra-thin liquid films. Convertech and e-Print Japan, Part 1: March/April, 96–103, Part 2: May/June, 4–8.

Chapter 13
Mixing of Adjacent Layers

Abstract This chapter addresses the question of interlayer mixing, be it by convection or diffusion. Convective mixing can be prevented if the multilayer film flows are kept laminar. Mixing by diffusion cannot be prevented, but ways are suggested to minimize adverse effects in the coated film.

13.1 Mixing Due to Convection

The question of whether or not adjacent layers in a simultaneous multilayer coating application will mix must be analyzed from two different points of view, i.e., convection-driven and diffusion-driven mixing. Mixing due to convection does not happen for adjacent layers in multilayer film or coating flows, even though they may flow around sharp corners and bends, and even though they may be highly accelerated in the boundary layer near the dynamic wetting line. The reason for the absence of convective mixing is shown in Fig. 13.1 and explained by the fact that these flows are truly laminar, with Reynolds numbers of individual layers and the entire layer pack typically being in the range of 0.1–10.

In Fig. 13.1a one layer is exiting a die slot, turning around a 90° corner and flowing down an inclined plane. A second layer is generated in the same way farther upstream, and this layer is subsequently flowing on top of the bottom layer. The interface between the two layers is visualized by injecting a thin filament of green dye inside the die slot of the lower layer. The form of the interface remains smooth and highly continuous, thus indicating that the interface is the dividing streamline between the two layers of this film flow. No fluid from the bottom layer can flow across the interface into the upper layer or vice versa. For mixing to occur, cross-flow relative to the main flow direction must exist, but the photo proves otherwise. In addition, the inertia of the bottom layer in this example is too high, such that the layer cannot turn around the sharp corner without separating from the die surface. Hence a vortex forms just downstream of the corner. The presence of a vortex does not at all indicate that the flow is turbulent. On the contrary, the fluid inside the vortex is slowly rotating clockwise and the flow is still laminar. This is visualized by a second green streamline generated inside the die slot, and this streamline separates

P. M. Schweizer, *Premetered Coating Methods*, Engineering Materials,
https://doi.org/10.1007/978-3-031-04180-8_13

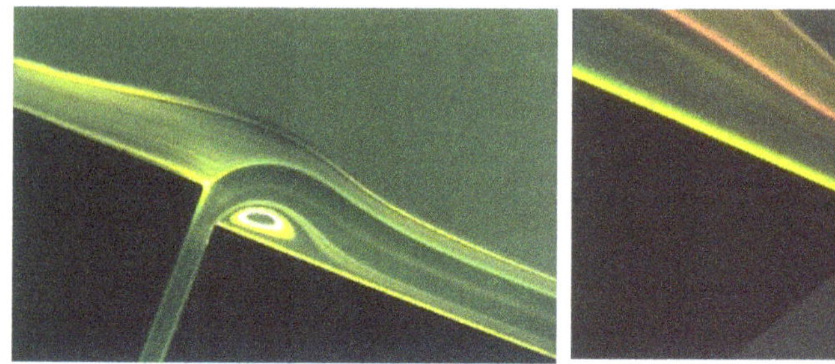

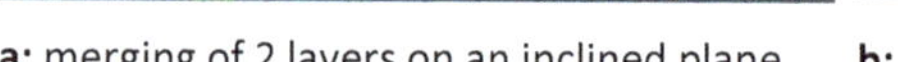

Fig. 13.1 Visualization of 2-layer film flow in slide coating. Photos taken from Schweizer (1997) and reproduced with permission from Springer Nature

the vortex from the film that flows around the vortex. Even if the flow inside the vortex were turbulent, it would not cause any mixing between the two layers on the die slide because mixing occurs at the interface between the two layers, and that interface is located far away from the vortex.

Following the interface between the two layers to the end of the die slide and into the bead of a slide coating process (Fig. 13.1b), the interface, now marked by a red dye filament, is still smooth and continuous. Inside the bead, the flow changes direction by about 90°, and, after being coated, it is accelerated by a factor of about 10 by the motion of the moving substrate. This can be verified by the thickness of the coated film, which is approximately 10 times smaller than the thickness of the 2-layer film on the die slide upstream of the bead. Despite these dramatic events imposed on the 2-layer film flow, the interface remains smooth and continuous throughout, thus indicating that the flow remains laminar and mixing does not occur.

13.2 Mixing Due to Diffusion

Taking a closer look at the interfaces between adjacent layers in Fig. 13.2, particularly between the red and white as well as the yellow and white layers, reveals that some small amount of diffusion-driven mixing did occur. Mixing due to diffusion cannot be prevented, if concentration differences of specific chemical species exist across the layer interface. The question is whether this diffusion effect negatively affects the performance of the final coated product, but this question is difficult to answer. On the one hand, the time for diffusion is short and does not exceed a few seconds. Diffusion starts on the die slide where adjacent layers merge, and it ends a short distance into the dryer, where enough solvent has evaporated such that the viscosity is high and the mobility of diffusing molecules is low. Should diffusion-driven mixing be a problem in a given product, then it can be reduced or suppressed by increasing the solids concentration of the coating fluid, which increases the viscosity, or by adding

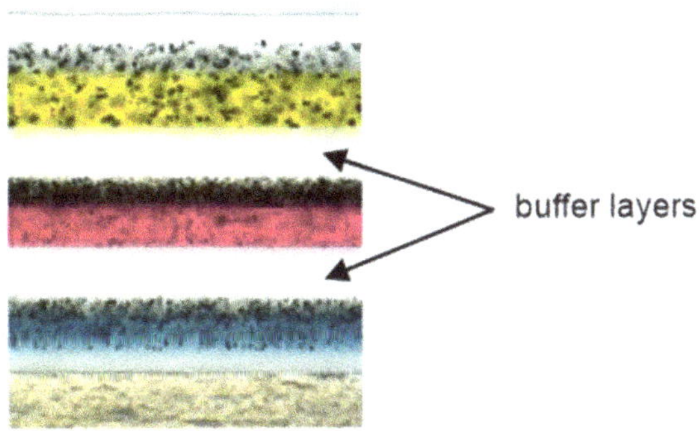

Fig. 13.2 Separation of functional layers by buffer layers to prevent unwanted effects from diffusion-driven mixing. Photo reproduced with permission from ILFORD Imaging Switzerland GmbH

chemically inactive buffer layers to physically separate adjacent functional layers, see white layers between colored layers in Fig. 13.2. This concept was often implemented in photographic materials as shown in the figure because a true rendition of natural colors is not possible by the photo, if interface diffusion effects are excessive. In this example, the photographic product is composed of eight functional layers, but the number of layers had to be increased by 2 to prevent detrimental diffusion effects. This complication, however, can easily be afforded because the product can still be coated in one or perhaps two passes, and thus at high productivity, instead of the three passes that would have been necessary without the buffer layers.

Raupp et al. (2017) used inverse micro-Raman spectroscopy for measuring the interdiffusion of a two-layer system. Their measurements confirmed that intermixing of the polymers does not occur in immiscible systems. However, interdiffusion could be measured in-situ when they investigated the miscible ternary model system PMMA—PVAc—toluene. They found that the average solvent content was more important than the molecular weight of the polymers for the interdiffusion kinetics. This kind of study is relevant for the manufacturing of multilayer OLEDs.

In a related paper, Raupp et al. (2019) developed and validated a theoretical model that was able to accurately predict the interdiffusion behavior of PMMA and PVAc in the presence of toluene.

Merklein et al. (2019) used slot coating to sequentially apply four layers in order to build a four-layered OLED device. The best product performance was obtained by suppressing interlayer diffusion, and this was the case when they worked with cross-linkable materials and when they applied the orthogonal solvent approach. Orthogonal means that the solvent used in one layer does not dissolve in any other layers.

While the results produced by Raupp et al. (2017, 2019) and Merklein et al. (2019) cannot directly be applied to simultaneous multilayer coating applications, they do provide valuable information that will allow emerging electronic products to be manufactured efficiently with simultaneous multilayer coating technology.

References

Merklein, L., Mink, M., Kourkoulos, D., Ulber, B., Raupp, S. M., Meerholz, K., Scharfer, P., & Schabel, W. (2019). Multilayer OLEDs with four slot die-coated layers. *Journal of Coatings Technology and Research, 16*(6), 1643–1652.

Raupp, S. M., Siebel, D. K., Kitz, P. G., Scharfer, P., & Schabel, W. (2017). Interdiffusion in polymeric multilayer systems studied by inverse micro-Raman spectroscopy. *Macromolecules, 50*, 6819–6828.

Raupp, S. M., Kitz, P. G., Siebel, D., Wunsch, A., Merklein, L., Scharfer, P., Schabel, W. (2019). Modeling of interdiffusion in poly(vinyl acetate)—poly(methyl methacrylate)—toluene multicomponent systems. *Journal of Applied Polymer Science.*

Schweizer, P. M. (1997). Experimental methods. In S. F. Kistler, P. M. Schweizer (Eds.), *Chapter 7 in Liquid Film Coating*.Chapman & Hall.

Chapter 14
Guidelines for Designing Single-Layer and Multilayer Films

Abstract Applying simultaneous multilayer coating technology, particularly slide and curtain coating, requires well-designed multilayer films that flow down on the inclined surface of the slide die and, in the case of curtain coating, through the curtain prior to being coated. These film and curtain flows must be optimized in terms of film thickness uniformity and fluid homogeneity because any flaws in these films will be transported through the coating process, and they will end up in the coated film where they deteriorate the uniformity of the final product. This chapter provides guidelines for designing singlelayer and multilayer liquid films.

14.1 Introduction

When designing singlelayer or multilayer films for premetered coating methods, the focus from a process point of view is on physical fluid properties, i.e., density, surface tension and rheological properties, as well as on how these properties should be structured across multilayer films. A comprehensive review of this subject with a focus on surfactants and static and dynamic surface tension was written by Tricot (1997).

Regarding the solids concentration, this parameter is not a physical property of the coating fluid. From a coating point of view, therefore, boundary conditions for the solids concentration do not exist. However, satisfying boundary conditions with regard to physical fluid properties may conveniently be achieved by altering the solids concentration of the fluid. Such an alteration is physical in nature and does not change the chemical composition of the product. Moreover, the solids concentration is tightly connected to the drying process. Therefore, satisfying boundary conditions relating to physical fluid properties may require a solids concentration of a specific value, which in turn would require—together with the dry coat weight and the desired coating speed—a specific drying capacity (dryer length).

P. M. Schweizer, *Premetered Coating Methods*, Engineering Materials,
https://doi.org/10.1007/978-3-031-04180-8_14

14.2 Guidelines for Single-layer Films

For single-layer films, guidelines and boundary conditions only apply to the viscosity and the surface tension, but not to the density.

14.2.1 Viscosity

Throughout the converting industry, the term "viscosity" is often assumed to be the low shear viscosity, such as measured with a Brookfield viscometer or DIN or FORD cup. However, as is explained in Sect. 4.4, the viscosity of most coating fluids depends on the solids concentration, the temperature, and the shear rate of the flow under consideration. Moreover, most polymer melts, polymer solutions, and dispersions exhibit a shear-thinning behavior, meaning that the viscosity decreases as the shear rate increases. Therefore, discussions involving the viscosity are often meaningless without knowing the dependence of the viscosity upon the shear rate, and without focusing on a particular flow field with a specified shear rate.

For premetered coating processes, the viscosity affects many quality- and productivity-related aspects, such as

- Degassing of the coating fluid.
- The pressure drop in solution supply lines, across filters, and the die
- the wall shear stress in pipe, duct, and slot flows, which is a measure for the tendency of a flow system to contaminate (sedimentation) or for the ability of a flow system to clean itself by way of fluid flow forces.
- The film thickness nonuniformity in the cross-web direction.
- The formation of waves in film flow on an inclined plane.
- The location of the dynamic wetting line in the curtain heel or the coating bead of slide and slot coating.
- The maximum speed of wetting (onset of air entrainment).
- The leveling of uneven films after coating.

As discussed in Chap. 5, many of the aspects in the above list can be modeled and optimized fairly well with simple analytical tools. Here, for reasons of simplicity, we focus on just a couple of general guidelines, one for the low shear viscosity μ_0, and one for the high shear viscosity μ_∞.

The guideline for the low shear viscosity μ_0 requires

$$\mu_0 = order\ (100\, mPas) \tag{14.1}$$

Ideally, the low shear viscosity should be on the order of a few hundred [mPas], say between 50 and 1,000 mPas. Such viscosities are low enough to facilitate degassing, to result in low pressure drops even for long pipes with small diameters and in die cavities. Moreover, such viscosities are high enough to generate high wall shear stresses, to prevent the formation of waves in film flows, to minimize flow after coating (leveling, formation of fish eyes), and to assure laminar flows throughout the coating process by keeping the Reynolds number low.

While premetered coating methods can successfully be implemented with coating fluids having a low shear viscosity in the range of 5,000–10,000 Pas, it is more comfortable to work with fluids having low shear viscosities in the range of 50–1,000 mPas. It is suggested, therefore, to aim for moderate values of the low shear viscosity, preferably by adjusting the solids concentration, or perhaps by adjusting the fluid temperature or by adding thickening agents.

In contrast, the guideline for the high shear viscosity μ_∞ requires

$$\mu_\infty = order\,(10\,mPas) \tag{14.2}$$

For premetered coating methods, the shear-thinning rheological behavior is very welcome, particularly for curtain coating regarding the flow field of the curtain impingement zone, which controls whether or not air is entrained at the dynamic wetting line. As summarized by Blake and Ruschak (1997), many scientific studies from the past show that the inception speed of air entrainment increases with decreasing viscosity near the dynamic wetting line. In concert with these findings, guideline 14.2 requires the viscosity at the prevailing shear rate near the dynamic wetting line to be on the order of a few 10 [mPas], say between 10 and 100 mPas. If the high shear viscosity is >100 mPas, then high web speeds without air entrainment may be difficult to achieve.

From a drying and energy consumption point of view, any kind of viscosity should always be maximized by maximizing the solids concentration, which in turn minimizes the solvent quantity that must be evaporated. The viscosity guidelines posted above, therefore, may conflict with guidelines for optimizing the drying process. Best viscosity values, or best values of the rheological parameters of the Carreau-Yasuda model, must always be determined by optimizing the various aspects of the various coating flows as explained throughout this book, and by finding a compromise between coating and drying requirements.

14.2.2 Surface Tension

As alluded to in Sect. 4.5.4, issues like curtain stability, resistance to repellencies, wetting of substrates at low speeds, capillary rise on a vertical surface, and edge withdrawal are all positively affected by low values of surface tension. On the other hand, leveling after coating, the formation of waves in film flow, and perhaps the resistance to film deformation due to ambient pressure disturbances are positively affected by high values of surface tension. Due to these conflicting requirements, it is difficult to find the best surface tension value for a given application. Optimizing the surface tension by weighing the importance of the various issues listed above might be the best or the only way forward to obtaining good results.

In our experience, good results with all premetered coating methods are often obtained if the following guideline is respected:

$$\sigma < 40\ mN/m \tag{14.3}$$

This operating boundary is particularly important for curtain coating to achieve acceptable curtain stability and to prevent the formation of holes in the curtain.

14.3 Guidelines for Multilayer Films

The guidelines for single-layer films presented in the previous chapter generally also apply to the individual layers of multilayer films, see Fig. 14.1. In addition, multilayer films should respect requirements concerning the **structuring** of density ρ, viscosity μ, and surface tension σ between adjacent layers of the layer pack.

14.3.1 Density Structuring

In a multilayer film, the densities of the various layers must satisfy the following requirement, see Fig. 14.2.

$$\rho_{i+1} \leq \rho_i \quad \mathrm{i} = 1 \ldots \mathrm{n} \tag{14.4}$$

Satisfying the density structuring requirement 8.7.4 prevents the multilayer film from becoming unstable and deformed by a heavier top layer pressing on a lighter bottom layer. In practice, however, it may often be difficult to satisfy this requirement.

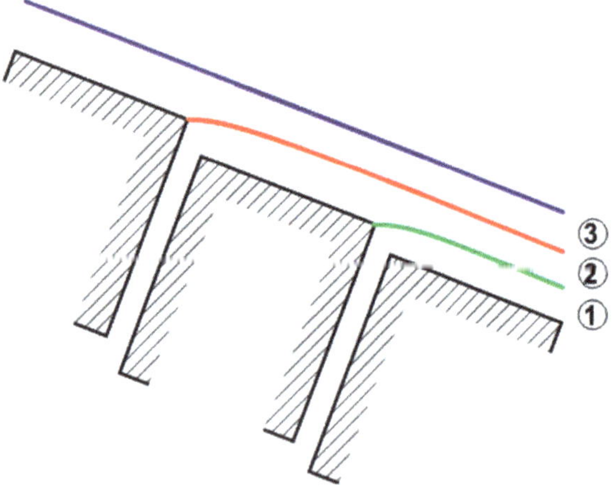

Fig. 14.1 Multilayer film on an inclined die slide

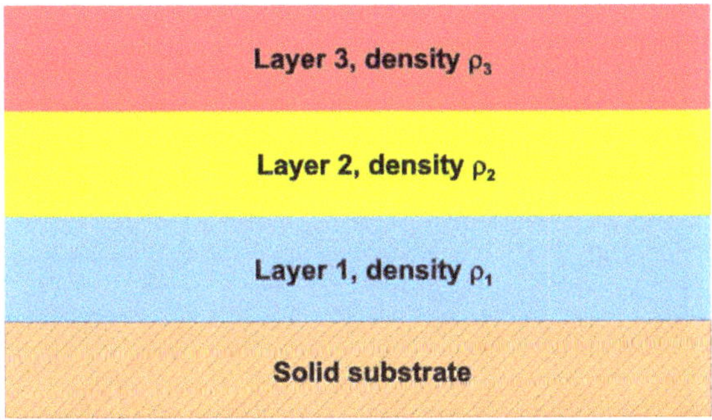

Fig. 14.2 Multilayer film with different densities

Adjusting the density of a given layer is usually not possible because the density results from the solid chemical materials contained in that layer, and those materials are in there to achieve the desired product function. Adding or removing solids from that layer would change the product function.

In practice, adverse effects from violating the density structuring requirement are rarely observed because typically, densities of coating fluids vary by less than a factor 2.

14.3.2 Viscosity Structuring

As explained in Sect. 5.7.4, interface waves in multilayer film flows are more damaging than surface waves. The instability originates at an interface between adjacent layers, and the resulting waves are transmitted across other interfaces towards the free film surface. Interface waves are of a much shorter wavelength than surface waves, even on the inclined surface of a slide die. In most cases, they dominate surface waves in terms of resulting film thickness nonuniformity. Since interfaces between miscible fluid layers are tension-free, the wave damping effect based on local surface tension differences is not available here.

There is no simple theory for predicting the onset of interface waves as a function of operating parameters and fluid properties. However, multilayer films with an inflection point in their velocity profile are especially susceptible to interface wave instability. The presence of an inflection point in the velocity profile is a necessary but not sufficient condition for the presence of interface waves.

As shown in the example of Fig. 14.3, velocity profiles with an inflection point are likely to occur, if the viscosity of the top layer is much lower than the one in the bottom layer, and if the flow rate in the top layer is much higher than that in the bottom layer.

Consequently, interface waves can be avoided by avoiding velocity profiles with inflection points in multilayer film flows, and this can be achieved by respecting the following requirement for structuring the viscosity of adjacent layers, see Fig. 14.4.

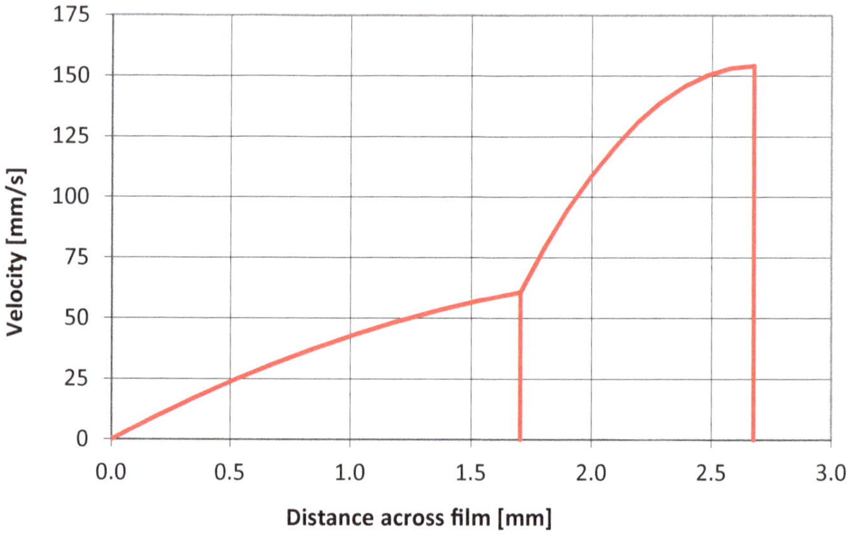

Fig. 14.3 Velocity profile of a 2-layer film; $\mu_2/\mu_1 = 0.1$; $Q_2/Q_1 = 2.0$

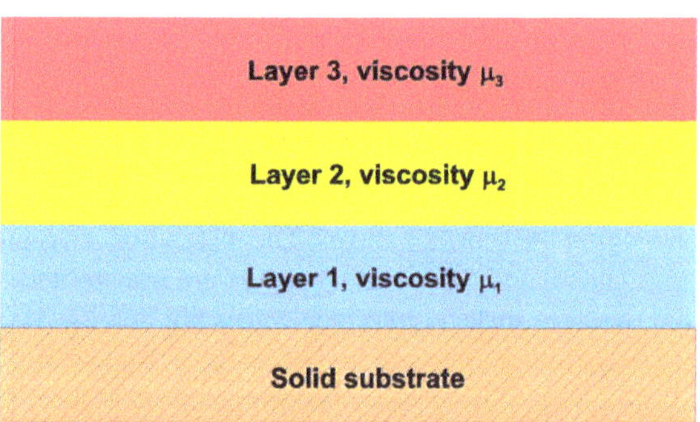

Fig. 14.4 Multilayer film with different viscosities

$$\mu_{i+1} \geq \mu_1 \quad i = 1 \ldots n \tag{14.5}$$

Satisfying the requirement for structuring the viscosities means that the bottom layer of a multilayer pack has the lowest of all viscosities. Often, this is also helpful for preventing air entrainment at the dynamic wetting line in high-speed coating applications, and it is necessary for implementing the carrier layer concept for preventing ribbing lines in slide coating, see Sect. 18.5.1 for more details.

14.3.3 Surface Tension Structuring

Apart from the aspects described for single-layer films, the relative wetting of adjacent layers is strongly affected by surface tension. It is essential that the upper layer of a multilayer film completely wets the adjacent lower layer, see Fig. 14.5a. If the upper layer is unable to properly wet the lower layer, then it will detach and withdraw from the confining side wall, contract, and form a heavy edge bead as depicted in Fig. 14.5b. At the same time, the lower layer becomes exposed to the surrounding atmosphere. This not only results in a reduced useable width of the coated product, but for curtain coating, it also negatively affects the flow of the curtain along the edge guide because the curtain thickness will be reduced by the missing top layer. Consequently, and worse yet in most cases, it will cause the curtain to detach from the edge guide, which is equivalent to a catastrophic process failure.

Wetting is related to spreading, and spontaneous spreading, i.e., proper wetting of the lower film by the upper film occurs, if the surface tensions of the various layers satisfy the following requirement, see also Sect. 4.6.1 and Fig. 14.6.

$$\sigma_{i+1} \leq \sigma_i \quad \mathrm{i} = 1 \ldots \mathrm{n} \tag{14.6}$$

Fig. 14.5 Relative wetting of adjacent layers in a multilayer film flow; **a** proper wetting, if $\sigma_2 < \sigma_1$; **b** improper wetting with edge withdrawal, if $\sigma_2 > \sigma_1$

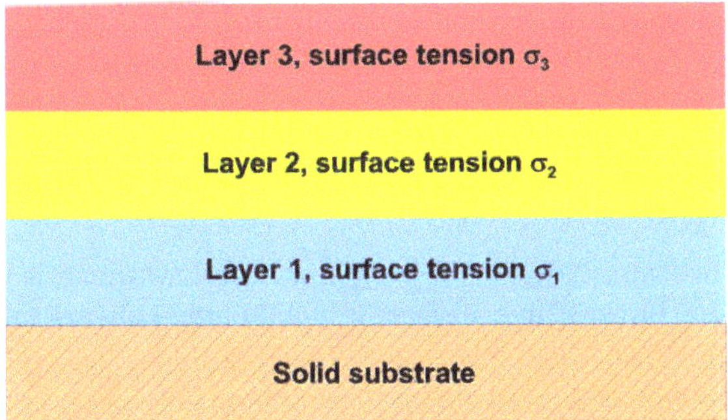

Fig. 14.6 Multilayer film with different surface tensions

Guideline 14.6 requires the surface tension to not increase from the bottom layer towards the top layer of a multilayer film. In other words, the surface tension should remain the same from one layer to the next layer above, or, preferably, it should slightly decrease. The surface tension difference between adjacent layers should be small, particularly if the liquid film contains many different layers. As mentioned in Sect. 4.5.1 for the ideal case, the admissible range for the surface tension is from about 20 mN/m to about 40 mN/m, which is rather small. Quantitatively, therefore, it is suggested that the surface tension difference $\Delta\sigma$ between adjacent liquid layers be also small, i.e.,

$$1 < \Delta\sigma < 3 \quad [\text{mN/m}] \tag{14.7}$$

Satisfying guideline 14.7 may be difficult, especially when trying to obtain dynamic surface tension curves that are parallel to each other over the entire range of relevant surface ages. Nevertheless, the surface tension difference between adjacent layers should not exceed 5 mN/m.

Adjusting the surface tension of adjacent layers of an aqueous multilayer film requires the addition of a proper amount of a suitable (fast diffusing) surfactant. This can be done by measuring the dynamic surface tension as a function of the surfactant concentration or by a simple visual approach. After adding a small amount of surfactant to an upper layer, its proper wetting behavior must be checked. This can be conveniently done by inserting a narrow plastic shim into the die slot of the lower layer, thereby blocking the flow of the upper layer on the lower layer, see left side in Fig. 14.7. Proper wetting is established if the width of the blocked-off upper layer (yellow) decreases upon flowing onto the lower layer (blue). Note in Fig. 14.7 that the superposition of a blue and yellow layer results in a green color.

For fluids containing an organic solvent, the procedure described above may be more difficult if not impossible to implement because most organic solvents have an inherently low surface tension (<20–40 mN/m). In such cases, adding a surfactant may simply remain ineffective in further lowering the surface tension.

In summary, the surface tension profile for a multilayer liquid film flowing down an inclined plane must either be constant or preferably, it must be asymmetrical, i.e., slightly decreasing from one layer to the next layer above. Note that the above issues do not apply if a multilayer slot die is used for implementing the curtain coating process.

14.3.4 Surface Tension Structuring for Multilayer Curtains

In contrast to a multilayer film, which flows down an inclined solid plane, and which has just one free surface, a liquid curtain always falls more or less vertically and always has two free surfaces, see Fig. 14.8. This difference is significant because it changes the requirement for structuring the surface tensions at the die lip, where the multilayer film changes into a multilayer curtain.

Fig. 14.7 Relative wetting of adjacent layers in a multilayer film flow; proper wetting, if the width of the blocked-off upper layer (yellow) decreases upon flowing onto the lower layer (blue or steel); photo reproduced with permission from Polytype Converting AG

Fig. 14.8 Multilayer curtain generated by a slide die

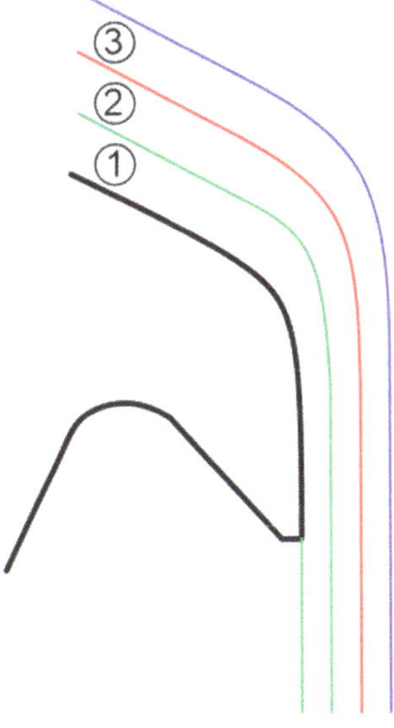

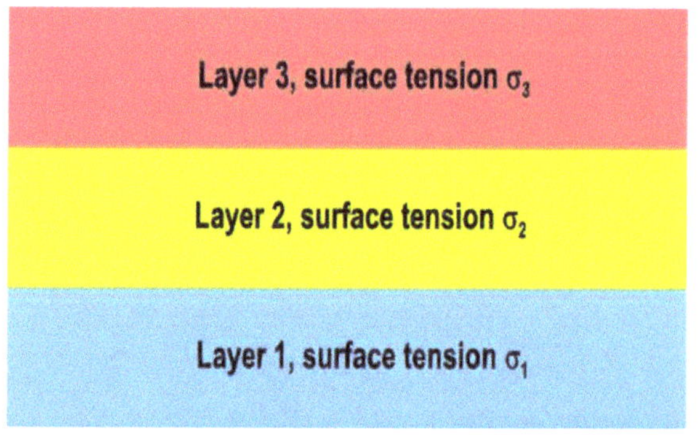

Fig. 14.9 Multilayer curtain with different surface tensions

Specifically, in a multilayer curtain, the surface tension of both outer layers (front and rear surface of the curtain) must be lower than the surface tensions of the adjacent inner layers. Hence, the surface tensions of the various layers must satisfy the following guideline, see also Fig. 14.9.

$$\sigma_{i+1} \geq \sigma_i \quad \mathrm{i} = 1 \ldots \mathrm{n}/2 \tag{14.8a}$$

$$\sigma_{i-1} \geq \sigma_i \quad \mathrm{i} = \mathrm{n} \ldots \mathrm{n}/2 \tag{14.8b}$$

Satisfying guideline 14.8 increases the robustness of the curtain to external disturbances such as particles and droplets with a low surface tension, which could trigger the formation of repellencies and ultimately holes in the curtain (Kheshgi, 1997).

Adjusting the surface tension of adjacent layers of an aqueous multilayer curtain requires the addition of a proper amount of a suitable, fast diffusing surfactant. The effect of this measure can be tested and visualized by measuring the dynamic surface tension with the Mach angle method as described in Sect. 4.5.3.

In summary, the surface tension profile for a multilayer curtain must be symmetrical, i.e., it must have a maximum in the middle layer of the curtain, and it must slightly decrease toward both outer layers. However, comparing the surface tension requirements for a multilayer film and a multilayer curtain reveals a conflict, that cannot be resolved perfectly, but that must nevertheless be managed in the best way possible. As schematically depicted in Fig. 14.10, the surface tension of a multilayer film should slightly decrease from the bottom toward the top layer, and the surface tension in a multilayer curtain should increase from the rear outer surface (which previously was the bottom layer of the film flow) toward the middle layer, and it should decrease from the middle layer toward the front surface. The conflict arises in the bottom layer of the film flow, which, upon leaving the die at its lip, turns into one of the free surfaces of the curtain. Obviously, however, this liquid cannot immediately change its surface tension.

The conflict and its associated problems can be minimized by keeping the surface tension of all layers below 40 mN/m, and by making the surface tension difference

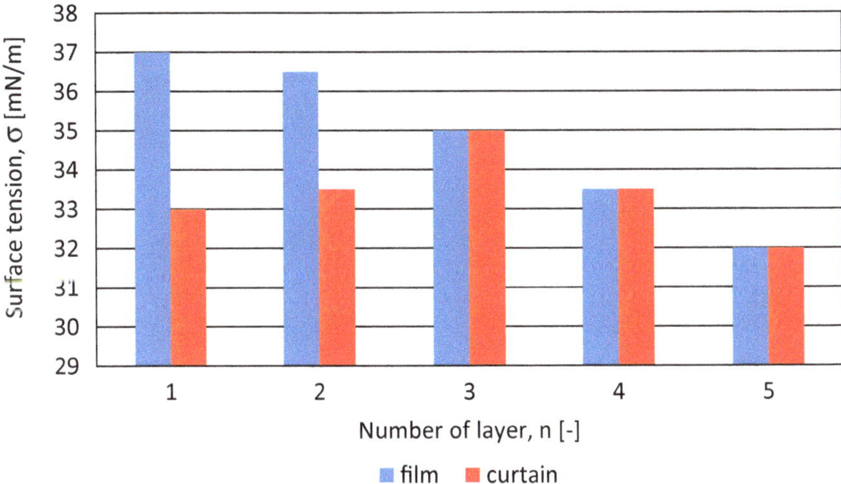

Fig. 14.10 Schematic surface tension structuring in a multilayer film (asymmetrical profile) and a multilayer curtain (symmetrical profile)

between adjacent layers small, see guideline 14.7. More stringently, the surface tension difference between the bottom and top layers of a multilayer film or curtain should be kept below a maximum value of, say

$$\Delta\sigma = \sigma_1 - \sigma_n < 5m \quad [\text{mN/m}] \tag{14.9}$$

Furthermore, attempting to make the surface tension of the bottom layer similar to the surface of the top layer will make it easier for the multilayer curtain to adhere to the auxiliary fluid of the curtain edge guide because the spreading behavior of both the rear and front surfaces of the curtain on the edge fluid will be similar.

Tricot (1997) proposed valuable recipe-like instructions for optimizing the surface tensions of a 3-layer film that is curtain coated. Tricot also pointed out that fast diffusing surfactants are required for increasing the resistance to the formation of repellencies in multilayer films and curtains because each layer must be able to reduce its surface tension as fast or faster than the layer underneath.

When optimizing the dynamic surface tension in a curtain, i.e., when attempting to make the two surface tensions equal, the different time scales of the two curtain surfaces must be considered, see schematic diagram in Fig. 14.11.

The front surface of the curtain emerges from the free surface of the top layer of the film flow, which starts at the exit of the uppermost slot of the slide die. This point marks time = 0 for the front surface of the curtain. If the fluid is water-based and contains a surfactant, then the surface tension is high at that point because the free surface of the film is just being created, and surfactant molecules must first diffuse to the free surface to lower the surface tension. This happens continuously as the film flows down on the straight portion of the die slide, and perhaps until the

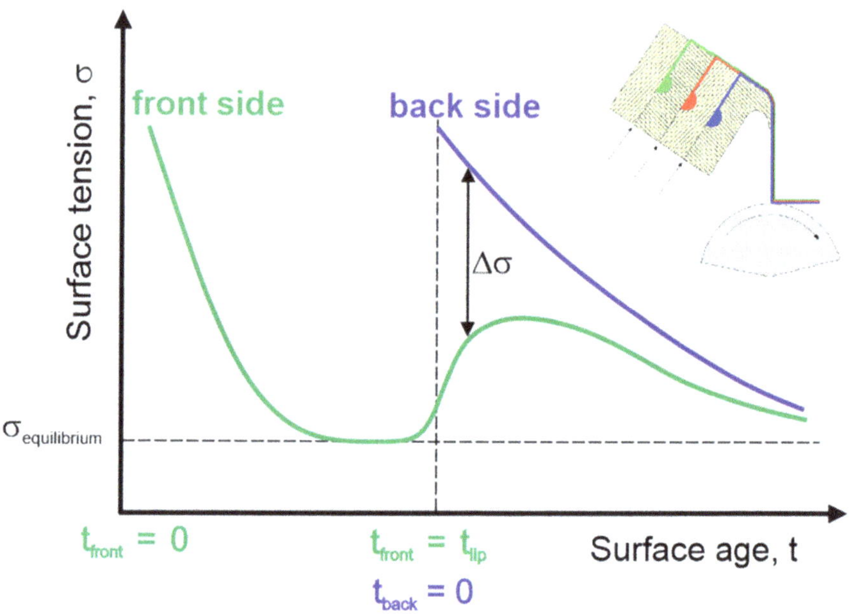

Fig. 14.11 Different time scales for dynamic surface tension in front and back surface of curtain

surface tension reaches its equilibrium value. When the slide starts to curve at the die lip, a new free film surface is created and the surface tension may start to increase again. When the film flow changes to curtain flow at the tip of the lip, i.e., when $t_{front} = t_{lip}$, a new free surface in the front side of the curtain is generated owing to the strong acceleration in the curtain flow, and the surface tension increases even further. After this transition zone, the curtain flow is steadily driven by gravity, which again generates a new curtain surface. Whether the surface tension further decreases, stays constant, or even increases along the curtain flow depends on the relative effects of the migration of surfactant molecules from the bulk to the free surface and the depletion of surfactant molecules in the curtain surface due to the generation of new curtain surface.

The rear curtain surface is created at the die lip, meaning that what is t_{lip} for the front surface is $t = 0$ for the back surface, and the surface tension of the back surface is high. At that point, therefore, it is difficult to match the surface tension values of both curtain surfaces. However, efforts must be undertaken to bring the two surface tension curves together in a parallel fashion as quickly and as much as possible as sketched in Fig. 14.11. Remember, though, that the actual changes of surface tension depend on the details of the flow field and the diffusional speed of the surfactant molecules. Moreover, the issue of different time scales in the curtain surfaces does not exist if the curtain is formed by an inverted slot die.

Coating fluids in the photographic industry are based on aqueous gelatin solutions of low solids concentration (typically of order 10% or less). Experience from that industry teaches that

- Anionic surfactants promote layer spreading and inhibit edge withdrawal. However, they tend to produce hydrophobic surfaces, which are difficult to wet during subsequent coating passes. Triton 770 is a good example of such a surfactant.
- Wetting of a coated and dried layer by a fluid of a subsequent coating pass, i.e., re-wetting, is promoted if the coated (top) layer contains a non-ionic surfactant. Olin 10G is a good example of such a surfactant.
- The formation of transient diffuse streaks or bands caused by ambient pressure disturbances can be inhibited if the potential for surface tension gradients in the top layer is maintained during the whole time that the liquid is mobile. One way of prolonging surface tension gradients is to mix a fast diffusing surfactant with a slowly diffusing surfactant that has a high surface tension lowering capability, e.g., a fluorinated surfactant. At young surface ages the fast surfactant dominates, and at longer times the slow surfactant dominates because of its ability to lower the surface tension still further. See also Ruschak (1987) and Sect. 5.10.4 for more details.
- The ability of a liquid film to resist disturbances is measured by its ability to maintain surface tension gradients, which in turn depends on the concentration and diffusivity of the surfactant. Changing the polymer (solids) concentration results in lower or higher viscosity, which affects the rate of diffusion of surfactant from the bulk to the free surface and the ability to maintain surface tension gradients. This fact must properly be accounted for when adjusting the surfactant concentration following a change in polymer concentration. Lower viscosities require less surfactant, and higher viscosities demand a higher surfactant concentration. Best values must be found experimentally by measuring the dynamic surface tension.
- The quantity of surfactant needed to obtain the desired surface tension depends on the nature of the fluid (solution, emulsion, dispersion, viscosity, etc.) and must be determined experimentally.

References

Blake, T. D., & Ruschak, K. J. (1997). Wetting: Static and dynamic contact lines. In: S. F. Kistler & P. M. Schweizer (Eds.), Chapter 3, *Liquid film coating*. Chapman & Hall.

Kheshgi, H. S. (1997). The fate of thin liquid films after coating. In: S. F. Kistler & P. M. Schweizer (Eds.), Chapter 6, *Liquid film coating*. Chapman & Hall.

Ruschak, K. J. (1987). Flow of a thin liquid layer due to ambient disturbances. *AIChE Journal, 33*(5), 801–807.

Tricot, Y. M. (1997). Surfactants: Static and dynamic surface tension. In S. F. Kistler & P. M. Schweizer (Eds.), Chapter 4, *Liquid film coating*. Chapman & Hall

Chapter 15
Modeling Economic Aspects of Coating Methods

Abstract Economic aspects of coating processes, i.e., the effect of an installed coating method on the process productivity and the resulting product unit costs, can be modeled easily and accurately. Many details of such models were presented previously by Schweizer (1997).

In this chapter, the effects of the coating speed, the size of the coating lot, and the non-productive time on the process productivity are discussed. In addition, cost savings from reduced coating passes and improved coat weight uniformity are illuminated and quantified. The purpose of these examples is to show that modeling financial aspects of coating processes are invaluable instruments for the management of industrial coating companies.

15.1 Introduction

Investment decisions in the converting industry are almost always driven by financial parameters, such as return on investment or payback time. It turns out that an accurate calculation of such parameters is difficult because investment costs into a new technology must be compared with operating savings when compared to existing technology. On the one hand, when evaluating several new technologies to an existing one, suppliers of coating methods and associated equipment would have to prepare several offers, which they are not eager to do at no cost for the interested customer. On the other hand, savings in manufacturing costs can be realized with new technology, if the efficiency of the coating method can be increased. This in turn can be achieved by reducing the consumption of raw materials, as well as by increasing the yield and the productivity of the coating process. These latter parameters can be modeled well as will be shown below. However, increasing the yield and productivity of a coating process also increases the capacity of a given coating machine, i.e., more product volume can be manufactured in a given time. Whether or not this extra coated surface can be sold and turned into revenue and profit cannot be predicted with certainty because it depends on the growth rate of the marketplace, and on

P. M. Schweizer, *Premetered Coating Methods*, Engineering Materials,
https://doi.org/10.1007/978-3-031-04180-8_15

whether or not it is possible to take market share from a competitor by implementing a successful marketing strategy.

One of the core parameters in financial modeling is the specific machine utilization SMU, which equals the time it takes for coating a unit substrate surface. This time is composed of productive manufacturing time, where the machine actually produces a usable product, and of non-productive time that can be allocated to a given product, such as time for changing the coating width or the product formulation, and time for maintenance of the equipment, etc. Small values of SMU are desirable. SMU is defined as

$$SMU = n\left[\frac{1}{UW} + \frac{t_{np}}{A_{gross}}\right] \tag{15.1}$$

SMU specific machine utilization [h/m^2],
A_{gross} gross lot size [m^2],
U coating speed [m/h],
W coating width [m],
n number of coating passes [−],
t_{np} non-productive time [h].

Product unit costs C_C associated with the coating process are obtained by dividing Eq. 15.1 by the substrate yield Y_s ($Y_s < 1$) and by multiplying it with the hourly cost of the coating machine C_{CM}. The latter parameter accounts for cost contributions from labor, energy, auxiliary material consumption, depreciation, etc., and it depends on the annual gross coating volume. The substrate yield accounts for the surface that cannot be sold due to coating defects, edge losses, start and finish losses, material needed for quality control, etc. Moreover, the annual product costs $C_{C,0}$ associated with the coating process are obtained by multiplying C_C with the annual production volume. Other cost factors that contribute to the product unit costs and should be considered as well are described by Schweizer (1997).

The following examples visualize the sort of information that can be obtained by modeling the effects of the installed coating technology and how this technology is operated on the resulting product unit costs.

15.2 Effect of the Coating Speed and the Size of the Coating Lot on the Specific Machine Utilization

In Fig. 15.1 SMU is computed as a function of the coating speed and the size of the coating lot. The other parameters in Eq. 15.1. are chosen arbitrarily and held constant; their values are listed in the figure caption. The main message is that it pays off to coat this product at a speed of at least 500 m/min because increasing the speed up to that value significantly lowers the product unit costs. Equally important is to avoid coating this product in many small lots. Increasing the lot size to 100,000 m^2 results

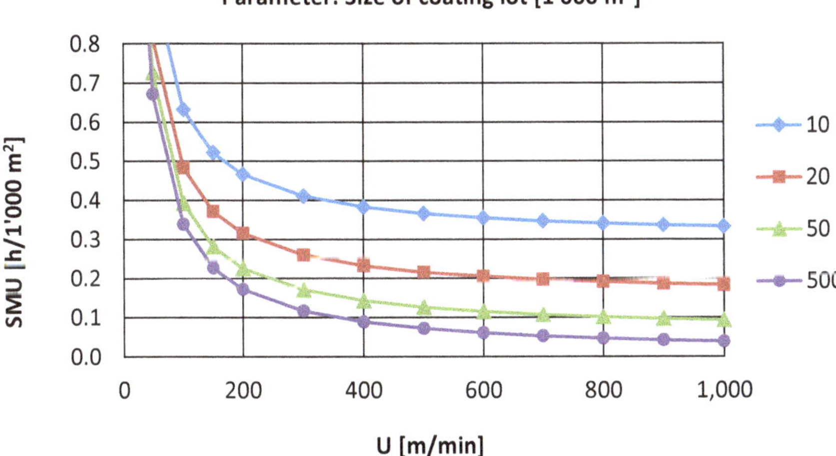

Fig. 15.1 Effect of the coating speed and the size of the coating lot on SMU. $t_{np} = 1.0$ h, W = 1500 mm, n = 3

in significantly lower unit costs. However, coating larger lots will generate additional storage or stock costs, if the product cannot be sold immediately after coating. How the lot size can be optimized by simultaneously considering raw material, coating, and storage costs is shown in an example by Schweizer (1997).

15.3 Effect of the Lot Size and the Non-productive Time on the Specific Machine Utilization

The example in Fig. 15.2 shows that SMU significantly depends on the non-productive time if the size of the coating lot is not large, say <1,000,000 m^2. For such situations, it pays off to invest in suitable measures for reducing the non-productive time.

Such measures include a switch from manually changing the coating width (about 45 min for a 3-layer slide die) to a motorized and automated system offered, for example, by TSE Troller AG (<10 min). Another idea is using a die that produces a high and constant wall shear stress throughout the internal fluid distribution system. Consequently, such a die does not contaminate and hence must not be cleaned during a product change. Instead, the old fluid will simply be pushed out of the delivery system and the die by the new fluid, which reduces the change-over-time considerably.

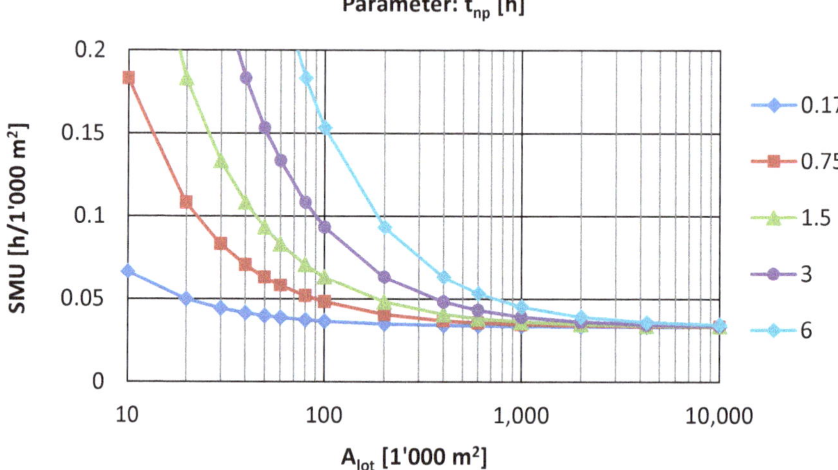

Fig. 15.2 Effect of the lot size and the non-productive time on SMU. n = 2, U = 500 m/min, W = 2,000 mm

15.4 Cost Savings from Reduced Coating Passes

One of the most attractive features of premetered coating processes is their ability to apply several different fluid layers simultaneously. Compared to single-layer coating methods, this feature allows the productivity of the manufacturing process for multi-layered products to be increased significantly through the reduction of coating passes. Increasing productivity is equivalent to reducing the time it takes for manufacturing a given volume of a given product. As a consequence, savings can be realized by reducing the specific costs of the coating machine (i.e., staffing, energy consumption, etc.), and/or revenues can be increased by manufacturing and selling additional products. Moreover, additional savings will be realized by an increased material yield because material losses of both substrate and coating liquid are proportional to the number of coating passes, e.g., waste generated at the start/stop procedures, during splice passages, by coating defects, etc. The experience from the photographic industry is a 2% loss of material yield per coating pass, which is significant.

According to Eq. 15.1 the specific machine utilization is proportional to the number of coating passes. The following example considers manufacturing a hypothetical 3-layer product, i.e., a thermal paper consisting of an undercoat UC that has the function of flattening and smoothing the rough surface of the raw paper, a thermal coat TC, which contains the thermo-reactive material, and an overcoat OC, that contains barrier properties.

It is further assumed that a simple coating machine is available, consisting of an unwinder, a coating station, a dryer, and a rewinder. Each liquid will be applied with the same blade coating process because this method is required for achieving the surface flattening effect of the undercoat.

Moreover, the following operational parameters are posted:

U_{UC} 1,000 m/min, coating speed of the undercoat;
U_{TC} 650 m/min, coating speed of the thermal coat;
U_{OC} 530 m/min, coating speed of the overcoat;
W 2,050 mm, coating width;
A_{gross} 8,000,000 m^2, size of coating lot;
A_0 100,000,000 m^2/year, annual production volume;
t_{np} 1 h, non-productive time between coating passes, for example for changing the coating liquid or the coating width.

For the situation described above, each paper roll must be sent 3 times through the coating machine, which is cumbersome. The interesting question is whether or not it is financially attractive to invest in a new coating technology, curtain coating for example, by which the number of coating passes can be reduced.

Note, however, that for the example at hand, the undercoat cannot be applied with the curtain coating method because curtain coating is a conformal method, which is not capable of flattening a rough paper surface. While the undercoat must still be applied with the blade coating method, the thermal coat and the overcoat can be applied simultaneously with the curtain coating method. Thus, a reduction of coating passes from 3 to 2 is possible.

The mathematical models presented below compare different manufacturing scenarios, whereby each multiple-pass scenario is characterized by a suitable average coating speed, which is properly defined as follows.

The time it takes for coating a surface corresponding to the desired lot size equals

$$\Delta t = \frac{A_{gross}}{UW} \tag{15.2}$$

For a multiple-pass operation the lot size and the coating width remain constant for each pass, but the coating speed, and so the coating time, may vary as is shown in the sample above. Therefore, as is schematically depicted in Fig. 15.3, the average coating speed must satisfy the following condition:

$$U_1\Delta t_1 + U_2\Delta t_2 + U_3\Delta t_3 = \overline{U}(\Delta t_1 + \Delta t_2 + \Delta t_3) \tag{15.3}$$

The general expression for a suitable average coating speed is given by Eq. 15.4:

$$\overline{U} = \frac{n}{\sum_{i=1}^{n} \frac{1}{U_i}} \tag{15.4}$$

U_i coating speed of pass i [m/min],
$\overline{U}$ average coating speed [m/min],
n number of coating passes [−].

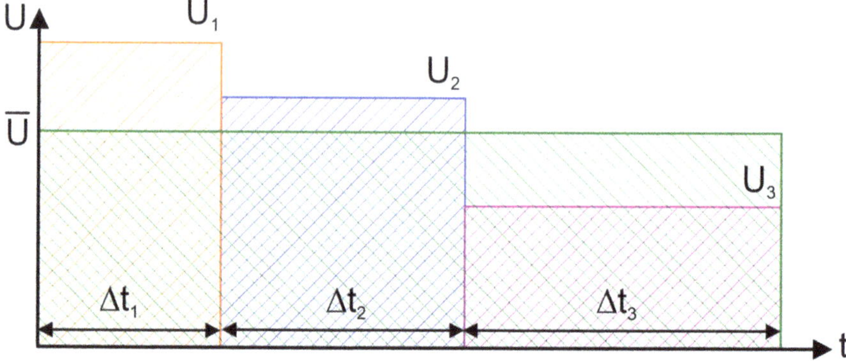

Fig. 15.3 Schematic speed diagram for multiple-pass coating operations

The average coating speed for the 3-pass sample process listed above is 678 m/min. This value is quite different from the inappropriate arithmetic mean value, which would amount to 727 m/min.

What is the average speed of the 2-pass improved situation involving curtain coating? For thermal paper, the maximum web temperature in the dryer must not exceed a critical value of about 60 °C to prevent the paper from turning black in the dryer. The coating speed of 650 m/min for applying the thermal coat alone corresponds to a maximum specific solvent evaporation rate expressed in units of [kg/h m^2]. If that same value is used as the upper limit for applying the thermal and the overcoat together, then the coating speed for the 2-layer curtain mode is estimated to be 360 m/min. Consequently, the average speed for the 2-pass situation is equal to 529 m/min.

All parameters are known for calculating the cost savings from reducing the number of coating passes by comparing two scenarios. The first one corresponds to the situation with 3 coating passes, and the second one corresponds to the same task but is now executed with only 2 coating passes. The corresponding equation is

$$\Delta C_C = A_0 C_{CM}(SMU_1 - SMU_2) \tag{15.5}$$

In Fig. 15.4 the annual cost savings are plotted as a function of the specific machine cost, C_{CM}, and of the average coating speed of the curtain coating process. Note that the same values for C_{CM} and Y_S were taken for both coating scenarios, which is an approximation. Moreover, it appears possible to increase the average speed of the curtain coating process by about 100 m/min while still respecting the upper limit of the specific solvent evaporation rate, if the average coat weight could be slightly decreased owing to an improved coat weight uniformity, and if the solids concentration could be slightly increased by reducing the variability of the fluid making process.

As mentioned above, cost savings from a reduction of coating passes can be very significant. For the example at hand, savings easily exceed 1 mio. $/year, which is

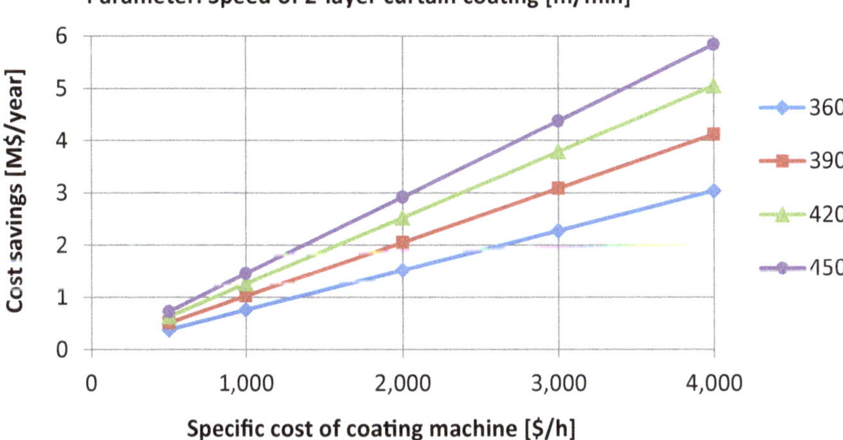

Fig. 15.4 Annual cost reduction resulting from a reduction of the number of coating passes; $A_{lot} = 5$ mio. m^2, $A_0 = 150$ mio. m^2, $W = 2{,}000$ mm, $t_{np} = 1.0$ h

extremely attractive because the investment cost for implementing the curtain coating technology, i.e., the cost for a curtain coating station including peripheral equipment for fluid conditioning (degassing) and delivery, are on the order of 1–2 million dollars, thus resulting in payback times for such an investment project of < 2 years.

15.5 Cost Savings from Improved Coat Weight Uniformity

Suppose that a given coating method produces coat weight variations of the magnitude dA_1, as schematically depicted in Fig. 15.5. Then, for most industrial applications, the average coat weight $A_{average,1}$ must be chosen such that the resulting minimum coat weight $A_{minimum} = A_{average,1} - dA_1$ meets the particular product requirement. Moreover, the fluid quantity corresponding to dA_1 can be considered as wasted because it is not required by the product performance, but it results from inadequacies of the coating method. However, it is impossible to run an industrial coating process with $dA = 0$.

Further suppose that an alternative coating method exists, which produces a smaller coat weight variation dA_2. Since product performance criteria require the minimum coat weight to be the same, the average coat weight $A_{average,2}$ of the alternative coating method can be reduced. The exact value of the reduced coat weight can be calculated with Eq. 15.6:

$$A_{average,2} = A_{average,1} - dA_1 + dA_2 \tag{15.6}$$

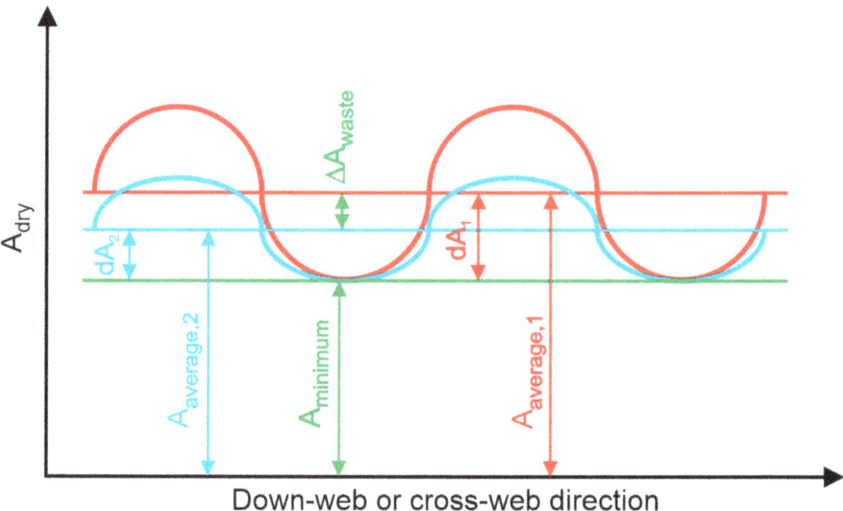

Fig. 15.5 Schematic representation of coat weight nonuniformities

or, alternatively:

$$A_{average,2} = A_{averag,1} \frac{1 - \epsilon_1}{1 - \epsilon_2} \tag{15.7}$$

ε_1 dA_1/A_1 [−]; relative coat weight variation of the old coating method 1 and
ε_2 dA_2/A_2 [−]; relative coat weight variation of the new coating method 2.

The difference between the two average coat weights $\Delta A = (A_{average,1} - A_{average,2})$ can be considered as true waste. While the coat weight difference ΔA may appear to be very small, it must be emphasized that premetered coating methods are particularly suitable for implementing this difference in an industrial coating process owing to the precise control of the flow rate that is pumped into the die (see Sect. 9.3). Therefore, avoiding the waste by resorting to an alternative coating method or by resorting to a die with higher mechanical precision (see Sect. 9.4), or by reducing the variability of the fluid mixing process will result in immediate savings by way of reduced raw material costs.

Moreover, being able to reduce the average coat weight reduces the solvent load in the dryer, which typically allows the coating speed to be increased accordingly. The resulting increase in productivity frees up additional manufacturing time, which in turn generates additional revenues, provided that the extra product volume can be sold in the marketplace.

Direct savings in raw material costs from reducing the coat weight variations, and consequently from reducing the average coat weight, can be calculated with Eq. 15.8:

$$\Delta C_{RM} = A_{average,1} A_0 \left(C_{M,1} - C_{M,2} \frac{1 - \varepsilon_1}{1 - \epsilon_2} \right) \tag{15.8}$$

ΔC_{RM} savings in raw material cost for a given coating fluid [$/year],
$C_{M,1}$ specific raw material cost of a given coating fluid associated with the old coating method 1 [$/kg],
$C_{M,2}$ specific raw material cost of a given coating fluid associated with the new coating method 2 [$/kg],
A_0 annual coated surface of a given product [m^2/year].

Note that the specific raw material cost will typically change a bit when switching from one coating method to another. A blade coating formulation, for example, may contain de-foaming agent, while this chemical component is not needed or desirable in a curtain coating formulation. On the other hand, a curtain coating formulation may require the addition of a surfactant, but, as is demonstrated in this example, the coated layer will most likely be applied at a reduced thickness compared to the blade coating case. Details have to be checked carefully for a given situation.

Calculating the savings of raw material costs from a reduced coat weight variation requires knowledge of sensitive information related to the current manufacturing operation at a given company. In particular, exact values for the current relative coat weight variation ε_1, the specific raw material cost $C_{M,1}$, and the annual production volume A_0 are typically not accessible to the public. Therefore, Fig. 15.6 contains a general diagram, which hopefully images the correct order of magnitude of a given situation, and into which a coating company can read its current values of sensitive parameters to estimate the possible savings. Following is an example that demonstrates how to work with this diagram.

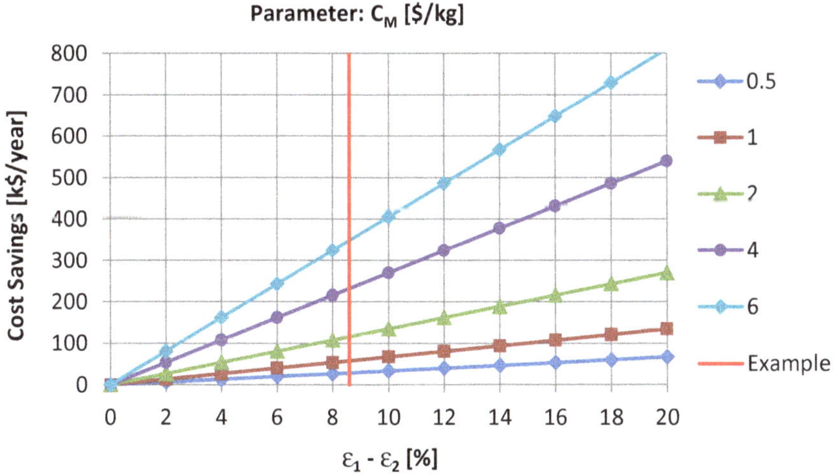

Fig. 15.6 Savings in raw material costs resulting from a reduction in coat weight variation; A_1 = 4.5 g/m^2, A_0 = 150 mio. m^2

Example: Calculation of raw material savings from reduced coat weight variation.

Current situation:

Coating method	flexible blade coating.
A_1	4.5 g/m^2.
dA_1	unknown.
ε_1	unknown.
A_0	150,000,000 m^2/year.
$C_{M,1}$	unknown.

Future situation:

Coating method	curtain coating.
ε_2	0.025 (2.5%), from experience.
A_0	150,000,000 m^2/year.
$C_{M,2}$	unknown but variable.

Assuming that the current value for dA_1 is 0.5 g/m^2, then ε_1 equals 0.111 (11.1%), and ($\varepsilon_1 - \varepsilon_2$) equals 0.086 (8.6%). Now, as is shown in Fig. 15.6, and depending on the specific raw material costs, which are assumed to be equal for both the old and new coating method in this example, annual savings in raw material costs amount to several hundred thousand dollars, which is quite significant.

In conclusion, many reasons for coat weight variation are known, some are related to the coating equipment, others to the processes and procedures for making the coated liquid. The choice of the coating method is an important factor as well.

For premetered coating methods, the origin of coat weight variations is well understood. Consequently, coat weight variations as small as $\pm 1\%$ can be achieved for dedicated applications, and values of 2–3% are typical for most ranges of applications. Reducing the coat weight variation allows the average coat weight to be reduced as well, and this not only generates immediate savings in raw material costs, but it also generates extra production capacity through improved process productivity. Depending on other parameter values, such as the annual coating volume and the specific raw material costs, annual savings in raw material costs can be significant. It is therefore worthwhile considering investing in alternative coating methods such as curtain coating, and/or investing in improved procedures and processes for making the coating liquid.

Reference

Schweizer, P. M. (1997). Minimization of product unit costs. Example on page 761 in Chapter 15 entitled control and optimization of coating processes. InS. F. Kistler & P.M. Schweizer (Eds.) *Liquid film coating*. Chapman & Hall.

Part III
Specific Properties of Premetered Coating Methods

Abstract

This part of the book describes properties that are specific to the three premetered coating methods. Each of the following chapters addresses the topics of process configuration, operating limits, and process optimization as well as other subjects that are particular to any of the methods.

Chapter 16
The Concept of the Coating Window

Abstract Coating windows are an instructive way to visualize various operating boundaries of a given coating process. In this chapter, suitable operating windows for premetered coating methods are presented and discussed. It is important to understand how the operating boundaries of the coating window can be moved such that a desired operating point comes to lie inside the window. Equally important is finding the optimum operating point inside the window that results in the best possible product uniformity and process productivity.

16.1 Introduction

The term *coating window* or *operating window* refers to a diagram, in which operating boundaries of a given coating process are plotted as a function of the two parameters that span the axes of the diagram, and possibly of one of the other relevant process parameters. Every coating process has its specific coating window. In addition, each specific coating process is characterized by more than one window because, typically, the performance of every process is determined by more than three relevant parameters. Using dimensionless numbers such as the Reynolds, the capillary, or the Weber number instead of original parameters such as the web speed, the flow rate, or the surface tension, is an elegant and efficient way for reducing the number of relevant process parameters. This approach, however, is not always possible or beneficial.

16.2 Coating Windows for Premetered Coating Methods

For premetered coating methods, an obvious and illustrative way for creating a coating window is by plotting the volumetric flow rate/width Q as a function of the web speed U, or vice versa because these two parameters are most relevant as they determine the wet thickness of the coated film.

P. M. Schweizer, *Premetered Coating Methods*, Engineering Materials,
https://doi.org/10.1007/978-3-031-04180-8_16

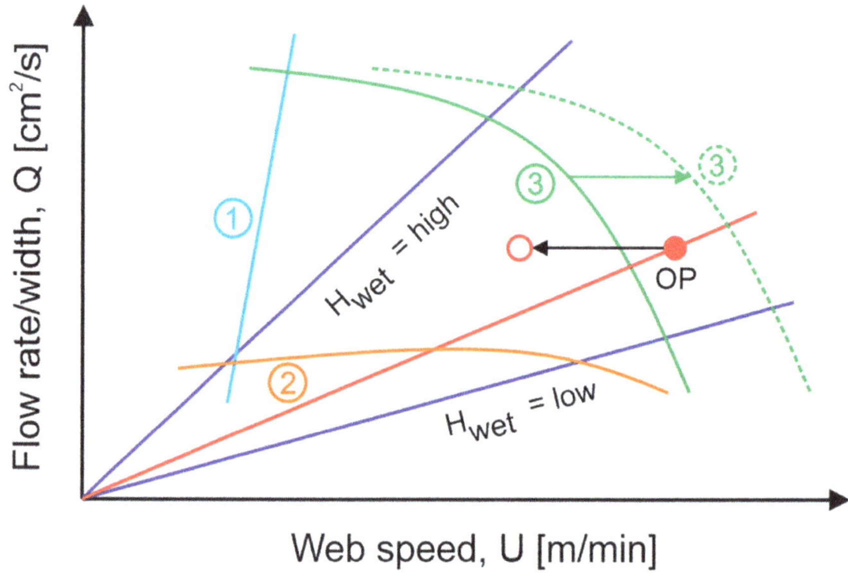

Fig. 16.1 Generic coating window for premetered coating methods

As generically depicted in Fig. 16.1, points of constant wet film thickness are located on straight lines through the origin of the coordinate system, with increasing film thickness resulting in a steeper slope of the line. Operating limits such as the onset of specific coating defects like ribbing lines or air entrainment at the dynamic wetting line can now conveniently be displayed in such a diagram as a function of physical fluid properties including viscosity and surface tension, geometric parameters such as the curtain height or the impingement angle, or other operating parameters like the pressure difference across the coating bead. In Fig. 16.1, three different but unspecified operating boundaries are drawn arbitrarily. The space framed by those boundaries is the coating window. Operating points lying inside the window produce acceptable coating quality, and operating points located outside the window do not.

In the example of Fig. 16.1, the operating point OP marked with a full red circle is outside the operating boundary 3. In order to move the operating point inside the window, the speed would have to be reduced, which would result in a higher wet film thickness (see open red circle). This may not be acceptable because it changes the product performance, and it increases the solvent content and hence the dryer load, which is neither good nor possible for a given dryer capacity. Also, reducing the coating speed reduces the productivity of the coating process, which is not helpful either. A better way is to reduce the speed and the flow rate such that the film thickness remains constant. An even better way is to move the operating boundary 3 to speeds beyond the desired speed of the operating point (see green arrow). This, however, requires knowledge as to which other parameter is capable of moving this boundary in the necessary direction. In this context, it is less important to know the exact location of a boundary as a function of other relevant process parameters than knowing qualitatively which parameter has to be changed in which direction to move the boundary in the desired direction.

Operating boundaries divide the space of the Q-U diagram into areas of good and bad coating, whereby the meaning of "good" is not always clearly defined. Often, it refers to operating conditions that result in a coherent wet film on the substrate. Chen and Scriven (1992) expanded the idea of the coating window by introducing an onion-like structure of several operating boundaries nested inside each other. In particular, they called the outermost boundary the *feasibility window*, where the coherent coating can be established and maintained. The next layer inside is called the *sensitivity window*, in which the coating is insensitive to spontaneous disturbances triggered by process fluctuations, and unresponsive to forced disturbances from the environment. The innermost layer is termed *quality window*, where all flow fields are free of vortices and static wetting lines are pinned to sharp edges of the flow boundaries. Schweizer (1997) illustrated this concept by using the example of curtain coating.

In Part III of this book, the discussion of coating windows focuses on the quality window because that is the goal of every industrial coating application. Based on our experience, moreover, we are convinced that flow fields containing vortices must be avoided and static wetting lines must be pinned to sharp corners whenever possible.

The presence of any operating boundary raises the question of where should the operating point of an industrial coating process be located relative to the location of the operating boundary? The answer is based on risk assessment. If the operating point is located very closely to the boundary, then increased process productivity and/or increased product uniformity may result but only at the cost of increased risk of process failure due to ever-present process fluctuations. The opposite is true if the operating point is located too far away from the process boundary. We believe that a good operating point is found when the Production Manager, i.e., the person responsible for generating a specified volume of a sellable product by the end of the day, can sleep well every night.

Presenting any given operating boundary as a function of the viscosity is difficult for most industrial coating fluids. Specifically, changing the flow rate and/or the web speed changes the characteristic shear rate of the flow under consideration, and a changing shear rate changes the viscosity if the shear rate is inside the shear-thinning part of the Carreau-Yasuda viscosity curve of a given fluid. For that reason, it is helpful when experimentally determining operating boundaries to use model fluids that exhibit Newtonian flow behavior for a wide shear rate range, such as aqueous glycerin solutions, and aqueous gelatin or PVA solutions at low solids concentrations.

As will be shown below in Chaps. 18 and 19, the approach of using the parameters Q and U for building coating windows works very well for slide and curtain coating. For slot coating, however, it is customary to plot important operating boundaries in terms of other process parameters, see Chap. 17.

Regarding the determination of operating boundaries, several authors of papers about coating technology, for example, Carvalho and Kheshgi (2000), argue that experimental approaches are more expensive than relying on theoretical models in combination with powerful numerical methods for solving complex differential equations. However, they admit that theoretical results should be verified and supported with experimental data, which they did in the paper just mentioned above. From a coating machine manufacturer's point of view, such as Polytype Converting AG, demonstrating process capabilities by way of pilot coating trials is even more important than claiming success based on theoretical predictions. When a customer is about to commit large investment funds on the order of 10^5–10^7 dollars for a new coating method installed on a coating station or even on a new machine, then performing experiments on a semi-industrial pilot coating machine with the fluid of the customer and with the customer standing by and observing whether this new process really works, or perhaps not, generates much more trust than presentations about theoretical results.

As already argued by Schweizer (1997), it is also proposed here not to focus too much on operating boundaries, but instead to concentrate on optimizing the coating process for a given set of operating conditions, i.e., for locating the best possible operating point inside the coating window. If implemented successfully, then knowing the exact location of the various operating boundaries becomes less important. To that extent, various optimization criteria are suggested in the following chapters describing slot, slide, and curtain coating.

References

Carvalho, M. S., & Kheshgi, H. S. (2000). Low-flow limit in slot coating: Theory and experiments. *AIChE Journal, 46*(10), 1907–1917.

Chen, K. S. A., & Scriven, L. E. (1992). Two views of multilayer coating beads. Ind Coat Res 2.

Schweizer, P. M. (1997). Control and optimization of coating processes. In: S. F. Kistler & P. M. Schweizer (Eds.), Chapter 15, *Liquid film coating*. Chapman & Hall.

Chapter 17
Specific Properties of Slot Coating

Abstract Slot coating is currently a popular application process, not the least for its capability to coat intermittently, which is a necessary feature for manufacturing electronic devices such as OPV's and OLED's, as well as electrodes for Li-ion batteries. After presenting possible process configurations and related equipment, the discussion turns to the operating window with the low flow limit as the major operating boundary. The onset of ribbing lines, air entrainment and vortices are other operating limits, and actions to prevent these flow defects are proposed. Then, the effect of the uniformity of the coating gap on the uniformity of the coated film, various operational aspects, and means to optimize the process are discussed. In the second part of the chapter, unique and attractive features of slot coating are presented, including tensioned-web coating, double-sided coating, stripe coating, intermittent, and pattern coating.

17.1 Introduction

The slot coating process has been used industrially for almost 70 years to explore its advantages regarding product quality and process productivity when compared to other coating methods. The first patent was granted in 1954 to Albert Beguin from Eastman Kodak. Starting in the 70ies of the last century, slot coating was developed to a high degree of sophistication by the research laboratories of large companies and by universities across the world. Durst and Wagner (1997) summarized the results of these activities up to the mid-90ies, and a comprehensive review of research results in slot coating up to about 2015 can be found in Ding et al. (2016).

Slot coating is applied for manufacturing very diverse products including medical imaging, medical diagnostics, electrical insulation, adhesive labels and tapes, holograms for security papers, and flexible organic photovoltaic cells, etc. Today, slot coating is still very popular for the manufacturing of new and emerging products. In particular, it is used for coating thick slurry layers onto metal foils for making anodes and cathodes of lithium-ion batteries (see Fig. 17.1a), and it is used for coating very thin layers of very high uniformity for optical films, organic photovoltaic modules (OPV), and organic light-emitting diodes (OLED). Some of these applications require

P. M. Schweizer, *Premetered Coating Methods*, Engineering Materials,
https://doi.org/10.1007/978-3-031-04180-8_17

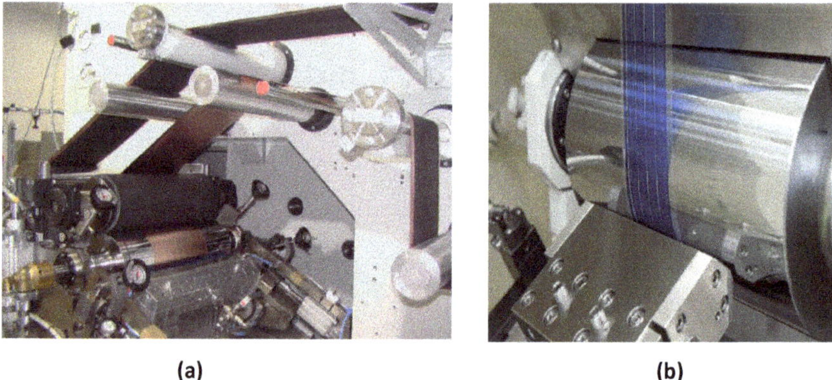

(a) (b)

Fig. 17.1 **a** Slot coating of a slurry onto copper foil (anode of Li-ion battery); photo reproduced with permission from Polytype Converting AG. **b** Slot coating of stripes for flexible, organic photovoltaic cells; photo reproduced with permission from Polytype Converting AG

stripe and/or intermittent coating, for which the slot coating process is very suitable, see Fig. 17.1b.

17.2 Process Configuration and Equipment

As illustrated in Fig. 17.2, slot coating can be implemented in different operating modes.

The most popular mode is characterized by a fixed geometry, meaning that the substrate is wrapped around a rigid backup or coating roll, and the coating gap, i.e., the distance between the die lip and the substrate surface, can accurately be adjusted to any desired value.

To answer the question as to which die position on the coating roll is the best one, the relevant forces, which govern coating processes, i.e., the inertia, viscous, surface tension and gravity forces, must be compared with each other. In particular, the gravity force is important, if the performance of the slot coating process depends on the position of the die on the backup roll. Taking the ratio of the gravity force to the other relevant forces results in three well known dimensionless numbers as follows:

Bond number = gravity/surface tension:

$$Bo = \frac{\rho_{liquid} g H_{gap}^2}{\sigma} \tag{17.2.1}$$

Froude number = inertia/gravity:

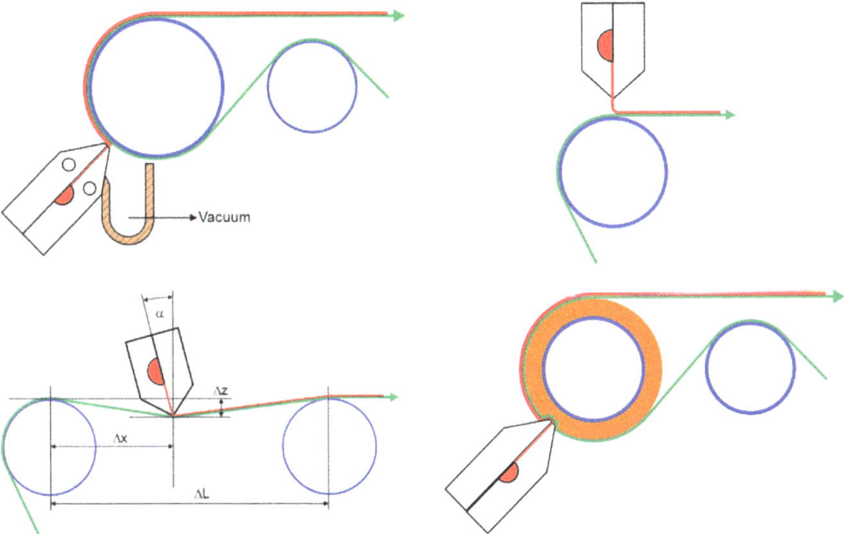

Fig. 17.2 Different operating modes for slot coating

$$Fr = \frac{U^2}{gH_{gap}} \quad (17.2.2)$$

Stokes number = viscous/gravity:

$$Sto = \frac{\mu U}{\rho_{liquid} g H_{gap}^2} \quad (17.2.3)$$

For most typical slot coating applications Bo << 1, Fr >> 1, and Sto >> 1. Thus, the gravity force is negligible compared to the other forces. Regarding the flow and coating behavior of the process, therefore, there would be no preferred location for the die around the coating roll. However, it is suggested that the preferred location of the die on the coating roll is either the 4:30 or the 7:30 position, see Figs. 17.1 and 17.3. These positions result in a die slot with an upward orientation toward the slot exit, which allows air pockets that may accumulate inside the die cavities during filling, coating, and cleaning to escape easily. It also prevents the die slot from draining spontaneously during long coating stops, such as overnight or over a weekend. Moreover, one of the problems in slot coating is the very small size of the coating bead such that visual observation of the bead flow field by the unaided eye is hardly possible. If the die is at the 7:30 position, then at least the flow at the exit of the coating gap can be seen with adequate illumination. Specifically, it can be seen if the static wetting line is attached to the downstream corner of the downstream lip, and if the wetting of the lip is uniform across the width of the coating. In contrast, the upstream meniscus of the bead cannot be seen, i.e., it cannot be observed if backflow or leakage occurs. On the other hand, mounting the die at the 4:30 position

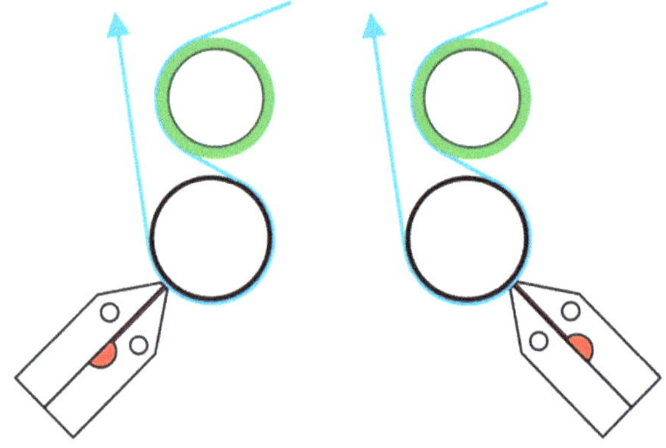

Fig. 17.3 7:30 and 4:30 configurations of the sot coating process

allows observation of the upstream meniscus, as long as the vacuum box is not in place. In addition, and as further explained in Sect. 17.10.6, experience from running numerous slot coating trials at the pilot center of Polytype Converting AG teaches that the propensity of generating lines and streaks in the coated film decreases if the die is located at the 4:30 position.

If slot coating is operated in the capillary mode (see below in Sect. 17.5), where large coating gaps compared to the wet film thickness are possible, then the 4:30 or 7:30 die position may result in undesired sagging of the coating bead. This situation can be avoided by mounting the die in the 12:00 position.

Finally, web paths resulting from process configurations shown in Fig. 17.3 are suitable if the dryer is located on the same level as or one level above the coating station.

Other slot coating configurations are referred to as elasto-hydrodynamic operating modes, where the die is pushed onto a stretch of unsupported substrate or the rubber cover of a backup roll, see Fig. 17.2. For these configurations, the coating gap cannot be set accurately before the coating start. Instead, the gap height results when equilibrium is reached between pressure forces generated by the flowing liquid, which push the substrate or the rubber cover away from the die lip and opposing elastic forces generated by the web tension or by the compression of the rubber cover. Pushing the die onto a free span of web is also called kiss coating or tensioned web coating. It is applied when attempting to coat very thin wet films of <20 μm and when the capillary number is not very small, i.e., Ca > 0.1, see Sect. 17.11. Furthermore, kiss coating is necessary when simultaneously coating both sides of the substrate, at least for coating one of the sides, for example for manufacturing electrodes of Li-ion batteries, see Sect. 17.12. Pushing the die onto the rubber cover of the coating roll is often done when coating highly viscous fluids such as hot melt adhesives or other polymer solutions. More information about elasto-hydrodynamic coating systems can be found in Pranckh and Coyle (1997).

Figure 17.4 shows typical examples of slot dies. In Fig. 17.4a, the side plate was removed to reveal the internal die geometry, which in this case consists of a dual

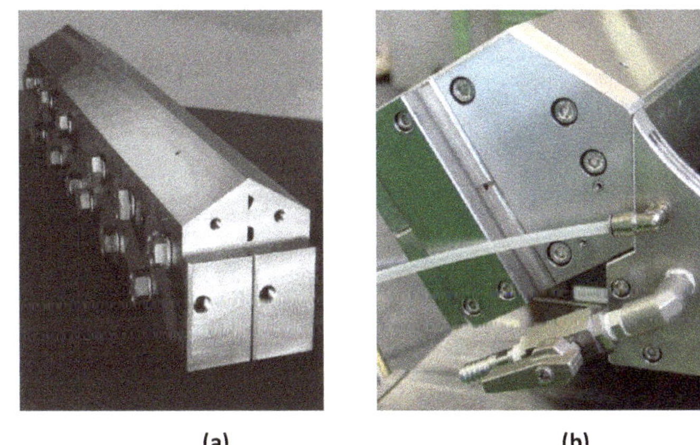

Fig. 17.4 **a** Dual cavity slot die; photo reproduced with permission from TSE Troller AG. **b** Slot die and dual chamber vacuum box assembly; photo reproduced with permission from TSE Troller AG

cavity design comprising two distribution systems in series. Figure 17.4b shows a slot die and vacuum box assembly.

The use of a vacuum box or vacuum chamber may allow the operating window of the slot coating process to be widened by applying a pressure difference across the coating bead, i.e., across the liquid bridge, which is suspended between the die lip and the substrate surface. In particular, vacuum is often required when operating in the capillary mode, see Sect. 17.5 for more details. The vacuum box also serves as a catch pan for the coating fluid exiting from the die slot before and after coating, which prevents the rest of the coating station from being excessively contaminated.

The coating gap must be very uniform across the coating width to guarantee an even thickness profile of the coated film. This in turn requires a highly precise die positioning system. Setting the coating gap to any desired value is best accomplished by pressing the positioning system with the die against adjustable mechanical stops that are designed with a pair of wedges allowing a positioning resolution in the micron range. Some positioning systems feature a design that allows the die to be tilted by a few degrees around an axis located at the exit of the die slot. Other systems include the ability to continuously change the position of the die around the coating roll. The latter two design features are not necessary from a process point of view, and they are not wanted from a design point of view because the additional movements are associated with mechanical play, which degrades the precision of the entire positioning system.

The direction of motion of the positioning system should be parallel to the orientation of the die slot, and the axis of the die slot should point to the center of the backup roll. This configuration corresponds to a perpendicular impingement of the flow leaving the slot, and hence to optimal implementation of the concept of hydrodynamic assist, see Chap. 7. If nevertheless, the possibility of a variable die tilt

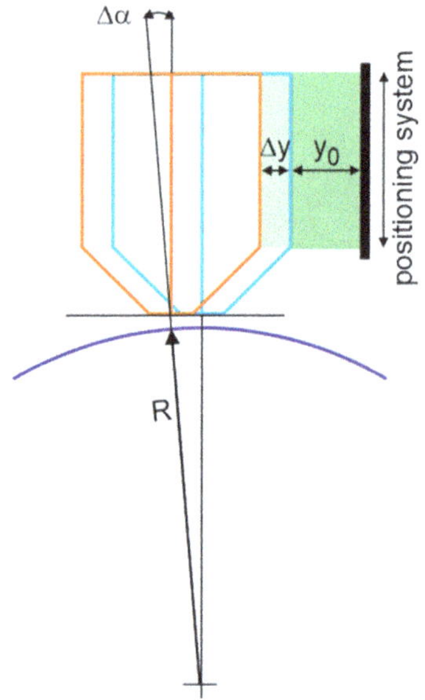

Fig. 17.5 Changing the die tilt angle by moving the die perpendicularly to the axis of the slot

angle is desirable, then it should not be designed with a die tilt feature but with an intermediate plate that is positioned between the die and the positioning system. As illustrated in Fig. 17.5, changing the thickness of that plate achieves the same effect as tilting the die but without reducing the mechanical precision because this approach does not introduce additional movable parts. In particular, the apparent die tilt angle $\Delta\alpha$ is related to the change in plate thickness Δy as follows:

$$\Delta\alpha = \arcsin\left(\frac{\Delta y}{R}\right) \tag{17.2.4}$$

A plate thickness of y_0 corresponds to an impingement angle of 0°, i.e., the flow exiting the slot impinges perpendicularly onto the substrate. Increasing the plate thickness ($\Delta y > 0$) results in a positive impingement angle, and vice versa.

As coating gaps are typically very narrow, i.e., <100 μm, the positioning system may have to provide a skip-out function to temporarily widen the coating gap during the passage of the splice.

The die, the vacuum box, and the positioning system are typically mounted into a frame structure called a coating station, which allows an ergonomic working position for the operator. The coating station can be permanently installed into a coating machine (see Fig. 17.6), or it can be built as an interchangeable trolley as shown in Fig. 17.1a. This latter design allows the trolley to be turned around in the coating machine, thereby providing easy access to the 4:30 and 7:30 die positions on the coating roll.

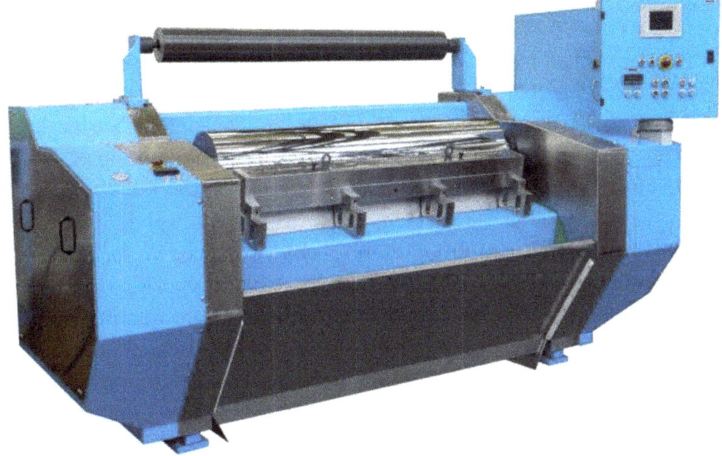

Fig. 17.6 Stationary slot coating station for adhesive coating; photo reproduced with permission from Polytype Converting AG

17.3 Geometry of the Coating Bead

The geometry of the coating bead, i.e., the space between the die lip and the substrate surface, is schematically depicted in Fig. 17.7.

Most of the time, when discussing slot coating, the coating bead is viewed in a simplified version as shown in Fig. 17.7a. In particular, the coating gap is assumed to be parallel, and the height and length of the downstream gap are equal to the height and length of the upstream gap. Typically, the axis of the die slot impinges perpendicularly onto the substrate. Furthermore, in illustrations like the ones in Fig. 17.7 as well as in plots generated by finite element calculations, the scale across the coating gap is amplified relative to the scale along the substrate surface, thus yielding a distorted view of the bead flow field. This, however, is necessary because details of the flow

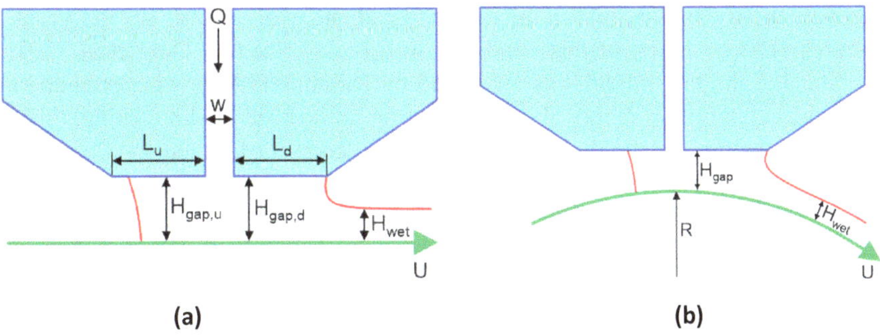

Fig. 17.7 **a** Simplified geometry for modeling purposes. **b** Realistic geometry

field, such as the presence of vortices, would barely be visible by the unaided eye if the two scales were equal, since the length of the die lips is typically much larger than the height of the gap. When modeling the flow field of the bead, however, the geometry may be much more flexible. In particular, the length and height of the downstream gap may be different from the values of the upstream gap, and the surface of both die lips may be angled positively or negatively relative to the orientation of the substrate. Also, the height of the die slot and the impingement angle, i.e., the orientation of the slot relative to the substrate surface, may vary. To discuss the relevant features of the bead flow field, the simplified view as shown in Fig. 17.7a, is sufficient and it keeps the mathematical efforts on a more affordable level. This is in contrast with a realistic bead geometry as encountered in industrial slot coating processes and as shown in Fig. 17.7b because, typically, the substrate surface assumes a circular shape as the substrate is wrapped around a backup roll. Consequently, the upstream gap converges from the upstream meniscus to the die slot, and the downstream gap diverges from the die slot to the downstream meniscus.

17.4 Operating Window

As mentioned in Chap. 16, the best way to represent the operating window of slot coating is not obvious. Durst and Wagner (1997) plotted the pressure difference ΔP across the coating bead, i.e., the bead vacuum, as a function of the capillary number Ca. In this diagram, the operating limit called *swelling* becomes visible if the bead vacuum is too high for a given value of Ca. Swelling refers to a flow condition where liquid at the upstream meniscus is leaking into the vacuum box. Another operating limit called *ribbing lines* appears if the bead vacuum is not high enough for a given value of the capillary number. Ribbing lines describe a situation where the thickness of the coated film is not uniform across the coating width but decays into a sinusoidal pattern with ridges and valleys spread evenly across the width.

In a rather popular version of the coating window, the bead vacuum is plotted as a function of the inverse of the wet coating thickness or the ratio of coating gap to wet thickness of the coated film. This view of the coating window is based on theoretical calculations of the flow field in the coating bead and evolved over several years of research. The evolution of this window is nicely described by Ding et al. (2016). Specifically, the first model was published by Ruschak (1976), who extended the Landau-Levich (1942) theory about dip coating to the bead of slot coating, thereby assuming that the flow field of the bead is controlled by surface tension forces only, and hence that viscous and inertia forces are negligible, i.e., both the capillary number Ca and the Reynolds number Re are very small. Consequently, this model is called the *capillary model*. This model provides operating limits for the bead vacuum and the wet film thickness, including the minimum film thickness, under the assumptions that the upstream and downstream menisci are circular and pinned to the upstream and downstream corners of the die lip. Ruschak's model was extended by Higgins and Scriven (1980), who allowed the surface tension in the upstream and downstream menisci as well as the height of the upstream and downstream coating gap to be different.

In the same paper, Higgins and Scriven also presented a purely *viscous model* for the operating limits of the slot coating process. They neglected surface tension and inertia effects, and they allowed the upstream meniscus to move freely between the upstream and the downstream corners of the upstream die lip. For these conditions, and further assuming that the coating gaps are parallel and the flow behavior of the coating fluid is Newtonian, the flow field of the coating bead is described by the flow rate pressure drop relationship in the downstream and upstream gap as given by Eqs. (5.6.26) and (5.6.30), respectively. For these assumptions, the minimum film thickness without bead vacuum is obtained by setting the pressure drop in the downstream gap equal to zero. The result says that the minimum film thickness equals half of the downstream coating gap, and it is independent of the capillary number:

$$H_0 = \frac{H_{gap,d}}{2} \tag{17.4.1}$$

The viscous model also allows one to estimate the position of the upstream meniscus L_m by setting the pressure drop in the upstream gap (Eq. 5.6.30) equal to the negative value of the pressure drop in the downstream gap (Eq. 5.6.26), and by solving the former equation for $L = L_m$ according to:

$$L_m = L_d \left(\frac{H_{gap,u}}{H_{gap,d}} \right)^2 \left(\frac{2H_{wet}}{H_{gap,d}} - 1 \right) \tag{17.4.2}$$

Equation (17.4.2) is not accurate because it does not account for the two-dimensional turn-around flow inside of the upstream meniscus nor the capillary pressure drop across the meniscus, but it may still provide valuable information in case of unexpected coating problems by indicating whether the upstream meniscus is likely to be located near the die slot or the upstream corner of the upstream gap.

Theoretical models that only account for capillary or viscous effects are insufficient. To overcome these shortcomings, Higgins and Scriven (1980) combined the two models to form the *visco-capillary* model of operating limits. The resulting coating window, when plotted in an axes system of bead vacuum ΔP versus thickness ratio H_{gap}/H_{wet} is shown in Fig. 17.8. The shape and size of the coating window depend upon the capillary number. Following is a qualitative description of the three operating limits based on explanations given by Romero et al. (2004) and Ding et al. (2016).

An acceptable uniformity of the coated film is obtained if the operating point lies inside the coating window as marked by the yellow dot in Fig. 17.8. However, if the thickness of the coated layer is larger than the thinnest that can be produced for a given coating gap and substrate speed, i.e., $H_{wet} > H_{min}$, then excessive vacuum pressure at the upstream meniscus causes the liquid of the bead to be drawn in the upstream direction and to leak into the vacuum chamber. The onset of this defect is called *high vacuum limit*. It also marks the loss of the feature of premetering.

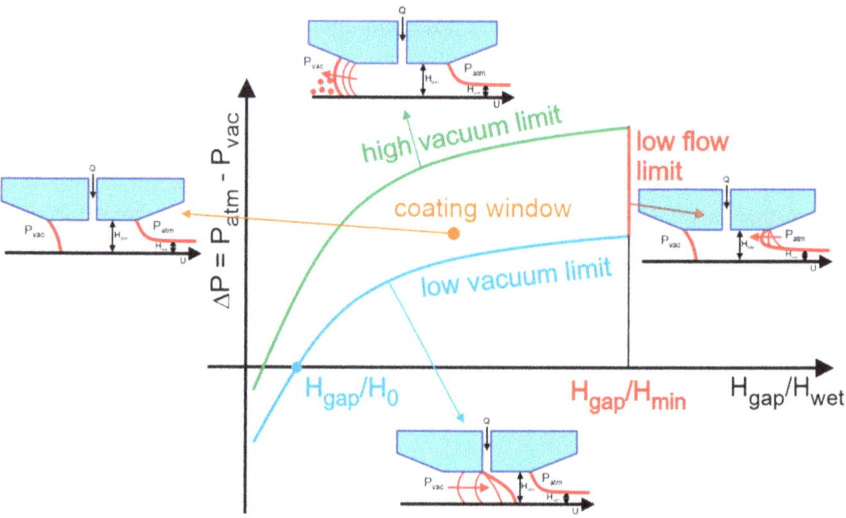

Fig. 17.8 Operating limits of slot coating predicted by the visco-capillary model

Reducing the vacuum at this point would prevent the defect and would allow thicker films to be coated.

In contrast, if not enough vacuum is applied at the upstream meniscus, then the viscous drag exerted on the liquid of the bead by the moving substrate moves the upstream meniscus in the downstream direction upon increasing the web speed. First, the upstream free surface is shifted to the downstream corner of the upstream lip, which is a limiting position for this meniscus. Upon further increasing the speed, the dynamic wetting is invading the flow field under the downstream lip until the dynamic contact angle approaches 180° and air is entrained between the substrate and the liquid. The onset of this defect is called the *low vacuum limit*. Increasing the vacuum at this point would prevent the defect, and would allow thinner films to be coated at higher speeds. Note that the thinnest film that can be applied without bead vacuum is denoted with H_0.

If at a given substrate speed, the flow rate/width is continuously decreased, then the downstream meniscus starts curving more and more, and it begins to invade the downstream gap until it can no longer bridge the gap between the die lip and the coated film. At that point, the meniscus adapts a three-dimensional form, which causes the initially uniform thickness of the coated film to decay into evenly spaced longitudinal disturbances known as ribbing lines, or into rivulets and dry lanes if the uniform film thickness is small enough. This defect is called the *low flow limit*. It marks the minimum film thickness that can be deposited uniformly for a given coating gap and web speed. Since slot coating is a premetered application method, it also marks the maximum substrate speed at which a uniform film can be coated for a given gap and flow rate/width. From this point of view, the operating limit is also called the *high-speed limit*. Moreover, the same defect is generated when, for a given flow rate/width and web speed, the downstream coating gap is increased beyond a critical value. This operating condition is called the *wide gap limit* (Romero et al., 2004). As indicated in Fig. 17.8 the low flow limit is independent of the bead vacuum, provided that the vacuum is high enough to draw the upstream meniscus away from the die slot.

Both the parameters H_0 and H_{min} stand for the minimum wet film thickness with and without vacuum. However, as explained above, the physical reasons for their existence are different. While H_0 describes the situation when the upstream meniscus excessively invades the downstream coating gap in the direction of the web motion, H_{min} occurs when the downstream meniscus excessively invades the downstream coating gap in the opposite direction to the web motion. Therefore, Lee et al. (1992) argued that the ratio of H_{min}/H_0 can be used for evaluating the effectiveness of bead vacuum on reducing the film thickness. Based on the visco-capillary model, it can be expected that bead vacuum becomes more effective as viscosity decreases. This conclusion was confirmed experimentally by Lee et al.

Chang et al. (2009) studied the effect of bead vacuum on the slot coating process both experimentally and theoretically. They worked with two aqueous glycerol solutions having viscosities of 3 and 120 mPas, respectively. Bead vacuum was not generated with the help of a vacuum chamber, but with a suction slot that was oriented parallel to and located just a short distance upstream of the feed slot for the coating fluid. They were able to visualize the form and the location of the upstream and downstream menisci. Their results clearly showed that applying bead vacuum reduced the minimum wet thickness of the coated film and increased the maximum coating speed for both tested viscosities, but the vacuum effect was stronger for the low-viscosity liquid. Furthermore, the low flow limit was reached at thinner films if vacuum was applied when compared to coating without vacuum. Visualizing the coating bead revealed that for the low-viscosity fluid, the upstream meniscus was pulled backward by the vacuum, while the form and location of the downstream meniscus were not changed. For the high-viscosity fluid, in contrast, the major effect of the vacuum was to reduce the dynamic contact angle, which in turn postponed the onset of air entrainment to higher web speeds. Theoretical modeling of this coating configuration confirmed the experimental results qualitatively, but the quantitative agreement was less helpful with theory predicting a stronger vacuum effect than was observed experimentally. It is important to note that the downstream meniscus always contained a vortex next to the static wetting line (vortex 1 in Fig. 17.26), which was not pinned to the lip corner but located somewhere along the inclined lip face. As a consequence, the downstream meniscus never invaded the coating gap for any level of bead vacuum.

The drawback of the visco-capillary model as derived by Higgins and Scriven is that predictions of the operating limits are not accurate because they depend on the surface tensions of the upstream and downstream menisci as well as on the dynamic contact angle, all of which are not known and must be assumed. In particular, surface tensions are difficult to assess if they are a dynamic property such as for aqueous coating fluids containing surfactants. Given the small dimensions of the menisci, i.e., on the order of the coating gap, the surface age may be estimated to be on the order of H_{gap}/U. Corresponding values of the surface age are very small, i.e., on the order of 1 ms or less, thus indicating that the surface tensions are close to the value of water. Lin et al. (2010) performed two-dimensional computations of the bead flow field and compared theoretically evaluated coating windows with experimental observations of the onset of ribbing lines and air entrainment. They found that the theoretically determined operating widow was much larger than the one determined experimentally in terms of web speed. On the other hand, they confirmed that theoretically predicted qualitative trends agreed with experimental observations.

17.5 Low Flow Limit

Based on the experience of running slot coating trials for many different applications over 15 years in the Polytype Converting pilot facility, the low flow operating limit discussed in the previous chapter is the most severe operating limit of this process. This can be explained by recent trends in the converting industry of exploring the benefits of the premetered slot coating process for applying thinner and thinner layers of high thickness uniformity, for example for electronic products such as OPVs and OLEDs. This trend not only tries to push fluid-mechanical operating limits to widen the coating window but, as explained later in this chapter, it also clashes with the mechanical limitations of the coating equipment.

Focusing on the low flow limit allows one to draw another type of operating window for the slot coating process. This view was first developed by Ruschak (1976). As illustrated in Fig. 17.9a, he resorted to findings from Landau-Levich (1942), who studied dip coating and derived the following relationship between the wet film thickness, the capillary number and the radius of curvature of the unbound meniscus:

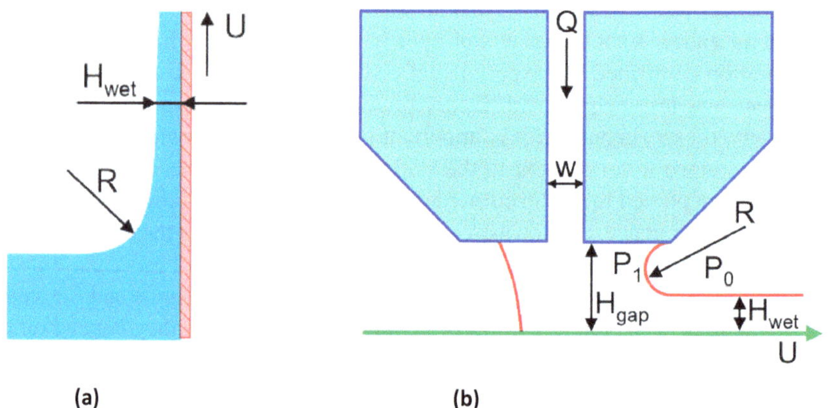

Fig. 17.9 **a** Radius of the film building meniscus in dip coating. **b** Minimum radius of the film building meniscus in slot coating

$$\Delta P = \frac{\sigma}{R} = 1.34Ca^{2/3}\frac{\sigma}{H_{wet}} \tag{17.5.1}$$

Ruschak then applied this result to the bound downstream meniscus of the coating bead (Fig. 17.9b) and argued that an operating limit is reached when the pressure drop across the downstream meniscus of the coating bead is at its maximum value. Based on the Young–Laplace equation, which says that surface tension supports a pressure difference across a curved meniscus (Weinstein & Palmer, 1997), this is the case when the radius of the meniscus is at its minimum. Assuming that the meniscus shape is circular and tangent to the lip face and the surface of the coated film, the minimum radius is given by a simple geometrical constraint as follows:

$$R_{min} = \frac{H_{gap} - H_{wet}}{2} \tag{17.5.2}$$

Combining Eqs. (17.5.1) and (17.5.2) results in an analytical expression for the low flow limit of slot coating:

$$Ca_c = \frac{\mu U}{\sigma} = 1.823\left[\frac{1}{H_{gap}/H_{wet} - 1}\right]^{3/2} \tag{17.5.3}$$

In concert with the work of Landau-Levich, this operating limit is only valid for vanishing viscous and inertia forces, i.e., for Ca << 1, and Re << 1. In that sense, Eq. (17.5.3) is a capillary model for the low flow limit. The corresponding coating window is obtained by plotting the critical capillary number Ca_c as a function of the ratio of coating gap to wet film thickness, see Fig. 17.10.

Stable and uniform coated films result for operating points that are located below the red curve in Fig. 17.10, i.e., for $Ca < Ca_c$. As the validity of Eq. (17.5.3) is limited to low values of the capillary number, which is not always typical of industrial applications, it is necessary to compare the theoretically predicted operating limit with experimental observations. Several authors have carried out such comparisons. Here, we took data from Lee et al. (1992), Romero et al. (2004), and Wengeler et al. (2014), and we compiled them in the same diagram shown in Fig. 17.11.

Lee et al. ran experiments with aqueous fluids at different coating gaps but without applying bead vacuum. All data sets follow Ruschak's low flow limit qualitatively, but the limit occurs at lower capillary numbers, thus suggesting a smaller coating window than predicted theoretically. The difference decreases with the increasing coating gap.

Wengeler et al. (2014) experimented with solvent-based fluids. They also changed the coating gap, and they did not apply bead vacuum. When performing a regression analysis on their data, they found that the dependence of Ca upon H_{gap}/H_{wet} is identical to Ruschak's theoretical predictions, but also shifted to lower values of Ca. Consequently, Eq. (17.5.3) must be modified by changing the constant in front of the bracket as follows:

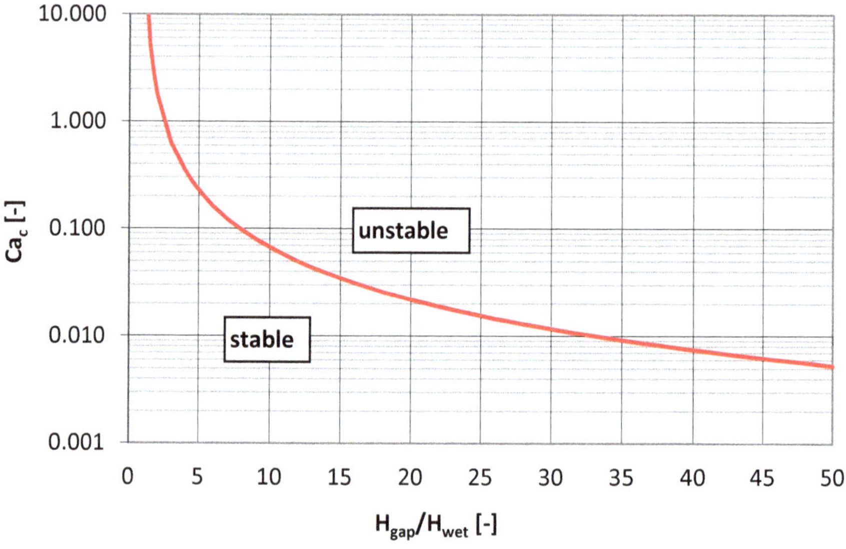

Fig. 17.10 Theoretical low flow limit for slot coating

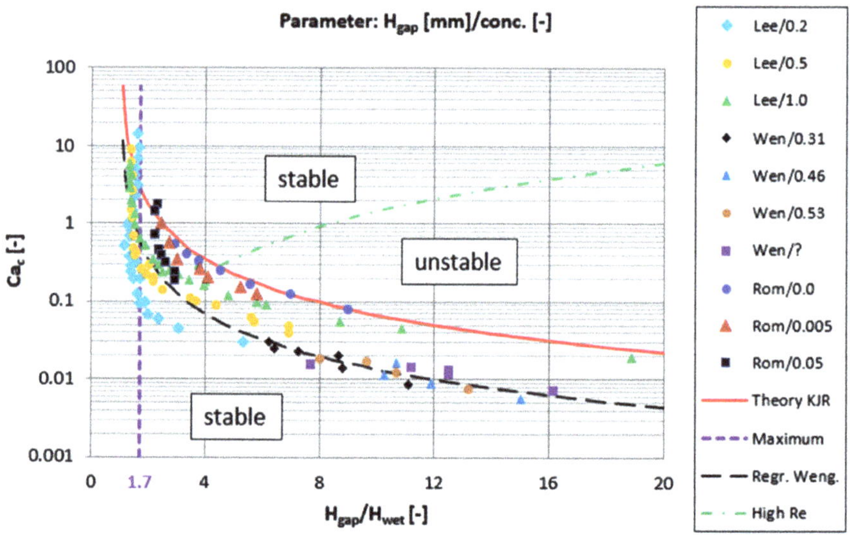

Fig. 17.11 Comparison of theoretical low flow limit with experimental data

$$Ca_c = \frac{\mu U}{\sigma} = 0.362\left[\frac{1}{H_{gap}/H_{wet} - 1}\right]^{3/2} \tag{17.5.4}$$

Romero et al. worked with aqueous polyethylene glycol solutions at different concentrations, and added polyethylene oxide at various concentrations to render the fluid moderately visco-elastic. For most of the trials, the coating gap was set to 100 μm. Bead vacuum was applied to keep the upstream meniscus at more or less the same position when decreasing the flow rate. Interestingly, the low flow limit for the viscous fluid, i.e., the fluid without polyethylene oxide, is very close to the limit predicted by Ruschak's model. However, with increasing visco-elasticity, the limit keeps its characteristic shape but is shifted to lower values of the capillary number.

Carvalho and Kheshgi (2000) carried out a theoretical and experimental study for operating conditions at high capillary and high Reynolds numbers. Their results showed that the low flow limit for such operating conditions is fundamentally different when compared to low values of Ca and Re. Specifically, the minimum film thickness that can be coated decreases with increasing web speed. Consequently, the operating window of the slot coating process can be expanded considerably for such operating conditions. Carvalho and Kheshgi explained that, when inertial forces are important, the low flow limit is no longer governed by viscous and surface tension forces alone, but that the shape of the downstream meniscus is controlled by the momentum of the liquid contained in the coating bead. The behavior of the flow field in the bead can be explained by boundary layer theory, much like the flow field of the bead in slide coating as visualized in Fig. 5.73 in Sect. 5.7.8, or the flow field of the impinging curtain as explained in Sect. 5.9.6. As illustrated in Fig. 17.12, the boundary layer starts at the dynamic wetting line, and it grows in length until the thickness of the boundary layer has reached the thickness of the coated film. Carvalho and Kheshgi show that the boundary layer length is proportional to the web speed. At low web speeds, therefore, the boundary layer may be shorter than the length of the bead as determined by the lengths of the upstream and downstream die lips, which in turn would allow the downstream meniscus to invade the downstream coating bead. At high web speeds, however, the boundary layer may be much longer than the bead geometry as defined by the die lips. Consequently, the drag force imposed by the moving web on the liquid inside the boundary layer pulls the downstream meniscus

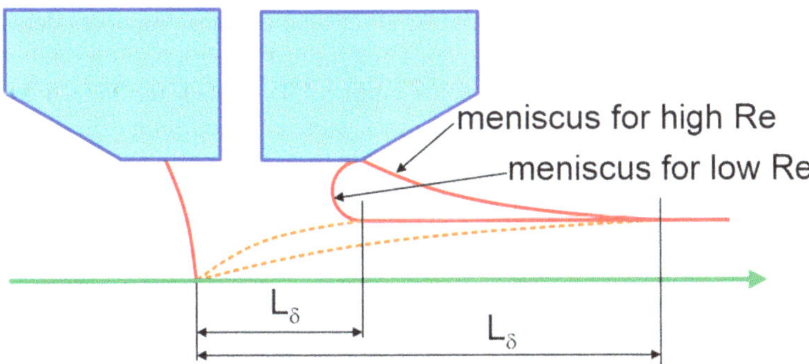

Fig. 17.12 Formation of a viscous boundary layer in the bead of slot coating at high Reynolds numbers

in the downstream direction such that an invasion into the bead area is no longer possible.

Carvalho and Kheshgi also show that the ratio of coating gap to film thickness is proportional to the square root of the web speed. The corresponding low flow limit for the high Reynolds number region is taken from Fig. 19 of the Carvalho and Kheshgi paper and drawn into Fig. 17.11. Specifically, H_{gap}/H_{wet} and Ca of a few points on their limiting curve were measured in their Fig. 19, and the following power law function was obtained by regression analysis:

$$Ca_c = a_0 \left(\frac{H_{gap}}{H_{wet}} \right)^{a_1} \tag{17.5.5}$$

$a_0 = 0.01329$, $a_1 = 2.05109$, $r^2 = 0.99945$.

While the boundary layer arguments presented above are theoretical, Carvalho and Kheshgi also confirmed their findings with experimental data, albeit with quite some scatter in the data. Consequently, a well-documented and experimentally supported regression equation such as Eq. (17.5.4) does not exist for the low flow limit at high values of Re. Moreover, boundary layer arguments explain that applying bead vacuum results in lower values of the ratio H_{gap}/H_{wet} because the position of the dynamic wetting line, i.e., the starting point of the boundary layer, is pulled in the upstream direction as the vacuum is increased. This trend works in the opposite direction compared to operating conditions at low values of Ca and Re, and hence it is not desirable. However, vacuum may have to be applied for high coating speeds to prevent air entrainment at the dynamic wetting line.

The following conclusions can be drawn from analyzing the experimental data shown in Fig. 17.11:

- The capillary model for the low flow limit derived by Ruschak captures the underlying physics of this operating limit if $Ca \ll 1$, and $Re \ll 1$, i.e., the minimum film thickness that can be coated decreases with decreasing coating speed.
- The experimental data in Fig. 17.11 scatter because the low flow limit depends on additional parameters, including the height of the coating gap and the degree of visco-elasticity. In addition, almost all critical capillary numbers determined experimentally are lower than the ones determined by Ruschak's model.
- For high capillary and low Reynolds numbers, i.e., when the flow field of the coating bead is controlled by viscous forces, there is a maximum value for the ratio of $H_{gap}/H_{wet} = H_{gap}/H_0$. Considering the data from Lee et al. (1992) and Wengeler et al. (2014), i.e., considering purely viscous fluids only that are coated without vacuum, this maximum value is about 1.7 as indicated by thc purple, vertical, dashed line in Fig. 17.11. Citing several other experimental investigations, Ding et al. (2016) concluded that the maximum value for H_{gap}/H_0 is in the range of about 1.4–2.0.
- As explained in Sect. 17.4, the physical reason for the existence of a minimum film thickness when coated without vacuum is different compared to coating with vacuum, i.e., invasion of the upstream meniscus into the downstream coating

gap versus invasion of the downstream meniscus into the downstream coating gap. It is surprising, therefore, that despite this physical difference, the qualitative behavior of these two low flow limits is identical and in line with Ruschak's model as illustrated in Fig. 17.11.

- The value of the capillary number that marks the transition from one operating range to the next is not well defined. Based on the data shown in Fig. 17.11, the transition from the capillary to the viscous region occurs at a capillary number in the range of 0.1–0.5, and the high inertia range stats at capillary numbers in the range of 0.3–0.7.
- Experimental data in support of the low flow limit scatters for all three operating ranges, i.e., the capillary range with Ca << 1 and Re << 1, the viscous range with Ca > 0.1 and Re << 1, and the high inertia range with Ca > 0.3 and Re > 1. Consequently, predictions of the low flow limit are associated with some degree of uncertainty. However, when running industrial slot coating processes, this uncertainty is not very important because relevant process parameters that affect the low flow limit, such as the coating gap, the bead vacuum, and the web speed, can easily be changed during coating. In addition, the process information contained in Fig. 17.11 readily shows which parameter must be changed in which direction to move an operating point from an unstable to a stable region.

17.6 Implications of the Low Flow Limit

Operating limits of the slot coating process in terms of the low flow limit are characterized by the dimensionless capillary and Reynolds numbers, which are defined as follows (Carvalho & Kheshgi, 2000):

$$Ca = \frac{\mu U}{\sigma} \tag{17.6.1}$$

$$Re = \frac{\rho U H_{gap}}{\mu} \tag{17.6.2}$$

Both numbers contain the viscosity μ, which is only clearly defined for Newtonian fluids having a viscosity that is independent of the prevailing shear rate of the flow field under consideration. Such situations are unrealistic for most industrial coating fluids because their viscosity is shear-dependent as described by the Carreau-Yasuda model given by Eq. (4.4.1). To obtain the correct viscosity, therefore, the shear rate of the flow field in the coating bead must be known. Unfortunately, an accurate equation for the shear rate of non-Newtonian slot flow with one wall at rest and one wall in motion as described by the power law model, for example, is not available. Therefore, the viscosity must be approximated by resorting to Newtonian flow equations.

As visualized by Eqs. (5.6.27) and (5.6.28) the shear rate in the downstream coating gap depends on the ratio of H_{wet}/H_{gap}, and it varies in magnitude and sign across the coating gap. Therefore, it is recommended to determine the viscosity for the average shear rate, the magnitude of which is given by Eq. (5.6.29) as follows:

$$\gamma_{average} = \frac{U}{H_{gap}} \tag{17.6.3}$$

Equation (17.6.3) is valid for both the downstream and upstream coating gap. Now, it is possible to determine an approximate viscosity value in combination with the Carreau-Yasuda model.

For the viscous operating region, i.e., for Ca > 0.1 and Re << 1, the ratio of H_{gap}/H_{wet} is limited to a maximum value of about 1.7, see Fig. 17.11. Consequently, coating thin films requires narrow coating gaps. However, the coating gap cannot be made narrower and narrower at will owing to the limited mechanical precision of the gap. As illustrated in Fig. 17.13, the mechanical precision of the gap is affected by the mechanical precision or uniformity of the die lip, the backup roller, and the substrate thickness. Since all of these three parameters are stochastic in nature, their effect on the precision of the gap can be quantified as follows:

$$\frac{dH_{gap}}{H_{gap}} = \pm\sqrt{\left(\frac{dH_{web}}{H_{gap}}\right)^2 + \left(\frac{dR}{H_{gap}}\right)^2 + \left(\frac{dZ_{lip}}{H_{gap}}\right)^2} \tag{17.6.4}$$

As shown at Polytype Converting AG, $dH_{web}/H_{web} < \pm 2.5\%$ and $dR \geq \pm 1\ \mu m$. The capability for manufacturing straight die lips depends on the die width. Specifically, values for dZ_{lip} can be quantified as $< \pm 1.0\ \mu m$ for $W < 1$ m and dZ_{lip} <

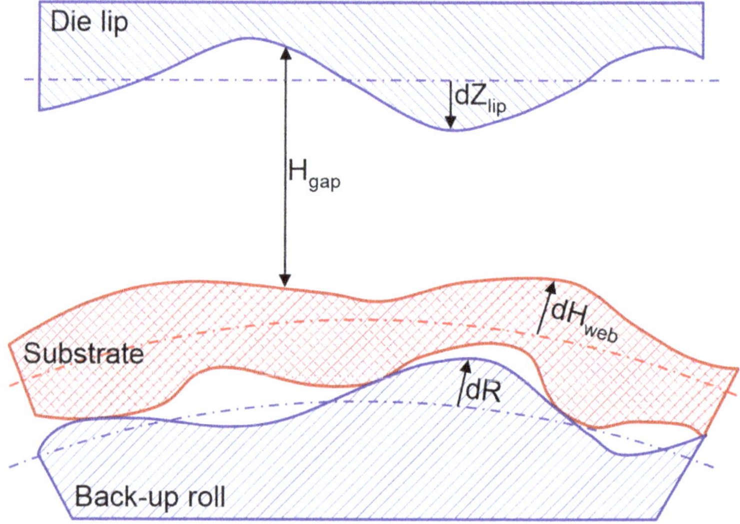

Fig. 17.13 Mechanical precision of coating gap

±2.5 μm for W > 1 m, with a gradual transition between the two width ranges. Note that the dominant contribution to gap variation stems from variations of the substrate thickness dH_{web}, if the substrate is thick. Moreover, mechanical engineers and designers of coating equipment at Polytype Converting AG prefer that coating gaps are not smaller than about 30 μm for processes involving close proximity between a stationary element, i.e., a slot die, and a moving element, i.e., a backup roll. This requirement applies in particular to industrial equipment with a width of >1 m.

Applying this mechanical limit of the coating gap to the low flow limit for the viscous operating mode means that the minimum film thickness that can be coated is about 30/1.7 = 18 μm. This result is in good agreement with my experience gained at Polytype Converting AG after years of running many varied slot coating trials in the viscous mode, mostly without bead vacuum. As explained above, applying bead vacuum would allow somewhat thinner films to be coated, but only if the viscosity of the coating fluid is not too high. This result also means that wet films that are thinner than about 10–15 μm cannot be applied with the slot coating process if the operating conditions lie within the viscous range.

Coating films that are thinner than about 10–15 μm requires the ratio of H_{gap}/H_{wet} to be larger than about 2. For the capillary operating mode, this in turn, requires the capillary number to be lower than about 0.1. Furthermore, thinner and thinner wet films require lower and lower capillary numbers. Using Wengeler's regression Eq. (17.5.4) for the low flow limit, and requiring the coating gap to assume its minimum value of 30 μm posted above, the maximum admissible capillary number can be calculated as a function of the desired wet coating thickness. The result is shown in Fig. 17.14.

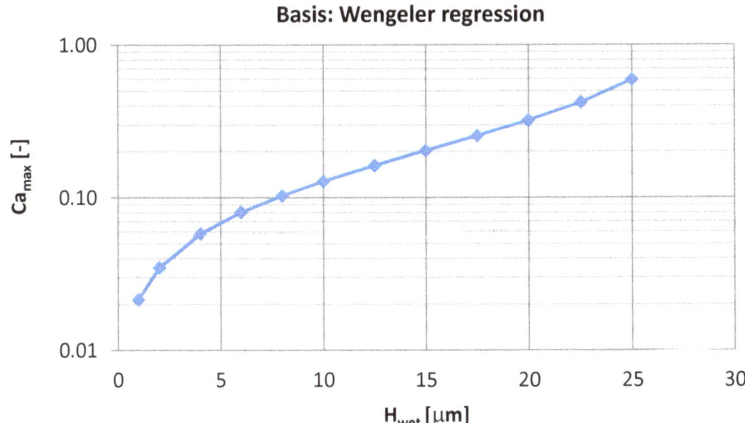

Fig. 17.14 Maximum capillary number as a function of the wet film thickness for a minimum coating gap of 30 μm and based on the regression Eq. (17.5.4) for the low flow limit by Wengeler et al. (2014)

Operating conditions requiring high values for H_{gap}/H_{wet} and low values for Ca are characterized by the following two disadvantages.

As illustrated in Fig. 5.47, a vortex is present under the downstream die lip if $H_{gap}/H_{wet} > 3$. This result is valid for Newtonian fluids, and the limiting gap ratio increases with decreasing power law index. However, achieving very low values of the capillary number requires low viscosities as well as low coating speeds and hence low shear rates. Consequently, the power law index is likely to be near 1.0 for many industrial coating fluids. As explained in Sect. 5.2, vortices in coating flows are not desirable because they are a major cause of lines and streaks in the coated film. A vortex under the downstream die lip belongs to this class of vortices. If the coating fluid is a pure and clean solution, then the presence of this vortex may not cause too many quality problems. But if the coating fluid is an emulsion or a suspension, lines and streaks are more likely to appear, therefore such operating conditions should be avoided. However, this would require considering alternative coating methods for coating very thin layers of high uniformity, such as curtain coating involving an inert bulk layer as explained in Chap. 12.

On the other hand, coating at low or even very low capillary numbers requires low web speeds, low viscosities and high surface tensions. The surface tension of a coating fluid is primarily determined by the type of solvent. Therefore, it seems easier to achieve low capillary numbers with aqueous fluids because their surface tensions are typically in the range of 60–70 mN/m if they do not contain surfactants. However, the surface tension of aqueous fluids must often be lowered to a value of around 40 mN/m by adding surfactants, for example to stabilize emulsion properties or to prevent coating defects such as repellencies. In contrast, the surface tension of most organic solvents is in the range of 20–40 mN/m. For that reason, the surface tension is not a good parameter for influencing the value of the capillary number.

In most cases, achieving low viscosities requires a high dilution of the coating fluid. As a result, such liquids contain large amounts of solvent, which increases the dryer load, which in turn increases the investment costs for longer dryers and it increases the operating costs due to higher energy consumption of the dryer, all of which is not desirable. Considering, however, that the wet film thicknesses are small in the context of this discussion, these monetary effects may not be all that significant.

Finally, coating at low speeds considerably decreases the productivity of the manufacturing process, which in turn increases the product unit costs and limits the annual production volume of a given coating machine, both of which are not desirable. Details regarding the economic aspects of coating processes are discussed in Chap. 15. Figure 17.15 shows possible coating speeds as a function of low values of the capillary number and as a function of selected viscosity and surface tension values. For $Ca < 0.1$, it seems, therefore, that coating speeds of >20 m/min are difficult to achieve if the viscosity is higher than 10 mPas and if the surface tension is lower than about 40 mN/m.

In summary, for $Ca \ll 1$ and $Re \ll 1$, it is possible to apply very thin films with the slot coating process without problems of mechanical precision, but only at the expense of low process productivities.

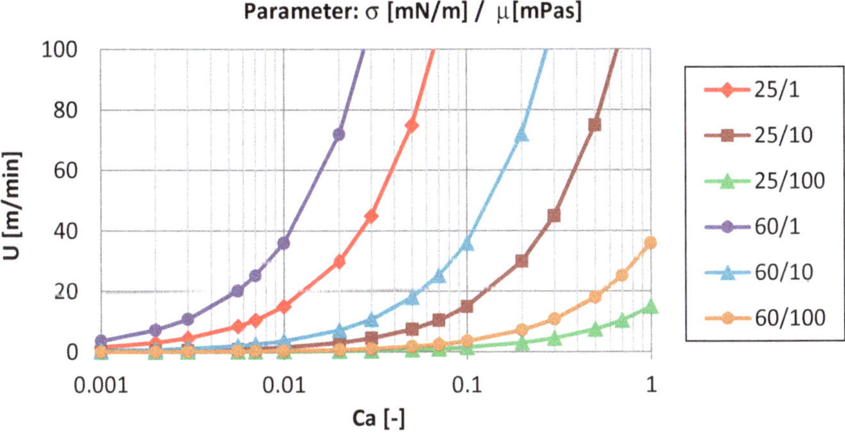

Fig. 17.15 Coating speed as a function of the capillary number and as a function of selected values of surface tension and viscosity

Regarding the high inertia operating range, the author of this book has no practical experience. Nevertheless, it would be interesting to know the range of Re-values that allow thin films to be applied at high web speeds with the slot coating process. To that extent, I examined the experimental work carried out by Carvalho and Kheshgi (2000). In particular, they ran trials with a fluid based on the organic solvent MEK and containing some unspecified but dissolved solid component. The surface tension of MEK is 25 mN/m, and the fluid density was assumed to be just a bit higher than that of MEK, i.e., 810 kg/m^3. The fluid viscosity was varied in the range of 13–75 mPas by varying the solids concentration. Moreover, the coating gap was held constant at 101.6 μm. Then, I measured the values of H_{gap}/H_{wet} and Ca of a few experimental points that are located on the low flow limit shown in their Fig. 19. Knowing the physical fluid properties and the coating gap, the corresponding web speed can be calculated from the capillary number according to Eq. (17.6.1), the Reynolds number is obtained from Eq. (17.6.2), and the resulting wet film thickness can be extracted from the ratio of gap/film thickness. The values of 6 experimental points are listed in Table 17.1.

It seems that wet film thicknesses of less than 10 μm can be achieved for Reynolds numbers in the range of about 1–10, with web speed ranging from about 50–100 m/min and viscosities in the range of about 20–100 mPas. It should not be too difficult to realize such parameter values for industrial coating applications. Moreover, the coating speeds for achieving very thin wet films in the high inertia region are more attractive from a process productivity point of view than the speed values for the low Ca and low Re operating range.

Table 17.1 Parameter values of 6 experimental points on the low flow limit for Ca > 0.3 and Re > 1

H_{gap}/H_{wet} (–)	Ca (–)	μ (mPas)	U (m/min)	Re (–)	H_{wet} (μm)
5.56	0.44	22	30.0	1.87	18.3
6.39	0.68	22	46.6	2.91	15.9
7.64	0.92	22	62.7	3.91	13.3
8.75	1.33	22	90.8	5.66	11.6
14.24	3.10	75	62.0	1.13	7.1
15.52	3.49	75	69.8	1.28	6.5

17.7 The Pressure Profile in the Coating Bead

Understanding the characteristic features of the pressure profile across the coating bead is important because it affects the behavior and the performance of the bead flow field. Figure 17.16 illustrates that the pressure profile is determined by fife relevant pressure values.

P_0 is the atmospheric pressure just outside of the downstream meniscus, and P_1 is the fluid pressure just inside the downstream meniscus. Since the downstream meniscus is always convex toward the fluid, P_1 is always lower than P_0. The pressure difference ΔP across the meniscus is given by the Young–Laplace equation, with σ being the surface tension and R the radius of the curved meniscus:

$$\Delta P = \frac{2\sigma}{R} \tag{17.7.1}$$

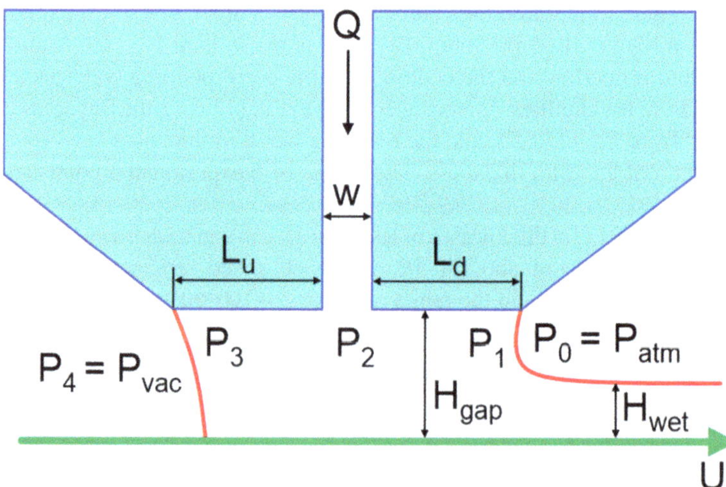

Fig. 17.16 Relevant pressure levels in the coating bead

The problem with Eq. (17.7.1) is that, typically, the meniscus radius is not known because it depends on the details of the two-dimensional bead flow field. In addition, the surface tension is also not known, if the fluid is aqueous and contains surfactants because then surface tension is a dynamic property, which also depends on the details of the flow field as well as on the viscosity and the nature of the surfactant molecules.

P_2 is the highest pressure in the bead. P_2 is located at the exit of the die slot. The pressure difference $\Delta P = P_1 - P_2$ is determined by the pressure drop of the flow under the downstream die lip according to Eq. (5.6.22) for Newtonian fluids. Unfortunately, a similar, simple analytical equation for the pressure drop of power law fluids does not exist.

Usually, it is assumed that the pressure across the exit of the die slot remains constant. Consequently, pressure P_3, which is located just inside of the upstream meniscus, is obtained by integrating Eq. (5.6.30) over the length L_u of the upstream gap. Equation (5.6.30) is also valid for Newtonian fluids. Finally, P_4 is the pressure just outside of the upstream meniscus. Its value is either ambient or below ambient, depending on whether or not bead vacuum is applied. The pressure difference across the meniscus is also given by the Young–Laplace Eq. (17.7.1). Since the shape of the upstream meniscus can be concave or convex, P_3 can be lower or higher than P_4. Moreover, and as with the downstream meniscus, the meniscus radius and the surface tension of aqueous fluids are not known, and neither is the dynamic contact angle at the intersection of the meniscus and the substrate surface.

Given these various unknowns for calculating the pressure drop across the coating bead, a simpler and approximate approach is proposed by only focusing on the viscous pressure drops in the downstream and upstream coating gaps. Since viscous pressure drops are often larger than capillary pressure drops across menisci, this approach still allows characteristic features of the pressure profile to be identified and visualized, specifically when working with a very variable gap geometry as presented in Fig. 17.17. The die can be tilted by an angle α_0 or, equivalently, displaced by a distance z_0 relative to the origin of the coordinate system. Both lip surfaces can be tilted by angles β and γ relative to a reference surface, and the length of each die lip can be adjusted individually. The two die plates can be offset relative to each other by a distance OS, and the die slot w is variable. The moving surface of the gap is not straight, but it is formed by a backup roll being of circular shape with radius R and rotating counter-clock-wise with surface speed U. Note, however, that this very variable gap geometry was chosen for the sake of playing "what-if" games on the computer. The gap geometry of most industrial slot dies is much simpler, with α, β, γ, OS $= 0$, and $L_u = L_d$. This simpler geometry is justified by more efficient die manufacturing and assembly procedures. Moreover, as is shown below, complex gap geometries are not necessary from a coating process point of view, at least for most industrial applications.

For calculating the pressure profile, the view of Wagner (1996) is adopted. He introduced a coordinate system with a reference plane that is perpendicular to the axis of the die slot and tangent to the roll at the origin for $\alpha = 0$, see Fig. 17.17. Moreover, the variable coating gap $H_{gap}(z)$ is measured from the die lip to the roll surface (substrate surface) in a direction perpendicular to the reference plane. Since the gap

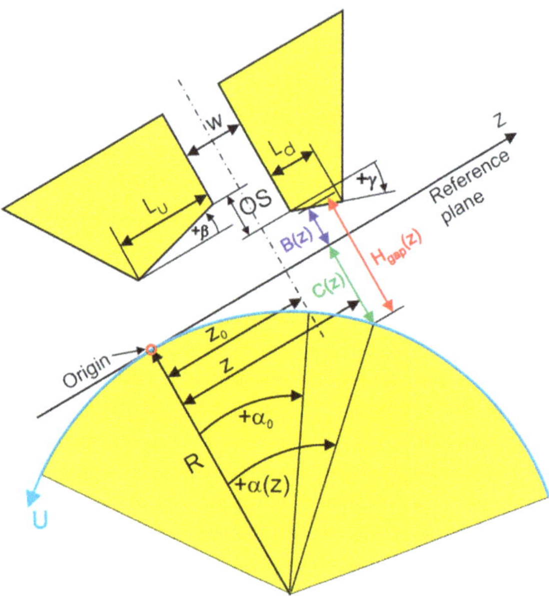

Fig. 17.17 Variable geometry of the coating gap

geometry may now be very complex, the pressure profile can no longer be calculated with Eqs. (5.6.22) and (5.6.30). Instead, Eqs. (5.6.20) and (5.6.30) for the pressure gradients in the downstream and upstream coating gaps are integrated numerically by using a 4-point Gauss quadrature. Therefore, the length of the downstream and upstream gaps is each divided into 50 short intervals of constant cross- section, and the pressure drop is calculated for each interval. Owing to the converging and diverging nature of the coating gap, each interval has a different but constant gap height. Therefore, the shear rate changes in each interval and so does the viscosity as it is determined by the shear rate according to Eq. (17.6.1) and the viscosity curve according to the Carreau-Yasuda Eq. (4.4.1). As explained above, an approximate position of the upstream meniscus is obtained by calculating the pressure drop in the upstream gap until P_3 equals either the ambient or the vacuum pressure as per Eq. (17.4.2).

Following are exemplary calculations of the pressure profile in the coating gap in an attempt to visualize how the profile is affected by characteristic geometrical features of the gap. The following operating conditions were assumed: $H_{wet} = 40\ \mu m$, $U = 50$ m/min, $\rho = 1{,}000$ kg/m^3, $\sigma = 40$ mN/m for both menisci, Carreau-Yasuda parameters: $\mu_0 = 2{,}000$ mPas, $\mu_\infty = 10$ mPas, $\gamma_c = 10\ s^{-1}$, $a = 0.7$, $b = 0.6$. The bead vacuum was set to 0. The die slot w was held constant at 300 μm, the roll radius R was held constant at 130 mm, and the ratio of H_{gap}/H_{wet} was held constant at 1.7. These values resulted in Ca = 2.61 and Re = 0.45, thus indicating a viscous operating region. Consequently, the maximum coating gap was 68 μm, which was placed at the upstream corner of the downstream gap because this is where the gap would be measured in most coating applications. Values for the other geometric parameters

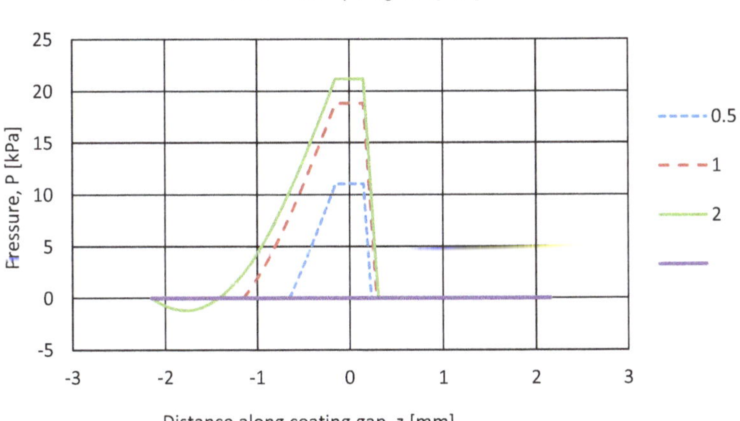

Fig. 17.18 Pressure profiles in the coating gap as a function of the lip length L; $\alpha = \beta = \gamma = \text{OS} = 0$

are listed below. The lengths of the downstream and upstream die lips are assumed to be equal. The substrate is moving from right to left in the following diagrams.

The effect of the lip length on the pressure profile is shown in Fig. 17.18.

The maximum pressure at the center of the gap ($z = 0$) increases with increasing lip length. As the gap is at its maximum value for the viscous operating mode, the upstream gap is almost completely free of liquid. The upstream meniscus is located at less than 8% into the upstream gap. The pressure profile in the upstream gap is linearly decreasing for all lip lengths. In contrast, the pressure profile in the downstream gap only decreases linearly for short lips, i.e., 0.5 mm in this example. As the lip length increases the profile becomes more and more curved, particularly toward the exit of the downstream gap. For lip lengths of longer than about 1.7 mm, the pressure drops below ambient and increases toward the gap exit. As explained below, this profile shape is highly undesirable because it promotes the onset of a coating defect called ribbing lines. Reducing the coating speed or reducing the viscosity by lowering γ_c or n cannot avoid the sub-ambient pressure in the downstream gap. However, tilting the die by a small amount, i.e., $\alpha - 0.5°$, or inclining the downstream lip by a small amount, i.e., $\beta = 0.5°$, eliminates the sub-ambient pressure and renders almost linearly decreasing pressures in the downstream gap.

Figure 17.19 shows the effect of the die tilt angle α on the pressure profile in the gap.

Without die tilt ($\alpha = 0°$) the pressure in the downstream gap drops below ambient toward the exit of the gap as shown in the previous figure, and the profile is not linear. However, as soon as α increases, or as the die is displaced by the corresponding z-value as depicted in Fig. 17.17, the downstream gap becomes less divergent. Consequently, the maximum pressure increases, the downstream profile becomes more linear but the upstream profile loses linearity, and the position of the upstream

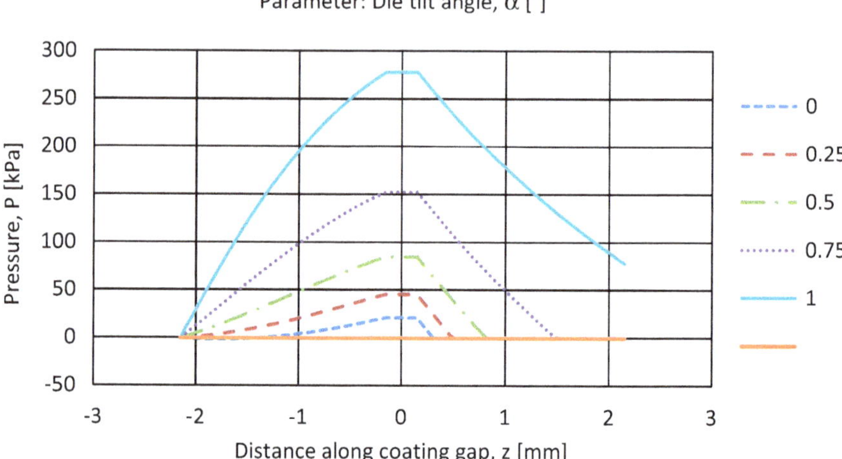

Fig. 17.19 Pressure profiles in the coating gap as a function of the die tilt angle α; $L = 2$ mm, $\beta = \gamma = OS = 0$

meniscus moves in the upstream direction. In fact, for $\alpha = 1°$, the pressure in the upstream gap has not reached ambient yet at the exit of the upstream gap, thus indicating that liquid is leaking into the vacuum box. The best profile is obtained for $\alpha = 0.5°$ in this example, with both profile branches being linear, the maximum pressure not being excessively high, and the upstream meniscus is located at about 34% of the upstream gap.

The effect of the coating gap on the pressure profile is depicted in Fig. 17.20. As

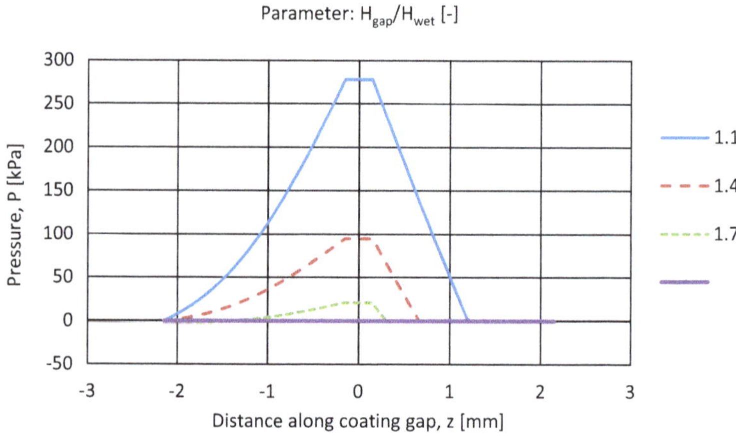

Fig. 17.20 Pressure profiles in the coating gap as a function of the gap ratio H_{gap}/H_{wet}; $L = 2$ mm, $\alpha = \beta = \gamma = OS = 0$

the operating conditions of this example indicate a viscous operating mode, the value of $H_{gap}/H_{wet} = 1.7$ is near the maximum value for this ratio following the low flow limit for this operating mode. Therefore, the gap can only be made narrower.

As seen previously, running the process near the low flow limit for the chosen operating conditions results in sub-ambient pressures near the exit of the downstream gap. Upon reducing the gap, the maximum pressure increases considerably, but negative pressures disappear. Moreover, the downstream pressure profile remains highly non-linear, and the position of the upstream meniscus moves toward the exit of the upstream gap. When reducing the gap, care must be taken that its value does not drop below the minimum value of 30 μm proposed for mechanical reasons in Sect. 17.6. In the example at hand, $H_{gap} = 44$ μm for $H_{gap}/H_{wet} = 1.1$.

For the viscous operating mode with Ca > 0.1 and Re << 1, the coating gap cannot exceed the value of about twice the wet film thickness. When coating at $H_{gap}/H_{wet} > 2$ is desirable, then either Ca and Re must be high (inertia mode) or Ca and Re must be small (capillary mode). Focusing on the pressure profile for applications in the capillary mode, we first must change the operating conditions. Specifically, in the example below the low shear viscosity is reduced from 2,000 to 100 mPas, and the web speed is lowered from 50 to 10 m/min. These changes result in Ca = 0.085 and Re = 0.65. Next, for the flow of Newtonian liquids in a parallel coating gap, Eq. (5.6.26) teaches that the pressure gradient in the downstream gap changes sign if $H_{gap}/H_{wet} = 2$. In particular, the pressure in the downstream gap decreases from the film building meniscus to the die slot for wide-gap applications in the capillary mode. Furthermore, since the flow is now controlled by capillary forces, the pressure jump across the downstream and upstream menisci must be accounted for. To that extent, we follow procedures outlined by Durst and Wagner (1997).

The pressure drop across the film building meniscus ΔP_d can be calculated with Eq. (17.7.2), which was derived for dip coating by Wilson (1982), and which is valid for Ca << 1.

$$\Delta P_d = (1.3376 Ca^{2/3} - 0.1551 Ca)\frac{\sigma_d}{H_{wet}} \tag{17.7.2}$$

σ_d is the surface tension in the downstream meniscus; it has a value of 40 mN/m in our example.

The pressure drop across the wetting meniscus is calculated with Eq. (17.7.3), which was derived by Higgins and Scriven (1980).

$$\Delta P_u = (\cos\theta + \cos\psi)\frac{\sigma_u}{H_{gap}} \tag{17.7.3}$$

σ_u is the surface tension in the upstream meniscus; it also has a value of 40 mN/m in our example. In addition, θ is the static contact angle between the coating fluid and the die material; it has a value of 30° in our example. ψ is the dynamic contact angle. Based on experimental investigations, Gutoff and Kendrik (1982) established the following empirical correlation for calculating ψ:

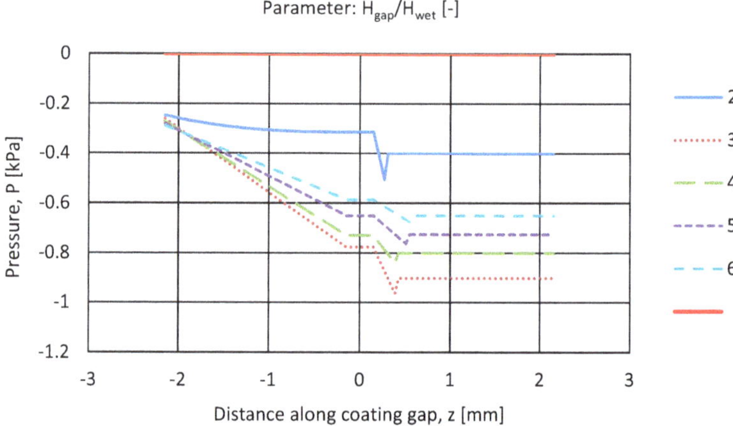

Fig. 17.21 Pressure profiles for the capillary operating mode as a function of the gap ratio H_{gap}/H_{wet}; Ca = 0.085, Re = 0.65, L = 1 mm, $\alpha = \beta = \gamma = OS = 0$, bead vacuum = variable

$$\psi = 541 Ca^{0.22} \sigma^{0.33} \mu^{-0.04} \tag{17.7.4}$$

Here, the viscosity μ and the surface tension σ must be taken in SI units of [Pas] and [N/m], respectively.

Figure 17.21 shows pressure profiles for the capillary operating mode as a function of H_{gap}/H_{wet}.

For all gap ratios, the capillary pressure drop across the downstream meniscus is in the range of 250–300 Pa, and it ranges from about 50–100 Pa at the upstream meniscus. For $H_{gap}/H_{wet} = 2$, the pressure gradient in the downstream gap has already changed sign from negative to positive due to the diverging gap geometry, i.e., the pressure is increasing from the die slot to the exit of the downstream gap. In addition, the maximum (negative) pressure in the gap is small because the pressure gradient is close to zero. However, as soon as H_{gap}/H_{wet} is larger than 2, the maximum pressure increases to a maximum value for H_{gap}/H_{wet} of about 3. Upon further increasing H_{gap}/H_{wet} the maximum pressure decreases again, which is expected. Since the pressure in the upstream gap decreases from the die slot to the exit of the gap, this operating mode must be run with a bead vacuum of a value that is a bit lower than the maximum negative pressure at the exit of the die slot. In this example, bead vacuum assumed values from 400–900 Pa, corresponding to 40–90 mm of water column. Such vacuum levels are comfortable because they can be held with a siphon of a reasonable length of, say, 100–300 mm, that can easily be connected to the vacuum chamber and attached to the side of the coating station.

As stated above, and as is further explained below in Sect. 17.8 about coating defects, the preferred pressure profile in the downstream gap is of linear shape because it prevents the onset of ribbing lines, at least for the viscous operating mode. For the inertia mode, a simple equation for calculating the pressure profile in the coating gap is not known. For the capillary mode, ribbing lines are not an important issue because the web speed and the viscosity must be low to achieve very low capillary numbers.

To assess the linearity of the pressure profile in the downstream gap, we have carried out many calculations by varying all the available geometric gap parameters as illustrated in Fig. 17.17. Then, we compared the shape of the downstream pressure profile with a straight line. The linearity of the profile was quantified by subjecting the profile data to a linear regression, and by calculating the r^2-value, which is a measure for the fit. The closer r^2 approaches the value of 1, the better is the fit. The calculations produced the following results:

- Reasonably linear pressure profiles are obtained for $r^2 > 0.995$.
- Linear pressure profiles require the divergence of the downstream gap to be small.
- Linear pressure profiles are obtained by varying the radius R of the backup roll, the angle β of the downstream die lip, and, as shown in Figs. 17.18, 17.19 and 17.20, the length of the downstream lip L_d, the die tilt angle α, and the coating gap H_{gap}.
- The preferred parameter for achieving a linear pressure profile is the length of the die lip if it is angled at 90° relative to the axis of the die slot because machining is easier than for lips with $\beta > 0$, and because it does not jeopardize the mechanical precision of the die positioning system.
- All attempts failed to correlate r^2 with various parameter combinations in an attempt to reduce the number of independent parameters, except for the parameter R/L_d^2. This empirical correlation is shown in Fig. 17.22. As can be seen, linear pressure profiles are obtained for $R/L_d^2 \geq 125$. Therefore, the maximum length for the downstream die lip can be estimated with the following correlation, which is graphically displayed in Fig. 17.23. Note that R and L_d are measured in units of [mm].

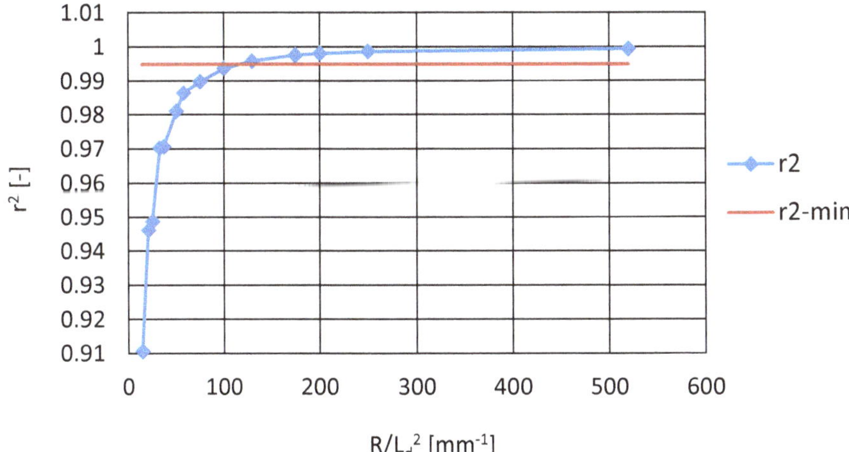

Fig. 17.22 r^2, i.e., the linearity of the downstream pressure profile as a function of the geometric group R/L_d^2

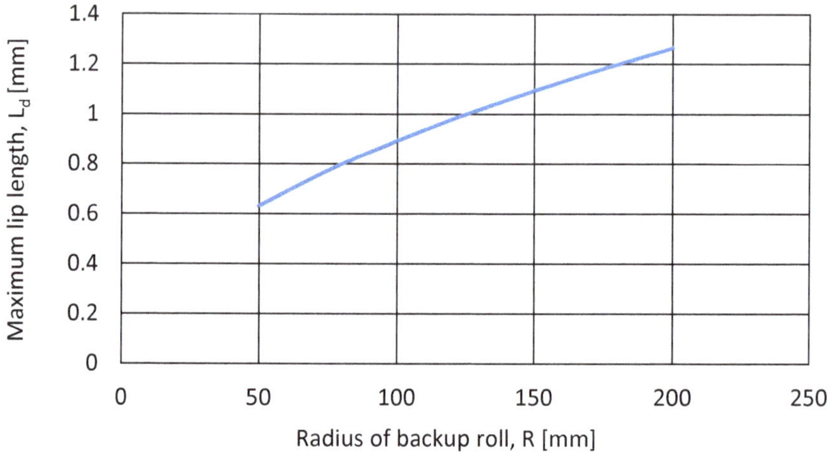

Fig. 17.23 Maximum length of the downstream die lip as a function of the roll radius

$$L_d \leq \sqrt{\frac{R}{125}} \tag{17.7.5}$$

As visualized by Figs. 17.18, 17.19, 17.20 and 17.21, the pressure profile in the upstream gap has much less of a tendency to become non-linear as various geometric parameters are varied. Moreover, the profile is not directly linked to the onset of a particular coating defect. Therefore, a requirement for a maximum lip length is not known or necessary. This is particularly so if the coating process is operated near the low flow limit, i.e., with the maximum admissible coating gap because then the upstream gap is mostly empty, and so the location of the upstream corner of the upstream gap is not relevant.

In addition to a maximum length, both die lips should respect criteria for a minimum length based on mechanical and fluid mechanical reasons. On the one hand, reducing the lip length makes grinding and lapping of the lip face more difficult. It is, however, very important for the lip face to be very uniform and straight across the coating width because this surface makes for one of the boundaries of the coating gap, and mechanical non-uniformities of this gap translate into non-uniformities of the coated film on the substrate. Even though both die plates are mounted next to each other during grinding and lapping to double the surface area to be machined, lip lengths of $L_d = L_u < 0.5$ mm are considered critical.

On the other hand, results from various theoretical models for calculating velocity and pressure profiles in the coating gap are only accurate and usable if the flow in the coating gap is fully developed as is assumed in these models. If the coating gap is too short, then the flow will consist of inlet and exit effects, and the fully developed flow area will be very short, if it exists at all.

The inlet effect, or the entrance length X_e, which marks the area of undeveloped slot flow can be calculated with Eq. (17.7.6) developed by Atkinson et al. (1969).

$$\frac{X_e}{H_{gap}} = 0.625 + 0.044Re \tag{17.7.6}$$

Here, the Reynolds number Re is calculated according to Eq. (17.6.2). For the operating conditions of the viscous example above and for a coating gap of 60 μm, $X_e = 0.04$ mm if Re = 1, or 0.06 mm if Re = 10, or 0.17 mm if Re = 50. Therefore, lip lengths in the range of 0.5–1.0 mm should be adequate for many slot coating applications.

17.8 Coating Defects

17.8.1 Ribbing Lines

Ribbing lines constitute a flow instability, and are found in several different coating methods, including slot and slide coating, as well as forward roll coating with a film split. When ribbing lines occur, the previously uniform surface of the coated film changes into a regular sinusoidal pattern of ridges and adjacent valleys in the machine direction, see Fig. 17.24. The amplitude and wave length of this surface deformation change with changing viscosity, surface tension, coating speed, wet film thickness, and geometrical process parameters. If the amplitude of the wave crest is large compared to the average film thickness, then it may happen that the minimum film thickness at the valley bottom becomes so small that the film ruptures, thereby generating dry stripes on the substrate, see Fig. 17.24. This form of ribbing lines is often called break lines.

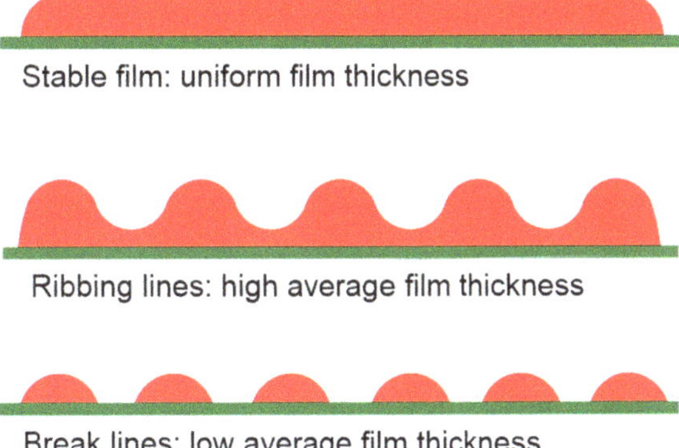

Fig. 17.24 Stable and unstable film with ribbing or break lines

In slot coating, ribbing lines are generated at the film-forming meniscus, just at the exit of the coating bead. Even though ribbing lines may level a bit between coating and drying, their presence usually destroys the uniformity of the final product, and hence they are not acceptable. The inception of ribbing lines is a process limitation and clearly marks one of the operating boundaries of this coating process, see Sect. 17.4.

Ribbing lines can be eliminated by increasing the wet film thickness, by decreasing the viscosity and/or the coating speed, and by applying or increasing the bead vacuum.

The fluid mechanical reason for ribbing lines in slot coating is not well understood, and no simple, analytical expression exists for visualizing and quantifying the inception of ribbing lines. Generally speaking, ribbing lines occur, if disturbing pressure forces in the flow field of the downstream coating bead cannot be controlled by surface tension forces. Using the analogy of the film splitting flow in forward roll coating, ribbing lines occur if the pressure gradient in flow direction is positive and exceed a critical value, which depends on the surface tension σ of the fluid and the radius of curvature of the meniscus R_c, see Savage (1977), Coyle (1997) and inequality (17.8.1).

$$\frac{dP}{dz} - \frac{\sigma}{2R_c^2}\frac{dR_c}{dz} > 0 \tag{17.8.1}$$

In other words, surface tension stabilizes the meniscus, while a positive pressure gradient is destabilizing.

In the absence of a useful and practical criterion for the onset of ribbing lines, it is suggested to take an alternative approach by not looking for a criterion, which describes one of the boundaries of the operating window, but by looking for a criterion, which describes a safe operating condition located away from this boundary of the operating window. In particular, we have learned that ribbing lines can be prevented, if the pressure gradient in the downstream coating gap is negative with a constant slope, i.e., if the pressure decreases linearly. As explained in the previous chapter, we have also learned that excessively diverging geometries of the coating gap prevent the pressure profile from decreasing linearly. In particular, we saw that it does not take much for a slightly non-linear pressure profile to generate locally negative pressures followed by a pressure increase in the flow direction, which conflicts with inequality (17.8.1). Even though we do not know the exact conditions that result in such a flow situation, we know that it must be avoided. Therefore, inequality (17.7.5) is a geometrical constraint that limits the geometrical divergence of the downstream gap, guarantees a linear pressure profile, and so prevents ribbing lines from occurring.

The arguments presented above were confirmed experimentally at Polytype Converting AG, where ribbing lines in slot coating were never an issue during more than 15 years of running trials on semi-industrial coating equipment for many different applications.

17.8.2 Air Entrainment

The term "air entrainment" refers to a situation where air, which is dragged along by the uncoated moving substrate, is entrained at the dynamic wetting line between the substrate surface and the coating fluid, see Fig. 17.25.

Air entrainment occurs when the hydrodynamic assist of dynamic wetting is insufficient for a given set of operating conditions (see Chap. 7 for more details), i.e., when the dynamic wetting line is located too far downstream (red shade in Fig. 17.25) or too far upstream (purple shade in Fig. 17.25) relative to the location of the impinging liquid, which is defined by the die slot.

If air entrainment occurs, then the uniformity of the coated film is destroyed, which is why air entrainment marks one of the boundaries of the operating window of the slot coating process, see Sect. 17.4.

For a given set of operating conditions, air entrainment typically occurs if the web speed is increased and exceeds a critical value. Blake and Ruschak (1997) provided a summary of many experimental studies on air entrainment. Unfortunately, most of these studies were done for tape plunging into a quiescent pool or for curtain coating, but not for slot coating. In one of these investigations, Blake et al. (1994) curtain-coated aqueous gelatin solutions onto a dry substrate and found that the maximum speed of wetting increases with decreasing viscosity taken at the shear rate near the dynamic wetting line, and with increasing impingement velocity normal to the substrate surface. The authors also confirmed that air entrainment is associated with a hysteresis, meaning that the web speed of air entrainment inception is higher than the speed, at which air entrainment disappears upon reducing the speed.

As explained in Sect. 17.9, a preferred way of operating the slot coating process is with the upstream gap being empty, or almost empty. Therefore, the red flow field in Fig. 17.25 is more likely to exist than the purple one. Hence, eliminating air entrainment calls for the following measures:

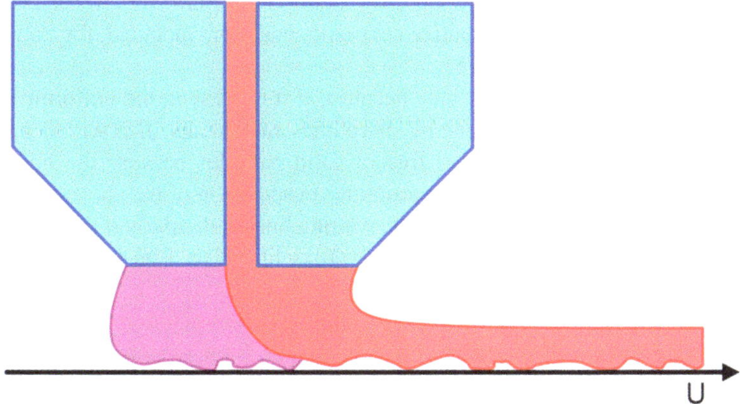

Fig. 17.25 Schematic representation of air entrainment in slot coating

- Reducing the web speed.
- Lowering the viscosity.
- Increasing the flow rate/width by increasing the wet film thickness without increasing the coating gap.
- Decreasing the height of the die slot to increase the momentum of the impinging fluid.
- Decreasing the coating gap.
- Increasing the bead vacuum.

Altering the viscosity requires detailed knowledge of the rheological behavior of the coating fluid under investigation. In particular, the viscosity must be known at the prevailing shear rate in the coating gap, and that can be estimated with Eq. (17.6.3).

Unfortunately, no theory or experimental data exist for predicting the onset of air entrainment in slot coating. If air entrainment is observed, it can be eliminated by resorting to the counter-measures listed above. Bhamidipati et al. (2012) carried out an experimental study about air entrainment in slot coating for highly viscous, shear-thinning fluids. Their goal was to study the defect, not the prevention of it. Specifically, they identified two bubble break-up mechanisms, and they investigated the qualitative impact of geometric conditions and viscosity of the coating solution on the sizes of the air bubbles.

17.8.3 Vortices

As described in Sect. 5.6, the slot coating process contains many potential locations for vortex formation. The questions are: Where exactly are these vortices located and how can they be prevented? Preventing vortices is particularly important as they are a major source for lines and streaks in the coated film (see Sect. 5.2). An overview of potential vortex locations is sketched in Fig. 17.26. In addition, computer-generated visualization of vortices can be found, for example, in Sartor (1990) or Durst and Wagner (1997), and vortices visualized experimentally are included, for example, in Sartor (1990) or Suszynski (2015).

Vortex 1 is particularly dangerous because, if it is present, the uniformity of the coated film is likely to be degraded by sharp lines and streaks. This will also happen for pure coating solutions that are free of solid particles because the flow inside the vortex is almost stagnant, which promotes evaporation of the solvent and so the formation of solid material near the static wetting line, particularly if the solvent is of organic nature. Vortex 1 exists if the static wetting line of the film-forming meniscus is not pinned to the downstream corner of the downstream die lip but is allowed to migrate along the inclined lip face to some unknown position. As explained in Sect. 5.6, therefore, vortex 1 can be prevented if the static wetting line sticks to the downstream lip corner, and that can be achieved if the lip corner approaches the form of a knife-edge, i.e., if the angle γ is small, preferably <90°, see Fig. 17.27a.

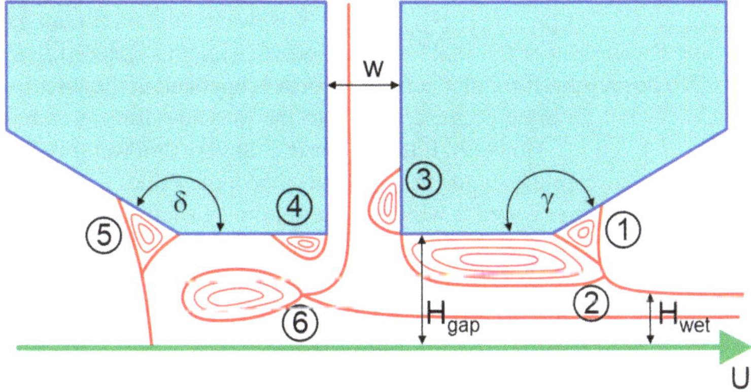

Fig. 17.26 Potential vortex locations in the bead of the slot coating process

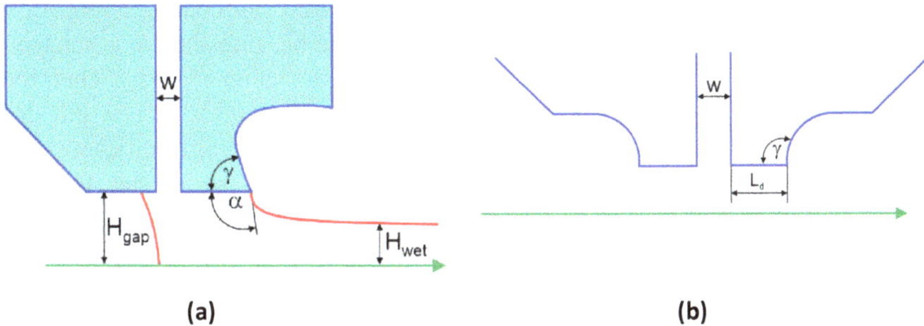

Fig. 17.27 **a** Preferred shape of the downstream lip corner. **b** Compromise of optimized lip shape

Corner angles of $\gamma < 90°$ are not possible for short lip lengths owing to manufacturing difficulties and loss of mechanical stability. Since we are promoting short lip lengths in the range of 0.5–1.0 mm, corner angles of <90° are not an option. Consequently, a design with identical downstream and upstream lip shapes and with $\gamma = 90°$ as depicted in Fig. 17.27b is a good and acceptable compromise. If, in addition, the coating process is run with a maximized coating gap, which we also recommend, then the separation angle α of the meniscus is minimized. This in turn promotes that the static wetting line of the film-building meniscus sticks to the lip corner under the Gibbs criterion described by inequality (5.6.66) and explained in Sect. 5.6.3.

Vortex 2 is also critical, particularly if vortex 1 is not present because it forms part of the film-building meniscus. Consequently, chances are high that this vortex degrades the uniformity of the coated film either by forming lines and streaks or by generating diffuse, longitudinal bands or cloudiness. This is particularly so if the coating fluid is a dispersion.

Figure 5.47 illustrates that the inception of vortex 2 depends on the gap ratio H_{gap}/H_{wet} and the power law index n. For Newtonian fluids ($n = 1.0$) and a parallel coating gap, vortex 2 is present if $H_{gap}/H_{wet} \geq 3$. However, the minimum gap ratio increases as n decreases. For $n = 0.2$, for example, H_{gap}/H_{wet} must exceed a value of 7 for vortex 2 to be present. If the slot coating process is operated in the viscous range, then vortex 2 is not an issue because, owing to the low flow limit, the maximum value for H_{gap}/H_{wet} is <2. However, if the goal is to coat very thin wet films with slot coating, either in the capillary or the inertia mode, then large values for H_{gap}/H_{wet} are necessary. Consequently, vortex 2 will be present in most such applications, and it becomes important to carefully check the uniformity of the coated film to see if it is still acceptable. If the uniformity is not acceptable, then alternative coating methods may have to be considered for applying ultra-thin films, such as curtain, 5-roll or gravure coating as described by Schweizer (2013).

Vortex 3 is located near the exit of the die slot. Conditions for the onset of this vortex were determined numerically by Durst and Wagner (1997). Specifically, they found that vortex 3 is present if the height w of the die slot is larger than twice the height H_{gap} of the coating gap, see inequality (5.6.65). As already elucidated in Sect. 5.6.3, avoiding vortex 3 is difficult for industrial slot coating applications, particularly if they are operated in the viscous regime because then w is related to H_{gap}, H_{gap} is related to H_{wet}, and H_{gap} cannot exceed the value of $2H_{wet}$. Therefore, coating thin wet films without the presence of vortex 3 requires narrow die slots, and they, in turn, may generate excessive film thickness non-uniformities owing to their susceptibility to the imperfect mechanical precision of the die slot. However, the consequences of the presence of vortex 3 on the uniformity of the coated film are not clear. They are expected to be less severe than those of vortices 1 and 2 because vortex 3 is located far away from either of the menisci of the coating bead and hence it is far away from any contact line.

Vortex 4 is present, if the momentum of the flow exiting the die slot is excessively high and if the coating gap is large, such that the flow cannot turn around the 90° corner into the upstream gap without separating from the flow boundary. Quantitative information about the onset of this vortex is not known. Theoretical calculations by Durst and Wagner (1997) indicate that vortex 4 is not present for any value of the coating gap if the die slot w is 6 times wider than the coating gap. However, this geometric ratio does not describe the inception of vortex 4.

Vortex 5 is similar to vortex 1, and these two vortices are similar to the vortex at the exit of the uppermost slot of a multilayer slide die as shown in Fig. 5.71d. Therefore, it is very likely that the wedge-shaped area of the separated flow field is not just filled with one vortex as sketched in Fig. 17.26, but with a series of vortices that diminish in size as they approach the static wetting line like shown in Fig. 5.71d. Vortex 5 only plays a role if the upstream meniscus is located at the upstream corner of the upstream gap. However, as we propose to operate the slot coating process near the low flow limit, at least for the viscous operating regime, i.e., with the largest possible coating gap, vortex 5 is likely to not be present because the upstream meniscus is located closer to the die slot than to the upstream corner of the upstream gap for such operating conditions.

Finally, vortex 6 is located in the upstream gap. It is similar to vortex 4 in slide coating as sketched in Fig. 5.70, and it is similar to the vortex in an excessively large heel of an impinging curtain as shown in Fig. 5.182 and visualized in Fig. 5.186. Vortex 6 is only present, if the upstream gap is full of liquid, and if the height of the coating gap is large. Again, as we propose to operate the slot coating process near the low flow limit, at least for the viscous operating regime, i.e., with the largest possible coating gap, vortex 6 is likely to not be present because, for such operating conditions, the upstream meniscus is located closer to the die slot than to the upstream corner of the upstream gap, and hence the upstream gap is hardly filled with liquid. Moreover, if the upstream gap is filled with fluid but vortex 6 is not present, then the flow field resembles the one in slide coating when the bead is pulled around the die lip and deep into the vacuum box as visualized in Fig. 5.71k. In this case, neighboring streamlines in the middle of the coating gap move in opposite direction. Flow visualization experiment with slide coating revealed that such flow fields are unstable and start to oscillate, which is not acceptable. Such observations underscore the need to operate the slot coating process with an almost empty upstream gap.

17.8.4 Heavy Edges

Section 5.6 describes the flow field in the coating gap of slot coating. In particular, this flow is characterized by a pressure drop in the downstream gap that can generate high pressures in the middle of the gap where the flow in the die slot merges with the flow in the coating gap. Since the flow in the coating gap is not bound by side walls, the fluid pressure generates cross flow, which causes the liquid in the gap to exit along its sides.

Schmitt et al. (2014) analyzed this flow and found that heavy coated edges may form that can cause problems in subsequent processing steps such as calendaring and winding, particularly when manufacturing electrodes for Li-ion batteries. Schmitt slot-coated an aqueous anode slurry and showed experimentally how the web speed and the height of the coating gap influence the height of the coated edge. Specifically, the height and the width of the edge did not significantly change with increasing coating speed, but the slope of the edge increased with increasing speed. The height and the slope of the edge are not affected by the dimensionless coating gap, but the dimensionless edge width clearly increases with increasing dimensionless gap.

Spiegel et al. (2021) continued the work by Schmitt et al. (2014) and found that the generation of heavy edges can be minimized by changing the geometry of the die slot near the end of the slot. In particular, Spiegel et al. (2021) increased the width of the die slot near the slot exit by altering the geometry of the U-shaped shim, which defines the coating width and the height of the die slot, see Fig. 17.28.

For an aqueous and shear-thinning anode slurry, the edge elevation was minimized when the diverging width W_d of the shim was about 1.5 mm, and when the diverging length L_d was about 10 mm. However, these results depend on the physical fluid properties, and this dependence must yet be determined experimentally. Also, the mechanism for generating heavy edges along both sides of the coated film is not yet completely understood, and that too calls for more experimental results.

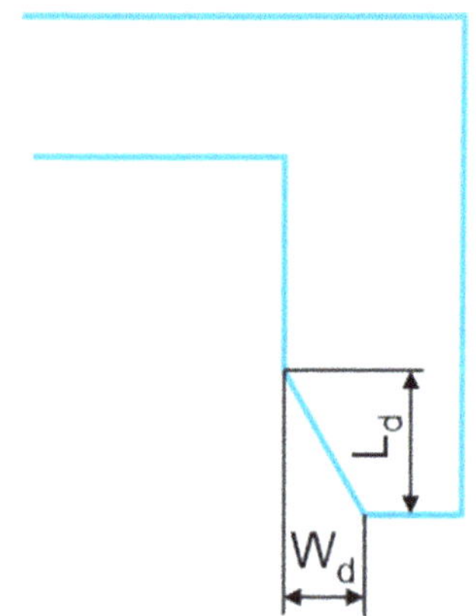

Fig. 17.28 Modification of the geometry of the U-shaped shim near the slot exit

17.8.5 Effect of the Uniformity of the Coating Gap on the Uniformity of the Wet Film Thickness

As explained in Sect. 17.6, the geometry and the mechanical uniformity of the coating gap are a complex matter. The question remains if and how the uniformity of the coated film, i.e., H_{wet}, is affected by the non-uniformity of the coating gap? The answers are explained by separately investigating effects in machine and cross-web direction.

17.8.5.1 Coating Gap Variation in Machine Direction

Looking at Fig. 17.29, the volumetric flow rate/width Q is uniformly supplied through

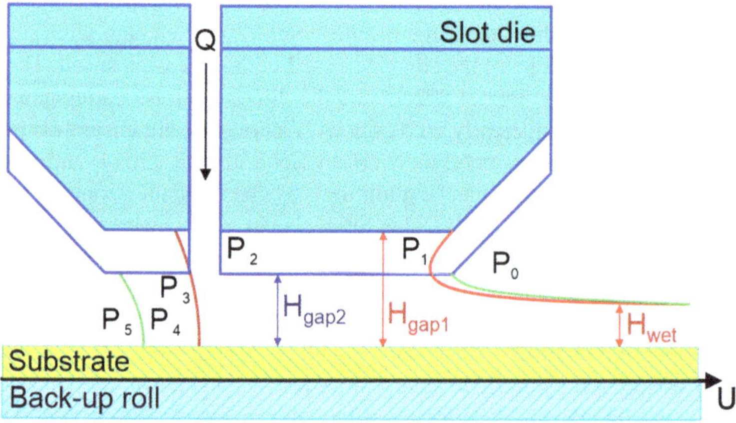

Fig. 17.29 Steady state change of the coating gap

the die slot. In a successful and **steady state** slot coating process, all this fluid quantity must flow through the downstream coating gap before it is carried away by the moving substrate. The characteristic features of the flow field in the downstream coating gap for Newtonian fluids are described in Sect. 5.6.1. In particular, Eq. (5.6.26) visualizes that the pressure drop across that gap depends on the wet film thickness H_{wet} or, by considering Eq. (5.6.22), on the volumetric flow rate/width Q, and on the coating gap H_{gap}. Given values of Q and U result in a wet film thickness H_{wet}, and the upstream and downstream menisci of the coating bead may look like depicted by the red lines in Fig. 17.29 if the coating gap has a value of H_{gap1}.

If the coating gap decreases to a new value of H_{gap2}, for example by moving the die closer to the substrate surface, then the pressure drop in the downstream gap increases, and so do the pressures P_2 and P_3. As a consequence, the location of the upstream meniscus moves further upstream. However, the entire flow rate Q is still transported through the downstream coating gap, thus not affecting the wet film thickness H_{wet}. Consequently, the upstream and downstream menisci of the coating bead may look like depicted by the green lines in Fig. 17.29. This is true as long as

- all the boundaries of the coating gap and the coating liquid are incompressible;
- the upstream meniscus is fully contained inside the upstream coating gap.

This latter requirement is fulfilled, if the position of the upstream meniscus, L_m (see Eq. 17.4.2) is smaller than the length of the upstream die lip, L_u. Combining Eqs. (5.6.26) and (17.4.2), and assuming a parallel gap, no vacuum pressure at the upstream meniscus, and negligible pressure drop across both menisci, the upstream meniscus remains confined under the upstream die lip, if H_{gap} satisfies the following expression:

$$H_{gap} \geq \frac{2H_{wet}}{1 + \frac{L_u}{L_d}} \tag{17.8.2}$$

Setting $L_u = L_d$, which is often the case in industrial applications, then inequality (17.8.2) simplifies to

$$H_{gap} \geq H_{wet} \tag{17.8.3}$$

Inequalities (17.8.2) and (17.8.3) set a lower bound for the coating gap. It turns out however, that inequality (17.8.3) is not acceptable in practice for the following reason. As illustrated in Fig. 17.27a, the separation angle α of the film building meniscus from the die lip would be 180° if $H_{gap} = H_{wet}$. In order to prevent vortex 1 located at the downstream lip face (see Fig. 17.26), which is a highly desirable process feature to prevent lines and streaks in the coated film, the separation angle α cannot be 180° according to the Gibbs inequality (5.6.66), if the die lip angle γ is 90°. This is because the advancing contact angle θ of most industrial coating fluids is <90°. Unfortunately, we cannot quantify the relationship between the separation angle α and the ratio of H_{gap}/H_{wet}. However, we believe that inequality (5.6.66) can be satisfied, and hence lines and streaks can be prevented for many industrial applications if $H_{gap}/H_{wet} > 1.2$.

On the other hand, the low flow limit of slot coating defines an upper bound for the coating gap, if the process is operated in the viscous mode. Specifically, and as documented in Fig. 17.11, $H_{gap}/H_{wet} < 1.7$. For the viscous operating mode, therefore, an operating window for the nominal value of the coating gap can be formulated as follows:

$$1.2 < \frac{H_{gap}}{H_{wet}} < 1.7 \tag{17.8.4}$$

In other words, optimized performance of the slot coating process operated in the viscous mode can only be expected if the admissible range for the coating gap respects inequality (17.8.5), i.e., the maximum variations of the coating gap should not exceed half of the wet film thickness:

$$\Delta H_{gap} < (1.7 - 1.2) H_{wet} \approx 0.5 H_{wet} \tag{17.8.5}$$

Note that the numbers in inequality (17.8.5) are associated with some uncertainty, particularly the factor of 1.2, as is explained above.

While inequality (17.8.5) is a process limitation for the admissible range of **nominal coating gaps**, it must also be applied to **temporary and random variations** of the gap as depicted in Fig. 17.13 and as quantified by Eq. (17.6.4). To that extent, Eqs. (17.6.4) and (17.8.5) are combined to yield the following process limitation:

$$2H_{gap}\sqrt{\left(\frac{dH_{web}}{H_{gap}}\right)^2 + \left(\frac{dR}{H_{gap}}\right)^2 + \left(\frac{dZ_{lip}}{H_{gap}}\right)^2} < 0.5 H_{wet} \tag{17.8.6}$$

Inequality (17.8.6) is graphically shown in Fig. 17.30. The terms on the left and right hand side of the < sign are each plotted as a function of the web thickness for exemplary values of the coating gap non-uniformity.

The left-hand term containing the square root function (blue curve) exceeds the right-hand term (red curve), thus violating inequality (17.8.6), for web thicknesses larger than about 180 μm. For such operating conditions the randomly and exceedingly varying coating gap (primarily as a result of web thickness fluctuations) results in

- the wetting meniscus to sporadically be pushed outside of the upstream coating gap;
- the vortex 1 along the downstream lip face to be formed;
- the low flow limit for the viscous operating mode to be violated.

To avoid such undesirable results, the coating gap should not be set to the maximum value as per the low flow limit of the viscous operating mode, i.e., 1.7 H_{wet}, but to a somewhat lower value of, say, 1.5 H_{wet}. This will allow the coating

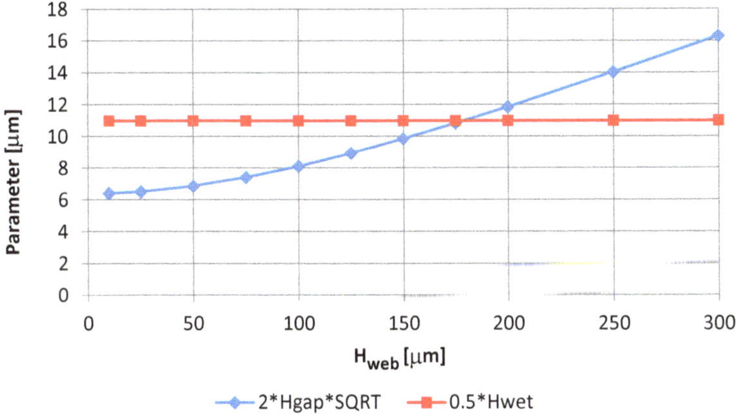

Fig. 17.30 Admissible web thickness as a function of various parameters defining the non-uniformity of the coating gap; $H_{wet} = 22\ \mu m$, $H_{gap}/H_{wet} = 1.5$, $dH_{web} = 0.025\ H_{web}$, $dR = 2.0\ \mu m$, $dZ_{lip} = 2.5\ \mu m$

gap to deviate in the positive and negative direction from the nominal value without causing the flow field to deteriorate.

If the coating gap varies randomly and temporarily as depicted in Fig. 17.13, then it is interesting to know how the uniformity of the coated wet film is affected by the variations of the coating gap. This information can be obtained for Newtonian fluids flowing through a parallel coating gap by integrating and rearranging Eq. (5.6.26), such that the wet film thickness is expressed as a function of the coating gap, the pressure drop across the downstream gap, and the other parameters as follows:

$$H_{wet} = H_{gap}\left(\frac{1}{2} - \frac{\Delta P_{12} H_{gap}^2}{12\mu L_d U}\right) \tag{17.8.7}$$

Then, the dependence of the thickness variation of the coated film upon the variation of the other parameters that control the flow in the gap is obtained by taking the total derivative of Eq. (17.8.7) and by normalizing it with the nominal wet film thickness. Moreover, since process-related fluctuations of these parameters are stochastic, the final expression takes the form:

$$\frac{dH_{wet}}{H_{wet}} = \pm\sqrt{\left(A\frac{dH_{gap}}{H_{gap}}\right)^2 + \left(B\frac{d\Delta P_{12}}{\Delta P_{12}}\right)^2 + \left(B\frac{dU}{U}\right)^2 + \left(B\frac{d\mu}{\mu}\right)^2} \tag{17.8.8}$$

The parameters A and B only depend on the ratio H_{gap}/H_{wet} as follows:

$$A = 3 - \frac{H_{gap}}{H_{wet}} \quad B = \frac{1}{2}\frac{H_{gap}}{H_{wet}} - 1 \tag{17.8.9}$$

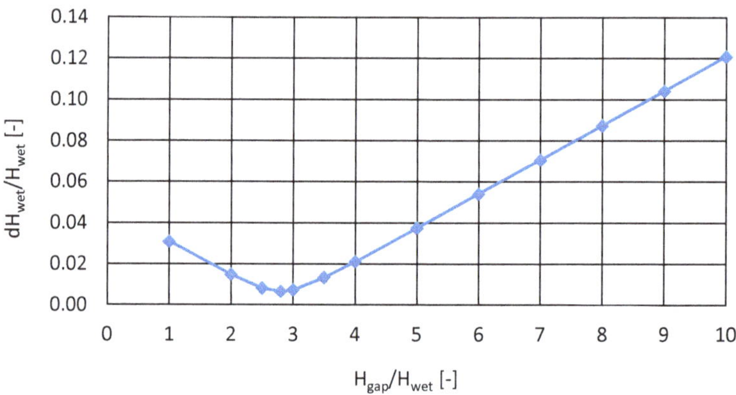

Fig. 17.31 Thickness non-uniformity of the coated film as a function of H_{gap}/H_{wet} and of process fluctuations of H_{gap}, ΔP_{12}, U and μ; $dH_{gap}/H_{gap} = 0.015$, $d\Delta P_{12}/\Delta P_{12} = 0.01$, $dU/U = 0.005$, $d\mu/\mu = 0.01$

The positive sign of Eq. (17.8.8) is plotted in Fig. 17.31 for an exemplary set of process fluctuations. As was demonstrated above for the viscous operating mode, the dimensionless coating gap typically assumes values in the range of 1.2–1.7. For such conditions, increasing (maximizing) the coating gap reduces (minimizes) the film thickness non-uniformity. In fact, increasing H_{gap}/H_{wet} to a value of 3, which marks the onset of backflow in the downstream gap, produces the absolute lowest film thickness variation. However, increasing H_{gap}/H_{wet} beyond a value of 3, which is easily possible for the capillary and inertia operating modes, increases the film thickness non-uniformity. To limit this undesirable effect, it is recommended to increase H_{gap}/H_{wet} not beyond a value of about 5.

Note that film thickness non-uniformities as a function of coating gap variations are higher when compared to the same effect in the die slot because coating gaps are typically narrower than die slots, and gap variations driven by variations of the substrate thickness are typically higher than variations of the slot height. Moreover, when comparing Eqs. (10.8), (10.12) and (17.8.8), it seems likely that the values of the parameters A and B (Eq. 17.8.9) increase with decreasing power law index n. Unfortunately, simple analytical expressions for quantifying this non-Newtonian effect are not available, but if true, it would worsen this issue. In any case, the arguments presented above emphasize the importance of

- Maximizing the coating gap when the process is operated in the viscous mode.
- Minimizing the gap variations by using highly precise equipment components such as
  - a slot die with a straight and uniform lip;
  - a precise backup roller with a minimized run-out;
  - high-precision roller bearings.
- Using thin and uniform substrates such as plastic films.

- Avoiding thick and non-uniform substrates such as board and non-woven webs.

17.8.5.2 Coating Gap Variation in Cross-Web Direction

Coating gap variations in the cross-web direction are a serious problem for the slot coating process because they immediately result in a variation of the wet film thickness as explained below. Such gap variations are often encountered in the tensioned-web operating mode (see Sect. 17.11) owing to deformations of the unsupported web as shown in Fig. 17.32. The geometrical situation of one such web deformation is schematically shown in Fig. 17.33, where the web is characterized by a local longitudinal deformation, for example owing to an excessive web tension, which results in a local reduction of the coating gap H_{gap2} when compared to the nominal uniform coating gap H_{gap1}.

If the web deformation extends all along the downstream coating gap, then the pressure drop across the gap at the location of the deformation will be higher than on either side of the deformation owing to the reduced gap height as visualized by Eq. (5.6.26). Therefore, the pressure P_2 at any location along the gap will be higher than the adjacent pressure values P_1, which in turn will induce local flows in the cross-web direction as depicted in Fig. 17.33. Consequently, less fluid will exit the coating gap at the location of the web deformation, and the film thickness will locally be reduced as schematically shown in Fig. 17.33. Based on the same reasons the local wet film thickness would be higher if the web deformation pointed in the opposite direction, such that the local coating gap would be higher than the neighboring values.

It is clear that web deformations as shown in Fig. 17.32 must be avoided by all means, if the uniformity of the coated film must be high, and this is particularly so for tensioned-web applications. This in turn is a non-trivial task, which may require, among other things:

Fig. 17.32 Example of excessive web deformations at the coating application point for tensioned-web slot coating; photo reproduced with permission from Polytype Converting AG

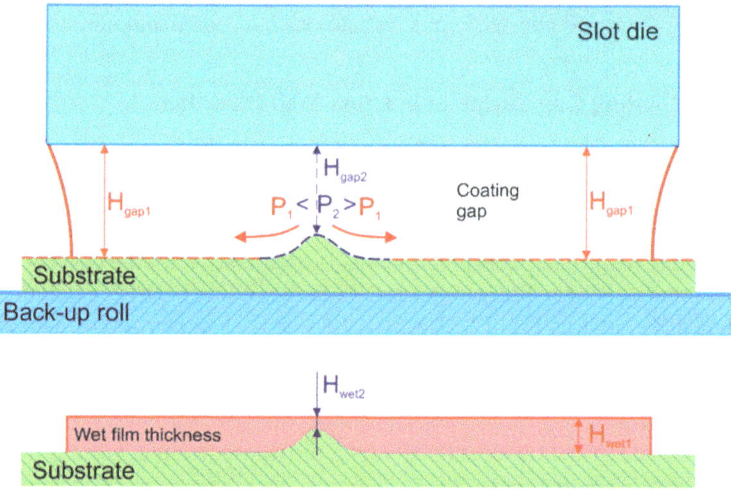

Fig. 17.33 Variation of the coating gap in cross-web direction owing, for example, to a corresponding deformation of the substrate

- perfectly aligning all idle rolls in the vicinity of the coating application point;
- minimizing the distance between the two idle rolls just upstream and downstream of the coating application point;
- minimizing the web tension across the coating process;
- working with substrates, which have uniform mechanical properties such as the modulus of elasticity;
- coating at an elevated temperature to avoid lateral web contractions due to excessive evaporative cooling.

In summary, the high sensitivity of the thickness of the coated film to variations of the coating gap is a drawback of the slot coating process, particularly if the requirements on the film thickness uniformity are high, and particularly for the tensioned-web operating mode. Consequently, high film thickness uniformities can only be achieved, if the mechanical precision of the slot die (straightness of the die lip) and the backup roll (run-out, cylindricity) are high, and if the thickness variations of the substrate are small compared to the nominal coating gap. The latter requirement is difficult to achieve, particularly if the substrate is thick and the desired wet film thickness is small. For such cases operating the slot coating process in the indirect reverse mode as shown in Fig. 17.34 might be an attractive option. With this approach, a uniform liquid film is first applied onto a uniform roller surface, and this film is then transferred onto the non-uniform substrate without having to pass through a narrow and non-uniform gap.

Care must be taken with this approach that the liquid transfer between rollers 1 and 2 is 100% complete, such that no left-over fluid on the return surface of roller 1 can negatively affect the application of fresh fluid by the slot die.

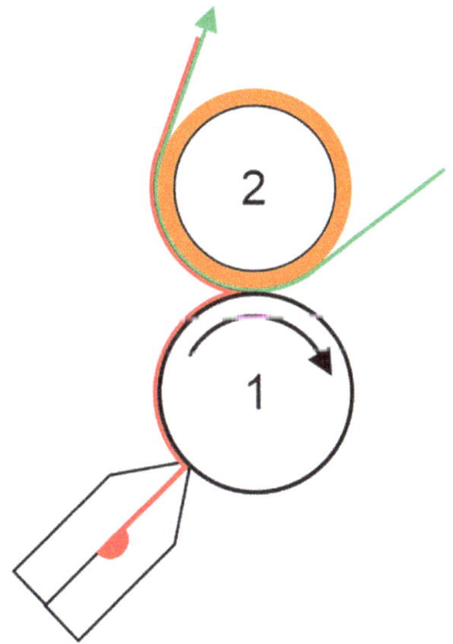

Fig. 17.34 Indirect reverse slot coating configuration

17.9 Process Optimization

Based on the information presented in Sect. 5.5 about characteristic features of slot flow, and based on the information discussed in Chap. 17 above, rules can be formulated for optimizing the slot coating process for any specific application. In particular, two sets of rules need to be formulated, one for the **capillary mode**, i.e., for Ca < 0.1 where large coating gaps relative to the wet film thickness can be achieved, and another one for the **viscous mode**, i.e., for Ca > 0.1, where the maximum coating gap is less than twice the wet film thickness. Some of the rules, however, apply to both operating modes and also to the inertia regime.

Note that the Ca value of 0.1, which distinguishes between the capillary and viscous operating modes, is not an exact value, but it gives the correct order of magnitude of this parameter.

17.9.1 Optimizing the Viscous Operating Mode

Admissible range for capillary number Ca and Reynolds number Re

$Ca = \mu U/\sigma > 0.1$ $Re < 1$.

Most industrial slot coating applications are operated in the viscous mode because neither the viscosity nor the web speed are very low, thus resulting in Ca > 0.1. Moreover, the width of most industrial slot coating applications is in the range of 1,000–2,000 mm.

Low flow limit

$H_{gap}/H_{wet} < 2$

The data presented in Fig. 17.11 suggests that $H_{gap}/H_{wet} < 1.7$. However, applying bead vacuum may allow this ratio to become as large as 2 or even higher. The effect of the bead vacuum on increasing the coating gap depends on the viscosity, and it is only significant for rather low viscosities of <<100 mPas.

Minimum coating gap

As all the fluid to be coated must flow through the downstream coating gap, this gap must be mechanically very uniform in the cross-web direction. Assuming the coating width to be in the range of 1,000 mm < W < 2,000 mm, and allowing all types of substrates to be admissible, i.e., films, papers, boards, and coils, then achieving sufficiently uniform coating gaps from a mechanical point of view is possible for.

$H_{gap} > 30\ \mu m$

Effective coating gap

The value of H_{gap} must be maximized to

- reduce the sensitivity to the mechanical imprecision of the die lip straightness, the variation of the web thickness, and the run-out of the backup roller;
- minimize web breaks;
- avoid a vortex in the upstream coating gap;
- facilitate the splice passage through the coating gap.

However, for the viscous operating mode, the effective coating gap is always related to the wet film thickness, and the maximum value for H_{gap} is in the range of 1.7–2.0 times the wet film thickness. Consequently, the coating gap must be adjusted as the wet film thickness changes from one application to the next.

Minimum wet film thickness

Based on the low flow limit and the minimum coating gap presented above, the minimum wet thickness of the coated film can be quantified as

$H_{wet} > 15–20\ \mu m$

Geometrical process configuration

The position of the slot die on the coating roller should be at –45° relative to the horizontal (7:30 position). In addition, the line through the die slot should pass through the center of the coating roller. This configuration allows for air bubbles, which might be contained in the coating liquid, to escape spontaneously from inside of the die. Moreover, the optical access to the film-building meniscus of the coating bead is open, which gives the operator good control over the process. For some applications, the 4:30 position may be preferable because it may reduce the propensity to form lines and streaks. The best position for a given application must be determined experimentally by trial and error.

The slot coating process shall be operated with a vacuum box, which has two functions, namely providing bead vacuum and serving as a catch pan. Bead vacuum may allow the coating gap to be increased and the onset of ribbing lines as well as the onset of air entrainment to be postponed to higher web speeds. The catch pan will collect coating liquid during filling of the die, and it will collect rinsing fluid during the cleaning phases.

Length of downstream lip

L_d must satisfy inequality (17.7.6) to assure a fully developed flow field in the downstream gap, and to facilitate proper grinding and lapping of the lip face surfaces. For manufacturing reasons, L_d should not be smaller than 0.5 mm. Moreover, L_d must be minimized to allow for a linearly decreasing pressure profile in the downstream gap, which in turn prevents the onset of ribbing lines. The best value for L_d depends on the value of the radius R of the backup roller.

Length of upstream lip

The value of L_2 is irrelevant as long as H_{gap} is maximized, thus causing the wetting meniscus to be located near the downstream corner of the upstream lip, i.e., near the die slot, and hence causing the upstream gap to be hardly filled with liquid. For reasons of simplicity, we recommend fabricating the downstream and upstream lips of equal length.

Radius of backup roller

The value of R must be >100 mm. Moreover, R must be large enough to allow for a linearly decreasing pressure profile in the downstream gap, which in turn prevents the onset of ribbing lines. The best value for R depends on the value of the length L_1 of the downstream die lip.

Height of die slot

The value of w should be <2 H_{gap} to prevent a vortex at the slot mouth. However, w will be very small, if H_{wet} is small, thus increasing the sensitivity of the slot height to the mechanical imprecision of the slot, which in turn increases the overall cross profile. Consequently, a compromise must be sought for small values of H_{wet}.

Angle of downstream corner of downstream die lip

The value of γ must be minimized to pin the static wetting line to the corner of the downstream lip, which in turn prevents the formation of a vortex along the downstream lip face. In particular, γ must be $\leq 90°$. The amount, by which γ can be less than 90° depends on the length L_d of the downstream die lip.

Bead vacuum

The value of the bead vacuum ΔP should be as small as possible to prevent the upstream coating gap from excessively filling up with liquid, and it should be as high as necessary to prevent the onset of ribbing lines and/or air entrainment. The necessary vacuum level, which will influence the flow behavior of the coating bead, is proportional to the pressure drop across the downstream coating gap. Typical vacuum pressures are on the order of [hPa], i.e., [cm of H_2O column].

Optimum coating bead

Accounting for all criteria listed above, the optimized flow field of the coating bead looks like shown in Fig. 17.35. In contrast to all illustrations of the coating bead shown above, the geometry of all relevant distances and dimensions are now drawn to scale in both the y and z directions.

Optimum operating point

The best operating point for a given application is obtained if the diameter of the coating roller is large, the downstream lip is short, the angle of the downstream corner of the downstream lip is $\leq 90°$, the height of the gap is maximized and the process is operated without vacuum assist because such conditions include desirable

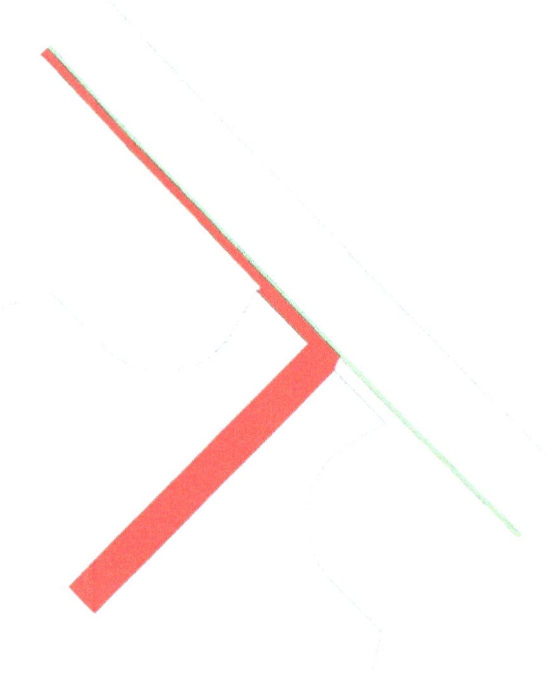

Fig. 17.35 Example of optimized coating bead and coating configuration drawn to scale for the viscous operating mode, i.e., Ca > 0.1; drawing reproduced with permission from Polytype Converting AG

and avoid undesirable features of the bead flow field. In particular, the bead is free of vortices, the pressure in the downstream gap decreases linearly and the upstream gap is mostly empty. These features result in the highest uniformity of the coated film because they prevent various coating defects. Referring to Chap. 16, the operating window shown in Fig. 17.8 is a feasibility window, which offers a large number of admissible operating conditions. However, these operating points only result in a coherent coated film with a uniformity that may not be as desired or required. In contrast, the optimized operating point as proposed above is representative of the quality coating window, which is much smaller than the feasibility window but results in a better film uniformity.

Should air entrainment be observed for these optimum operating conditions, then it is very likely to eliminate this defect by applying bead vacuum because this measure pulls the upstream meniscus in the upstream direction, thereby reducing the dynamic contact angle as explained by Chang et al. (2009). However, the vacuum should not much exceed the level that is necessary to clear the air entrainment. The problem with applying vacuum is that its effect on the bead flow field cannot be seen in an industrial coating process. Applying too much vacuum fills the upstream gap, which in turn may result in neighboring streamlines that move in opposite directions, or worse, it may form a vortex in the upstream gap. Both effects are not desirable because they degrade the uniformity of the coated film.

While coating without vacuum produces the best film uniformity, this desirable process feature comes at a cost. Specifically, the minimum wet film thickness is not as low as it could be with vacuum. Therefore, if thinner films must be obtained, then vacuum is required, but the resulting film uniformity must be examined carefully to assure that it still meets specifications. Alternatively, other coating methods might be considered that can coat very thin films, such as 5-roll coating, gravure coating, or multilayer curtain coating as described by Schweizer (2013).

Finally, as described by Schweizer (1997), the performance of any coating process is determined by three classes of parameters, namely

- operating conditions, such as coating speed, flow rate, level of vacuum;
- physical fluid properties, such as density, surface tension, and rheological properties;
- geometrical parameters, such as slot height, coating gap height, and die lip length.

Therefore, operating a coating process at its optimum point requires changing at least one parameter contained in two of the classes to be changed as soon as one of the parameters of the third class is changed. If, for example, the web speed is increased a bit, then the vacuum level may also have to be increased a bit to prevent air entrainment to occur at that higher speed. This effort is a small price to pay if the uniformity of the coated film is always optimal for any combination of the operating parameters.

17.9.2 Optimizing the Capillary Operating Mode

Admissible range for capillary number Ca and Reynolds number Re

$Ca = \mu U/\sigma < 0.1$ $Re << 1$

Very low values of the capillary number require the viscosity and the web speed to be very low. In addition, very low viscosities require the solids concentration to be very low as well. This, in turn, allows the dry coat weight to become very low, which explains why the capillary operating mode of the slot coating process is often used for coating extremely thin, dry films with high uniformity at the same time, such as components of FPD's, OPV's, and OLED's. Slot coating applications in the capillary mode are most often carried out in conjunction with smooth substrate surfaces, such as TAC, PET, PP, and PA films. Moreover, the width of most capillary mode applications is in the range of 100–1,000 mm.

Low flow limit

$H_{gap}/H_{wet} > 2$

The data presented in Fig. 17.11 suggests that H_{gap}/H_{wet} can assume values of >> 2. In fact, values of >50 have recently been reported by TSE Troller AG. It seems, however, that such very low values of H_{gap}/H_{wet} would require extremely low Ca on the order of about 0.01, which commercially could not be of much interest because the coating speed would have to be extremely low as well.

Minimum coating gap

As all the fluid to be coated must flow through the downstream coating gap, this gap must be mechanically very uniform in both the machine and the cross-web direction. Assuming the coating width W to be <1,000 mm, and limiting the types of substrates to include plastic films only, Polytype Converting AG achieves sufficiently uniform coating gaps for

$H_{gap} > 15\ \mu m$.

Such small coating gaps require an extremely high mechanical precision of the slot coating equipment, particularly concerning the straightness of the die lip and the total indicated runout (TIR) of the backup roller. Precision rollers require precision bearings and grinding in their bearings. TIR values of $<\pm 1.0\ \mu m$ are possible (see for example LKS Kronenberger GmbH). The straightness of the die lip depends on the width of the die; values of $<\pm 1\ \mu m$ are possible if W < 1,000 mm. Moreover, to achieve such high mechanical precision, the die and the backup roller should be tempered and the entire coating station should be placed in a cleanroom environment.

Minimum wet film thickness

In the capillary mode, very small wet film thicknesses can be achieved, and values of less than 5 μm seem possible. There is, however, no well-defined lower limit. To be on the safe side, the minimum wet film thickness is postulated to be

$H_{wet} > 5\ \mu m$

Note that the ratio of H_{gap}/H_{wet} would have to be at least 4 for the coating gap to assume a mechanically sound value for a wet film thickness of 5 μm.

Geometrical process configuration

If the values of the Bond, Froude or Stokes numbers (see Eqs. 17.2.1–17.2.3) indicate that gravitational forces are significant, then sagging of the coating bead as a result of gravitational pull can be prevented by moving the position of the slot die on the coating roller to the 12 o'clock position, see Fig. 17.2. In addition, the line through the center of the die slot must pass through the center of the coating roller. This configuration does not allow for air bubbles, which might be contained in the coating liquid, to escape spontaneously from inside of the die, which is a disadvantage. However, the optical access to the coating bead is still fairly open, which gives the operator good control over the process.

The die mounting and positioning system should allow the die to be rotated such that the die slot can assume an upward orientation during filling and cleaning operations. This position allows the die to be filled properly before coating. Moreover, the die slot must be quite narrow to prevent the coating fluid from draining out of the die slot owing to gravity, when the die is turned upside down into coating position. This is particularly important when the viscosity is low.

Preventing vortices

The vortex underneath the downstream die lip cannot be prevented if $H_{gap}/H_{wet} > 3$, which is the case for most capillary mode applications. Therefore, the uniformity of the coated film must be checked carefully for cloudiness, diffuse bands, or other quality-degrading defects generated by this vortex. This is particularly important if the coating fluid is not a clear solution but an emulsion or a suspension.

17.10 Operational Aspects

17.10.1 Coating Start

Starting the slot coating process is most crucial for obtaining a highly uniform and defect-free coated film. We strongly recommend building a fluid delivery system as discussed in Sect. 8.1 and sketched in Fig. 8.1. In particular, the delivery system must have a recirculation loop with the return valve located just upstream of the die entrance. This way, the delivery system can be operated in a recirculation mode before and after coating. Moreover, the system can be properly filled and the pump can run at the correct speed for delivering the desired flow rate when a coating is started.

We have learned that return valves must open and close as quickly as possible. Therefore, they must not be driven pneumatically but electrically. If a 3-way ball valve, which is a convenient piece of equipment, is driven pneumatically, then the rotation of the ball is slow, and it takes a relatively long time for the flow cross-section leading toward the die to open completely. Consequently, the initial flow rate leaving the die slot and entering the coating gap is too low for the gap height that is set for steady and continuous coating. As such operating conditions violate the low flow limit, the initially coated wet film is highly non-uniform, most likely bothered by ribbing or break lines as sketched in Fig. 17.24. However, if ribbing or break lines are present, then the static wetting line of the film-building meniscus cannot stick to the corner of the downstream gap. Instead, the wetting line assumes a sinusoidal form across the die lip, which contaminates the lip accordingly. Very often we have observed that this contamination does not disappear once the flow rate reaches its nominal value and that the uniformity of the coated film is degraded by lines and streaks. Also, we have seen that it is very difficult to clean the die lip while coating is in progress, for example by moving a thin wooden stick across the lip. Interrupting coating, cleaning the die lip, and re-starting coating are often of no help because the same thing will happen again.

The best way for improving this situation is to build a more rigid fluid delivery system that can establish steady-state coating conditions in the shortest time possible. Suitable problem solutions can be found under the subject of *intermittent coating* (see Sect. 17.14), where it is necessary to obtain close to perfect leading edges for every coated patch. When starting the slot coating process, the difference is that the patch length is not of order centimeter, but it corresponds to the area of the coating campaign. However, the requirements for a high-quality leading edge at the coating start are identical.

An easy approach to better coating starts is to replace the 3-way ball valve with two magnetically driven, fast-acting and computer-controlled ball valves located in the return branch and in the pipe toward the die just downstream of the branch point, see Fig. 17.36.

The optimum timing between closing one and opening the other valve must be found experimentally. It might be advantageous to close the valve in the return line a short moment before opening the valve toward the die while the pump is running. This way, the pressure in the delivery line increases momentarily, which helps to push the fluid out of the die and to establish steady coating conditions. Moreover, the return line must be equipped with a pressure-regulating valve. In particular, it is important that the pressure drop in the return line is equal to the pressure drop through the die to avoid pressure jumps upon switching the valves because adjusting pressure differences, except perhaps for a very short pressure peak when switching the valves, takes time and prevents the formation of high-quality leading edges.

Alternatively, starting the coating process at a speed, which is lower than the nominal value for achieving the desired film thickness, reduces the problem of contaminating the die lip at the start because the lower speed increases the wet film thickness, hopefully to the value corresponding to the pre-set coating gap. Exactly by how much the speed has to be lowered, and for how long this lower speed has to be maintained, must be found by trial-and-error.

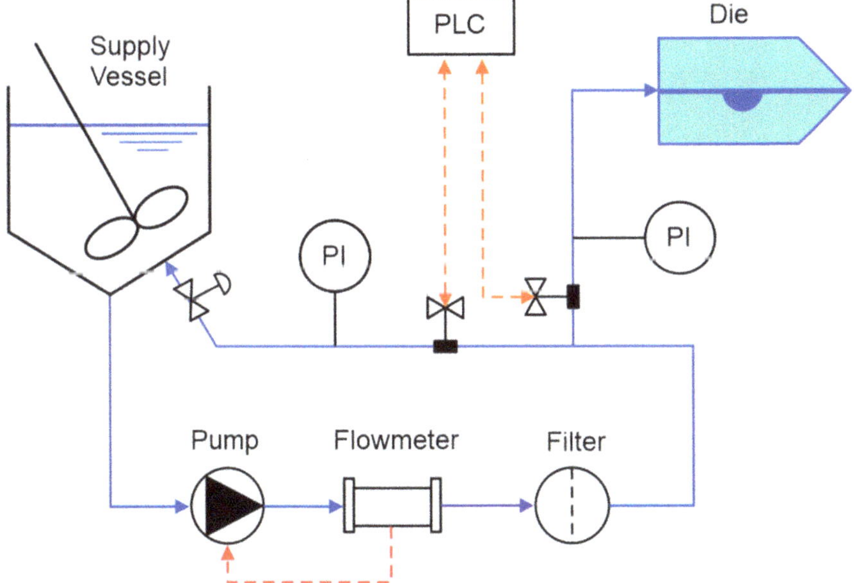

Fig. 17.36 P&ID for achieving good slot coating starts

17.10.2 Coating Stop

The procedure for stopping the coating process is simpler than the start procedure. For most applications, it is not necessary to implement the same stop sequence as is used in an intermittent coating. If coating resumes with the same fluid, then it suffices to clean the die lip thoroughly with a wet towel until the die lip is completely free of liquid or solidified material and to apply an adhesive tape over the die lip to protect the liquid inside the die from oxygen.

17.10.3 Cleaning the Die and Changing the Coating Fluid

The issue of cleaning the inside of the slot die is best handled by working with a die having an internal geometry that is not only optimized regarding the cross profile, but also the level and the cross-web distribution of the wall shear stress in the distribution chambers, see Sects. 6.3 and 8.2.4. The big advantage of this design and operating concept is the fact that the die does not have to be cleaned with water or solvent between subsequent coating fluids. Instead, the old fluid is simply pushed out of the die by the new fluid. This procedure results in a very short change-over time, and a very small quantity of wasted fluid, provided that the viscosity of the new fluid is not much lower than the viscosity of the old liquid.

If the viscosity of the new fluid is much lower than the viscosity of the old fluid, then the new fluid will "carve" itself a narrow passage of least resistance through the internal die geometry. Consequently, the new fluid will not be able to efficiently replace the old fluid in both end zones of the distribution system, owing to an insufficient shear force associated with this low viscosity. For such situations, the cleaning efficiency can be increased and the change-over time can be reduced by temporarily plugging the center 80% of the die slot with a suitable strip of plastic shim. In particular, the operator must take a plastic shim having a thickness that is a bit lower than the height of the die slot and having a length of about 80% of the coating width, and he must insert this shim into the center portion of the die slot while the pump is running at low speed. Consequently, the new liquid is forced to flow toward the ends of the distribution cavity, thereby pushing out the old fluid even if the viscosity ratio is ill-designed. Sliding the plastic shim from one end to the other end of the die slot may help to accelerate this cleaning process.

If the die is equipped with deckles for changing the coating width, then the deckles must be designed such as to not plug the inner distribution cavity. Moreover, the side plates of the die must be equipped with bleeding valves located at both ends of the inner distribution cavity. These bleeding valves must be opened during cleaning and product change-over phases. Note that a die equipped with deckles for changing the coating width can only be used for non-reactive fluids, as the deckled sections of the inner cavity will be filled with stagnant fluid during the coating phase.

17.10.4 Changing the Coating Width

Changing the coating width of a slot die is not a trivial job because it takes time and special equipment. The best way for slot coating is to avoid the need for changing the coating width. If this is not possible, then the second-best way is to use several dies, each one dedicated to a different but fixed width. This approach may be interesting from a financial point of view, if the number of coating widths is small, say ≤ 3. Other design concepts for changing the width of slot dies are known and used in the converting industry. A detailed description of this topic is given in Chap 10.

17.10.5 Splice Passage

The passage of splices through the coating bead presents a severe disturbance to the local flow field. The total thickness of the splice may be higher than the coating gap for many applications. Such splices cannot pass through the coating gap without causing the web to break and potentially causing mechanical damage to the die lip. Details of how to handle the splice passage through the bead of slot coating are discussed in Chap. 11.

17.10.6 Prevention of Lines and Streaks

In addition to preventing vortices in the coating bead as described in Sect. 17.8.3, we noticed that the propensity for forming streaks for slot coating operated in the viscous mode is lowered by mounting the slot die at the 4:30 position on the backup roll instead of at the 7:30 position, see Fig. 17.3.

At the expense of losing optical access to the downstream lip, we argue that the direction of gravity helps to reduce the contamination of the upstream lip during the coating start. We further conclude that the static wetting line at the downstream lip tends to remain pinned at the downstream corner of the downstream lip.

Finally, Bruno Kaeser, one of the machine operators in the Pilot Center of Polytype Converting, discovered that streaks generated at the die lip can quite often be eliminated by temporarily and locally applying a wet paper towel to the uncoated substrate just upstream of the slot die. The water, or solvent, which is released from the towel, seems to alter the flow field in the bead in such a way that contaminations at the lip are washed away. However, care must be taken as to not apply too much solvent from the towel to prevent flooding of the upstream lip, which in turn may then worsen the streak problem.

17.11 Tensioned-Web Coating

Tensioned web coating, also known as kiss coating or as tensioned-web-over-slot-die coating (TWOSD), is a version of slot coating, whereby the die is not positioned against a rigid backup roll but a flexible, unsupported section of the substrate. In many applications, the die is pressed into the web, which is supported by two rollers positioned not too far apart from each other as schematically depicted in Fig. 17.2, and as shown in Fig. 17.37.

The advantage of this configuration is that the rigid die is not positioned near the surface of a rigid roller. Therefore, the danger of a mechanical collision between a stationary and a moving rigid surface does not exist. This in turn allows narrower coating gaps to be installed and thinner wet films to be coated, particularly also at high web speeds, when compared to the standard fixed-gap slot coating configuration. Another advantage is that tensioned-web coating is operated without vacuum assist. The disadvantage of this approach is that the exact geometry of the coating gap is not known because one of the gap surfaces is deformable. Moreover, the uniformity of the coated film may be severely degraded if web deformations in the cross-web direction extend to the coating point.

The popularity of tensioned-web coating increased some 40 years ago when magnetic storage media such as videotapes and floppy discs were in high demand, and when the manufacturing of optical films started to be industrialized. Nowadays, tensioned-web coating is necessary when both sides of the substrate are coated simultaneously, for example for the manufacturing of electrodes for Li-ion batteries.

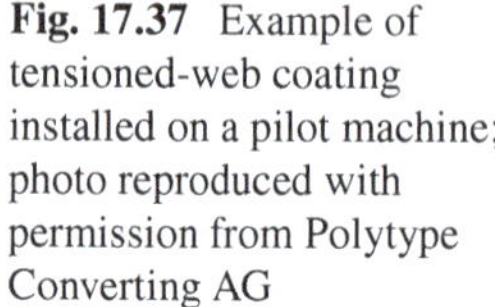
Fig. 17.37 Example of tensioned-web coating installed on a pilot machine; photo reproduced with permission from Polytype Converting AG

Many papers have been published on tensioned-web coating, and an overview can be found in Ding et al. (2016).

Modeling of the flow field of the coating bead is much more difficult than for rigid-gap slot coating because, as explained in Sect. 17.2, tensioned-web coating is an elasto-hydrodynamic process. Consequently, elaborate numerical calculations are required as analytical equations are insufficient for predicting relevant properties of the bead flow field. Most publications examined the situation where the slot die is pushed either perpendicularly or at an angle α (see Fig. 17.2) into the unsupported substrate. This approach leads to a very complex and awkward geometry of the coating gap as schematically sketched in Fig. 17.38a.

From analyzing the performance of rigid-gap slot coating we have learned that a parallel gap is the best geometry because an excessively divergent downstream gap leads to ribbing lines, and an excessively divergent upstream gap in combination

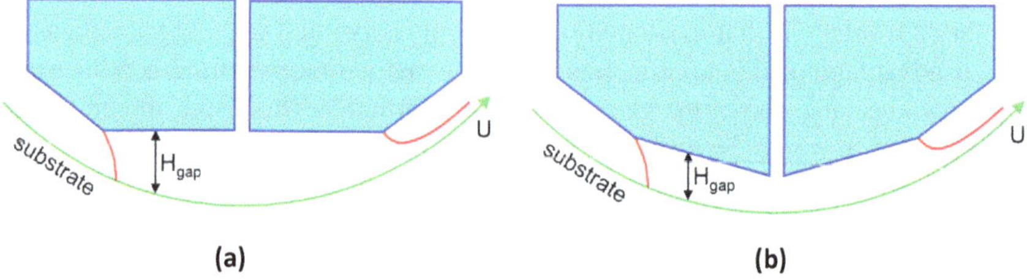

Fig. 17.38 **a** Complex gap geometry in combination with a simple lip geometry in tensioned-web coating. **b** Simple gap geometry in combination with a complex lip geometry in tensioned-web coating

with an excessively convergent downstream gap as shown in Fig. 17.38a increases the pressure at the slot exit considerably, such that leakage at the upstream meniscus will occur eventually. To avoid such an undesirable flow behavior and to render both coating gaps more parallel, the geometry of the die lip would have to be altered, for example as shown in Fig. 17.38b. Even worse in the ideal case, such geometrical alterations would have to be changed as soon as the operating conditions are changed, and this is not an attractive proposition for industrial applications.

In an attempt to avoid the problems addressed above we are proposing a different process configuration for tensioned-web coating that has proven successful in various applications on the pilot machine of Polytype Converting AG. The goal of this approach is to generate a similar or even a better bead flow field as in the rigid-gap case, and so to produce coated films with a high uniformity.

In particular, and as sketched in Fig. 17.39, the die is positioned perpendicularly to the direction where the coated web needs to go. This could be an idle roller in a fixed position, or, if the backside of the substrate is coated simultaneously, the entrance to the flotation dryer as shown in the illustration. In such a configuration, the die is not pushed into the substrate. Instead, the downstream lip is parallel to the substrate, thus always forming a coating gap of uniform height, which is perfect, see Fig. 17.40. The die can be moved up or down, but the die lip is never positioned beyond the departing web. Furthermore, an idle roll that can be moved up or down is mounted just upstream of the die. The position of this roll relative to the position of the die determines the angle α of the incoming web at the coating point and determines the normal force N that the substrate can generate to balance the fluid pressure P in the coating bead, see Fig. 17.40. The overall orientation of this configuration can be changed, i.e., the coated web does not have to depart in a horizontal direction.

A simple analytical formula for calculating the force N is not known. However, based on the analysis of the tensile stress in a thin wall of a round vessel under pressure (boiler formula), it can be deduced that N increases with increasing web tension Z

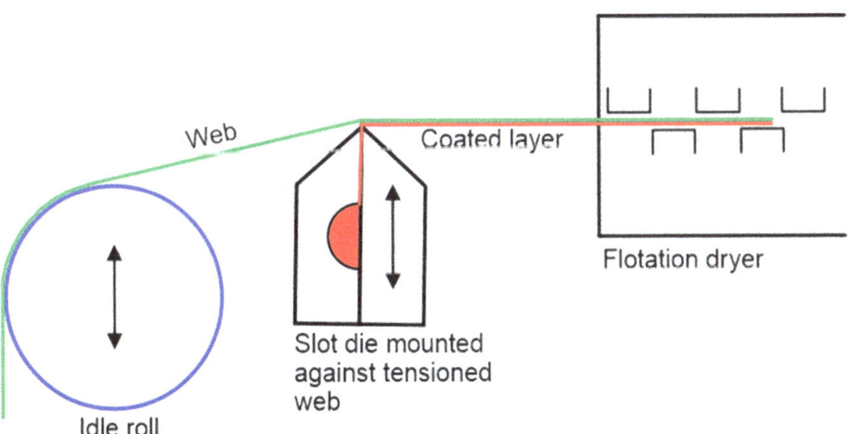

Fig. 17.39 Configuration for tensioned-web coating

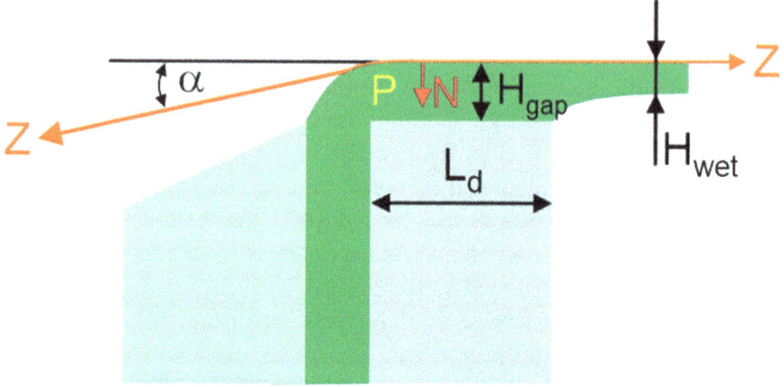

Fig. 17.40 Optimum bead geometry for tensioned-web coating

and increasing web angle α. Normally, the web tension at the coating station can be adjusted continuously and while coating. If the substrate enters a flotation dryer immediately after coating as shown in Fig. 17.39, then the maximum admissible web tension is limited because excessive tension prevents the web from floating properly. The angle α of the incoming web can also be changed continuously during coating. In our experience, this angle should not be too large, say <20°. To prevent a collision between the web and the upstream die lip, the lip should be beveled with an angel that is a bit larger than the maximum web angle. If the thickness of the coated film is high, then it is advantageous to configure the die with an offset OS, see Fig. 17.17. Care must be taken, though, that the incoming web is riding on the liquid film exiting the die slot, and that it does not touch the upstream lip, particularly if the web is a metal foil. It might be advisable, therefore, to use a die with replaceable lips such as shown in Fig. 10.5b.

As discussed in Sect. 17.8.4, it is important to avoid any sort of web deformation in the cross-web direction at the coating point because such deformations always generate thickness variations in the coated film. Web deformations can be avoided, or at least reduced by

- avoiding excessive web tension;
- perfectly aligning all idle rollers around the coating station;
- minimizing the distance between the die and idle rolls upstream and, if applicable, downstream of the die;
- using spreader rolls, if necessary.

Figure 17.41 shows the back-side of an aluminum foil that is slot-coated in the tensioned-web mode. The web is moving vertically from bottom to top, and the web arriving at the coating point is free of any deformations. However, immediately downstream of the coating point the foil is severely deformed in the cross-web direction. Tests have shown that these deformations immediately disappear when the coating is temporarily interrupted, but they immediately reappear upon resuming

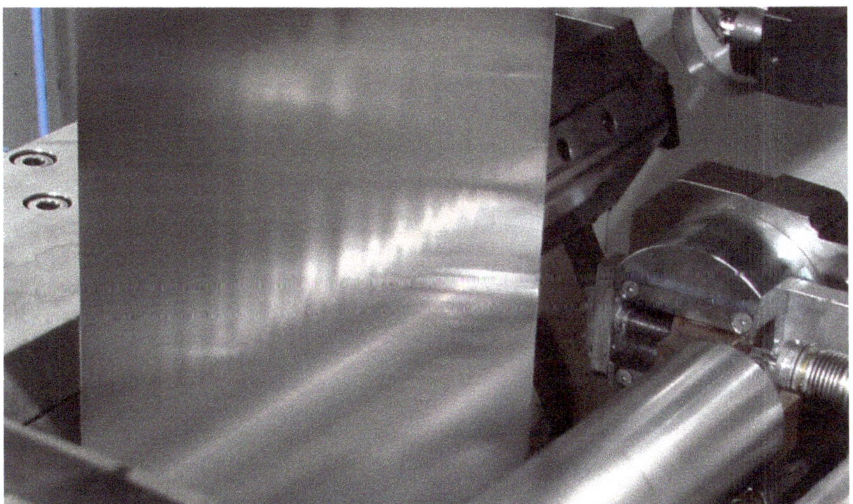

Fig. 17.41 Severe web deformations in an aluminum foil that is slot-coated in the tensioned-web mode; photo reproduced with permission from Polytype Converting AG

coating. We assume that these deformations are generated by the flow in the coating gap, but the exact cause is not known.

If the substrate is simultaneously coated on both sides, then chances are high that web flutter generated by the air currents inside the flotation dryer propagate upstream and possibly to the coating point. If that happens, the uniformity of the coated film is degraded by cross lines. Tests have shown that reducing the airspeed in the first few nozzles inside the dryer is not of much help. Therefore, we tried to calm the web by squeezing it between several pairs of roller discs mounted in the dry edges of the coated foil as illustrated in Figs. 17.42 and 17.43. If stripes are coated, then additional disc pairs cold be mounted in the dry areas between stripes.

The discs can be rotated a bit about a vertical axis, which provides a slight spreading effect, and therefore helps to prevent web deformations. The effect of this approach is modest, particularly if adjacent disc pairs are far apart from each other. While the disc pair manage to stabilize the edges of the substrate, web flutter still has a chance to move upstream in the unsupported areas of the substrate.

Tensioned-web slot coating involves more parameters that must be set properly and controlled than slot coating with a fixed coating gap. Moreover, modeling of the flow in the coating bead is more difficult owing to the elasto-hydrodynamic nature of the bead flow field. Nevertheless, all relevant parameters that affect the bead are known and many of them can be adjusted during coating. With a bit of experience and with a bit of trial-and-error, therefore, successful tensioned-web coating can be implemented and a uniform film can be obtained even simultaneously on both sides of the substrate as shown in Fig. 17.44.

Fig. 17.42 Pair of roller discs to dampen web flutter generated in the flotation dryer and propagating upstream toward the application point of tensioned-web slot coating, side view; photo reproduced with permission from Polytype Converting AG

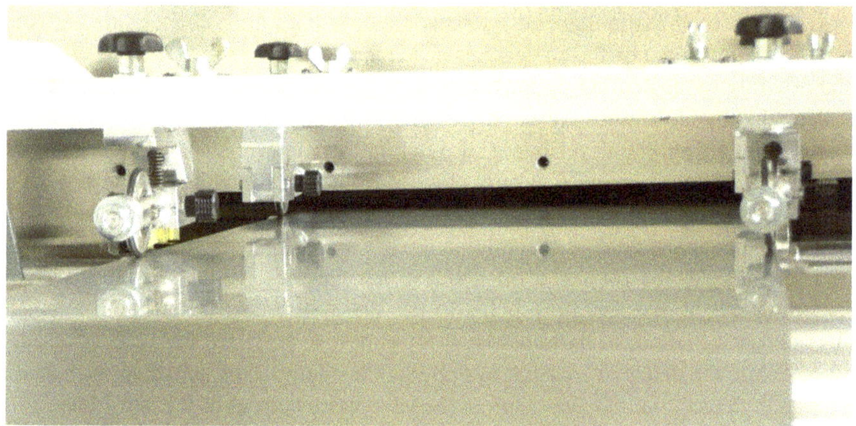

Fig. 17.43 Pair of roller discs to dampen web flutter generated in the flotation dryer and propagating upstream toward the application point of tensioned-web slot coating, front view; photo reproduced with permission from Polytype Converting AG

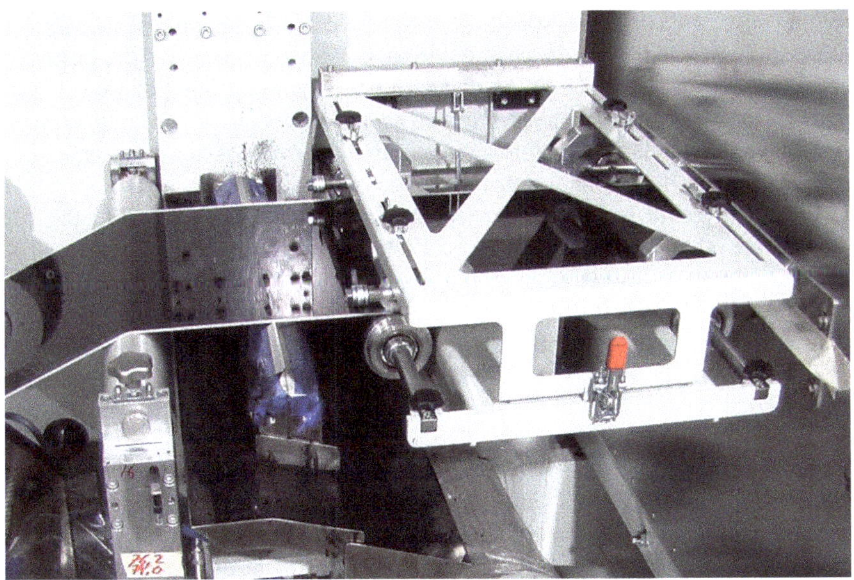

Fig. 17.44 Simultaneous double-sided slot coating in the tensioned-web operating mode; photo reproduced with permission from Polytype Converting AG

17.12 Double-Sided Coating

Coating both sides of the substrate is currently in high demand for the manufacturing of electrodes for Li-ion batteries. Of the three premetered coating methods only slot coating can be used if both sides are coated simultaneously before entering the dryer. In this case, typically, the back side of the substrate is coated upside down as shown in Fig. 17.44.

Three different operating modes and machine designs are known as depicted in Fig. 17.45. Picture a illustrates the simplest version from a coating machine point of view because only one side can be coated at a time. Therefore, two passes are necessary. If an intermittent coating is necessary, then some companies require the

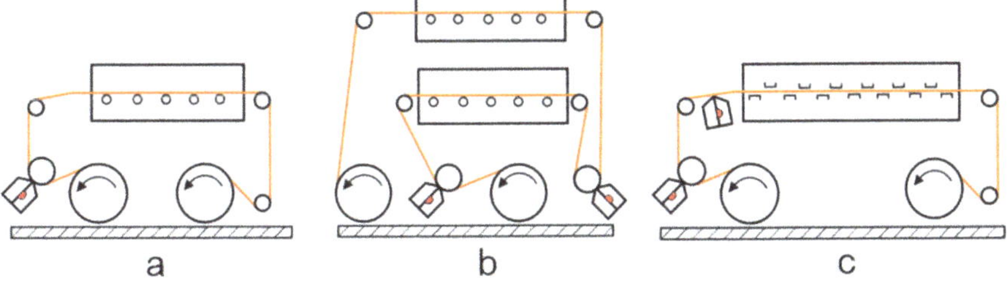

Fig. 17.45 Different machine layouts for coating both sides of the substrate

leading edge of the patches on both sides to be located in the same place, which in turn requires rewinding before coating the backside. Consequently, three passes are necessary, which is the worst case from a product unit cost point of view. On the other hand, the coating machine is simple and relatively short; hence, investments costs are relatively low. In addition, tensioned-web coating is not necessary, which is an advantage when an intermittent coating is required.

Illustration b shows two coating machines in series, i.e., one for coating the front side and the other one for coating the backside. The advantages of this approach are no need for tensioned-web coating, the possibility for intermittent coating, leading patch edges on both sides in the right place, only one pass, and low product unit costs. However, the investment costs are relatively high. Versions a and b allow the dryer to be of support roll type or flotation type.

In picture c, both sides of the substrate are coated simultaneously. Hence, only one coating machine and one coating pass are needed, but the dryer is almost twice as long as the dryers of the other two versions, and it must be of the flotation type. While the front side can be coated with fixed gap slot coating, tensioned-web coating is necessary for applying the backside. An intermittent coating is much more difficult, if not impossible with the tensioned-web mode, owing to the difficulties of achieving concise leading and trailing edges for each patch.

In configurations a and b, the front side of the coated substrate must pass twice through the dryer, which may negatively affect the final product properties.

From our point of view, version b is most attractive despite the relatively high investment costs because tensioned-web coating is not needed, which makes coating easier. Moreover, version b is necessary if intermittent coating is required. Version c might be attractive from an investment and product unit cost point of view despite the need for tensioned-web coating, but it is only possible if intermittent coating is not necessary. Version a is not attractive from an economic point of view.

17.13 Stripe Coating

Nowadays, stripe coating is an important issue for the manufacturing of products such as electrodes for Li-ion batteries, specifically when coated onto wide substrates, optical products like OPV's or OLED's, and security features for credit cards.

Many papers have been published about stripe coating, but most of this literature focuses on manufacturing electronic devices and not on the fundamentals of the coating process. Krebs (2009) reviewed various coating processes that are suitable for stripe coating. He pointed out that slot coating is a preferred method. Papers focusing on issues related to stripe coating with a slot die include Wen and Liu (1995), Han et al. (2014), Schmitt et al. (2014), Lin et al. (2014), Abbel et al. (2017), Raupp et al. (2018), Parsekian et al. (2019), and Lee and Park (2020).

Figure 17.46 visualizes that stripe coating can easily be implemented into the slot coating process by mounting a comb-like shim between the die plates. The

Fig. 17.46 Stripe coating accomplished with a slot die; photo reproduced with permission from Polytype Converting AG

stripe pattern can be adjusted at will by changing the comb geometry as depicted in Fig. 17.47.

The major challenge in stripe coating is controlling the width of the coated stripe. In general, the width of the coated stripe differs from the stripe width defined by the shim. On the one hand, the fluid pressure in the coating gap pushes liquid sideways into both adjacent blocked areas (Raupp et al., 2018), and on the other hand, the coated stripe will either spread or withdraw on the substrate. The difficulty of the latter issue may be amplified if the product contains several layers that are coated one after the other because the substrate of any given layer is the dried surface of the previously coated layer. Figure 17.48 is a graphic rendition of what happens in stripe coating.

The width of the coated stripe depends on

- Fluid viscosity.
- Fluid density.
- Fluid surface tension.
- Substrate surface energy.
- Coated film's wet thickness.
- Time for spreading or withdrawing.

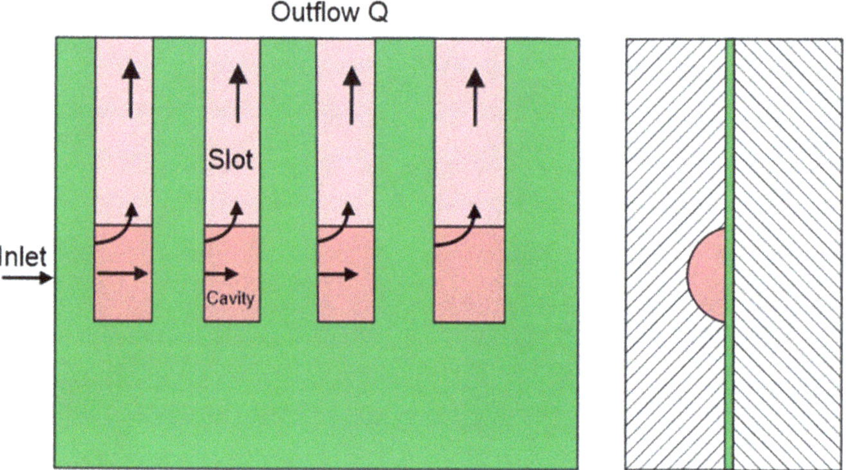

Fig. 17.47 Comb shim used for stripe coating with a slot die

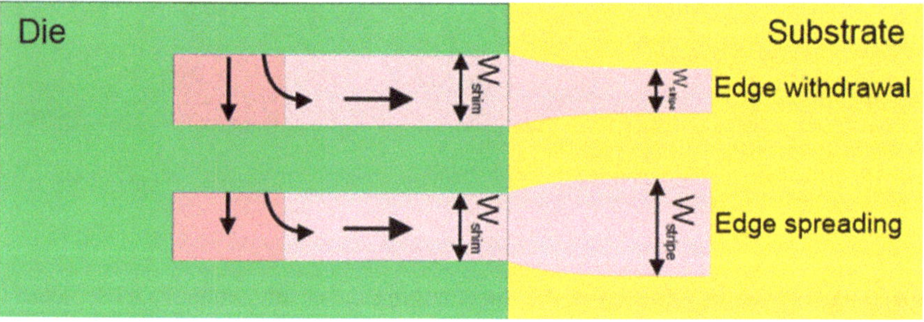

Fig. 17.48 Increasing or decreasing width of the coated stripe relative to the stripe width defined by the comb shim

  - Web speed.
  - Distance between coating and drying.
- Fluid pressure in coating gap:
  - Height of coating gap.
  - Length of coating gap.

In particular, Raupp et al. (2018) carried out an experimental stripe coating study with a focus on low-viscosity fluids. They showed that

- air entrainment-induced film breakup determines the minimum wet film thickness in the coatings;
- thick films limit the process window due to merging stripes;
- decreasing the coating gap did not significantly affect the process window;

- a larger process window was observed for higher spacer widths between the stripes;
- increasing the viscosity and decreasing the surface tension resulted in larger process windows with surface tension being more important than viscosity.

Controlling the stripe width would be easier if the spreading coefficient between the coated layer and the substrate surface was zero, see Sect. 4.6.1. This, however, is difficult to implement in practice. Sondergaard et al. (2013) proposed to work with so-called meniscus guides, which are an extension beyond the end of the die slot of the flow-blocking portions of the comb shim as illustrated in Fig. 17.49. As such shims protrude beyond the die slot, they may touch the substrate and therefore should be made of plastic instead of metal. The purpose of these shims is to guide the meniscus across the coating gap and to maintain the stripe width as defined by the comb shim. However, these shims cannot prevent the subsequent spreading or withdrawal of the coated stripe on the substrate.

Another approach to controlling the stripe width for low viscosity and low surface tension fluids was presented by de Vries (2018), who used a special slot die with notched lip inserts in an attempt to reduce capillary effects between adjacent stripes. De Vries reported that the difference between the width of the coated stripes and the width specified by the comb shim was less than 0.5 mm.

Based on our experience, it is difficult to coat accurate stripes with very narrow dry stripes in between as it is desirable, for example, for OPV's and OLED's. The reason is edge spreading and pressure-driven lateral flow in the coating gap, which may cause adjacent stripes to connect and thereby to eliminate the dry lanes between stripes.

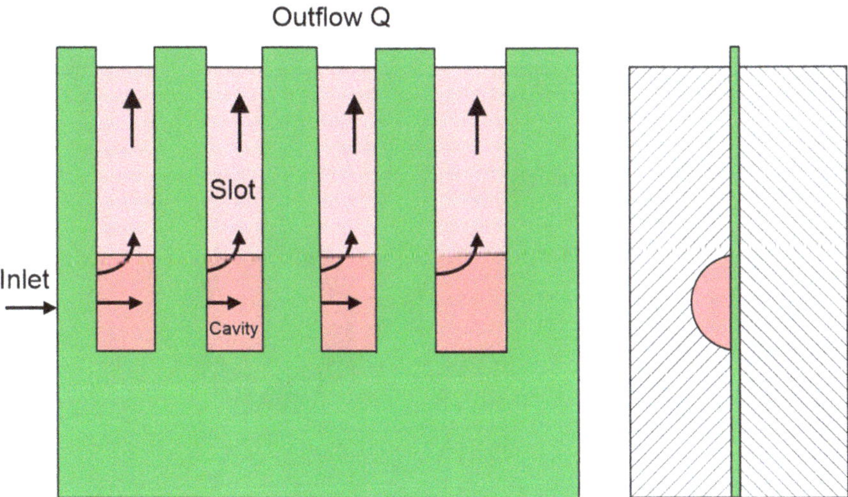

Fig. 17.49 Comb shim with meniscus guides

17.14 Intermittent Coating

Intermittent coating is currently in demand for the manufacturing of battery electrodes and optical devices such as OPS's and OLED's. Intermittent coating refers to the temporary interruption of a continuous coating process. The result of interrupting coating are patches that are as wide as the substrate, or, if stripes are being coating, as wide as the stripes, see Fig. 17.50.

Among premetered coating methods, slot coating is the only process suitable for intermittent coating. In fact, it is the best process of all for this purpose. In principle, patches can also be produced with indirect roll processes, for example with an indirect comma coater as shown in Fig. 17.51. Here, coating interruption is achieved by temporarily lifting the coating roll away from the applicator roll. This concept can also be used in combination with tensioned web slot coating, i.e., by temporarily lifting the substrate away from the die lip. The disadvantage of this approach is that few parameters, if any, are available for controlling the quality of the leading and trailing edges of the patch.

Fig. 17.50 Patches generated by interrupting a continuous coating process; photo reproduced with permission from Polytype Converting AG

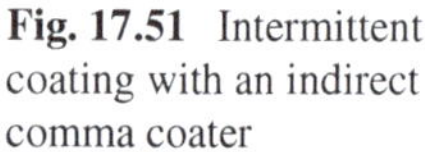

Fig. 17.51 Intermittent coating with an indirect comma coater

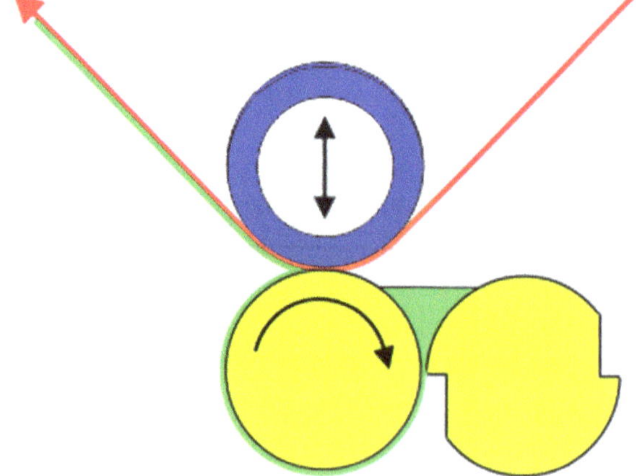

Focusing on slot coating, various design concepts and operating principles have been presented and patented over the past 25 years. To date, nevertheless, the subject is not well understood from a scientific and technological point of view. The reason is that the performance of technological concepts for intermittent coating is controlled by time-dependent flow phenomena, i.e., by starting and stopping of a steady-state coating process, and such processes are difficult to model or visualize experimentally.

Generally speaking, intermittent coating concepts for the slot coating process can be implemented as a die-external or die-internal solution. Some concepts interrupt the coating process by temporarily interrupting the liquid flow to the slot die. In an early example (Hidetoshi et al., 2001), the liquid supply is stopped by a valve located upstream of the die, while a second valve opens a fluid path between the die and an auxiliary vessel under vacuum, which causes the liquid in the die to be sucked back into the vessel. In another example (Noboru & Masaru, 2001), the liquid supply to the die is temporarily recirculated into the supply vessel while the liquid in the die is sucked back with the help of a piston. This concept is schematically depicted in Fig. 17.52.

Gabel and de Vries (2014) obtained intermittent coating by using a substrate surface comprising a pre-patterned layer of high surface energy areas and low surface energy areas. The fluid supply to the slot die is periodically interrupted such that only the high surface energy areas are coated. In addition, the uniformity of the patches is improved by periodically moving the die away from the substrate by a short distance when the die slot is moving over areas of low surface energy.

An example of a die-internal design concept was published by Janssen (2011), who uses a rotating bar that is incorporated into the outer die cavity to block the fluid flow through the die at the end of each patch. One argument for following die-internal design concepts is that equipment manipulations should be carried out

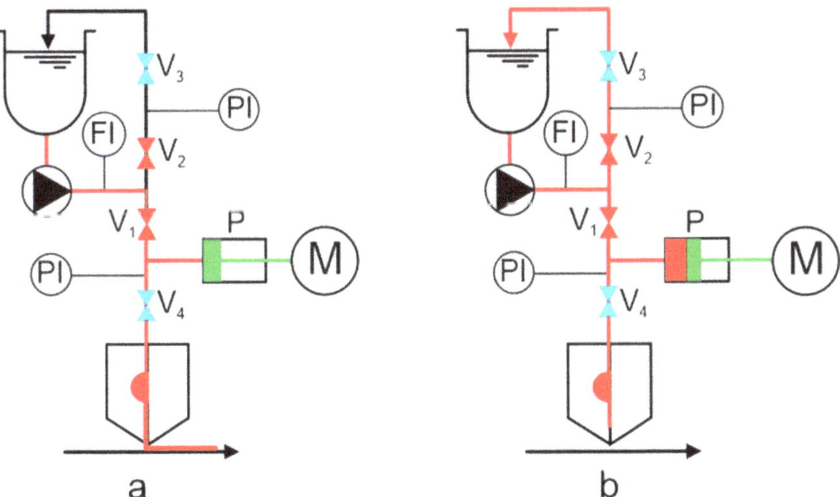

Fig. 17.52 Example of die-external design concept for intermittent coating

as closely as possible to the coating point. However, the longer distance between equipment manipulations and coating point for die-external designs is irrelevant, as long as the coating fluid is incompressible. This assumption is correct here because the flow inside the die is confined, and because the coating fluid must be free of gas bubbles to obtain uniformly coated films.

In our view, die-external concepts are preferable because the internal die design remains simple. In addition, two or more such systems can be installed in parallel, which becomes more and more important as switching frequencies increase in concert with increasing coating speeds. Also, concepts with stationary dies are preferable over designs with an oscillating die position because the coating width for battery electrodes is about to increase from about 600 mm to about 1,500 mm. Consequently, the weight of the die is about to increase accordingly, and moving a heavy die in and out for a short distance at ever-increasing frequencies is a non-trivial task that should be avoided.

In contrast, other concepts do not interrupt the liquid flow to the slot die. An example of this approach is given by Scharfer et al. (2014), and the invention is schematically illustrated in Fig. 17.53. Specifically, a portion of the die slot is formed by a flexible membrane that is communicating with a series of pressure-regulating valves. At the end of each patch, the membrane is deformed in such a way that a cavity is generated that takes up all the fluid supplied to the die. Consequently, no fluid is exiting the die slot. Upon resuming coating, the fluid temporarily stored in the cavity is combined with the fluid supplied by the pump in such a way that the resulting flow rate of the fluid exiting the die always equals the desired value for achieving the desired film thickness. This concept is rather interesting because, as was demonstrated experimentally, it can generate high-quality patches at speeds of up to 100 m/min.

The goal of every design concept is to generate patches that have a perfectly rectangular thickness profile as depicted in Fig. 17.54. In reality, the perfect form

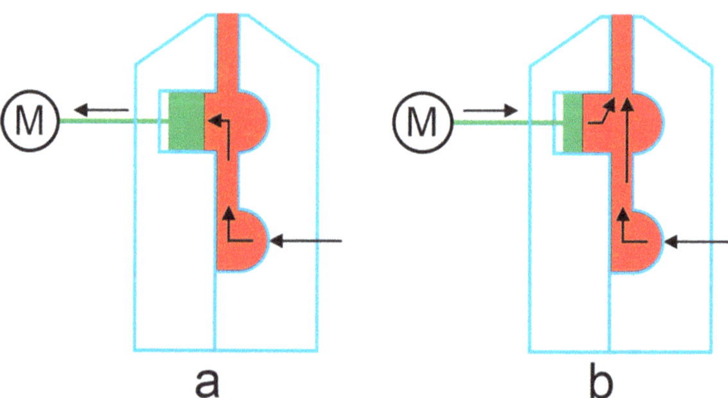

Fig. 17.53 Example of die-internal design concept for intermittent coating; patented by Scharfer et al. (2014)

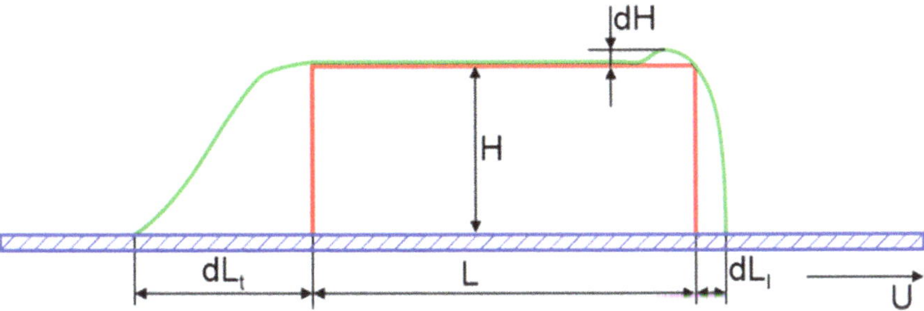

Fig. 17.54 Ideal (red) and real (green) thickness profile of a patch

cannot be reached because the leading edge is smeared over a length dL_l and the patch thickness may locally overshoot by a value of dH. In the same way, the trailing edge extends over a length dL_t, which typically is longer than dL_l. The goal, therefore, is to avoid the overshoot and to minimize the extent of both edge spreads. Reaching this goal becomes more and more difficult as the coating speed increases. Maza and Carvalho (2017) investigated how operating conditions, die geometry, and physical fluid properties affect the break-up of the coating bead. They found that the uniformity of the trailing edge of each patch improves if the static wetting line at the film-forming meniscus is pinned to the downstream corner of the downstream gap, i.e., if the vortex at the corner of the die lip is avoided, see Figs. 17.26 and 17.27.

Another issue that becomes more challenging with increasing web speed is the switching of hardware components that are needed for generating the patches. As depicted in Fig. 17.55, the switching frequency f depends on the web speed and the length of the patch or the length of the dry area between subsequent patches. The latter dimension is more crucial because it should be minimized to minimize raw material losses. For battery electrodes, a typical range for the length of the uncoated area is $\Delta L = 25$–30 mm.

For $\Delta L = 25$ mm, switching frequencies are in the range of 16–66 s^{-1} for web speeds between 25 and 100 m/min. Such high frequencies are challenging from a mechanical point of view.

To learn how to minimize the size of the leading and trailing edges as well as to avoid the thickness over-shoot at the leading edge, we carried out an experimental study at the Pilot Center of Polytype Converting AG. Specifically, we chose a die-external system as shown in Fig. 17.52. The fluid flow was either directed toward the die for coating or recirculated during coating interruption by opening and closing a pair of fast-acting magnetic ball valves V_1 and V_2. Additionally, after V_1 has closed, the liquid from the coating bead was pulled back into the die slot by a piston that was driven by a fast-acting linear motor M. The function of the piston was carried out by a commercial membrane pump, and the diameter of the membrane was chosen such that only a few millimeters of piston stroke were needed to pull the fluid from the bead just a few millimeters into the die slot. The purpose of this type of liquid

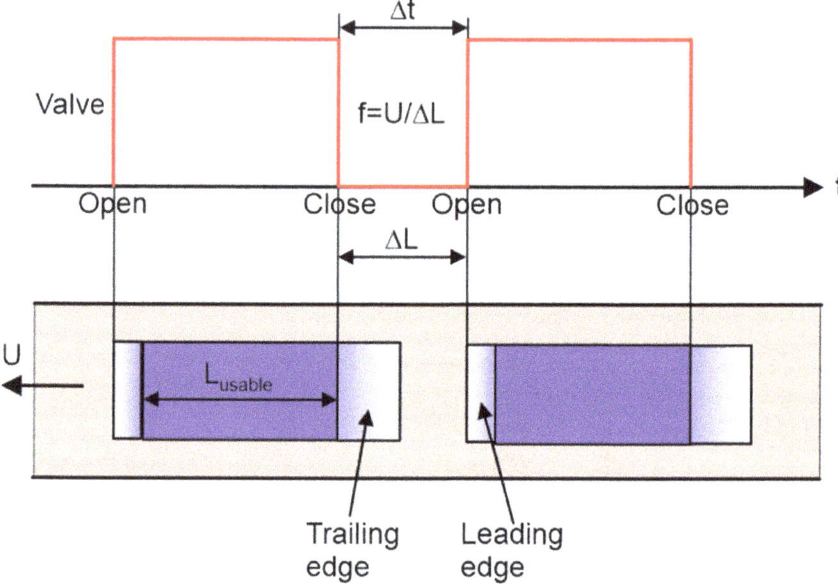

Fig. 17.55 Sketch of valve switching and resulting patches during intermittent coating

withdrawal was to facilitate coating interruption and to form short trailing edges of the coated patches.

At about the time of re-opening the valve V_1 the piston was moved forward to re-supply the liquid, which was previously pulled back, thereby facilitating a sharp leading edge of the patch. V_3 and V_4 are manually operated needle valves. V_3 may be needed for keeping the pressure around the magnetic valves above a minimum value of about 2 bar to guarantee proper and fast functioning of the magnetic valves. V_4 is used to adjust the pressure drop in the return line to be equal to the pressure drop in the feed line to prevent a varying load on the pump motor during switching of the valves V_1 and V_2.

As

- the slot coating process is a premetered process, meaning that the wet thickness of the coated film is solely determined by the ratio of volumetric flow rate/width to web speed;
- the coating liquid can be considered to be incompressible;
- the flow rate of confined flows is strictly related to the pressure in the flow system by the so-called flow rate-pressure drop relationship;

the fluid pressure measured just upstream of the slot die can be used for visualizing the film thickness of the coated film, and in particular for visualizing the shape of the leading and trailing edges of the coated patches. Therefore, we used fast responding

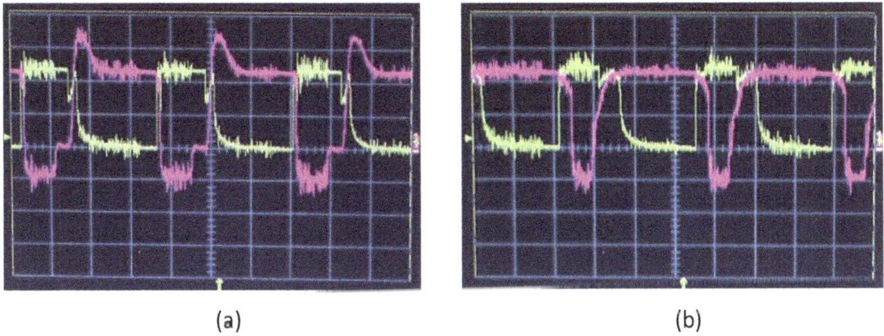

Fig. 17.56 Fluid pressure as a function of time; pink trace: measured just upstream of die entrance; yellow trace: measured in return pipe; **a** large pressure gradients at both patch edges but large overshoot; **b** no overshoot but still acceptable pressure gradients

piezo-electric pressure sensors, and we displayed the sensor signals on the screen of an oscilloscope as displayed in Fig. 17.56.

The shape of the pressure curve of a patch is a fair representation of the true form of the patch, including its leading and trailing edges and the overshoot, if present. If the slope of the increasing pressure is small, then the leading edge is smeared over a large distance in web direction, and dL_l is large, see Fig. 17.54. In addition, the pressure at the leading edge often overshoots, thus leading to an over-thick edge, and hence dH is large. At the trailing edge, the pressure rapidly decreases. Again, if the slope of the decreasing pressure is small, then the edge is smeared over a wide distance, and dL_t is large. Such a non-ideal situation with a pronounced overshoot is visualized in Fig. 17.56a. However, if the intermittent coating process is well mastered, then dL_l, dL_t and dH are close to zero, or at least minimized to an acceptable level for any given set of operating conditions. Such a situation with no overshoot is shown in Fig. 17.56b.

Figure 17.56 also confirms that the pressure in the return pipe (yellow curve) is equal to the pressure in the feed line to the die (pink curve) and that the pressure in the feed line during coating interruption is negative, thus indicating suction generated by the action of the piston.

We used the following mathematical tools provided by the oscilloscope for analyzing and quantifying the signal from the pressure sensor:

Rise time: measures the time it takes for the signal to rise from 10 to 90% of a steep step change; is indicative of the slope of the pressure curve at the leading edge, and hence is a measure of dL_l.

Fall time: measures the time it takes for the signal to fall from 90 to 10% of a steep step change; is indicative of the slope of the pressure curve at the trailing edge, and so is a measure of dL_t.

Overshoot: measures the amplitude difference between the peak and the upper level after a steep step change; is indicative of the overshoot of the pressure curve at the leading edge, and is a measure of dH.

Moreover, the process concept shown in Fig. 17.52 offers several equipment parameters, by which the values of dL_l, dL_t and dH can be influenced, namely:

- The closing time of V_1 relative to the starting time of the pulling motion of the piston.
- The opening time of V_2 relative to the starting time of the pulling motion of the piston.
- The opening time of V_1 relative to the starting time of the pushing motion of the piston.
- The closing time of V_2 relative to the starting time of the pushing motion of the piston.
- The travel distance of the piston.
- The pulling speed of the piston.
- The pushing speed of the piston.

We ran many trials, first off-line, then on a 500 mm wide coating machine at the Pilot Center of Polytype Converting AG (see Fig. 17.50), to learn how the properties of the leading and trailing edges of a patch can be influenced by the equipment parameters described above. We also equipped the slot die with a plexiglass plate, which allowed us to visually verify that the fluid was indeed pulled back uniformly into the die slot upon moving the piston backward. The main results are summarized as follows.

17.14.1 Trailing Edge

Regarding the trailing edge of the patch, time "zero" in the following graphs refers to the starting time of the piston withdrawal. Sharp trailing edges, i.e., short fall times of the pressure in the feed line, can be obtained if

- V_1 closes after the piston is moved (negative time values), or before the piston is moved but not more than about 100 ms, and if V_2 does not open before the piston moves (avoid negative times), see Fig. 17.57.
- The piston is withdrawn far, i.e., >4 mm for our experimental set-up, and at a high speed, i.e., >400 mm/s, see Fig. 17.58.
- The fluid pressure in the feed line reaches sufficiently negative values during the piston withdrawal, see Figs. 17.59 and 17.60.

All results from such types of trials show a clear way regarding desirable values of various process parameters for obtaining short trailing edges. It seems, therefore, that the equipment concept under investigation is suitable for obtaining sharp trailing edges.

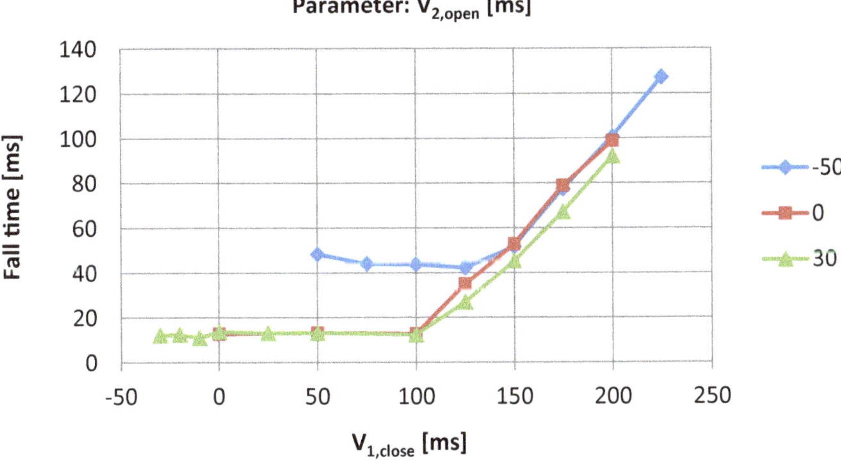

Fig. 17.57 Fall time as a function of the closing time of V_1 and the opening time of V_2

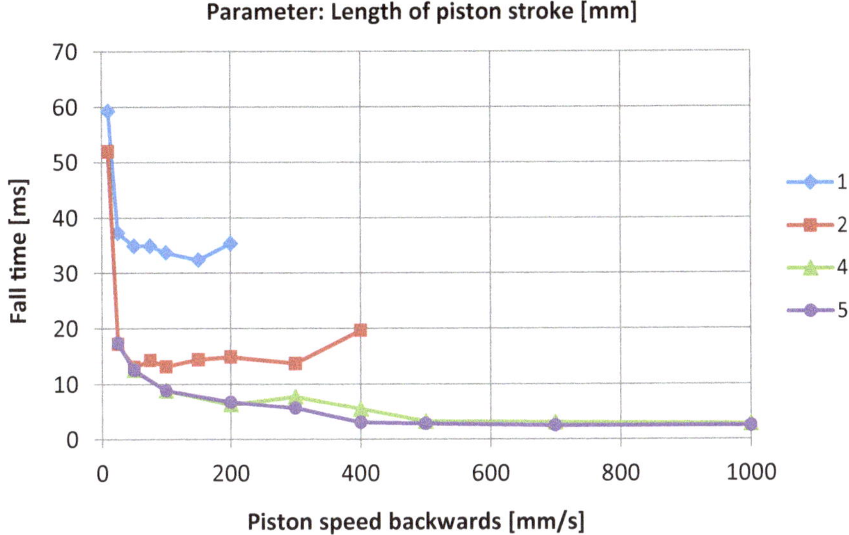

Fig. 17.58 Fall time as a function of the piston speed and the piston travel distance

17.14.2 Leading Edge

Regarding the leading edge of the patch, time "zero" in the following graphs refers to the starting time of the forward motion of the piston. Sharp leading edges, i.e., short rise times of the pressure in the feed line, can be obtained if.

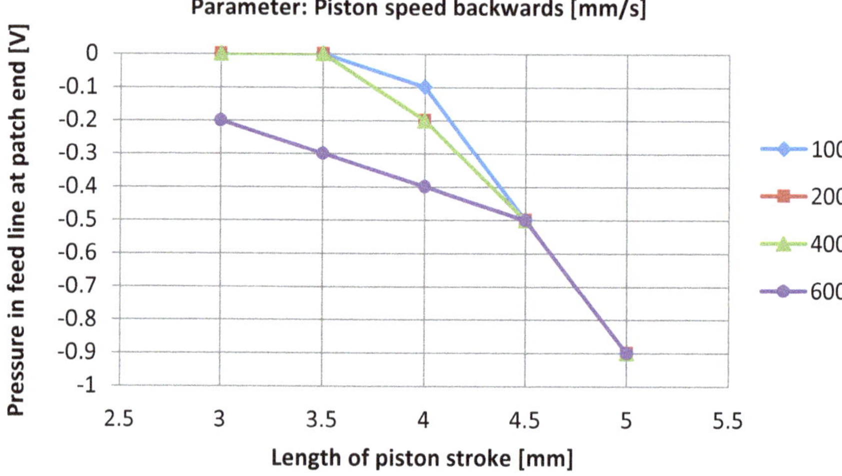

Fig. 17.59 Fluid pressure in feed line as a function of piston speed and piston stroke

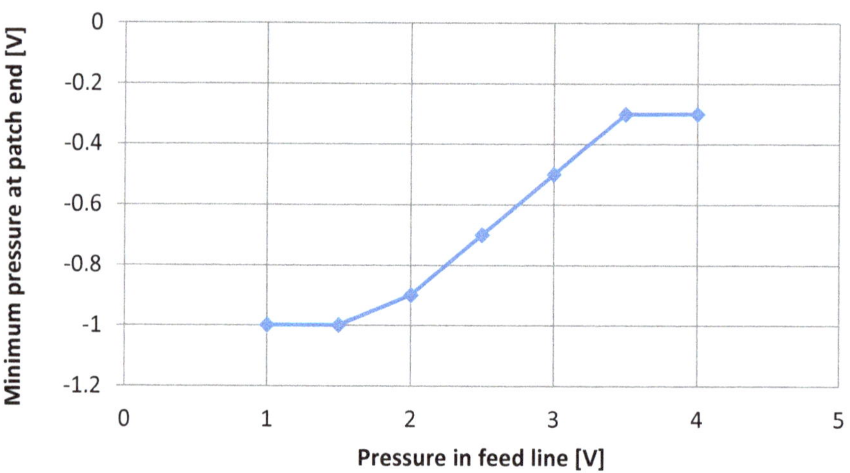

Fig. 17.60 Minimum fluid pressure in feed line during piston withdrawal as a function of the overall fluid pressure in the feed line

- V_1 opens before the piston is moved forward (positive time values of >100 ms), and this is so for positive and negative values of the closing time for V_2, see Fig. 17.61. However, such parameter values result in excessive overshoots, which is not at all desirable.
- The forward piston speed is >300 mm/s for any piston stroke. However, achieving low overshoots requires short strokes, which is in contradiction with obtaining short rise times, see Fig. 17.62. Moreover, there is an optimum piston speed for obtaining small overshoots, and that optimum value depends on the piston stroke.

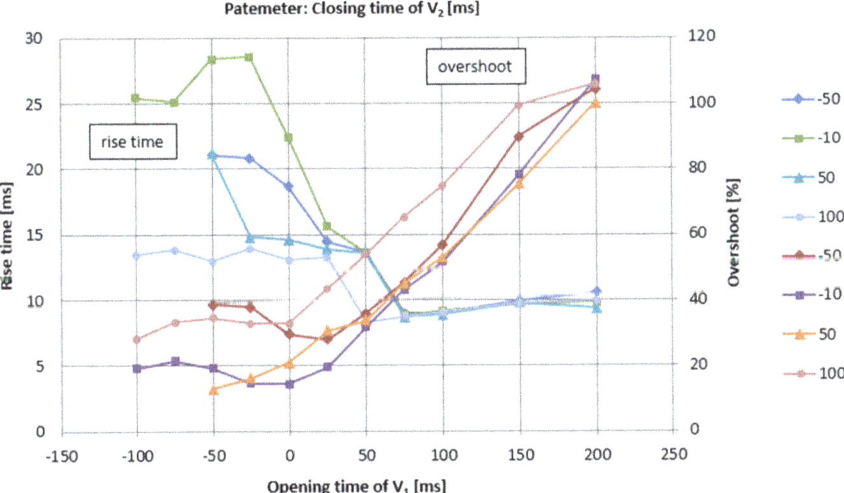

Fig. 17.61 Rise time as a function of the opening time of V_1 and the closing time of V_2

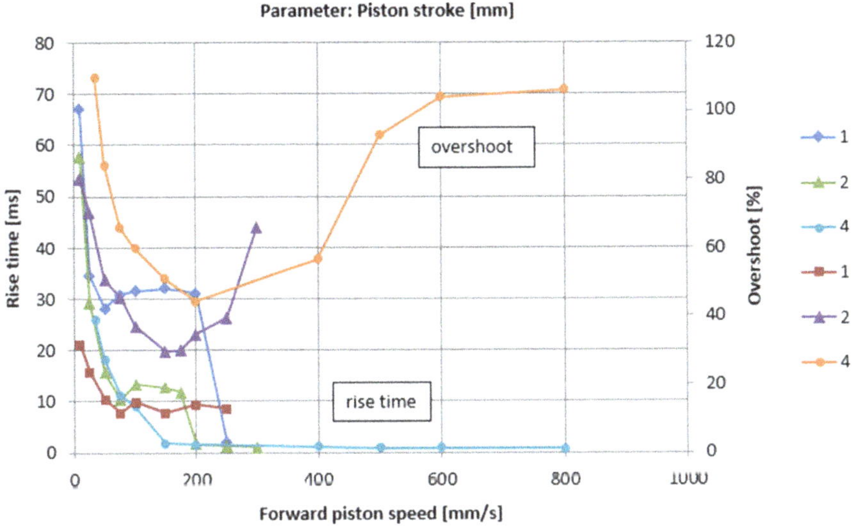

Fig. 17.62 Rise time as a function of the forward piston speed and the piston stroke distance

Summarizing the findings for obtaining good leading and good trailing edges, it can be concluded that the parameters necessary for obtaining good trailing edges are not suitable for obtaining good leading edges with a short rise time and a small overshoot. This is particularly so for the speed and the stroke distance of the piston motion. However, the piston position (stroke) after obtaining a good trailing edge is the starting position for generating the leading edge, and this position is not suitable for getting good results, which is a dilemma. As the time between the trailing and the next leading edge is typically small, there is no chance to move the piston into a better position, which would allow us to generate a good leading edge. However, this dilemma can be resolved by installing a second piston as well as additional valves as depicted in Fig. 17.63.

All equipment used in the initial concept (see Fig. 17.52) is still in place. In addition, a second piston is added, and the pipelines between each piston and the main feed line to the die are equipped with two additional magnetic ball valves. Now, the switching sequence of the valves and the moving action of the pistons are illustrated in Fig. 17.64.

Generating a good trailing edge requires one of the pistons to be moved backward with a high speed and for a long distance. This position is not suitable for shortly thereafter generating a good leading edge. Instead, the second piston, which, during the coating of the previous patch, was moved from the end position of the previous trailing edge to the starting position for making a good next leading edge, is activated forward at the best possible speed. Thereafter, the second piston is in the correct position for making the next trailing edge. In the meantime, i.e., during the coating

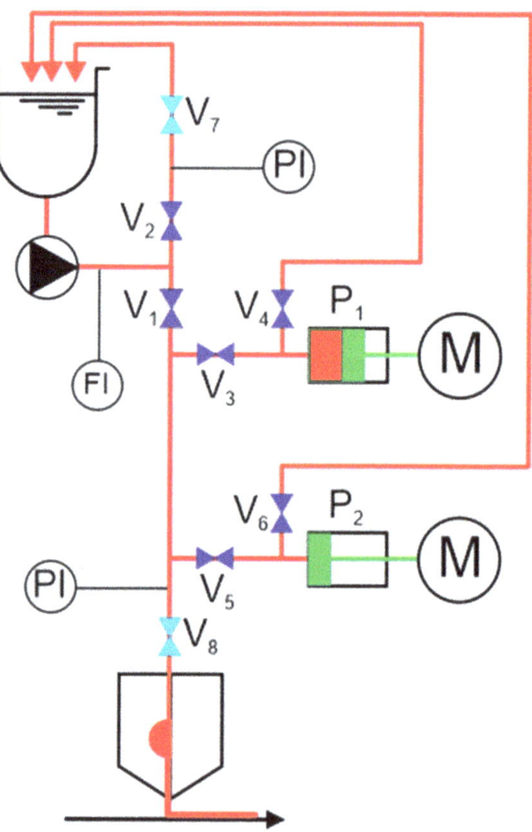

Fig. 17.63 Improved concept for obtaining good leading and trailing patch edges without overshoot

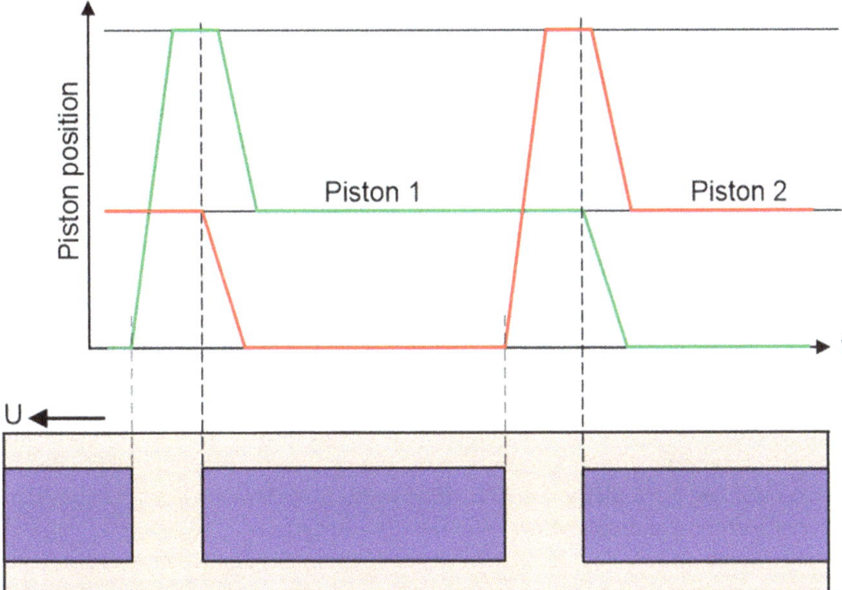

Fig. 17.64 Sequence of tandem piston motion

of the next patch, the first piston is moved forward a bit to the position, which allows a good formation of the next leading edge. During this forward motion of the first piston, the liquid pushed forward cannot easily be fed into the feed line going to the die. Instead, it must be diverted back to the fluid supply vessel by temporarily closing valve V_3 and opening valve V_4. Thereafter, this sequence of switching valves and moving pistons is repeated until coating is stopped.

Unfortunately, due to financial pressures, Polytype Converting AG stopped this project before we were able to experimentally test the improved design concept and a short time later, I retired. Nor did we have time to quantifying the width of the patch edges and the overshoot as a function of relevant operating parameters such as the web speed, the wet film thickness, the coating gap, and the fluid viscosity. Such investigations were not published either for most of the patented design concepts, which is unfortunate because it leaves industrial coating companies that are interested in intermittent coating in the dark about the performance of available systems. One exception is the paper by Diehm et al. (2020a, 2020b), in which they slot-coated a water-based and shear-thinning anode slurry and examined the influence of coating speeds up to 50 m/min and wet film thicknesses up to 400 μm on the quality of the leading and trailing edges of the patch. Interestingly, they found that the length of both the rising and falling edges remained constant for any of the tested speeds. In addition, they observed that the rising edge was always shorter than the falling edge for any of the tested speeds and wet film thicknesses. However, when increasing the wet film thickness, the length of the falling edge increases proportionally but with a gradient that depends on the coating speed. In contrast, the length of the rising edge increases proportionally to the wet film thickness but the effect is independent of the speed.

Regarding other scientific literature about intermittent coating, useful technical information is still sparse, but some insight about this process can be found in Diehm (2017), Diehm et al. (2017, 2019), who report about electrode coating for Li-ion batteries, and in de Vries et al. (2017), who applied this technology to OLED's and OPV's.

17.15 Pattern Coating

Slot coating is capable of coating patterns, although only in the form of rectangles or squares, see Fig. 17.65. Currently, potential applications include electrodes for Li-ion batteries, particularly if the substrate width is >600 mm, OPV's, OLED's and security features for credit cards.

Patterns are achieved by simultaneously combining stripe and intermittent coating. Therefore, pattern coating requires knowledge and experience in these technologies as described in the previous chapters.

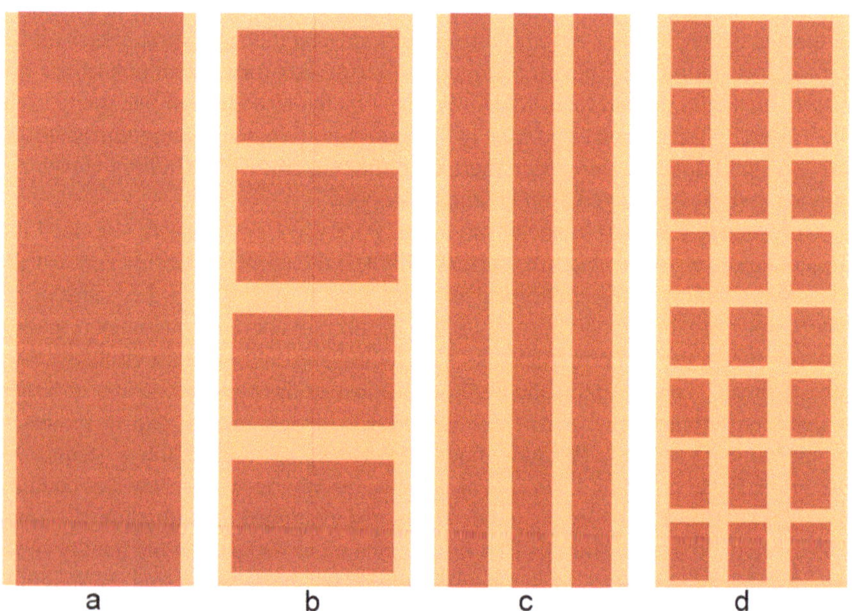

Fig. 17.65 Pattern coating by combining stripe and intermittent coating with the slot die. **a** continuous coating, **b** intermittent coating, **c** stripe coating, **d** pattern coating

17.16 Simultaneous Multilayer Coating

All premetered coating methods are predestined for simultaneous multilayer applications. In contrast to slide and curtain coating, however, the maximum number of layers that can be coated at the same time with the slot coating method is limited to 3, if a die design as depicted in Fig. 2.1b is used. In this design concept, the slot of each individual layer exits into the coating gap with only a short distance between adjacent slots, see sketch in Fig. 17.66. In addition, neighboring slots, i.e., adjacent fluid distribution systems, are separated from each other by a wedge-shaped die plate. Consequently, the opening angle of the entire die becomes excessively large, and hence the use of the die is excessively cumbersome if more than 3 slots and two wedges are assembled.

From a fluid mechanics point of view, this design concept may be problematic, particularly if the thickness of the bottom layer is small, and/or if the viscosity of the bottom layer at the shear rate in the middle gap is low. As explained in Sect. 17.6 and for mechanical reasons, there is a minimum value of approximately 30 μm for any coating gap. If this gap is too wide for a thin and low-viscosity bottom layer, then the interface between the two layers is pushed toward the substrate surface when the bottom layer enters the downstream gap. Consequently, the interface near the die lip blocks the flow of the bottom layer in the middle gap, which causes a vortex to form, and that is not desirable, see Fig. 17.66a. On the other hand, enlarging the downstream gap as illustrated in Fig. 17.66b prevents the interface from being pushed toward the substrate, and hence prevents the vortex formation, but this geometrical configuration is only possible if the bottom layer thickness is in concert with the requirement of a minimum coating gap.

In the example presented in Fig. 5.53, the interface is located at 28.9 μm above the substrate surface, which just about satisfies the mechanical requirement of a minimum gap of 30 μm. So, it seems possible to obtain a gap geometry as depicted in Fig. 17.66b with optimum flow conditions in every gap. However, if the web speed were to be increased from 100 to 200 m/min without changing any of the other operating conditions listed in Table 5.13, then the interface would move closer to the substrate, i.e., to a distance of 20.6 μm. Since the coating gap could not be reduced

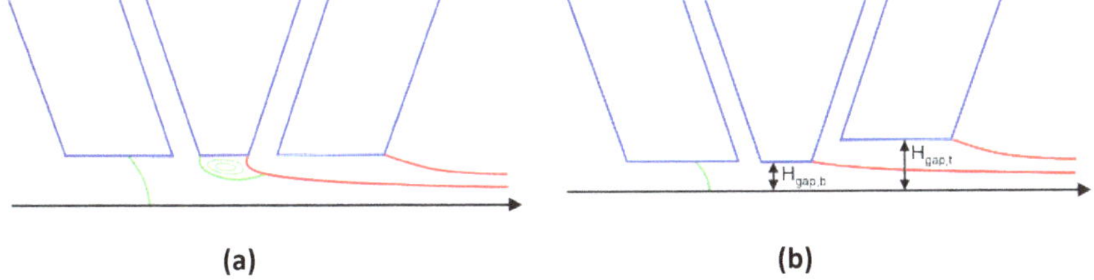

Fig. 17.66 **a** Coating bead with uniform gap height for both layers. **b** Coating bead with different gap heights for the bottom and top layer

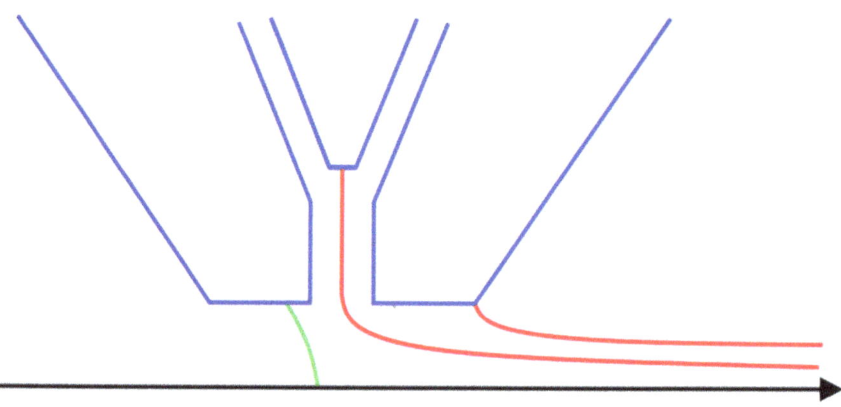

Fig. 17.67 Merging of the two layers inside the die

to the same value, a flow situation as illustrated in Fig. 17.66a would result. Such a flow field is not desirable owing to the presence of a vortex.

Potentially critical and unwanted situations as described here could be prevented, if the die was designed such that both layers merge into one slot before that slot exits the die, see Fig. 17.67. Pressure-driven two-layer slot flow is discussed in Sect. 5.6.3, and Fig. 5.52 shows an example of a typical velocity profile.

Both layers exit the die together, and they flow together through the downstream gap. Consequently, and owing to the higher flow rate in that gap, it is easier to respect the mechanical requirement of a minimum gap height. To our knowledge, most if not all die manufacturers offer 2-layer slot dies according to the slot design shown in Fig. 17.66.

Quite a few papers about 2-layer slot coating have been published in the scientific literature, for example from the group of Skip Scriven in Minnesota, i.e., Cohen (1993) and Musson (2001), from the group of Ta-Jo Liu in Taiwan, i.e., Lu et al. (2001) and Lin et al. (2005), and from the group of Marcio Carvalho in Brazil, i.e., Nam and Carvalho (2009, 2010) and Horiuchi et al. (2015). More recently, Wilhelm Schabel's group at the Karlsruhe Institute of Technology investigated 2-layer slot coating for manufacturing Li-ion battery electrodes, i.e., Schmitt et al. (2015), Diehm (2018), Diehm and Kumberg (2019) and Diehm et al. (2020a, 2020b). Note that I have no experience with 2-layer slot coating.

References

Abbel, R., de Vries, I., Langen, A., Kirchner, G., t'Mannetje, H., Gorter, H., Wilson, J., & Groen, P. (2017). Toward high volume solution based roll-to-roll processing of OLED's. *Journal of Materials Research, 32*(12), 2219–2229.

Atkinson, B., Brocklebank, M. P., Card, C. C. H., & Smith, J. M. (1969). Low Reynolds number Developing Flow. *AIChE Journal, 1*(6).

Blake, T. D., & Ruschak, K. J. (1997). Wetting: Static and dynamic contact lines. In S. F. Kistler & P. M. Schweizer (Eds.), *Chapter 3 in Liquid film coating*. Chapman & Hall.

Blake, T. D., Clarke, A., & Ruschak, K. J. (1994). Hydrodynamic assist of dynamic wetting. *AIChE Journal, 40*, 229–242.

Beguin, A. E. (1954). Method of coating strip material, US 2,681,694.

Bhamidipati, K., Didari, S., & Harris, T. A. L. (2012). Experimental study on air entrainment in slot die coating of high-viscosity, shear-thinning fluids. *Chemical Engineering Science, 80*, 195–204.

Carvalho, M. S., & Kheshgi, H. S. (2000). Low-flow limit in slot coating: Theory and experiments. *AIChE Journal, 46*(10), 1907–1917.

Chang, H.-M., Lee, C.-C., & Liu, T.-J. (2009). The effect of bead vacuum on slot die coating. *Intern. Polymer Processing, 24*(2), 157–165.

Cohen, D. (1993). *Two-layer slot coating flow: Visualization and modeling*. Thesis, University of Minnesota.

Coyle, D. J. (1997). Knife and roll coating. In S.F. Kistler & P.M. Schweizer (Eds.), Chapter 12a in Liquid film coating. Chapman & Hall.

De Vries, I. (2018). Advances in organic & printed electronic processing. In *Proceedings of the 3rd Thin Film Technology Forum*, Karlsruhe Institute of Technology, June 7–8, 2018.

De Vries, I., Gorter, H., Kirchner, G., Kolbush, T., Crone, K. P., Pankratz, A., & Groen, P. (2017). Recent developments on intermittent coating. In *Proceedings of the 12th European Coating Symposium*, Fribourg, Switzerland, November 8–10, 2017.

Diehm, R. (2017). Limitations in industrial coating of battery electrodes. In *Proceedings of the 2nd Thin Film Technology Forum*, Karlsruhe Institute of Technology, June 1–2, 2017.

Diehm, R. (2018). Advances in coating of Li-ion battery electrodes. In *Proceedings of the 3rd Thin Film Technology Forum*, Karlsruhe Institute of Technology, June 7–8, 2018.

Diehm, R., & Kumberg, J. (2019). Advances in coating and drying of multilayer Li-ion battery electrodes. In *Proceedings of the 4th Thin Film Technology Forum*, Karlsruhe Institute of Technology, May 13–17, 2019.

Diehm, R., Kumberg, J., Dörrer, C., Müller, M., Bauer, W., Scharfer, P., & Schabel, W. (2020a). In situ investigations of simultaneous two-layer slot die coating of component-graded anodes for improved high-energy Li-ion batteries. *Energy Technology, 8*.

Diehm, R., Weinmann, H., Kumberg, J., Schmitt, M., Fleischer, J., Scharfer, P., & Schabel, W. (2020b). Edge formation in high-speed intermittent slot-die coating of disruptively stacked thick battery electrodes. *Energy Technology, 8*.

Diehm, R., Kumberg, J., Scharfer, P., & Schabel, W. (2019). Advantages and limitations of structured coatings of lithium-ion battery electrodes. In *Proceedings of the 13th European Coating Symposium*, Heidelberg, Germany, September 8–11, 2019.

Diehm, R., Schmitt, M., Scharfer, P., & Schabel, W. (2017). Intermittent slot die coating for Li-ion battery electrodes. In *Proceedings of the 12th European Coating Symposium*, Fribourg, Switzerland, November 8–10, 2017.

Ding, X., Liu, J., & Harris, T. A. L. (2016). A review of the operating limits in slot coating processes. *AIChE Journal, 62*(7), 2508–2524.

Durst, F., & Wagner, H.-G. (1997). Slot coating. In S. F. Kistler & P. M. Schweizer (Eds.), *Chapter 11a in Liquid film coating*. Chapman & Hall.

Gabel, S. J., & de Vries, I. G. (2014). Slot-die coating method, apparatus, and substrate. EP 2 799 154 A1.

Gutoff, E. B., & Kendrik, C. E. (1982). Dynamic contact angles. *AIChE Journal, 28*, 456–466.

Han, G. H., Lee, S. H., Ahn, W. G., Nam, J., & Jung, H. W. (2014). Effect of shim configuration on flow dynamics and operability windows in stripe slot coating process. *Journal of Coatings Technology and Research, 11*(1), 19–29.

Hidetoshi, Y., Yoshitoshi, M., Kazuto, U., Eiji, M., & Eishin, M. (2001). Discharge type coating device. JP 2001179156.

Higgins, B. G., & Scriven, L. E. (1980). Capillary pressure and viscous pressure drop set bounds on coating bead operability. *Chemical Engineering Science, 35*(3), 673–682.

Horiuchi, R., Suszynski, W. J., & Carvalho, M. S. (2015). Simultaneous multilayer coating of water-based and alcohol-based solutions. *Journal of Coating Technology and Research, 12*, 819–826.

Janssen, F. (2010). Coating nozzle for intermittent application. DE 10 2010 017 965 A1.

Krebs, F. C. (2009). Fabrication and processing of polymer solar cells: A review of printing and coating techniques. *Solar Energy Mater. Solar Cells, 93*(4), 394–412.

Landau, L., & Levich, B. (1942). Dragging of a liquid by a moving plate. *Acta Physicochimica URSS, 17*, 42–54.

Lee, K.-Y., Liu, L.-D., & Liu, T.-J. (1992). Minimum wet thickness in extrusion slot coating. *Chemical Engineering Science, 47*, 1703–1713.

Lee, J., & Park, J. (2020). Increased stripe density of slot-coated PEDOT: PSS using a meniscus guide with linearly tapered μ-tips for OLED's. *Organic Electronics, 83.*

Lin, C.-F., Hill, W., David, S., Liu, T.-J., & Wu, P.-Y. (2010). Operating windows of slot die coating: Comparison of theoretical predictions with experimental observations. *Advances in Polymer Technology, 29*(1), 31–44.

Lin, Y. N., Liu, T.-J., & Hwang, S. J. (2005). Thickness of double-layer slide-slot coating of polyvinyl-alcohol solutions. *Polymer Engineering Science, 45*(12), 1590–1599.

Lin, C. F., Wang, B. K., Lo, S. H., Wong, D. S. H., & Liu, T. J. (2014). Operating window of stripe coating. *Asia-Pacific Journal of Chemical Engineering, 9*(1).

Lu, S. Y., Lin, Y. P., & Liu, T.-J. (2001). Coating window for double layer extrusion slot coating of polyvinyl-alcohol solutions. *Polymer Engineering Science, 41*(10), 1823–1829.

Maza, D., & Carvalho, M. S. (2017). Trailing edge formation during slot coating of rectangular patches. *Journal of Coatings Technology and Research, 14*, 1003–1013.

Musson, L. C. (2001). *Two-layer slot coating.* Ph.D. Thesis, University of Minnesota.

Nam, J., & Carvalho, M. S. (2009). Mid-gap invasion in two-layer slot coating. *Journal of Fluid Mechanics, 631*, 397–417.

Nam, J., & Carvalho, M. S. (2010). Linear stability analysis of two-layer rectilinear flow in slot. *AIChE Journal, 56*, 2503–2512.

Noboru, M., & Masaru, W. (2001). Intermittent coating device and intermittent coating method. JP 2001191005.

Parsekian, A., Jeong, T. J., & Harris, T. A. L. (2019). A process model for slot coating of narrow stripes. *Journal of Coatings Technology and Research, 16*(41).

Pranckh, F. R., & Colye, D. J. (1997). Elasto-hydrodynamic coating systems. In S. F. Kistler & P. M. Schweizer (Eds.), *Chapter 12c in Liquid film coating.* Chapman & Hall.

Raupp, S. M., Schmitt, M., Walz, A. L., Diehm, R., Hummel, H., Scharfer, P., & Schabel, W. (2018). Slot die stripe coating of low viscous fluids. *Journal of Coatings Technology and Research, 5*, 899–911.

Romero, O. J., Suszynski, W. J., Scriven, L. E., & Carvalho, M. S. (2004). Low-flow limit in slot coating of dilute solutions of high molecular weight polymer. *Journal of Non-Newtonian Fluid Mechanics, 118*, 137–156.

Ruschak, K. J. (1976). Limiting flow in a pre-metered coating device. *Chemical Engineering Science, 31*(11), 1057–1060.

Sartor, L. (1990). *Slot coating: Fluid mechanics and die design.* Ph.D. Thesis, University of Minnesota.

Savage, M. D. (1977). Cavitation in lubrication. Part 2. Analysis of wavy interfaces. *Journal of Fluid Mechanics* **80**, 757-767.

Scharfer, P., Schmitt, M., Schabel, W., & Diehm, R. (2014). Verfahren und Vorrichtung zur intermittierenden Beschichtung. DE 10 2014 118 524 A1.

Schmitt, M., Raupp, S., Wagner, D., Scharfer, P., & Schabel, W. (2015). Analytical determination of process windows for bilayer slot die coating. *Journal of Coating Technology and Research, 12*, 877–887.

Schmitt, M., Scharfer, P., & Schabel, W. (2014). Slot die coating of lithium-ion battery electrodes: Investigations on edge effect issues for stripe and pattern coatings. *Journal of Coatings Technology and Research, 11*(1), 57–63.

Schweizer, P. M. (1997). Control and optimization of coating processes. In S. F. Kistler & P. M. Schweizer (Eds.), *Chapter 15 in Liquid film coating*. Chapman & Hall.

Schweizer, P. M. (2013). Coating of ultra-thin liquid films. Convertech and e-Print Japan, Part 1: March/April, 96–103, Part 2: May/June, 4–8.

Sondergaard, R. R., Hösel, M., & Krebs, F. C. (2013). Roll-to roll fabrication of large area functional organic materials. *Journal of Polymer Science, Part B: Polymer Physics, 51*(1), 16–34.

Spiegel, S., Heckmann, T., Altvater, A., Diehm, R., Scharfer, P., & Schabel, W. (2021). *Study of edge formation during slot die coating of Li-ion battery electrodes*. Paper presented at the 14th European Coating Symposium, September 6–9, 2021, Brussels, Belgium.

Suszynski, W. J. (2015). *Converting Quarterly, Quarter 4.*

Wagner, H.-G. (1996). *Die Schlitzfliesserbeschichtung und Laminartrocknung als Grundverfahren optimierter Beschichtungsanlagen*. Ph.D. Thesis, University of Erlangen.

Weinstein, S. J., & Palmer, H. J. (1997). Capillary hydrodynamics and interfacial phenomena. In S. F. Kistler & P. M. Schweizer (Eds.), *Chapter 2 in Liquid film coating*. Chapman & Hall.

Wen, S. H., & Liu, T. J. (1995). Extrusion die design for multiple stripes. *Polymer Engineering & Science, 9*(35), 759–767.

Wengeler, L., Peters, K., Schmitt, M., Wenz, T., Scharfer, P., & Schabel, W. (2014). Fluid-dynamic properties and wetting behavior of coating inks for roll-to-roll production of polymer-based solar cells. *Journal of Coatings Technology and Research, 11*(1), 65–73.

Wilson, S. D. R. (1982). The drag-out problem in film coating theory. *Journal of Engineering Mathematics, 16*, 209–221.

Chapter 18
Specific Properties of Slide Coating

Abstract Slide coating was invented in the photographic industry by Eastman Kodak. Owing to its capability to apply several layers at the same time, slide coating resulted in a large productivity increase when compared to slot coating. The chapter begins with a description of the process configuration and the associated equipment, followed by a discussion of the flow field of the coating bead. The main operating limits of slide coating are the onset of ribbing lines and several vortices. The onset of ribbing lines is documented with experimental data. The chapter closes with suggestions for optimizing the process.

18.1 Introduction

The slide coating process has been used industrially for about 65 years. The first important patent was granted in 1956 to Mercier, Torpey and Russell, who worked for Eastman Kodak. Slide coating is an advancement over slot coating, with the main difference and main advantage being the possibility of applying many different layers at the same time. It was reported that Polaroid Corporation applied 19 different layers under production conditions, and one of the Japanese photographic companies managed to coat 25 layers simultaneously, although "only" in a laboratory environment. In contrast to curtain coating, which is also capable of multilayer applications, slide coating has less of a need for low surface tensions, and hence for adding surfactants, particularly if the fluid is aqueous. In addition, there is no need for a minimum web speed, a requirement associated with the stability of liquid curtains. On the other hand, slide coating is characterized by rather severe process limitations such as a relatively high minimum wet film thickness and a relatively low maximum web speed, both related to the onset of a flow instability called *ribbing lines*.

For several decades in the second half of the last century, slide coating was the workhorse for all photographic companies across the world. These companies gained much knowledge and experience about this coating technology. Unfortunately, however, not much of this know-how and know-why was ever published. A noteworthy exception is the comprehensive report by Hens and van Abbenyen (1997), who worked for Agfa Gevaert at that time.

P. M. Schweizer, *Premetered Coating Methods*, Engineering Materials,
https://doi.org/10.1007/978-3-031-04180-8_18

Nowadays, and particularly since the significant decline of the photographic industry, slide coating has lost most of its former popularity, not only relative to slot and curtain coating but also when compared to the self-metering coating processes. The main reason is that slide coating works well for low viscosities and high wet film thicknesses, and these conditions were given in the photographic industry, specifically when simultaneously applying several layers. Today, only a few new products have emerged that meet these requirements. Nevertheless, slide coating deserves to be presented in as much detail as possible, and that is the purpose of the chapter at hand.

18.2 Process Configuration and Equipment

Figure 18.1 illustrates the two process configurations that are known from the photographic industry. The difference is related to the location of the dryer relative to the location of the coating roll. If the substrate leaves the roll in a vertical direction up toward the dryer, then the coating or application point must be located in the lower quadrant of the coating roll to prevent web flutter at the coating point. In contrast, if the web departs horizontally toward the dryer, then the application point is preferably located in the upper quadrant. As is explained in Chap. 7, the hydrodynamic assist to dynamic wetting can best be exploited if the prolongation of the slide surface of the die coincides with the center of the coating roll as depicted in Fig. 18.1b. The application angle α, and hence the slide angle β, should be maximized to maximize the momentum of the impinging liquid film, but it must be kept low enough to prevent waves in the film flow on the die slide. In concert with the relatively low viscosities of gelatin-based photographic materials, slide angles in the photographic industry were in the range of 15°–25°.

Regardless of the location of the application point, slide coating is typically operated with bead vacuum. The pressure difference across the bead is generated with a vacuum box that is mounted below the bead between the die and the coating roll. The

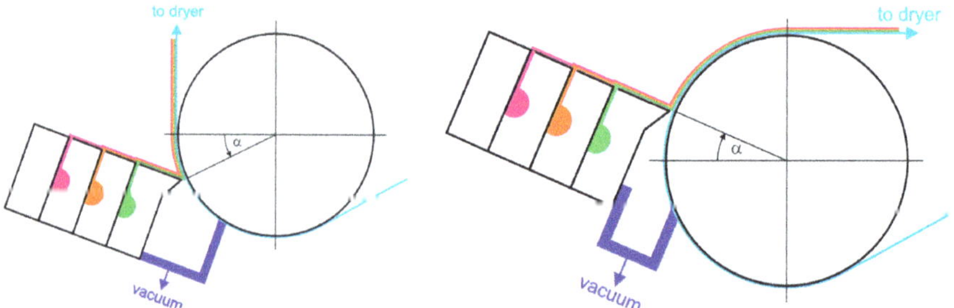

Fig. 18.1 **a** Process configuration with the web departing vertically from the coating roll. **b** Process configuration with the web departing horizontally from the coating roll

vacuum box also serves as a catch pan before and after coating, which is a convenient feature because it prevents the liquids flowing down the die slide from contaminating the coating station. The design of the box can be simple such as sketched in Fig. 18.1, or it can be sophisticated by incorporating two chambers as shown in Fig. 18.2.

The first chamber serves as a catch pan, which is connected to a drain, and which must be sealed with a siphon. The second chamber is connected to the vacuum source and it remains free of liquid. The wall between the chambers forms a narrow gap with the substrate that is wrapped around the coating roll. Consequently, the sub-ambient pressure in the first chamber, which affects the coating bead, is very uniform in the cross-web direction. Vacuum pressures are typically low, just a few centimeters of water column, i.e., <1,000 Pa.

Regarding the geometry of the die lip, many different forms are known from the photographic industry. Examples of patented lip designs from the photographic industry include Jackson from Kodak (1976) and Choinski from Polaroid (1981). Figure 18.3 shows a sketch of the geometric parameters that may be varied. α is the application angle. It determines the application point where the die slide (or the

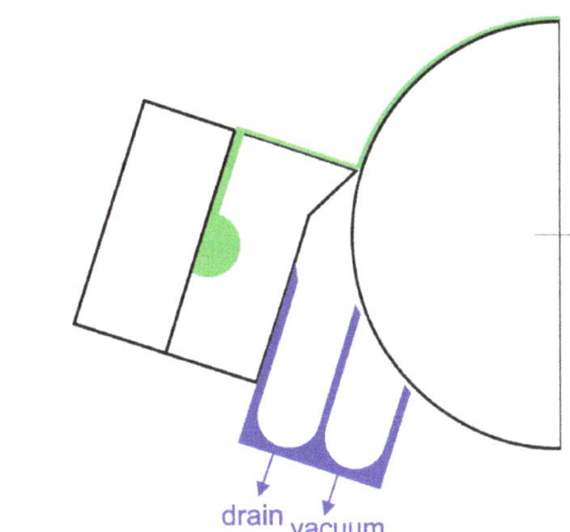

Fig. 18.2 Dual-chamber vacuum box

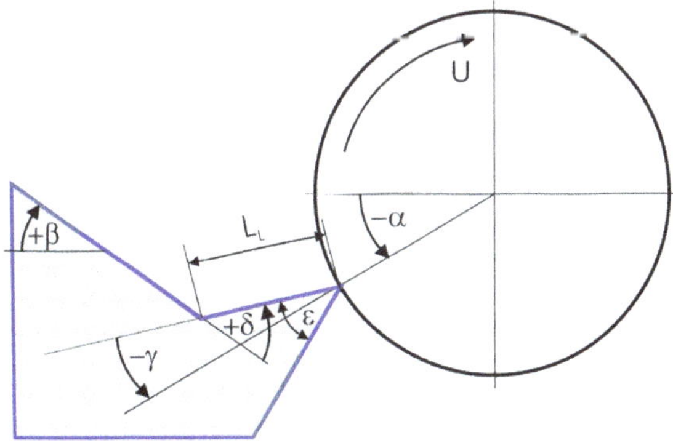

Fig. 18.3 Sketch of general lip geometry

extension of the slide in case of a modified lip geometry) meets the coating roll, and it is measured relative to the horizontal line through the center of the coating roll. α is positive if the application point is in the upper quadrant, and negative if the coating point is in the lower quadrant. β is the slide angle of the die surface. γ is the impingement angle; it is measured between the lip surface at the tip of the lip and the reference line that connects the application point and the center of the coating roll. γ is positive if the reference line lies above the lip surface. δ is the upsweep or down-sweep angle; it is measured between the lip surface at the tip of the lip and the slide surface. δ is negative if the slope of the lip surface is more negative than the slope of the die slide. ε is the lip angle as indicated in Fig. 18.3.

From a fluid mechanics point of view, the impingement angle γ is important because it determines the relevant features of the bead flow field, whereas, from an equipment point of view, the application angle α is relevant because it determines the location of the die lip on the coating roll, and it determines admissible web paths around the coating station. γ depends on α as well as the slide angle β and the lip up-sweep angle δ according to Eq. (18.2.1).

$$\gamma = \alpha - \beta + \delta \tag{18.2.1}$$

The form shown in Fig. 18.3 is called "ski jump" with the angle δ being positive. It was used for coating points in the lower quadrant in an attempt to improve the direction of the impinging liquid film. However, this design was only effective if the lip length L_L was correctly related to the total thickness of the arriving liquid film. If L_L was too short, the effect was not significant, and if L_L was too long, the impingement momentum of the approach film was reduced. Upsweep geometries that exceed horizontal are not desirable because gravity acts against the impingement momentum.

If the angle δ is negative, the lip shape is characterized by a down-sweep, which may lead to significantly negative impingement angles γ, particularly if the application angle α is not very positive. Such an undesirable situation is shown in Fig. 18.4b. For the impingement angle γ to be zero, which is the preferred value, the upsweep angle δ must satisfy the following criterion:

$$\delta = \beta - \alpha \tag{18.2.2}$$

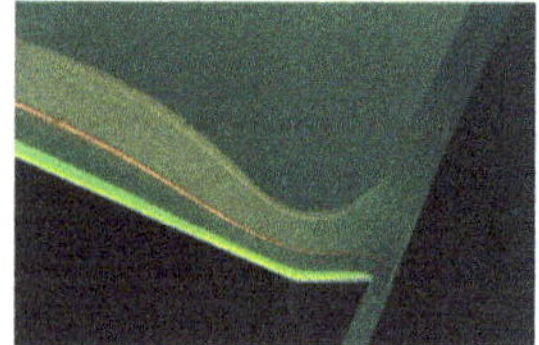

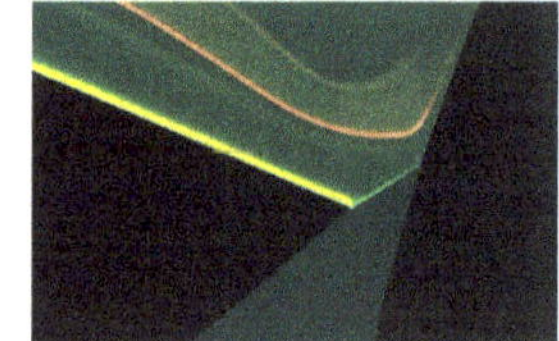

Fig. 18.4 **a** Flow field of the coating bead; $\alpha = 23°$, $\beta = 23°$, $\gamma = 23°$, $\delta = 23°$, $\varepsilon = 67°$, $L_l = 1.0$ mm. **b** Flow field of the coating bead; $\alpha = 23°$, $\beta = 23°$, $\gamma = -45°$, $\delta = --45°$, $\varepsilon = 135°$, $L_l = 1.41$ mm. **c** Flow field of the coating bead; $\alpha = 23°$, $\beta = 23°$, $\gamma = 0°$, $\delta = 0°$, $\varepsilon = 60°$, $L_l = 0$ mm

Fig. 18.5 **a** 3-layer slide die with dual-chamber vacuum box for slide coating; photo reproduced with permission from Polytype Converting AG; slide die and vacuum box manufactured by TSE Troller AG. **b** Slide coating of a single layer with a 3-layer slide die; photo reproduced with permission from Polytype Converting AG; slide die manufactured by TSE Troller AG

If the coating point is located in the preferred upper quadrant, then the best lip is the one with the simplest shape, i.e., with $\delta = 0°$ and $\varepsilon < 90°$, say 60°, see Fig. 18.4c. Based on the Gibbs inequality (5.6.82), $\varepsilon < 90°$ is necessary to promote the pinning of the static contact line to the sharp corner of the die lip, which in turn prevents the formation of vortices below the die lip, see Figs. 5.71k and 5.76.

Schweizer (1997) visualized the flow field of the coating bead as a function of the pressure difference across the bead (bead vacuum), the application angle and the geometry of the lip. The latter photos are reproduced here in Fig. 18.4 with permission from Springer Nature.

Figure 18.5a shows a 3-layer slide die with a dual-chamber vacuum box on the pilot machine of Polytype Converting, and in Fig. 18.5b one layer is being applied with that same die. The application point is located in the upper quadrant.

Figure 18.6 shows a typical multilayer slide die from the photographic industry. Number of layers = 7, die width approximately 1,600 mm.

18.3 Flow Field of the Coating Bead

The term "coating bead" refers to the quantity of liquid that bridges the gap between the tip of the die lip and the substrate surface. In contrast to slot coating, the liquid

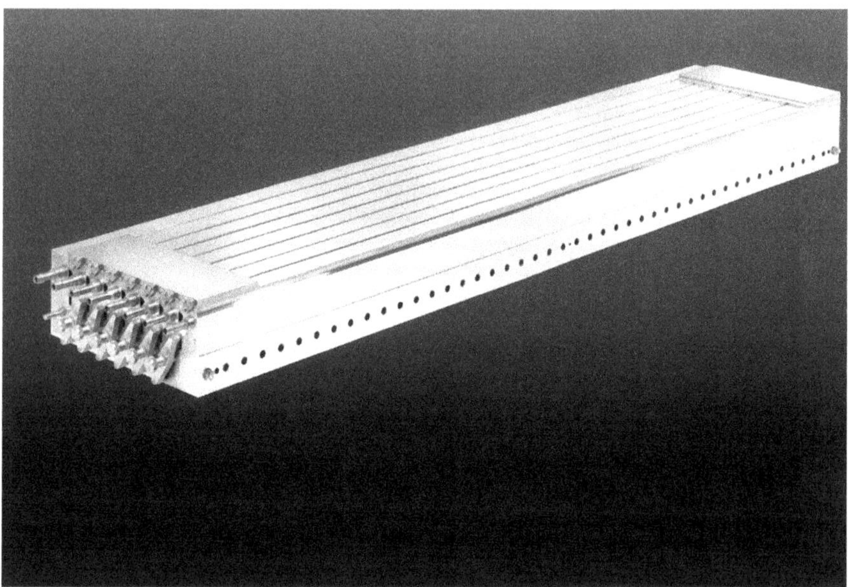

Fig. 18.6 7-layer slide die used for slide coating of photographic materials; photo reproduced with permission from TSE Troller AG

arriving at the coating point does not have to pass through a narrow gap that is formed between the die lip and the substrate surface. Instead, the liquid film simply bridges the gap between the tip of the die lip and the substrate. This is an advantage of slide coating over slot coating. In addition, the gap is not so strictly related to the wet thickness of the coated film as it is in slot coating. In the photographic industry, where several layers were coated simultaneously, the total wet film thickness was on the order of 100 μm, and typical coating gaps ranged from 50 to 150 μm.

In that sense, slide coating is more similar to curtain than to slot coating. One might say that slide coating is like curtain coating, but with a curtain length that is close to zero, and with a liquid film that does not approach the web in the vertical direction of gravity but at a relatively small slide angle, and that, consequently, has an impingement momentum that is much smaller, i.e., by a factor of about 20. All these differences are in favor of curtain coating, see also Chap. 7 on the hydrodynamic assist for more discussions about the differences between the premetered coating methods.

Figure 18.7b visualizes how a uniform liquid film on the inclined die slide approaches the coating point. Upon impinging on the substrate, the flow changes direction by more or less 90°, and it is accelerated by the drag force exerted by the moving web. A characteristic feature of this flow is the formation of standing waves in the free surface at the end of the film flow and in the adjacent film-forming meniscus of the coating bead. The flow conditions in Fig. 18.7b lead to just one wave valley at the transition from film to bead flow, and one adjacent wave mountain at

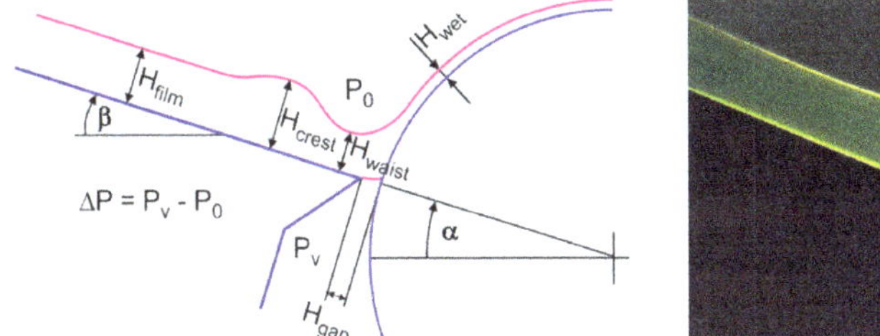

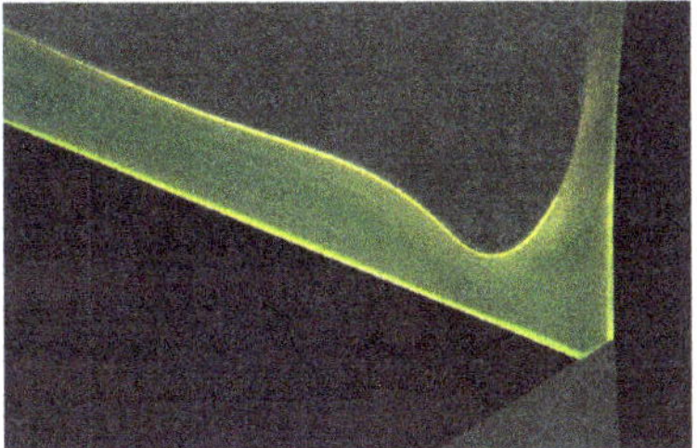

Fig. 18.7 **a** Flow field of the coating bead: sketch of relevant geometric parameters **b** Flow field of the coating bead: flow visualization. $\alpha = 0°$, $\beta = 23°$, $\gamma = -23°$, $\delta = 0°$, $\varepsilon = 53°$, $L_1 = 0$ mm, $H_{gap} = 508$ μm, $\Delta P = 0$ Pa, $\mu = 24.7$ mPas, $\sigma = 67.6$ mN/m, $\rho = 1184$ kg/m^3, $Q = 3.49$ cm^2/s, $U = 0.78$ m/s; photo taken from Schweizer (1997) and reproduced with permission from Springer Nature

the end of the film flow. The valley and mountain are measured by their respective film thickness H_{waist} and H_{crest}.

The phenomenon of standing waves forming in a liquid film flowing into a quiescent pool was first described and analyzed by Ruschak (1978). Applying these findings to slide coating, Ruschak first referred to the work of Landau and Levich (1942) and assumed that the film-forming meniscus in the coating bead is circular and that the resulting sub-atmospheric pressure in the bead can be calculated with Eq. (17.5.1) presented in Sect. 17.5. He argued that if the uniform thickness of the film flow upstream on the die slide is slightly disturbed such that the thickness decreases in flow direction, then the film velocity increases, which results in larger velocity gradients, and hence the viscous energy dissipation is larger than the energy input from gravity. Consequently, the pressure falls below atmospheric, which produces a concave film surface, i.e., the curvature of the film surface turns positive, which causes the film thickness to increase. Similarly, if the film thickness increases, the viscous energy dissipation is smaller than the energy input from gravity, and the pressure rises. Therefore, the curvature of the film surface turns negative, resulting in a convex surface, and the film thickness decreases. The combination of these two phenomena results in a series of standing waves. Since the pressure has to relax from atmospheric upstream in the film to sub-atmospheric in the bead, the sinusoidal process recurs with increasing amplitude until the curvature of the film profile at a trough (valley or waist) becomes so large that the meniscus of the film blends with the meniscus of the bead, which curves away from the die slide.

Additional standing wave forms associated with slide coating are visualized and discussed in Sect. 5.7.5. Regardless of the number of standing waves, the deepest valley is always located at the entrance to the bead. Consequently, the momentum of the impinging film is not determined by the equilibrium thickness of the film far upstream of the bead, but by H_{waist} as indicated in Fig. 18.7a. Moreover, as discussed in Sect. 18.5.1, the onset of ribbing lines is also affected by the thinning of the liquid film just upstream of the bead. Regrettably, however, the height of the associated mountain and valley cannot be calculated accurately with simple mathematical models.

Just like in slot or curtain coating, the acceleration of the bead fluid in web direction takes place inside the boundary layer, which starts at the dynamic wetting line and ends where all fluid supplied by the film flow has reached web speed. Unfortunately, however, and unlike curtain coating, slide coating cannot offer any convenient geometric reference for the position of the dynamic wetting line or the end point of the boundary layer. As explained in Sect. 5.9.6, the front surface of the curtain is assumed to coincide with the end point of the boundary layer in curtain coating. This concept cannot be transferred to slide coating as can be verified in Fig. 18.7b. Specifically, if the straight line coinciding with the free surface of the undisturbed film far upstream of the bead is extended until it meets the film-forming meniscus in the bead, then the wet thickness of the coated film is still greater than the uniform equilibrium thickness far downstream of the bead. The same argument holds if a similar line is placed on the crest of the valley bottom of the first standing wave.

The location of the dynamic wetting line cannot serve as a geometric reference either because it is unknown. Moreover, as visualized by Schweizer (1988, 1997), the wetting line can easily be moved along the substrate surface by applying a vacuum. Additional flow visualization studies revealed that the position of the wetting line also depends on the coating gap, i.e., it moves downstream as the gap is enlarged. As a result, a comprehensive model like Eq. (5.9.37) for curtain coating that describes how operating and geometric parameters as well as physical fluid properties affect the position of the dynamic wetting line cannot be formulated for slide coating.

On the other hand, Blake and Ruschak (1997) presented a model that explains the importance of bead vacuum in slide coating. Specifically, the upper, film-forming meniscus is curved such that the pressure in the bead, P_b, is sub-atmospheric, see Fig. 18.8. P_b decreases with decreasing thickness of the coated film. Therefore, obtaining equilibrium conditions requires a capability, i.e., a vacuum box, for setting the pressure P_v below the wetting meniscus to values below atmospheric and of the same magnitude as P_b.

As analyzed by Ruschak (1976), the film-forming meniscus is circular with radius R if the capillary, Weber and Bond numbers are small. To a first approximation, the meniscus radius R, and P_b, are determined by the web speed, the coating thickness, and the physical fluid properties, see Sect. 17.5 for more details. The wetting meniscus is assumed to be pinned to the tip of the die lip and it ends at the dynamic wetting line. Furthermore, it intersects the substrate surface at a dynamic contact angle ψ, which is measured across the liquid.

The dynamic contact angle indicates the side of the meniscus that is liquid. Therefore, $P_b < P_v$, if the meniscus is concave as viewed from the liquid side, and $P_b > P_v$, if the meniscus is convex. For these conditions, Blake and Ruschak derived inequalities for the largest possible coating gap as a function of the meniscus radius and the dynamic contact angle. Then, by replacing the meniscus radius with the relationship between pressure drop and meniscus curvature according to the Young–Laplace

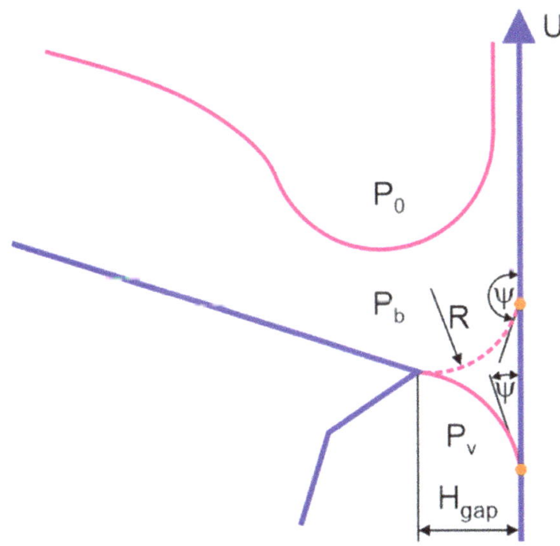

Fig. 18.8 Sketch of the coating bead

equation (Weinstein & Palmer, 1997), they arrived at an inequality for the pressure drop across the wetting meniscus as a function of the coating gap, the surface tension and the dynamic contact angle as follows:

$$-1 - \cos\psi \leq \frac{(P_b - P_v)H_{gap}}{\sigma} \leq 1 - \cos\psi \qquad (18.3.1)$$

If this relationship is violated, then a circular wetting meniscus is not possible and process failure such as bead break-up occurs. However, for a given dynamic contact angle, a range of pressure drops is admissible because the meniscus can adopt different curvatures. The application of a vacuum at the wetting meniscus allows a broader range of bead pressures and operating conditions, owing to the various geometrical configurations available to the wetting meniscus. This is important because otherwise, P_v would have to be set to one precise value. Inequality (18.3.1) further visualizes that small gaps are necessary in slide coating because the latitude for P_v to vary around P_b decreases and approaches zero as H_{gap} increases. Moreover, the possibility to adjust P_v to values less than P_b is particularly important when the dynamic contact angle approaches 180°, i.e., an operating condition that is close to air entrainment during high-speed coating. Without vacuum ($P_v = P_0$), $P_b - P_0 < 0$, but the configurations available for the wetting meniscus limit P_b to values just slightly below P_0. Indeed, no values for P_b are possible without vacuum if ψ approaches 180° and the process fails.

In summary, applying vacuum across the bead in slide coating can not only shift the onset of air entrainment (wetting failure as ψ approaches 180°) to higher web speeds but, as discussed in Sect. 18.5.1, it also shifts the onset of ribbing lines to higher speeds. The extent, to which P_v can be reduced for preventing air entrainment is limited, and the dependence of P_v on other process parameters is visualized by inequality (18.3.1), at least qualitatively. In reality, unfortunately, i.e., in an industrial slide coating process, the best value for the vacuum pressure P_v in combination with a given set of operating conditions cannot be predicted accurately, nor can the effect of slightly changing P_v on the overall process performance be visualized owing to a lack of optical access to the coating bead. Consequently, the optimum value for P_v must be determined experimentally by a trial-and-error procedure, i.e., by finding operating conditions that lead to the onset of process failures such as ribbing lines or air entrainment, and then by increasing the bead vacuum, if possible, or by reducing the coating speed until the failure disappears, all while considering a desired process safety margin.

18.4 Operating Window

The operating window of slide coating is affected by at least 9 independent process parameters, see Table 18.1 and Fig. 18.7a. More realistically, the viscosity should be replaced by at least four of the fife Carreau-Yasuda parameters (Eq. 4.4.1), and the surface tension should be replaced by the dynamic surface tension measured at the surface age of the film flow at the end of the die slide.

The slide angle expresses the importance of the gravitational acceleration g, and the application angle defines the location of the die lip on the coating roll. This large number of parameters makes it difficult to condense the relevant process information into a simple form, even if dimensionless numbers are used. Therefore, we chose the plane of Q versus U to depict the operating window. In this coordinate system, straight lines through the origin mark points of constant wet film thickness H_{wet}, whereby the thickness decreases with an increasing slope of the line, see Fig. 18.9. Note that the coordinate axes Q and U are switched relative to the generic operating window proposed in Chap. 16, the reason being that the onset of ribbing lines is the most important operating boundary of slide coating, and this representation allows one to readily read the maximum possible web speed at the inception of ribbing lines as a function of other relevant process parameters.

Besides ribbing lines, air entrainment and various vortices constitute the major boundaries of the operating window. At a given flow rate/width various vortices may be present in the film flow on the die slide or in the coating bead, if the web speed is too low, as visualized by Chen (1992), or in Sect. 5.7.8 and Fig. 5.71. When the web speed is increased these vortices disappear and good coating quality is possible.

Table 18.1 Relevant operating parameters of slide coating

Density ρ	Slide angle β	Bead vacuum ΔP
Viscosity μ	Application angle α	Web speed U
Surface tension σ	Coating gap H_{gap}	Volumetric flow rate/width Q

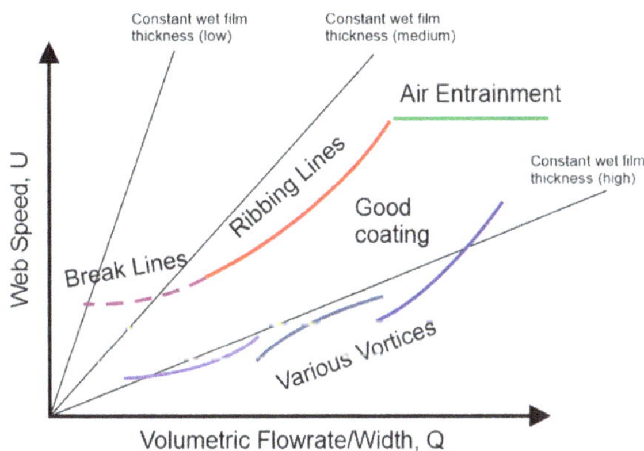

Fig. 18.9 Generic operating window of slide coating

Upon further increasing the web speed, however, ribbing lines appear, which degrade the uniformity of the coated film. If the flow rate/width is sufficiently low, then the liquid film will be thin enough such that the thin portions of the ribbed film will rupture, thus leading to a uniform pattern of dry and wet stripes. This defect is called "break lines", see Fig. 17.24. On the other hand, if the flow rate/width is sufficiently high, air entrainment will occur before the onset of ribbing lines. The web speed at the onset of air entrainment cannot be increased, even if the flow rate/width is further increased.

Like with other coating processes, it is difficult to predict and quantify the boundaries of the operating window with simple theoretical models (Hens & van Abbenyen, 1997). This can be explained, at least in part, by the lack of a suitable geometrical reference frame for the fluid flow in the coating bead. Therefore, the onset of ribbing lines, i.e., the main operating boundary of slide coating, is shown below by way of experimental data. These data were first published by Schweizer and Rossier (2001, 2003).

Figure 18.10 shows an example of such a coating window. Specifically, it visualizes the onset of ribbing lines and the onset of vortices 1 and 2 (see Fig. 5.70) for the same fluid, i.e., an aqueous glycerol solution. Unfortunately, however, the vortex data was taken for a slide angle β of 15°, whereas the ribbing line data are related to a slide angle of 23°. Vortex and ribbing line data for the same slide angle are not known. Nevertheless, Fig. 18.10 provides an idea of how a realistic operating window for slide coating may look like. In particular, vortices 1 and 2 are present, if the operating point lies below the respective line in the diagram. Ribbing lines are present if the operating point is located above the inception line. Consequently, a bead flow field that is free of vortices and ribbing lines is rather small for the chosen operating conditions, i.e., it is located above the inception curve for vortex 1 and below the onset of ribbing. Moreover, the admissible wet film thicknesses are relatively high, i.e., above about 150 μm. Increasing the viscosity would lower the inception speed for vortex 1, which is desirable, but it would also lower the speed at the onset of ribbing lines, thereby offsetting the gain obtained with reference to

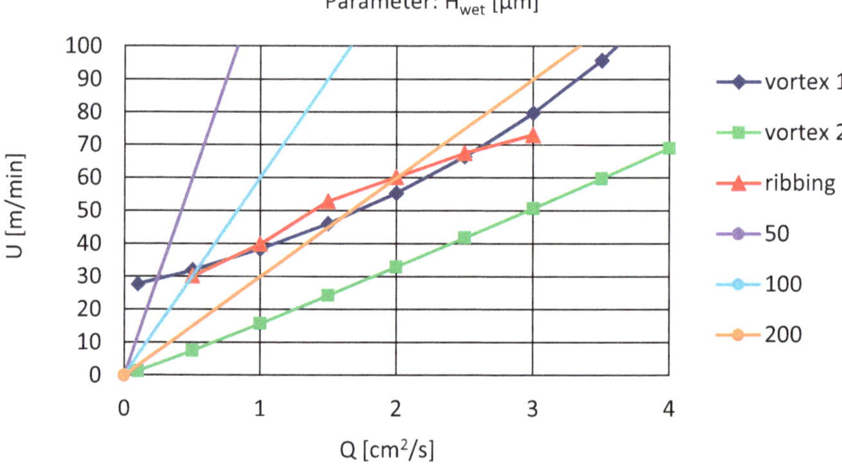

Fig. 18.10 Operating window of slide coating; vortex data for $\beta = 15°$, ribbing line data for $\beta = 23°$; $\rho = 1146\ \mathrm{kg/m^3}$, $\mu = 8.8$ mPas, $\sigma = 69.2$ mN/m, $\Delta P = 500$ Pa, $H_{gap} = 200\ \mu$m, $\alpha = 23°$

vortex 1. Increasing the pressure difference across the bead, i.e., lowering the pressure in the vacuum box, might postpone the onset of ribbing to higher speeds, but experimental data to confirm this idea do not exist.

18.5 Coating Defects

18.5.1 Ribbing Lines

Ribbing lines are the most severe limitation of slide coating, particularly with respect to the minimum film thickness and the maximum web speed. The mechanisms leading to the onset of ribbing (and break) lines in slide coating are the same as described for other coating processes (Chen 1992 and Sect. 17.8.1). In simple terms, ribbing lines occur if disturbing pressure forces in the coating bead can no longer be balanced by restoring surface tension forces. Evaluation of ribbing line data in combination with flow visualization has shown that such conditions are potentially present in the flow fields of rapidly thinning films. In slide coating, a pattern of standing waves including a thinning film is found at the end of the die slide, just upstream of the bead as visualized in Fig. 18.7b. The amount of local film thinning depends on the geometry of the flow field, the physical fluid properties, and the operating conditions. In particular, film thinning becomes excessive if the web speed is above a critical value, thus leading to the onset of the ribbing instability.

The onset of ribbing lines in slide coating was extensively investigated experimentally by Schweizer and Rossier (2001, 2003). They used a 260 mm wide two-layer

Table 18.2 Physical properties of aqueous glycerol solutions

Fluid	Concentration	Density	Viscosity	Surface tension	Comment
[#]	[–]	[kg/m^3]	[mPas]	[mN/m]	
1	0.785	1203	52.0	66.8	
2	0.690	1178	20.5	67.9	
3	0.568	1146	8.8	69.2	
4	0.318	1080	2.6	71.8	
5	0.330	1083	2.8	71.7	Carrier-layer fluid
6	0.763	1197	41.7	66.6	Top layer for carrier layer experiments
7	0.680	1175	18.6	67.0	Base fluid for surface tension experiments
8	0.685	1176	20.5	32.5	With addition of 0.3% Aerosol OT-75

slide coating table-top apparatus provided by TSE Troller AG. The slide angle β was fixed at 23°, the roller diameter was 350 mm, and the lip angle ε was 90°. As no substrate was involved in the coating, the roll surface was cleaned with a Teflon and a rubber squeegee. Single and two-layer experiments were conducted at room temperature using various concentrations of aqueous glycerol solutions as Newtonian model fluids. The viscosity was changed by altering the concentration of the glycerol solution, and the resulting physical fluid properties are listed in Table 18.2.

To determine the onset of ribbing lines in slide coating, a specific geometric process configuration was prepared and the volumetric flow rate was set at a constant value. For a given fluid, the web speed was then slowly increased until the defect became visible. The ribbing inception point was found when the reflection of a straight line object in the coated film just downstream of the bead, e.g., a fluorescent light tube, became wavy. The above procedure was carried out for the following sets of independent process parameters:

Volumetric flow rate/width Q: 0.5, 1.0, 1.5, 2.0, 2.5, 3.0 cm^2/s
Application angle α: −23°, 0°, 23°, 37°, 40°
Impingement angle γ: −46°, −23°, 0°, 14°, 17°
Coating gap H_{gap}: 100, 200, 300, 500 μm
Bead vacuum ΔP: 0, −100, −250, −500 Pa

Results are presented in terms of the maximum attainable web speed, U_{max}, at the onset of ribbing lines. Consequently, ribbing lines are present if the operating point lies above the respective inception curve. In the Q-U coordinate system, this information is also indicative of the minimum attainable wet film thickness, $H_{wet,min}$. Figure 18.11 shows the experimental rig with ribbing and break lines on the coating roll.

Fig. 18.11 **a** Example of ribbing lines in slide coating; photo reproduced with permission from TSE Troller AG. **b** Example of break lines in slide coating; photo reproduced with permission from TSE Troller AG

18.5.1.1 Effect of Flow Rate

As visualized in Fig. 18.12, U and $H_{wet,min}$ increase with increasing Q for all investigated values of viscosity, surface tension, coating gap, bead vacuum, and application angle. The sensitivity of U upon Q changes only slightly with changing the coating gap or application angle, but it decreases with increasing viscosity and decreasing vacuum.

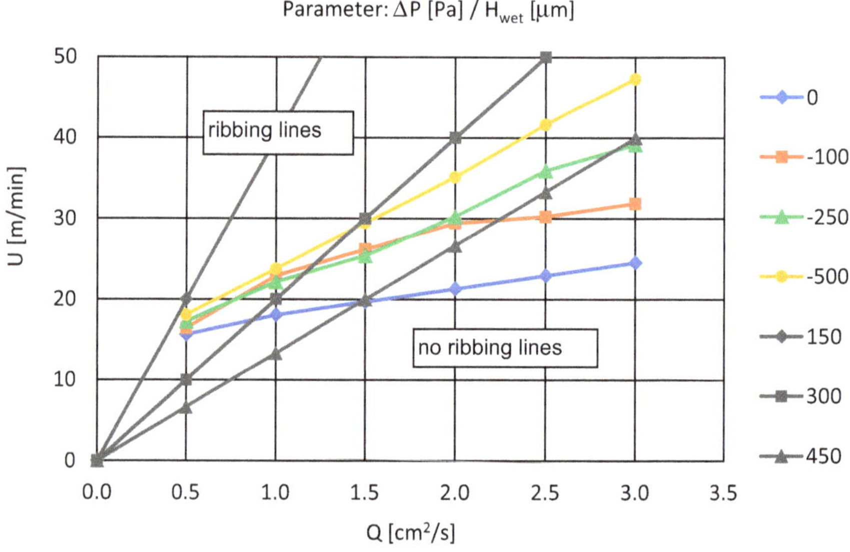

Fig. 18.12 Onset of ribbing lines; effect of volumetric flow rate/width and bead vacuum on maximum web speed and minimum wet film thickness; μ = 52.0 mPas, σ = 66.8 mN/m, α = 0°, H_{gap} = 200 μm

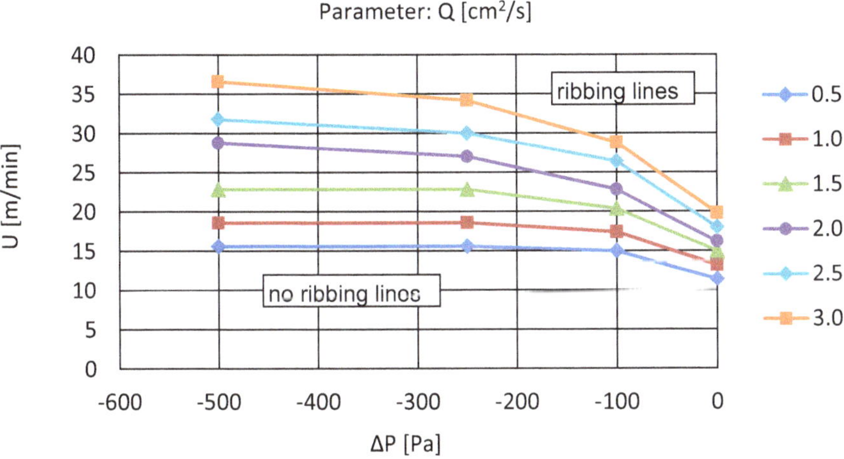

Fig. 18.13 Onset of ribbing lines; effect of bead vacuum and volumetric flow rate/width on maximum web speed; $\mu = 52.0$ mPas, $\sigma = 66.8$ mN/m, $\alpha = -23°$, $H_{gap} = 200$ µm

18.5.1.2 Effect of Bead Vacuum

U increases with increasing ΔP. This effect is stronger at high Q, α and μ values than at low values for these parameters. Above a certain vacuum level, i.e., –300 to –500 Pa, U is no longer affected by ΔP. For low μ and high H_{gap} values, high vacuum levels (e.g., < –300 Pa) lead to an unwanted effect called "pull-through", where some of the liquid is pulled into the vacuum box instead of being coated onto the web. $H_{wet,min}$ decreases with increasing ΔP if Q is high (about >2 cm^2/s); however, for smaller Q values, $H_{wet,min}$ becomes independent of ΔP. An example of this behavior is shown in Fig. 18.13.

18.5.1.3 Effect of Coating Gap

U and $H_{wet,min}$ do not depend on H_{gap} for higher μ-values (>50 mPas), but U increases with decreasing H_{gap} for lower μ-values (<10 mPas). This effect becomes stronger with decreasing Q (Fig. 18.14).

18.5.1.4 Effect of Application Angle

Except for low μ (<5 mPas), there is an optimum impingement angle in the range of 0 to slightly negative values (application angle = 23° or slightly less) leading to a maximum value for U, see Fig. 18.15. In contrast, strongly positive and strongly negative values for α or γ result in lower web speeds at the onset of ribbing lines. This pronounced optimum value for α or γ is lost for low viscosities (<5 mPas), where it

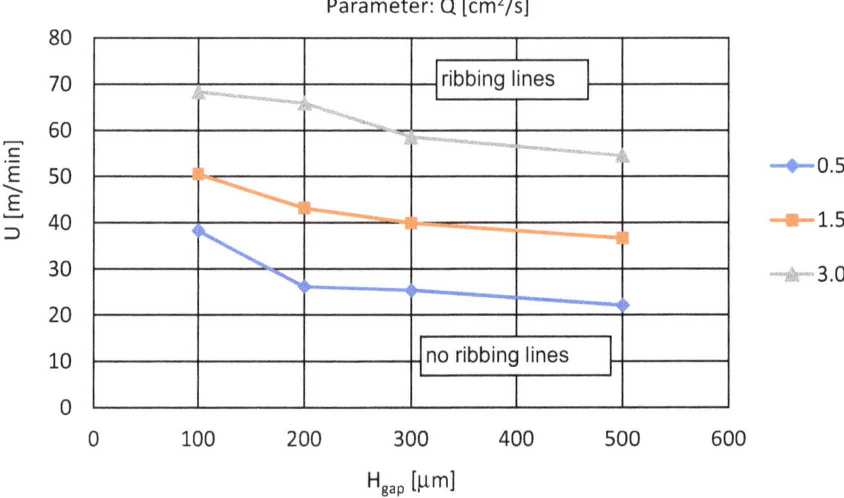

Fig. 18.14 Onset of ribbing lines; effect of coating gap and volumetric flow rate/width on maximum web speed; $\mu = 8.8$ mPas, $\sigma = 69.2$ mN/m, $\alpha = -23°$, $\Delta P = 0$ Pa

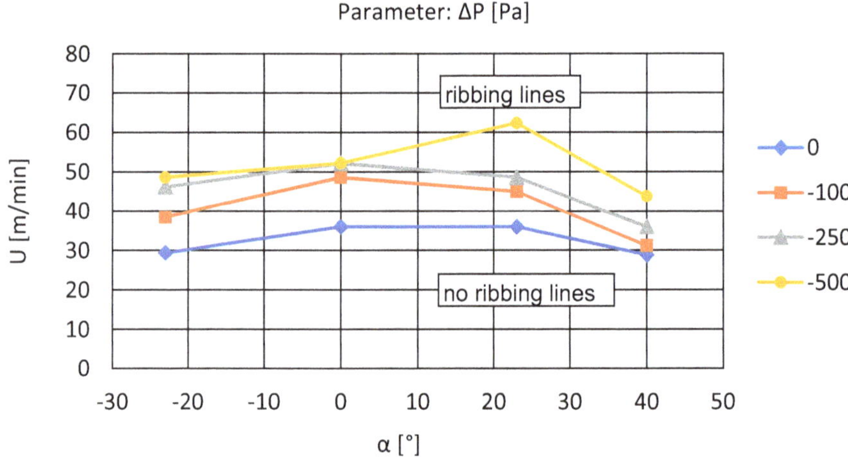

Fig. 18.15 Onset of ribbing lines; effect of application angle and bead vacuum on maximum web speed; $\mu = 20.5$ mPas, $\sigma = 67.9$ mN/m, $H_{gap} = 200$ µm, $Q = 3.0$ cm^2/s

is shifted to more negative angle values. The dependence of U upon α or γ increases with increasing Q if μ is low to moderate, but becomes independent of Q for high μ values (>50 mPas).

18.5.1.5 Effect of Viscosity

Figures 18.16 and 18.17 visualize that U decreases with increasing μ for all values of σ, Q, ΔP, H_{gap}, and α or γ. The sensitivity of U upon μ is large for μ values below about 10 mPas, it decreases a bit for viscosities of up to about 20 mPas, and the dependence of U upon μ becomes weak for μ > 20 mPas.

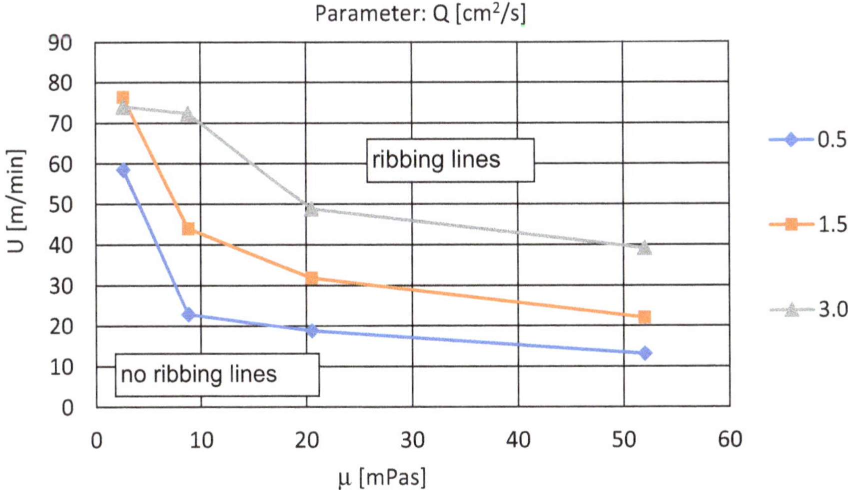

Fig. 18.16 Onset of ribbing lines; effect of viscosity and volumetric flow rate/width on maximum web speed; $\sigma_{average} = 68.9$ mN/m, $H_{gap} = 200$ μm, α = 23°, ΔP = –250 Pa

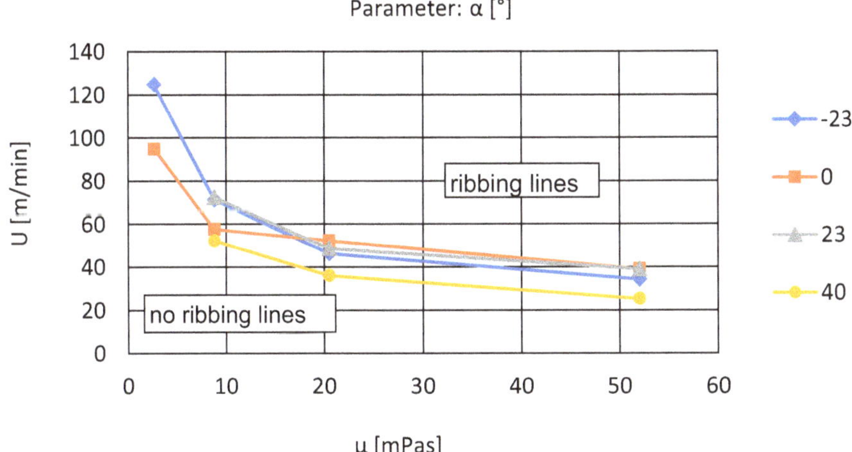

Fig. 18.17 Onset of ribbing lines; effect of viscosity and application angle on maximum web speed; $\sigma_{average} = 68.9$ mN/m, $H_{gap} = 200$ μm, Q = 3.0 cm^2/s, ΔP = –100 Pa

18.5.1.6 Effect of Flow Rate, Viscosity and Bead Vacuum

Figure 18.18 summarizes the combined effects of flow rate, viscosity and bead vacuum on the inception speed of ribbing lines. Accordingly, increasing the flow rate, decreasing the viscosity, and applying a small amount of bead vacuum all result in higher web speeds before the ribbing lines set in. In particular, increasing the volumetric flow rate/width from 3.0 to 5.0 cm^2/s just about doubles the web speed window. On the other hand, the minimum wet film thickness is not affected by higher flow rates. Depending on viscosity, it remains high, i.e., >100 μm, and thus unattractive for many industrial applications.

In contrast, Fig. 18.19 shows the situation for a glycerol solution having a low viscosity of 3.1 mPas. For low viscosity fluids, bead vacuum is no longer effective in postponing the onset of ribbing lines to higher web speeds. At lower flow rates (up to about 2.5 cm^2/s), slide coating is free of ribbing lines up to a critical speed, which is the same behavior as for more viscous fluids. At a flow rate of about 3.0 cm^2/s, some sort of bifurcation seems to take place. Specifically, the flow field is characterized by ribbing lines that moved laterally across the coating bead if the web speed was low. Upon increasing the web speed, an operating window without ribbing lines was detected, the size of which increased with an increasing vacuum.

Above a second critical web speed, ribbing lines were present again. For flow rates >3.5 cm^2/s, the flow field was again free of ribbing lines for all low web speeds and irrespective of the vacuum level. Between a lower and an upper critical web speed, an island of ribbing lines was detected, and above the second critical speed, another window without ribbing lines was seen. Upon further increasing the web speed, air was entrained between the substrate and the coated film. The speed of air entrainment strongly depended on the applied vacuum. Note that in the operating

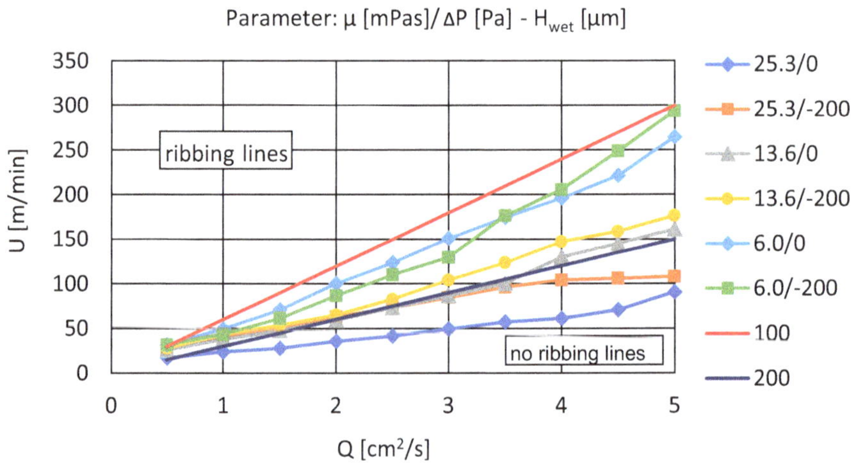

Fig. 18.18 Onset of ribbing lines as a function of flow rate, viscosity and bead vacuum; H_{gap} = 150 μm, α = 23°

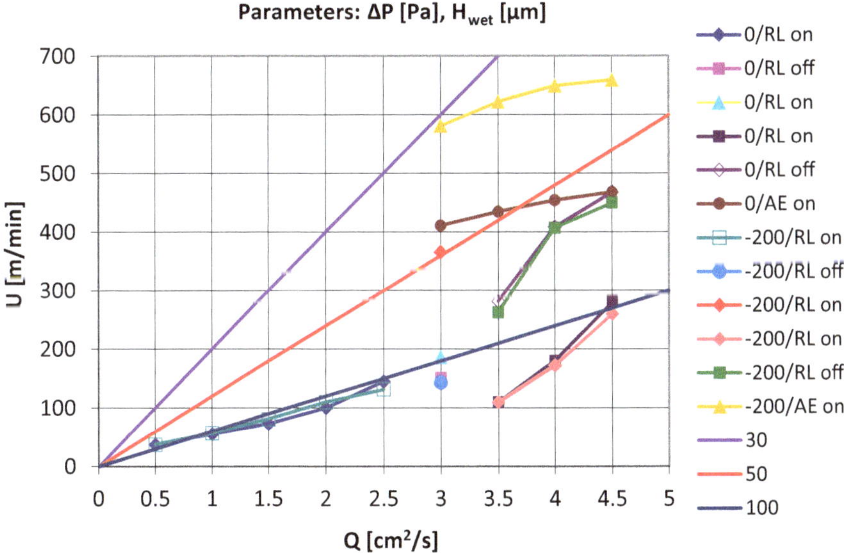

Fig. 18.19 Onset of ribbing lines for low viscosity fluids; $\mu = 3.1$ mPas, $\rho = 1088$ kg/m^3, $\sigma =$ 71.5 mN/m, $H_{gap} = 150$ μm, $\alpha = 23°$

window between ribbing lines and air entrainment, web speeds of up to 11 m/s (660 m/min) and wet film thicknesses as low as 30 μm could be realized.

18.5.1.7 Effect of Surface Tension

In our experiments, the surface tension was lowered by adding surfactant. This process is diffusion-controlled and depends on the features of the local flow field, i.e., expansion and contraction of the free surface where ribbing lines are generated. To estimate the real surface tension in the upper meniscus of the bead, the dynamic surface tension was measured with the maximum bubble pressure method, see Fig. 18.20. The surface age was calculated by using the film flow equation for the inclined die plane between the slot exit and the tip of the lip, see Sect. 5.7.6 and Figs. 18.1 and 18.21. Accordingly, a high flow rate value results in a shorter surface age, thus yielding a higher surface tension compared to low flow rate values. In Fig. 18.22, the indicated surface tension values (32.5 and 67.0 mN/m) are averaged over the flow rate range.

The inception speed for ribbing lines, U, decreases considerably with decreasing surface tension. Moreover, the optimum α or γ value, which is pronounced at high surface tension, practically vanishes at low surface tension, which is true for all levels of bead vacuum.

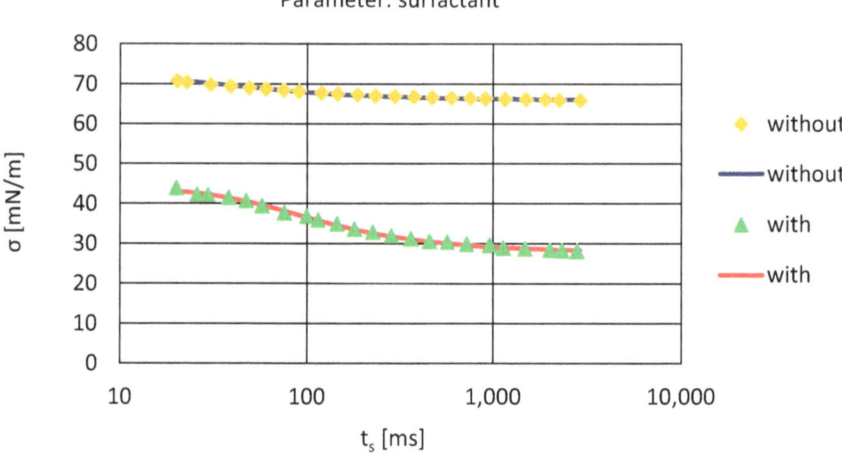

Fig. 18.20 Surface tension of aqueous glycerol solutions; **a** without surfactant; **b** with 0.3% surfactant Aerosol OT-75

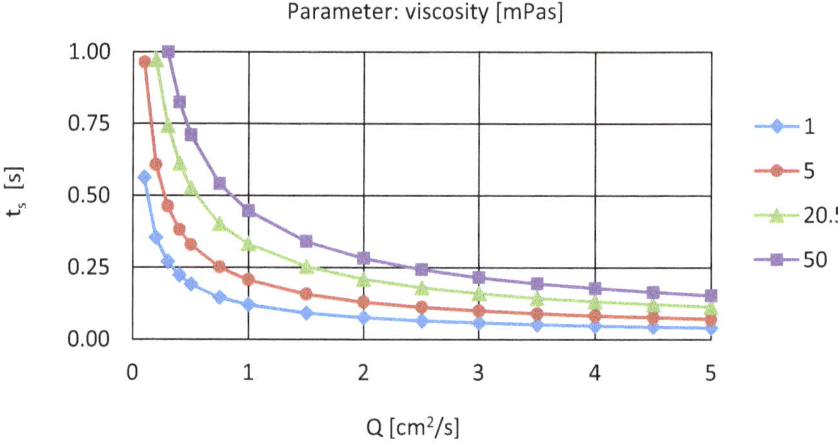

Fig. 18.21 Film flow on inclined plane: age of free liquid surface t_s as a function of volumetric flow rate/width and viscosity; $\beta = 23°$, $\rho = 1175$ kg/m^3, $L_{slide} = 45$ mm

18.5.1.8 Effect of Carrier Layer

Following Figs. 18.16 and 18.17, high web speeds in slide coating can only be achieved for low viscosities. However, by implementing the concept of the "carrier layer" (Dittman & Rozzi, 1977), see also Chap. 12, where one or several moderately viscous layers of arbitrary thickness are combined with a thin but low-viscosity bottom layer, it becomes possible to also coat reasonably viscous fluids at high speeds without incurring ribbing lines. The data in Fig. 18.23 show that a speed

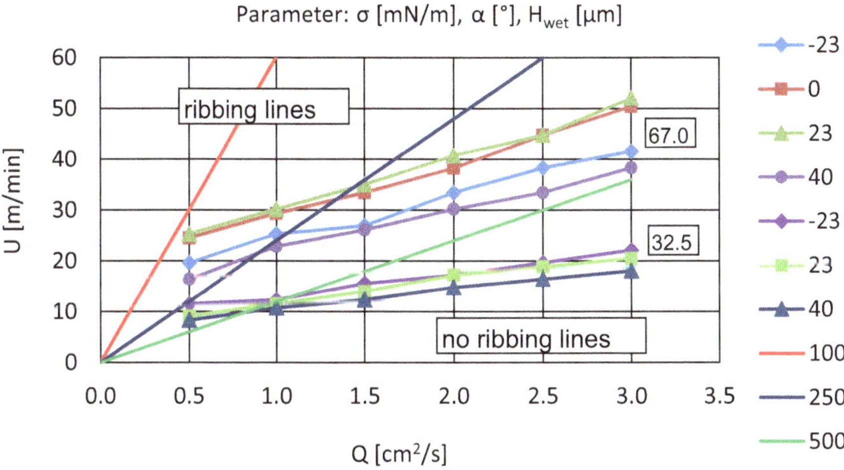

Fig. 18.22 Onset of ribbing lines; effect of volumetric flow rate/width, application angle and surface tension on maximum web speed; $\mu = 20.5$ mPas, $H_{gap} = 200$ µm, $\Delta P = 0$ Pa

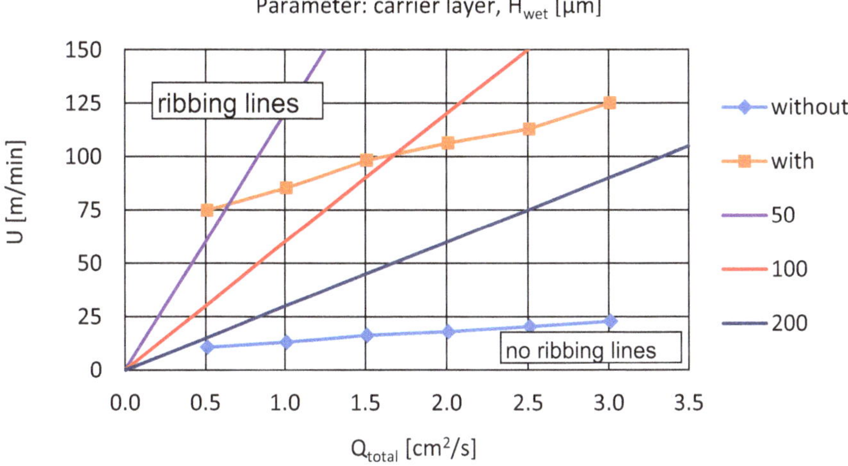

Fig. 18.23 Onset of ribbing lines; effect of carrier layer on maximum web speed; $H_{gap} = 200$ µm, $\Delta P = 0$ Pa, $\alpha = 23°$, $\mu = 2.8/41.7$ mPas, $\sigma = 66.6$ mN/m (with carrier), $\mu = 52.0$ mPas, $\sigma = 66.8$ mN/m (without carrier)

increase by a factor of >5 is possible for a top layer viscosity of >40 mPas by using a carrier layer having a flow rate/width of 0.21 cm^2/s and a viscosity of 2.8 mPas. These data, which were taken for aqueous glycerol solutions, also show that the features of the bead flow field are mainly determined by the (local) viscosity of the bottom layer.

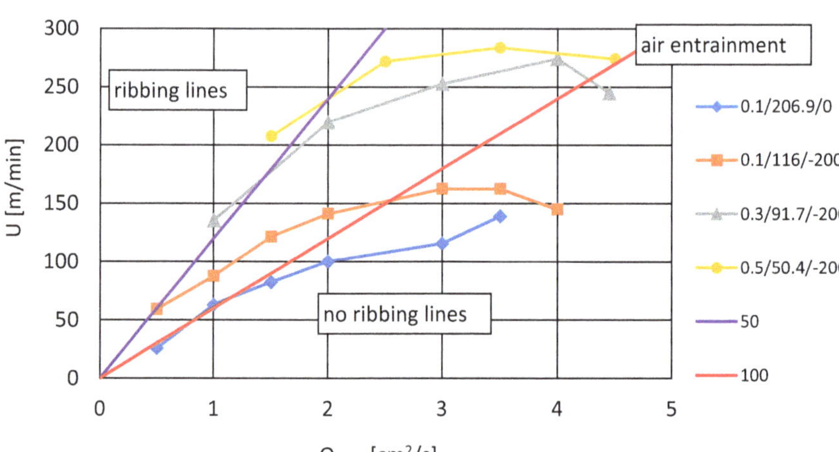

Fig. 18.24 Onset of ribbing lines and air entrainment; effect of carrier layer on maximum web speed and minimum wet film thickness; $H_{gap} = 150\ \mu m$, $\alpha = 23°$, $\mu_{bottom} = 3.1$ mPas

More ribbing line and air entrainment inception data for glycerin solutions are shown in Fig. 18.24. In particular, the carrier layer has a viscosity of 3.1 mPas. Attractive web speeds or thin layers of <50 μm can be obtained by increasing the flow rate of the carrier layer, by increasing the bead vacuum, and by limiting the viscosity of the top layer to approximately <100 mPas. If the carrier layer flow rate is high enough, then process failure is not caused by ribbing lines but by air entrainment, see the yellow curve with round symbols in the figure.

Finally, Fig. 18.25 shows ribbing line data for aqueous polyvinyl alcohol (PVA) solutions. As depicted in Fig. 4.11, PVA fluids are shear-thinning in principle. However, the solids concentrations of the carrier and top layers for all experiments were 0.016 and 0.093, respectively. Moreover, the shear rates of the two-layer film flow at the end of the die slide, where ribbing lines are formed, were considered lower than the respective critical shear rates according to the Carreau-Yasuda model. Therefore, the flow behavior was considered to be Newtonian, and all viscosity values correspond to the constant low shear viscosity. Specifically, the carrier layer had a viscosity of 3.0 mPas, and the value for the top layer was 255 mPas. The densities were 1002 kg/m^3 and 1012 kg/m^3, respectively, and the top layer had a surface tension of 45 mN/m.

Compared to the glycerin data, the ribbing line inception speed for PVA depends much less on the flow rate, particularly if $Q_{total} > 2$ cm^2/s. Otherwise, the two fluids behave similarly regarding the onset of ribbing lines, namely: the inception speed increases with increasing flow rate of the carrier layer and with increasing bead vacuum. The lower maximum speeds and the higher minimum film thicknesses

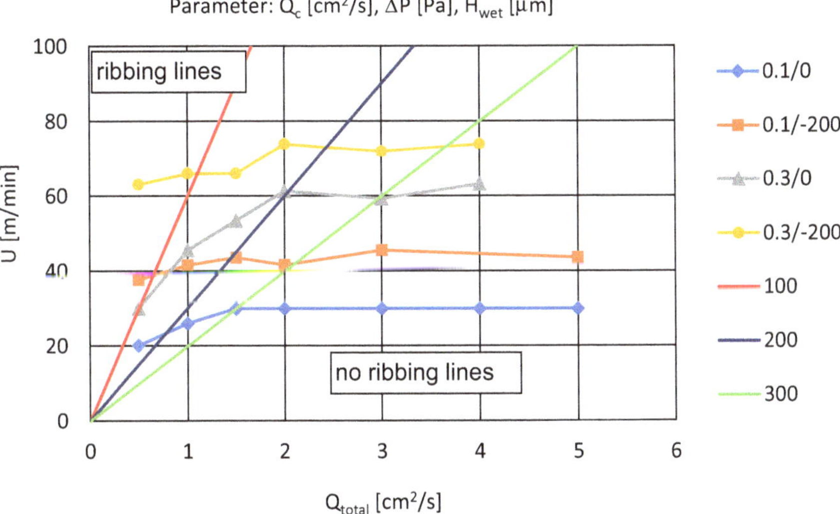

Fig. 18.25 Onset of ribbing lines for PVA solutions; effect of carrier layer on maximum web speed and minimum wet film thickness; $H_{gap} = 150\ \mu m$, $\alpha = 23°$, $\mu_{carrier} = 3.0$ mPas

obtained for the PVA experiments can be explained by the higher viscosity of the top layer when compared to the glycerin experiments.

The experimental data presented in this chapter confirm that the carrier layer concept is an interesting compromise for postponing the onset of ribbing lines to higher web speeds without incurring excessive dryer loads due to fluid dilution when compared to applying a single layer only.

18.5.1.9 Dimensionless Representation of Ribbing Line Data

For flow fields that are characterized by film splitting, which is typical for forward roll coating operations, all the data depicting the inception of ribbing lines can be reduced to a master curve involving the capillary number as a function of the ratio of nip gap to roll radius (Coyle, 1997). For slide coating, a similar master curve involving the capillary number does not exist, one reason being the lack of a suitable geometrical reference point.

However, using a trial-and-error approach in combination with flow visualization of the coating bead, we were able to condense all ribbing line data onto a master curve involving the Weber number and either the ratio of wet film thickness on the substrate H_{wet} to bead waist H_{waist}, or the ratio of bead waist to film thickness on the die slide H_{film}, see Figs. 18.7a and 18.26. The Weber number compares inertia to surface tension forces, and for slide coating, it is defined according to Eq. (18.5.1).

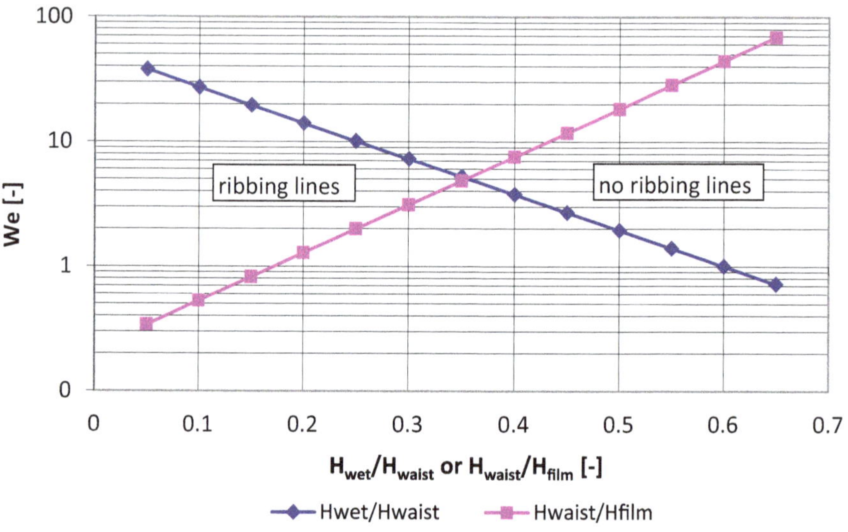

Fig. 18.26 Dimensionless master curve for the onset of ribbing lines in slide coating

$$We = \frac{\rho Q U}{\sigma} = \frac{\rho H_{wet} U^2}{\sigma} \tag{18.5.1}$$

In slide coating, the liquid film is thinned in two steps from the flow on the die slide to the flow on the substrate. H_{waist}/H_{film} is a measure for the film thinning on the slide, and H_{wet}/H_{waist} is a measure for the thinning in the bead. As shown in Fig. 18.26, ribbing lines can be avoided if the operating point is on the right side of the respective curve. Therefore, high Weber numbers (high web speeds) require high values of H_{waist}/H_{film} or low values of H_{wet}/H_{waist}. In other words, small film thinning on the slide seems to favor high web speeds, and such conditions seem to be given for Weber numbers larger than 5 or so.

In Fig. 18.27, bead flow fields for a low and a high Weber number are compared. Both photos are taken from Schweizer (1988) and reproduced with permission from Cambridge University Press. The high Weber number flow field (Fig. 18.27b) indeed shows little film thinning on the slide, and it is characterized by several standing waves, whose amplitudes decrease in the upstream direction.

From a practical point of view, the dimensionless master curve in Fig. 18.26 is of little value because the height of the bead waist is not known and can only be quantified with the help of flow visualization/image analysis or two-dimensional numerical computations, both of which are difficult and time-consuming. As a compromise, an alternative dimensionless representation of the ribbing line data can be plotted, this time involving the Weber number and the ratio of the wet thickness of the coated film to the average film thickness on the die slide, see Fig. 18.28. This film thickness ratio is now accessible (see Eqs. 2.1 and 5.7.2) and can be calculated according to Eq. (18.5.2):

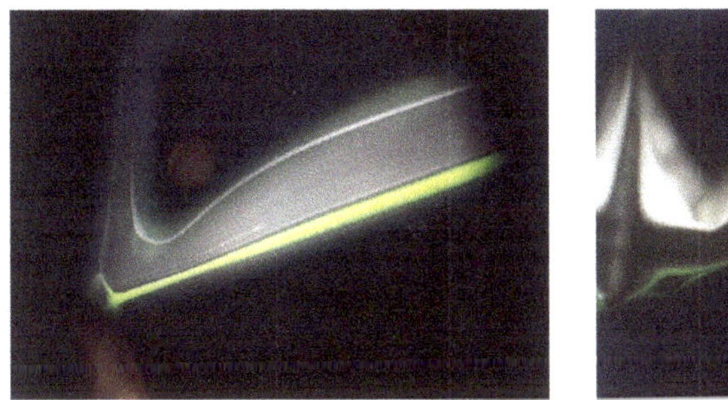
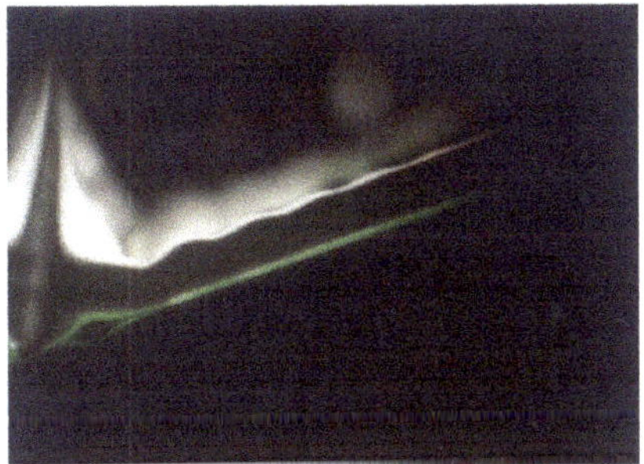

Fig. 18.27 **a** Bead flow field characterized by a low Weber number; We = 2.0, Re = 6.36, Ca = 0.32, $H_{waist}/H_{film} = 0.49$. **b** Bead flow field characterized by a high Weber number; We = 31.9, Re = 162.6, Ca = 0.20, $H_{waist}/H_{film} = 0.79$

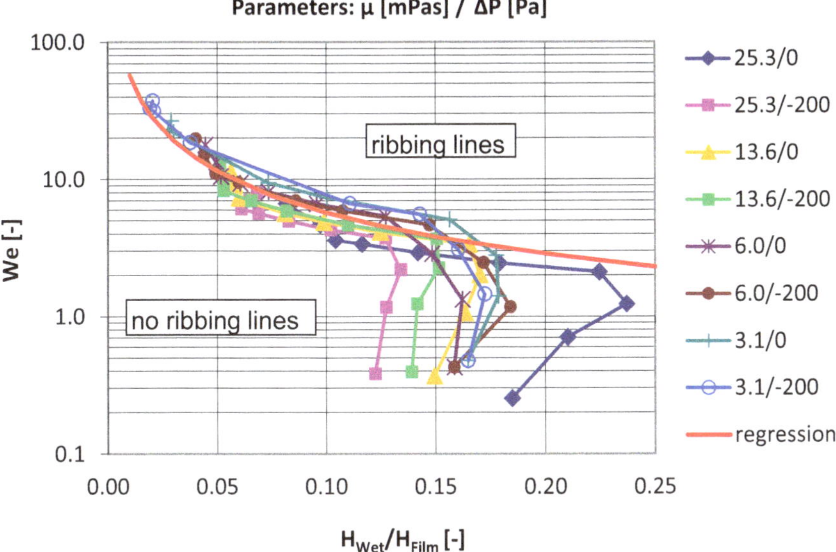

Fig. 18.28 Alternative dimensionless master curve for the onset of ribbing lines in slide coating

$$\frac{H_{wet}}{H_{film}} = \left(\frac{\rho g \sin\beta Q^2}{3\mu U^3}\right)^{1/3} \tag{18.5.2}$$

As can be seen in Fig. 18.28, ribbing lines are present, if a desired operating point in terms of We and H_{wet}/H_{film} lies above or to the right of the respective curve. Moreover, the data no longer collapse onto a single master curve if the Weber number is less than about 5. In this low We-regime, the film thinning on the slide is larger than the film thinning in the bead, the onset of ribbing lines is strongly affected by the value of the viscosity, and increasing the surface tension tends to stabilize the flow. For the high-inertia regime ($We > \approx 5$), however, all data collapse onto a single curve, the viscosity is no longer discriminating much, and decreasing the surface tension tends to promote the onset of ribbing lines.

The empirical regression equation for the master curve (red line in Fig. 18.28) is given by:

$$We = a_0 \left(\frac{H_{wet}}{H_{film}} \right)^{a_1} \tag{18.5.3}$$

$a_0 = 0.5754$, $a_1 = -0.9983$.

18.5.1.10 Summary

The onset of ribbing lines is the most severe operating boundary of the slide coating process. This process limitation is visualized and quantified by way of experimental data. In line with the range of parameters covered in this study, and as a rule of thumb, it can be concluded that slide coating is a suitable process for web speeds of less than about 100–150 m/min, and wet film thicknesses of more than about 50 μm. These limits apply to typical operating conditions and agree with values observed in industrial applications.

The inception speed of ribbing lines increases with.

- increasing flow rate/width Q,
- decreasing viscosity μ,
- increasing bead vacuum ΔP,
- increasing surface tension σ,
- optimized application angle α or impingement angle γ.

In particular, good results are obtained if the impingement angle assumes values near 0°, i.e., if the liquid film on the die slide impinges perpendicularly onto the substrate. For simple lip geometries without up- or down-sweep, this implies that the impingement angle and the application angle must be equal. This in turn requires the application point to be located in the upper quadrant of the coating roll, and thus the coated web cannot depart vertically from the coating roll toward the dryer.

Moreover, higher ribbing line inception speeds can be obtained for rather high flow rates/width in the range of $Q = 3$–$5\ cm^2/s$, and rather low viscosities in the range of $\mu = 2$–10 mPas. A high-speed operating range for slide coating was detected, which is characterized by high inertia forces ($We > 5$), the absence of ribbing lines, a maximum web speed of 650 m/min (determined by the onset of air entrainment), and a minimum wet film thickness of 30 μm. These latter two values are much more attractive than the general values reported above. However, operating points in this high-speed regime require a low low-shear viscosity, which in turn requires a low solids concentration, and a high volumetric flow rate/width. In combination with a high web speed, such conditions result in generally poor uniformity of the coated film and a rather long dryer. Therefore, such high-speed operating conditions are not really of industrial interest after all.

A compromise solution allowing to achieve the benefits of high web speeds and thin wet films without the need for an excessive dryer capacity is offered by the concept of the **carrier layer**. The use of a carrier layer is a temporary means for improving the performance of the coating process. Typically, the carrier layer is a thin layer (as thin as possible, with Q being in the range of 0.1–0.5 cm^2/s) of low viscosity and hence of low solids concentration (as low as needed with μ being on the order of 5 mPas), which is combined with at least one thicker layer of higher viscosity and hence higher solids concentration. Ideally, the carrier layer and its adjacent layer are of the same chemical composition. Thus, the use of a carrier layer requires a multilayer coating process, even if the final product consists of only one layer. If the viscosity of the layer above the carrier layer is not too high, say <200 mPas, then web speeds of up to about 250 m/min and wet film thicknesses of <50 μm can be obtained without ribbing lines, and such values are attractive for industrial applications.

18.5.2 Air Entrainment

Experimental data about the onset of air entrainment in slide coating are scarce. Figures 18.19 and 18.24 show examples for a low-viscosity single layer and a carrier layer application, respectively.

This lack of experimental evidence for quantifying the onset of air entrainment in slide coating is not truly cumbersome because, for many operating conditions, process failure occurs due to the onset of ribbing lines, not air entrainment, and so at lower web speeds. Moreover, if achieving high web speeds is a dominant goal for a given coating application, then curtain coating should be chosen over slide coating because no experimental evidence about the onset of ribbing lines in curtain coating has ever been reported, and because the momentum of the impinging curtain is much higher than the momentum of the impinging liquid film in slide coating, thus allowing much higher web speeds before air entrainment set in. As is explained in Chap. 19, curtain coating can be chosen over slide coating, if the volumetric flow rate/width summed over all layers applied simultaneously exceeds the value of about 1.0 cm^2/s, and if no objections exist for lowering the surface tensions of all layers to be coated simultaneously to a value of less than about 40 mN/m.

18.5.3 Vortices

As sketched in Fig. 5.70, slide coating, i.e., the multilayer film flow on the die slide and the continuation of that flow through the coating bead, offers at least eleven potential locations for the occurrence of vortices. These vortices are visualized and discussed in more detail in Sect. 5.7.8. Specifically, empirical equations quantifying the inception of many of these eddies were determined with the help of flow visualization experiments, and they are presented in the same chapter.

Operating the slide coating process without any vortices at all may be challenging, and it may require a complex and difficult geometry of the die slide, involving steps between adjacent die plates, see Figs. 5.70 and 5.71. Vortices that must be avoided above all include those attached to the free film and bead surface, i.e., vortices #1, 8, and 10, as well as those below the die lip, i.e., vortices #4 and 5, see Fig. 5.70.

18.6 Process Optimization

Regarding the geometrical configuration of slide coating, the optimum is achieved if both the impingement angle γ and the up-sweep angle δ are set to 0°, see Eq. (18.2.1) and Fig. 18.3. This configuration maximizes the momentum of the impinging film flow, which is desirable. Furthermore, setting the lip angle ε to a value of <90°, say to 60°, results in a simple but effective lip geometry because it promotes pinning of the static wetting line to the sharp corner of the die lip, and this, in turn, prevents the coating bead from flowing around the die lip, and hence the formation of vortices # 4 and 5, see Fig. 5.70. In addition, this configuration requires the application angle α to be equal to the slide angle β. On the other hand, β should be maximized to maximize the momentum of the impinging liquid film, but β must be kept small enough to prevent the formation of surface and/or interface waves in the (multilayer) film flow on the die slide, see Sect. 5.7.4.

Ribbing lines constitute the major operating boundary of slide coating. Maximizing the inception speed of this defect is best accomplished by incorporating the concept of the carrier layer. Accordingly, the bottom layer of the final product must temporarily be split into two layers of equal chemical composition. Specifically, the bottom portion must be diluted such that the low shear viscosity does not exceed a value of 10 mPas, preferably less. What is more, the flow rate/width of this bottom portion should be in the range of 0.1–0.5 cm^2/s. The best values for the viscosity and the flow rate of the carrier layer must be found experimentally by a trial-and-error procedure.

Maximizing the inception speed of ribbing lines also requires the surface tension of the top layer to be maximized. However, this surface tension may need to be lowered to promote wetting of the layer below, and to prevent the onset of vortices or other defects occurring between the coating and drying such as repellencies.

Of all the vortices that might occur in slide coating, vortices #1, 4, 5, 8, and 10 are the worst and must be prevented because they all are located in the free surface of the film flow on the die slide or in the film-forming or the wetting meniscus of the coating bead. Detailed information about how to prevent these vortices is given in Sect. 5.7.8.

Regarding the coating gap, the inception speed for ribbing lines, U, and hence the minimum wet film thickness does not depend on H_{gap} for higher μ-values (>50 mPas), but U increases with decreasing H_{gap} for lower μ-values (<10 mPas). This effect becomes stronger with decreasing Q. Generally speaking, thinner coated films require narrower coating gaps. A good starting value for the gap is on the order of 100 μm, but the best value must be found experimentally by trial-and-error. This, however, can be carried out easily, if the die positioning system is well designed. Moreover, depending on the design of the splice and on the value of the coating gap, a die skip out must be performed during the splice passage through the coating station. From that point of view, therefore, it might be interesting to test if the coating process works well with a gap so big that a skip out is not necessary.

Finally, bead vacuum may or may not be necessary. Applying bead vacuum generates uncertainty concerning the exact effect of the pressure difference across the bead on the behavior of the bead flow field. Specifically, applying too much vacuum may cause the approaching film to flow around the die lip, which would form at least one, but possibly two vortices, below the die lip, and that must be avoided. On the other hand, applying bead vacuum postpones the onset of ribbing lines and air entrainment to higher web speeds. In search of the best vacuum level for a new coating formulation, it is suggested, therefore, to start with no vacuum, or with only a weak level of, say, −100 Pa. If the desired coating speed cannot be reached without triggering ribbing lines or air entrainment, then the vacuum level can gently be increased. However, vacuum levels of <−500 Pa increase the chances of pulling the liquid below the die lip, which must be avoided. Instead of further increasing the pressure difference across the bead, it might be wiser to lower the viscosity of the carrier layer in order to achieve the desired coating speed, or switch from slide to curtain coating.

References

Blake, T. D., & Ruschak, K. J. (1997). Wetting: Static and dynamic contact lines. In S. F. Kistler & P. M. Schweizer (Eds.), *Chapter 3 in Liquid film coating*. Chapman & Hall.

Chen, K. S. A. (1992). *Studies of multilayer slide coating and related processes*. Ph.D. Thesis, University of Minnesota.

Choinski, E. J. (1981). Method and apparatus for coating webs. US 4,283,443.

Coyle, D. J. (1997). Knife and roll coating. In S. F. Kistler & P. M. Schweizer (Eds.), *Chapter 12a in Liquid film coating*. Chapman & Hall.

Dittman, D. A., & Rozzi, F. A. (1977). US 4,001,024.

Hens, J., & van Abbenyen, W. (1997). Slide coating. In S. F. Kistler & P. M. Schweizer (Eds.), *Chapter 11b in Liquid film coating*. Chapman & Hall.

Jackson, B. W. (1976). Apparatus for coating a substrate. US 3,993,019.

Landau, L., & Levich, B. (1942). Dragging of a liquid by a moving plate. *Acta Physicochimica URSS, 17*, 42–54.

Mercier, J. A., Torpey, W. A., & Russell, T. A. (1956). Multiple coating apparatus. US 2,761,419.

Ruschak, K. J. (1976). Limiting flow in a pre-metered coating device. *Chemical Engineering Science, 31*, 1057–1060.

Ruschak, K. J. (1978). Flow of a falling film into a pool. *AIChE Journal, 24*(4), 705–709.

Schweizer, P. M. (1988). Visualization of coating flows. *Journal of Fluid Mechanics, 193*, 285–302.

Schweizer, P. M. (1997). Experimental methods. In S. F. Kistler & P. M. Schweizer (Eds.), *Chapter 7 in Liquid film coating*. Chapman & Hall.

Schweizer, P. M., & Rossier, P. A. (2001). The operating window of slide coating. In *Proceedings of the 4th European Coating Symposium*, October 1–4, 2001, Brussels, Belgium.

Schweizer, P. M., & Rossier, P. A. (2003). High-speed operating range for slide coating. In *Proceedings of the 5th European Coating Symposium*, September 17–19, 2003, Fribourg, Switzerland.

Weinstein, S. J., & Palmer, H. J. (1997). Capillary hydrodynamics and interfacial phenomena. In S. F. Kistler & P. M. Schweizer (Eds.), *Chapter 2 in Liquid film coating*. Chapman & Hall.

Chapter 19
Specific Properties of Curtain Coating

Abstract This chapter starts out by listing all process-technological advantages that curtain coating has over any other application methods. The discussion concerning process configuration and equipment addresses issues such as the web path through and the design of the coating station, whether a slot or slide die should be used, the suction baffle for removing the air boundary layer on the uncoated substrate, the inboard and overboard operating modes, and how to start and stop the coating process. The coating window including the dominant operating limits is presented for both long and short curtains. Finally, guidelines for optimizing the coating process are proposed.

19.1 Introduction

A simple version of curtain coating was already known in the middle of the last century for applying paint to wooden boards and other plate-like structures. The high-tech version was formally introduced in 1970 for manufacturing multilayer photographic films and paper, when Don Hughes applied for a broad patent that was granted to Eastman Kodak. Multilayer curtain coating can be considered as a further development of multilayer slide coating with the main difference being the distance between the die lip and the substrate surface. While this distance is on the order of 100 μm in slide coating, it is on the order of 100 mm in curtain coating. This difference of a factor of 1'000 generates the following major advantages in favor of curtain coating:

- Absence of narrow gaps at the application point through which the coating fluid has to flow. As a consequence, curtain-coated products are virtually free of line and streak defects.
- For long curtains, the thickness of the curtain decreases gradually from the die lip downwards such that the thickness of the curtain at the impingement point approaches the thickness of the coated film. As a result, excessive thinning of the curtain just upstream of coating, as it is observed in the approaching film in slide coating, does not occur. This explains why data describing the onset of ribbing lines in curtain coating have never been published.

P. M. Schweizer, *Premetered Coating Methods*, Engineering Materials,
https://doi.org/10.1007/978-3-031-04180-8_19

- The momentum of the impinging curtain is higher by a factor of 10–100 compared to the momentum of the liquid film in slide coating. Therefore, air entrainment at the dynamic wetting line can be postponed to high web speeds. Nowadays, state-of-the-art coating machines in the converting industry for applying acrylic emulsions such as pressure-sensitive adhesives (PSA) run at speeds of up to 1'200 m/min. In the paper industry, fast machines for making thermal paper approach 2'000 m/min.

It is the opinion of this author that curtain coating is by far the best coating method with regard to product uniformity and process productivity when compared to any other application technique. This observation is based on some 35 years of practical experience in the laboratory, pilot plants and production environments as well as during the commissioning of slot dies and curtain coaters in industrial companies. Furthermore, this observation is supported by the following facts:

- Excellent control of the nominal film thickness owing to the feature of premetering.
- Excellent film thickness uniformity in both machine and cross-web direction.
- Absence of air entrainment allowing high production speeds.
- Absence of major coating defects.
- Capability of simultaneous multilayer applications.
- Capability of implementing the concept of high wall shear stress through smart equipment design.
- Possibility for process optimization with geometrical parameters only, thereby leaving more freedom for optimizing product design.

Like with multilayer slide coating, most of the research and development work carried out in the last third of the twentieth century was done by large photographic companies. Not many results were published because these companies wanted to protect their competitiveness. Nevertheless, a comprehensive summary of all published work relating to curtain coating up to the late nineties was given by Miyamoto and Katagiri (1997). The purpose of this chapter is to provide additional information about the application process. Note that much detailed information about the various flow fields that lead up to the coating point are presented throughout this book.

19.2 Process Configuration and Equipment

19.2.1 Web Path

Curtain coating can be implemented with different web paths through the coating station, depending on the design and the location of the dryer relative to the location of the coating station. Figure 19.1 shows examples of possible web paths if the coating roll is located at more or less the same level as the entrance to the dryer.

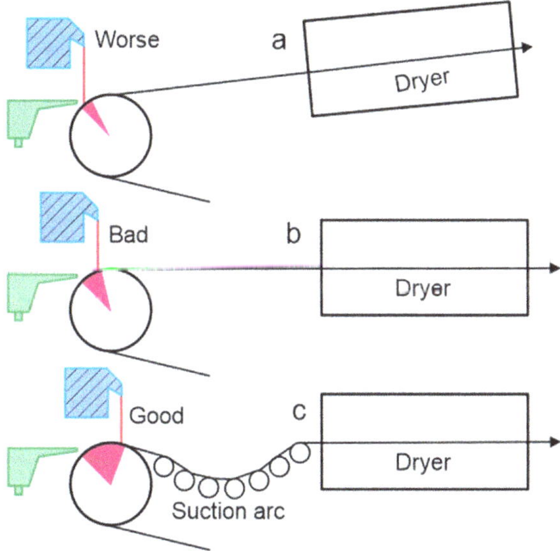

Fig. 19.1 Variations of web path through the curtain coating station

In Fig. 19.1a, the dryer has a curved design, which is often found in the paper industry. The substrate leaves the coating roll in an upward direction. Therefore, only negative application angles covering a relatively small range as indicated by the purple wedge are possible. In Fig. 19.1b, the dryer is straight and horizontal, thus increasing the range for possible but still only negative application angles. This range is widened to include slightly positive application angles by inserting a suction arc between the coating roll and the dryer as illustrated in Fig. 19.1c. This configuration was successfully implemented by ILFORD Imaging Switzerland GmbH in the late 1980's. Design c is preferred over design b, and b is better than a owing to the larger range of possible applications angles. However, all three configurations suffer from the fact that the catch pan, over which the die is prepared prior to the coating start, is located to the left of the coating roll in Fig. 19.1. Therefore, when the die is moved from the pan to the substrate for multilayer applications, it is the top layer in the curtain that first contacts the web. This causes a temporary mess on the web as the order of the layers in the multilayer film must first be inverted. As this mess can likely not be dried completely, it should be removed immediately to prevent contamination of rollers inside the dryer or between the dryer and the re-winder. This in turn requires some kind of full-width suction device, which is difficult to operate and so highly undesirable.

Another disadvantage of the configurations shown in Fig. 19.1 is the fact that, for most coating machines, the dryer is located at least one floor level above the main floor of the machine. Therefore, the coating station must also be placed on that upper level, often associated with ergonomically uncomfortable working positions for the operators, particularly if the curtain coating process is installed into an existing machine.

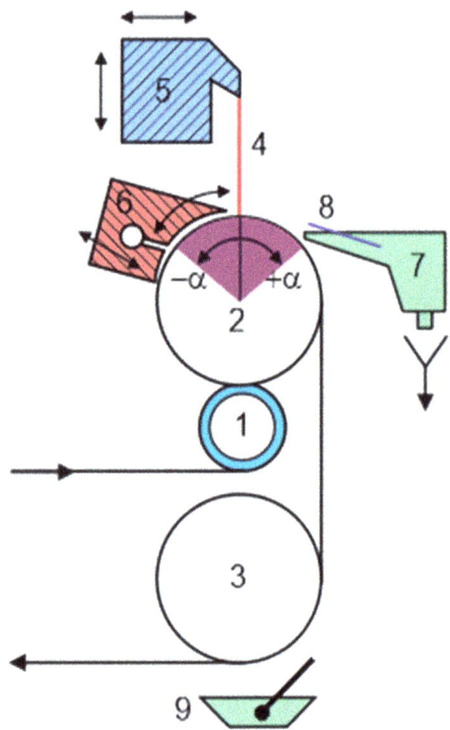

Fig. 19.2 Preferred web path through the curtain coating station

A much more attractive web path without any of the disadvantages discussed above is shown in Fig. 19.2. A coating station of this design is preferably located on the main floor of the coating machine. Moreover, the height of the coating roll and associated equipment above the floor level can be chosen such that it results in a comfortable working position for the operator.

The substrate arriving horizontally at the coating station is guided around a rubber-covered roller 1 before it wraps around the coating roll 2, from where it departs downward in a vertical direction toward roller 3, which guides the web back into a horizontal position before leaving the station. Curtain 4, which is formed by die 5, is then able to drop onto the substrate over a wide range of negative and positive application angles α as visualized by the purple wedge in the figure. Roller 1 presses the substrate against coating roll 2, thereby eliminating the thick air boundary layer carried along by the uncoated web, and minimizing the distance between the nip between rolls 1 and 2 and the upstream edge of the suction baffle 6. This distance is reduced to about 1/4 of the circumference of the coating roll. Along this distance, however, the thickness of the air boundary layer on the uncoated web can grow again before it is eliminated for the most part by the suction baffle. In principle, it is possible to operate the coating process with or without engaging roller 1 against the coating roll. If the substrate has sticky surfaces, it might be advantageous to leave an open gap to prevent the formation of wrinkles.

The suction baffle is mounted in such a way that it can rotate around the center of the coating roll, which allows the distance between the front tip of the baffle and the back side of the curtain to be minimized. This optimization feature is important because it is over this distance that the air boundary layer grows again in thickness and momentum, before impinging onto the curtain. The suction baffle is also mounted such that the narrow concentric gap that is formed between the baffle and the substrate surface can be adjusted continuously.

The die, be it of slot or multilayer slide type, is mounted on a two-dimensional positioning system. Adjusting the horizontal position allows the die to move from the preparation location over catch pan 7 to the coating position at the desired application point, or even farther to the left in Fig. 19.2 to a park position in case the coating roll must be removed for maintenance purposes. Changing the vertical position of the die allows the curtain length to be optimized for any given set of operating conditions.

Catch pan 7 also houses a so-called sliding shield 8. The purpose of this piece of equipment is to facilitate a perfect coating start as well as a perfect and quick coating stop. The operating principle of the sliding shield is explained in detail in Sect. 19.2.4.

The configuration shown in Fig. 19.2 causes the coated web to first move vertically down before it changes to a horizontal direction, but now with the coated side being upside down. Furthermore, the coated web is normally guided upward to at least one level above the main floor before it reaches the entrance to the dryer. Experience at Polytype Converting has shown time after time that this web path can be operated without any problems, even at speeds of up to 1’200 m/min for the application of pressure-sensitive adhesives. The coated liquid does not fling off the substrate, even if the web changes direction by 90°. Care must be taken, though, that the thickness of the coated layer is uniform across the entire width. In particular, the coated edges must not be over-thick, but this can be achieved with well-designed curtain edge guides, see Sect. 5.9.5.

This configuration, and in particular the location of the catch pan relative to the location of the coating roll, does not require the order of layers in the curtain to be reversed after the coating start because it is always the bottom layer on the die slide or the rear layer in the curtain, that first comes in contact with the substrate. Moreover, should splashes or other contaminations occur during the coating start, the related mess on the substrate can easily be removed by the full-width scraper blade 9 that is mounted against idle roll 3, and that can be engaged temporarily during the coating start until the operator signals that the uniformity of the coated film is perfect. Note that the performance of the scraper blade does not need to be perfect. It must only remove over-thick portions on the coated web that could not be dried properly.

The coating station as described above was developed by Polytype Converting AG. It works well, and it has proven industrially viable in many curtain coating projects.

Figure 19.3 shows how the location of the curtain coating station is related to the location of the dryer in a typical coating machine.

The roller located just upstream of the entrance to the dryer is a vacuum roll. It serves to establish different web tensions across the coating station and across the

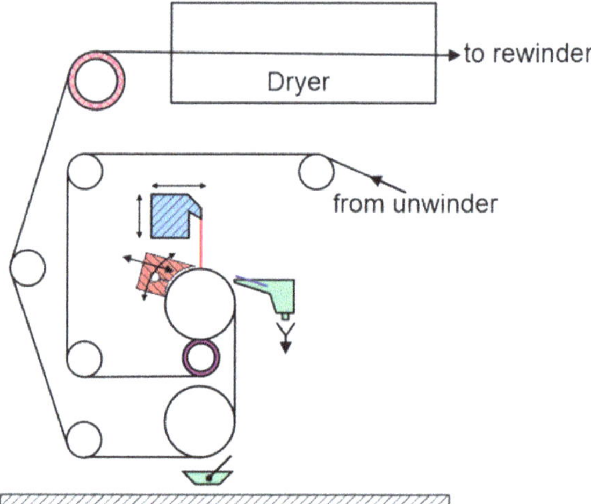

Fig. 19.3 Location of the curtain coating station relative to the location of the dryer in a typical coating machine

dryer. Slightly modified web paths may be necessary depending on the location of the unwinder relative to the location of the coater. If, for example, the adhesive coating is done in-line after silicone coating for manufacturing adhesive labels, then the web would arrive from the left side of the coater in Fig. 19.3. In that case, the curtain coater could be raised a bit above the main floor to create space for the web to pass underneath the coater and beyond the platform for the operator before turning upward and around the coater as shown in the sketch. Other web paths involving suctions arcs or air turns in combination with pre-drying are conceivable as illustrated in Fig. 19.4.

Suction arcs and air turns are convenient machine components for changing the web direction when contact between the web and the substrate, particularly the coated side of the substrate, must be avoided. These elements have successfully been used in

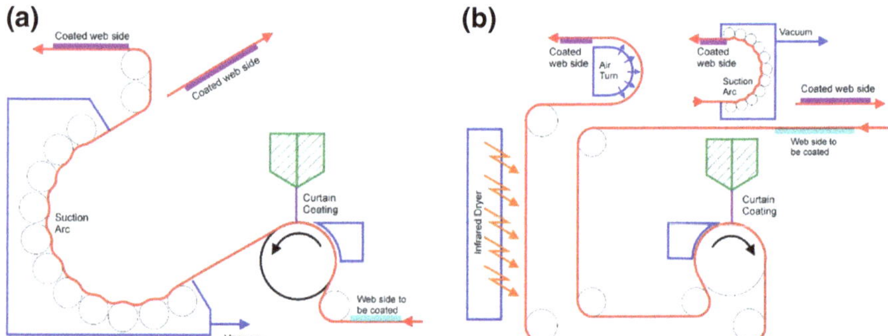

Fig. 19.4 **a** Web path around curtain coater involving a 180° suction arc. **b** Web path around curtain coater involving a 180° suction arc or air turn with pre-drying

Fig. 19.5 180° suction arc in coating machine of ILFORD Imaging Switzerland GmbH. Photo reproduced with permission from ILFORD Imaging Switzerland GmbH

the photographic industry and also in coating machines made by Polytype Converting AG. An example of a 180° suction arc that connected the chill section with the dryer on the coating machine of ILFORD Imaging Switzerland is shown in Fig. 19.5.

If the web enters the dryer with the coated side upside-down, then, obviously, a flotation dryer is required. The best web path solution for a given situation depends on the relative location of the unwinder, curtain coater, and dryer.

It may be advisable to install a fast-acting, full-width web cutting device just upstream of the curtain coating station to prevent a collision between the coating roll and the curtain edge guides in case of a web break, particularly if the coating speed is high. The issue here is the very narrow but fixed gap between the bottom of the curtain edge guides and the substrate surface. One time, when curtain coating onto aluminum foil at high speed on the Polytype pilot machine, we experienced a web break. This caused the suddenly loose foil to immediately be wrapped around the coating roll, thereby increasing the roll diameter so much that it collided with the edge guides within a few seconds. The forces involved in this incident were so high that the sturdy aluminum mounting brackets of the edge guides were severely bent, thus causing serious damage to the equipment. This prompted us to change the design of the mounting brackets by implementing a pivot point such that the edge guides can turn away in a forward direction upon only a small force acting on them from behind. A web cutting device is an additional means for preventing such collisions.

19.2.2 Design of Coating Station

Regarding equipment, the curtain coating process concept depicted in Fig. 19.2 can be designed and built in three different ways. The simplest and least costly approach uses a trolley, i.e., a cart that can easily be moved in and out of the coating machine, and that houses all the elements of the process. The idea is that application methods other than curtain coating can also be installed on such a machine. The drawback of this design is the limited range, over which the relevant process parameters can be varied, and the limited working comfort for the operators because all the equipment components are tightly packed onto the trolley.

A more generously designed and more comfortable version is the so-called compact design shown in Fig. 19.6. The catch pan can be lowered for easy cleaning and for providing full access to the coating roll. In addition, this particular version is equipped with a full-width widow that can be lowered and placed in front of the curtain and the coating roll during coating. The purpose of the window is to provide safety for the operator and to protect the curtain from excessive air currents in the environment. In practice, we have learned, however, that such a window is

Fig. 19.6 Compact curtain coating station. Photo reproduced with permission from Polytype Converting AG, multilayer slide die with variable coating width from TSE Troller AG

not convenient because it prevents the operator from intervening quickly in case of an emergency, for example when the curtain detaches from the edge guide. A more practical way of handling these issues is to guarantee operator safety by way of an optical barrier instead of a window and to install shields against air currents, if necessary, at a larger perimeter around the coating station, i.e., outside the operator platform.

The preferred and best but most costly station design is based on a solid overhead frame, such as the one sketched in Fig. 8.11. The frame is anchored to an isolated and vibration-free floor segment of the coating machine. Attached to this frame is only the die with the curtain edge guides and the two-dimensional positioning system. Underneath the die, a second frame is placed on the floor. This frame houses all driven and idle rollers as well as other mechanical components that move, such as the catch pan, the suction baffle, and the full-width scraper blade. The big advantage of this design concept is the mechanical separation of the die from all parts that move, thereby preventing vibrations from disturbing the liquid curtain. In addition, the overhead frame is large enough to include the operator platform. Therefore, if air shields are needed, they can conveniently be attached to that frame.

19.2.3 *Slide or Slot Die*

As illustrated in Fig. 2.1, curtain coating offers a choice regarding the type of die being used. When choosing the best die for a given application, the various features of these dies must be considered and the related advantages and disadvantages have to be evaluated.

19.2.3.1 Number of Layers

Each die type can be designed as a single-layer or multilayer device. However, slot dies with more than 3 layers are difficult to build and so are not recommended. For applications with more than 3 layers, slide dies are the only choice.

19.2.3.2 Cross Profile

Slot dies, particularly single-layer slot dies, can be built with either a fixed or a deformable slot. The internal geometry of dies with a fixed slot must be designed correctly for the intended range of application because such dies have no mechanism for adjusting the die geometry during operation. In contrast, dies with a deformable slot have screws or thermal bolts or other mechanisms located a few centimeters apart in the cross-web direction to adjust the resulting cross profile in response to changing rheological fluid properties or operating conditions. Slide dies with several slots are typically built with fixed slots as it is difficult if not impossible to implement independent slot deforming mechanisms for each slot. Generally speaking, fixed slot designs are preferred over deformable slots because they are much easier to operate. Nowadays, moreover, the understanding of the physical die design concepts is so high that a need for adjusting the geometry of the die slot after the die is built does normally not exist, see Sect. 9.4.

19.2.3.3 Orientation of Die Slot

As shown in Fig. 2.1, the orientation of the slot for inverted slot dies used in curtain coating, and hence the direction of the slot flow, is vertically downward. This is a big disadvantage with regard to the effect of air bubbles in the coating fluid. Depending on the performance of degassing equipment, if installed in the first place, coating fluids will always carry some amount, albeit often small, of bubbles into the die. Owing to the buoyancy effect, bubbles on their own can only rise, which is in the opposite direction of the slot flow. As a consequence, small bubbles tend to accumulate and form bigger bubbles on top of the distribution chamber during long periods of production. This is particularly so if the die is designed without respecting the requirement of a minimum wall shear stress. Big bubbles will then negatively affect the cross-web distribution of the coating fluid because their presence effectively modifies the internal die geometry.

Undesirable effects of air pockets inside the die can only be avoided if air bubbles can be purged from the die by the flowing liquid and against the buoyancy effect. This in turn requires a high wall shear stress everywhere and at all times inside the die. High wall shear stress is facilitated by high flow rates, high viscosities, and small flow cross-sections, see Sect. 6.3 for details. In contrast, high wall shear stress is difficult to achieve for low flow rate and low viscosity applications. This is the same for dies allowing a variable coating width, owing to the necessary constant cross-section of the distribution chamber in the regions of the variable width. For such situations, the accumulation of air bubbles inside the die can be avoided by periodically opening bleeding valves located in the sidewalls of the die and at the end of the distribution chamber. This, however, is an operational nuisance.

Furthermore, owing to the constant action of gravity, inverted slot dies tend to leak when the pump is stopped after a coating run. This is particularly so for wide die slots and low fluid viscosities, and again is an operational nuisance.

All the problems described above do not exist for slide dies owing to the general upward orientation of the die slots, with a typical slot angle against vertical on the order of 30°.

19.2.3.4 Design of Die Lip

A proper design of the die lip is crucial for the overall performance of the curtain coating process because the details of the micro-flow field around the die lip determine whether the curtain is clean and homogeneous or full of lines and streaks. As explained in Sect. 5.8.7, most of the lines and streaks in the curtain are generated by vortices that are present near the die lip, see Fig. 5.39.

The curtain should be formed at the die lip without any vortices present. In the case of the slot die, this requires both static wetting lines to be located at the inside corners of the die lip as shown in Fig. 5.39b. In our experience, this is only possible, if the height of the die slot at its exit is wider than the curtain thickness. Moreover, since the curtain thickness depends on the volumetric flow rate/width, the slot height would have to be adjusted as a function of the flow rate, which is not possible, or at least not practical in industrial applications. In addition, the slot height of the die is primarily controlled by cross profile requirements; therefore, it cannot be controlled by requirements related to the location of the static wetting lines. Finally, if the coating liquid exhibits non-Newtonian flow behavior, in particular viscoelastic behavior, then the so-called die swell effect will occur, which tends to move the static wetting lines to the outside corners of the die lip. In most industrial applications, therefore, the static wetting lines are attached to the outside corners of the die lip, which in turn results in a pair of vortices located underneath the flatlands of the die lip, see Fig. 5.39b. Normally, these vortices cannot be avoided, but their size can be minimized by minimizing the length of the lip land L_L.

Regarding the propensity for line and streak formation, slot dies are twice as bad as slide dies because due to the symmetrical design of the die lip, two vortices as opposed to one vortex are present.

In the case of the slide die, the static wetting line should be located at the right corner of the die lip as shown in Fig. 5.39a, however, this is not possible either. The main reason is that there is a flow phenomenon called the tea-pot effect, or curtain bend back, or curtain deflection, see Sect. 5.8.6 for details. Consequently, the static wetting line is pinned at the left corner of the die lip, and a vortex is formed underneath the die lip.

Again, this vortex cannot be avoided, but its size can be minimized by minimizing the length of the lip land. The minimum lip length depends on the width of the die and the manufacturing capabilities of the die supplier. Lip length values of ≤ 0.25 mm should be possible and are preferred, see Fig. 19.7a. In addition, it is possible to further reduce this land length by attaching a thin plastic film or a commercially available metal blade to the outside of the die lip as sketched in Fig. 19.7b. This way, the vortex size can virtually be reduced to zero.

Many slide dies used for curtain coating have a lip form that is sometimes referred to as *bull nose* because the end of the slide gradually curves from the inclination β to a vertical orientation, see Fig. 19.7a. Several die manufacturers offer slide dies for curtain coating that look like slide dies for slide coating, i.e., without a bull nose lip, particularly if the slide angle β is large, see Fig. 19.7c. However, the issue of vortex formation at the die lip does not change. In fact, it is slightly worse with a straight lip because gravity helps to push the static wetting line to the lower lip corner, which is not desirable. In addition, the curtain trajectory follows a ballistic form as illustrated in Fig. 5.17. Consequently, the location of the curtain impingement point on the substrate is not known, which is not desirable either.

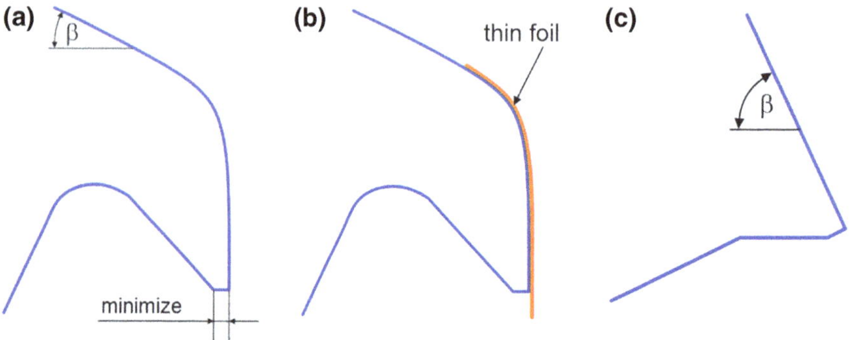

Fig. 19.7 **a** Optimized lip geometry for slide die with a bull nose lip. **b** Extension of die lip by thin, non-wetting plastic sheet, or thin, flexible steel foil. **c** Slide die for curtain coating without a bull nose lip

19.2.3.5 Visual Access to Die Lip

Unobstructed visual access to critical points in the characteristic flow fields of the curtain coating process is important for the operator to control the quality and resulting uniformity of the curtain and the coated film on the substrate. Two of these critical points are the locations of the static wetting lines. In the case of slot dies, both static wetting lines are located at the die lip as depicted in Fig. 5.39b. In the case of slide dies, one static wetting line is located at the die lip, while the other wetting line is located on the inclined surface of the die, at the exit of the uppermost slot, see Fig. 4.1. We have noted that visual access to these points is better for slide dies than for slot dies. This is particularly so if slot dies from TSE Troller AG are used. These slot dies are equipped with additional auxiliary plates that obscure the visual access to the die lip, but that also protect the die lip in case the die must be removed from the coating station and placed onto a stone table for maintenance purposes, see Fig. 19.8.

19.2.3.6 Film Flow on Inclined Surface of Slide Die

In the case of slot dies, the liquid exiting the die slot is directly transferred into the curtain. In contrast for the case of slide dies, liquid exiting the die slot is first transferred into a film flow on an inclined plane (slide). As mentioned above, this film flow comprises one of the static wetting lines and is guided along both sides by slide edge guides. The curtain is formed at the end of the slide (the die lip), where the film flow changes into the curtain flow.

During film flow, many undesirable things can happen to the coating fluid. In fact, the presence of film flow is a disadvantage of slide dies when compared to slot dies. Therefore, it is important to control all details of this film flow, particularly along the static wetting line and the slide edge guides, to prevent any undesirable flow

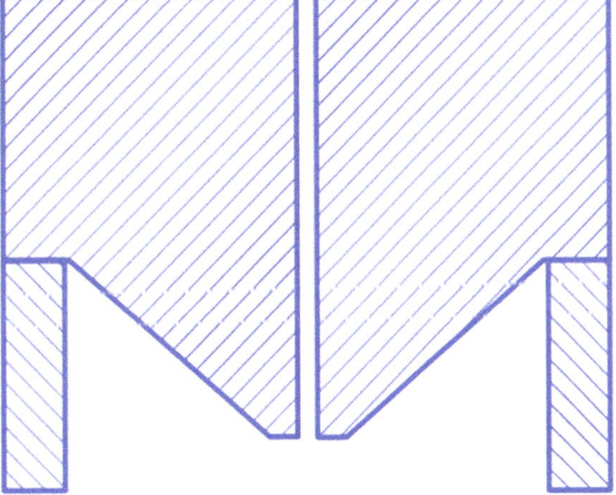

Fig. 19.8 Auxiliary plates obscuring the visual access to the die lip; die design by TSE Troller AG

effects that might reduce the quality and uniformity of the subsequent curtain flow. Section 5.7 provides ample information about how to optimize all details of the film flow for application-specific operating conditions.

19.2.3.7 Curtain Deflection (Bend-Back, Tea-Pot Effect)

In the case of slide dies, it is well known that the trajectory of the falling liquid curtain is not always vertically downward, but, depending on the values of several relevant parameters, bent backward. The reason for curtain deflection is found in the transition from the film flow on the die slide to the free-falling curtain flow. On the die slide, the velocity profile of the film flow is non-symmetrical and has the form of a half-parabola. In fluid-mechanical terms, this flow is said to have a component of rotation, i.e., the flow wants to curve toward the slide surface. However, the very presence of the slide surface prevents this from happening. At the die lip, however, the slide surface is suddenly removed, and the rotational component, which is still present in a short, transitional zone between film and curtain flow, but which is no longer present in the fully developed curtain flow, makes the curtain bending backward. It is known that the amount of curtain deflection depends on the physical fluid properties, the volumetric flow rate/width, the inclination angle of the die slide, and the distance along the falling curtain.

In the case of single-layer slot dies, curtain deflection does not exist because the velocity profile in the slot flow preceding the curtain flow is fully parabolic, and hence fully symmetrical. Consequently, slot flow contains two rotational components of opposite signs, which cancel each other. In the contrasting case of multi-layer slot dies, curtain deflection exists if the flow rates/width and the physical properties of the fluids, notably the viscosity, differ from one layer to the next.

Curtain deflection is undesirable because it negatively affects the flow of the curtain along the curtain edge guide. Moreover, it alters the curtain impingement point on the substrate, which in turn may affect the quality and uniformity of the coated film on the substrate. If significant curtain deflection is present for single-layer applications, then slide dies are at a disadvantage when compared to slot dies.

Experimental data that quantify the amount of curtain deflection for slide dies as a function of all relevant process parameters are presented in Sect. 5.8.6. Furthermore, to absorb excessive curtain deflection, it is advantageous to use surface-type rather than line-type curtain edge guides, see Sect. 5.9.5 for details. Experimental data describing curtain deflection for multilayer slot dies are not known.

19.2.3.8 Price Difference Between Slot and Slide Die

Most converting companies do not manufacture their dies. Therefore, they must rely on and purchase dies at suitable suppliers. Regarding slot dies with fixed geometry there is a choice of suppliers including Cloeren, Nordson, Mitsubishi Materials, FMP Technology, and TSE Troller. Regarding slide dies, particularly multilayer slide dies, the choice is more limited.

Usually, slide dies are 10–20% more expensive than slot dies for the same number of slots.

19.2.3.9 Conclusion

Both slot and slide dies have advantages and disadvantages. Summing up the various aspects, slide dies with fixed slot geometries are preferred over slot dies for most single-layer and all multilayer applications. The main reasons are the flexibility for expanding the number of layers, the correct orientation of the die slot relative to gravity, the reduced propensity for forming lines and streaks in the curtain, and the good visual access to fluid-mechanically relevant areas. Slot dies with fixed slot geometries might be considered for single-layer applications involving fluids based on low boiling point solvents such as acetone to prevent problems related to solvent evaporation and solidification along the slide edge guides.

19.2.4 Suction Baffle

Reducing the energy of the air boundary layer that is dragged along the uncoated substrate and that impinges onto the backside of the curtain is necessary. Without suitable equipment for achieving this goal, the curtain would be disturbed in an uncontrolled manner by pressure forces generated by the airflow of the boundary layer. In particular, the curtain would bulge in the downstream direction, which would prevent the dynamic wetting line from extending in a straight line across the substrate.

This in turn would cause the curtain to meet the substrate at different impingement angles across the substrate if the web was wrapped around the coating roll as depicted in Fig. 19.2, and that would result in a laterally changing hydrodynamic assist, which is highly undesirable. All these negative effects are amplified with increasing web speed.

A pinch point such as the contact line between rollers 1 and 2 in Fig. 19.2 would eliminate the air boundary layer on the uncoated web. However, such a pinch point cannot be located right next to the dynamic wetting line. Consequently, there is always a certain length of uncoated substrate just upstream of the curtain, over which a new air boundary layer will develop. The challenge, therefore, is to design a process and associated equipment, here called suction baffle, in such a way that the negative effects described above do not cause excessive harm to the impinging curtain and the dynamic wetting process at the desired coating speed.

Many baffle designs are known from the patent literature. Examples include Schweizer and Troller (1997), Frediani et al. (2001) and Horbach and Reinhard (2004). Most of these devices explore the possibilities of combining mechanical scraping with suction. Figure 19.9 shows a suction baffle developed by Polytype Converting AG. The operating concept and an optimization procedure are described in detail in Sect. 5.6.1.

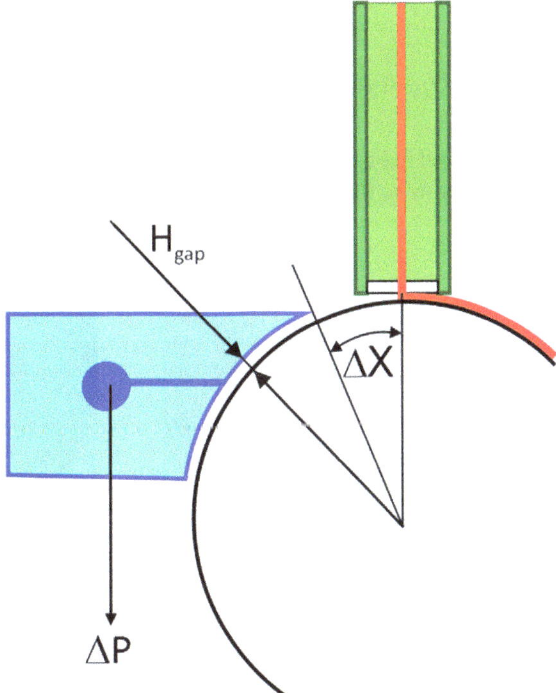

Fig. 19.9 Suction baffle developed by Polytype Converting AG

Most of the air boundary layer arriving at the suction baffle is blocked and diverted away from the substrate by the body of the baffle. A small amount of air enters a narrow and concentric gap between the baffle and the substrate. Most of this air, as well as air from the space between the tip of the baffle and the curtain, is sucked away through a narrow gap in the baffle body. In fact, the baffle works like a slot die operated in reverse. However, a small amount of air exits the concentric gap in form of a thinner boundary layer. This boundary layer grows in thickness by drawing in air from the space behind the curtain until it impinges onto the rear curtain surface.

The challenge is to keep the energy of the impinging air boundary layer low enough, such that it does not excessively disturb the falling curtain and that air entrainment at the dynamic wetting line can be prevented for a given set of operating conditions. Important features of this device for achieving this goal include the possibilities for

- adjusting the height of the concentric gap H_{gap};
- adjusting the distance between the tip of the baffle and the impinging curtain ΔX;
- adjusting the vacuum pressure ΔP.

Experience with industrial curtain coaters has shown that this baffle concept is viable and effective for coating width of >2'000 mm and coating speeds of >1'000 m/min.

19.2.5 Operating Modes

As visualized in Fig. 19.10, curtain coating can be operated in two different ways, i.e., in the *inboard* or the *overboard* mode. Based on the experience of TSE Troller and

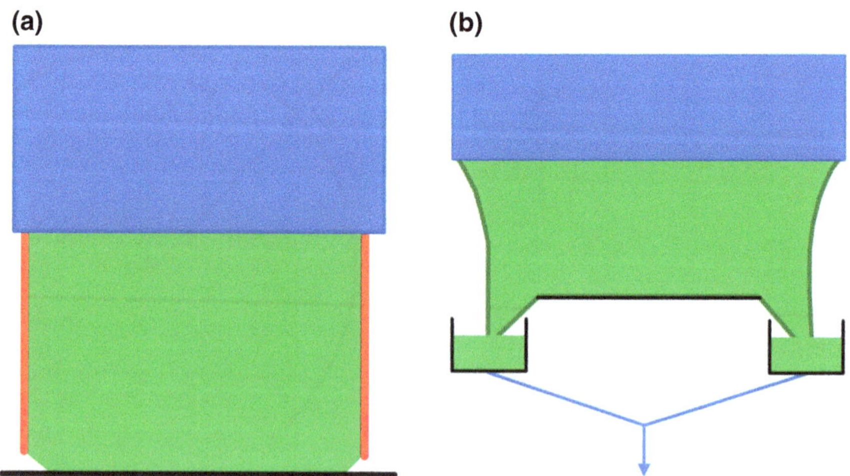

Fig. 19.10 **a** Inboard mode for curtain coating. **b** Overboard mode for curtain coating

Polytype Converting, the inboard mode is more often implemented in the converting industry than the overboard mode.

19.2.5.1 Inboard Operating Mode

The distinctive feature of coating inboard is the fact that the curtain is narrower than the substrate. In addition, the curtain must be confined between a pair of edge guides to prevent excessive curtain contraction due to surface tension forces, and thus to prevent excessively over-thick edges on the coated substrate. Consequently, the substrate is characterized by narrow but dry zones along both edges. Finally, the substrate at the curtain impingement point is usually supported by a coating or backup roll. The best process configuration of the inboard operating mode is discussed in Sect. 19.2.1 and illustrated in Fig. 19.2. A characteristic feature of this design is that the coating roll is in a fixed position and the die moves between a position over the catch pan and the coating position over the backup roll during coating start and end.

19.2.5.2 Overboard Operating Mode

In the classical overboard configuration, the horizontal position of the die is fixed such that the curtain impinges onto an unsupported substrate between rollers 2 and 10. The die is considerably wider than the maximum substrate width, such that the excess fluid can shear off the substrate edges and falls into catch pan 7 located underneath the substrate. For starting and stopping the coating process, catch pan 11 is moved horizontally between a park position over drip pan 12 and a position above catch pan 7, which allows the curtain to be intercepted. The width of catch pan 11 is similar to the maximum substrate width. This allows the excess fluid to shear off the side edges of the pan and to flow along the edge guides 14 past the substrate into catch pan 7, see Fig. 19.11.

For this configuration, the curtain edge guides are usually of a simple kind, i.e., a solid body without auxiliary fluid, perhaps with an inward inclination toward the bottom to prevent the curtain from detaching from the guides. The edge guides are typically longer than the height of the curtain to guide the excess liquid past the substrate into the catch pan, see Fig. 19.11b. Moreover, the distance between the curtain edge guide and the edge of the substrate with the maximum width is on the order of several centimeters to assure that undesired effects from the viscous boundary layer in the curtain along the edge guide will not negatively affect the coating performance.

If the curtain does not shear off cleanly from the substrate edges, then the underside of the substrate will be contaminated, which in turn will contaminate idle rollers downstream of the coating station. Such roller contamination may be reduced or even eliminated by mounting narrow suction nozzles (position number 15 in Fig. 19.12) at the underside of the substrate edges.

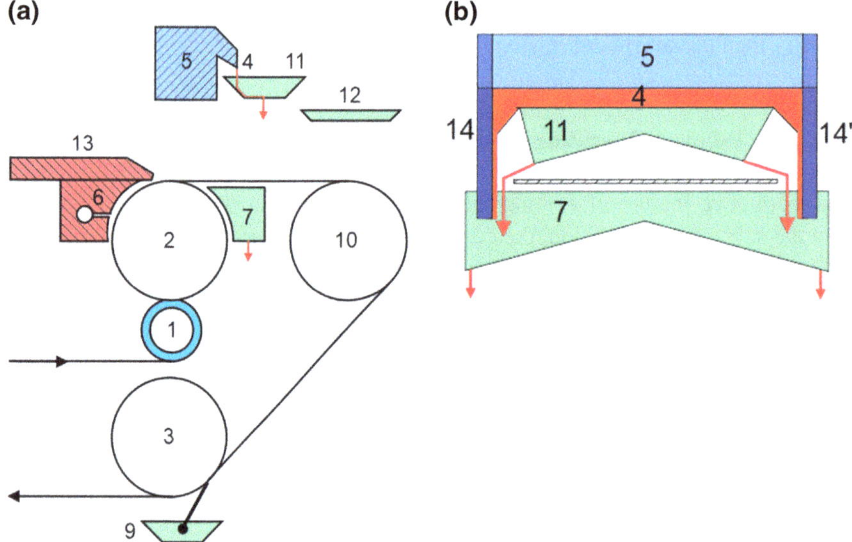

Fig. 19.11 **a** Concept for overboard curtain coating, side view of start/stop configuration. **b** Concept for overboard curtain coating, front view of start/stop configuration

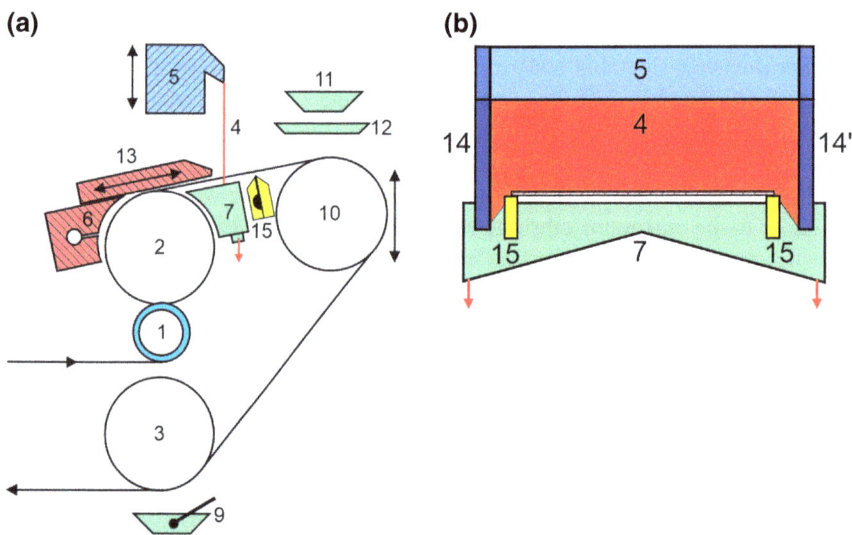

Fig. 19.12 **a** Concept for overboard curtain coating, side view of coating configuration. **b** Concept for overboard curtain coating, front view of coating configuration

Fig. 19.13 **a** Curtain coating in over-board mode: slide die 5 with curtain, catch pan 11 intersecting curtain prior to the coating start. Photo reproduced with permission from Polytype Converting AG. **b** Curtain coating in over-board mode: slide die 5 with curtain, catch pan 11 and drip pan 12 in front of the curtain, coated substrate, roller 10 with a protective shield in foreground; photo reproduced with permission from Polytype Converting AG

As the performance of the suction baffle (position number 6 in Fig. 19.11) is best when it is operated with a fixed gap geometry, the best position for the suction baffle is against backup roll 2, even if the coating position (the curtain impingement position) is between rolls 2 and 10, which is far away from the location of the suction baffle. However, good performance of the baffle can be maintained, if the gap between the baffle and the substrate surface is extended from roll 2 to near the curtain impingement point. This is achieved by plate 13, the horizontal position of which can be adjusted to optimize the distance between the tip of plate 13 and the rear surface of the curtain, see Fig. 19.12.

The curtain height can be changed by simply moving the die vertically up or down. Of course, the length of the curtain edge guides must be longer than the maximum curtain height.

The curtain impingement angle can be changed from negative to positive values by simply changing the vertical position of the idle roll 10, see Fig. 19.12. As the position of roll 10 is changed, the suction baffle 6 and the associated cover plate 13 must be rotated accordingly around roll 2.

Polytype Converting AG has developed an equipment module containing a frame with side plates that can be attached to the side plate of the curtain coating station designed for inboard applications as shown in Fig. 19.3. The module comprises roller 10 with a mechanism for vertical roll displacement, catch pan 11 and drip pan 12. Figure 19.13 shows photographs of this overboard module.

19.2.5.3 Comparison Between In-Board and Over-Board Operating Modes

In an attempt to compare the two operating modes, several criteria relating to investment costs, downtime, material waste, and product uniformity were defined as follows. The in-board process configuration shown in Fig. 19.2 is taken as the base case. It is compared to the over-board configuration depicted in Fig. 19.12a.

Investment costs: The over-board die must be wider than the maximum intended substrate width by a distance on the order of 100 mm. The over-board mode requires an extension of the coating station, which includes wider side plates that hold an extra roller with bearing blocks and a mechanism for moving the roller vertically up and down to change the curtain impingement angle, as well as an extra catch pan and a drip pan located above the substrate. In addition, the suction baffle must be modified for over-board applications by providing a longer cover plate and a mechanism for sliding the cover plate to optimize the distance between the tip of the plate and the rear surface of the curtain. On the other hand, the over-board mode works with just one long set of simple solid-body curtain edge guides. Furthermore, a precision, two-dimensional positioning system for the die is not necessary, but it may be convenient to provide a possibility for moving the die horizontally out of the way to have access for removing the rollers 2 and 10 in case of maintenance needs. In contrast, the inboard mode requires a sophisticated edge guide design and an infrastructure for preparing and supplying optimized edge fluid. Moreover, the length of the curtain edge guides must be changed every time the curtain height is changed.

Downtime: The over-board mode generates no downtime related to equipment change when the substrate width is changed. As described in Chap. 10, a width change can be accomplished in just a couple of minutes for single-layer applications in the in-board mode, if movable curtain edge guides are used as depicted in Fig. 10.4. Otherwise, the die width must be changed as per the substrate width by changing the length of the deckles or the geometry of the U-shaped shim. A deckle change takes about 10 min, if a motorized system from TSE Troller AG is used. A manual change for a 3-layer die can be completed in about 45 min. Changing the shim geometry takes more than 1 h as the die must be removed from the coating station, opened, and cleaned before replacing the shim, as well as closing and re-mounting the die.

Apart from re-threading the coating machine, no downtime results in the case of a web break during over-board coating because the running curtain simply drops into the catch pan located underneath the web. In in-board coating, it is generally not possible to remove the curtain from the coating roll or to stop the fluid supply to the die fast enough in case of a web break. Consequently, the still running curtain contaminates the coating roll and other equipment such as the bearing blocks of the coating roll and the suction baffle. The subsequent cleaning time depends on the type of coating fluid and on the fluid amount that caused contamination. It is particularly long if the fluid is an adhesive.

Film thickness uniformity: Owing to web flutter in the over-board mode, the position of the substrate at the curtain impingement point is not fixed. Web flutter causes the length of the curtain to continuously change both in machine and cross-web direction. However, since curtain coating is a premetered application method, the wet thickness of the coated film does not depend on geometric parameters such as the curtain length. Curtain coating is also a conformal application method meaning that web flutter should not significantly affect the uniform thickness of the coated film.

In contrast, web flutter affects the position of the dynamic wetting line. If the web moves upward, the dynamic wetting line moves upstream relative to the position of the impinging curtain and vice versa. The same phenomenon is observed in slot coating as a result of a changing coating gap, see Sect. 17.8.5.1 and Fig. 17.6.

The changing curtain length also affects the momentum of the impinging curtain, which affects the dynamic wetting of curtain coating (see Eq. 7.1.1). Neglecting the initial curtain velocity for simplicity reasons (see Eq. 5.8.1), the curtain velocity at the impingement point is given by $V_c = (2gL_c)^{1/2}$. Now, the effect of the changing curtain length on the curtain impingement momentum can be quantified by taking the derivative of the momentum with respect to the curtain length. The result in normalized form is:

$$\frac{dM_c}{M_c} = \frac{1}{2}\frac{dL_c}{L_c} \tag{19.2.1}$$

dL_c is a measure for the web flutter, and Eq. 19.2.1 visualizes that the curtain momentum decreases if the web flutter moves the substrate upward (dL_c = negative). It is estimated that the amplitude of web flutter is in the range of 1–10 mm, particularly along the edges of the substrate. With typical curtain lengths being in the range of 100–300 mm, the curtain momentum varies by less than 5%. This number is small enough such that web flutter does not significantly affect the wetting behavior of the impinging curtain.

Dry edges and contamination of the web underside: When coating in-board, both edges of the coated substrate are always dry. The width of the dry stripe depends on the quality of the coated edge (over-thickness, spreading or withdrawal of the coated edge, etc.) and on the quality of the web steering system. It also depends on how well the width of the substrate matches the sum of the formats of the final product after cutting. When coating over-board, in contrast, there are usually no dry edges on the coated side of the substrate. However, this does allow for the possibility of contaminating the edge regions of the underside of the substrate, particularly if the curtain does not shear off properly from the substrate edges. A comparative example of these situations is shown in Fig. 19.14. Contamination of the web-underside may have to be removed with a suitable suction device.

In addition, innovations have been presented that allow dry edges to be formed while coating in the over-board mode, for example by using mechanical means for temporarily forcing the substrate edges to curl downward so much so that the impinging curtain starts to shear off on the curled section leaving a dry edge, see illustration of the concept and real coating in Fig. 19.15, as well as a concept patented by Bernert and Ueberschär (2000) who worked for Voith Paper. Experience has shown that this idea works in practice, but the location of the coated edge is less well defined, and the quality of the edge is not as good when compared to the in-board edge generated by a well-designed curtain edge guide, see Fig. 19.15b.

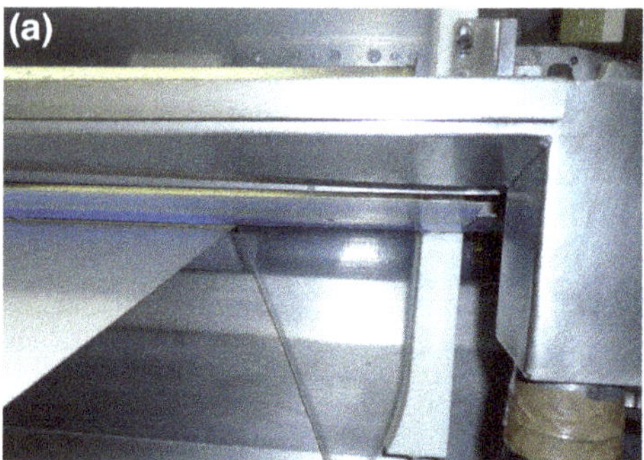

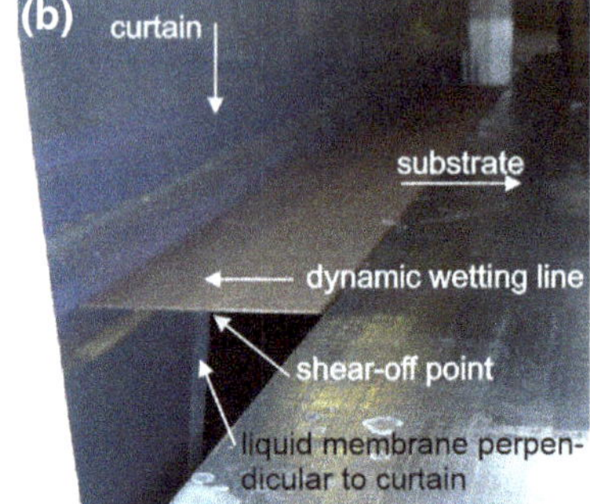

Fig. 19.14 **a** Curtain coating in over-board mode: catch pan 11 and drip pan 12, curtain edge guide, curtain properly shearing off the edge of the substrate; photo reproduced with permission from Polytype Converting AG. **b** Curtain coating in the over-board mode with improper curtain shear-off from the substrate edge; photo reproduced with permission from Polytype Converting AG

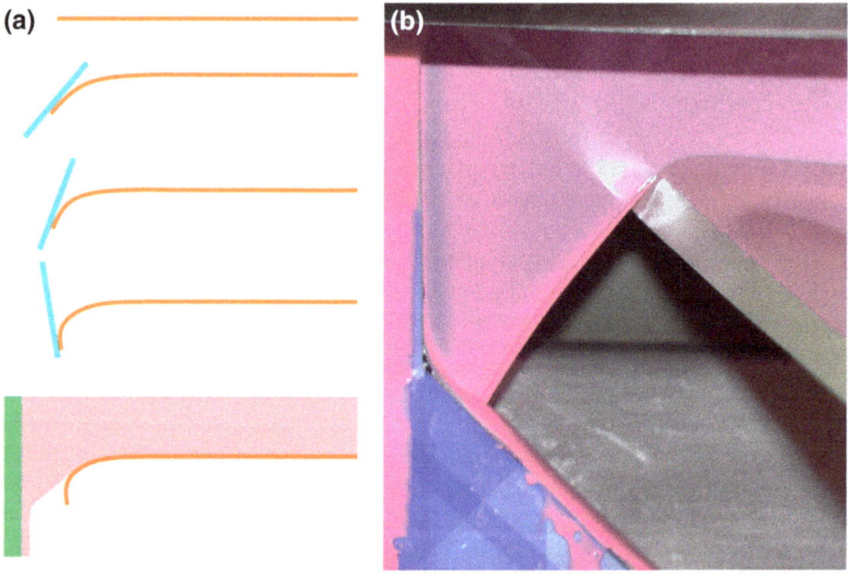

Fig. 19.15 **a** Concept sketch showing successive deformation of the substrate edge until the impinging curtain shears off the curled substrate edge. **b** Over-board coating with a dry edge; photo reproduced with permission from Polytype Converting AG

Excess fluid: Excess fluid is always present in over-board coating because the die is always wider than the substrate width. In contrast, in-board coating does not generate excess fluid, except for single-layer applications where the coating width can be changed instantly by laterally moving a special curtain edge guide—catch pan assembly depicted in Fig. 10.4.

Excess fluid from single-layer applications can be reused, but requires appropriate handling, i.e., collecting it in the catch pan located underneath the substrate and pumping it back into the supply vessel that contains the fresh fluid. Excess fluid may contain air bubbles generated when the extra-wide curtain and the sheared-off fluid flowing along the edge guide impinge onto the bottom of the catch pan. However, experience shows that such bubbles are of little to no concern because they are large and therefore collapse spontaneously after a short time. Micro-bubbles are not generated by this type of flow.

Simultaneous multilayer coating: Applying several layers at the same time with curtain coating operated in the over-board mode is normally not a good proposition because the layers of the excess fluid will mix in the catch pan and must be discarded. This can be costly depending on the specific costs of the raw materials involved and on the width of the excess fluid.

Voith Paper has patented an idea where the fluid waste in simultaneous multilayer curtain coating in the over-board mode can be minimized. Specifically, only the bottom layer of a multilayer film is designed to be wider than the substrate width. All other layers are a bit narrower than the substrate, and are applied in the in-board mode. Consequently, only the bottom layer generates excess fluid. This idea may be viable if the bottom layer is a dummy fluid or a functional fluid consisting of low-cost raw materials. On the down-side, the width of the narrower, functional layers is not well defined because these layers are not guided laterally during the film and curtain flow.

In summary, weighing the relative importance of the various criteria must be done in view of any specific coating application. Nevertheless, we prefer and recommend the in-board operating mode, particularly for simultaneous multilayer applications and when uniformity requirements of the coated film are high.

19.2.5.4 Extrusion Coating

Another operating mode called extrusion coating is depicted in Fig. 19.16. The term "extrusion" refers to the fact that the coating fluid is not drawn from a vessel and supplied by a "normal" pump at or near room temperature, but it is delivered by an extruder at temperatures much above ambient such as 150 °C. Typically, the fluids so applied are highly viscous polymer melts.

In contrast to the in-board and over-board operating modes presented in the previous chapters, extrusion curtain coating is an in-board mode operated without edge guides. In addition, the curtain is not impinging more or less perpendicularly onto the substrate where the liquid is accelerated by a viscous boundary layer, but it is falling vertically into a nip between two rollers, and the fluid is accelerated upstream of the nip by extensional forces. Also, high molecular weight polymers are typically more viscoelastic than fluids coated at room temperature.

The lack of curtain edge guides can be explained by the geometrical configuration of this application mode, because it is practically impossible to design edge guides that reach all the way to the nip. Consequently, the curtain contracts laterally as it

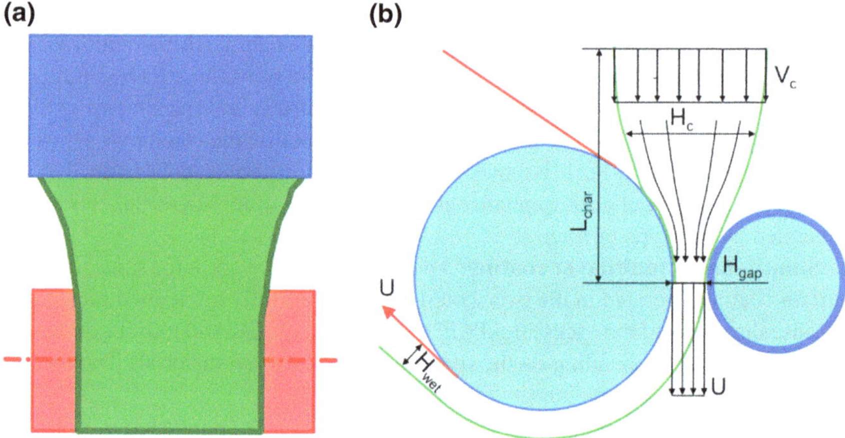

Fig. 19.16 **a** Plain view of extruded curtain. **b** Cross-section of extruded curtain between a pair of rollers

falls to the nip, which results in heavy edges that cannot be avoided. Normally, such edges are cut off before further processing the coated film.

Another difference between extrusion and "normal" curtain coating is the way multilayer films are generated. Figure 2.1f shows the latter configuration where each fluid of a multilayer system is distributed across the coating width by its own individual slot die. In contrast, a so-called external feed block is used for combining all individual fluids upstream of the slot die, and only one common distribution system (cavity/slot) is used for the lateral distribution. Moreover, the die is equipped with a slot that can be deformed during coating, which allows the resulting cross profile to be adjusted. Figure 19.17 shows an extrusion curtain coating station with the slot die being laterally moved out of the station for maintenance purposes.

Since the author of this book has no experience with this curtain coating operating mode, it is recommended to consult with experts for further assistance, such as suppliers of these processes and the related equipment, including Nordson (www.nordson.com), Cloeren (www.cloeren.com), and Polytype Converting (www.polytype-converting.com).

19.2.6 Coating Start and Stop

19.2.6.1 Introduction

Starting a curtain coating process firstly involves generating a homogeneous and uniform curtain and secondly moving the flowing curtain from a stationary catch pan onto the moving substrate, or alternatively, withdrawing a catch pan from the flowing but stationary curtain, such that it drops onto the moving substrate. In any case, if not

Fig. 19.17 Extruder and slot die of an extrusion curtain coating process; photo reproduced with permission from Polytype Converting AG

executed properly, this start-up sequence often produces imperfect curtains and big splashes of liquid and locally over-thick coated films that contaminate the coating and drying equipment, that cannot be dried, and that cause winding problems as well as unwinding problems for subsequent processing steps. To avoid such problems, the start-up sequence must be optimized as is described in the following sections.

19.2.6.2 Forming Uniform Curtains

Suppose that, in preparation for a coating start, a (multilayer) slide die for curtain coating is filled with a coating liquid, and the pump is then turned off, so that the inclined die surface including the die lip can be cleaned and dried with soft towels. If now the pump is turned back on, apart from boundary layer effects along both slide edge guides, the initial liquid front exiting from the die slot is either straight as shown in Fig. 19.18a, or it is very non-uniform with a finger pattern as shown in Fig. 19.18b.

Occasionally it has been observed that a liquid front that is straight on the inclined die surface becomes fingered on the vertical die surface just upstream of the die lip. This is undesirable, because when a fingered liquid front arrives at the die lip, it causes non-uniform wetting along the static wetting line on the backside of the die lip, which subsequently may cause bands in the curtain that can then be seen in the coated film. This effect seems to be stronger for solvent-based liquids than for

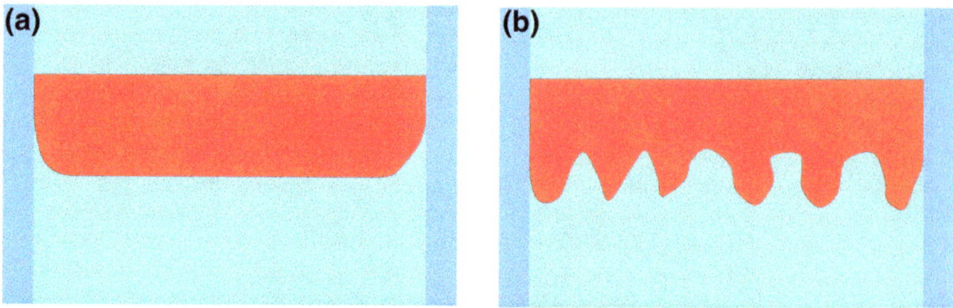

Fig. 19.18 **a** Straight initial liquid front. **b** Fingered initial liquid front

aqueous liquids, particularly if the solvent has a low boiling point. The effect is also amplified, if the horizontal land of the die lip L_L is large rather than small because the size of the vortex at the lip is proportional to L_L, see Fig. 5.39. Moreover, the dimensions and the spatial frequency of the finger pattern in the liquid front seem to be transferred through the curtain onto the substrate.

We also observed that since it takes some time for a pump motor to reach its final speed, the fingering effect is amplified if the pump is started from a state of rest because the initial flow rate/width is lower than the intended equilibrium value. However, fingering could not be prevented entirely if the pump was running at the desired speed, and if the fluid supply to the die was controlled (turned on or off) by way of a 3-way ball valve.

The solution of this problem is based on repeated observations that streaks in the curtain can be prevented during the start-up phase, if the die slide is generously wetted just prior to turning on the pump, for example with a solvent-soaked paper towel. A streak-free curtain was obtained this way even if fingering still occurred in the film flow on the die slide.

Good results were also obtained for high or low nominal flow rates, with a long horizontal lip land of 1.0 mm, and whether the pump was started from rest or was running with the fluid supply to the die being controlled by a 3-way valve or a pair of fast-acting magnetic valves.

Reducing the length of the horizontal lip land and minimizing the time it takes to establish the correct flow rate in the curtain is still helpful for reducing negative effects caused by fingering in the initial film flow on the die slide, but these measures are far less effective than wetting the die slide before feeding the liquid to the die.

19.2.6.3 Generating Uniform Coated Films

A liquid jet (curtain) impinging on a stationary wall produces two liquid films, that flow away from the jet in directions perpendicular to the line of impingement. If the impingement angle α is 0°, then both films have the same thickness. Otherwise, the two wall jets flowing away from the impinging curtain are of different thickness, see Fig. 19.19a.

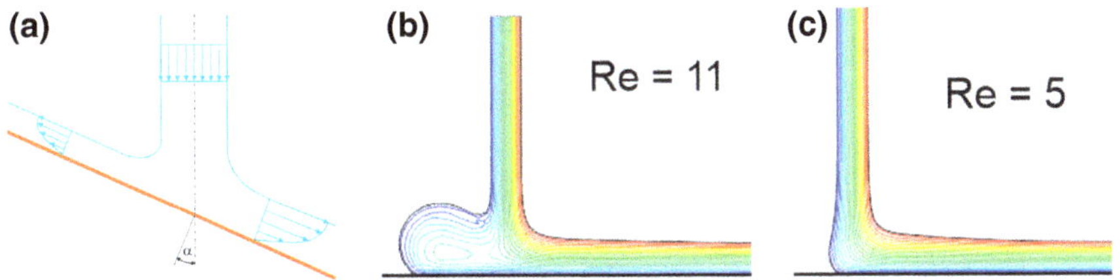

Fig. 19.19 **a** Liquid jet impinging on a stationary wall. **b** Curtain impinging onto moving wall: big heel; graph reproduced with permission from Sunderhaut (1997). **c** Curtain impinging onto moving wall: small heel; graph reproduced with permission from Sunderhaut (1997)

If the wall, or the substrate in a coating process, is moving to the right side (downstream) with a speed U, then the liquid film on the wall, which initially moved to the left side (upstream) with a speed W, is turned around and pulled with the substrate in the downstream direction, see Fig. 19.19b. This happens if the velocity of the substrate U is higher than the velocity W of the upstream film on the wall. Depending on the relative magnitudes of the velocities U and W, a heel of a certain size is formed in the jet impingement zone. If the velocities U and W are optimized relative to each other and in combination with other relevant process parameters, then the heel size will also be optimized (not too big nor too small), as shown in Fig. 19.19c. If U and W are not optimized, then the heel may turn out to be too small or too big, as shown in Fig. 19.19b.

The same qualitative results are obtained, if the wall is stationary and if the curtain is moving to the left (upstream) with a speed V. Still the same qualitative result is obtained, if both the substrate is moving downstream with a speed U and the curtain is moving upstream with a speed V. In all cases, the relevant velocity parameter is the difference in speed between substrate and curtain (U-V). Note that the sign of V is the opposite of the sign of U.

Based on these observations, Polytype Converting AG developed a start procedure for the curtain coating process that uses equipment as depicted in Figs. 19.20, 19.21, 19.22 and 19.23, and that is carried out according to the following process steps.

The size and the location of catch pan 7 are designed such that its edge adjacent to the substrate allows the top portion of the coating roll to be free of equipment, which in turn allows the curtain to impinge on the moving substrate in the range of, say, 11:00–1:00 o'clock on the coating roll, i.e., inside the application angle range of $-30° < \alpha < +30°$.

The reach of the catch pan is extended by the so-called sliding shield 8. The sliding shield is moved out of the catch pan such that its leading edge is located close to the intended impingement point of the curtain on the coating roll, for example, the 12 o'clock position ($\alpha = 0°$) in Fig. 19.20a. The motion of the sliding shield is pneumatically driven and computer-controlled, so that the distance, the speed, and the timing of the motion can be changed and optimized.

If the 12 o'clock position is the optimum impingement point for a given coating application, thus leading to a perfect size of the curtain heel (not too big and not too

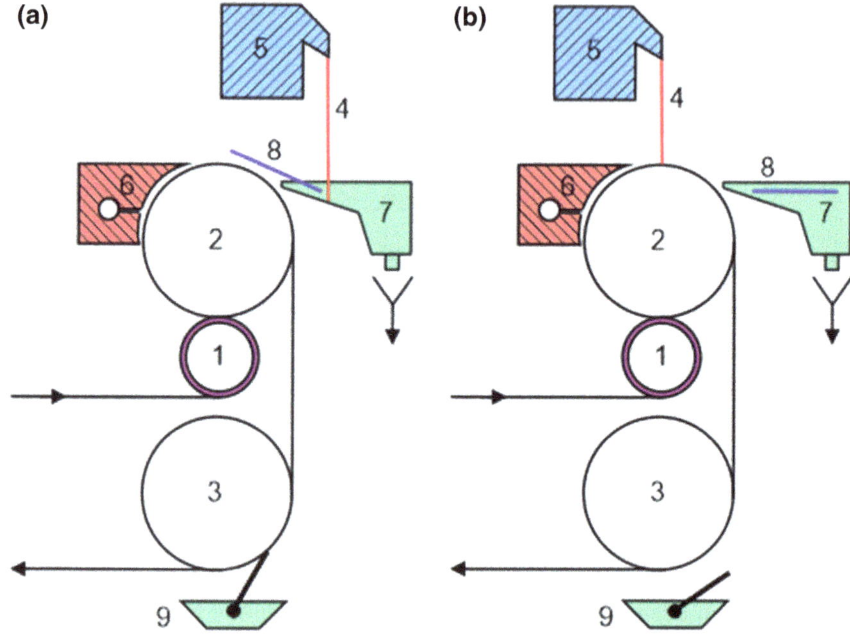

Fig. 19.20 **a** Equipment configuration during the start-up phase. **b** Equipment configuration during the coating phase

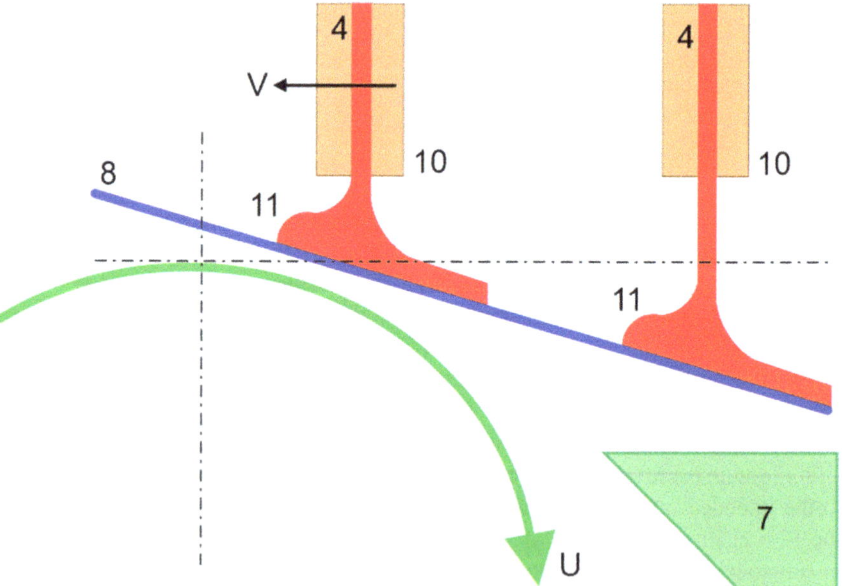

Fig. 19.21 First step of the coating start: the curtain is moving over the stationary sliding shield

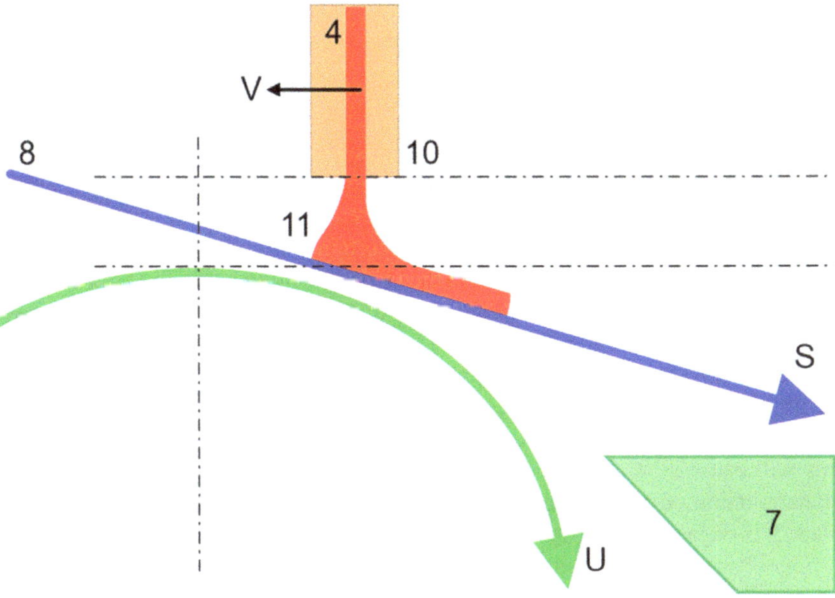

Fig. 19.22 Second step of the coating start: the sliding shield is withdrawn from underneath the moving curtain

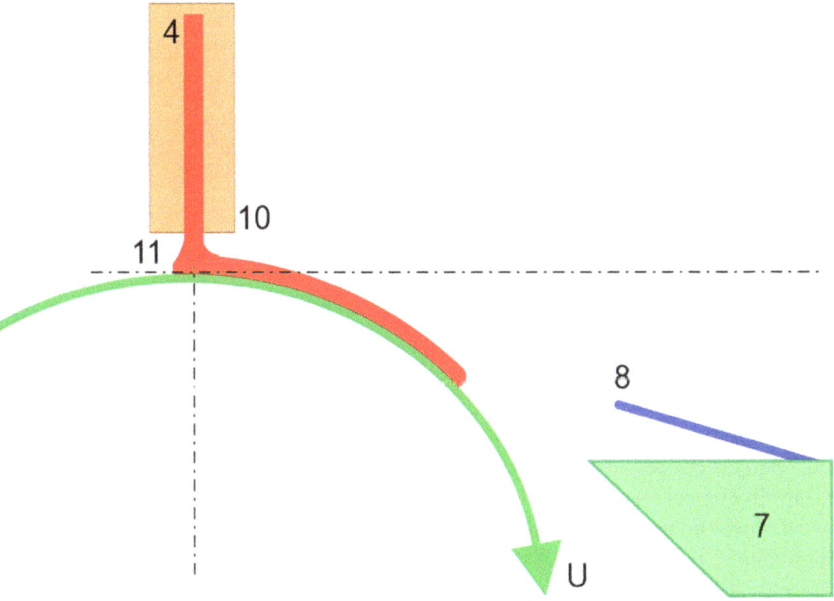

Fig. 19.23 Third step of the coating start: the curtain is falling onto the moving substrate, and the sliding shield is parked over the catch pan

small), then the sliding shield prevents the curtain from initially dropping onto the moving substrate at about 1:30 h (position of the edge of the catch pan), at which point the curtain heel would be ill-shaped, and would cause problems during the start-up, such as splashing, machine contamination, etc.

The beginning of the start procedure is illustrated in Fig. 19.21. A stable curtain 4 is first established so it falls along the curtain edge guides 10 into the catch pan 7. The substrate is then accelerated to the desired speed (production speed, or lower start speed). Then, the die with the flowing curtain is moved with speed V from the catch pan toward the coating point on the backup roll, first over the bottom surface of the catch pan, then over the sliding shield, until it drops onto the substrate very near the intended coating point.

During this phase, according to our observations, a big heel 11 is formed downstream of the moving curtain as shown in Figs. 19.19b and 19.21. This heel will drop onto the moving substrate just before the curtain falls onto the substrate, and this will cause a big splash. Adverse effects from this splash, such as subsequent contamination of the coating equipment, can be prevented if the full-width scraper blade 9 is engaged during this process step, see Fig. 19.20a.

Our observations have shown that the best possible coating start is achieved if the heel size is small as depicted in Figs. 19.19c and 19.22. This is why the sliding shield must be withdrawn with speed S at the same time as the curtain moves with speed V from the park position over the catch pan to the coating position over the substrate, see Fig. 19.22. Pulling the sliding shield through the curtain in the same direction as the substrate motion simulates a coating situation that will be reached momentarily, and it reduces the size of the heel prior to the actual coating. The length of the sliding shield, the starting time of the sliding shield motion relative to the motion of the curtain, as well as the speed of the sliding shield S, and the speed of the horizontal curtain motion V are parameters available for optimizing the curtain heel size during start-up. The best values for these parameters must be found by trial and error during the commissioning of each new product to be coated. In our experience, the sliding shield must be withdrawn just before the curtain drops over the leading edge of the shield.

Long curtains and high flow rates lead to large heels on the sliding shield. High viscosity fluids lead to thick heels, but the precursor length of the heel tends to be short. In contrast, low viscosity fluids produce thin but long heels. Which of the heel shapes is worse is not clear because it depends on yet other parameters. If the heel is too big (too long and/or too thick) during start-up, its size can be reduced by increasing the speed of the sliding shield, and/or by increasing the time during which the sliding shield is in motion. If the motion time of the sliding shield must be long to be effective, then the length of the sliding shield must probably be extended beyond the intended coating point on the backup roll, see Fig. 19.22. In this case, however, the start of the sliding shield motion must be timed such as to avoid a collision with the curtain edge guide 10, which moves with speed V toward the coating point. In other words, at the time of their crossing, the point marking the leading edge of the sliding shield 8 must be below the horizontal line indicated by the bottom of the moving edge guide 10.

The start procedure is terminated when the leading edge of the sliding shield comes to a halt over the catch pan, see Fig. 19.23. In this way, sporadic liquid drops detaching from the edge of the shield will not fall onto the coated substrate but into the catch pan. If this start procedure is well-executed, and if all relevant process parameters are optimized for a particular coating application, then a coherent and good coating will be established in a very short time (typically less than 1 s), overly thick films on the substrate are prevented, and the coating equipment will not be contaminated by splashes. Also, if the start procedure is well optimized, then the full-width scraper blade 9 is no longer needed because the leading edge of the coated film is of good quality.

19.3 Operating Window

This chapter discusses the operating windows for long and short curtains. Almost all industrial curtain coating processes work with long curtains, i.e., the curtain length is in the range of 50–300 mm. In contrast, short curtains that were investigated by Eggerath (2012) have a typical length of <2 mm. Short curtains are introduced in Sect. 5.8.5, which discusses curtain stability.

19.3.1 Operating Window for Long Curtains

As explained in Chap. 16, the operating window of the curtain coating process is best drawn in a diagram with the web speed U on the x-axis and the volumetric flow rate/width Q on the y-axis. A generic version of such a window for curtain coating is shown in Fig. 19.24.

Specifically, the operating window is bound by three process limitations, i.e., curtain stability, air entrainment, and puddling. Operating points located inside these boundaries result in good uniformity of the coated film. One of these points is called the optimum operating point because it produces the best possible uniformity for a given set of operating conditions. Moreover, straight lines through the origin of the coordinate system mark the points of a constant wet thickness of the coated film, with the value of the film thickness being proportional to the slope of the line. Therefore, achieving a lower film thickness requires reducing the flow rate or increasing the web speed.

19.3.1.1 Curtain Stability

Curtain stability, or the break-up of a stable liquid curtain, is discussed in detail in Sect. 5.8.5. The loss of curtain stability is quantified by a critical, or a minimum value of the volumetric flow rate/width Q. For viscous fluids, the minimum flow rate

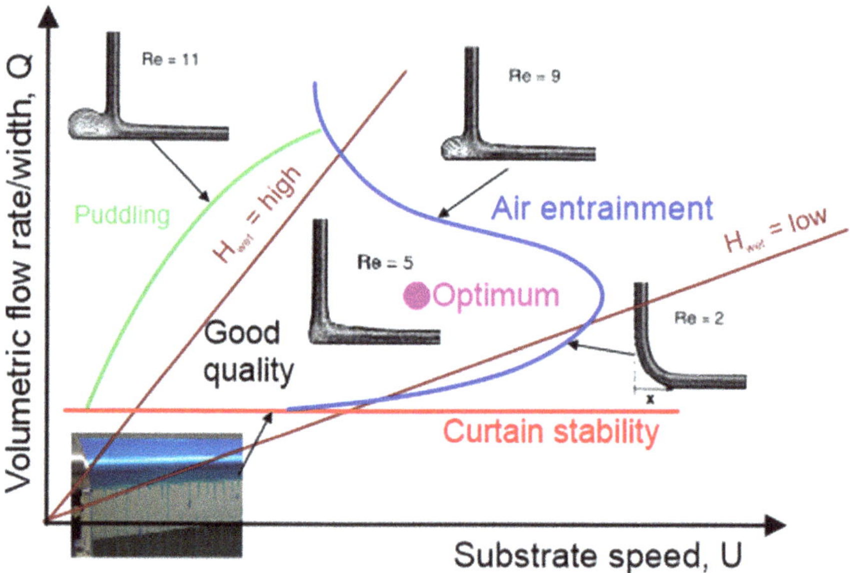

Fig. 19.24 Generic operating window of curtain coating; illustrations of curtain heel from Sünderhauf (1997)

depends primarily on the surface tension, i.e., the flow rate decreases with decreasing surface tension. For Reynolds numbers of <10 and Ohnesorge numbers of >0.1, the minimum flow rate/width also depends on the viscosity. Furthermore, the minimum flow rate decreases substantially with increasing visco-elasticity.

For aqueous fluids and surface tensions of <40 mN/m, a recommended maximum tension value for curtain coating, the minimum flow rate/width is often in the range of 0.7–1.0 cm^2/s. For fluids based on organic solvents, minimum flow rate values of around 0.5 cm^2/s are not uncommon.

In the operating window, the loss of curtain stability is marked with a straight, horizontal line located at the minimum flow rate value. In practice, the minimum flow rate is easily and quickly determined experimentally by starting with a stable curtain and by slowly reducing the flow rate until the curtain breaks up into rivulets. In the pilot facility of Polytype Converting, determining the minimum flow rate was always the first order of business when experimenting with a new coating formulation.

19.3.1.2 Air Entrainment

The onset of air entrainment, or dynamic wetting failure, is a difficult issue. To date, research is ongoing in an attempt to fully understand the physical concepts that lead to this operating boundary of curtain coating. Simple theoretical models based on material properties that are easily accessible in an industrial coating plant for predicting the onset of air entrainment do not exist. Therefore, determining the onset of air entrainment for a given set of operating conditions is best done experimentally. However, this may be challenging because air entrainment in curtain coating quite often only begins at high web speeds of several hundred meters per minute, and because not many pilot coating machines exist across the world that can run that fast. In any case, to find the onset of air entrainment for a given fluid and a given geometry of the coating process, it is recommended to fix a flow rate value and start with a low enough web speed that results in a good coating. Then, the web speed should be continuously increased until wetting failure is observed. Blake et al. (1994) observed a hysteresis for this operating limit. Specifically, the maximum speed at the onset of air entrainment is higher than the speed, at which air entrainment clears again upon reducing the speed from that maximum value.

On the other hand, air entrainment does not seem to be a major problem in curtain coating processes in the converting industry. In fact, during more than 15 years of running curtain coating trials on the pilot machines of Polytype Converting, air entrainment was encountered less than a handful of times even though the maximum speed of the two machines was 1’000 and 1’500 m/min, respectively, and many vastly different fluids and operating conditions were tested.

In the operating window, the air entrainment boundary is marked by a complicated curve, which consists of two branches, see blue line in Fig. 19.24. For lower flow rates, the speed at the air entrainment inception increases with increasing flow rate until a turning point is reached. Along this branch of the curve, air entrainment is triggered by a dynamic wetting line that moves too far downstream relative to the position of the impinging curtain. In other words, the value of the relative wetting line position ℓ as calculated by Eq. 5.9.37 is too small. The turning point marks the maximum possible coating speed without entraining air for a given set of operating conditions, see also Blake et al. (1994). Upon further increasing the flow rate beyond the value of the turning point, the speed at the onset of air entrainment decreases, and ℓ increases with increasing flow rate, thus indicating that the heel of the impinging curtain becomes too big. It can be concluded from these observations that, for a given set of operating conditions, air entrainment is triggered if the dynamic wetting line is located either too far upstream or too far downstream relative to the position of the impinging curtain.

Blake et al. (1994) carried out experiments to investigate air entrainment in curtain coating. Their data focused on the clearance of air entrainment, and two examples of aqueous gelatin solutions coated onto PET substrate are shown here. In Fig. 19.25, the air entrainment process boundary is plotted as a function of the low-shear viscosity μ_0, and the same limitation is depicted in Fig. 19.26 as a function of the curtain length L_c.

The following conclusions can be drawn:

- The shape and location of the air entrainment boundary in the Q-U diagram changes as other relevant process parameters are changed, notably the viscosity, the curtain length, and the curtain impingement angle.

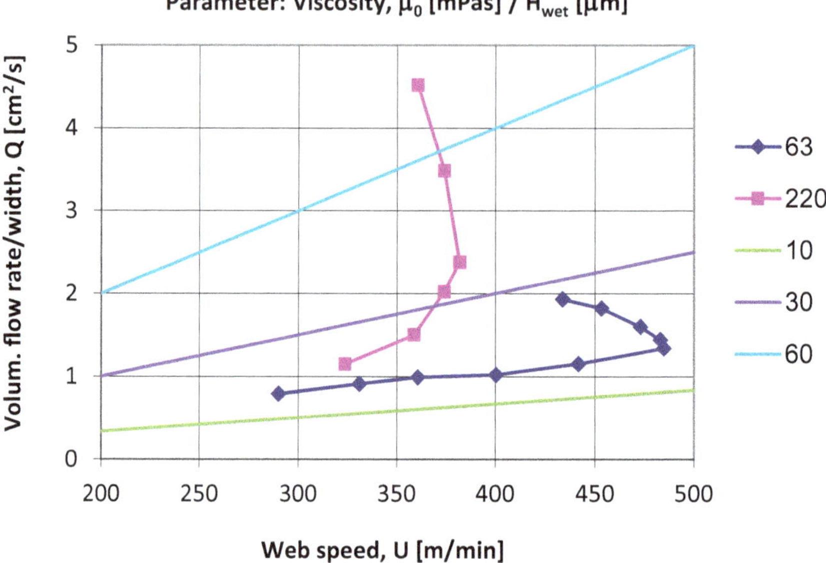

Fig. 19.25 Clearance of air entrainment as a function of the low-shear viscosity; $\alpha = 0°$, $L_c = 254$ mm. Data taken from Blake et al. (1994) and reproduced with permission from AIChE Journal

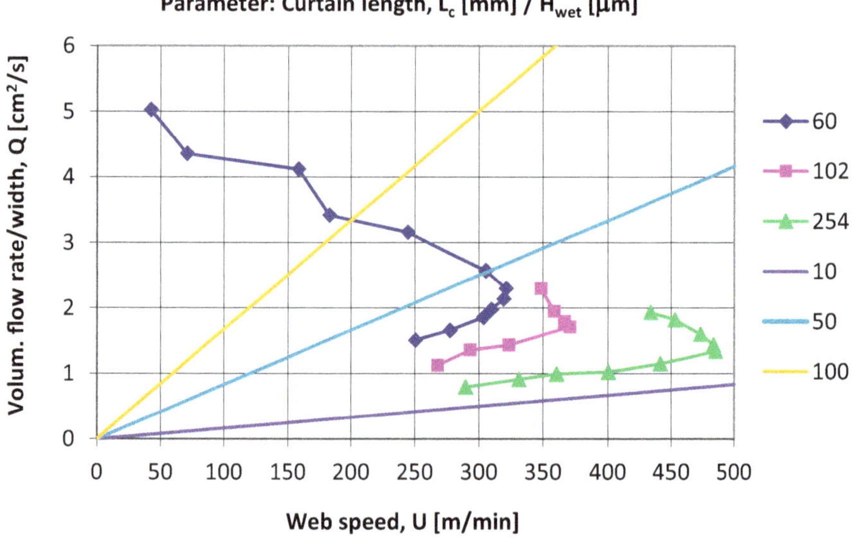

Fig. 19.26 Clearance of air entrainment as a function of the curtain length; $\alpha = 0°$, $\mu_0 = 63$ mPas. Data taken from Blake et al. (1994) and reproduced with permission from AIChE Journal

- Decreasing the viscosity and increasing the curtain length allows faster coating speeds before air entrainment sets in.
- The curtain length and the curtain impingement angle are two geometric parameters, by which the onset of air entrainment can be postponed to higher speeds.
- Increasing the impingement angle reduces the size of the curtain heel; it may be an appropriate measure for preventing air entrainment if the flow rate is higher than the flow rate of the turning point.
- Decreasing the impingement angle increases the size of the curtain heel; it may be an appropriate measure for preventing air entrainment if the flow rate is lower than the flow rate of the turning point.
- The concept of the carrier layer is suitable for lowering the viscosity of the bottom layer of a multilayer film, which also postpones the onset of air entrainment to higher speeds.
- For flow rates of $Q < 1\ cm^2/s$, it is difficult to prevent air entrainment at web speeds that are attractive for the converting industry.

The air entrainment data shown in Figs. 19.25 and 19.26 are only valid for the gelatin solutions investigated by Blake et al. (1994). For any other fluid, similar data would have to be generated, which is labor-intensive and hence expensive. A better way to avoid air entrainment is to ensure that a curtain coating station such as the one schematically depicted in Fig. 19.2 is used, which allows the curtain height and the impingement angle to be changed over wide ranges. Also, it is important to understand in which direction these parameters must be changed if air entrainment is present for a given set of operating conditions. More information on this topic can be found in Sect. 19.4.

19.3.1.3 Puddling

Puddling is a flow condition that results in an excessively big curtain heel, see Figs. 5.58 and 5.62. High flow rates and low web speeds favor the onset of puddling. However, continuously decreasing the speed or increasing the flow rate increases the size of the heel. So, it is difficult in an experiment to detect the onset of puddling, which may be defined as the flow condition that generates one or two vortices in the heel. In our experience, puddling is likely to be present, if the heel can readily be seen by the unaided eye.

Air entrainment does not occur if puddling is present. However, the flow in the cross-web direction inside the vortex-laden and large heel is nonuniform, which causes coating defects such as diffuse, longitudinal bands and cloudiness, both of which are undesirable. Puddling can be prevented by taking the known measures that reduce the size of the curtain heel, see Sects. 5.9.6 and 19.3.1.2.

19.3.1.4 Minimum Coating Speed

Curtain coating is often called a high-speed coating method. This statement is not true per se because the speed is not a primary operating limit of this process. Instead, it is the volumetric flow rate/width, which is a limiting parameter, and which depends on the coating speed and the wet thickness of the coated film according to Eq. 2.1. Curtain coating is characterized by two different, limiting flow rates, one of which is related to the stability of liquid curtains as described in Sect. 5.8.5. This parameter primarily depends on the surface tension and the amount of visco-elasticity. Typical values for industrial curtain coating formulations are in the range of 0.5–1.0 cm^2/s. The other minimum flow rate marks the onset of air entrainment as described in Sect. 19.3.1.2. Based on limited data for the onset of air entrainment such as those shown in Figs. 19.25 and 19.26, and on our experience, air entrainment cannot be prevented for industrially attractive coating speeds, if the flow rate/width is much smaller than 1.0 cm^2/s. If experimental data show that the inception of air entrainment occurs at a flow rate of, say, 0.8 cm^2/s for a given set of operating conditions, then an industrial curtain coating process cannot be run at that flow rate. Instead, it must be operated with a safety margin of about 0.1 cm^2/s, i.e., with a minimum flow rate value that is close to 1.0 cm^2/s.

We have noted that the minimum flow rate related to the onset of air entrainment normally overrides the minimum flow rate related to curtain stability. Therefore, assuming a minimum flow rate/width of $Q_{min} = 1.0\ cm^2/s$ is a conservative but appropriate value for discussing the issue of the minimum coating speed in industrial curtain coating processes. If, after experimental verification and the inclusion of a safety factor, lower speeds than those predicted with a minimum flow rate of 1.0 cm^2/s are possible, then these lower speeds can be accepted as a welcome benefit.

In effect, the discussion about a minimum coating speed is misleading. Since it is not the speed but the flow rate/width that is a process limitation, running the curtain coating process at a lower minimum flow rate not only allows the process to be operated at a lower speed for a given film thickness, but also allows a smaller film thickness to be obtained for a given speed.

Taking Eq. 2.1, requiring the flow rate/width Q to be larger than the minimum flow rate/width Q_{min}, and replacing the wet film thickness by the industrially more interesting dry coat weight according to Eq. 3.5.3, results in the following expressions for the minimum coating speed for single-layer and simultaneous multilayer applications, respectively.

$$\text{Singlc} - \text{layer}: U > \frac{Q_{min} C_{W,solid} \rho_{liqid}}{A_{dry}} \tag{19.3.1a}$$

$$\text{Multilayer}: U > \frac{Q_{min}}{\sum_{i=1}^{k} \frac{A_{dry,i}}{C_{W,solid,i} \rho_{liquid,i}}} \tag{19.3.1b}$$

A diagram of Eq. 19.3.1a is shown in Fig. 19.27.

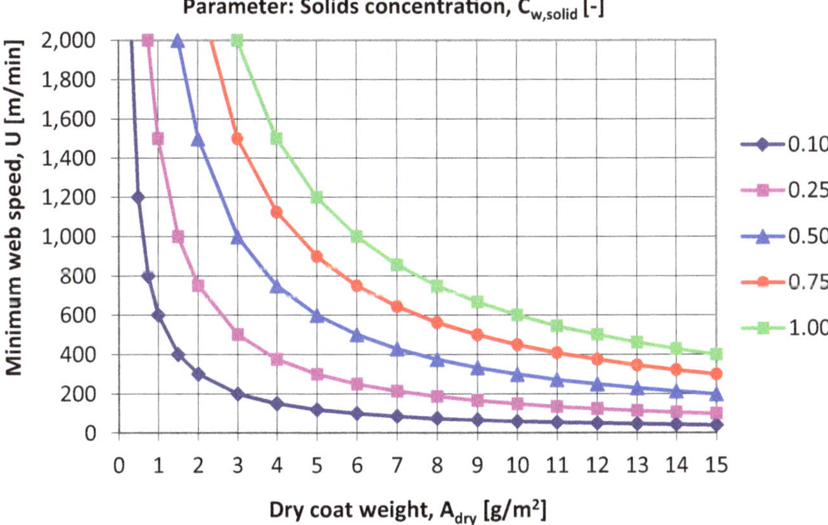

Fig. 19.27 Minimum web speed as a function of the dry coat weight and the solids concentration; $Q_{min} = 1.0\ cm^2/s$, $\rho = 1'000\ kg/m^3$

All curves have a hyperbolic form, thus indicating that decreasing the coat weight requires higher speeds, and this effect is amplified as the solids concentration is increased. It is obvious now that curtain coating can be operated at rather low speeds if only the dry coat weight is high enough and/or the solids concentration is low enough. On the other hand, it is also obvious that thin films of highly concentrated fluids cannot be curtain-coated because the required minimum speed would be unrealistically high. An example thereof is the manufacturing of a release layer for adhesive labels, where, often, solventless silicone ($C_{w,solid} = 1.0$) is applied at a coat weight of about 1 g/m^2.

19.3.1.5 Other Coating Defects

In most industrial applications, curtain coated products are free of any defects if the following requirements are satisfied:

- Clean and homogeneous fluids (filtered, de-gassed, well mixed).
- Low surface tension.
- Optimized position of the dynamic wetting line.
- Prevention of vortices.
- Optimized geometrical details.

Occasional holes in the curtain, as well as diffuse longitudinal bands, may occur if the above requirements are not satisfied. Specifically, sharp lines, streaks, and ribbing lines are absent. However, an open question remains regarding **ribbing lines in curtain coating**. In our experience, ribbing lines in curtain coating have never been observed or reported, at least for speeds of up to 1'000 m/min. On the other hand, investigations of the ribbing line defect in slide coating teach us that this operating limit is triggered if the impinging film shows excessive thinning just upstream of the impingement point, see Sect. 18.5.1.9 for more details. Sünderhauf (2001) and Dragomirescu et al. (2001) used the finite element method for calculating the flow field of the impinging curtain for high-speed applications of the sort that are common in the paper industry, i.e., for web speeds in the range of 1'000–2'000 m/min. The curtain thickness just upstream of the impingement point increases with increasing web speed. If at the same time, the wet thickness of the coated film is small, Sünderhauf showed that the curtain thins considerably in the transition region between curtain flow and the solid body motion of the coated film on the substrate. He also found that curtain thinning increases with decreasing high-shear viscosity.

An example of his calculations is presented in Fig. 19.28. The ratio of curtain waist to curtain thickness just upstream of the impingement area is about 0.43, and the ratio of the thickness of the coated film to curtain thickness just upstream of the impingement area is about 0.1. For $\rho = 1'500\ \mathrm{kg/m^3}$ and $\sigma = 35$ mN/m, the resulting Weber number amounts to nearly 386, which is very high. Projecting these numbers into Fig. 18.16, which shows the onset of ribbing lines in slide coating, would suggest that this flow field of the impinging curtain is subject to ribbing lines.

Comparing the bead flow field of slide coating at moderate speeds with the flow field of an impinging curtain at a very high coating speed may not be appropriate. Therefore, resolving the open question of ribbing lines in high-speed curtain coating requires convincing experimental evidence, which must yet be generated.

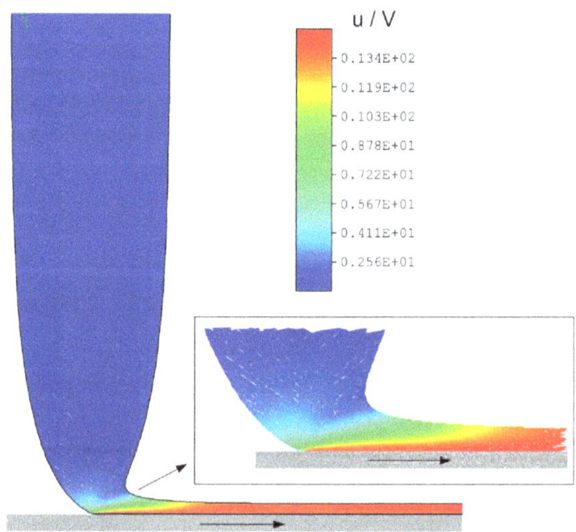

Fig. 19.28 Curtain thinning just upstream of the impingement point; U = 1'800 m/min, H_{wet} = 10 μm, μ_∞ = 20 mPas; V = curtain velocity, u = web velocity. Graph reproduced with permission from Sünderhauf (2001)

Fig. 19.29 Curtain coating with a short curtain; distance between die lip and substrate = 800 μm. Photo reproduced with permission from Eggerath (2012)

19.3.2 Operating Window for Short Curtains

Coating with a short curtain is not unlike slot coting in the capillary operating mode, where the ratio of H_{gap}/H_{wet} can reach values of >> 2. Eggerath (2012) investigated the coating behavior of short curtains. An example thereof is visualized in Fig. 19.29, where $H_{gap} = 800$ μm and $H_{gap}/H_{wet} \approx 2$. If H_{gap} were 1'500 μm, thus corresponding to a value that is close to the maximum length of a short curtain, and if H_{gap}/H_{wet} were 20, a value that can be reached in slot coating, then the wet film thickness would amount to 75 μm, which does not appear to be unrealistic.

Unfortunately, Eggerath's experimental results regarding the onset of air entrainment and puddling are not conclusive. On the pilot machine of Polytype Converting AG we ran short curtain trials for two customers, who were interested in exploring the possibilities of this new application method. In both cases, we were unable to obtain coherent and uniform coated films for the specified operating conditions. Therefore, more experiments are needed for visualizing and quantifying the operating limits of the short curtain coating method.

19.4 Process Optimization

Optimizing the curtain coating process is the ultimate and most important task when developing this process for a new product application because it produces the best possible product uniformity and the best possible process productivity for a given set of operating conditions.

During this optimization procedure, when a chosen combination of operating parameters is not yet optimal, then either the operating point has to be moved inside the coating window, or the boundaries of the operating window have to be moved further away from the coating point. The former step is often not desirable nor acceptable because industrial coating companies typically specify goals for the product performance in terms of the thickness of the coated film, and the process productivity in terms of the web speed. Therefore, moving the boundaries of the operating widow is a more attractive procedure, and knowing how to accomplish this is crucial.

Optimizing the curtain coating process was described in detail by Schweizer (1997) and also in Sect. 7.2 of this book. The available theoretical tools are

- Eqs. 5.9.36 and 5.9.37 or 5.9.39, which describe the position of the dynamic wetting line, ℓ, relative to the position of the impinging curtain based on the concept of hydrodynamic assist developed by Blake et al. (1994).
- Eq. 7.1.1, which quantifies the momentum M of the impinging curtain.

Different optimum values for the relative wetting line position ℓ have been proposed in the literature. Specifically, Van Abbenyen et al. (1992) suggested $\ell = 1.0$ based on the maximum speed of wetting. Blake et al. (1994) proposed $\ell = 2.0$ by maximizing the impingement pressure of the curtain. Schweizer (1997) argued for $\ell \approx 1.5$ by maximizing the robustness of the wetting line position with respect to disturbing forces acting on the wetting line in the direction of the web motion. Experience gained at Polytype Converting AG over more than 15 years from running many different coating trials and commissioning several industrial coating machines at speeds of up to >1’000 m/min confirms good process performances for $0.5 < \ell < 2.0$. The preferred ℓ-value is about 1.0, and ℓ-values being slightly below 1.0 work well for high-speed applications.

Regarding the momentum of the impinging curtain, Eq. 7.1.1 can be re-written in terms of the dry coat weight(s) of the (multilayer) film, which define(s) the performance of the end product. Neglecting the initial curtain velocity and the exact effect of the densities, the impingement momentum can be approximated by

$$M_c = \sum_{i=1}^{k} \left(\frac{A_{dry,i}}{C_{w,solid,i}} \right) U \sqrt{2gL_c} \cos\alpha \tag{19.4.1}$$

Using these tools, the curtain coating process can be optimized by carrying out the following steps:

1. Maximize the desired effect of the hydrodynamic assist by adjusting the parameters in Eq. 5.9.36 such that the location of the dynamic wetting line coincides more or less with the location of the rear curtain surface, i.e., such that the value of ℓ is close to 1.0.
2. Optimize the effect of the impingement momentum of the curtain to prevent air entrainment by adjusting the parameters in Eq. 19.4.1. If air entrainment is observed even though the dynamic wetting line is located near the optimum position, then the momentum must be increased. In contrast, air entrainment

may be avoided by reducing the momentum if $\ell >> 1$, or by increasing the momentum if $\ell << 1$, i.e., if the dynamic wetting line is located too far upstream or too far downstream relative to the location of the curtain.

3. Steps 1 and 2 must be carried out in parallel as both the wetting line position ℓ and the curtain momentum M depend upon the same parameters. Therefore, changing ℓ also changes M and vice versa. Specifically:

$$\ell = \ell(L_c, \alpha, U, A_{dry}, C_w, \rho, \mu, \sigma) \tag{19.4.2}$$

$$M = M(L_c, \alpha, U, A_{dry}, C_w) \tag{19.4.3}$$

Note, however, that the operating parameters U and A_{dry} as well as the physical fluid properties in terms of their chemical composition are usually specified and should not be changed. This leaves the curtain length L_c, the curtain impingement angle α and the solids concentration of the layer(s) $C_{w,i}$ as free parameters that can be changed. Therefore, it is important to design the curtain coating station such that the geometric parameters L_c and α can be changed. A viable example of such a design is shown in Fig. 19.2.

The curtain length is a relatively weak parameter because the impingement momentum depends on the square root of L_c. Thus, doubling the curtain height increases the momentum only by a factor of 1.41.

Regarding the impingement angle, the best, most convenient, and preferred value is $\alpha = 0°$ because it not only maximizes the effect of the hydrodynamic assist, but also eliminates the chances for a mechanical collision between the curtain edge guides and the coating roll during the start procedure when the die is moved from the park position over the catch pan to the coating position over the backup roll. If $\alpha > 0°$, this die displacement is crucial because the curtain edge guides must be positioned very close to the substrate surface such as illustrated in Fig. 5.45. For a station design shown in Fig. 19.2, if $\alpha < 0°$, the die movement is even more cumbersome because the die must not only be displaced horizontally, but it also must be lowered to the desired application point after the die and the edge guides have passed the 12 o'clock position on the coating roll. Moreover, if $\alpha << 0°$, the start and stop procedures as described in Sect. 19.2.6.3 cannot be carried out effectively.

The reservations mentioned above regarding the length and the impingement angle of the curtain as being convenient parameters for optimizing the curtain coating process might just increase the attractiveness of the solids concentration for accomplishing this task, particularly if several layers are applied simultaneously. Changing the solids concentration not only affects the impingement momentum by changing the flow rate in the curtain, but it also affects the wet thickness of the coated film and the shear rate in the boundary layer of the impinging curtain, as well as all the parameters of the Carreau-Yasuda rheological model and the resulting viscosity in the boundary layer. However, reducing the solids concentration is not attractive, and it might not even be possible because it increases the dryer load. Nevertheless, if the solids concentration should be reduced, then the measure should be limited to the lowest layer of a multilayer film because it is the viscosity of the lowest layer that has the strongest effect on the location of the dynamic wetting line. For single-layer applications, and for thick bottom layers of multilayer films, the concept of the carrier layer as described in Chap. 12 should be used for changing the viscosity of the boundary layer without excessively increasing the dryer load.

In addition to the optimization steps 1 and 2 listed above, several other aspects of the curtain coating process must be optimized as described throughout this book, notably

4. Avoiding vortices in the film flow on the die slide and at the die lip.
5. Optimizing the design of the curtain edge guides.
6. Optimizing the design of the internal and external die geometry.
7. Optimizing all aspects of the fluid conditioning and delivery system.

Guidelines for optimizing steps 4–7 are discussed throughout this book. When all steps are optimized as suggested, then curtain coating produces very pleasing results in terms of the uniformity of the coated film and the productivity of the application process. In fact, the results are second to none compared to the results that can be produced by any other coating method.

References

Blake, T. D., Clarke, A., & Ruschak, K. J. (1994). Hydrodynamic assist of dynamic wetting. *AIChE Journal, 40*(2), 229–242.

Bernert, R., & Ueberschär, M. (2000). Vorhang-Auftragsvorrichtung. DE 100 47 167.

Dragomirescu, Sünderhauf, G., Raszillier, H., & Durst, F. (2001). Curtain coating at high speed—Fluid dynamical limitations. In *Proceedings of the 4th European Coating Symposium*, October 1–4, Brussels, Belgium.

Eggerath, D. (2012). Wissenschaftliche Beiträge zur Vorhangbeschichtung und ihrem industriellen Einsatz. Ph.D. thesis, University of Erlangen-Nürnberg.

Frediani, L., Hardegger, H., Holtmann, B., & Metzger, R. (2001). Absaugung einer Luftgrenzschicht von einer laufenden Materialbahn. DE 101 17 667.

Horbach, H., & Reinhard, F. (2004). Anordnung und Verfahren zur Vorhangbeschichtung bewegter Substrate. EP 1 817 116.

Hughes, D. J. (1970). Method for simultaneously applying a plurality of coated layers by forming a stable multilayer free-falling vertical curtain. US 3,508,947.

Miyamoto, K., & Katagiri, Y. (1997). Curtain coating. In S. F. Kistler & P. M. Schweizer (Eds.), *Chapter 11c in Liquid Film Coating*. Chapman & Hall.

Schweizer, P. M. (1997). Position of the dynamic wetting line for premetered coating processes. In S. F. Kistler, & P.M. Schweizer (Eds.), *Example 15.3.3.1 in Chapter 15 in Liquid Film Coating*. Chapman & Hall

Schweizer, P.M., & Troller U. (1997). Verfahren und Vorrichtung zur Vorhangbeschichtung eines bewegten Trägers. EP 0 906 789.

Sünderhauf, G. (1997). Vorhangbeschichtung. Kurzlehrgang über Grundlagen und Verfahren der Beschichtungstechnik, Lehrstuhl für Strömungsmechanik, Universität Erlangen-Nürnberg, Germany.

Sünderhauf, G. (2001). Strömungsuntersuchung des Vorhangstreichens von Papier. Ph.D. thesis, University of Erlangen-Nürnberg, Germany.

Van Abbenyen, W., Mues, W., & Goetmaeckers, B. (1992). Explaining limits of operability in curtain coating. In Paper 43a presented at the 6th International Coating Process Science and Technology Symposium of the AIChE Spring National Meeting, New Orleans, March 29–April 2.

Index

P. M. Schweizer, *Premetered Coating Methods*, Engineering Materials,
https://doi.org/10.1007/978-3-031-04180-8

T

U

www.ingramcontent.com/pod-product-compliance
Ingram Content Group UK Ltd.
Pitfield, Milton Keynes, MK11 3LW, UK
UKHW021832270726
14058UKWH00001B/102
* 9 7 8 3 0 3 1 0 4 1 8 2 2 *